Engineering Surveying

Engineering Surveying

Sixth Edition

W. Schofield
Former Principal Lecturer, Kingston University

M. Breach
Principal Lecturer, Nottingham Trent University

AMSTERDAM • BOSTON • HEIDELBERG • LONDON • NEW YORK • OXFORD
PARIS • SAN DIEGO • SAN FRANCISCO • SINGAPORE • SYDNEY • TOKYO
Butterworth-Heinemann is an imprint of Elsevier

Butterworth-Heinemann is an imprint of Elsevier
Linacre House, Jordan Hill, Oxford OX2 8DP, UK
30 Corporate Drive, Suite 400, Burlington, MA 01803, USA

First edition 1972
Second edition 1978
Third edition 1984
Fourth edition 1993
Reprinted 1995, 1997, 1998
Fifth edition 2001
Reprinted 2002
Sixth edition 2007

British Library Cataloguing in Publication Data
Schofield, W. (Wilfred)
Engineering surveying
– 6th ed.
1. Surveying
I. Title II. Breach, Mark
526. 9'02462

Library of Congress Cataloging-in-Publication Data
Library of Congress Control Number: 2006933363

ISBN–13: 978-0-7506-6949-8
ISBN–10: 0-7506-6949-7

For information on all Butterworth-Heinemann publications visit our website at www.books.elsevier.com

Transferred to digital printing in 2009.

Contents

Wilf Schofield
1933–2006

This is the sixth edition of Wilf Schofield's book which he invited me to share with him. A few weeks after we started work, Wilf was diagnosed with cancer. In spite of this he worked tirelessly on the project but, sadly, was overcome before this edition came to press.

Wilf brought to this, and previous editions, a wealth of experience as a mine surveyor and as an academic. He was, at heart, a teacher with a passion for his subject and this comes over in the clarity of his writing. He will be remembered by colleagues and former pupils not only for his technical ability but also for his incisive wit, his storytelling and his distinctive and commanding appearance.

Mark Breach

Preface to the sixth edition

The subject of engineering surveying continues to develop at a rapid pace and this has been reflected in the many and substantial changes that have been made in updating and revising the previous edition. The authors have taken the opportunity to examine in detail all the previous material making both minor and major changes throughout. As always, decisions have to be made as to what should be retained that is still current and relevant and to identify the material that needs to be cut to make way for new text to describe the emerging technologies.

The subject of survey control is now treated in much greater depth. The chapter on traditional methods still in current practice is followed by a whole new chapter on rigorous methods of control, that is, the application of the technique of least squares in the determination of coordinates and their quality. This topic was dropped from the fifth edition of this book but now reappears in a completely rewritten chapter which reflects modern software applications of a technique that underlies much of satellite positioning and inertial navigation as well as rigorous survey control.

Satellite positioning brings up to date the many advances that have been made in the development of GPS and its applications, as well as looking to the changes now taking place with GLONASS and the European GALILEO systems.

The chapter on underground surveying includes an enlarged section on gyrotheodolites which reflects new techniques that have been developed and the application of automation in modern instrumentation. The final chapter on mass data methods brings together substantial sections on simple applications of photogrammetry with the revolutionary new technology of laser scanning by aerial and terrestrial means. Inertial technology, once seen as an emerging standalone surveying technology, now reappears in a completely new guise as part of aircraft positioning and orientation systems used to aid the control of aerial photogrammetry and laser scanners.

In spite of all this new material the authors have been able to keep the same level of worked examples and examination questions that have been so popular in previous editions. We are confident that this new edition will find favour with students and practitioners alike in the areas of engineering and construction surveying, civil engineering, mining and in many local authority applications. This book will prove valuable for undergraduate study and professional development alike.

Mark Breach

Preface to the fifth edition

Since the publication of the fourth edition of this book, major changes have occurred in the following areas:

- surveying instrumentation, particularly Robotic Total Stations with Automatic Target Recognition, reflectorless distance measurement, etc., resulting in turnkey packages for machine guidance and deformation monitoring. In addition there has been the development of a new instrument and technique known as laser scanning
- GIS, making it a very prominent and important part of geomatic engineering
- satellite positioning, with major improvements to the GPS system, the continuance of the GLONASS system, and a proposal for a European system called GALILEO
- national and international co-ordinate systems and datums as a result of the increasing use of satellite systems.

All these changes have been dealt with in detail, the importance of satellite systems being evidenced by a new chapter devoted entirely to this topic.

In order to include all this new material and still retain a economical size for the book, it was necessary but regrettable to delete the chapter on Least Squares Estimation. This decision was based on a survey by the publishers that showed this important topic was not included in the majority of engineering courses. It can, however, still be referred to in the fourth edition or in specialised texts, if required.

All the above new material has been fully expounded in the text, while still retaining the many worked examples which have always been a feature of the book. It is hoped that this new edition will still be of benefit to all students and practitioners of those branches of engineering which contain a study and application of engineering surveying.

W. Schofield
February 2001

Acknowledgements

The authors wish to acknowledge and thank all those bodies and individuals who contributed in any way to the formation of this book.

For much of the illustrative material thanks are due to:

ABA Surveying

Air Force Link

AGTEK Development

Delta Aerial Surveys Ltd

Deutsche Montan Technologie GmbH

GOCA – GPS-based online Control and Alarm Systems

ISP (Integrated Software Products)

Leica Geosystems (UK) Ltd

Ordnance Survey of Great Britain

Riegl UK Ltd

terra international surveys ltd

Topcon

Trimble

Wehrli & Associates Inc.

We must also acknowledge the help received from the many papers, seminars and conferences, which have been consulted in the preparation of this edition.

1
Basic concepts of surveying

The aim of this chapter is to introduce the reader to the basic concepts of surveying. It is therefore the most important chapter and worthy of careful study and consideration.

1.1 DEFINITION

Surveying may be defined as the science of determining the position, in three dimensions, of natural and man-made features on or beneath the surface of the Earth. These features may be represented in analogue form as a contoured map, plan or chart, or in digital form such as a *digital ground model* (DGM).

In engineering surveying, either or both of the above formats may be used for planning, design and construction of works, both on the surface and underground. At a later stage, surveying techniques are used for dimensional control or setting out of designed constructional elements and also for monitoring deformation movements.

In the first instance, surveying requires management and decision making in deciding the appropriate methods and instrumentation required to complete the task satisfactorily to the specified accuracy and within the time limits available. This initial process can only be properly executed after very careful and detailed reconnaissance of the area to be surveyed.

When the above logistics are complete, the field work – involving the capture and storage of field data – is carried out using instruments and techniques appropriate to the task in hand.

Processing the data is the next step in the operation. The majority, if not all, of the computation will be carried out with computing aids ranging from pocket calculator to personal computer. The methods adopted will depend upon the size and precision of the survey and the manner of its recording; whether in a field book or a data logger. Data representation in analogue or digital form may now be carried out by conventional cartographic plotting or through a totally automated computer-based system leading to a paper- or screen-based plot. In engineering, the plan or DGM is used when planning and designing a construction project. The project may be a railway, highway, dam, bridge, or even a new town complex. No matter what the work is, or how complicated, it must be set out on the ground in its correct place and to its correct dimensions, within the tolerances specified. To this end, surveying procedures and instrumentation of varying precision and complexity are used depending on the project in hand.

Surveying is indispensable to the engineer when planning, designing and constructing a project, so all engineers should have a thorough understanding of the limits of accuracy possible in the construction and manufacturing processes. This knowledge, combined with an equal understanding of the limits and capabilities of surveying instrumentation and techniques, will enable the engineer to complete the project successfully in the most economical manner and in the shortest possible time.

1.2 PRINCIPLES

Every profession must be founded upon sound practice and in this engineering surveying is no different. Practice in turn must be based upon proven principles. This section is concerned with examining the principles of survey, describing their interrelationship and showing how they may be applied in practice. Most of the principles below have an application at all stages of a survey and it is an unwise and unprofessional surveyor who does not take them into consideration when planning, executing, computing and presenting the results of the survey work. The principles described here have application across the whole spectrum of survey activity, from field work to photogrammetry, mining surveying to metrology, hydrography to cartography, and cadastral to construction surveying.

1.2.1 Control

A control network is the framework of survey stations whose coordinates have been precisely determined and are often considered definitive. The stations are the reference monuments, to which other survey work of a lesser quality is related. By its nature, a control survey needs to be precise, complete and reliable and it must be possible to show that these qualities have been achieved. This is done by using equipment of proven precision, with methods that satisfy the principles and data processing that not only computes the correct values but gives numerical measures of their precision and reliability.

Since care needs to be taken over the provision of control, then it must be planned to ensure that it achieves the numerically stated objectives of precision and reliability. It must also be complete as it will be needed for all related and dependent survey work. Other survey works that may use the control will usually be less precise but of greater quantity. Examples are setting out for earthworks on a construction site, detail surveys of a greenfield site or of an as-built development and monitoring many points on a structure suspected of undergoing deformation.

The practice of using a control framework as a basis for further survey operations is often called 'working from the whole to the part'. If it becomes necessary to work outside the control framework then it must be extended to cover the increased area of operations. Failure to do so will degrade the accuracy of later survey work even if the quality of survey observations is maintained.

For operations other than setting out, it is not strictly necessary to observe the control before other survey work. The observations may be concurrent or even consecutive. However, the control survey must be fully computed before any other work is made to depend upon it.

1.2.2 Economy of accuracy

Surveys are only ever undertaken for a specific purpose and so should be as accurate as they need to be, but not more accurate. In spite of modern equipment, automated systems, and statistical data processing the business of survey is still a manpower intensive one and needs to be kept to an economic minimum. Once the requirement for a survey or some setting out exists, then part of the specification for the work must include a statement of the relative and absolute accuracies to be achieved. From this, a specification for the control survey may be derived and once this specification has been achieved, there is no requirement for further work.

Whereas control involves working from 'the whole to the part' the specification for all survey products is achieved by working from 'the part to the whole'. The specification for the control may be derived from estimation based upon experience using knowledge of survey methods to be applied, the instruments to be used and the capabilities of the personnel involved. Such a specification defines the expected quality of the output by defining the quality of the work that goes into the survey. Alternatively a statistical analysis of the proposed control network may be used and this is the preferable approach. In practice a good specification will involve a combination of both methods, statistics tempered by experience. The accuracy of any survey work will never be better than the control upon which it is based. You cannot set out steelwork to 5 mm if the control is only good to 2 cm.

1.2.3 Consistency

Any 'product' is only as good as the most poorly executed part of it. It matters not whether that 'product' is a washing machine or open heart surgery, a weakness or inconsistency in the endeavour could cause a catastrophic failure. The same may apply in survey, especially with control. For example, say the majority of control on a construction site is established to a certain designed precision. Later one or two further control points are less well established, but all the control is assumed to be of the same quality. When holding-down bolts for a steelwork fabrication are set out from the erroneous control it may require a good nudge from a JCB to make the later stages of the steelwork fit.

Such is the traditional view of consistency. Modern methods of survey network adjustment allow for some flexibility in the application of the principle and it is not always necessary for all of a particular stage of a survey to be of the same quality. If error statistics for the computed control are not to be made available, then quality can only be assured by consistency in observational technique and method. Such a quality assurance is therefore only second hand. With positional error statistics the quality of the control may be assessed point by point. Only least squares adjustments can ensure consistency and then only if reliability is also assured. Consistency and economy of accuracy usually go hand in hand in the production of control.

1.2.4 The Independent check

The independent check is a technique of quality assurance. It is a means of guarding against a blunder or gross error and the principle must be applied at all stages of a survey. Failure to do so will lead to the risk, if not probability, of 'catastrophic failure' of the survey work. If observations are made with optical or mechanical instruments, then the observations will need to be written down. A standard format should be used, with sufficient arithmetic checks upon the booking sheet to ensure that there are no computational errors. The observations should be repeated, or better, made in a different manner to ensure that they are in sympathy with each other. For example, if a rectangular building is to be set out, then once the four corners have been set out, opposite sides should be the same length and so should the diagonals. The sides and diagonals should also be related through Pythagoras' theorem. Such checks and many others will be familiar to the practising surveyor.

Checks should be applied to ensure that stations have been properly occupied and the observations between them properly made. This may be achieved by taking extra and different measurements beyond the strict minimum required to solve the survey problem. An adjustment of these observations, especially by least squares, leads to misclosure or error statistics, which in themselves are a manifestation of the independent check.

Data abstraction, preliminary computations, data preparation and data entry are all areas where transcription errors are likely to lead to apparent blunders. Ideally all these activities should be carried out by more than one person so as to duplicate the work and with frequent cross-reference to detect errors. In short, wherever there is a human interaction with data or data collection there is scope for error.

Every human activity needs to be duplicated if it is not self-checking. Wherever there is an opportunity for an error there must be a system for checking that no error exists. If an error exists, there must be a means of finding it.

1.2.5 Safeguarding

Since survey can be an expensive process, every sensible precaution should be taken to ensure that the work is not compromised. Safeguarding is concerned with the protection of work. Observations which are written down in the field must be in a permanent, legible, unambiguous and easily understood form so that others may make good sense of the work. Observations and other data should be duplicated at the earliest possible stage, so that if something happens to the original work the information is not lost. This may be by photocopying field sheets, or making backup copies of computer files. Whenever the data is in a unique form or where all forms of the data are held in the same place, then that data is vulnerable to accidental destruction.

In the case of a control survey, the protection of survey monuments is most important since the precise coordinates of a point which no longer exists or cannot be found are useless.

1.3 BASIC MEASUREMENTS

Surveying is concerned with the fixing of position whether it be control points or points of topographic detail and, as such, requires some form of reference system.

The physical surface of the Earth, on which the actual survey measurements are carried out, is not mathematically definable. It cannot therefore be used as a reference datum on which to compute position.

Alternatively, consider a level surface at all points normal to the direction of gravity. Such a surface would be closed and could be formed to fit the mean position of the oceans, assuming them to be free from all external forces, such as tides, currents, winds, etc. This surface is called the geoid and is defined as the equipotential surface that most closely approximates to mean sea level in the open oceans. An equipotential surface is one from which it would require the same amount of work to move a given mass to infinity no matter from which point on the surface one started. Equipotential surfaces are surfaces of equal potential; they are not surfaces of equal gravity. The most significant aspect of an equipotential surface going through an observer is that survey instruments are set up relative to it. That is, their vertical axes are in the direction of the force of gravity at that point. A level or equipotential surface through a point is normal, i.e. at right angles, to the direction of gravity. Indeed, the points surveyed on the physical surface of the Earth are frequently reduced, initially, to their equivalent position on the geoid by projection along their gravity vectors.

The reduced level or elevation of a point is its height above or below the geoid as measured in the direction of its gravity vector, or plumb line, and is most commonly referred to as its height above or below mean sea level (MSL). This assumes that the geoid passes through local MSL, which is acceptable for most practical purposes. However, due to variations in the mass distribution within the Earth, the geoid, which although very smooth is still an irregular surface and so cannot be used to locate position mathematically.

The simplest mathematically definable figure which fits the shape of the geoid best is an ellipsoid formed by rotating an ellipse about its minor axis. Where this shape is used by a country as the surface for its mapping system, it is termed the reference ellipsoid. *Figure 1.1* illustrates the relationship between these surfaces.

The majority of engineering surveys are carried out in areas of limited extent, in which case the reference surface may be taken as a tangent plane to the geoid and the principles of plane surveying applied. In other words, the curvature of the Earth is ignored and all points on the physical surface are orthogonally projected onto a flat plane as illustrated in *Figure 1.2*. For areas less than 10 km square the assumption of a flat Earth is perfectly acceptable when one considers that in a triangle of approximately 200 km^2, the difference between the sum of the spherical angles and the plane angles would be 1 second of arc, or that the difference in length of an arc of approximately 20 km on the Earth's surface and its equivalent chord length is a mere 8 mm.

The above assumptions of a flat Earth, while acceptable for some positional applications, are not acceptable for finding elevations, as the geoid deviates from the tangent plane by about 80 mm at 1 km or 8 m at 10 km from the point of contact. Elevations are therefore referred to the geoid, at least theoretically, but usually to MSL practically.

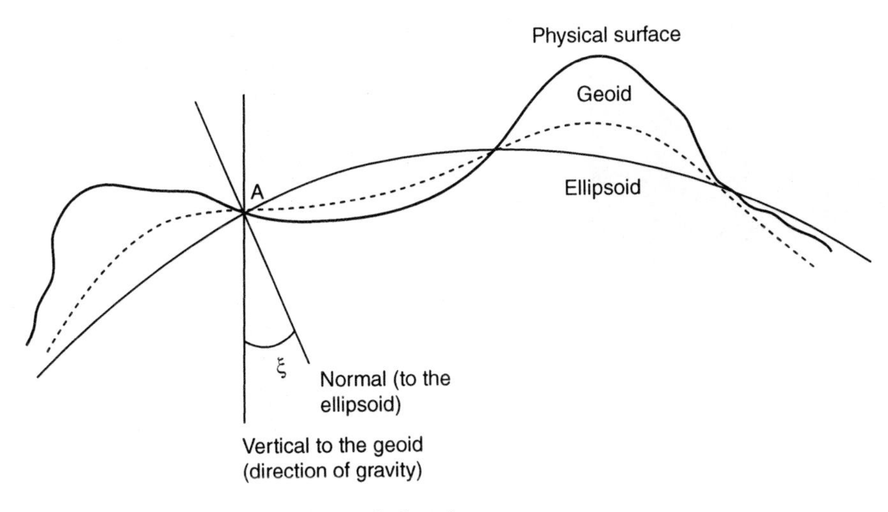

Fig. 1.1 *Geoid, ellipsoid and physical surface*

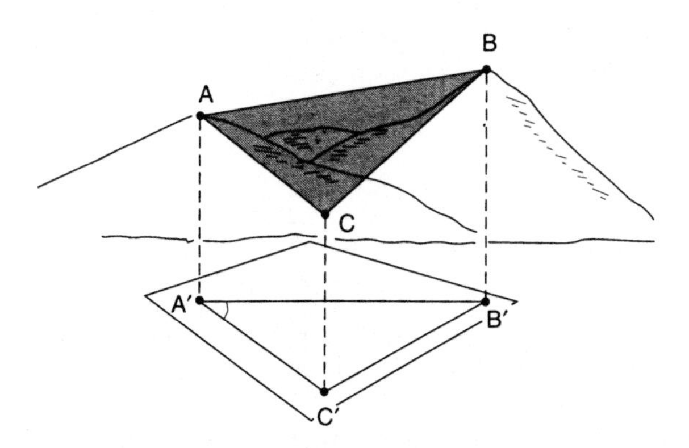

Fig. 1.2 *Projection onto a plane surface*

An examination of *Figure 1.2* clearly shows the basic surveying measurements needed to locate points A, B and C and plot them orthogonally as A', B' and C'. Assuming the direction of B from A is known then the measured *slope distance* AB and the *vertical angle* to B from A will be needed to fix the position of B relative to A. The *vertical angle* to B from A is needed to reduce the slope distance AB to its equivalent horizontal distance $A'B'$ for the purposes of plotting. Whilst similar measurements will fix C relative to A, it also requires the *horizontal angle* at A measured from B to C ($B'A'C'$) to fix C relative to B. The *vertical distances* defining the relative elevation of the three points may also be obtained from the slope distance and vertical angle or by direct levelling (Chapter 3) relative to a specific reference datum. The five measurements mentioned above comprise the basis of plane surveying and are illustrated in *Figure 1.3*, i.e. AB is the slope distance, AA' the horizontal distance, $A'B$ the vertical distance, BAA' the vertical angle (α) and $A'AC$ the horizontal angle (θ).

It can be seen from the above that the only measurements needed in plane surveying are angle and distance. Nevertheless, the full impact of modern technology has been brought to bear in the acquisition and processing of this simple data. Angles may now be resolved with single-second accuracy using optical and electronic theodolites; electromagnetic distance measuring (EDM) equipment can obtain distances up

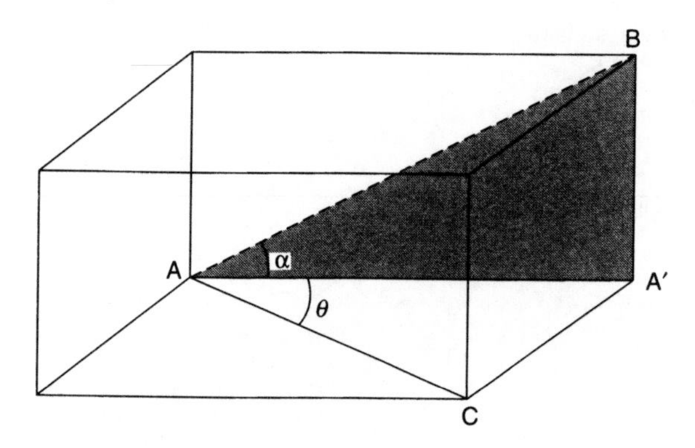

Fig. 1.3 *Basic measurements*

to several kilometres with millimetre precision, depending on the distance measured; lasers and north-seeking gyroscopes are virtually standard equipment for tunnel surveys; orbiting satellites are being used for position fixing offshore as well as on; continued improvement in aerial and terrestrial photogrammetric and scanning equipment makes mass data capture technology an invaluable surveying tool; finally, data loggers and computers enable the most sophisticated procedures to be adopted in the processing and automatic plotting of field data.

1.4 CONTROL NETWORKS

The establishment of two- or three-dimensional control networks is the most fundamental operation in the surveying of large or small areas of land. Control networks comprise a series of points or positions which are spatially located for the purpose of topographic surveying, for the control of supplementary points, or dimensional control on site.

The process involved in carrying out the surveying of an area, the capture and processing of the field data, and the subsequent production of a plan or map, will now be outlined briefly.

The first and obvious step is to know the purpose and nature of the project for which the surveys are required in order to assess the accuracy specifications, the type of equipment required and the surveying processes involved.

For example, a major construction project may require structures, etc., to be set out to subcentimetre accuracy, in which case the control surveys will be required to an even greater accuracy. Earthwork volumes may be estimated from the final plans, hence contours made need to be plotted at 2 m intervals or less. If a plan scale of 1/500 is adopted, then a plotting accuracy of 0.5 mm would represent 0.25 m on the ground, thus indicating the accuracy of the final process of topographic surveying from the supplementary control and implying major control to a greater accuracy. The location of topographic data may be done using total stations, GPS satellites, or, depending on the extent of the area, aerial photogrammetry. The cost of a photogrammetric survey is closely linked to the contour interval required and the extent of the area. Thus, the accuracy of the control network would define the quality of the equipment and the number of observations required.

The duration of the project will affect the design of survey stations required for the control points. A project of limited duration may only require a long, stout wooden peg, driven well into solid, reliable ground and surrounded by a small amount of concrete. A fine nail in the top defines the geometrical position to be located. A survey control point designed to be of longer duration is illustrated in Chapter 6, *Figure 6.15*.

The next stage of the process is a detailed reconnaissance of the area in order to establish the best positions for the control points.

Initially, data from all possible sources should be studied before venturing into the field. Such data would comprise existing maps and plans, aerial photographs and any previous surveying data of that area. Longitudinal sections may be drawn from the map contours to ensure that lines of sight between control points are well above ground level and so free of shimmer or refraction effects. If the surveys are to be connected into the national surveys of the country (Ordnance Survey National Grid in the UK), then the position of as many national survey points as possible, such as (in the UK) those on the GPS Active or Passive Network, should be located. These studies, very often referred to as the 'paper survey', should then be followed up with a detailed field reconnaissance.

This latter process locates all existing control in the area of interest, both local and national, and establishes the final positions for all the new control required. These final positions should be chosen to ensure clear, uninterrupted lines of sight and the best observing positions. The location of these points, and the type of terrain involved, would then influence the method of survey to be used to locate their spatial position.

Figure 1.4 indicates control points $A, B, \ldots, F$, in the area to be surveyed. It is required to obtain the coordinate positions of each point. This could be done using any of the following methods:

(a) Intersection or resection
(b) Traversing
(c) Networks
(d) GPS satellites

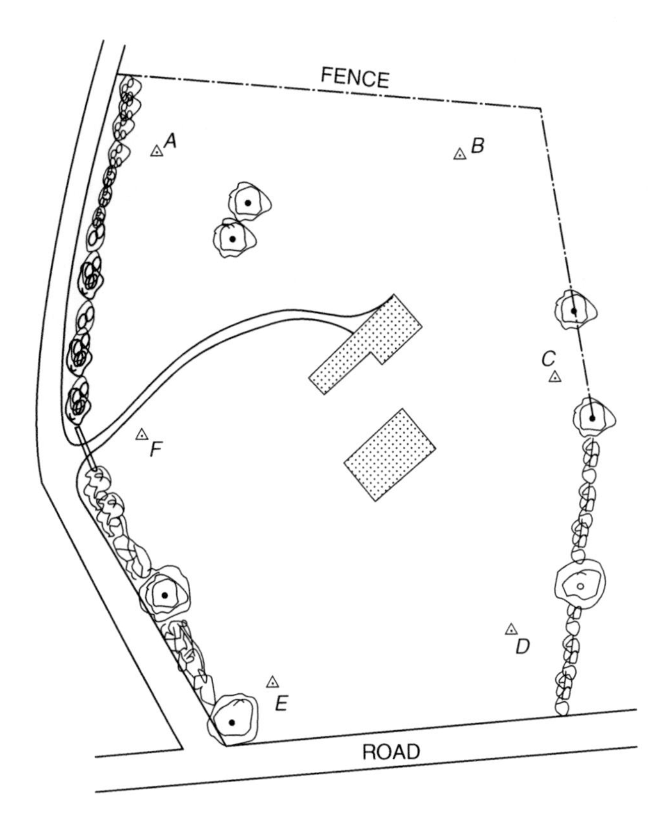

Fig. 1.4 *Control points*

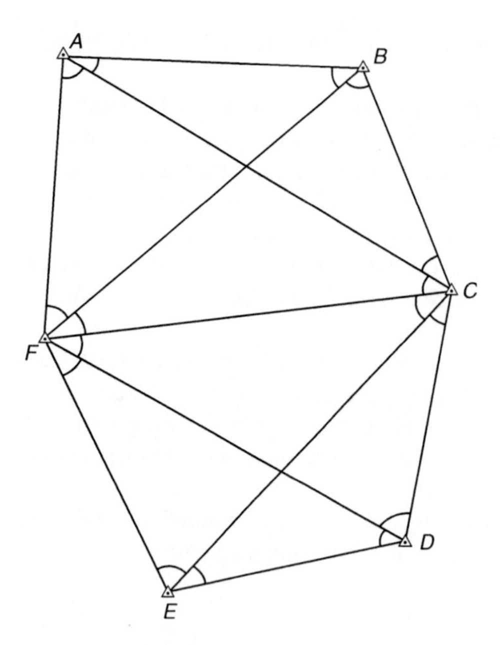

Fig. 1.5 *A network*

Figure 1.5 illustrates possible lines of sight. All the horizontal angles shown would be measured to the required accuracy to give the shape of the network. At least one side would need to be measured, say *AB*, to give the scale or size of the network. By measuring a check baseline, say *ED*, and comparing it with its value, computed through the network, the scale error could be assessed. This form of survey is classical triangulation and although forming the basis of the national maps of many countries, is now regarded as obsolete because of the need for lines of sight between adjacent points. Such control would now be done with GPS.

If the lengths of all the sides were measured in the same triangular configuration without the angles, this technique would be called 'trilateration'. Although giving excellent control over scale error, swing errors may occur. For local precise control surveys the modern practice therefore is to use a combination of angles and distance. Measuring every angle and every distance, including check sights wherever possible, would give a very strong network indeed. Using sophisticated least squares software it is now possible to optimize the observation process to achieve the accuracies required.

Probably the most favoured simple method of locating the relative coordinate positions of control points in engineering and construction is traversing. *Figure 1.6* illustrates the method of traversing to locate the same control points *A* to *F*. All the adjacent horizontal angles and distances are measured to the accuracies required, resulting in much less data needed to obtain coordinate accuracies comparable with the previous methods. Also illustrated are minor or supplementary control points *a*, *b*, *c*, *d*, located with lesser accuracy by means of a link traverse. The field data comprises all the angles as shown plus the horizontal distances *Aa*, *ab*, *bc*, *cd*, and *dB*. The rectangular coordinates of all these points would, of course, be relative to the major control.

Whilst the methods illustrated above would largely supply a two-dimensional coordinate position, GPS satellites could be used to provide a three-dimensional position.

All these methods, including the computational processes, are dealt with in later chapters of this book.

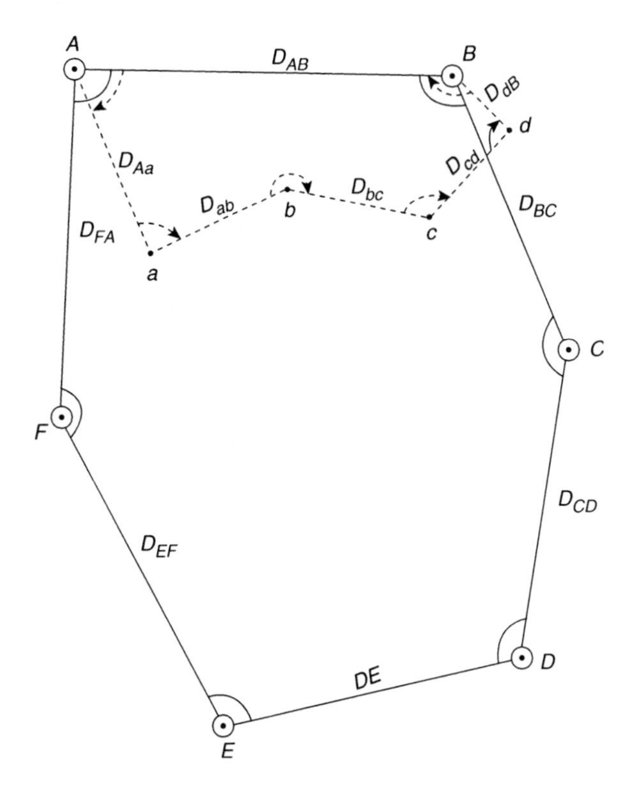

Fig. 1.6 *Major traverse A to F. Minor link traverse A to B*

1.5 LOCATING POSITION

Establishing control networks and their subsequent computation leads to an implied rectangular coordinate system over the area surveyed. The minor control points in particular can now be used to position topographic data and control the setting out of a construction design.

The methods of topographic survey and dimensional control will most probably be:

(a) by polar coordinates (distance and bearing) using a total station; or
(b) by GPS using kinematic methods.

Considering method (a), a total station would be set up over a control point whose coordinates are known as '*a*', and back-sighted to another control point '*b*' whose coordinates are also known. Depending on the software on board, the coordinates may be keyed into the total station. Alternatively, the bearing of the line '*ab*', computed from the coordinates, may be keyed in. Assuming the topographic position of a road is required, the total station would be sighted to a corner cube prism fixed to a detail pole held vertically at (P_1), the edge of the road as shown in *Figures 1.7* and *1.8*. The field data would comprise the horizontal angle $baP_1(\alpha_1)$ and the horizontal distance D_1 (*Figure 1.8*). Depending on the software being used, the angle α_1 would be used to compute the bearing aP_1 relative to '*ab*' and, with the horizontal distance D_1, compute the rectangular coordinates of P_1 in the established coordinate system. This process is repeated for the remaining points defining the road edge and any other topographic data within range of the total station. The whole area would be surveyed in this way by occupying pairs of control points situated throughout the area.

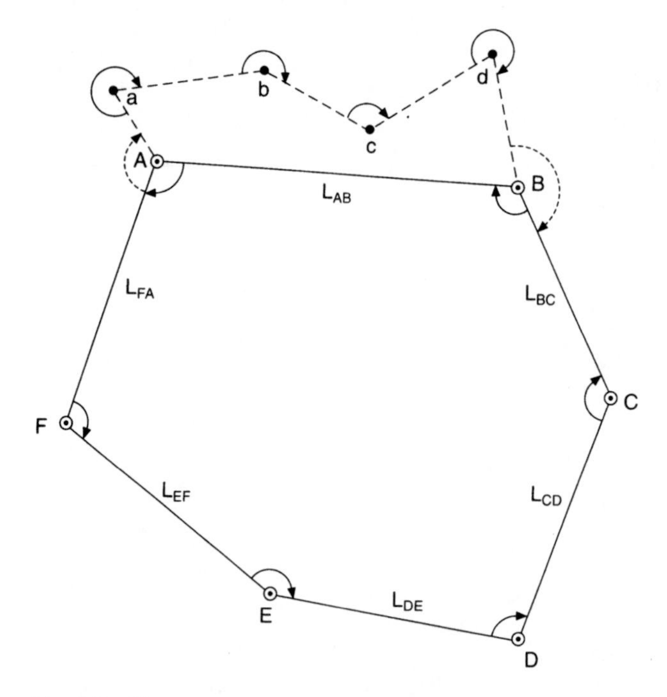

Fig. 1.7 *Traverse*

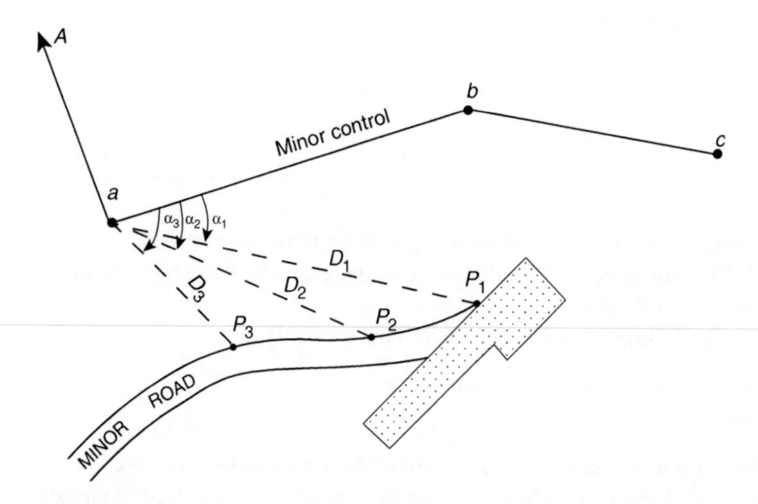

Fig. 1.8 *Detailing by polar coordinates*

For method (b), using GPS equipment, the methods are dealt with in detail in Chapter 9: Satellite positioning.

A further development is the integration of a total station with GPS. This instrument, produced by Leica Geosystems and called SmartStation, provides the advantages of both the systems (a) and (b).

If using existing control, the local coordinate data is copied into the SmartStation and used in the usual manner on existing pairs of control points. Where the GPS cannot be used because of excessive tree cover for instance, then the total station may be used in the usual way.

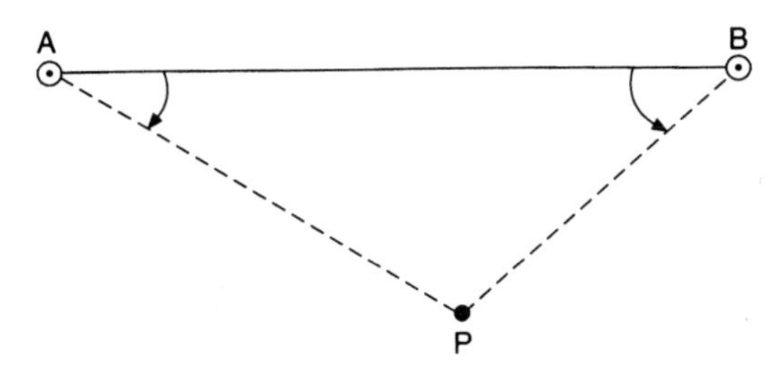

Fig. 1.9 *Intersection*

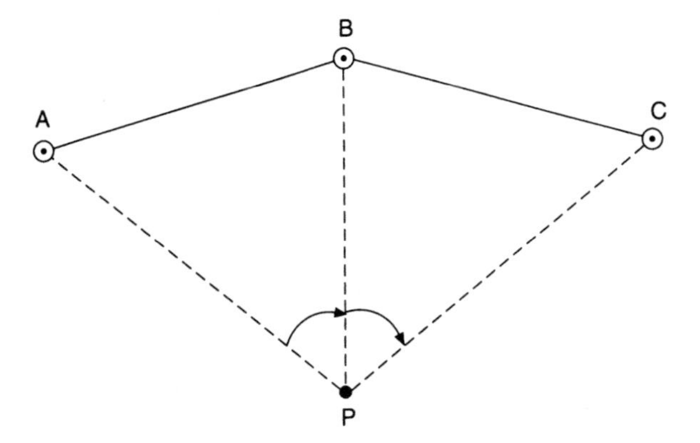

Fig. 1.10 *Resection*

Perhaps the most significant aspect of this instrument is that pairs of points for orientation as well as position could be established by GPS thereby eliminating the need to establish a prior control network, with great savings on time and money.

Alternative methods used very often for the location of single points are intersection, where *P* is fixed by measuring the horizontal angles *BAP* and *PBA* as shown in *Figure 1.9*, and resection (*Figure 1.10*). This method forms the basis of triangulation. Similarly, *P* could be fixed by the measurement of horizontal distances *AP* and *BP*, and forms the basis of the method of trilateration. In both these instances there is no independent check, as a position for *P* (not necessarily the correct one) will always be obtained. Thus at least one additional measurement is required, either by combining the angles and distances, by measuring the angle at *P* as a check on the angular intersection, or by producing a trisection from an extra control station.

Resection (*Figure 1.10*) is done by observing the horizontal angles at *P* to at least three control stations of known position. The position of *P* may be obtained by a mathematical solution as illustrated in Chapter 6.

It can be seen that all the above procedures simply involve the measurement of angles and distances.

1.6 PLOTTING DETAIL

In the past, detail would have been plotted on paper or a more stable medium such as plastic film. However, today all practical 'plotting' is computer based and there is now an abundance of computer plotting software available that will not only produce a contour plot but also supply three-dimensional views, digital ground models, earthwork volumes, road design, drainage design, perspective views, etc.

1.6.1 Computer systems

To be economically viable, practically all major engineering/surveying organizations use an automated plotting system. Very often the total station and data logger are purchased along with the computer hardware and software as a total operating system. In this way interface and adaptation problems are precluded. *Figure 1.11* shows a computer driven plotter which is networked to the system and located separately.

The essential characteristics of such a system are:

(1) Capability to accept, store, transfer, process and manage field data that is input manually or directly from an interfaced data logger (*Figure 1.12*).
(2) Software and hardware are in modular form for easy access.
(3) Software will use all modern facilities, such as 'windows', different colour and interactive screen graphics, to make the process user friendly.
(4) Continuous data flow from field data to finished plan.
(5) Appropriate database facility for the storage and management of coordinate and cartographic data necessary for the production of DGMs and land/geographic information systems.
(6) Extensive computer storage facility.
(7) High-speed precision flat-bed or drum plotter.

To be truly economical, the field data, including appropriate coding of the various types of detail, should be captured and stored by single-key operation, on a data logger interfaced to a total station. The computer

Fig. 1.11 *Computer driven plotter*

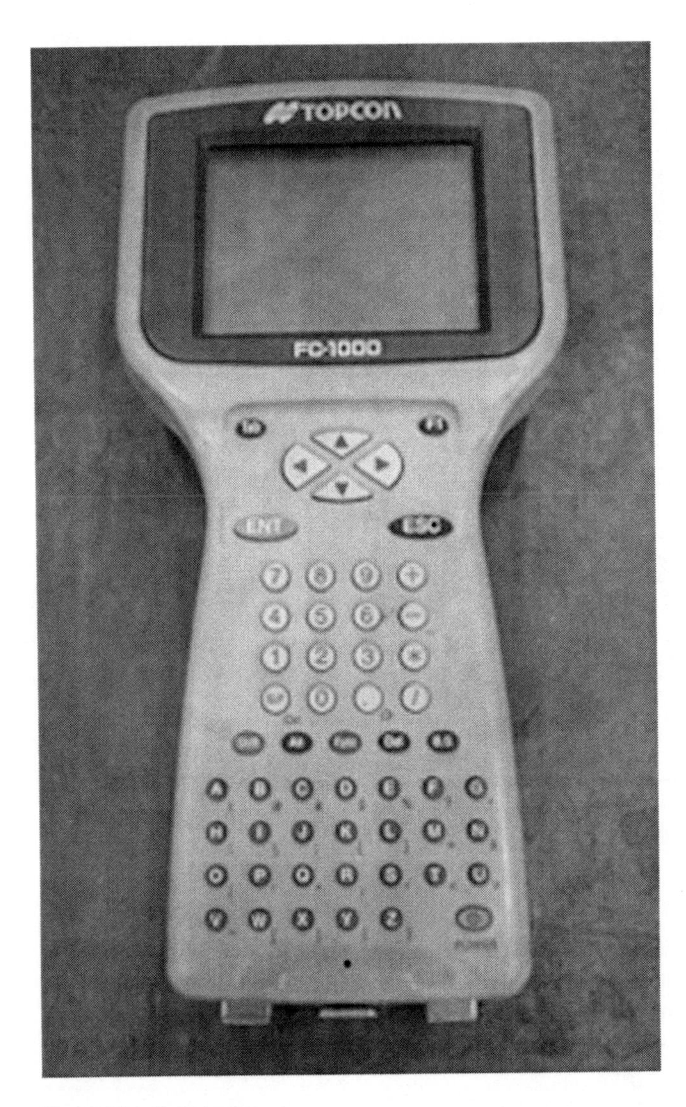

Fig. 1.12 *Data logger*

system should then permit automatic transfer of this data by direct interface between the logger and the system. The software should then: store and administer the data; carry out the mathematical processing, such as network adjustment, produce coordinates and elevations; generate data storage banks; and finally plot the data on completion of the data verification process.

Prior to plotting, the data can be viewed on the screen for editing purposes. This can be done from the keyboard or touch screen using interactive graphics routines. The plotted detail can be examined, moved, erased or changed as desired. When the examination is complete, the command to plot may then be activated. *Figure 1.13* shows an example of a computer plot.

1.6.2 Digital ground model (DGM)

A DGM is a three-dimensional, mathematical representation of the landform and all its features, stored in a computer database. Such a model is extremely useful in the design and construction process, as it permits

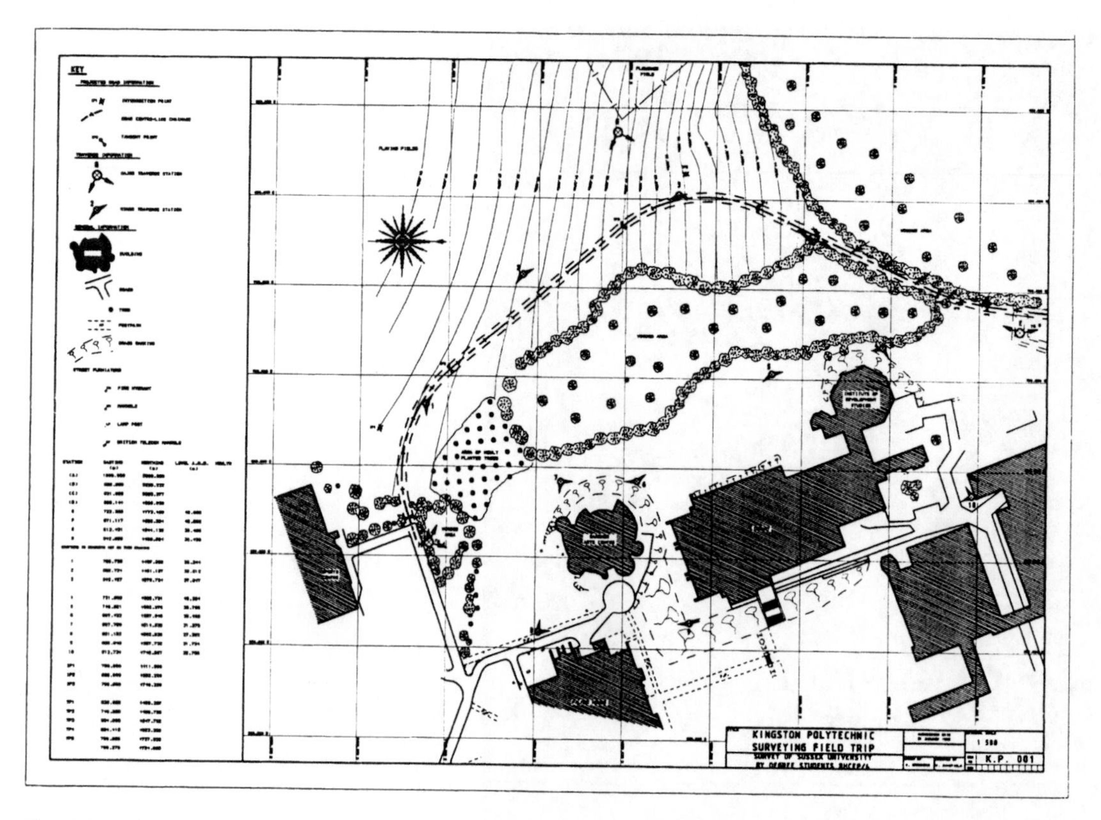

Fig. 1.13 *Computer plot*

quick and accurate determination of the coordinates and elevation of any point. The DGM is formed by sampling points over the land surface and using appropriate algorithms to process these points to represent the surface being modelled. The methods in common use are modelling by 'strings', 'regular grids' or 'triangulated irregular networks'. Regardless of the methods used, they will all reflect the quality of the field data.

A 'string' comprises a series of points along a feature and so such a system stores the position of features surveyed. The system is widely used for mapping purposes due to its flexibility, accuracy along the string and the ability to process large amounts of data very quickly. However, as the system does not store the relationship between strings, a searching process is essential when the levels of points not included in a string are required. Thus the system's weakness lies in the generation of accurate contours and volumes.

The 'regular grid' method uses appropriate algorithms to convert the sampled data to a regular grid of levels. If the field data permits, the smaller the grid interval, the more representative of landform it becomes. Although a simple technique, it only provides a very general shape of the landform, due to its tendency to ignore vertical breaks of slope. Volumes generated also tend to be rather inaccurate.

In the 'triangulated irregular networks' (TIN) method, 'best fit' triangles are formed between the points surveyed. The ground surface therefore comprises a network of triangular planes at various inclinations (*Figure 1.14*(*a*)). Computer shading of the model (*Figure 1.14*(*b*)) provides an excellent indication of the landform. In this method vertical breaks are forced to form the sides of triangles, thereby maintaining correct ground shape. Contours, sections and levels may be obtained by linear interpolation through

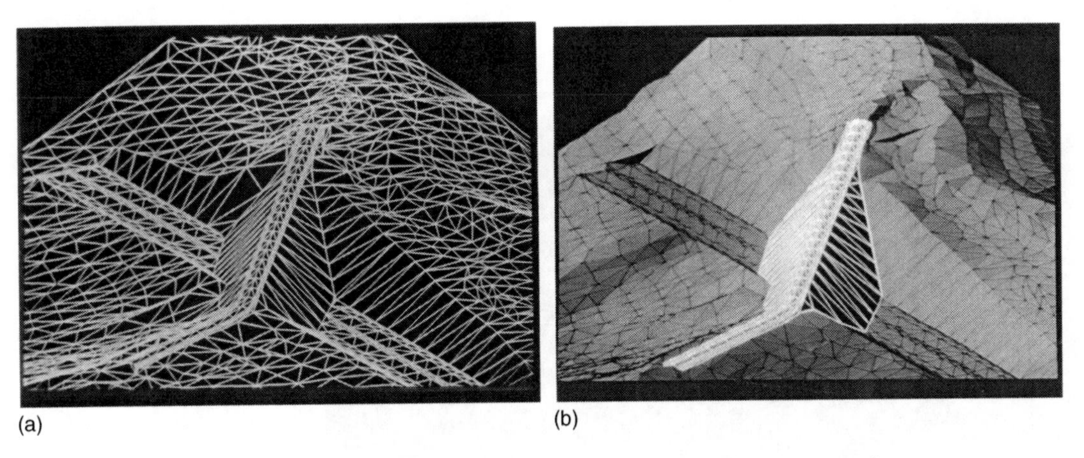

Fig. 1.14 *(a) Triangular grid model, and (b) Triangular grid model with computer shading*

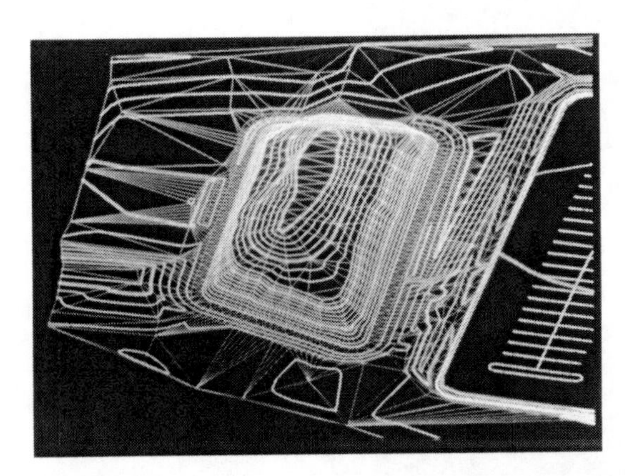

Fig. 1.15 *Computer generated contour model*

the triangles. It is thus ideal for contour generation (*Figure 1.15*) and computing highly accurate volumes. The volumes may be obtained by treating each triangle as a prism to the depth required; hence the smaller the triangle, the more accurate the final result.

1.6.3 Computer-aided design (CAD)

In addition to the production of DGMs and contoured plans, a computer-based surveying system permits the finished plan to be easily related to the designed structure. The three-dimensional information held in the database supplies all the ground data necessary to facilitate the finished design. *Figure 1.16* illustrates its use in road design.

The environmental impact of the design can now be more readily assessed by producing perspective views as shown in *Figure 1.17*. Environmental impact legislation makes this latter tool extremely valuable.

Fig. 1.16 *Computer aided road design – courtesy of ISP (Integrated Software Products)*

Fig. 1.17 *Perspectives with computer shading – courtesy of ISP (Integrated Software Products)*

1.7 SUMMARY

In the preceding sections the basic concepts of surveying have been outlined. Because of their importance they will now be summarized as follows:

(1) *Reconnaissance* is the first and most important step in the surveying process. Only after a careful and detailed reconnaissance of the area can the surveyor decide upon the techniques and instrumentation required to complete the work economically and meet the accuracy specifications.

(2) *Control networks* not only form a reference framework for locating the position of topographic detail and setting out constructions, but may also be used for establishing minor control networks containing a greater number of control stations at shorter distances apart and to a lower order of accuracy, i.e. *a, b, c, d* in *Figure 1.6*. These minor control stations may be better placed for the purpose of locating the topographic detail.

This process of establishing the major control to the highest order of accuracy, as a framework on which to connect the minor control, which is in turn used as a reference framework for detailing, is known as *working from the whole to the part* and forms the basis of all good surveying procedure.

(3) *Errors* are contained in all measurement procedures and a constant battle must be waged by the surveyor to minimize and evaluate their effect.

It follows from this that the greater the accuracy specifications the greater the cost of the survey because it results in more observations, taken with greater care, over a longer period of time and using more precise (and therefore more expensive) equipment. It is for this reason that major control networks contain the minimum number of stations necessary and surveyors adhere to the economic principle of working to an accuracy neither greater than nor less than that required.

(4) *Independent checks* should be introduced not only into the field work, but also into the subsequent computation and reduction of field data. In this way, errors can be quickly recognized and dealt with. Data should always be measured more than once and preferably in different ways. Examination of several measurements will generally indicate if there are blunders in the measuring process. Alternatively, close agreement of the measurements is indicative of high precision and generally acceptable field data, although, as shown later, high precision does not necessarily mean high accuracy, and further data processing may be necessary to remove any systematic error that may be present.

(5) *Commensurate accuracy* is advised in the measuring process, i.e. the angles should be measured to the same degree of accuracy as the distances and vice versa. For guidance: 1″ of arc subtends 1 mm at 200 m. This means that if distance is measured to, say, 1 in 200 000, the angles should be measured to 1″ of arc, and so on.

In the majority of engineering projects, sophisticated instrumentation such as 'total stations' interfaced with electronic data recording is the norm. In some cases the recorded data can be used to produce screen plots in real time.

GPS and other satellite systems are used to fix three-dimensional position. Such is the accuracy and speed of positioning using satellites that they may be used to establish control points, fix topographic detail, set out position on site and carry out continuous deformation monitoring. However, they cannot be used to solve every positioning problem and conventional survey techniques continue to have about equal importance.

However, regardless of the technological advances in surveying, attention must always be given to instrument calibration and checking, carefully designed projects and meticulous observation. As surveying is essentially the science of measurement, it is necessary to examine the measured data in more detail, as discussed in the next chapter.

2

Error and uncertainty

In surveying nothing is ever absolutely certain.

The product of surveying may be thought of as being in two parts, that is, the derivation of the desired quantities such as coordinates of, or distances between, points, and the assessment and management of the uncertainty of those quantities. In other words not only must the survey results be produced, but there should be numerical statements of the quality of the results for them to be meaningful.

Survey results can never be exactly true for a number of reasons. Surveying equipment, like any other piece of equipment in the real world can only be manufactured to a certain level of precision. This means that there is a limit upon the quality of a measurement that can be made by any instrument. Although survey measuring procedures are designed to remove as many errors as possible there will always be some sources of error that cannot be compensated for. Whatever the scale on the instrument, be it digital or analogue, there is a limit to the number of significant digits that it can be read to. Surveyors are trained to get the most out of their instrumentation, but no observer can make perfect measurements. There is a limit to the steadiness of the hand and the acuity of the eye. All survey measurements are subject to external factors, for example all observed angles are subject to the effects of refraction, and observed distances, whether EDM or tape, will vary with temperature. The process of getting from observations to coordinates involves reductions of, and corrections to, observed data. Some mathematical formulae are rigorous, others are approximate. These approximations and any rounding errors in the computations will add further error to the computed survey results.

The surveyor's task is to understand the source and nature of the errors in the survey work and appreciate how the observing methods and the computing process may be designed to minimize and quantify them. It is important to understand the nature of the measurement process. Firstly, the units in which the measurement is to take place must be defined, for example distances may be measured in metres or feet and angles may be in degrees, gons or mils. Next, the operation of comparing the measuring device with the quantity to be measured must be carried out, for example laying a tape on the ground between two survey stations. A numerical value in terms of the adopted units of measure is then allocated to the measured quantity. In one of the examples already quoted the readings of the tape at each station are taken and the difference between them is the allocated numerical value of the distance between the stations. The important point is that the true value of the interstation distance is never known, it can only be estimated by an observational and mathematical process.

Since the true value of a measurement or coordinate can never be known it is legitimate to ask what is the accuracy or the precision of, or the error in, the estimate of that measurement or coordinate. *Accuracy*, *precision* and *error* have specific meanings in the context of surveying. Accuracy is a measure of reliability. In other words

$$\text{Accuracy} = \text{True value} - \text{Most probable value}$$

where the 'most probable value' is derived from a set of measurements. In the example above the most probable value might be the arithmetic mean of a number of independent measurements. Since the true value is never known then it is also impossible for the accuracy to be known. It can only be estimated.

Accuracy can be estimated from 'residuals', for example, in the two sets of measurements below, which mean is the more accurate, that of the measurements of line *AB* or line *XY*?

Line	*AB*		*XY*	
	measure	*residuals*	*measure*	*residuals*
	25.34 m	+0.02 m	25.31 m	−0.01 m
	25.49 m	+0.17 m	25.33 m	+0.01 m
	25.12 m	−0.20 m	25.32 m	0.00 m
	25.61 m	+0.29 m	25.33 m	+0.01 m
	25.04 m	−0.28 m	25.31 m	−0.01 m
Mean	25.32 m		25.32 m	

The residuals in this case are differences between the individual observations and the best estimate of the distance, that is the arithmetic mean. It is clear from inspection of the two sets of residuals that the length of line *XY* appears to be more accurately determined than that of line *AB*.

Precision is a measure of repeatability. Small residuals indicate high precision, so the mean of line *XY* is more precisely determined than the mean of line *AB*. High precision does not necessarily indicate high accuracy. For example, if the tape used to measure line *XY* was in decimals of a yard and the surveyor assumed it was in metres, then the computed mean of line *XY* would be very precise but also very inaccurate. In general

Precision > Accuracy

but in practice the computed precision is often taken as the assessed accuracy.

Coordinates and their accuracy and precision may be stated as being 'relative' or 'absolute'. Absolute values are with respect to some previously defined datum. Relative values are those with respect to another station. For example, the Ordnance Survey (OS) coordinates of a GPS passive network station might be assumed to be absolute coordinates since they are with respect to the OSTN02 datum of UK. The coordinates of a new control station on a construction site may have been determined by a series of observations including some to the GPS station. The precision of the coordinates of the new station may better be expressed with respect to the OSTN02 datum, or alternatively with respect to the coordinates of another survey station on site. In the former case they may be considered as absolute and in the latter as relative. The difference between absolute and relative precisions is largely one of local definition and therefore of convenience. In general

Relative precision > Absolute precision

Accuracy and precision are usually quoted as a ratio, or as parts per million, e.g. 1:100 000 or 10 ppm, or in units of the quantity measured, e.g. 0.03 m.

Error is the difference between an actual true valve and an estimate of that true value. If the estimate is a bad one, then the error will be large.

Of these three concepts, accuracy, precision and error, only precision may be numerically defined from appropriate computations with the observations. Accuracy and error may be assumed, sometimes erroneously, from the precision but they will never be known for sure. The best estimate of accuracy is usually the precision but it will usually be overoptimistic.

2.1 UNITS OF MEASUREMENT

The system most commonly used for the definition of units of measurement, for example of distance and angle, is the 'Système Internationale', abbreviated to SI. The basic units of prime interest are:

Length in metres (m)

from which we have:

$$1 \text{ m} = 10^3 \text{ millimetres (mm)}$$
$$1 \text{ m} = 10^{-3} \text{ kilometres (km)}$$

Thus a distance measured to the nearest millimetre would be written as, say, 142.356 m.

Similarly for areas we have:

$$1 \text{ m}^2 = 10^6 \text{ mm}^2$$
$$10^4 \text{ m}^2 = 1 \text{ hectare (ha)}$$
$$10^6 \text{ m}^2 = 1 \text{ square kilometre (km}^2\text{)}$$

and for volumes, m^3 and mm^3.

There are three systems used for plane angles, namely the sexagesimal, the centesimal and radians (arc units).

The sexagesimal units are used in many parts of the world, including the UK, and measure angles in degrees (°), minutes (′) and seconds (″) of arc, i.e.

$$1° = 60'$$
$$1' = 60''$$

and an angle is written as, say, 125° 46′ 35″.

The centesimal system is quite common in Europe and measures angles in gons (g), i.e.

$$1 \text{ gon} = 100 \text{ cgon (centigon)}$$
$$1 \text{ cgon} = 10 \text{ mgon (milligon)}$$

A radian is that angle subtended at the centre of a circle by an arc on the circumference equal in length to the radius of the circle, i.e.

$$2\pi \text{ rad} = 360° = 400 \text{ gon}$$

Thus to transform degrees to radians, multiply by $\pi/180°$, and to transform radians to degrees, multiply by $180°/\pi$. It can be seen that:

$$1 \text{ rad} = 57.2957795° \ (= 57° \ 17' \ 44.8'') = 63.6619972 \text{ gon}$$

A factor commonly used in surveying to change angles from seconds of arc to radians is:

$$\alpha \text{ rad} = \alpha''/206\,265$$

where 206 265 is the number of seconds in a radian.

Other units of interest will be dealt with where they occur in the text.

2.2 SIGNIFICANT FIGURES

Engineers and surveyors communicate a great deal of their professional information using numbers. It is important, therefore, that the number of digits used, correctly indicates the accuracy with which the field data were measured. This is particularly important since the advent of pocket calculators, which tend to present numbers to as many as eight places of decimals, calculated from data containing, at the most, only three places of decimals, whilst some eliminate all trailing zeros. This latter point is important, as 2.00 m is an entirely different value to 2.000 m. The latter number implies estimation to the nearest millimetre as opposed to the nearest 10 mm implied by the former. Thus in the capture of field data, the correct number of significant figures should be used.

By definition, the number of significant figures in a value is the number of digits one is certain of plus one, usually the last, which is estimated. The number of significant figures should not be confused with the number of decimal places. A further rule in significant figures is that in all numbers less than unity, the zeros directly after the decimal point and up to the first non-zero digit are not counted. For example:

Two significant figures: 40, 42, 4.2, 0.43, 0.0042, 0.040

Three significant figures: 836, 83.6, 80.6, 0.806, 0.0806, 0.00800

Difficulties can occur with zeros at the end of a number such as 83 600, which may have three, four or five significant figures. This problem is overcome by expressing the value in powers of ten, i.e. 8.36×10^4 implies three significant figures, 8.360×10^4 implies four significant figures and 8.3600×10^4 implies five significant figures.

It is important to remember that the accuracy of field data cannot be improved by the computational processes to which it is subjected.

Consider the addition of the following numbers:

155.486
7.08
2183.0
42.0058

If added on a pocket calculator the answer is 2387.5718; however, the correct answer with due regard to significant figures is 2387.6. It is rounded off to the most extreme right-hand column containing all the significant figures, which in the example is the column immediately after the decimal point. In the case of $155.486 + 7.08 + 2183 + 42.0058$ the answer should be 2388. This rule also applies to subtraction.

In multiplication and division, the answer should be rounded off to the number of significant figures contained in that number having the least number of significant figures in the computational process. For instance, $214.8432 \times 3.05 = 655.27176$, when computed on a pocket calculator; however, as 3.05 contains only three significant figures, the correct answer is 655. Consider $428.4 \times 621.8 = 266379.12$, which should now be rounded to $266\,400 = 2.664 \times 10^5$, which has four significant figures. Similarly, $41.8 \div 2.1316 = 19.609683$ on a pocket calculator and should be rounded to 19.6.

When dealing with the powers of numbers the following rule is useful. If x is the value of the first significant figure in a number having n significant figures, its pth power is rounded to:

$n - 1$ significant figures if $p \leq x$

$n - 2$ significant figures if $p \leq 10x$

For example, $1.5831^4 = 6.28106656$ when computed on a pocket calculator. In this case $x = 1, p = 4$ and $p \leq 10x$; therefore, the answer should be quoted to $n - 2 = 3$ significant figures $= 6.28$.

Similarly, with roots of numbers, let x equal the first significant figure and r the root; the answer should be rounded to:

n significant figures when $rx \geq 10$

$n - 1$ significant figures when $rx < 10$

For example:

$36^{\frac{1}{2}} = 6$, because $r = 2$, $x = 3$, $n = 2$, thus $rx < 10$, and answer is to $n - 1 = 1$ significant figure.
$415.36^{\frac{1}{4}} = 4.5144637$ on a pocket calculator; however, $r = 4$, $x = 4$, $n = 5$, and as $rx > 10$, the answer is rounded to $n = 5$ significant figures, giving 4.5145.

As a general rule, when field data are undergoing computational processing which involves several intermediate stages, one extra digit may be carried throughout the process, provided the final answer is rounded to the correct number of significant figures.

2.3 ROUNDING NUMBERS

It is well understood that in rounding numbers, 54.334 would be rounded to 54.33, whilst 54.336 would become 54.34. However, with 54.335, some individuals always round up, giving 54.34, whilst others always round down to 54.33. Either process creates a systematic bias and should be avoided. The process which creates a more random bias, thereby producing a more representative mean value from a set of data, is to round to the nearest even digit. Using this approach, 54.335 becomes 54.34, whilst 54.345 is 54.34 also.

2.4 ERRORS IN MEASUREMENT

It should now be apparent that position fixing simply involves the measurement of angles and distance. However, all measurements, no matter how carefully executed, will contain error, and so the true value of a measurement is never known. It follows from this that if the true value is never known, the true error can never be known and the position of a point known only with a certain level of uncertainty.

The sources of error fall into three broad categories, namely:

(1) Natural errors caused by variation in or adverse weather conditions, refraction, unmodelled gravity effects, etc.
(2) Instrumental errors caused by imperfect construction and adjustment of the surveying instruments used.
(3) Personal errors caused by the inability of the individual to make exact observations due to the limitations of human sight, touch and hearing.

2.4.1 Classification of errors

(1) Mistakes are sometimes called gross errors, but should not be classified as errors at all. They are blunders, often resulting from fatigue or the inexperience of the surveyor. Typical examples are omitting a whole tape length when measuring distance, sighting the wrong target in a round of angles, reading '6' on a levelling staff as '9' and vice versa. Mistakes are the largest of the errors likely to arise, and therefore great care must be taken to obviate them. However, because they are large they are easy to spot and so deal with.

(2) Systematic errors can be constant or variable throughout an operation and are generally attributable to known circumstances. The value of these errors may often be calculated and applied as a correction to the measured quantity. They can be the result of natural conditions, examples of which are: refraction of light rays, variation in the speed of electromagnetic waves through the atmosphere, expansion or contraction of steel tapes due to temperature variations. In all these cases, corrections can be applied to reduce their effect. Such errors may also be produced by instruments, e.g. maladjustment of the theodolite or level, index error in spring balances, ageing of the crystals in EDM equipment. One form of systematic error is the constant error, which is always there irrespective of the size of the measurement or the observing conditions. Examples of constant errors in tape measurement might be due to a break and join in the tape or tape stretch in the first metre of the tape. In this case the remedy is to ensure that the tape is undamaged and also not to use the first metre of the tape. Examples of constant errors in the repeated observations of a horizontal angle with a theodolite to elevated targets might be miscentring over the station or dislevelment of the theodolite. In this case, the remedy is to ensure that the theodolite is correctly set up.

There is the personal error of the observer who may have a bias against setting a micrometer or in bisecting a target, etc. Such errors can frequently be self-compensating; for instance, a person observing a horizontal angle to a cylindrical target subject to phase, the apparent biased illumination by the sun when shining on one side will be subject to a similar bias on a similar target nearby and so the computed angle between will be substantially correct.

Systematic errors, in the main, conform to mathematical and physical laws; thus it is argued that appropriate corrections can be computed and applied to reduce their effect. It is doubtful, however, whether the effect of systematic errors is ever entirely eliminated, largely due to the inability to obtain an exact measurement of the quantities involved. Typical examples are: the difficulty of obtaining group refractive index throughout the measuring path of EDM distances; and the difficulty of obtaining the temperature of the steel tape, based on air temperature measurements with thermometers. Thus, systematic errors are the most difficult to deal with and therefore they require very careful consideration prior to, during, and after the survey. Careful calibration of all equipment is an essential part of controlling systematic error.

(3) Random errors are those variates which remain after all other errors have been removed. They are beyond the control of the observer and result from the human inability of the observer to make exact measurements, for reasons already indicated above.

Random errors should be small and there is no procedure that will compensate for or reduce any one single error. The size and sign of any random error is quite unpredictable. Although the behaviour of any one observation is unpredictable the behaviour of a group of random errors is predictable and the larger the group the more predictable is its behaviour. This is the basis of much of the quality assessment of survey products.

Random variates are assumed to have a continuous frequency distribution called normal distribution and obey the law of probability. A random variate, x, which is normally distributed with a mean and standard deviation, is written in symbol form as $N(\mu, \sigma^2)$. Random errors alone are treated by statistical processes.

2.4.2 Basic concept of errors

The basic concept of errors in the data captured by the surveyor may be likened to target shooting.

In the first instance, let us assume that a skilled marksman used a rifle with a bent sight, which resulted in his shooting producing a scatter of shots as at A in *Figure 2.1*.

That the marksman is skilled (or consistent) is evidenced by the very small scatter, which illustrates excellent precision. However, as the shots are far from the centre, caused by the bent sight (systematic error), they are completely inaccurate. Such a situation can arise in practice when a piece of EDM equipment produces a set of measurements all agreeing to within a few millimetres (high precision) but, due to an

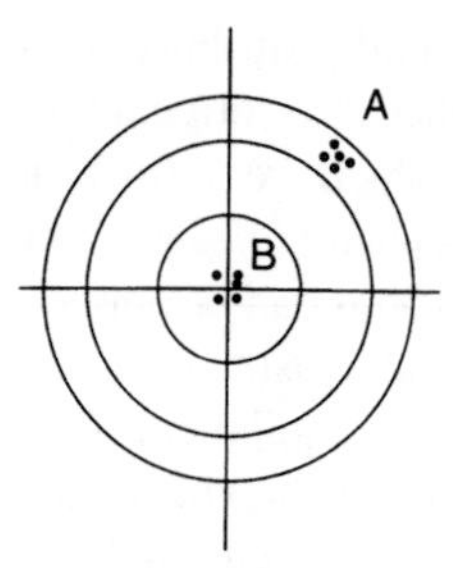

Fig. 2.1 *Scatter*

operating fault and lack of calibration, the measurements are all incorrect by several centimetres (low accuracy). If the bent sight is now corrected, i.e. systematic errors are minimized, the result is a scatter of shots as at *B*. In this case, the shots are clustered near the centre of the target and thus high precision, due to the small scatter, can be related directly to accuracy. The scatter is, of course, due to the unavoidable random errors.

If the target was now placed face down, the surveyors' task would be to locate the most probable position of the centre based on an analysis of the position of the shots at *B*. From this analogy several important facts emerge, as follows.

(1) Scatter is an 'indicator of precision'. The wider the scatter of a set of results about the mean, the less repeatable the measurements are.
(2) Precision must not be confused with accuracy; the former is a relative grouping without regard to nearness to the truth, whilst the latter denotes absolute nearness to the truth.
(3) Precision may be regarded as an index of accuracy only when all sources of error, other than random errors, have been eliminated.
(4) Accuracy may be defined only by specifying the bounds between which the accidental error of a measured quantity may lie. The reason for defining accuracy thus is that the absolute error of the quantity is generally not known. If it were, it could simply be applied to the measured quantity to give its true value. The error bound is usually specified as symmetrical about zero. Thus the accuracy of measured quantity x is $x \pm \varepsilon_x$ where ε_x is greater than or equal to the true but unknown error of x.
(5) Position fixing by the surveyor, whether it is the coordinate position of points in a control network, or the position of topographic detail, is simply an assessment of the most probable position and, as such, requires a statistical evaluation of its precision.

2.4.3 Further definitions

(1) The true value of a measurement can never be found, even though such a value exists. This is evident when observing an angle with a one-second theodolite; no matter how many times the angle is read, a slightly different value will always be obtained.
(2) True error (ε_x) similarly can never be found, for it consists of the true value (X) minus the observed value (x), i.e.

$$X - x = \varepsilon_x$$

(3) Relative error is a measure of the error in relation to the size of the measurement. For instance, a distance of 10 m may be measured with an error of ± 1 mm, whilst a distance of 100 m may also be measured to an accuracy of ± 1 mm. Although the error is the same in both cases, the second measurement may

clearly be regarded as more accurate. To allow for this, the term relative error (R_x) may be used, where

$$R_x = \varepsilon_x / x$$

Thus, in the first case $x = 10$ m, $\varepsilon_x = \pm 1$ mm, and therefore $R_x = 1/10\,000$; in the second case, $R_x = 1/100\,000$, clearly illustrating the distinction. Multiplying the relative error by 100 gives the percentage error. 'Relative error' is an extremely useful definition, and is commonly used in expressing the accuracy of linear measurement. For example, the relative closing error of a traverse is usually expressed in this way. The definition is clearly not applicable to expressing the accuracy to which an angle is measured, however.

(4) Most probable value (MPV) is the closest approximation to the true value that can be achieved from a set of data. This value is generally taken as the arithmetic mean of a set, ignoring at this stage the frequency or weight of the data. For instance, if A is the arithmetic mean, X the true value, and ε_n the errors of a set of n measurements, then

$$A = X - \frac{\Sigma \varepsilon_n}{n}$$

where $\Sigma \varepsilon_n$ is the sum of the errors. As the errors are equally as likely to be positive as negative, then for a finite number of observations $\Sigma \varepsilon_n / n$ will be very small and $A \approx X$. For an infinite number of measurements, it could be argued that $A = X$.

(5) The residual is the difference between the MPV of a set, i.e. the arithmetic mean, and the observed values. Using the same argument as before, it can be shown that for a finite number of measurements, the residual r is approximately equal to the true error ε.

2.4.4 Probability

Consider a length of 29.42 m measured with a tape and correct to ±0.05 m. The range of these measurements would therefore be from 29.37 m to 29.47 m, giving 11 possibilities to 0.01 m for the answer. If the next bay was measured in the same way, there would again be 11 possibilities. Thus the correct value for the sum of the two bays would lie between $11 \times 11 = 121$ possibilities, and the range of the sum would be $2 \times \pm 0.05$ m, i.e. between −0.10 m and +0.10 m. Now, the error of −0.10 m can occur only once, i.e. when both bays have an error of −0.05 m; similarly with +0.10. Consider an error of −0.08; this can occur in three ways: (−0.05 and −0.03), (−0.04 and −0.04) and (−0.03 and −0.05). Applying this procedure through the whole range can produce *Table 2.1*, the lower half of which is simply a repeat of

Table 2.1 *Probability of errors*

Error	*Occurrence*	*Probability*
−0.10	1	1/121 = 0.0083
−0.09	2	2/121 = 0.0165
−0.08	3	3/121 = 0.0248
−0.07	4	4/121 = 0.0331
−0.06	5	5/121 = 0.0413
−0.05	6	6/121 = 0.0496
−0.04	7	7/121 = 0.0579
−0.03	8	8/121 = 0.0661
−0.02	9	9/121 = 0.0744
−0.01	10	10/121 = 0.0826
0	11	11/121 = 0.0909
0.01	10	10/121 = 0.0826

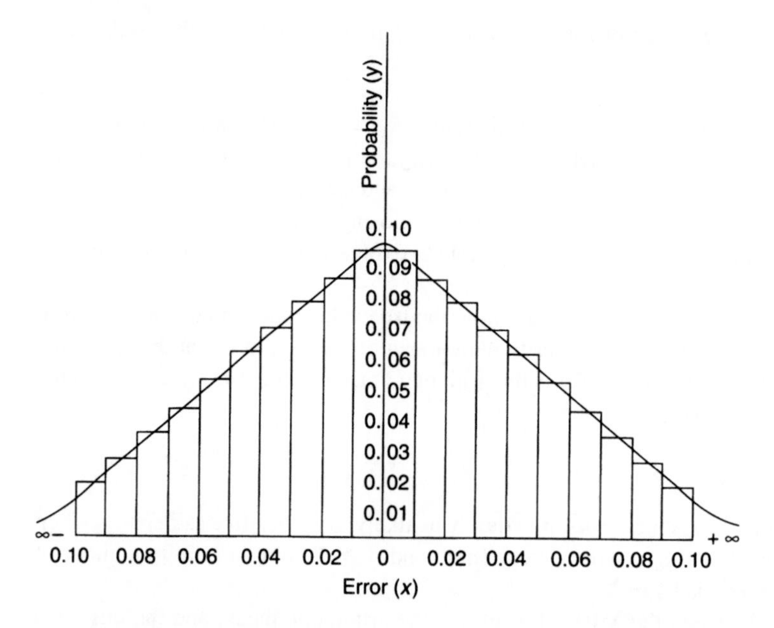

Fig. 2.2 *Probability histogram*

the upper half. If the decimal probabilities are added together they equal 1.0000. If the above results are plotted as error against probability the histogram of *Figure 2.2* is obtained, the errors being represented by rectangles. Then, in the limit, as the error interval gets smaller, the histogram approximates to the superimposed curve. This curve is called the normal probability curve. The area under it represents the probability that the error must lie between ±0.10 m, and is thus equal to 1.0000 (certainty) as shown in *Table 2.1*.

More typical bell-shaped probability curves are shown in *Figure 2.3*; the tall thin curve indicates small scatter and thus high precision, whilst the flatter curve represents large scatter and low precision. Inspection of the curve reveals:

(1) Positive and negative errors are equal in size and frequency; they are equally probable.
(2) Small errors are more frequent than large; they are more probable.
(3) Very large errors seldom occur; they are less probable and may be mistakes or untreated systematic errors.

The equation of the normal probability distribution curve is:

$$y = \frac{e^{-\frac{1}{2}(x-\mu)^2\sigma^{-2}}}{\sigma(2\pi)^{\frac{1}{2}}}$$

where y = probability of the occurrence of $x - \mu$, i.e. the probability that x the variate deviates this far from the central position of the distribution μ, σ is the spread of the distribution and e = the base of natural logarithms. If $\mu = 0$, i.e. the centre of the distribution is at zero and $\sigma = 1$, i.e. the spread is unity, the formula for the probability simplifies to:

$$y = \frac{e^{-\frac{1}{2}x^2}}{(2\pi)^{\frac{1}{2}}}$$

As already illustrated, the area under the curve represents the limit of relative frequency, i.e. probability, and is equal to unity. Thus a table of Normal Distribution curve areas (*Table 2.2*) can be used to calculate

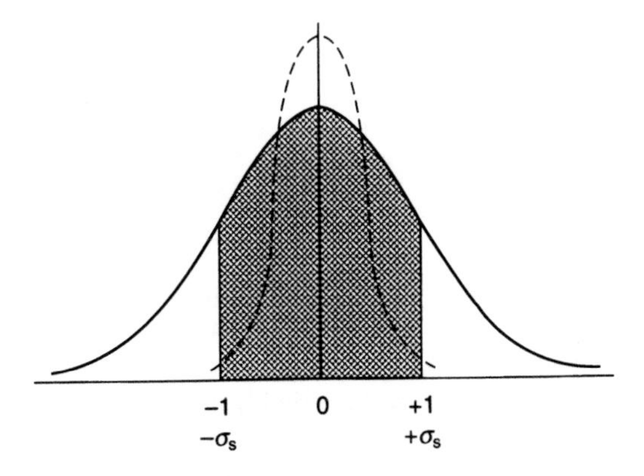

Fig. 2.3 *Probability curve*

Table 2.2 *Area under the Normal Distribution curve*

z	*0.00*	*0.01*	*0.02*	*0.03*	*0.04*	*0.05*	*0.06*	*0.07*	*0.08*	*0.09*
0.0	0.5000	0.5040	0.5080	0.5120	0.5160	0.5199	0.5239	0.5279	0.5319	0.5359
0.1	0.5398	0.5438	0.5478	0.5517	0.5557	0.5596	0.5636	0.5675	0.5714	0.5753
0.2	0.5793	0.5832	0.5871	0.5910	0.5948	0.5987	0.6026	0.6064	0.6103	0.6141
0.3	0.6179	0.6217	0.6255	0.6293	0.6331	0.6368	0.6406	0.6443	0.6480	0.6517
0.4	0.6554	0.6591	0.6628	0.6664	0.6700	0.6736	0.6772	0.6808	0.6844	0.6879
0.5	0.6915	0.6950	0.6985	0.7019	0.7054	0.7088	0.7123	0.7157	0.7190	0.7224
0.6	0.7257	0.7291	0.7324	0.7357	0.7389	0.7422	0.7454	0.7486	0.7517	0.7549
0.7	0.7580	0.7611	0.7642	0.7673	0.7704	0.7734	0.7764	0.7794	0.7823	0.7852
0.8	0.7881	0.7910	0.7939	0.7967	0.7995	0.8023	0.8051	0.8078	0.8106	0.8133
0.9	0.8159	0.8186	0.8212	0.8238	0.8264	0.8289	0.8315	0.8340	0.8365	0.8389
1.0	0.8413	0.8438	0.8461	0.8485	0.8508	0.8531	0.8554	0.8577	0.8599	0.8621
1.1	0.8643	0.8665	0.8686	0.8708	0.8729	0.8749	0.8770	0.8790	0.8810	0.8830
1.2	0.8849	0.8869	0.8888	0.8907	0.8925	0.8944	0.8962	0.8980	0.8997	0.9015
1.3	0.9032	0.9049	0.9066	0.9082	0.9099	0.9115	0.9131	0.9147	0.9162	0.9177
1.4	0.9192	0.9207	0.9222	0.9236	0.9251	0.9265	0.9279	0.9292	0.9306	0.9319
1.5	0.9332	0.9345	0.9357	0.9370	0.9382	0.9394	0.9406	0.9418	0.9429	0.9441
1.6	0.9452	0.9463	0.9474	0.9484	0.9495	0.9505	0.9515	0.9525	0.9535	0.9545
1.7	0.9554	0.9564	0.9573	0.9582	0.9591	0.9599	0.9608	0.9616	0.9625	0.9633
1.8	0.9641	0.9649	0.9656	0.9664	0.9671	0.9678	0.9686	0.9693	0.9699	0.9706
1.9	0.9713	0.9719	0.9726	0.9732	0.9738	0.9744	0.9750	0.9756	0.9761	0.9767
2.0	0.9772	0.9778	0.9783	0.9788	0.9793	0.9798	0.9803	0.9808	0.9812	0.9817
2.1	0.9821	0.9826	0.9830	0.9834	0.9838	0.9842	0.9846	0.9850	0.9854	0.9857
2.2	0.9861	0.9864	0.9868	0.9871	0.9875	0.9878	0.9881	0.9884	0.9887	0.9890
2.3	0.9893	0.9896	0.9898	0.9901	0.9904	0.9906	0.9909	0.9911	0.9913	0.9916
2.4	0.9918	0.9920	0.9922	0.9925	0.9927	0.9929	0.9931	0.9932	0.9934	0.9936
2.5	0.9938	0.9940	0.9941	0.9943	0.9945	0.9946	0.9948	0.9949	0.9951	0.9952
2.6	0.9953	0.9955	0.9956	0.9957	0.9959	0.9960	0.9961	0.9962	0.9963	0.9964
2.7	0.9965	0.9966	0.9967	0.9968	0.9969	0.9970	0.9971	0.9972	0.9973	0.9974
2.8	0.9974	0.9975	0.9976	0.9977	0.9977	0.9978	0.9979	0.9979	0.9980	0.9981
2.9	0.9981	0.9982	0.9982	0.9983	0.9984	0.9984	0.9985	0.9985	0.9986	0.9986
3.0	0.9987	0.9987	0.9987	0.9988	0.9988	0.9989	0.9989	0.9989	0.9990	0.9990

How to use the table: If $(x - \mu)/\sigma = 1.75$ *look down the left column to 1.7 and across the row to the element in the column headed 0.05; the value for the probability is 0.9599, i.e. the probability is 95.99%.*

probabilities provided that the distribution is the standard normal distribution, i.e. $N(0, 1^2)$. If the variable x is $N(\mu, \sigma^2)$, then it must be transformed to the standard normal distribution using $Z = (x - \mu)/\sigma$, where Z has a probability density function equal to $(2\pi)^{-\frac{1}{2}} e^{-Z^2/2}$.

For example, when $x = N(5, 2^2)$ then $Z = (x - 5)/2$
When $x = 9$ then $Z = 2$

Thus the curve can be used to assess the probability that a variable x will fall between certain values. There are two new terms, μ and σ which are usually found from the data, rather than being assumed before the data is investigated. In the normal distribution, the variable x may take values from minus infinity to plus infinity. μ is estimated as the mean value of x and σ gives a measure of the spread or width of the distribution. σ is estimated as the square root of the sum of the squares of the individual differences of x from μ divided by the number of differences less 1, that is the degrees of freedom. Note that whatever the population, for example a set of repeated measurements, μ and σ can only be estimated from the population, they are never known exactly. σ is known as the standard deviation or more often in surveying terminology, the standard error.

An assumption here is that the population of x is large, in fact theoretically infinite. Clearly sets of measurements are never infinite but providing the sets are large then this theoretical problem does not become a practical problem. When the population is small then the 't' distribution, described later, is more appropriate.

Any single observation will deviate from its expected value for two reasons. There will be random errors in the individual observation including rounding errors and there may be systematic errors that unknowingly invalidate the expectation. In practical survey statistics the process of testing whether the difference between an individual observation and the mean of the set is significant is first to estimate the parameters of the appropriate normal distribution, μ and σ, and then test whether the individual observation is an outlier.

In *Figure 2.3*, the area under the curve represents the sum total of all possibilities and therefore is 1. The greater the value of σ, the flatter and hence broader is the curve. The maximum gradients of the curve are where

$$x = \mu \pm \sigma$$

For example, the probability that x will fall between 0.5 and 2.4 is represented by area A on the normal curve (*Figure 2.4(a)*). This statement can be written as:

$P(0.5 < x < 2.4) =$ area A

Now Area A = Area B − Area C (*Figure 2.4(b)* and *(c)*)

where Area B represents $P(x < 2.4)$

and Area C represents $P(x < 0.5)$

i.e. $P(0.5 < x < 2.4) = P(X < 2.4) - P(X < 0.5)$

From the table of the Normal Distribution (*Table 2.2*):

When $x = 2.4$, Area $= 0.9918$

When $x = 0.5$, Area $= 0.6915$

$\therefore\ P(0.5 < x < 2.4) = 0.9918 - 0.6195 = 0.3723$

That is, there is a 37.23% probability that x will lie between 0.5 and 2.4.

If verticals are drawn from the points of inflexion of the normal distribution curve (*Figure 2.5*) they will cut that base at $-\sigma_x$ and $+\sigma_x$, where σ_x is the standard deviation. The area shown indicates the probability that x will lie between $\pm\sigma_x$ and equals 0.683 or 68.3%. This is a very important statement.

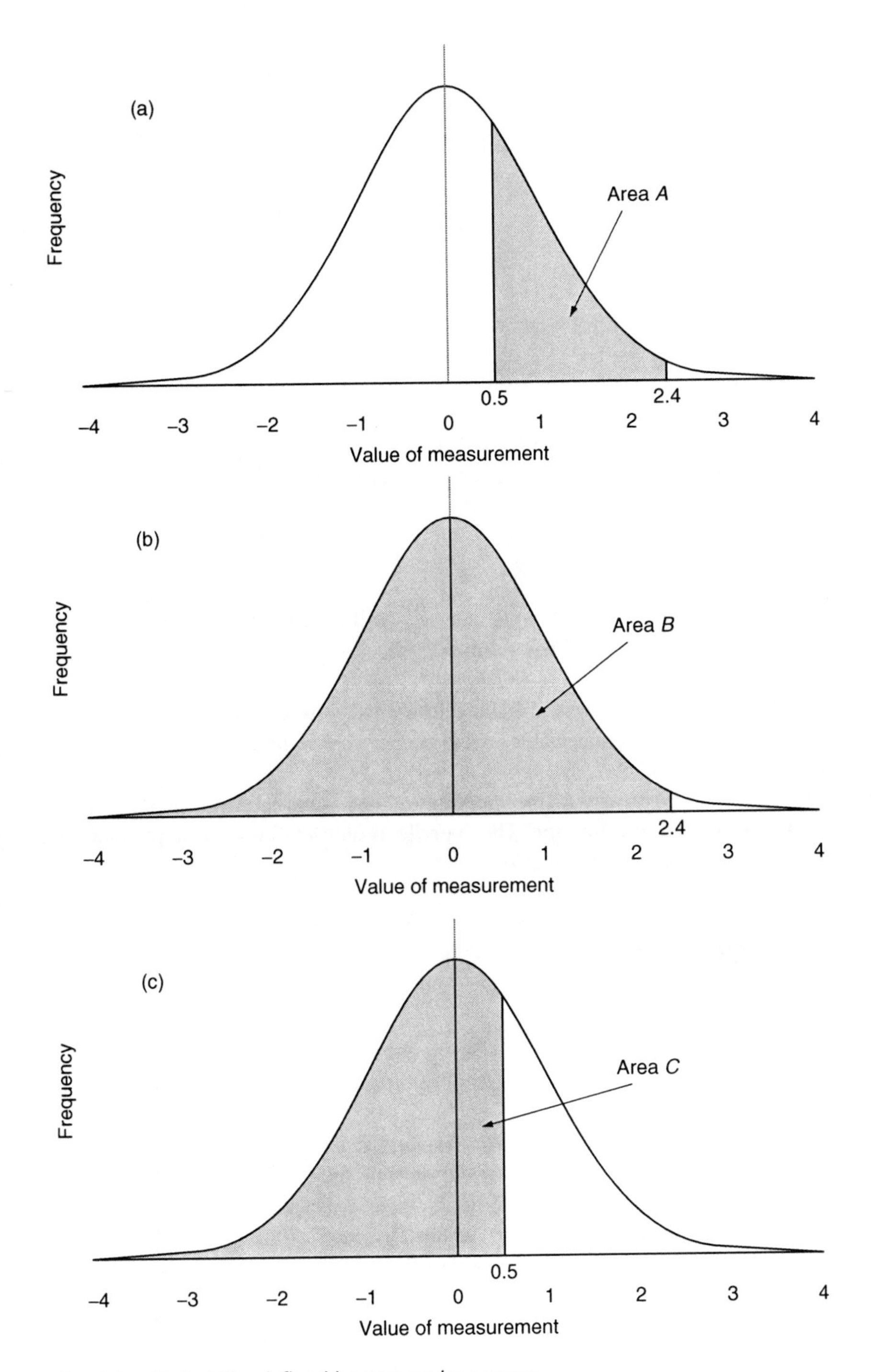

Fig. 2.4 *Probability defined by area under a curve*

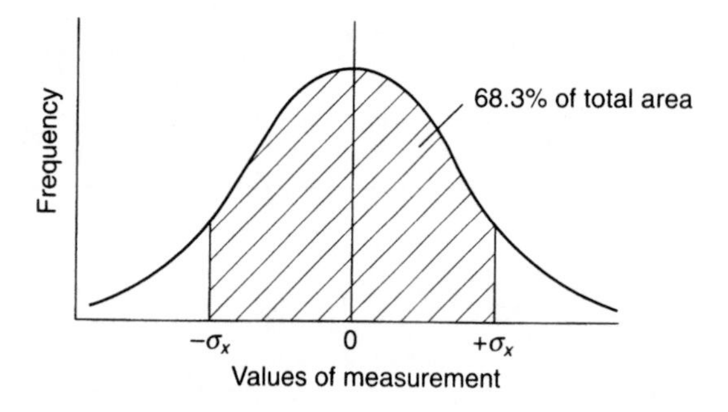

Fig. 2.5 *Normal distribution curve*

The standard deviation (σ_x), if used to assess the precision of a set of data, implies that 68% of the time, the arithmetic mean ($\bar{x}$) of that set should lie between ($\bar{x} \pm \sigma_x$). Put another way, if the sample is normally distributed and contains only random variates, then 7 out of 10 should lie between ($\bar{x} \pm \sigma_x$). It is for this reason that two-sigma or three-sigma limits are preferred in statistical analysis:

$$\pm 2\sigma_x = 0.955 = 95.5\% \text{ probability}$$

$$\text{and } \pm 3\sigma_x = 0.997 = 99.7\% \text{ probability}$$

Thus using two-sigma, we can be 95% certain that a sample mean ($\bar{x}$) will not differ from the population mean μ by more than $\pm 2\sigma_x$. These are called 'confidence limits', where $\bar{x}$ is a point estimate of μ and ($\bar{x} \pm 2\sigma_x$) is the interval estimate.

If a sample mean lies outside the limits of $\pm 2\sigma_x$ we say that the difference between $\bar{x}$ and μ is statistically significant at the 5% level. There is, therefore, reasonable evidence of a real difference and the original null hypothesis ($H_0 \cdot \bar{x} = \mu$) should be rejected.

It may be necessary at this stage to more clearly define 'population' and 'sample'. The 'population' is the whole set of data about which we require information. The 'sample' is any set of data from population, the statistics of which can be used to describe the population.

2.5 INDICES OF PRECISION

It is important to be able to assess the precision of a set of observations, and several standards exist for doing this. The most popular is standard deviation (σ), a numerical value indicating the amount of variation about a central value.

In order to find out how precision is determined, one must first consider a measure which takes into account *all* the values in a set of data. Such a measure is the deviation from the mean ($\bar{x}$) of each observed value (x_i), i.e. ($x_i - \bar{x}$), and one obvious consideration would be the mean of these values. However, in a normal distribution the sum of the deviations would be zero because the sum of the positive deviations would equal the sum of the negative deviations. Thus the 'mean' of the squares of the deviations may be used, and this is called the *variance* (σ^2).

$$\sigma^2 = \sum_{i=1}^{i=n} (x_i - \bar{x})^2 / n \tag{2.1}$$

Theoretically σ is obtained from an infinite number of variates known as the *population*. In practice, however, only a *sample* of variates is available and S is used as an unbiased estimator. Account is taken of the small number of variates in the sample by using $(n - 1)$ as the divisor, which is referred to in statistics as the *Bessel correction*; hence, variance is:

$$S^2 = \sum_{i=1}^{n} (x_i - \bar{x})^2 / n - 1 \tag{2.2}$$

As the deviations are squared, the units in which variance is expressed will be the original units squared. To obtain an index of precision in the same units as the original data, therefore, the square root of the variance is used, and this is called *standard deviation* (S), thus:

$$\text{Standard deviation} = S = \left\{ \sum_{i=1}^{n} (x_i - \bar{x})^2 / n - 1 \right\}^{\frac{1}{2}} \tag{2.3}$$

Standard deviation is represented by the shaded area under the curve in *Figure 2.5* and so establishes the limits of the error bound within which 68.3% of the values of the set should lie, i.e. seven out of a sample of ten.

Similarly, a measure of the precision of the mean ($\bar{x}$) of the set is obtained using the *standard error* ($S_{\bar{x}}$), thus:

$$\text{Standard error} = S_{\bar{x}} = \left\{ \sum_{i=1}^{n} (x_i - \bar{x})^2 / n(n - 1) \right\}^{\frac{1}{2}} = S / n^{\frac{1}{2}} \tag{2.4}$$

Standard error therefore indicates the limits of the error bound within which the 'true' value of the mean lies, with a 68.3% certainty of being correct.

It should be noted that S and $S_{\bar{x}}$ are entirely different parameters. The value of S will *not* alter significantly with an increase in the number (n) of observations; the value of $S_{\bar{x}}$, however, will alter significantly as the number of observations increases. It is important therefore that to describe measured data both values should be used.

2.6 WEIGHT

Weights are expressed numerically and indicate the relative precision of quantities within a set. The greater the weight, the greater the precision of the observation to which it relates. Thus an observation with a weight of two may be regarded as more precise than an observation with a weight of one. Consider two mean measures of the same angle: $A = 50° \, 50' \, 50''$ of weight one, and $B = 50° \, 50' \, 47''$ of weight two. This is equivalent to three observations, $50'', 47'', 47''$, all of equal weight, and having a mean value of

$$(50'' + 47'' + 47'')/3 = 48''$$

Therefore the mean value of the angle $= 50° \, 50' \, 48''$.

Inspection of this exercise shows it to be identical to multiplying each observation a by its weight, w, and dividing by the sum of the weights Σw, i.e.

$$\text{Weighted mean} = A_m = \frac{a_1 w_1 + a_2 w_2 + \cdots + a_n w_n}{w_1 + w_2 + \cdots + w_n} = \frac{\Sigma aw}{\Sigma w} \tag{2.5}$$

Weights can be allocated in a variety of ways, such as: (a) by personal judgement of the prevailing conditions at the time of measurement; (b) by direct proportion to the number of measurements of the quantity, i.e. $w \propto n$; (c) by the use of variance and co-variance factors. This last method is recommended and in the case of the variance factor is easily applied as follows. *Equation (2.4)* shows

$$S_{\bar{x}} = S/n^{\frac{1}{2}}$$

That is, error is inversely proportional to the square root of the number of measures. However, as $w \propto n$, then

$$w \propto 1/S_{\bar{x}}^2$$

i.e. weight is proportional to the inverse of the variance.

2.7 REJECTION OF OUTLIERS

It is not unusual, when taking repeated measurements of the same quantity, to find at least one which appears very different from the rest. Such a measurement is called an *outlier*, which the observer intuitively feels should be rejected from the sample. However, intuition is hardly a scientific argument for the rejection of data and a more statistically viable approach is required.

As already indicated, standard deviation σ represents 68.3% of the area under the normal curve and is therefore representative of 68.26% confidence limits. This leaves 31.74% of the area under the tails of the curve, i.e. 0.1587 or 15.87% on each side. In *Table 2.2* the value of z at $1.00 \times \sigma$ is 0.8413 ($0.8413 = 1 - 0.1587$) and indicates that the table is only concerned with the tail on one side of μ not both. Therefore to calculate confidence limits for a variate both tails must be considered, so if 95% confidence limits are required that implies that each tail will be 2.5% and so look for 97.5% or 0.9750 in the table. The value of $z = (\mu - x)/\sigma$ associated with 0.9750 in *Table 2.2* is 1.96. This indicates that for a Normal Distribution 95% of the population lies within $1.96 \times \sigma$ of μ.

Thus, any random variate x_i, whose residual error $(x_i - \bar{x})$ is greater than $\pm 1.96S$, must lie in the extreme tail ends of the normal curve and might therefore be ignored, i.e. rejected from the sample. In the Normal Distribution the central position of the distribution is derived from the theoretical infinite population. In practice, in survey, it is derived from a limited data set. For example, the true value of a measurement of a particular distance could only be found by averaging an infinite number of observations by an infinite number of observers with an infinite number of measuring devices. The best one could hope for in practice would be a few observations by a few observers with very few instruments. Therefore the computed mean value of the observations is an estimate, not the true value of the measurement. This uncertainty is taken into account by using the 't' distribution (Table 2.3) rather than the Normal Distribution.

Worked example

Example 2.1. The following observations of an angle have been taken:

30° 42′ 24″ 30° 42′ 22″ 30° 42′ 23″
30° 42′ 25″ 30° 42′ 22″ 30° 42′ 40″

What is the probability that the last observed angle is an outlier?

Compute the mean of the sample 30° 42′ 26″
Find the deviation, d, of the last observation 30° 42′ 40″ − 30° 42′ 26″ = 14″
Compute s, the estimate of σ, from the mean and the residuals 6.78″
Find $t = d/s = 14''/6.78'' = 2.064$
Number of degrees of freedom, N = number of observations $- 1 = 5$

Table 2.3 *Area under the 't' distribution curve*

Area = probability	*0.800*	*0.900*	*0.950*	*0.980*	*0.990*	*0.995*	*0.998*	*0.999*
degrees of freedom N	't'							
1	1.376	3.078	6.314	15.895	31.821	63.657	159.153	318.309
2	1.061	1.886	2.920	4.849	6.965	9.925	15.764	22.327
3	0.978	1.638	2.353	3.482	4.541	5.841	8.053	10.215
4	0.941	1.533	2.132	2.999	3.747	4.604	5.951	7.173
5	0.920	1.476	2.015	2.757	3.365	4.032	5.030	5.893
6	0.906	1.440	1.943	2.612	3.143	3.707	4.524	5.208
7	0.896	1.415	1.895	2.517	2.998	3.499	4.207	4.785
8	0.889	1.397	1.860	2.449	2.896	3.355	3.991	4.501
9	0.883	1.383	1.833	2.398	2.821	3.250	3.835	4.297
10	0.879	1.372	1.812	2.359	2.764	3.169	3.716	4.144
12	0.873	1.356	1.782	2.303	2.681	3.055	3.550	3.930
14	0.868	1.345	1.761	2.264	2.624	2.977	3.438	3.787
16	0.865	1.337	1.746	2.235	2.583	2.921	3.358	3.686
18	0.862	1.330	1.734	2.214	2.552	2.878	3.298	3.610
20	0.860	1.325	1.725	2.197	2.528	2.845	3.251	3.552
25	0.856	1.316	1.708	2.167	2.485	2.787	3.170	3.450
30	0.854	1.310	1.697	2.147	2.457	2.750	3.118	3.385
40	0.851	1.303	1.684	2.123	2.423	2.704	3.055	3.307
60	0.848	1.296	1.671	2.099	2.390	2.660	2.994	3.232
100	0.845	1.290	1.660	2.081	2.364	2.626	2.946	3.174
1000	0.842	1.282	1.646	2.056	2.330	2.581	2.885	3.098

Find the appropriate value in the row $N = 5$ in the t table (Table 2.3). At a probability of 0.95 the value of t is 2.015 therefore the computed value of 2.064 indicates that there is slightly more than a 95% chance that the last observation contains a non-random error.

It should be noted that *successive* rejection procedures should not be applied to the sample.

2.8 COMBINATION OF ERRORS

Much data in surveying is obtained indirectly from various combinations of observed data, for instance the coordinates of the ends of a line are a function of its length and bearing. As each measurement contains an error, it is necessary to consider the combined effect of these errors on the derived quantity.

The general procedure is to differentiate with respect to each of the observed quantities in turn and sum them to obtain their total effect. Thus if $a = f(x, y, z, \ldots)$, and each independent variable changes by a small amount (an error) δx, δy, δz, ..., then a will change by a small amount equal to δa, obtained from the following expression:

$$\delta a = \frac{\partial a}{\partial x} \cdot \delta x + \frac{\partial a}{\partial y} \cdot \delta y + \frac{\partial a}{\partial z} \cdot \delta z + \cdots \qquad (2.6)$$

in which $\partial a/\partial x$ is the partial derivative of a with respect to x, etc.

Consider now a set of measurements and let the residuals δx_i, δy_i, and δz_i, be written as x_i, y_i, and z_i and the error in the derived quantity δa_I is written as a_i:

$$a_1 = \frac{\partial a}{\partial x} \cdot x_1 + \frac{\partial a}{\partial y} \cdot y_1 + \frac{\partial a}{\partial z} \cdot z_1 + \cdots$$

$$a_2 = \frac{\partial a}{\partial x} \cdot x_2 + \frac{\partial a}{\partial y} \cdot y_2 + \frac{\partial a}{\partial z} \cdot z_2 + \cdots$$

$$\vdots \quad \vdots \quad \vdots \quad \vdots$$

$$a_n = \frac{\partial a}{\partial x} \cdot x_n + \frac{\partial a}{\partial y} \cdot y_n + \frac{\partial a}{\partial z} \cdot z_n + \cdots$$

Now squaring both sides gives

$$a_1^2 = \left(\frac{\partial a}{\partial x}\right)^2 \cdot x_1^2 + 2\left(\frac{\partial a}{\partial x}\right)\left(\frac{\partial a}{\partial y}\right) x_1 y_1 + \cdots + \left(\frac{\partial a}{\partial y}\right)^2 y_1^2 + \cdots$$

$$a_2^2 = \left(\frac{\partial a}{\partial x}\right)^2 \cdot x_2^2 + 2\left(\frac{\partial a}{\partial x}\right)\left(\frac{\partial a}{\partial y}\right) x_2 y_2 + \cdots + \left(\frac{\partial a}{\partial y}\right)^2 y_2^2 + \cdots$$

$$\vdots \quad \vdots \quad \vdots \quad \vdots$$

$$a_n^2 = \left(\frac{\partial a}{\partial x}\right)^2 \cdot x_n^2 + 2\left(\frac{\partial a}{\partial x}\right)\left(\frac{\partial a}{\partial y}\right) x_n y_n + \cdots + \left(\frac{\partial a}{\partial y}\right)^2 y_n^2 + \cdots$$

In the above process many of the square and cross-multiplied terms have been omitted for simplicity. Summing the results gives

$$\sum a^2 = \left(\frac{\partial a}{\partial x}\right)^2 \sum x^2 + 2\left(\frac{\partial a}{\partial x}\right)\left(\frac{\partial a}{\partial y}\right) \sum xy + \cdots + \left(\frac{\partial a}{\partial y}\right)^2 \sum y^2 + \cdots$$

As the measured quantities may be considered independent and uncorrelated, the cross-products tend to zero and may be ignored.

Now dividing throughout by $(n - 1)$:

$$\frac{\sum a^2}{n-1} = \left(\frac{\partial a}{\partial x}\right)^2 \frac{\sum x^2}{n-1} + \left(\frac{\partial a}{\partial y}\right)^2 \frac{\sum y^2}{n-1} + \left(\frac{\partial a}{\partial z}\right)^2 \frac{\sum z^2}{n-1} + \cdots$$

The sum of the residuals squared divided by $(n - 1)$, is in effect the variance σ^2, and therefore

$$\sigma_a^2 = \left(\frac{\partial a}{\partial x}\right)^2 \sigma_x^2 + \left(\frac{\partial a}{\partial y}\right)^2 \sigma_y^2 + \left(\frac{\partial a}{\partial z}\right)^2 \sigma_z^2 + \cdots \tag{2.7}$$

which is the general equation for the variance of any function. This equation is very important and is used extensively in surveying for error analysis, as illustrated in the following examples.

2.8.1 Uncertainty of addition or subtraction

Consider a quantity $A = a + b$ where a and b have standard errors σ_a and σ_b, then

$$\sigma_A^2 = \left\{\frac{\partial(a+b)}{\partial a}\sigma_a\right\}^2 + \left\{\frac{\partial(a+b)}{\partial b}\sigma_b\right\}^2 = \sigma_a^2 + \sigma_b^2 \quad \therefore \sigma_A = \left(\sigma_a^2 + \sigma_b^2\right)^{\frac{1}{2}} \tag{2.8}$$

As subtraction is simply addition with the signs changed, the above also holds for the error in a *difference*:

$$\text{If } \sigma_a = \sigma_b = \sigma, \quad \text{then} \quad \sigma_A = \sigma(n)^{\frac{1}{2}} \tag{2.9}$$

Equation (2.9) should not be confused with *equation (2.4)* which refers to the *mean*, not the *sum* as above.

Worked examples

Example 2.2. Three angles of a triangle each have a standard error of 2″. What is the total error (σ_T) in the triangle?

$$\sigma_T = (2^2 + 2^2 + 2^2)^{\frac{1}{2}} = 2(3)^{\frac{1}{2}} = 3.5''$$

Example 2.3. In measuring a round of angles at a station, the third angle c closing the horizon is obtained by subtracting the two measured angles a and b from 360°. If angle a has a standard error of 2″ and angle b a standard error of 3″, what is the standard error of angle c?

$$\text{since } c = 360° - a - b$$

$$\text{then } \sigma_c = (\sigma_a^2 + \sigma_b^2)^{\frac{1}{2}} = (2^2 + 3^2)^{\frac{1}{2}} = 3.6''$$

Example 2.4. The standard error of a *mean* angle derived from four measurements is 3″; how many measurements would be required, using the same equipment, to halve this uncertainty?

$$\text{From } \textit{equation (2.4)} \quad \sigma_m = \frac{\sigma_s}{n^{\frac{1}{2}}} \qquad \therefore \sigma_s = 3'' \times 4^{\frac{1}{2}} = 6''$$

i.e. the instrument used had a standard error of 6″ for a single observation; thus for $\sigma_m = 1.5''$, when $\sigma_s = 6''$:

$$n = \left(\frac{6}{1.5}\right)^2 = 16$$

Example 2.5. If the standard error of the sum of independently observed angles in a triangle is to be not greater than 6.0″, what is the permissible standard error per angle?

$$\text{From } \textit{equation (2.9)} \quad \sigma_T = \sigma_p(n)^{\frac{1}{2}}$$

where σ_T is the triangular error, σ_p the error per angle, and n the number of angles.

$$\therefore \sigma_p = \frac{\sigma_T}{(n)^{\frac{1}{2}}} = \frac{6.0''}{(3)^{\frac{1}{2}}} = 3.5''$$

2.8.2 Uncertainty of a product

Consider $A = (a \times b \times c)$ where a, b and c have standard errors σ_a, σ_b and σ_c. The variance

$$\sigma_A^2 = \left\{\frac{\partial(abc)}{\partial a}\sigma_a\right\}^2 + \left\{\frac{\partial(abc)}{\partial b}\sigma_b\right\}^2 + \left\{\frac{\partial(abc)}{\partial c}\sigma_c\right\}^2$$

$$= (bc\sigma_a)^2 + (ac\sigma_b)^2 + (ab\sigma_c)^2$$

$$\therefore \sigma_A = abc\left\{\left(\frac{\sigma_a}{a}\right)^2 + \left(\frac{\sigma_b}{b}\right)^2 + \left(\frac{\sigma_c}{c}\right)^2\right\}^{\frac{1}{2}} \tag{2.10}$$

The terms in brackets may be regarded as the relative errors R_a, R_b, R_c giving

$$\sigma_A = abc\,(R_a^2 + R_b^2 + R_c^2)^{\frac{1}{2}} \tag{2.11}$$

2.8.3 Uncertainty of a quotient

Consider $A = a/b$, then the variance

$$\sigma_A^2 = \left\{\frac{\partial(ab^{-1})}{\partial a}\sigma_a\right\}^2 + \left\{\frac{\partial(ab^{-1})}{\partial b}\sigma_b\right\}^2 = \left(\frac{\sigma_a}{b}\right)^2 + \left(\frac{\sigma_b a}{b^2}\right)^2 \tag{2.12}$$

$$\therefore \sigma_A = \frac{a}{b}\left\{\left(\frac{\sigma_a}{a}\right)^2 + \left(\frac{\sigma_b}{b}\right)^2\right\}^{\frac{1}{2}}$$

$$= \frac{a}{b}(R_a^2 + R_b^2)^{\frac{1}{2}} \tag{2.13}$$

2.8.4 Uncertainty of powers and roots

The case for the power of a number must not be confused with the case for multiplication, for example $a^3 = a \times a \times a$, with each term being exactly the same.

Thus if $A = a^n$, then the variance

$$\sigma_A^2 = \left(\frac{\partial a^n}{\partial a}\sigma_a\right)^2 = \left(na^{n-1}\sigma_a\right)^2 \quad \therefore \sigma_A = \left(na^{n-1}\sigma_a\right) \tag{2.14}$$

$$\text{Alternatively} \quad R_A = \frac{\sigma_A}{a^n} = \frac{na^{n-1}\sigma_a}{a^n} = \frac{n\sigma_a}{a} = nR_a \tag{2.15}$$

Similarly for roots, if the function is $A = a^{1/n}$, then the variance

$$\sigma_A^2 = \left(\frac{\partial a^{1/n}}{\partial a}\sigma_a\right)^2 = \left(\frac{1}{n}a^{1/n-1}\sigma_a\right)^2 = \left(\frac{1}{n}a^{1/n}a^{-1}\sigma_a\right)^2 \tag{2.16}$$

$$= \left(\frac{a^{1/n}}{n}\frac{\sigma_a}{a}\right)^2 \qquad \therefore \sigma_A = \left(\frac{a^{1/n}}{n}\frac{\sigma_a}{a}\right) \tag{2.17}$$

The same approach is adapted to general forms which are combinations of the above.

Worked examples

Example 2.6. The same angle was measured by two different observers using the same instrument, as follows:

Observer A			Observer B		
°	′	″	°	′	″
86	34	10	86	34	05
	33	50		34	00
	33	40		33	55
	34	00		33	50
	33	50		34	00
	34	10		33	55
	34	00		34	15
	34	20		33	44

Calculate: (a) The standard deviation of each set.
(b) The standard error of the arithmetic means.
(c) The most probable value (MPV) of the angle. (KU)

	Observer A			r	r^2	Observer B			r	r^2
	°	′	″	″	″	°	′	″	″	″
	86	34	10	10	100	86	34	05	7	49
		33	50	−10	100		34	00	2	4
		33	40	−20	400		33	55	−3	9
		34	00	0	0		33	50	−8	64
		33	50	−10	100		34	00	2	4
		34	10	10	100		33	55	−3	9
		34	00	0	0		34	15	17	289
		34	20	20	400		33	44	−14	196
Mean =	86	34	00	0	$1200 = \Sigma r^2$	86	33	58	0	$624 = \Sigma r^2$

(a) (i) Standard deviation $(\Sigma r^2 = \Sigma(x_i - \bar{x})^2)$

$$S_A = \left(\frac{\sum r^2}{n-1}\right)^{\frac{1}{2}} = \left(\frac{1200}{7}\right)^{\frac{1}{2}} = 13.1''$$

(b) (i) Standard error $S_{\bar{x}A} = \dfrac{S_A}{n^{\frac{1}{2}}} = \dfrac{13.1}{8^{\frac{1}{2}}} = 4.6''$

(a) (ii) Standard deviation $S_B = \left(\dfrac{624}{7}\right)^{\frac{1}{2}} = 9.4''$

(b) (ii) Standard error $S_{\bar{x}B} = \dfrac{9.4}{8^{\frac{1}{2}}} = 3.3''$

(c) As each arithmetic mean has a different precision exhibited by its $S_{\bar{x}}$ value, the arithmetic means must be weighted accordingly before they can be averaged to give the MPV of the angle:

$$\text{Weight of } A \propto \frac{1}{S_{\bar{x}A}^2} = \frac{1}{21.2} = 0.047$$

$$\text{Weight of } B \propto \frac{1}{10.9} = 0.092$$

The ratio of the weight of A to the weight of B is 0.047:0.092

$$\therefore \text{MPV of the angle} = \frac{(0.047 \times 86^\circ\, 34'\, 00'' + 0.092 \times 86^\circ\, 33'\, 58'')}{(0.047 + 0.092)}$$

$$= 86^\circ\, 33'\, 59''$$

As a matter of interest, the following point could be made here: any observation whose residual is greater than 2.998S should be rejected at the 98% level (see *Section 2.7*). Each data set has 8 observations and therefore the mean has 7 degrees of freedom. This is a 2-tailed test therefore the 0.99 column is used. As $2.998S_A = 39.3''$ and $2.998S_B = 28.2''$, all the observations should be included in the set. This test should normally be carried out at the start of the problem.

Example 2.7. Discuss the classification of errors in surveying operations, giving appropriate examples.

In a triangulation scheme, the three angles of a triangle were measured and their mean values recorded as 50° 48′ 18″, 64° 20′ 36″ and 64° 51′ 00″. Analysis of each set gave a standard deviation of 4″ for each of these means. At a later date, the angles were re-measured under better conditions, yielding mean values of 50° 48′ 20″, 64° 20′ 39″ and 64° 50′ 58″. The standard deviation of each value was 2″. Calculate the most probable values of the angles. (KU)

The angles are first adjusted to 180°. Since the angles within each triangle are of equal weight, then the angular adjustment within each triangle is equal.

50° 48′ 18″ + 2″ =	50° 48′ 20″	50° 48′ 20″ + 1″ =	50° 48′ 21″
64° 20′ 36″ + 2″ =	64° 20′ 38″	64° 20′ 39″ + 1″ =	64° 20′ 40″
64° 51′ 00″ + 2″ =	64° 51′ 02″	64° 50′ 58″ + 1″ =	64° 50′ 59″
179° 59′ 54″	180° 00′ 00″	179° 59′ 57″	180° 00′ 00″

$$\text{Weight of the first set} = w_1 = 1/4^2 = \frac{1}{16}$$

$$\text{Weight of the second set} = w_2 = 1/2^2 = \frac{1}{4}$$

Thus $w_1 = 1$, when $w_2 = 4$.

$$\therefore \text{MPV} = \frac{(50^\circ\,48'\,20'') + (50^\circ\,48'\,21'' \times 4)}{5} = 50^\circ\,48'\,20.8''$$

Similarly, the MPVs of the remaining angles are:

64° 20′ 39.6″ 64° 50′ 59.6″

The values may now be rounded off to single seconds.

Example 2.8. A base line of ten bays was measured by a tape resting on measuring heads. One observer read one end while the other observer read the other – the difference in readings giving the observed length of the bay. Bays 1, 2 and 5 were measured six times, bays 3, 6 and 9 were measured five times and the remaining bays were measured four times, the means being calculated in each case. If the standard errors of single readings by the two observers were known to be 1 mm and 1.2 mm, what will be the standard error in the whole line due only to reading errors? (LU)

$$\text{Standard error in reading a bay} \quad S_s = (1^2 + 1.2^2)^{\frac{1}{2}} = 1.6\,\text{mm}$$

Consider bay 1. This was measured six times and the mean taken; thus the standard error of the mean is:

$$S_{\bar{x}} = \frac{S_s}{l^{\frac{1}{2}}} = \frac{1.6}{6^{\frac{1}{2}}} = 0.6\,\text{mm}$$

This value applies to bays 2 and 5 also. Similarly for bays 3, 6 and 9:

$$S_{\bar{x}} = \frac{1.6}{5^{\frac{1}{2}}} = 0.7\,\text{mm}$$

For bays 4, 7, 8 and 10 $\quad S_{\bar{x}} = \dfrac{1.6}{4^{\frac{1}{2}}} = 0.8\,\text{mm}$

These bays are now summed to obtain the total length. Therefore the standard error of the whole line is

$$(0.6^2 + 0.6^2 + 0.6^2 + 0.7^2 + 0.7^2 + 0.7^2 + 0.8^2 + 0.8^2 + 0.8^2 + 0.8^2)^{\frac{1}{2}} = 2.3\,\text{mm}$$

Example 2.9.

(a) A base line was measured using electronic distance-measuring (EDM) equipment and a mean distance of 6835.417 m recorded. The instrument used has a manufacturer's quoted accuracy of 1/400 000 of the length measured 20 mm. As a check the line was re-measured using a different type of EDM equipment having an accuracy of 1/600 000 ± 30 mm; the mean distance obtained was 6835.398 m. Determine the most probable value of the line.

(b) An angle was measured by three different observers, A, B and C. The mean of each set and its standard error is shown below.

Observer	*Mean angle*			$S_{\bar{x}}$
	°	′	″	″
A	89	54	36	0.7
B	89	54	42	1.2
C	89	54	33	1.0

Determine the most probable value of the angle. (KU)

(a) Standard error, 1st instrument $\quad S_{\bar{x}_1} = \left\{\left(\dfrac{6835}{400\,000}\right)^2 + (0.020)^2\right\}^{\frac{1}{2}}$

$= 0.026\,\text{m}$

Standard error, 2nd instrument $\quad S_{\bar{x}_2} = \left\{\left(\dfrac{6835}{600\,000}\right)^2 + (0.030)^2\right\}^{\frac{1}{2}}$

$= 0.032\,\text{m}$

These values can now be used to weight the lengths and find their weighted means as shown below.

	Length, L (m)	$S_{\bar{x}}$	*Weight ratio*	*Weight, W*	$L \times W$
1st instrument	0.417	0.026	$1/0.026^2 = 1479$	1.5	0.626
2nd instrument	0.398	0.032	$1/0.032^2 = 977$	1	0.398
				$\Sigma W = 2.5$	$1.024 = \Sigma LW$

$$\therefore \text{MPV} = 6835 + \frac{1.024}{2.5} = 6835.410\,\text{m}$$

(b)

Observer	Mean angle			$S_{\bar{x}}$	Weight ratio	Weight, W	L × W
	°	′	″	″			
A	89	54	36	0.7	$1/0.7^2 = 2.04$	2.96	$6'' \times 2.96 = 17.8''$
B	89	54	42	1.2	$1/1.2^2 = 0.69$	1	$12'' \times 1 = 12''$
C	89	54	33	1.0	$1/1^2 = 1$	1.45	$3'' \times 1.45 = 4.35''$
						$\Sigma W = 5.41$	$34.15 = \Sigma LW$

$$\therefore \text{MPV} = 89^\circ\, 54'\, 30'' + \frac{34.15''}{5.41} = 89^\circ\, 54'\, 36''$$

Example 2.10. In an underground correlation survey, the sides of a Weisbach triangle were measured as follows:

$$W_1W_2 = 5.435 \text{ m} \qquad W_1W = 2.844 \text{ m} \qquad W_2W = 8.274 \text{ m}$$

Using the above measurements in the cosine rule, the calculated angle $WW_1W_2 = 175^\circ\, 48'\, 24''$. If the standard error of each of the measured sides is 1/20 000 of its length, find the standard error of the calculated angle in seconds of arc. (KU)

From *Figure 2.6*, by the cosine rule $c = a^2 + b^2 - 2ab \cos W_1$.
Differentiating with respect to each variable in turn:

$$2c\delta c = 2ab \sin W_1 \delta W_1 \quad \text{thus} \quad \delta W_1 = \frac{c\delta c}{ab \sin W_1}$$

Similarly: $a = c - b^2 + 2ab \cos W_1$

$$2a\delta a = 2b \cos W_1 \delta a - 2ab \sin W_1 \delta W_1$$

$$\therefore \delta W_1 = \frac{2b \cos W_1 \delta a - 2a\delta a}{2ab \sin W_1} = \frac{\delta a(b \cos W_1 - a)}{ab \sin W_1}$$

but, since angle $W_1 \approx 180^\circ$, $\cos W_1 \approx -1$ and $(a + b) \approx c$

$$\therefore \delta W_1 = \frac{-c\delta a}{ab \sin W_1}$$

now $\quad b^2 = a^2 - c^2 + 2ab \cos W_1$

and $\quad 2b\delta b = 2a \cos W_1 \delta b - 2ab \sin W_1 \delta W_1$

$$\therefore \delta W_1 = \frac{\delta b(a \cos W_1 - b)}{ab \sin W_1} = \frac{-c\delta b}{ab \sin W_1}$$

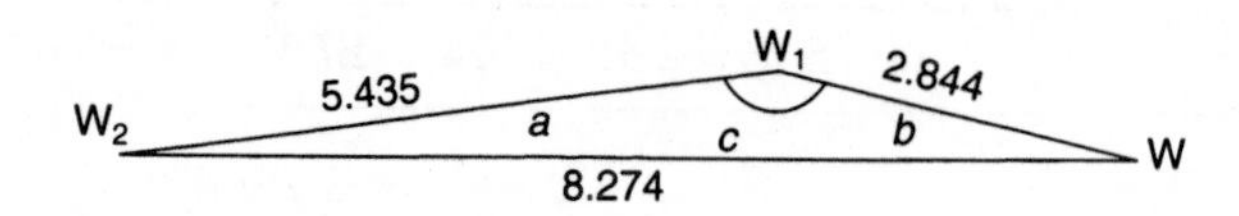

Fig. 2.6 *Cosine rule*

Since W_1 is a function of a, b and c:

$$\sigma_{W_1} = \frac{c}{ab \sin W_1}(\sigma_a^2 + \sigma_b^2 + \sigma_c^2)^{\frac{1}{2}}$$

where $$\sigma_a = \frac{5.435}{20\,000} = 2.7 \times 10^{-4}$$

$$\sigma_b = \frac{2.844}{20\,000} = 1.4 \times 10^{-4}$$

$$\sigma_a = \frac{8.274}{20\,000} = 4.1 \times 10^{-4}$$

$$\therefore \sigma_{W_1} = \frac{8.274 \times 206\,265 \times 10^{-4}}{5.435 \times 2.844 \sin 175^\circ 48' 24''}(2.7^2 + 1.4^2 + 4.1^2)^{\frac{1}{2}}$$

$$= 56''$$

This is a standard treatment for small errors. There are numerous examples of its application throughout the remainder of this book.

Exercises

(2.*1*) Explain the meaning of the terms *random error* and *systematic error*, and show by example how each can occur in normal surveying work.

A certain angle was measured ten times by observer A with the following results, all measurements being equally precise:

$74^\circ\, 38'\, 18'', 20'', 15'', 21'', 24'', 16'', 22'', 17'', 19'', 13''$

(The degrees and minutes remained constant for each observation.)
The same angle was measured under the same conditions by observer B with the following results:

$74^\circ\, 36'\, 10'', 21'', 25'', 08'', 15'', 20'', 28'', 11'', 18'', 24''$

Determine the standard deviation for each observer and relative weightings. (ICE)

(*Answer*: 3.4″; 6.5″. A:B is 9:2)

(2.2) Derive from first principles an expression for the standard error in the computed angle W_1 of a Weisbach triangle, assuming a standard error of σ_w in the Weisbach angle W, and equal proportional standard errors in the measurement of the sides. What facts, relevant to the technique of correlation using this method, may be deduced from the reduced error equation? (KU)

(*Answer*: see Chapter 13)

3
Vertical control

3.1 INTRODUCTION

This chapter describes the various heighting procedures used to obtain the elevation of points of interest above or below a reference datum. The most commonly used reference datum is mean sea level (MSL). There is no such thing as a common global MSL, as it varies from place to place depending on local conditions. It is important therefore that MSL is clearly defined wherever it is used.

The engineer is, in the main, more concerned with the relative height of one point above or below another, in order to ascertain the difference in height of the two points, rather than a direct relationship to MSL. It is not unusual, therefore, on small local schemes, to adopt a purely arbitrary reference datum. This could take the form of a permanent, stable position or mark, allocated such a value that the level of any point on the site would not be negative. For example, if the reference mark was allocated a value of 0.000 m, then a ground point 10 m lower would have a negative value, minus 10.000 m. However, if the reference value was 100.000 m, then the level of the ground point in question would be 90.000 m. As minus signs in front of a number can be misinterpreted, erased or simply forgotten about, they should, wherever possible, be avoided.

The vertical height of a point above or below a reference datum is referred to as the reduced level or simply the level of a point. Reduced levels are used in practically all aspects of construction: to produce ground contours on a plan; to enable the optimum design of road, railway or canal gradients; to facilitate ground modelling for accurate volumetric calculations. Indeed, there is scarcely any aspect of construction that is not dependent on the relative levels of ground points.

3.2 LEVELLING

Levelling is the most widely used method for obtaining the elevations of ground points relative to a reference datum and is usually carried out as a separate procedure from that used for fixing planimetric position.

Levelling involves the measurement of vertical distance relative to a horizontal line of sight. Hence it requires a graduated staff for the vertical measurements and an instrument that will provide a horizontal line of sight.

3.3 DEFINITIONS

3.3.1 Level line

A level line or level surface is one which at all points is normal to the direction of the force of gravity as defined by a freely suspended plumb-bob. As already indicated in *Chapter 1* in the discussion of the geoid, such surfaces are ellipsoidal in shape. Thus in *Figure 3.1* the difference in level between A and B is the distance $A'B$, provided that the non-parallelism of level surfaces is ignored.

3.3.2 Horizontal line

A horizontal line or surface is one that is normal to the direction of the force of gravity at a particular point. *Figure 3.1* shows a horizontal line through point C.

3.3.3 Datum

A datum is any reference surface to which the elevations of points are referred. The most commonly used datum is that of mean sea level (MSL).

In the UK the MSL datum was measured and established by the Ordnance Survey (OS) of Great Britain, and hence it is often referred to as Ordnance Datum (OD). It is the mean level of the sea at Newlyn in Cornwall calculated from hourly readings of the sea level, taken by an automatic tide gauge over a six-year period from 1 May 1915 to 30 April 1921. The readings are related to the Observatory Bench Mark, which is 4.751 m above the datum. Other countries have different datums; for instance, Australia used 30 tidal observatories, interconnected by 200 000 km of levelling, to produce their national datum, whilst just across the English Channel, France uses a different datum, rendering their levels incompatible with those in the UK.

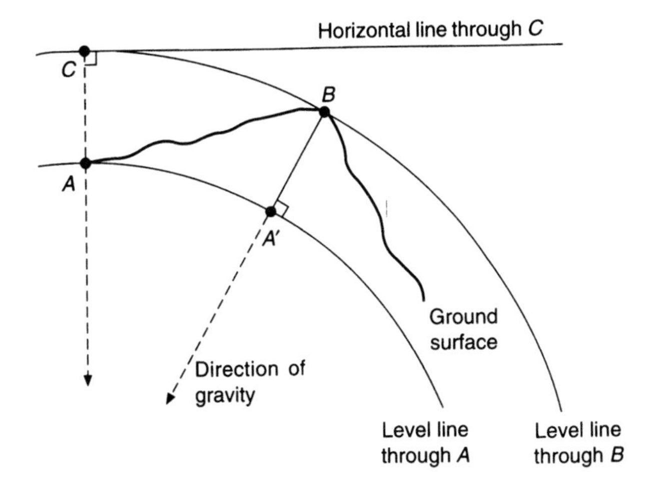

Fig. 3.1 *Horizontal and level lines*

3.3.4 Bench mark (BM)

In order to make OD accessible to all users throughout the country, a series of permanent marks were established, called bench marks. The height of these marks relative to OD has been established by differential levelling and until about 1970 was regularly checked for any change in elevation.

(1) Cut bench marks

The cut bench mark is the most common type of BM and is usually cut into the vertical surface of a stable structure (*Figure 3.2*).

(2) Flush brackets

Flush brackets are metal plates, 180 mm × 90 mm, cemented into the face of buildings and were established at intervals of about 2 km (*Figure 3.3*).

(3) Bolt bench marks

Bolt bench marks are 60-mm-diameter brass bolts set in horizontal surfaces and engraved with an arrow and the letters OSBM (*Figure 3.4*).

(4) Fundamental bench marks (FBM)

In the UK, FBMs were established by precise geodetic levelling, at intervals of about 50 km. Each mark consists of a buried chamber containing two reference points, whilst the published elevation is to a brass bolt on the top of a concrete pillar (*Figure 3.5*).

Rivet and pivot BMs are also to be found in horizontal surfaces.

Details of BMs within the individual's area of interest may be obtained in the form of a Bench Mark List from the OS. Their location and value are currently also shown on OS plans at the 1/2500 and 1/1250 scales. Their values are quoted as precise to the nearest 12 mm relative to neighbouring bench marks at the time of the original observation only.

Bench marks established by individuals, other than the OS, such as engineers for construction work, are called temporary bench marks (TBM).

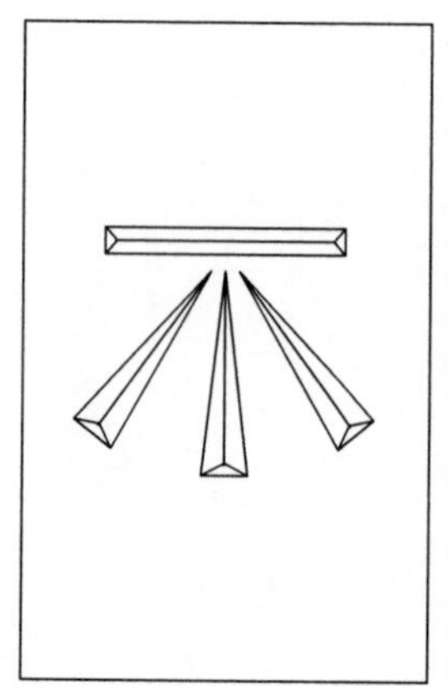

These are to be found on the vertical faces of buildings, bridges, walls, milestones, gate posts, etc.

The mark is approximately 0.1 m × 0.1 m and cut to a depth of about 6 mm and placed about 0.5 m above ground level.

Some very old marks may be considerably larger than this and Initial Levelling marks 1840–60 may have a copper bolt set in the middle of the horizontal bar or offset to one side of the mark.

The exact point of reference is the centre of the V shaped horizontal bar.

Fig. 3.2 *Cut bench mark*

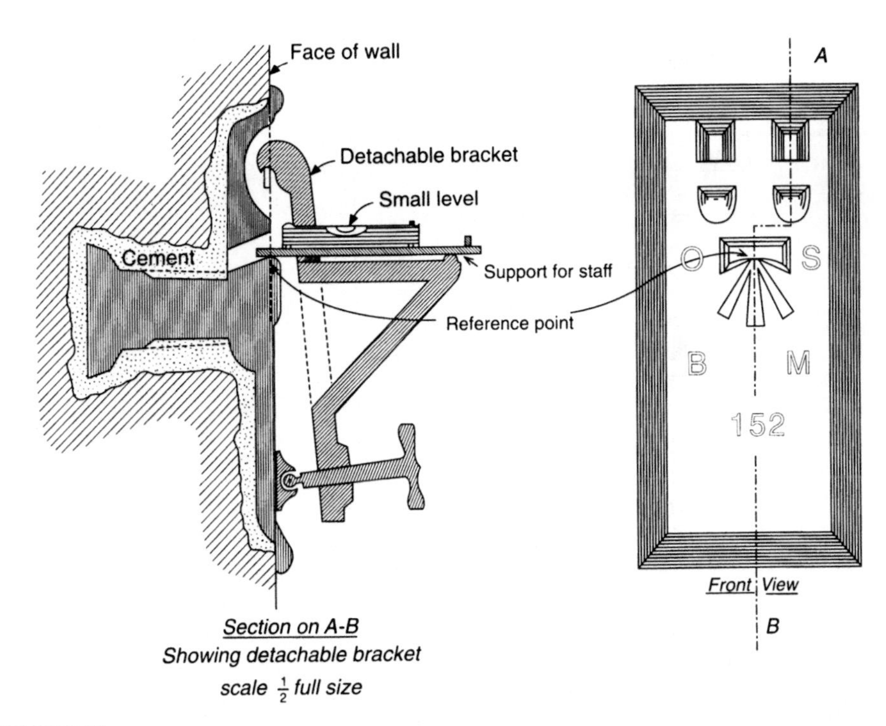

DESCRIPTION

These are normally placed on vertical walls of buildings and in the sides of triangulation pillars.

They are cast in brass and rectangular in shape (180 mm × 90 mm) with a large boss at the rear of the plate. The boss is cemented into a prepared cavity, so that the face of the bracket is vertical and in line with the face of the object on which it is placed. These marks in precise levelling necessitate the use of a special fitting as above.

Each flush bracket has a unique serial number and is referred to in descriptions as F.I. Br. No........

They are sited at approximately 2 kilometre intervals along Geodetic lines of levels and at 5 to 7 kilometre intervals on Secondary lines of levels.

Fig. 3.3 *Flush bracket (front and side view)*

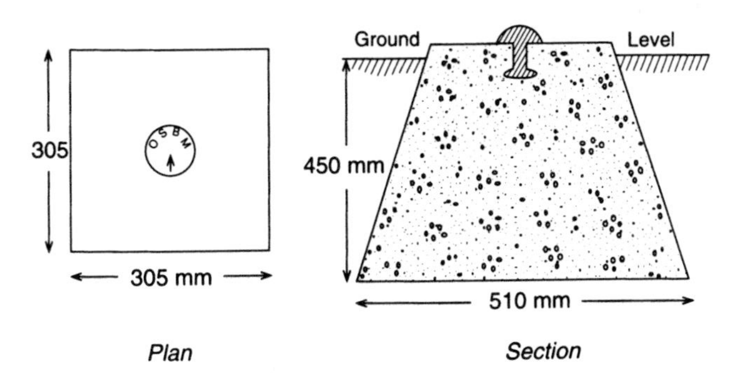

O.S.B.M.bolts are established on horizontal surfaces where no suitable site exists for the emplacement of a flush bracket or cut bench mark.

They are made of brass and have a mushroom shaped head. The letters O.S.B.M. and an arrow pointing to the centre are engraved on the head of the bolt. Typical sites for the bolts are:

(a) Living rock.
(b) Foundation abutments to buildings, etc.
(c) Steps, ledges, etc.
(d) Concrete blocks (as in fig.)

Fig. 3.4 *Bolt bench mark*

3.3.5 Reduced level (RL)

The RL of a point is its height above or below a reference datum.

FUNDAMENTAL BENCH MARK

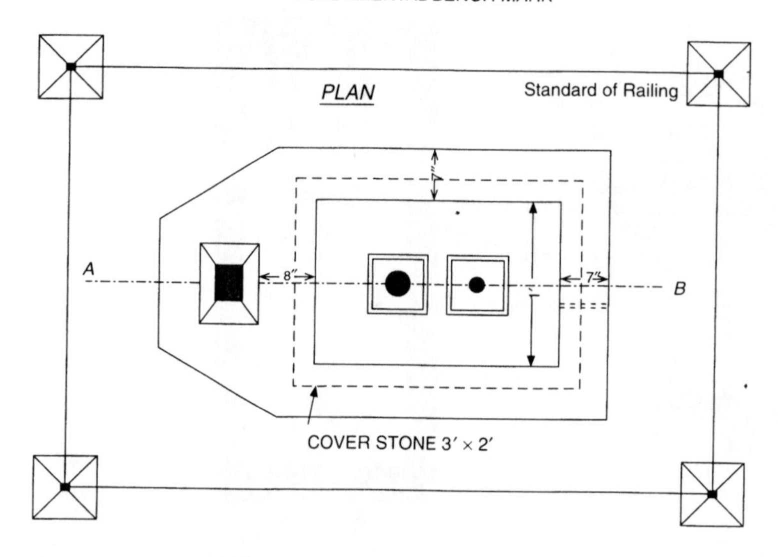

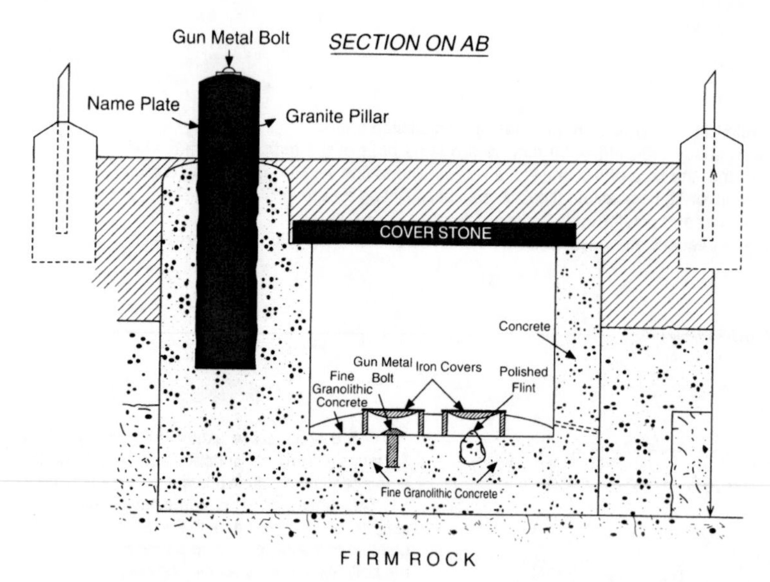

DESCRIPTION

The sites are specially selected with reference to the geological structure, so that they may be placed on sound strata clear of areas liable to subsidence. They are established along the Geodetic lines of levels throughout Great Britain at approximately 50 kilometre intervals. They have three reference points, two of which, a gun metal bolt and a flint are contained in a buried chamber. The third point is a gun metal bolt set in the top of a pillar projecting about I foot above ground level.

The piller bolt is the reference point to be used by Tertiary Levellers and other users.

The buried chamber is only opened on instructions from Headquarters.

Some Fundamental Bench Marks are enclosed by iron railings, this was done where necessary, as a protective measure. These marks are generally referred to as F.B.M's.

Fig. 3.5 *Fundamental bench mark*

3.4 CURVATURE AND REFRACTION

Figure 3.6 shows two points A and B at exactly the same level. An instrument set up at X would give a horizontal line of sight through X'. If a graduated levelling staff is held vertically on A the horizontal line would give the reading A'. Theoretically, as B is at the same level as A, the staff reading should be identical (B'). This would require a level line of sight; the instrument, however, gives a horizontal line and a reading at B'' (ignoring refraction). Subtracting vertical height AA' from BB'' indicates that point B is lower than point A by the amount $B'B''$. This error (c) is caused by the curvature of the Earth and its value may be calculated as follows:

With reference to *Figure 3.7*, in which the instrument heights are ignored and the earth is assumed to be sperical with a radius of R:

$$(X'B'')^2 = (OB'')^2 - (OX')^2 = (R+c)^2 - R^2 = R^2 + 2Rc + c^2 - R^2 = \left(2Rc + c^2\right)$$

As both c and the instrument heights have relatively small values the distance $X'B''$ may be assumed equal to the arc distance $XB = D$. Therefore

$$D = \left(2Rc + c^2\right)^{\frac{1}{2}}$$

Now as c is very small compared with R, c^2 may be ignored, giving

$$c = D^2/2R \tag{3.1}$$

Taking the distance D in kilometres and an average value for R equal to 6370 km, we have

$$c = (D \times 1000)^2/2 \times 6370 \times 1000$$
$$c = 0.0785D^2 \tag{3.2}$$

with the value of c in metres, when D is in kilometres.

In practice the staff reading in *Figure 3.6* would not be at B'' but at Y due to refraction of the line of sight through the atmosphere. In general it is considered that the effect is to bend the line of sight down,

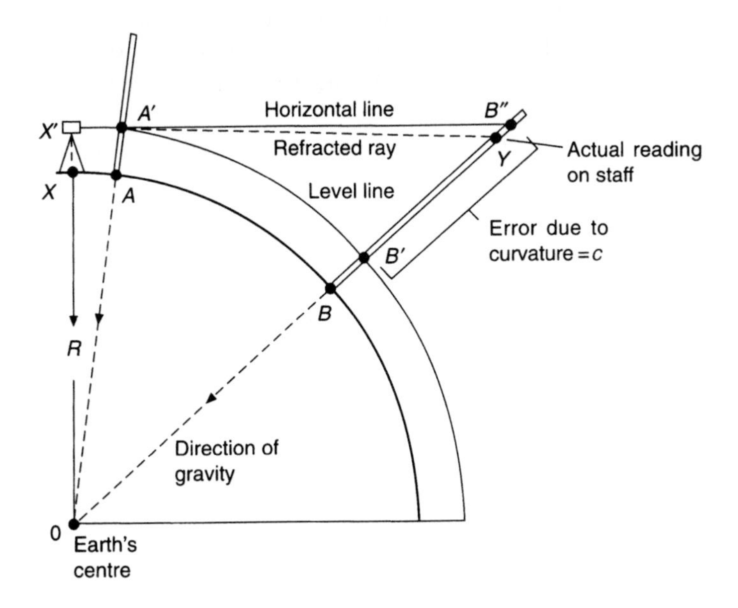

Fig. 3.6 *Horizontal and level lines*

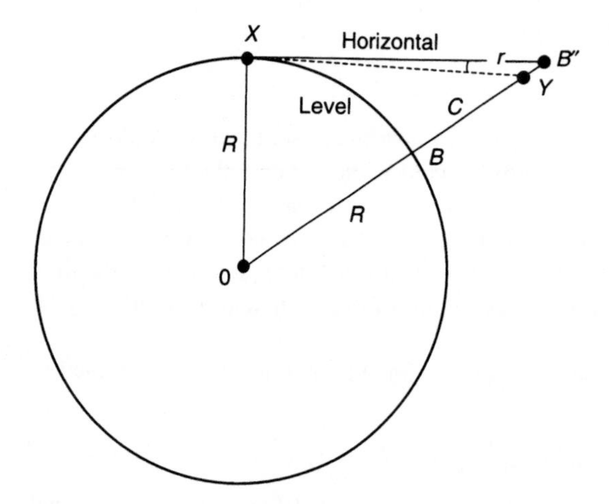

Fig. 3.7 *Refraction effect*

reducing the effect of curvature by 1/7th. Thus the combined effect of curvature and refraction $(c - r)$ is $(6/7)(0.0785D^2)$, i.e.

$$(c - r) = 0.0673D^2 \tag{3.3}$$

Thus if D is 122 m the value of $(c - r)$ is only 1 mm. So in tertiary levelling, where the length of sights are generally 25–30 m, the effect may be ignored.

It should be noted that although the effect of refraction has been shown to bend the line of sight down by an amount equal to 1/7th that of the effect of curvature, this is a most unreliable assumption for precise levelling.

Refraction is largely a function of atmospheric pressure and temperature gradients, which may cause the bending to be up or down by extremely variable amounts.

There are basically three types of temperature gradient ($\mathrm{d}T/\mathrm{d}h$):

(1) *Absorption*: occurs mainly at night when the colder ground absorbs heat from the atmosphere. This causes the atmospheric temperature to increase with distance from the ground and $\mathrm{d}T/\mathrm{d}h > 0$.
(2) *Emission*: occurs mainly during the day when the warmer ground emits heat into the atmosphere, resulting in a negative temperature gradient, i.e. $\mathrm{d}T/\mathrm{d}h < 0$.
(3) *Equilibrium*: no heat transfer takes place ($\mathrm{d}T/\mathrm{d}h = 0$) and occurs only briefly in the evening and morning.

The result of $\mathrm{d}T/\mathrm{d}h < 0$ is to cause the light ray to be convex to the ground rather than concave as generally shown. This effect increases the closer to the ground the light ray gets and errors in the region of 5 mm/km have occurred.

Thus, wherever possible, staff readings should be kept at least 0.5 m above the ground, using short observation distances (25 m) equalized for backsight and foresight.

3.5 EQUIPMENT

The equipment used in the levelling process comprises optical levels and graduated staffs. Basically, the optical level consists of a telescope fitted with a spirit bubble or automatic compensator to ensure long horizontal sights onto the vertically held graduated staff (*Figure 3.8*).

3.5.1 Levelling staff

Levelling staffs are made of wood, metal or glass fibre and graduated in metres and centimetres. The alternate metre lengths are usually shown in black and red on a white background. The majority of staffs are telescopic or socketed in three or four sections for easy carrying. Although the graduations can take various forms, the type adopted in the UK is the British Standard (BS 4484) E-pattern type as shown in *Figure 3.9*. The smallest graduation on the staff is 0.01 m and readings are estimated to the nearest millimetre. As the staff must be held vertical during observation it should be fitted with a circular bubble.

3.5.2 Optical levels

The types of level found in general use are the tilting, the automatic level, and digital levels.

(1) Tilting level

Figure 3.10 shows the telescope of the tilting level pivoted at the centre of the tribrach; an attachment plate with three footscrews. The footscrews are used to centre the circular bubble, thereby setting the telescope approximately in a horizontal plane. After the telescope has been focused on the staff, the line of sight is set more precisely to the horizontal using the highly sensitive tubular bubble and the tilting screw that raises or lowers one end of the telescope.

The double concave internal focusing lens is moved along the telescope tube by its focusing screw until the image of the staff is brought into focus on the cross-hairs. The Ramsden eyepiece, with a magnification of about 35 diameters, is then used to view the image in the plane of the cross-hairs.

Fig. 3.8 *Levelling procedure – the observer uses an automatic level to view a staff held vertically on a levelling plate*

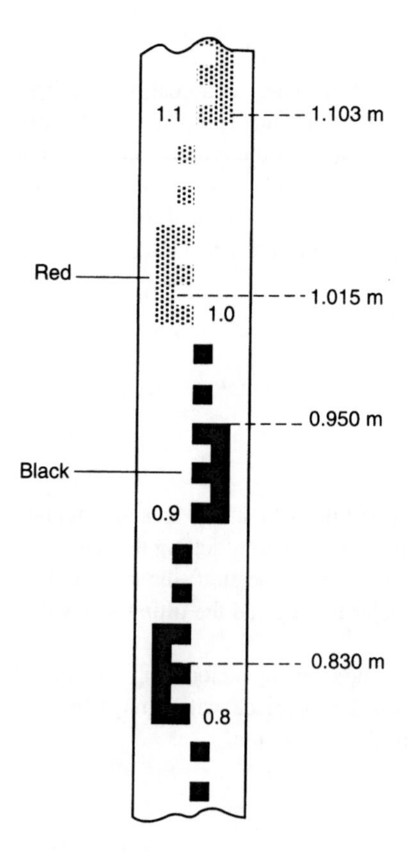

Fig. 3.9 *Levelling staff*

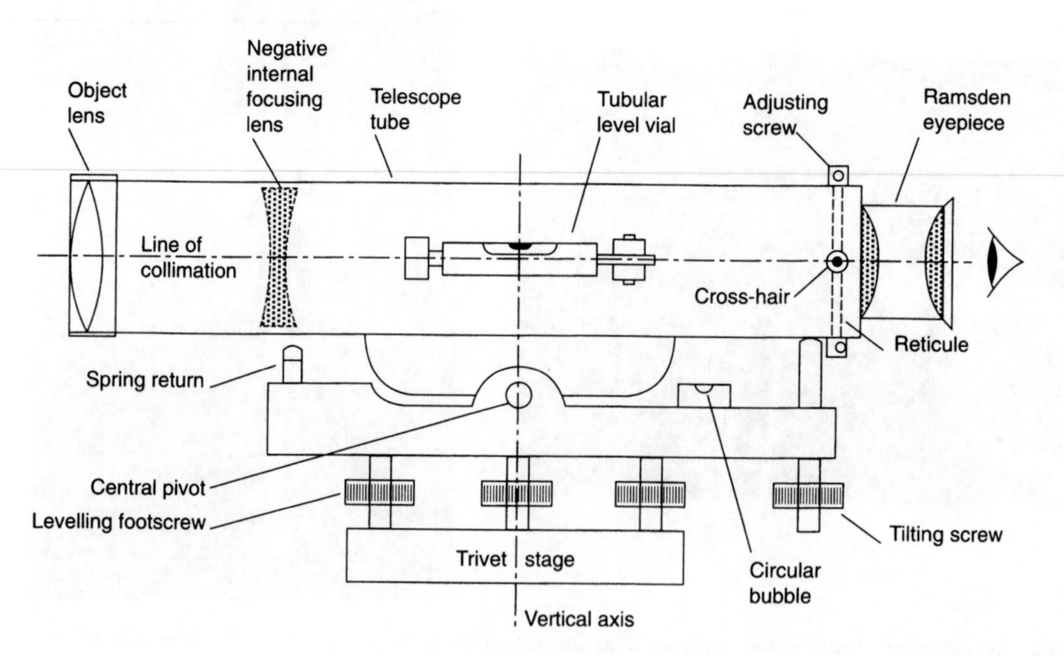

Fig. 3.10 *Tilting level*

The cross-hairs which are etched onto a circle of fine glass plate called a reticule must be brought into sharp focus by the eyepiece focusing screw prior to commencing observations. This process is necessary to remove any cross-hair parallax caused by the image of the staff being brought to a focus in front of or behind the cross-hair. The presence of parallax can be checked by moving the head from side to side or up and down when looking through the telescope. If the image of the staff does not coincide with the cross-hair, movement of the observer's head will cause the cross-hair to move relative to the staff image. The adjusting procedure is therefore:

(1) Using the eyepiece focusing screw, bring the cross-hair into very sharp focus against a light background such as a sheet of blank paper held in front of the object lens.
(2) Now focus on the staff using the main focusing screw until a sharp image is obtained without losing the clear image of the cross-hair.
(3) Check by moving your head from side to side several times. Repeat the whole process if necessary.

Different types of cross-hair are shown in *Figure 3.11*. A line from the centre of the cross-hair and passing through the centre of the object lens is the line of sight or line of collimation of the telescope.

The sensitivity of the tubular spirit bubble is determined by its radius of curvature (R) (*Figure 3.12*); the larger the radius, the more sensitive the bubble. It is filled with sufficient synthetic alcohol to leave a small air bubble in the tube. The tube is graduated generally in intervals of 2 mm.

If the bubble moves off centre by one such interval it represents an angular tilt of the line of sight of 20 seconds of arc. Thus if 2 mm subtends $\theta = 20''$, then:

$$R = (2 \text{ mm} \times 206\,265)/20'' = 20.63 \text{ m}$$

The bubble attached to the tilting level may be viewed directly or by means of a coincidence reading system (*Figure 3.13*). In this latter system the two ends of the bubble are viewed and appear as shown at (a) and (b). (a) shows the image when the bubble is centred by means of the tilting screw; (b) shows the image when the bubble is off centre. This method of viewing the bubble is four or five times more accurate than direct viewing.

The main characteristics defining the quality of the telescope are its powers of magnification, the size of its field of view, the brightness of the image formed and the resolution quality when reading the staff.

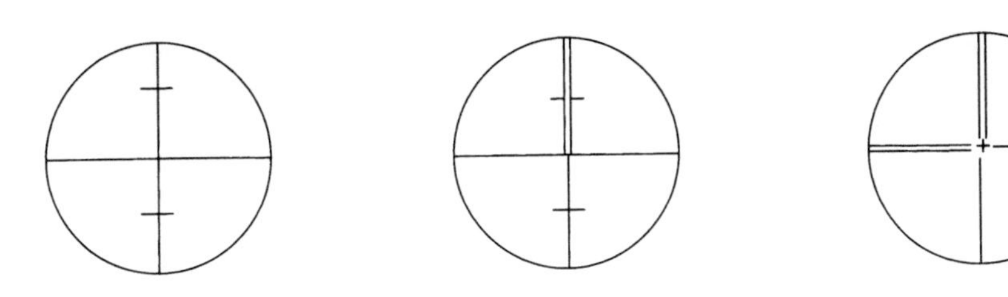

Fig. 3.11 *Cross-hairs*

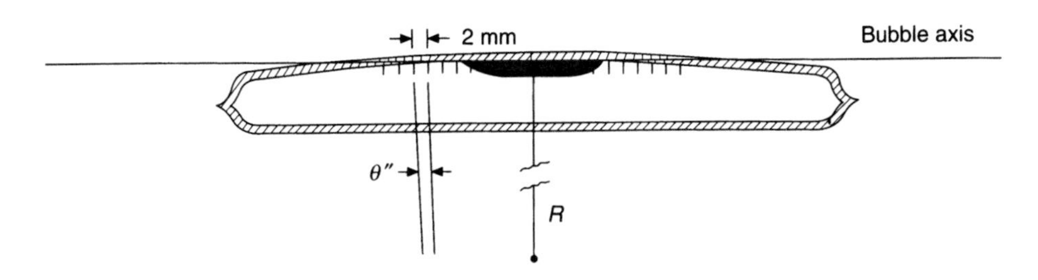

Fig. 3.12 *Tubular bubble*

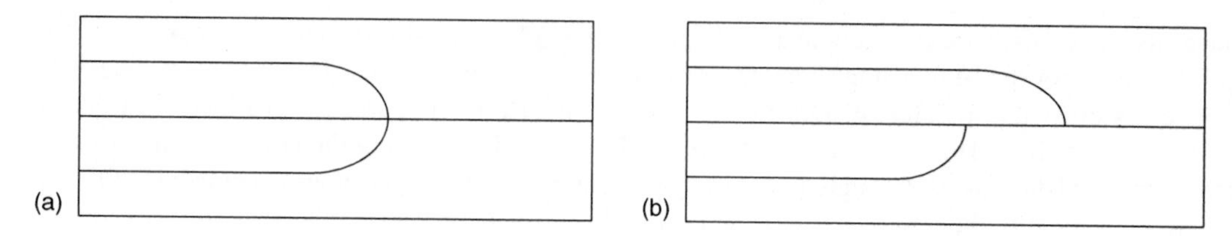

Fig. 3.13 *Bubble coincidence reading system: (a) bubble centred, (b) bubble off centre*

All these are a function of the lens systems used and vary accordingly from low-order builders' levels to very precise geodetic levels.

Magnification is the ratio of the size of the object viewed through the telescope to its apparent size when viewed by the naked eye. Surveying telescopes are limited in their magnification in order to retain their powers of resolution and field of view. Also, the greater the magnification, the greater the effect of heat shimmer, on-site vibration and air turbulence. Telescope magnification lies between 15 and 50 times.

The field of view is a function of the angle of the emerging rays from the eye through the telescope, and varies from 1° to 2°. Image brightness is the ratio of the brightness of the image when viewed through the telescope to the brightness when viewed by the naked eye. It is argued that the lens system, including the reticule, of an internal focusing telescope loses about 40% of the light. The resolution quality or resolving power of the telescope is its ability to define detail and is independent of magnification. It is a function of the effective aperture of the object lens and the wavelength (λ) of light and is represented in angular units. It can be computed from P radians $= 1.2\lambda/(\text{effective aperture})$.

(2) Using a tilting level

(1) Set up the instrument on a firm, secure tripod base.
(2) Centralize the circular bubble using the footscrews or ball and socket arrangement.
(3) Eliminate parallax.
(4) Centre the vertical cross-hair on the levelling staff and clamp the telescope. Use the horizontal slow-motion screw if necessary to ensure exact alignment.
(5) Focus onto the staff.
(6) Carefully centre the tubular bubble using the tilting screw.
(7) With the staff in the field of view as shown in *Figure 3.14* note the staff reading (1.045) and record it.

Operations (4) to (7) must be repeated for each new staff reading.

(3) Automatic levels

The automatic level is easily recognized by its clean, uncluttered appearance. It does not have a tilting screw or a tubular bubble as the telescope is rigidly fixed to the tribrach and the line of sight is made horizontal by a compensator inside the telescope.

The basic concept of the automatic level can be likened to a telescope rigidly fixed at right angles to a pendulum. Under the influence of gravity, the pendulum will swing into the vertical, as defined by a suspended plumb-bob and the telescope will move into a horizontal plane.

As the automatic level is only approximately levelled by means of its low-sensitivity circular bubble, the collimation axis of the instrument will be inclined to the horizontal by a small angle α (*Figure 3.15*) so the entering ray would strike the reticule at a with a displacement of ab equal to $f\alpha$. The compensator

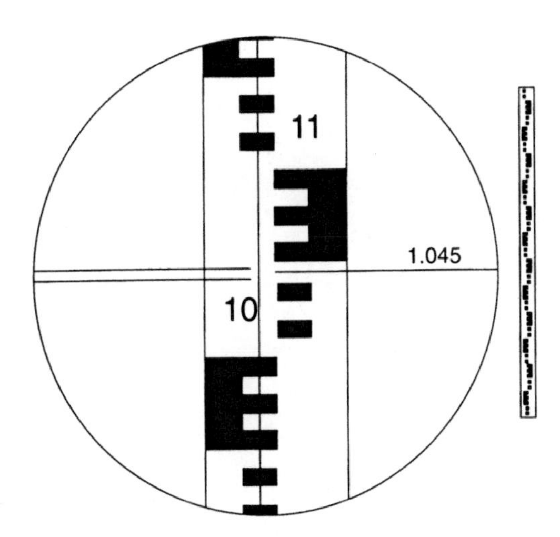

Fig. 3.14 *Staff and cross-hairs*

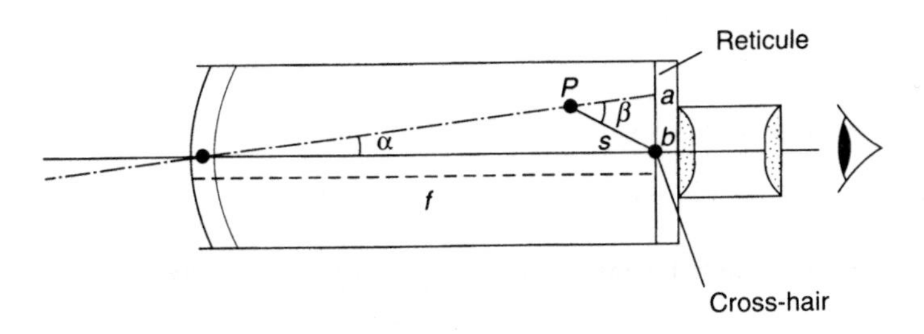

Fig. 3.15 *Principle of compensator*

situated at P would need to redirect the ray to pass through the cross-hair at b. Thus

$$f\alpha = ab = s\beta$$

and $$\beta = \frac{f\alpha}{s} = n\alpha$$

It can be seen from this that the positioning of the compensator is a significant feature of the compensation process. For instance, if the compensator is fixed halfway along the telescope, then $s \approx f/2$ and $n = 2$, giving $\beta = 2\alpha$. There is a limit to the working range of the compensator, about 20′; hence the need of a circular bubble.

In order, therefore, to compensate for the slight residual tilts of the telescope, the compensator requires a reflecting surface fixed to the telescope, movable surfaces influenced by the force of gravity and a dampening device (air or magnetic) to swiftly bring the moving surfaces to rest and permit rapid viewing of the staff. Such an arrangement is illustrated in *Figure 3.16*.

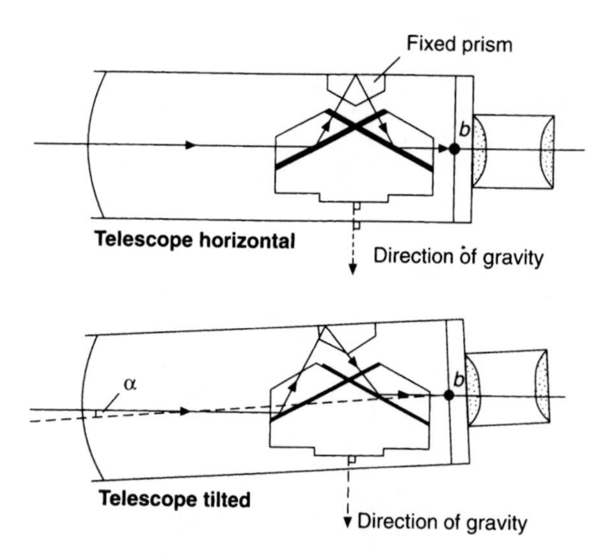

Fig. 3.16 *Suspended compensation*

The advantages of the automatic level over the tilting level are:

(1) Much easier to use, as it gives an erect image of the staff.
(2) Rapid operation, giving greater productivity.
(3) No chance of reading the staff without setting the bubble central, as can occur with a tilting level.
(4) No bubble setting error.

A disadvantage is that it is difficult to use where there is vibration caused by wind, traffic or, say, piling operations on site, resulting in oscillation of the compensator. Improved damping systems have, however, greatly reduced this defect. During periods of vibration it may be possible to reduce the effect by lightly touching a tripod leg.

(4) Using an automatic level

The operations are identical to those for the tilting level with the omission of operation (6). Some automatic levels have a button, which when pressed moves the compensator to prevent it sticking. This should be done just prior to reading the staff, when the cross-hair will be seen to move. Another approach to ensure that the compensator is working is to move it slightly off level and note if the reading on the staff is unaltered, thereby proving the compensator is working.

3.6 INSTRUMENT ADJUSTMENT

For equipment to give the best possible results it should be frequently tested and, if necessary, adjusted. Surveying equipment receives continuous and often brutal use on construction sites. In all such cases a calibration base should be established to permit weekly checks on the equipment.

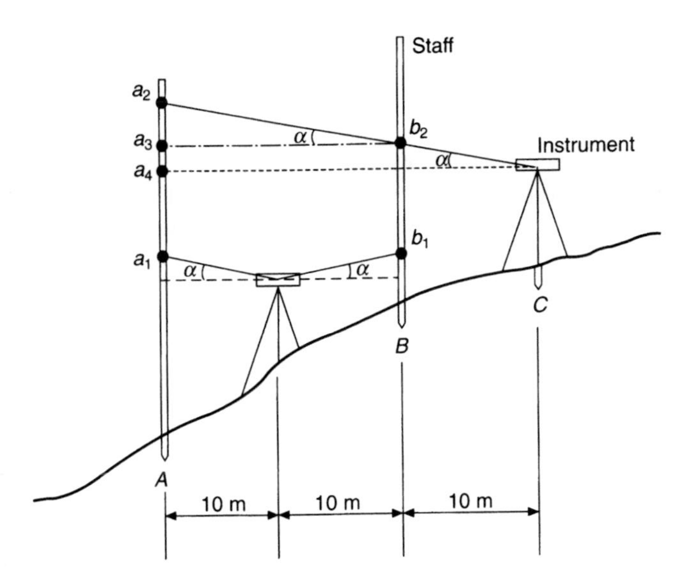

Fig. 3.17 *Two-peg test*

3.6.1 Tilting level

The tilting level requires adjustment for collimation error only. Collimation error occurs if the line of sight is not truly horizontal when the tubular bubble is centred, i.e. the line of sight is inclined up or down from the horizontal. A check known as the 'Two-Peg Test' is used, the procedure being as follows (*Figure 3.17*):

(a) Set up the instrument midway between two pegs A and B set, say, 20 m apart and note the staff readings, a_1 and b_1, equal to, say, 1.500 m and 0.500 m respectively.

Let us assume that the line of sight is inclined up by an angle of α; as the lengths of the sights are equal (10 m), the error in each staff reading will be equal and so cancel out, resulting in a 'true' difference in level between A and B.

$$\Delta H_{\text{TRUE}} = (a_1 - b_1) = (1.500 - 0.500) = 1.000 \text{ m}$$

Thus we know that A is truly lower than B by 1.000 m. We do not at this stage know that collimation error is present.

(b) Move the instrument to C, which is 10 m from B and in the line AB and observe the staff readings a_2 and b_2 equal to, say, 3.500 m and 2.000 m respectively. Then

$$\Delta H = (a_2 - b_2) = (3.500 - 2.000) = 1.500 \text{ m}$$

Now as $1.500 \neq$ the 'true' value of 1.000, it must be 'false'.

$$\Delta H_{\text{FALSE}} = 1.500 \text{ m}$$

and it is obvious that the instrument possesses a collimation error the amount and direction of which is as yet still unknown, but which has been revealed by the use of unequal sight lengths CB (10 m) and CA (30 m). Had the two values for ΔH been equal, then there would be no collimation error present in the instrument.

(c) Imagine a horizontal line from reading b_2 (2.000 m) cutting the staff at A at reading a_3. Because A is truly 1.000 m below B, the reading at a_3 must be $2.000 + 1.000 = 3.000$ m. However, the actual reading was 3.500 m, and therefore the line of sight of the instrument was too high by 0.500 m in 20 m (the distance between the two pegs). This is the amount and direction of collimation error.

(d) Without moving the instrument from C, the line of sight must be adjusted down until it is horizontal. To do this one must compute the reading (a_4) on staff A that a horizontal sight from C, distance 30 m away, would give.

By simple proportion, as the error in 20 m is 0.500, the error in 30 m = (0.500 × 30)/20 = 0.750 m. Therefore the required reading at a_4 is 3.500 − 0.750 = 2.750 m.

(e) (i) Using the 'tilting screw', tilt the telescope until it reads 2.750 m on the staff. (ii) This movement will cause the tubular bubble to go off centre. Re-centre it by means of its adjusting screws, which will permit the raising or lowering of one end of the bubble.

The whole operation may be repeated if thought necessary.

The above process has been dealt with in great detail, as collimation error is one of the main sources of error in the levelling process.

The diagrams and much of the above detail can be dispensed with if the following is noted:

(1) $(\Delta H_{FALSE} - \Delta H_{TRUE})$ = the amount of collimation error.
(2) If $\Delta H_{FALSE} > \Delta H_{TRUE}$ then the line of sight is inclined *up* and *vice versa*.

An example follows which illustrates this approach.

Worked example

Example 3.1. Assume the same separation for A, B and C. With the instrument midway between A and B the readings are A(3.458), B(2.116). With the instrument at C, the readings are A(4.244), B(2.914).

(i) From 'midway' readings, ΔH_{TRUE} = 1.342
(ii) From readings at 'C', ΔH_{FALSE} = 1.330

Amount of collimation error = 0.012 m in 20 m

(iii) $\Delta H_{FALSE} < \Delta H_{TRUE}$, therefore direction of line of sight is *down*
(iv) With instrument at C the reading on A(4.244) must be raised by (0.012 × 30)/20 = 0.018 m to read 4.262 m

Some methods of adjustment advocate placing the instrument close to the staff at B rather than a distance away at C. This can result in error when using the reading on B and is not suitable for precise levels. The above method is satisfactory for all types of level.

For very precise levels, it may be necessary to account for the effect of curvature and refraction when carrying out the above test. See *Equation 3.3*. A distance of 50 m would produce a correction to the staff readings of only −0.17 mm so it can be ignored for all but the most precise work.

An alternative two-peg test that removes the need for placing the instrument exactly midway between the staffs is as follows (*Figure 3.18*). The staffs at pegs A and B are observed by the instrument from positions C and D. The readings are as shown. In this case the collimation of the instrument is given by

$$\tan \alpha = (a_2 - a_1 - b_2 + b_1)/2d$$

3.6.2 Automatic level

There are two tests and adjustments necessary for an automatic level:

(1) To ensure that the line of collimation of the telescope is horizontal, within the limits of the bubble, when the circular bubble is central.
(2) The two-peg test for collimation error.

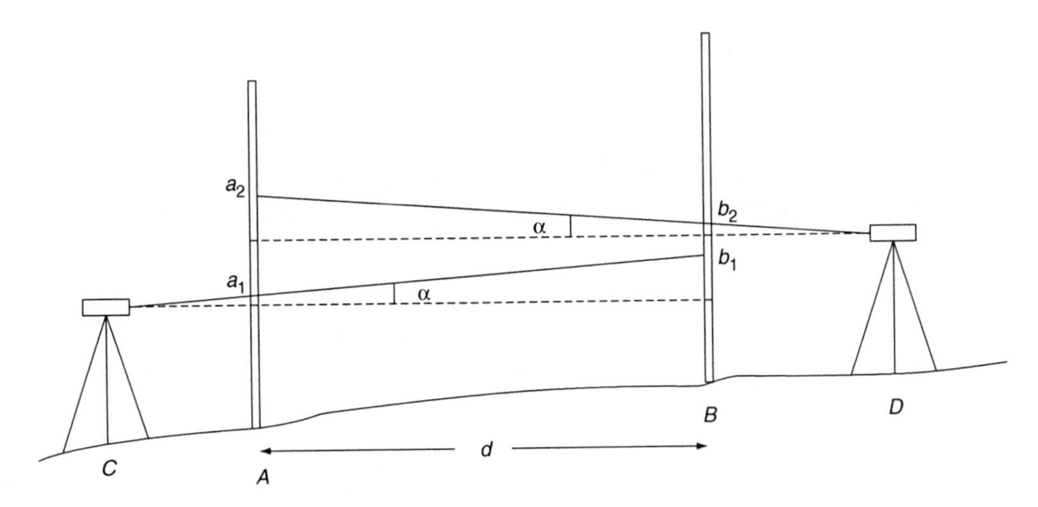

Fig. 3.18 *Alternative two-peg test*

(1) Circular bubble

Although the circular bubble is relatively insensitive, it nevertheless plays an important part in the efficient functioning of the compensator:

(1) The compensator has a limited working range. If the circular bubble is out of adjustment, thereby resulting in excessive tilt of the line of collimation (and the vertical axis), the compensator may not function efficiently or, as it attempts to compensate, the large swing of the pendulum system may cause it to stick in the telescope tube.
(2) The compensator gives the most accurate results near the centre of its movement, so even if the bubble is in adjustment, it should be carefully and accurately centred.
(3) The plane of the pendulum swing of the freely suspended surfaces should be parallel to the line of sight, otherwise over- or under-compensation may occur. This would result if the circular bubble were in error transversely. Any residual error of adjustment can be eliminated by centring the bubble with the telescope pointing backwards, whilst at the next instrument set-up it is centred with the telescope pointing forward. This alternating process is continued throughout the levelling.
(4) Inclination of the telescope can cause an error in automatic levels, which does not occur in tilting levels, known as 'height shift'. Due to the inclination of the telescope the centre of the object lens is displaced vertically above or below the centre of the cross-hair, resulting in very small reading errors, but which cannot be tolerated in precise work.

From the above it can be seen that not only must the circular bubble be in adjustment but it should also be accurately centred when in use.

To adjust the bubble, bring it exactly to centre using the footscrews. Now rotate the instrument through 180° about the vertical axis. If the bubble moves off centre, bring it halfway back to centre with the footscrews and then exactly back to the centre using its adjusting screws.

(2) Two-peg test

This is carried out exactly as for the tilting level. However, the line of sight is raised or lowered to its correct reading by moving the cross-hair by means of its adjusting screws.

If the instrument is still unsatisfactory the fault may lie with the compensator, in which case it should be returned to the manufacturer.

3.7 PRINCIPLE OF LEVELLING

The instrument is set up and correctly levelled in order to make the line of sight through the telescope horizontal. If the telescope is turned through 360°, a horizontal plane of sight is swept out. Vertical measurements from this plane, using a graduated levelling staff, enable the relative elevations of ground points to be ascertained. Consider *Figure 3.19* with the instrument set up approximately midway between ground points *A* and *B*. If the reduced level (RL) of point *A* is known and equal to 100.000 m above OD (AOD), then the reading of 3.000 m on a vertically held staff at *A* gives the reduced level of the horizontal line of sight as 103.000 m AOD. This sight onto *A* is termed a backsight (BS) and the reduced level of the line of sight is called the height of the plane of collimation (HPC). Thus:

$$RL_A + BS = HPC$$

The reading of 1.000 m onto the staff at *B* is called a foresight (FS) and shows the ground point *B* to be 1.000 m below HPC; therefore its RL = (103.000 − 1.000) = 102.000 m AOD.

An alternative approach is to subtract the FS from the BS. If the result is positive then the difference is a *rise* from *A* to *B*, and if negative a *fall*, i.e.

$$(3.000 - 1.000) = +2.000 \text{ m rise from } A \text{ to } B;$$

$$\text{therefore, } RL_B = 100.000 + 2.000 = 102.000 \text{ m AOD}$$

This then is the basic concept of levelling which is further developed in *Figure 3.20*.

The field data are entered into a field book that is pre-drawn into rows and columns. An example of levelling observations from a practical project is shown in *Figure 3.21*. Observations are booked using either the rise and fall or the HPC method.

It should be clearly noted that, in practice, the staff readings are taken to three places of decimals, that is, to the nearest millimetre. However, in the following description only one place of decimals is used and the numbers kept very simple to prevent arithmetic interfering with an understanding of the concepts outlined.

The field procedure for obtaining elevations at a series of ground points is as follows.

The instrument is set up at *A* (as in *Figure 3.20*) from which point a horizontal line of sight is possible to the TBM at 1*A*. The first sight to be taken is to the staff held vertically on the TBM and this is called a

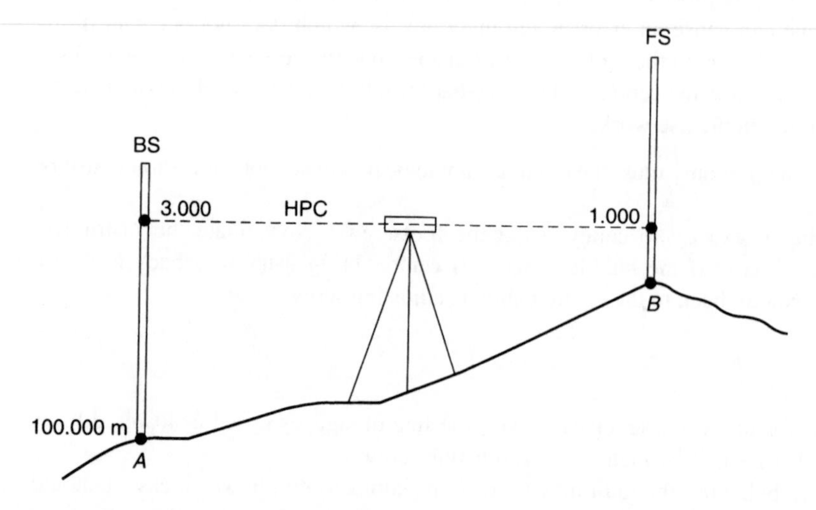

Fig. 3.19 *Basic principle of levelling*

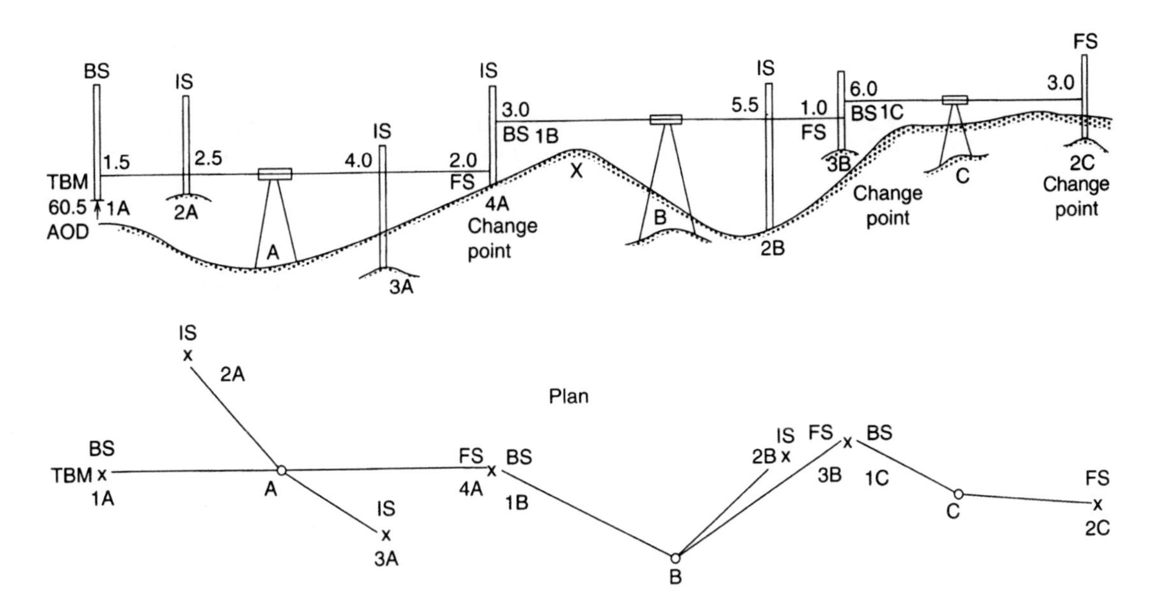

Fig. 3.20 *A levelling line*

Date 26/07/06 Levels taken for Road Alignment
From Chn. 2040 at Bridge N°6 To Chn 3040 at Bridge N°8

Back sight	Inter-mediate	Fore sight	Rise	Fall	Reduced level	Distance	Remarks
2.856					35.688	0m	TBM on Bridge Abut. (N°6)
	1.432		1.424		37.112	20m	Chn. 2060m
		3.543		2.111	35.001	30m	C.P. at Chn. 2070m

Date 26/07/06 Levels taken for Road Alignment
From Chn 2040 at Bridge N°6 To Chn 3040 at Bridge N°8

Back sight	Inter-mediate	Fore sight	Collimation or H.P.C.	Reduced level	Distance	Remarks
2.856			38.544	35.688	0m	TBM on Bridge Abut. (N°6)
	1.432			37.112	20m	Chn 2060m
		3.543		35.001	30m	C.P. at Chn 2070m

Fig. 3.21 *Project levelling observations*

backsight (BS), the value of which (1.5 m) would be entered in the appropriate column of a levelling book. Sights to points 2*A* and 3*A* where further levels relative to the TBM are required are called intermediate sights (IS) and are again entered in the appropriate column of the levelling book. The final sight from this instrument is set up at 4*A* and is called the foresight (FS). It can be seen from the figure that this is as far as one can go with this sight. If, for instance, the staff had been placed at *X*, it would not have been visible and would have had to be moved down the slope, towards the instrument at *A*, until it was visible. As foresight 4*A* is as far as one can see from *A*, it is also called the change point (CP), signifying a change of instrument position to *B*. To achieve continuity in the levelling the staff must remain at *exactly* the same point 4*A* although it must be turned to face the instrument at *B*. It now becomes the BS for the new instrument set-up and the whole procedure is repeated as before. Thus, one must remember that all levelling commences on a BS and finishes on a FS with as many IS in between as are required; and that CPs are always FS/BS. Also, it must be closed back into a known BM to ascertain the misclosure error.

3.7.1 Reduction of levels

From *Figure 3.20* realizing that the line of sight from the instrument at *A* is truly horizontal, it can be seen that the higher reading of 2.5 at point 2*A* indicates that the point is lower than the TBM by 1.0, giving 2*A* a level therefore of 59.5. This can be written as follows:

$$1.5 - 2.5 = -1.0, \text{ indicating a } \textit{fall} \text{ of } 1.0 \text{ from } 1A \text{ to } 2A$$

$$\text{Level of } 2A = 60.5 - 1.0 = 59.5$$

Similarly between 2*A* and 3*A*, the higher reading on 3*A* shows it is 1.5 below 2*A*, thus:

$$2.5 - 4.0 = -1.5 \text{ (fall from } 2A \text{ to } 3A)$$

$$\text{Level of } 3A = \text{level of } 2A - 1.5 = 58.0$$

Finally the *lower* reading on 4*A* shows it to be *higher* than 3*A* by 2.0, thus:

$$4.0 - 2.0 = +2.0, \text{ indicating a } \textit{rise} \text{ from } 3A \text{ to } 4A$$

$$\text{Level of } 4A = \text{level of } 3A + 2.0 = 60.0$$

Now, knowing the *reduced level* (RL) of 4*A*, i.e. 60.0, the process can be repeated for the new instrument position at *B*. This method of reduction is called the *rise-and-fall (R-and-F) method*.

3.7.2 Methods of booking

(1) Rise-and-fall

The following extract of booking is largely self-explanatory. Note that:

(a) Each reading is booked on a separate line except for the BS and FS at change points. The BS is booked on the same line as the FS because it refers to the same point. As each line refers to a specific point it should be noted in the remarks column.
(b) Each reading is subtracted from the previous one, i.e. 2*A* from 1*A*, then 3*A* from 2*A*, 4*A* from 3*A* and stop; the procedure recommencing for the next instrument station, 2*B* from 1*B* and so on.

BS	*IS*	*FS*	*Rise*	*Fall*	*RL*	*Distance*	*Remarks*	
1.5					60.5	0	TBM (60.5)	1*A*
	2.5			1.0	59.5	30		2*A*
	4.0			1.5	58.0	50		3*A*
3.0		2.0	2.0		60.0	70	CP	4*A* (1*B*)
	5.5			2.5	57.5	95		2*B*
6.0		1.0	4.5		62.0	120	CP	3*B* (1*C*)
		3.0	3.0		65.0	160	TBM (65.1)	2*C*
10.5		6.0	9.5	5.0	65.0		Checks	
6.0			5.0		60.5		Misclosure	0.1
4.5			4.5		4.5		*Correct*	

(c) Three very important checks must be applied to the above reductions, namely:

The sum of BS − the sum of FS = sum of rises − sum of falls

= last reduced level − first reduced level

These checks are shown in the above table. It should be emphasized that they are nothing more than checks on the arithmetic of reducing the levelling results; they are in no way indicative of the accuracy of fieldwork.

(d) It follows from the above that the first two checks should be carried out and verified before working out the reduced levels (RL).

(e) Closing error = 0.1, and can be assessed only by connecting the levelling into a BM of known and proved value or connecting back into the starting BM.

(2) Height of collimation

This is the name given to an alternative method of booking. The reduced levels are found simply by subtracting the staff readings from the reduced level of the line of sight (plane of collimation). In *Figure 3.20*, for instance, the *height of the plane of collimation* (HPC) at *A* is obviously (60.5 + 1.5) = 62.0; now 2*A* is 2.5 below this plane, so its level must be (62.0 − 2.5) = 59.5; similarly for 3*A* and 4*A* to give 58.0 and 60.0 respectively. Now the procedure is repeated for *B*. The tabulated form shows how simple this process is:

BS	*IS*	*FS*	*HPC*	*RL*	*Remarks*	
1.5			62.0	60.5	TBM (60.5)	1*A*
	2.5			59.5		2*A*
	4.0			58.0		3*A*
3.0		2.0	63.0	60.0	Change pt	4*A* (1*B*)
	5.5			57.5		2*B*
6.0		1.0	68.0	62.0	Change pt	3*B* (1*C*)
		3.0		65.0	TBM (65.1)	2*C*
10.5	12.0	6.0		65.0	Checks	
6.0				60.5	Misclosure	0.1
4.5				4.5	*Correct*	

Thus it can be seen that:

(a) BS is added to RL to give HPC, i.e. $1.5 + 60.5 = 62.0$.
(b) Remaining staff readings are *subtracted* from HPC to give the RL.
(c) Procedure repeated for next instrument set-up at *B*, i.e. $3.0 + 60.0 = 63.0$.
(d) Two checks same as R-and-F method, i.e.:

$$\text{sum of BS} - \text{sum of FS} = \text{last RL} - \text{first RL}$$

(e) The above two checks are not complete; for instance, if when taking 2.5 from 62.0 to get RL of 59.5, one wrote it as 69.5, this error of 10 would remain undetected. Thus the *intermediate* sights are *not* checked by those procedures in (d) above and the following cumbersome check must be carried out:

$$\text{sum of all the RL except the first} = (\text{sum of each HPC multiplied by the number of IS or FS taken from it}) - (\text{sum of IS and FS})$$

e.g. $362.0 = 62.0 \times 3 + 63.0 \times 2 + 68.0 \times 1 - (12.0 + 6.0)$

3.7.3 Inverted sights

Figure 3.22 shows inverted sights at *B*, *C* and *D* to the underside of a structure. It is obvious from the drawing that the levels of these points are obtained by simply adding the staff readings to the HPC to give $B = 65.0$, $C = 63.0$ and $D = 65.0$; *E* is obtained in the usual way and equals 59.5. However, the problem of inverted sights is completely eliminated if one simply treats them as *negative* quantities.

BS	*IS*	*FS*	*Rise*	*Fall*	*HPC*	*RL*	*Remarks*	
2.0					62.0	60.0	TBM	*A*
	−3.0		5.0			65.0		*B*
	−1.0			2.0		63.0		*C*
	−3.0		2.0			65.0		*D*
		2.5		5.5		59.5	Misclosure	*E* (59.55)
2.0	−7.0	2.5	7.0	7.5		60.0	Checks	
		2.0		7.0		59.5	Misclosure	0.05
		0.5		0.5		0.5	*Correct*	

R-and-F method			*HPC method*
$2.0 - (-3.0) = +5.0$	=	Rise	$62.0 - (-3.0) = 65.0$
$-3.0 - (-1.0) = -2.0$	=	Fall	$62.0 - (-1.0) = 63.0$
$-1.0 - (-3.0) = +2.0$	=	Rise	$62.0 - (-3.0) = 65.0$
$-3.0 - 2.5 = -5.5$	=	Fall	$62.0 - (+2.5) = 59.5$

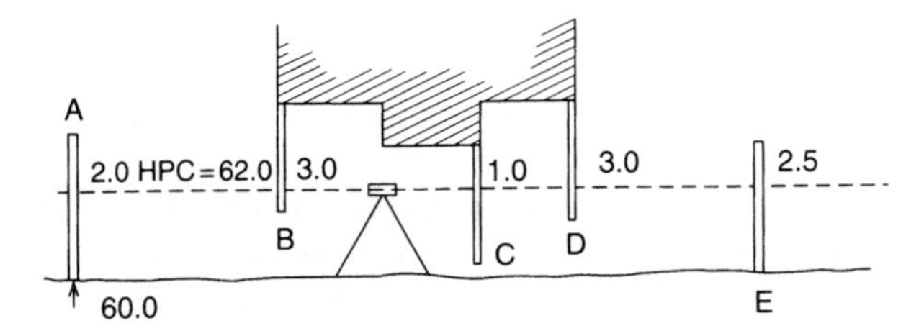

Fig. 3.22 *Inverted sights*

In the checks, inverted sights are treated as negative quantities; for example check for IS in HPC method gives:

$$252.5 = (62.0 \times 4.0) - (-7.0 + 2.5)$$

$$= (248.0) - (-4.5) = 248.0 + 4.5 = 252.5$$

3.7.4 Comparison of methods

The rise-and-fall method of booking is recommended as it affords a complete arithmetical check on all the observations. Although the HPC method appears superior where there are a lot of intermediate sights, it must be remembered that there is no simple straightforward check on their reduction.

The HPC method is useful when setting out levels on site. For instance, assume that a construction level, for setting formwork, of 20 m AOD is required. A BS to an adjacent TBM results in an HPC of 20.834 m; a staff reading of 0.834 would then fix the bottom of the staff at the required level.

3.8 SOURCES OF ERROR

All measurements have error. In the case of levelling, these errors will be instrumental, observational and natural.

3.8.1 Instrumental errors

(1) The main source of instrumental error is residual collimation error. As already indicated, keeping the horizontal lengths of the backsights and foresights at each instrument position equal will cancel this error. Where the observational distances are unequal, the error will be proportional to the difference in distances.

    The easiest approach to equalizing the sight distances is to pace from backsight to instrument and then set up the foresight change point the same number of paces away from the instrument.

(2) Parallax error has already been described.

(3) Staff graduation errors may result from wear and tear or repairs and the staffs should be checked against a steel tape. Zero error of the staff, caused by excessive wear of the base, will cancel out on backsight and foresight differences. However, if two staffs are used, errors will result unless calibration corrections are applied.

(4) In the case of the tripod, loose fixings will cause twisting and movement of the tripod head. Overtight fixings make it difficult to open out the tripod correctly. Loose tripod shoes will also result in unstable set-ups.

3.8.2 Observational errors

(1) Levelling involves vertical measurements relative to a horizontal plane so it is important to ensure that the staff is held strictly vertical.

It is often suggested that one should rock the staff back and forth in the direction of the line of sight and accept the minimum reading as the truly vertical one. However, as shown in *Figure 3.23*, this concept is incorrect when using a flat-bottomed staff on flat ground, due to the fact that the staff is not being tilted about its face. Thus it is preferable to use a staff bubble, which should be checked frequently with the aid of a plumb-bob.

(2) There may be errors in reading the staff, particularly when using a tilting level which gives an inverted image. These errors may result from inexperience, poor observation conditions or overlong sights. Limit the length of sight to about 25–30 m, to ensure the graduations are clearly defined.

(3) Ensure that the staff is correctly extended or assembled. In the case of extending staffs, listen for the click of the spring joint and check the face of the staff to ensure continuity of readings. This also applies to jointed staffs.

(4) Do not move the staff off the CP position, particularly when turning it to face the new instrument position. Always use a well defined and stable position for CPs. Levelling plates (*Figure 3.24*) should be used on soft ground.

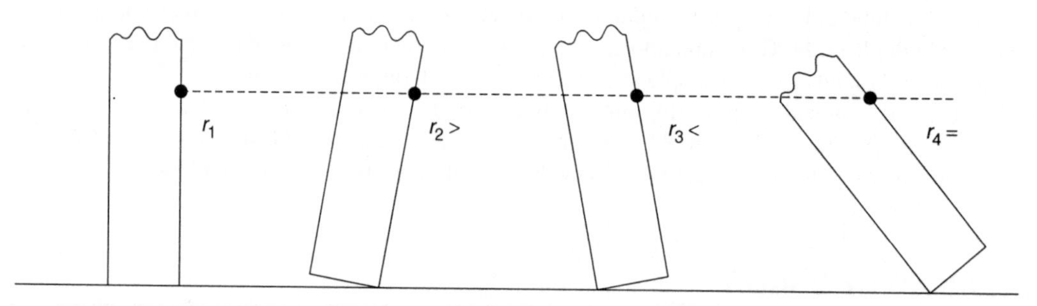

Fig. 3.23 *The effect of rocking the staff on staff readings r_1, r_2, r_3, r_4*

Fig. 3.24 *Levelling plate*

(5) Avoid settlement of the tripod, which may alter the height of collimation between sights or tilt the line of sight. Set up on firm ground, with the tripod feet firmly thrust well into the ground. On pavements, locate the tripod shoes in existing cracks or joins. In precise levelling, the use of two staffs helps to reduce this effect.

Observers should also refrain from touching or leaning on the tripod during observation.

(6) Booking errors can, of course, ruin good field work. Neat, clear, correct booking of field data is essential in any surveying operation. Typical booking errors in levelling are entering the values in the wrong columns or on the wrong lines, transposing figures such as 3.538 to 3.583 and making arithmetical errors in the reduction process. Very often, the use of pocket calculators simply enables the booker to make the errors quicker.

To avoid this error source, use neat, legible figures; read the booked value back to the observer and have them check the staff reading again; reduce the data as it is recorded.

(7) When using a tilting level remember to level the tubular bubble with the tilting screw prior to each new staff reading. With the automatic level, carefully centre the circular bubble and make sure the compensator is not sticking.

Residual compensator errors are counteracted by centring the circular bubble with the instrument pointing backwards at the first instrument set-up and forward at the next. This procedure is continued throughout the levelling.

3.8.3 Natural errors

(1) Curvature and refraction have already been dealt with. Their effects are minimized by equal observation distances to backsight and foresight at each set-up and readings more than 0.5 m above the ground.
(2) Wind can cause instrument vibration and make the staff difficult to hold in a steady position. Precise levelling is impossible in strong winds. In tertiary levelling keep the staff to its shortest length and use a wind break to shelter the instrument.
(3) Heat shimmer can make the staff reading difficult if not impossible and may make it necessary to delay the work to an overcast day. In hot sunny climes, carry out the work early in the morning or in the evening.

Careful consideration of the above error sources, combined with regularly calibrated equipment, will ensure the best possible results but will never preclude random errors of observation.

3.9 CLOSURE TOLERANCES

It is important to realize that the amount of misclosure in levelling can only be assessed by:

(1) Connecting the levelling back to the BM from which it started, or
(2) Connecting into another BM of known and proved value.

When the misclosure is assessed, one must then decide if it is acceptable or not.

In many cases the engineer may make the decision based on his/her knowledge of the project and the tolerances required.

Alternatively the permissible criteria may be based on the distance levelled or the number of set-ups involved.

A common criterion used to assess the misclosure (E) is:

$$E = m(K)^{\frac{1}{2}} \tag{3.4}$$

where K = distance levelled in kilometres, m = a constant with units of millimetres, and E = the allowable misclosure in millimetres.

The value of m may vary from 2 mm for precise levelling to 12 mm or more for engineering levelling.

In many cases in engineering, the distance involved is quite short but the number of set-ups quite high, in which case the following criterion may be used:

$$E = m(n)^{\frac{1}{2}} \tag{3.5}$$

where n = the number of set-ups, and m = a constant in millimetres.

As this criterion would tend to be used only for construction levelling, the value for m may be a matter of professional judgement. A value frequently used is ±5 mm.

3.10 ERROR DISTRIBUTION

In previous levelling examples in this chapter misclosures have been shown. The misclosure cannot be ignored and the error must be distributed among the points concerned. In the case of a levelling circuit, a simple method of distribution is to allocate the error in proportion to the distance levelled. For instance, consider a levelling circuit commencing from a BM at A, to establish other BMs at B, C, D and E (*Figure 3.25*) for which the heights have been computed without taking the misclosure into account.

On completing the circuit the observed value for the BM at A is 20.018 m compared, with its known and hence starting value of 20.000 m, so the misclosure is 0.018 m. The distance levelled is 5.7 km. Considering the purpose of the work, the terrain and observational conditions, it is decided to adopt a value for m of 12 mm. Hence the acceptable misclosure is $12(5.7)^{\frac{1}{2}} = 29$ mm, so the levelling is acceptable.

The difference in heights is corrected by (0.018/5.7) × distance in kilometres travelled. Therefore correction to $AB = -0.005$ m, to $BC = -0.002$ m, to $CD = -0.003$ m, to $DE = -0.006$ m and to $EA = -0.002$ m. The values of the BMs will then be $B = 28.561$ m, $C = 35.003$ m, $D = 30.640$ m, $E = 22.829$ m and $A = 20.000$ m.

In many instances, a closing loop with known distances is not the method used and each reduced level is adjusted in proportion to the cumulative number of set-ups to that point from the start. Consider the table

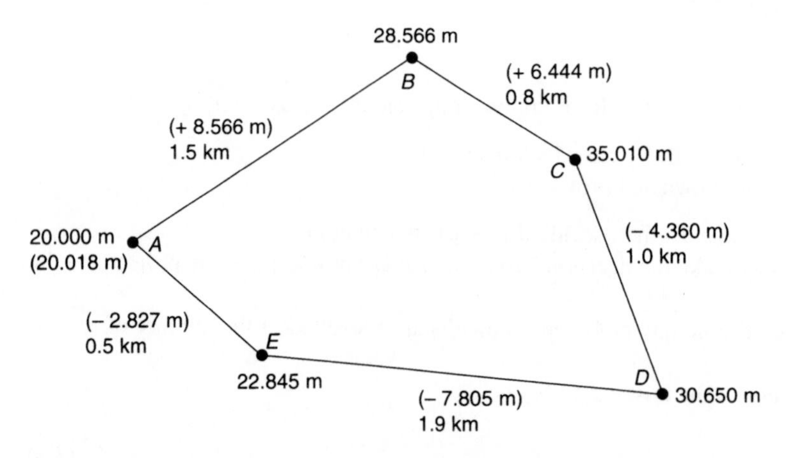

Fig. 3.25 *Levelling circuit*

below which shows the observations for a short section of levelling between two bench marks of known height:

BS	*IS*	*FS*	*Rise*	*Fall*	*R.L.*	*Adj.*	*Final R.L.*	*Remarks*
1.361					20.842		20.842	TBM 'A'
	2.844			1.483	19.359	–0.002	19.357	
	2.018		0.826		20.185	–0.002	20.183	
0.855		3.015		0.997	19.188	–0.002	19.186	C.P.
	0.611		0.244		19.432	–0.004	19.428	
2.741		1.805		1.194	18.238	–0.004	18.234	C.P.
2.855		1.711	1.030		19.268	–0.006	19.262	C.P.
	1.362		1.493		20.761	–0.008	20.753	
	2.111			0.749	20.012	–0.008	20.004	
	0.856		1.255		21.267	–0.008	21.259	
		2.015		1.159	20.108	–0.008	20.100	TBM 'B' (20.100)
7.812		8.546	4.848	5.582	20.842			
		7.812		4.848	20.108			
		0.734		0.734	0.734			Arith. check

(1) There are four set-ups, and therefore $E = 5(4)^{\frac{1}{2}} = 0.010$ m. As the misclosure is only 0.008 m, the levelling is acceptable.
(2) The correction per set-up is $(0.008/4) = -0.002$ m and is cumulative as shown in the table.

3.11 LEVELLING APPLICATIONS

Of all the surveying operations used in construction, levelling is the most common. Practically every aspect of a construction project requires some application of the levelling process. The more general are as follows.

3.11.1 Sectional levelling

This type of levelling is used to produce ground profiles for use in the design of roads, railways and pipelines.

In the case of such projects, the route centre-line is set out using pegs at 10 m, 20 m or 30 m intervals. Levels are then taken at these peg positions and at critical points such as sudden changes in ground profiles, road crossings, ditches, bridges, culverts, etc. A plot of these elevations is called a longitudinal section. When plotting, the vertical scale is exaggerated compared with the horizontal, usually in the ratio of 10 : 1. The longitudinal section is then used in the vertical design process to produce formation levels for the proposed route design (*Figure 3.26*).

Whilst the above process produces information along a centre-line only, cross-sectional levelling extends that information at 90° to the centre-line for 20–30 m each side. At each centre-line peg the levels are taken to all points of interest on either side. Where the ground is featureless, levels at 5 m intervals or less are taken. In this way a ground profile at right angles to the centre-line is obtained. When the design template

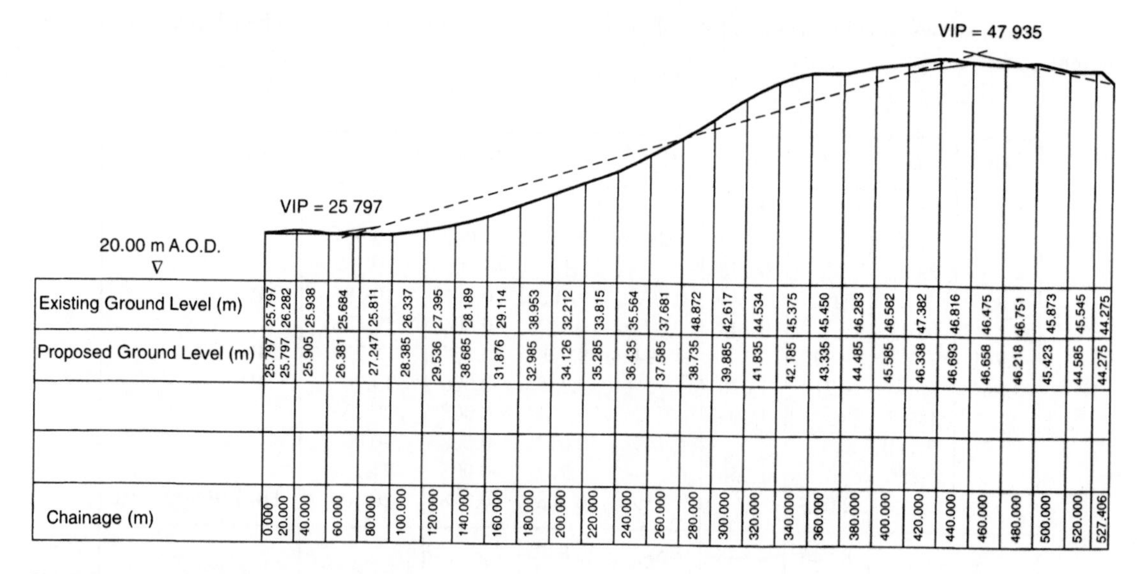

Fig. 3.26 *Longitudinal section of proposed route*

showing the road details and side slopes is plotted at formation level, a cross-sectional area is produced, which can later be used to compute volumes of earthwork. When plotting cross-sections the vertical and horizontal scales are the same, to permit easy scaling of the area and side slopes (*Figure 3.27*).

From the above it can be seen that sectional levelling also requires the measurement of horizontal distance between the points whose elevations are obtained. As the process involves the observation of many points, it is important to connect to existing BMs at regular intervals. In most cases of route construction, one of the earliest tasks is to establish BMs at 100 m intervals throughout the area of interest.

Levelling which does not require the measurement of distance, such as establishing BMs at known positions, is sometimes called 'fly levelling'.

3.11.2 Contouring

A contour is a horizontal curve connecting points of equal elevation. Contours graphically represent, in a two-dimensional format on a plan or map, the shape or morphology of the terrain. The vertical distance between contour lines is called the contour interval. Depending on the accuracy required, they may be plotted at 0.1 m to 0.5 m intervals in flat terrain and at 1 m to 10 m intervals in undulating terrain. The interval chosen depends on:

(1) The type of project involved; for instance, contouring an airstrip requires an extremely small contour interval.
(2) The type of terrain, flat or undulating.
(3) The cost, for the smaller the interval the greater the amount of field data required, resulting in greater expense.

Contours are generally well understood so only a few of their most important properties will be outlined here.

(1) Contours are perpendicular to the direction of maximum slope.
(2) The horizontal separation between contour lines indicates the steepness of the ground. Close spacing defines steep slopes, wide spacing gentle slopes.
(3) Highly irregular contours define rugged, often mountainous terrain.

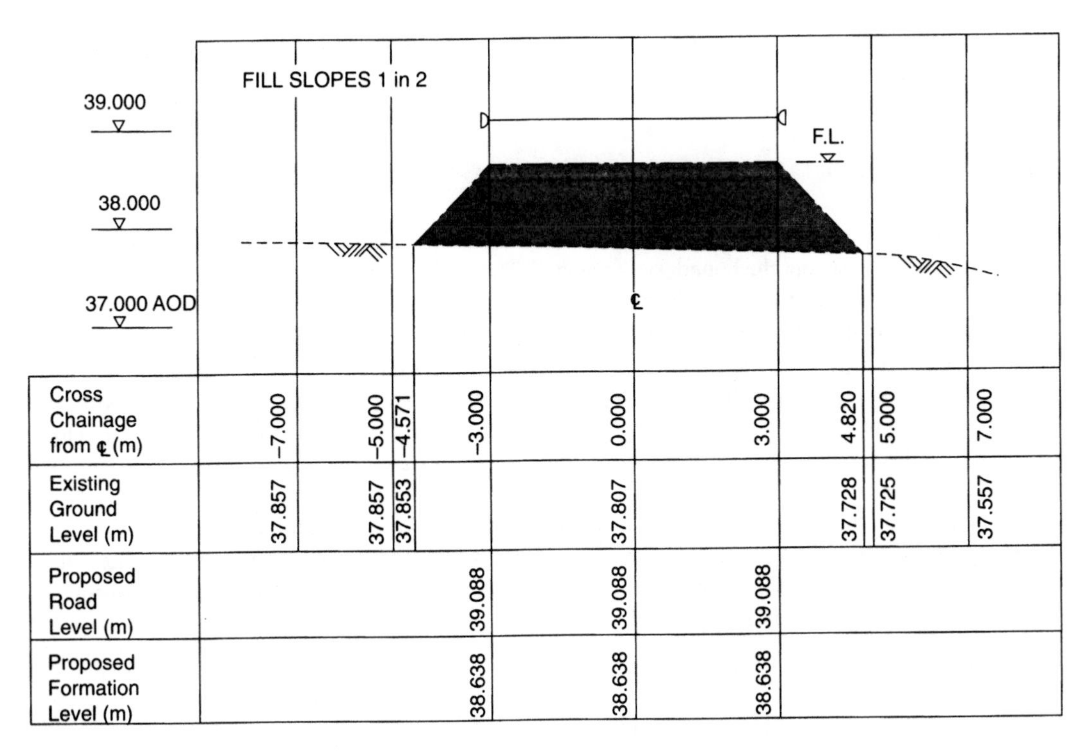

Cross Chainage from ℄ (m)	−7.000	−5.000	−4.571	−3.000	0.000	3.000	4.820	5.000	7.000
Existing Ground Level (m)	37.857	37.857	37.853		37.807		37.728	37.725	37.557
Proposed Road Level (m)				39.088	39.088	39.088			
Proposed Formation Level (m)				38.638	38.638	38.638			

Fig. 3.27 *Cross-section No. 3 at chainage 360.000 m*

(4) Concentric closed contours represent hills or hollows, depending on the increase or decrease in elevation.
(5) The slope between contour lines is assumed to be regular.
(6) Contour lines crossing a stream form V's pointing upstream.
(7) The edge of a body of water forms a contour line.

Contours are used by engineers to:

(1) Construct longitudinal sections and cross-sections for initial investigation.
(2) Compute volumes.
(3) Construct route lines of constant gradient.
(4) Delineate the limits of constructed dams, road, railways, tunnels, etc.
(5) Delineate and measure drainage areas.

If the ground is reasonably flat, the optical level can be used for contouring using either the *direct* or *indirect* methods. In steep terrain it is more economical to use other heighting, as outlined later.

(1) Direct contouring

In this method the actual contour is pegged out on the ground and its planimetric position located. A backsight is taken to an appropriate BM and the HPC of the instrument is obtained, say 34.800 m AOD. A staff reading of 0.800 m would then place the foot of the staff at the 34 m contour level. The staff is then moved throughout the terrain area, with its position pegged at every 0.800 m reading. In this way the 34 m contour is located. Similarly a staff reading of 1.800 m gives the 33 m contour and so on. The planimetric position of the contour needs to be located using an appropriate survey technique.

This method, although quite accurate, is tedious and uneconomical and could never be used over a large area. It is ideal, however, in certain construction projects that require excavation to a specific single contour line.

(2) Indirect contouring

This technique requires establishing a grid of intersecting evenly spaced lines over the site. A theodolite and steel tape may be used to set out the boundary of the grid. The grid spacing will depend upon the roughness of the ground and the purpose for which the data are required. All the points of intersection throughout the grid may be pegged or shown by means of paint from a spray canister. Alternatively ranging rods at the grid intervals around the periphery would permit the staff holder, with the aid of an optical square, to align himself with appropriate pairs and thus fix each grid intersection point, for example, alignment with rods *B*-*B* and 2-2 fixes point *B*2 (*Figure 3.28*). Alternatively assistants at ranging rods *B* and 2 could help to line up the staff holder. When the RLs of all the intersection points are obtained, the contours are located by linear interpolation between the levels, on the assumption of a uniform ground slope between each pair of points. The interpolation may be done arithmetically using a pocket calculator, or graphically.

Consider grid points *B*2 and *B*3 with reduced levels of 30.20 m and 34.60 m respectively and a horizontal grid interval of 20 m (*Figure 3.29*). The height difference between *B*2 and *B*3 is 4.40 m and the 31 m contour is 0.80 m above *B*2. The horizontal distance of the 31 m contour from $B2 = x_1$

where $(20/4.40) = 4.545 \text{ m} = K$

and $x_1 = K \times 0.80 \text{ m} = 3.64 \text{ m}$

Similarly for the 32 m contour:

$x_2 = K \times 1.80 \text{ m} = 8.18 \text{ m}$

and so on, where (20/4.40) is a constant *K*, multiplied each time by the difference in height from the reduced level of *B*2 to the required contour value. For the graphical interpolation, a sheet of transparent paper (*Figure 3.30*) with equally spaced horizontal lines is used. The paper is placed over the two points and rotated until *B*2 obtains a value of 30.20 m and *B*3 a value of 34.60 m. Any appropriate scale can be used for the line separation. As shown, the 31, 32, 33 and 34 m contour positions can now be pricked through onto the plan.

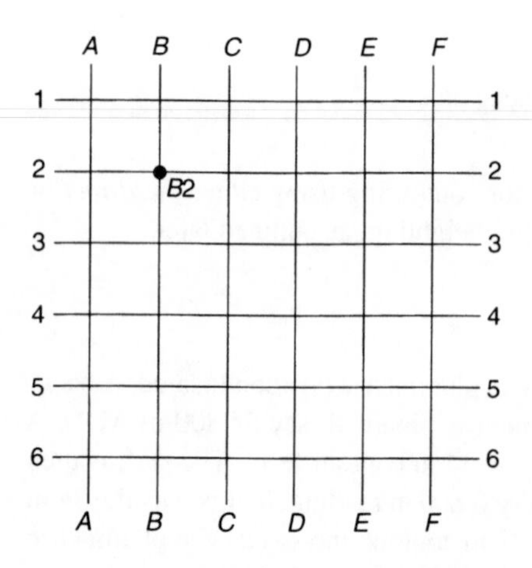

Fig. 3.28 *Grid layout for contouring*

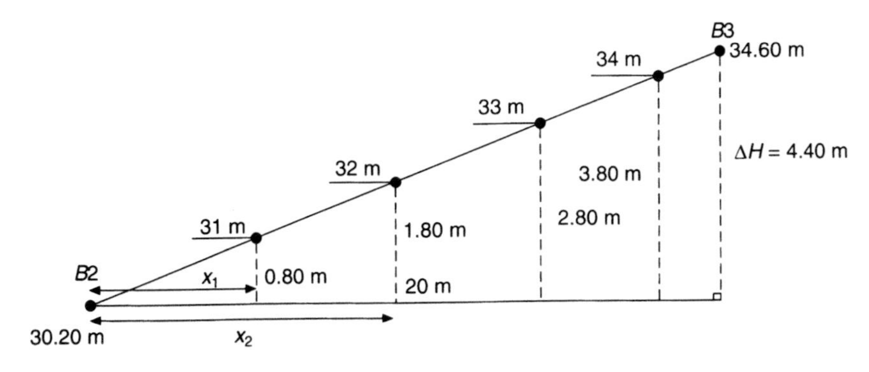

Fig. 3.29 *Contour calculations*

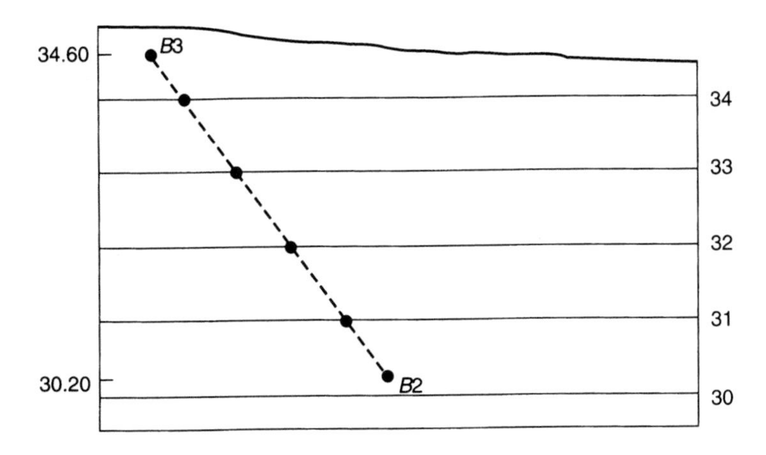

Fig. 3.30 *Graphical contour plotting*

This procedure is carried out on other lines and the equal contour points joined up to form the contours required.

An alternative way of creating the grid intersections that does not require the use of an optical square is to set out the ranging rods as in *Figure 3.31*. In this case it is important that the pairs of ranging rods at A, B, ... 1, 2, etc are set out precisely. However once set out, the staff holder can find position much more easily.

3.12 RECIPROCAL LEVELLING

When obtaining the relative levels of two points on opposite sides of a wide gap such as a river, it is impossible to keep the length of sights short and equal. Collimation error, Earth curvature and refraction affect the longer sight much more than the shorter one. In order to minimize these effects, the method of reciprocal levelling is used, as illustrated in *Figure 3.32*.

If the instrument near A observes a backsight onto A and a foresight onto B, the difference in elevation between A and B is:

$$\Delta H_{AB} = x_2 - x_1 - (c - r)$$

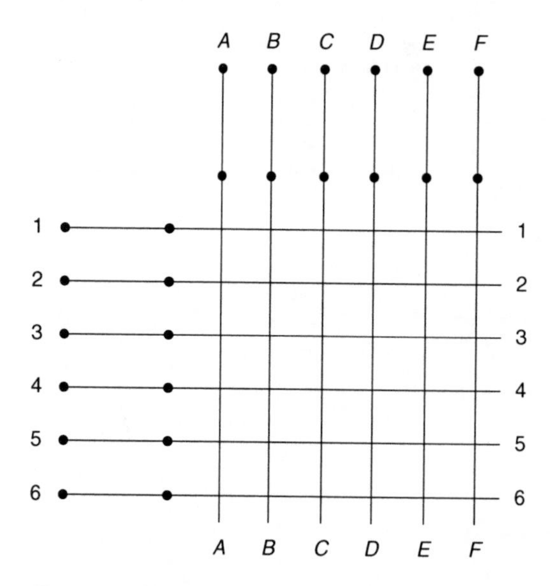

Fig. 3.31 *Alternative Grid layout for contouring*

where:

$$x_1 = \text{BS on } A$$
$$x_2 = \text{FS on } B$$
$$(c - r) = \text{the combined effect of curvature and refraction}$$
(with collimation error intrinsically built into r)

Similarly with the instrument moved to near B:

$$\Delta H_{AB} = y_1 - y_2 + (c - r)$$

where

$$y_1 = \text{BS on } B$$

$$y_2 = \text{FS on } A$$

then

$$2\Delta H_{AB} = (x_2 - x_1) + (y_1 - y_2)$$

and

$$\Delta H_{AB} = \frac{1}{2}[(x_2 - x_1) + (y_1 - y_2)] \tag{3.6}$$

This proves that the mean of the difference in level obtained with the instrument near A and then with the instrument near B is free from the errors due to curvature, refraction and collimation error. Random errors of observation will still be present, however.

Equation (3.6) assumes the value of refraction is equal in both cases. Refraction is a function of temperature and pressure and so varies with time. Thus refraction may change during the time taken to transport the instrument from side A to side B. To preclude this it is advisable to use two levels and take simultaneous reciprocal observations. However, this procedure creates the problem of each instrument having a different residual collimation error. The instruments should therefore be interchanged and the whole procedure repeated. The mean of all the values obtained will then give the most probable value for the difference in level between A and B.

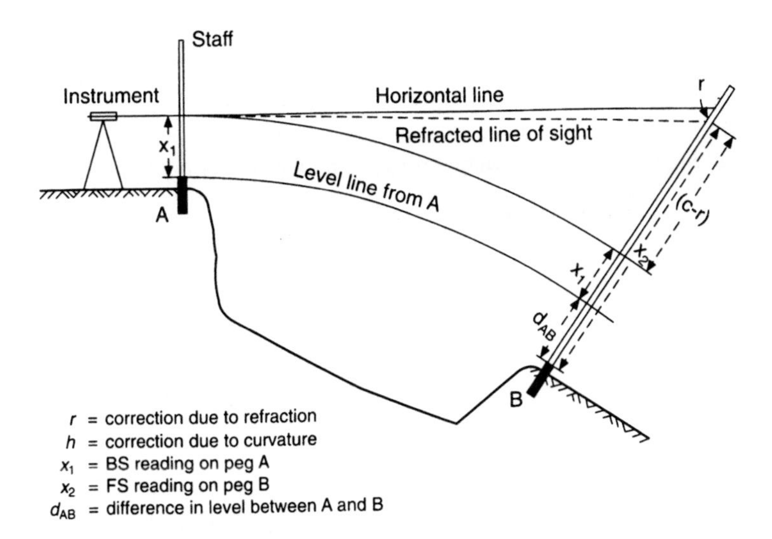

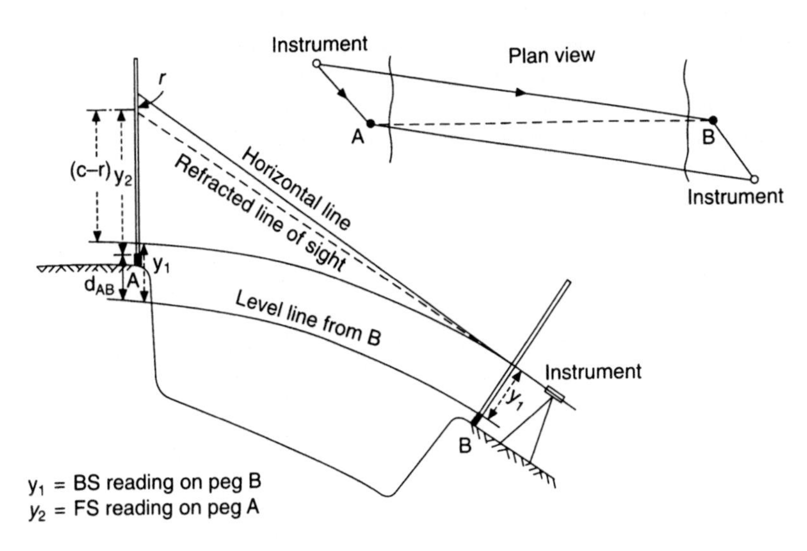

Fig. 3.32 *Reciprocal levelling*

3.13 PRECISE LEVELLING

Precise levelling may be required in certain instances in construction such as in deformation monitoring, the provision of precise height control for large engineering projects such as long-span bridges, dams and hydroelectric schemes and in mining subsidence measurements. For example, a dam that has been in place for many years is unlikely to be moving. However, should the dam fail the results would be catastrophic for those on the downstream side. Being under the pressure of water when full, the dam may be liable to distortion. The behaviour of the dam must therefore be monitored. One way of monitoring any vertical movement along the dam is by levelling. Since early warning of small movement is required, and since conclusions about movement must be made with statistical confidence, the levelling must be very precise.

There is more to precise levelling than precise levels. High quality equipment is very important, but so is the method by which it is used. Indeed the two components of precise levelling are precise equipment and precise procedures. Precise levelling uses the same principles as ordinary levelling but with:

(1) Higher quality instruments and more accurate staves
(2) More rigorous observing techniques
(3) Restricted climatic and environmental conditions
(4) Refined booking and reduction
(5) Least squares adjustment for a levelling net

3.13.1 Precise invar staff

The precise levelling staff has its graduations precisely marked (and checked by laser interferometry) on invar strips, which are attached to wooden or aluminium frames. The strip is rigidly fixed to the base of the staff and held in position by a spring-loaded tensioning device at the top. This arrangement provides support for the invar strip without restraining it in any way.

Usually there are two scales on each staff, offset from each other by a fixed amount (*Figure 3.33*). The staff is placed upon a change plate at intermediate stations. A conventional levelling change plate is small and light and is designed to give a firm platform for the staff on soft ground. Precise levelling should only ever take place on firm ground and the precise levelling change plate is designed to be unmoving on a hard surface. It is therefore heavy. The feet are rounded so that they do not slowly sink or heave when placed on tarmac. The top is smooth, round and polished. The change plate in *Figure 3.34* is made from a solid piece of steel and weighs about 10 kilograms.

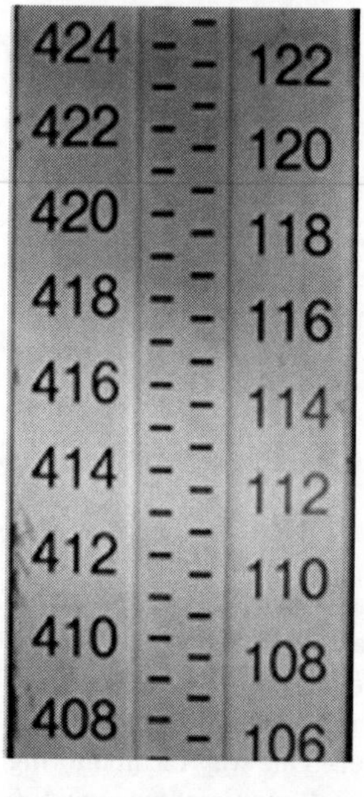

Fig. 3.33 *Segment of a precise levelling staff*

Fig. 3.34 *Precise levelling change plate*

For the most precise work, two staffs are used; in which case they should be carefully matched in every detail. A circular bubble built into the staff is essential to ensure verticality during observation. The staff should be supported by means of steadying poles or handles.

(1) The staff should have its circular bubble tested at frequent intervals using a plumb-bob.
(2) Warping of the staff can be detected by stretching a fine wire from end to end.
(3) Graduation and zero error can be counteracted by regular calibration.
(4) For the highest accuracy a field thermometer should measure the temperature of the strip in order to apply scale corrections.

3.13.2 Instruments

The instruments used should be precise levels of the highest accuracy. They should provide high-quality resolution with high magnification (×40) and be capable of being adjusted to remove any significant collimation. This may be achieved with a highly sensitive tubular bubble with a large radius of curvature that gives a greater horizontal bubble movement per angle of tilt. In the case of the automatic level a highly refined compensator would be necessary.

In either case a parallel plate micrometer, fitted in front of the object lens, would be used to obtain submillimetre resolution on the staff.

The instrument's cross-hairs may be as shown in *Figure 3.35(a)*. The distance that the staff is away from the instrument will affect which side of the cross-hairs is to be used. If the staff is far away use the normal horizontal hair, right-hand side of diagram. If the staff is close, the mark on the staff will appear too large to be bisected accurately. By comparing the two white wedges formed by the sloping cross-hairs and the mark on the staff, see the arrows in *Figure 3.35(b)* where the mark on the staff is not correctly aligned and *Figure 3.35(c)* where it is, a more precise setting of the micrometer can be made.

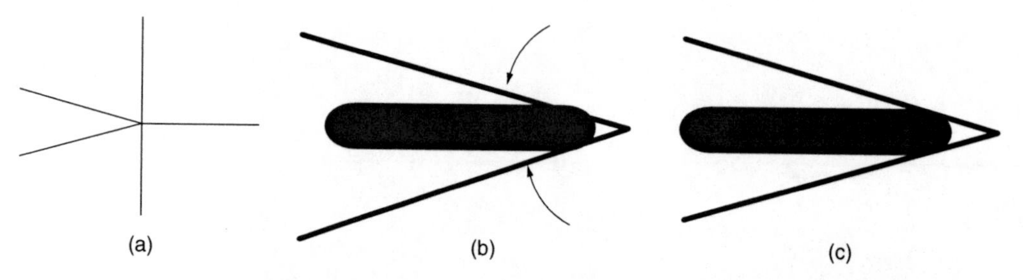

Fig. 3.35 *Precise level cross-hairs*

3.13.3 Parallel plate micrometer

For precise levelling, the estimation of 1 mm is not sufficiently accurate. A parallel plate glass micrometer in front of the object lens enables readings to be made direct to 0.1 mm, and estimated to 0.01 mm. The parallel plate micrometer works by refracting the image of a staff graduation to make it coincident with the cross-hair. There is, therefore, no estimation of the position of the cross-hair with respect to the graduation. The principle of the attachment is seen from *Figure 3.36*. Had the parallel plate been vertical the line of sight would have passed through without deviation and the reading would have been 1.026 m, the final figure being estimated. However, by manipulating the micrometer the parallel plate is tilted until the line of sight is displaced to the nearest division marked on the staff, which is 1.02 m. The rotation of the micrometer drum is proportional to the displacement of the image of the staff. The amount of displacement s is measured on the micrometer and added to the exact reading to give 1.02647 m, only the last decimal place is estimated.

It can be seen from *Figure 3.36* that the plate could have been moved in the opposite direction, displacing the line of sight up. Since the parallel plate micrometer run is normally equal to the gap between two successive divisions on the staff it will not be possible to gain coincidence on more than one division.

The displacement is related to the rotation of the parallel plate as follows. In *Figure 3.37* the plate pivots about A. The displacement is BC and the rotation is equal to the angle of incidence i. The thickness of the plate is t and the ray of light from the staff is refracted by an angle r. μ is the refractive index of the glass of the plate.

$$\text{Displacement} = BC = AB \sin(i - r)$$

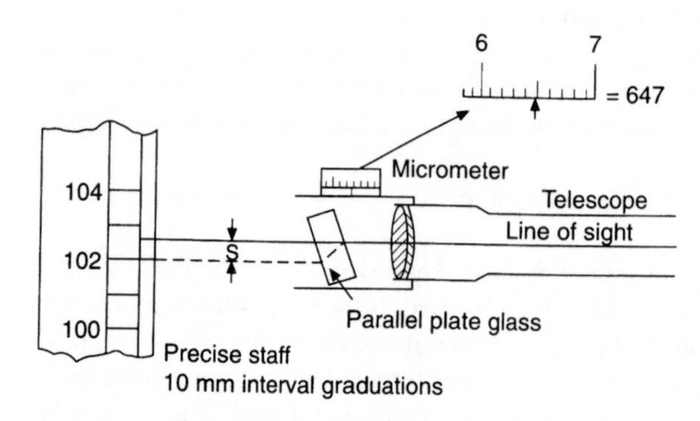

Fig. 3.36 *Parallel plate micrometer*

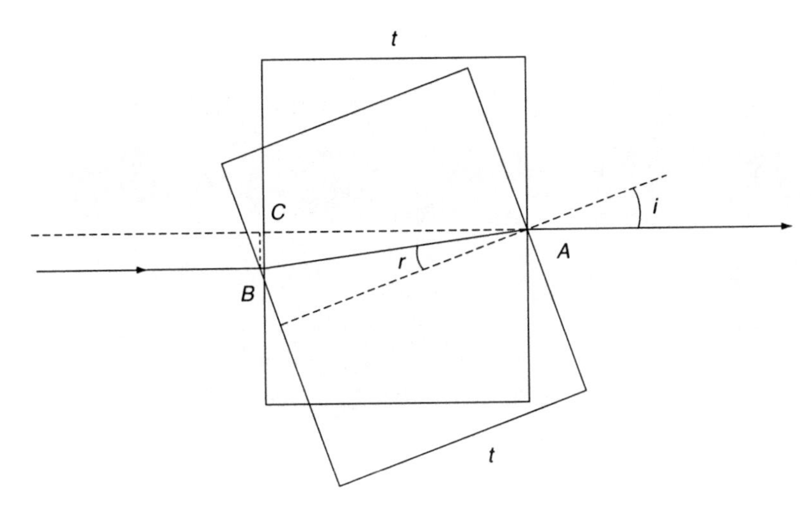

Fig. 3.37 *Parallel plate displacement*

But $AB = t \sec r$ so

$$\text{Displacement} = t \sin(i - r) \sec r$$
$$= t(\sin i \cos r - \cos i \sin r) \sec r$$
$$= t(\sin i - \cos i \sin r \sec r)$$

But from Snell's Law of refraction $\sin i = \mu \sin r$. So upon substitution and rearrangement the equation becomes

$$\text{Displacement} = t \sin i \left\{ 1 - (1 - \sin^2 i)^{\frac{1}{2}} (\mu^2 - \sin^2 i)^{-\frac{1}{2}} \right\}$$

If i is small then $\sin^2 i$ is negligible compared with 1 or μ^2 and $\sin i = i$ radians. So

$$\text{Displacement} = t(1 - \mu^{-1})i$$

Since t and μ are fixed properties of the plate then displacement is directly proportional to rotation. Parallel plate micrometers are also manufactured for use with 5 mm graduations.

3.13.4 Field procedure

At the beginning and end of each levelling run a stable and precise benchmark is required. Intermediate points are not observed. To avoid accidental damage or vandalism wall mounted benchmarks can be removed from the wall leaving the barrel, which has been fixed with epoxy resin, capped for protection (*Figure 3.38*).

The size of the levelling team depends upon the observing conditions and the equipment available. In ordinary levelling an observer and staff holder are required. In precise levelling there are two staves and therefore two staff holders are required. If a programmed data logger is available then the observer can also do the booking. If the observations are to be recorded on paper a booker should also be employed. The booker's task, other than booking, is to do a series of quality control checks at the end of each set of observations, before moving to the next levelling bay. Finally, in sunny weather, an umbrella holder is required because it is necessary to shield the instrument and tripod from the heating effects of the sun's rays.

Fig. 3.38 *Wall mounted benchmark*

Just as with ordinary levelling, a two-peg test is required to confirm that the instrumental collimation is acceptable. Precise levelling procedures are designed to minimize the effect of collimation, but even so, only a well-adjusted instrument should be used.

Precise level lines should follow communication routes where possible because they generally avoid steep gradients; they are accessible and have hard surfaces. However, there may be vibration caused by traffic, especially if using an automatic level.

The following procedures should be adhered to when carrying out precise levelling:

(1) Precise levelling can be manpower intensive, and therefore expensive to undertake. It is important to carry out a full reconnaissance of the proposed levelling route prior to observations being taken to ensure that the best possible route has been chosen.
(2) End and intermediate benchmarks should be constructed well before levelling starts to prevent settling during levelling operations.
(3) Steep slopes are to be avoided because of the unequal and uncertain refraction effects on the tops and bottoms of staves.
(4) Long lines should be split into workable sections, usually each section will not be more than about 3 km, because that is about as much as a team can do in one day. There must be a benchmark at each end of the line to open and close on. The length of each line will depend upon terrain, transport, accommodation and other logistical considerations.
(5) Each section is to be treated as a separate line of levelling and is checked by forward and backward levelling. This will isolate errors and reduce the amount of re-levelling required in the case of an unacceptable misclosure.
(6) On each section, if the forward levelling takes place in the morning of day 1, then the backward levelling should take place in the afternoon or evening of day 2. This will ensure that increasing refraction on one part of the line in one direction will be replaced by decreasing refraction when working in the other direction. This will help to compensate for errors due to changing refraction effects.
(7) On bright or sunny days an observing umbrella should be held over the instrument and tripod to avoid differential heating of the level and of the tripod legs.
(8) Take the greatest care with the base plate of the staff. Keep it clean. Place it carefully onto the change plate and do not drop the staff. This will avoid any change in zero error of the staff. When the staff is not being used, it should be rested upon the staff-man's clean boot.

(9) The distances of foresight and backsight must be as nearly equal as possible so as to limit the effect of the Earth's curvature, refraction and bad instrumental collimation. This will also avoid the need to re-focus the level between sightings.

(10) Take care when levelling along roads or railways. Stop levelling when traffic or vibrations are heavy. When the staff is not being used, it should be rested upon the staff-man's clean boot. Vibration may damage the staff base plate and so change its zero error.

(11) On tarmac and soft ground the instrument or staff may rise after it has been set up. This may be apparent to the observer but not by the staff person.

(12) In gusty or windy conditions stop levelling because there will be uncertainty in the readings. In variable weather conditions consider levelling at night.

(13) The bottom 0.5 m of the staff should not to be used because of unknown and variable refraction effects near the ground.

(14) If a precise automatic level is to be used, it should be lightly tapped and rotated before each reading to ensure that the compensator is freely operative. This will reduce errors by ensuring that the compensator always comes from the same direction. Some automatic levels have a press button for this purpose.

(15) The rounded centre on the change plate should be kept polished and smooth to ensure that the same staff position is taken up each time it is used.

(16) The change plate must be firmly placed and not knocked or kicked between foresight and backsight readings. Remember there is no check on the movement of a change plate between these observations. The staff holder should stand clear between observations.

(17) The observation to the back staff must be followed immediately by an observation to the forward staff, both on one scale. This is to ensure that refraction remains constant during the forward and back observations of one bay. Then, an observation to the forward staff is followed immediately by an observation to the back staff on the other scale. This procedure helps to compensate for unknown changes in refraction, by balancing the errors. Using two double scale rods the sequence of observation would be:

    (1) BS left-hand scale on staff *A*
    (2) FS left-hand scale on staff *B*
    (3) FS right-hand scale on staff *B*
    (4) BS right-hand scale on staff *A*

    Then (1)$-$(2) $= \Delta H_1$ and (4)$-$(3) $= \Delta H_2$; if these differences agree within the tolerances specified, the mean is accepted. Staff *A* is now leapfrogged to the next position and the above procedure repeated starting with staff *A* again (*Figure 3.39*).

(18) If the back staff is observed first at one set-up, then the forward staff is observed first at the next set-up. This ensures that changing refraction will affect each successive bay in an equal and

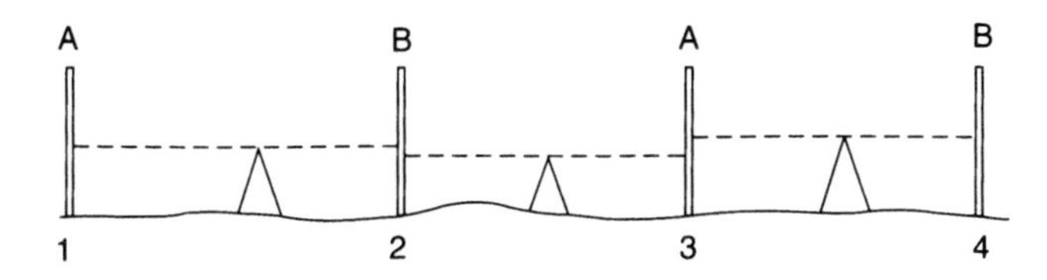

Fig. 3.39 *Staff leapfrog*

opposite manner. The order of observations this time will be:

(1) FS left-hand scale on staff *A*
(2) BS left-hand scale on staff *B*
(3) BS right-hand scale on staff *B*
(4) FS right-hand scale on staff *A*

Note that in each case the first observation of a bay is to the same staff, which is alternately the back and then the forward staff.

(19) The same staff that was used for the opening backsight must also be used for the closing foresight. This will eliminate the effect of different zero errors on the two staves. This means that there must always be an even number of set-ups on any line.
(20) Levelling should always be carried out in both directions, forward and back. If, on the forward levelling, the *A* staff was used to open and close the line, then the *B* staff should be used to open and close the line on the backward levelling. This will equalize the number of readings on each staff.
(21) Lines of sight should not exceed 50 m, especially in haze, or on sloping ground. This will minimize the effects of refraction, curvature of the Earth and difficulty of reading the staff. A good average length of sight is 35 m.
(22) Use the procedure already outlined for levelling the circular bubble on automatic levels. This will happen as a matter of course if the telescope is aimed at staff *A* each time when centring the circular bubble.

3.13.5 Booking and computing

Figure 3.40 shows a sample of precise levelling observations. Note the order in which the observations are made and that it agrees with paragraphs (17) and (18) above. Once the observations are complete and

Back Staff A/B	*Back Dist*	*Back Left*	*Back Right*	*Back R – L*	*Back R + L*	*Fwd Dist*	*Fwd Left*	*Fwd Right*	*Fwd R – L*	*Fwd R + L*	*Run Diff Dist*	*Run Diff Height*	*Δ (R–L) $<\pm$ 0.00070 m?*	*Distances $<$50 m?*	*Δ Distances $<\pm$0.3 m?*
A	30.1	0.85419	4.01683	3.16264	4.87102	30.2	2.44140	5.60412	3.16272	8.04552	+0.1	−1.58725	✓	✓	✓
B	30.5	0.21760	3.38033	3.16273	3.59793	30.3	2.09982	5.26243	3.16261	7.36225	−0.1	−3.46941	✓	✓	✓
A	29.7	1.10329	4.26617	3.16288	5.36946	29.8	1.71900	4.88162	3.16262	6.60062	0.0	−4.08499	✓	✓	✓
B	30.1	1.42299	4.58562	3.16263	6.00861	30.2	1.45819	4.62135	3.16316	6.07954	+0.1	−4.12046	✓	✓	✓
	The order of making the observations														
A	1	3	6			2	4	5							
B	1	4	5			2	3	6							

Fig. 3.40 *Precise levelling observations*

before the back staff and the instrument leapfrog forward the following reductions and quality checks are made. Examples from the fourth line of the observations are shown.

(1) Compute the Right minus Left readings for the back staff and the forward staff. Because these are constants of the invar strips they should be the same. Check that they are within 0.00070 m of each other.

$$\text{Back R} - \text{L} = 4.58562 - 1.42299 = 3.16263$$

$$\text{Forward R} - \text{L} = 4.62135 - 1.45819 = 3.16316$$

$$\text{Check } 0.0007 > (3.16263 - 3.16316 = -0.00053) > -0.0007. \text{ It is.}$$

(2) Check that the forward and back distances are both less than 50 m and that they agree to within 0.3 m of each other.

Distances are 30.1 and 30.2.

(3) Compute the Running Difference Distance as the forward − back distance + the previous value. And make sure that it stays close to 0.0 m by adjusting future forward or back distances as appropriate.

$$30.2 - 30.1 + 0.0 = +0.1$$

(4) Compute the Right plus Left readings for the back staff. This is equivalent to twice a back staff reading plus a large but constant offset. Do likewise for the forward staff readings.

$$\text{Back R} + \text{L} = 4.58562 + 1.42299 = 6.00861$$

$$\text{Forward R} + \text{L} = 4.62135 + 1.45819 = 6.07954$$

(5) Compute the Running Difference Height as sum of its previous value + $\frac{1}{2}$ (Forward R + L) − $\frac{1}{2}$ (Back R + L).

$$\text{Running Difference Height} = -4.08499 + \tfrac{1}{2}\,6.07954 - \tfrac{1}{2}\,6.00861 = -4.12046$$

3.14 DIGITAL LEVELLING

The digital level is an instrument that uses electronic image processing to evaluate the staff reading. The observer is in effect replaced by a detector which derives a signal pattern from a bar-code type levelling staff. A correlation procedure within the instrument translates the pattern into the vertical staff reading and the horizontal distance of the instrument from the staff. Staff-reading errors by the observer are thus eliminated.

The basic field data are automatically stored by the instrument thus further eliminating booking errors (*Figure 3.41*).

3.14.1 Instrumentation

The design of both the staff and instrument are such that it can be used in the conventional way as well as digitally.

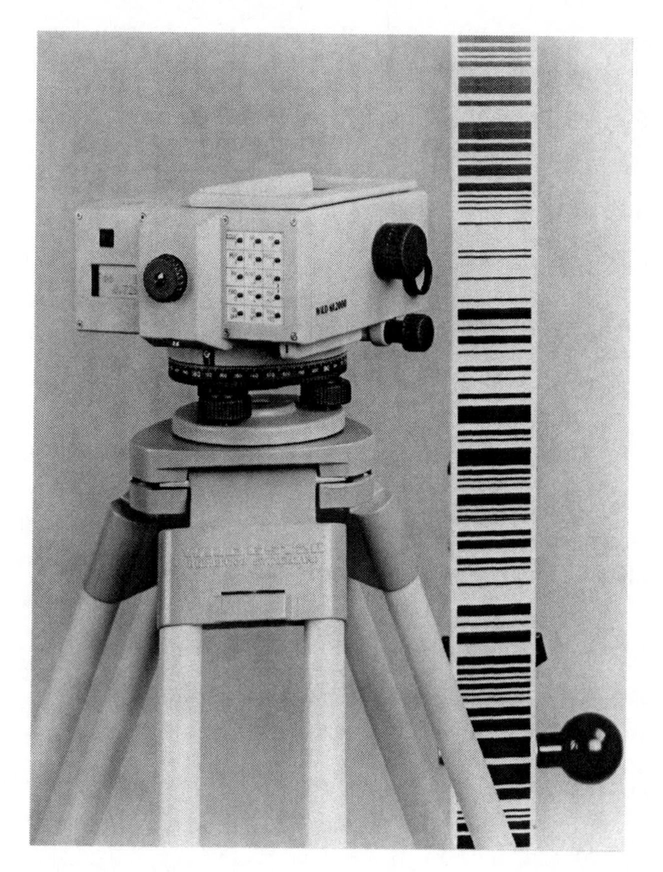

Fig. 3.41 *Digital level and staff*

(1) The levelling staff

The staff is usually made from a synthetic material, which has a small coefficient of expansion. The staff may be in one or more sections. There are precise invar staves for precise levelling. On one side of the staff is a binary bar code for electronic measurement, and on the other side there are often conventional graduations in metres. The black and white binary code comprises many elements over the staff length. The scale is absolute in that it does not repeat along the staff. As the correlation method is used to evaluate the image, the elements are arranged in a pseudo-random code. The code pattern is such that the correlation procedure can be used over the whole working range of the staff and instrument. Each manufacturer uses a different code on their staffs therefore an instrument will only work with a staff from the same manufacturer.

(2) The digital level

The digital level has the same optical and mechanical components as a normal automatic level. However, for the purpose of electronic staff reading a beam splitter is incorporated which transfers the bar code image to a detector. Light reflected from the white elements of the bar code is divided and sent to the observer and to the detector. The detector is a form of charge couple device (CCD) which turns the black and white staff pattern into a binary code. The angular aperture of the instrument is quite small, of the order of 1°–2°, resulting in a short section of the staff being imaged at the minimum range and up to the whole staff at the maximum range. The bar code image is compared with a stored reference code to find the height collimation on the staff. The instrument may not need to see the part of the staff where the cross-hairs lie. The distance from instrument to staff is dependent on the image scale of the code.

The data processing is carried out within the instrument and the data are displayed in a simple format.

The measurement process is initiated by a very light touch on a measure button. A keypad on the eyepiece face of the instrument permits the entry of further numerical data and pre-programmed commands. The data can be stored and transferred to a computer when required. The instrument may have an interface, which permits external control, data transfer and power supply.

3.14.2 Measuring procedure

There are two external stages to the measuring procedure; pointing and focusing on the staff and triggering the digital measurement. The whole process takes a few seconds.

Triggering the measurement determines the focus position, from which the distance to the staff is measured, and initiates monitoring of the compensator.

A coarse correlation approximately determines the target height and the image scale and a fine correlation using calibration constants produces the final staff reading and instrument to staff distance.

For best results a number of observations are taken automatically and the result averaged. This reduces biases due to oscillations of the compensator and air turbulence within the instrument.

The results may be further processed within the instrument, displayed and recorded. The programs incorporated will vary from instrument to instrument but typically may include those for:

(1) A single measurement of staff reading and horizontal distance.
(2) The start of a line of levelling and its continuation including intermediate sights. Automatic reduction of data. Setting out of levels.
(3) Calibration and adjustment of the instrument (two-peg test).
(4) Data management.
(5) Recognition of an inverted staff.
(6) Set the parameters of the instrument; a process similar to the initializing procedures used when setting up electronic theodolites.

3.14.3 Factors affecting the measuring procedure

Every operation in a measurement procedure is a possible error source and as such requires careful consideration in order to assess the effect on the final result.

(1) Pointing and focusing

Obviously the instrument will not work if it is not pointed at the staff. The amount of staff that needs to be read depends on the range of the instrument to the staff. However, there will be a minimum amount necessary at short ranges. It may not be critical to have the staff pointing directly at the instrument.

The precision of the height measurement may be independent of sharpness of image; however, a clear, sharply focused image reduces the time required for the measurement. If the image is too far out of focus then the instrument may not read at all. Some instruments have an auto-focus function to avoid potential focusing problems.

(2) Vibrations and heat shimmer

Vibration of the compensator caused by wind, traffic, etc., has a similar effect on the bar code image as that of heat shimmer. However, as digital levelling does not require a single reading, but instead is dependent on a section of the code, the effects of shimmer and vibration may not be critical.

Similarly, scale errors on the staff are averaged.

(3) Illumination

As the method relies on reflected light from the white intervals of the bar code, illumination of the staff is important. During the day, this illumination will be affected by cloud, sun, twilight and the effects of shadows. Up to a point these variations are catered for by the instrument but under adverse conditions there may be an increase in the measuring time.

(4) Staff coverage

In some conditions part of the bar code section being interrogated by the instrument may be obscured. Consult the manufacturer's handbook to ensure that sufficient of the staff is showing to the instrument.

(5) Collimation

The collimation value is set in the instrument but can be checked and changed as required. The method of determining the collimation is based upon one of the two peg methods described earlier. Once the collimation value has been determined it is applied to subsequent readings thereby minimizing its effect. Note, however, that it can never be completely removed and appropriate procedures according to the precision required must still be applied.

(6) Physical damage

It is likely that the instrument will be seriously damaged if it is pointed directly at the sun.

3.14.4 Operating features

The resolution for most instruments is 0.1 mm for height and 10 mm for distance or better with instrumental ranges up to 100 m. At such distances the effects of refraction and curvature become significant. The effect of curvature can be precisely calculated, the effect of refraction cannot. Most digital levels can also be used as conventional optical automatic levels but in that case the standard error of 1 km of double-run levelling becomes less. Although the digital level can also measure distance, the precision of the distance measurement is only of the order of a few centimetres.

3.14.5 Advantages of digital levelling

One advantage claimed for digital levelling is that there is less fatigue for the observer. While it is true that the observer does not have to make observations the instrument still needs to be set up, pointed at the target and focused. The digital display needs no interpretation such as reading the centimetre from the *E* on a conventional staff and estimating the millimetre. Measurements are of consistent quality, subject to the observer taking the same care with the instrument to ensure consistency of target distances and illumination of the staff. Also the staff holder must not move the staff between the forward reading in one bay and the back reading in the next, and that the staff must be kept vertical.

There is an acceptable range of illumination, but too much or too little light may make observations impossible. Some, but not all, digital levels will recognize when staffs are inverted, others will indicate an error if not told that the staff is inverted. Like any automatic level, the digital level will need to be at least coarsely levelled for the compensator to be in range.

Although exact focusing may not be required, the instrument will not work if the focusing is too far out but if the instrument has automatic focusing this would not be a problem. Automatic data storage eliminates the need for manual booking and its associated errors, and automatic reduction of data to produce ground levels eliminates arithmetical errors. However, checks for levelling circuit misclosure need to be made or at least checked and an adjustment to the intermediate points for misclosure needs to be made.

As with all surveying instruments the digital level should be allowed to adapt to the ambient air temperature.

The scale of the height measurements is primarily fixed by the scale of the staff. An invar staff will vary less with change of temperature. The scale will also be dependent on the quality of the CCD. How the dimensional stability of CCDs may vary with time is not well known.

There are a number of menus and functions that can be called on to make the levelling process easier, in particular the two-peg test for collimation error and calibration.

Overall, digital levelling is generally a faster process than levelling with an automatic level. Data can be directly downloaded to a suitable software package to enable computation and plotting of longitudinal sections and cross-sections. The digital level can be used in just about every situation where a conventional level can be used, and should the batteries fail it can be used as a conventional level if necessary.

Worked examples

Example 3.2. The positions of the pegs which need to be set out for the construction of a sloping concrete slab are shown in the diagram. Because of site obstructions the tilting level which is used to set the pegs at their correct levels can only be set up at station X which is 100 m from the TBM. The reduced level of peg A is to be 100 m and the slab is to have a uniform diagonal slope from A towards J of 1 in 20 downwards.

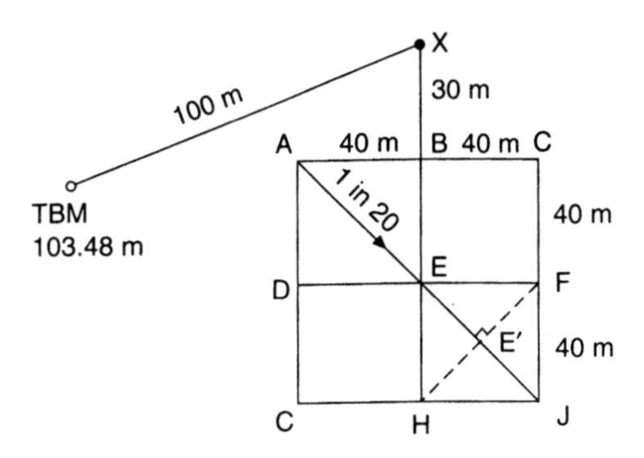

To ensure accuracy in setting out the levels it was decided to adjust the instrument before using it, but it was found that the correct adjusting tools were missing from the instrument case. A test was therefore carried out to determine the magnitude of any collimation error that may have been present in the level, and this error was found to be 0.04 m per 100 m downwards.

Assuming that the backsight reading from station X to a staff held on the TBM was 1.46 m, determine to the nearest 0.01 m the staff readings which should be obtained on the pegs at A, F and H, in order that they may be set to correct levels.

Describe fully the procedure that should be adopted in the determination of the collimation error of the tilting level. (ICE)

The simplest approach to this question is to work out the true readings at A, F and H and then adjust them for collimation error. Allowing for collimation error the true reading on TBM = 1.46 + 0.04 = 1.50 m.

HPC = 103.48 + 1.50 = 104.98 m

True reading on A to give a level of 100 m = 4.98 m
Distance AX = 50 m (ΔAXB = 3, 4, 5)
∴ Collimation error = 0.02 m per 50 m
Allowing for this error, actual reading at A = 4.98 − 0.02 = 4.96 m
Now referring to the diagram, line HF through E' will be a strike line
∴ H and F have the same level as E'
Distance $AE' = (60^2 + 60^2)^{\frac{1}{2}} = 84.85$ m
Fall from A to E' = 84.85 ÷ 20 = 4.24 m
∴ Level at E' = level at F and H = 100 − 4.24 = 95.76 m
Thus true staff readings at F and H = 104.98 − 95.76 = 9.22 m
Distance $XF = (70^2 + 40^2)^{\frac{1}{2}} = 80.62$ m
Collimation error ≈ 0.03 m
Actual reading at F = 9.22 − 0.03 = 9.19 m
Distance XH = 110 m, collimation error ≈ 0.04 m
Actual reading at H = 9.22 − 0.04 = 9.18 m

Example 3.3. The following readings were observed with a level: 1.143 (BM 112.28), 1.765, 2.566, 3.820 CP; 1.390, 2.262, 0.664, 0.433 CP; 3.722, 2.886, 1.618, 0.616 TBM.

(1) Reduce the levels by the R-and-F method.
(2) Calculate the level of the TBM if the line of collimation was tilted upwards at an angle of 6′ and each BS length was 100 m and FS length 30 m.
(3) Calculate the level of the TBM if in all cases the staff was held not upright but leaning backwards at 5° to the vertical. (LU)

(1) The answer here relies on knowing once again that levelling always commences on a BS and ends on a FS, and that CPs are always FS/BS (see table below)
(2) Due to collimation error

the BS readings are too great by 100 tan 6′
the FS readings are too great by 30 tan 6′

net error on BS is too great by 70 tan 6′

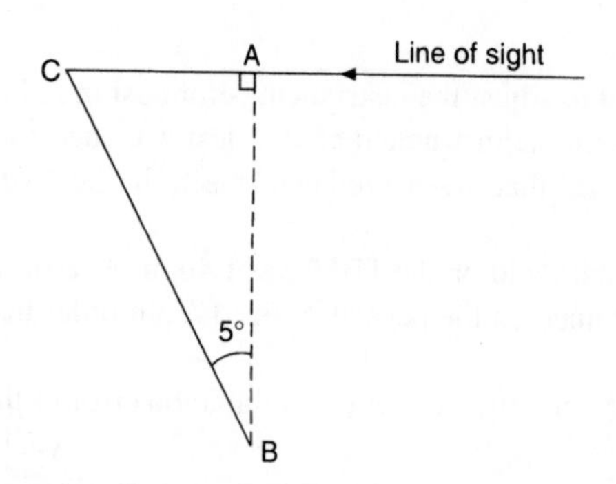

BS	*IS*	*FS*	*Rise*	*Fall*	*RL*	*Remarks*
1.143					112.280	BM
	1.765			0.622	111.658	
	2.566			0.801	110.857	
1.390		3.820		1.254	109.603	
	2.262			0.872	108.731	
	0.664		1.598		110.329	
3.722		0.433	0.231		110.560	
	2.886		0.836		111.396	
	1.618		1.268		112.664	
		0.616	1.002		113.666	TBM
6.255		4.869	4.935	3.549	113.666	
4.869			3.549		112.280	
1.386			1.386		1.386	Checks

Note that the intermediate sights are unnecessary in calculating the value of the TBM; prove it for yourself by simply covering up the IS column and calculating the value of TBM using BS and FS only.

There are three instrument set-ups, and therefore the *total net error* on BS $= 3 \times 70 \tan 6' = 0.366$ m (too great).

$$\text{level of TBM} = 113.666 - 0.366 = 113.300 \text{ m}$$

(3) From the diagram it is seen that the true reading AB = actual reading $CB \times \cos 5°$ Thus each BS and FS needs to be corrected by multiplying it by $\cos 5°$; however, this would be the same as multiplying the $\sum$BS and $\sum$FS by $\cos 5°$, and as one subtracts BS from FS to get the difference, then

$$\textit{True} \text{ difference in level} = \text{actual difference} \times \cos 5°$$

$$= 1.386 \cos 5° = 1.381 \text{ m}$$

$$\text{level of TBM} = 112.28 + 1.381 = 113.661 \text{ m}$$

Example 3.4. One carriageway of a motorway running due N is 8 m wide between kerbs and the following surface levels were taken along a section of it, the chainage increasing from S to N. A concrete bridge 12 m in width and having a horizontal soffit, carries a minor road across the motorway from SW to NE, the centre-line of the minor road passing over that of the motorway carriageway at a chainage of 1550 m.

Taking crown (i.e. centre-line) level of the motorway carriageway at 1550 m chainage to be 224.000 m:

(a) Reduce the above set of levels and apply the usual arithmetical checks.
(b) Assuming the motorway surface to consist of planes, determine the minimum vertical clearance between surface and the bridge soffit. (LU)

The HPC method of booking is used because of the numerous intermediate sights.

Intermediate sight check

$$2245.723 = [(224.981 \times 7) + (226.393 \times 3) - (5.504 + 2.819)]$$

$$= 1574.867 + 679.179 - 8.323 = 2245.723$$

BS	*IS*	*FS*	*Chainage (m)*	*Location*
1.591			1535	West channel
	1.490		1535	Crown
	1.582		1535	East channel
	−4.566			Bridge soffit*
	1.079		1550	West channel
	0.981		1550	Crown
	1.073		1550	East channel
2.256		0.844		CP
	1.981		1565	West channel
	1.884		1565	Crown
		1.975	1565	East channel

**Staff inverted*

BS	*IS*	*FS*	*HPC*	*RL*	*Remarks*
1.591				223.390	1535 West channel
	1.490			223.491	1535 Crown
	1.582			223.399	1535 East channel
	−4.566			229.547	Bridge soffit
	1.079			223.902	1550 West channel
	0.981		224.981*	224.000	1550 Crown
	1.073			223.908	1550 East channel
2.256		0.844	226.393	224.137	CP
	1.981			224.412	1565 West channel
	1.884			224.509	1565 Crown
		1.975		224.418	1565 East channel
3.847 2.819	5.504	2.819		224.418 223.390	
1.028				1.028	Checks

**Permissible to start here because this is the only known RL; also, in working back to 1535 m one still subtracts from HPC in the usual way.*

Now draw a sketch of the problem and add to it all the pertinent data as shown.

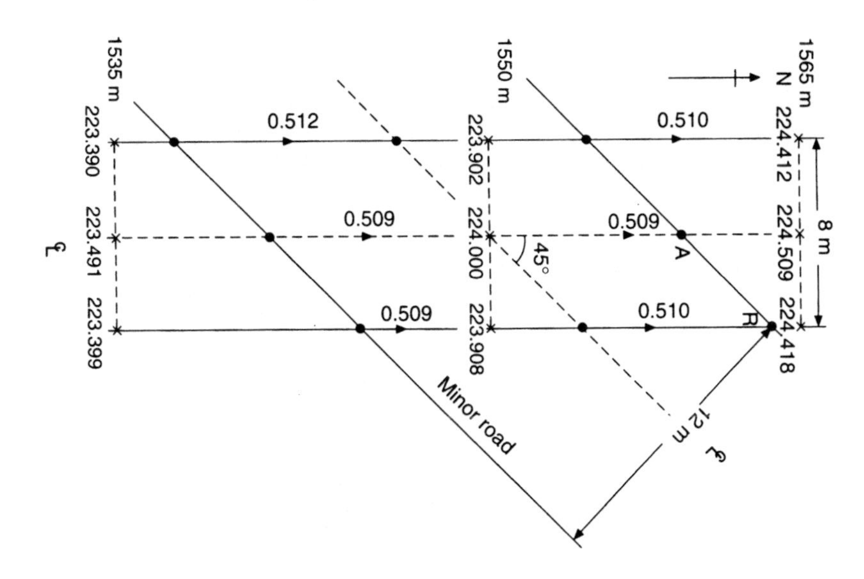

Examination of the sketch shows the road to be rising from S to N at a regular grade of 0.510 m in 15 m. This implies then, that the most northerly point (point *B* on east channel) should be the highest; however, as the crown of the road is higher than the channel, one should also check point *A* on the crown; all other points can be ignored. Now, from the illustration the distance 1550 to *A* on the centre-line:

$$= 6 \times (2)^{\frac{1}{2}} = 8.5\,\text{m}$$

$\therefore$ Rise in level from 1550 to $A = (0.509/15) \times 8.5 = 0.288$ m

$\therefore$ Level at $A = 224.288$ m giving a clearance of $(229.547 - 224.288) = 5.259$ m

Distance 1550 to *B* along the east channel $8.5 + 4 = 12.5$ m

$\therefore$ Rise in level from 1550 to $B = (0.510/15) \times 12.5 = 0.425$ m

$\therefore$ Level at $B = 223.908 + 0.425 = 224.333$ m

$\therefore$ Clearance at $B = 229.547 - 224.333 = 5.214$ m

$\therefore$ Minimum clearance occurs at the most northerly point on the east channel, i.e. at *B*

Example 3.5. In extending a triangulation survey of the mainland to a distant off-lying island, observations were made between two trig stations, one 3000 m and the other 1000 m above sea level. If the ray from one station to the other grazed the sea, what was the approximate distance between stations, (a) neglecting refraction, and (b) allowing for it? ($R = 6400$ km) (ICE)

Refer to *equation (3.1)*.

(a) $D_1 = (2Rc_1)^{\frac{1}{2}} = (2 \times 6400 \times 1)^{\frac{1}{2}} = 113$ km

$D_2 = (2Rc_2)^{\frac{1}{2}} = (2 \times 6400 \times 3)^{\frac{1}{2}} = 196$ km

Total distance = 309 km

(b) With refraction: $D_1 = (7/6 \times 2Rc_1)^{\frac{1}{2}}, \quad D_2 = (7/6 \times 2Rc_2)^{\frac{1}{2}}.$

By comparison with the equation in (a) above, it can be seen that the effect of refraction is to increase distance by $(7/6)^{\frac{1}{2}}$:

$$\therefore D = 309(7/6)^{\frac{1}{2}} = 334 \text{ km}$$

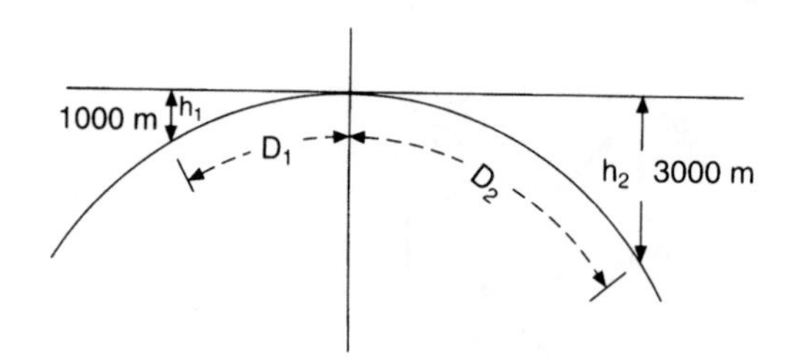

Example 3.6. Obtain, from first principles, an expression giving the combined correction for the Earth's curvature and atmospheric refraction in levelling, assuming that the Earth is a sphere of 12740 km diameter. Reciprocal levelling between two points Y and Z 730 m apart on opposite sides of a river gave the following results:

Instrument at	*Height of instrument* (m)	*Staff at*	*Staff reading* (m)
Y	1.463	Z	1.688
Z	1.436	Y	0.991

Determine the difference in level between Y and Z and the amount of any collimation error in the instrument. (ICE)

(1) $(c - r) = \dfrac{6D^2}{14R} = 0.0673D^2$ m

(2) With instrument at Y, Z is lower by $(1.688 - 1.463) = 0.225$ m
With instrument at Z, Z is lower by $(1.436 - 0.991) = 0.445$ m

$$\textit{True} \text{ height of } Z \text{ below } Y = \frac{0.225 + 0.445}{2} = 0.335 \text{ m}$$

Instrument height at $Y = 1.463$ m; knowing now that Z is lower by 0.335 m, then a truly horizontal reading on Z should be $(1.463 + 0.335) = 1.798$ m; it was, however, 1.688 m, i.e. -0.11 m too low ($-$ indicates low). This error is due to curvature and refraction $(c - r)$ *and* collimation error of the instrument (e).

Thus: $\quad (c - r) + e = -0.110$ m

Now $\quad (c - r) = \dfrac{6D^2}{14R} = \dfrac{6 \times 730^2}{14 \times 6370 \times 1000} = 0.036$ m

$\therefore \quad e = -0.110 - 0.036 = -0.146$ m in 730 m

$\therefore \quad$ Collimation error $e = 0.020$ m *down* in 110 m

Example 3.7. A and B are 2400 m apart. Observations with a level gave:

A, height of instrument 1.372 m, reading at B 3.359 m
B, height of instrument 1.402 m, reading at A 0.219 m

Calculate the difference of level and the error of the instrument if refraction correction is one seventh that of curvature. (LU)

Instrument at A, B is lower by $(3.359 - 1.372) = 1.987$ m
Instrument at B, B is lower by $(1.402 - 0.219) = 1.183$ m

True height of B below $A = 0.5 \times 3.170$ m $= 1.585$ m

Combined error due to curvature and refraction

$$= 0.0673D^2 \text{ m} = 0.0673 \times 2.4^2 = 0.388 \text{ m}$$

Now using same procedure as in *Example 3.6*:

Instrument at $A = 1.372$, thus true reading at $B = (1.372 + 1.585)$
$= 2.957$ m
Actual reading at $B = 3.359$ m

Actual reading at B too high by $+ 0.402$ m

Thus: $(c - r) + e = +0.402$ m

$$e = +0.402 - 0.388 = +0.014 \text{ m in } 2400 \text{ m}$$

Collimation error $e = +0.001$ m *up* in 100 m

Exercises

(*3.1*) The following readings were taken with a level and a 4.25-m staff:

0.683, 1.109, 1.838, 3.398 [3.877 and 0.451] CP, 1.405, 1.896, 2.676 BM (102.120 AOD), 3.478 [4.039 and 1.835] CP, 0.649, 1.707, 3.722

Draw up a level book and reduce the levels by

(a) R-and-F,
(b) height of collimation.

What error would occur in the final level if the staff had been wrongly extended and a plain gap of over 12 mm occurred at the 1.52-m section joint? (LU)

Parts (a) and (b) are self checking. Error in final level = zero.
(Hint: all readings greater than 1.52 m will be too small by 12 mm. Error in final level will be calculated from BM only.)

(*3.2*) The following staff readings were observed (in the order given) when levelling up a hillside from a TBM 135.2 m AOD. Excepting the staff position immediately after the TBM, each staff position was higher than the preceding one.

1.408, 2.728, 1.856, 0.972, 3.789, 2.746, 1.597, 0.405, 3.280, 2.012, 0.625, 4.136, 2.664, 0.994, 3.901, 1.929, 3.478, 1.332

Enter the readings in level-book form by both the R-and-F and collimation systems (these may be combined into a single form to save copying). (LU)

(*3.3*) The following staff readings in metres were obtained when levelling along the centre-line of a straight road *ABC*.

BS	*IS*	*FS*	*Remarks*
2.405			Point *A* (RL = 250.05 m AOD)
1.954		1.128	CP
0.619		1.466	Point *B*
	2.408		Point *D*
	−1.515		Point *E*
1.460		2.941	CP
		2.368	Point *C*

D is the highest point on the road surface beneath a bridge crossing over the road at this point and the staff was held inverted on the underside of the bridge girder at *E*, immediately above *D*. Reduce the levels correctly by an approved method, applying the checks, and determine the headroom at *D*. If the road is to be regraded so that *AC* is a uniform gradient, what will be the new headroom at *D*?
The distance *AD* = 240 m and *DC* = 60 m. (LU)

(*Answer*: 3.923 m, 5.071 m)

(*3.4*) Distinguish, in construction and method of use, between dumpy and tilting levels. State in general terms the principle of an automatic level.

(*3.5*) The following levels were taken with a metric staff on a series of pegs at 100-m intervals along the line of a proposed trench.

BS	*IS*	*FS*	*Remarks*
2.10			TBM 28.75 m
	2.85		Peg *A*
1.80		3.51	Peg *B*
	1.58		Peg *C*
	2.24		Peg *D*
1.68		2.94	Peg *E*
	2.27		
	3.06		
		3.81	TBM 24.07 m

If the trench is to be excavated from peg *A* commencing at a formation level of 26.5 m and falling to peg *E* at a grade of 1 in 200, calculate the height of the sight rails in metres at *A*, *B*, *C*, *D* and *E*, if a 3-m boning rod is to be used.

Briefly discuss the techniques and advantages of using laser beams for the control of more precise work. (KU)

(*Answer*: 1.50, 1.66, 0.94, 1.10, 1.30 m)

(3.6) (a) Determine from first principles the approximate distance at which correction for curvature and refraction in levelling amounts to 3 mm, assuming that the effect of refraction is one seventh that of the Earth's curvature and that the Earth is a sphere of 12740 km diameter.
(b) Two survey stations A and B on opposite sides of a river are 780 m apart, and reciprocal levels have been taken between them with the following results:

Instrument at	*Height of instrument* (m)	*Staff at*	*Staff reading* (m)
A	1.472	B	1.835
B	1.496	A	1.213

Compute the ratio of refraction correction to curvature correction, and the difference in level between A and B.

(*Answer*: (a) 210 m (b) 0.14 to 1; B lower by 0.323 m)

3.15 TRIGONOMETRICAL LEVELLING

Trigonometrical levelling is used where difficult terrain, such as mountainous areas, precludes the use of conventional differential levelling. It may also be used where the height difference is large but the horizontal distance is short such as heighting up a cliff or a tall building. The vertical angle and the slope distance between the two points concerned are measured. Slope distance is measured using electromagnetic distance measurers (EDM) and the vertical (or zenith) angle using a theodolite.

When these two instruments are integrated into a single instrument it is called a 'total station'. Total stations contain algorithms that calculate and display the horizontal distance and vertical height, This latter facility has resulted in trigonometrical levelling being used for a wide variety of heighting procedures, including contouring. However, unless the observation distances are relatively short, the height values displayed by the total station are quite useless, if not highly dangerous, unless the total station contains algorithms to apply corrections for curvature and refraction.

3.15.1 Short lines

From *Figure 3.42* it can be seen that when measuring the angle

$$\Delta h = S \sin \alpha \tag{3.7}$$

When using the zenith angle z

$$\Delta h = S \cos z \tag{3.8}$$

If the horizontal distance is used

$$\Delta h = D \tan \alpha = D \cot z \tag{3.9}$$

The difference in elevation (ΔH) between ground points A and B is therefore

$$\begin{aligned}\Delta H &= h_i + \Delta h - h_t \\ &= \Delta h + h_i - h_t \end{aligned} \tag{3.10}$$

where h_i = vertical height of the measuring centre of the instrument above A
h_t = vertical height of the centre of the target above B

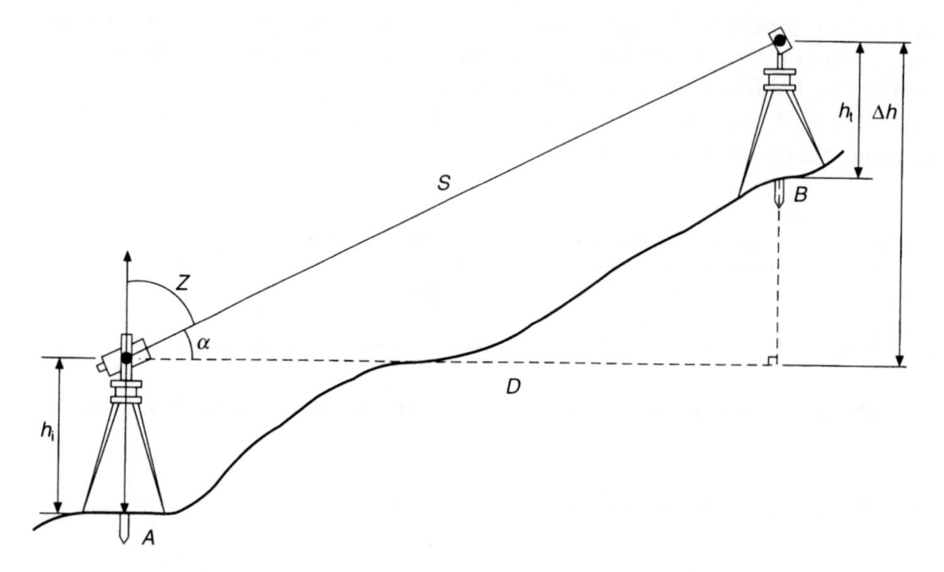

Fig. 3.42 *Trigonometric levelling – short lines*

This is the basic concept of trignometrical levelling. The vertical angles are positive for angles of elevation and negative for angles of depression. The zenith angles are always positive, but naturally when greater than 90° they will produce a negative result.

What constitutes a short line may be derived by considering the effect of curvature and refraction compared with the accuracy expected. The combined effect of curvature and refraction over 100 m = 0.7 mm, over 200 m = 3 mm, over 300 m = 6 mm, over 400 m = 11 mm and over 500 m = 17 mm.

If we apply the standard treatment for small errors to the basic equation we have

$$\Delta H = S \sin\alpha + h_i - h_t \tag{3.11}$$

and then

$$\delta(\Delta H) = \sin\alpha \cdot \delta S + S \cos\alpha \; \delta\alpha + \delta h_i - h_t \tag{3.12}$$

and taking standard errors:

$$\sigma^2_{\Delta H} = (\sin\alpha \cdot \sigma_s)^2 + (S \cos\alpha \cdot \sigma_\alpha)^2 + \sigma_i^2 + \sigma_t^2$$

Consider a vertical angle of $\alpha = 5°$, with $\sigma_\alpha = 5''$ (= 0.000024 radians), $S = 300$ m with $\sigma_s = 10$ mm and $\sigma_i = \sigma_t = 2$ mm. Substituting in the above equation gives:

$$\sigma^2_{\Delta H} = 0.9^2 \text{ mm}^2 + 7.2^2 \text{ mm}^2 + 2^2 \text{ mm}^2 + 2^2 \text{ mm}^2$$

$$\sigma_{\Delta H} = 7.8 \text{ mm}$$

This value is similar in size to the effect of curvature and refraction over this distance and indicates that short sights should never be greater than 300 m. It also indicates that the accuracy of distance S is not critical when the vertical angle is small. However, the accuracy of measuring the vertical angle is very critical and requires the use of a theodolite, with more than one measurement on each face.

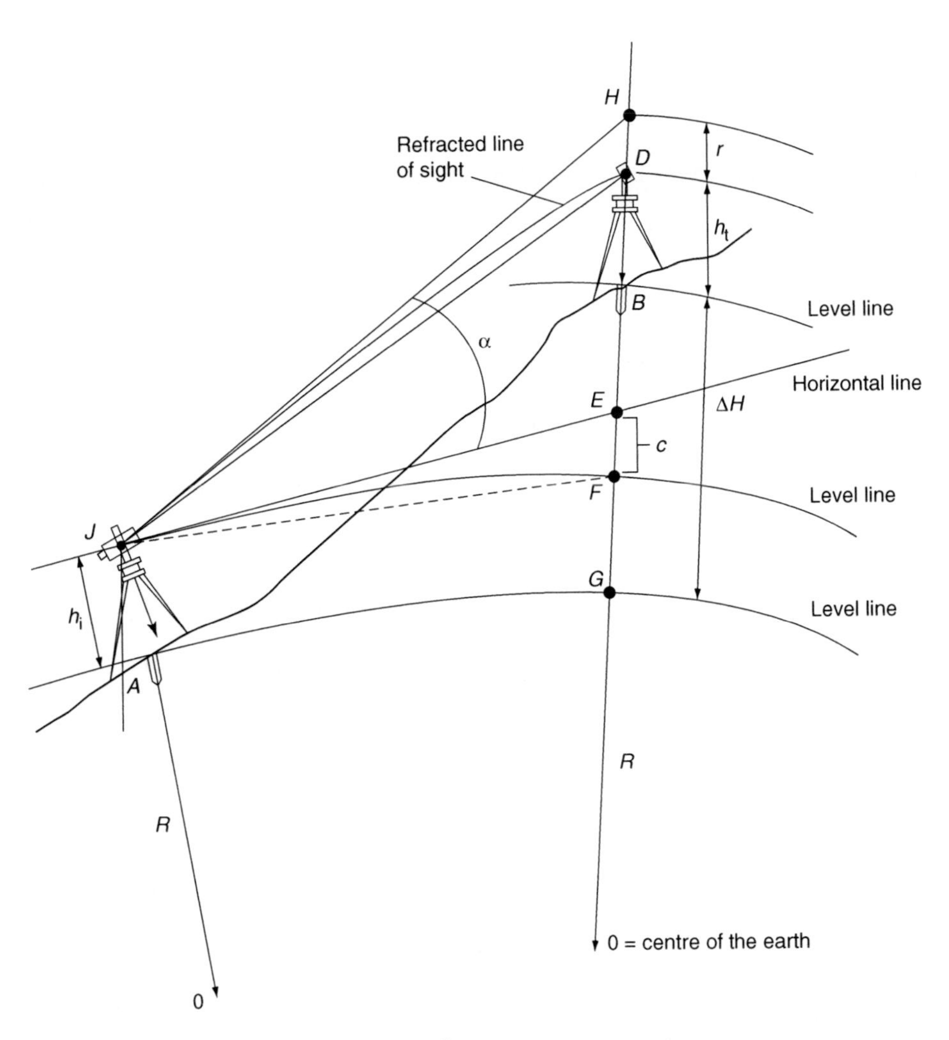

Fig. 3.43 *Trigonometric levelling – long lines*

3.15.2 Long lines

For long lines the effect of curvature (c) and refraction (r) must be considered. From *Figure 3.43*, it can be seen that the difference in elevation (ΔH) between A and B is:

$$\Delta H = GB = GF + FE + EH - HD - DB$$

$$= h_i + c + \Delta h - r - h_t$$

$$= \Delta h + h_i - h_t + (c - r) \quad (3.13)$$

Thus it can be seen that the only difference from the basic equation for short lines is the correction for curvature and refraction ($c - r$).

Although the line of sight is refracted to the target at D, the telescope is pointing to H, thereby measuring the angle α from the horizontal. It follows that $S \sin \alpha = \Delta h = EH$ and requires a correction for refraction equal to HD.

The correction for refraction is based on a quantity termed the 'coefficient of refraction' (K). Considering the atmosphere as comprising layers of air which decrease in density at higher elevations, the line of sight from the instrument will be refracted towards the denser layers. The line of sight therefore approximates to a circular arc of radius R_s roughly equal to $8R$, where R is the radius of the Earth. However, due to the uncertainty of refraction one cannot accept this relationship and the coefficient of refraction is defined as

$$K = R/R_s \tag{3.14}$$

An average value of $K = 0.15$ is frequently quoted but, as stated previously, this is most unreliable and is based on observations taken well above ground level. Recent investigation has shown that not only can K vary from -2.3 to $+3.5$ with values over ice as high as $+14.9$, but it also has a daily cycle. Near the ground, K is affected by the morphology of the ground, by the type of vegetation and by other assorted complex factors. Although much research has been devoted to modelling these effects, in order to arrive at an accurate value for K, the most practical method still appears to be by simultaneous reciprocal observations.

As already shown, curvature (c) can be approximately computed from $c = D^2/2R$, and as $D \approx S$ we can write

$$c = S^2/2R \tag{3.15}$$

Now considering *Figures 3.43* and *3.44* the refracted ray JD has a radius R_s and a measured distance S and subtends angles δ at its centre, then

$$\delta = S/R_s$$

$$\delta/2 = S/2R_s$$

As the refraction $K = R/R_s$ we have

$$\delta/2 = SK/2R$$

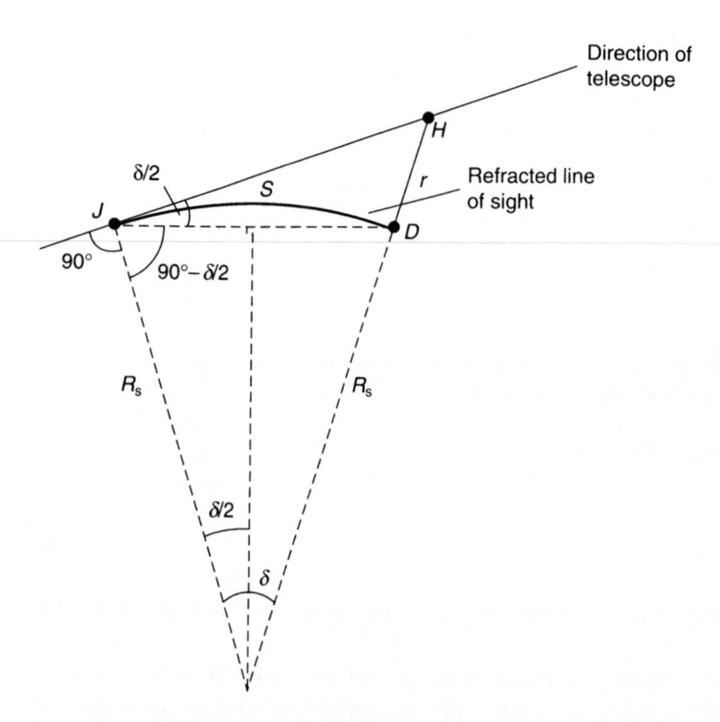

Fig. 3.44 *Curvature and refraction*

Without loss of accuracy we can assume $JH = JD = S$ and treating the HD as the arc of a circle of radius S:

$$HD = S \cdot \delta/2 = S^2K/2R = r \tag{3.16}$$

$$(c - r) = S^2(1 - K)/2R \tag{3.17}$$

All the above equations express c and r in linear terms. To obtain the angles of curvature and refraction, EJF and HJD in *Figure 3.43*, reconsider *Figure 3.44*. Imagine JH is the horizontal line JE in *Figure 3.43* and JD the level line JF of radius R. Then δ is the angle subtended at the centre of the Earth and the angle of curvature is half this value. To avoid confusion let $\delta = \theta$ and as already shown:

$$\theta/2 = S/2R = \hat{c} \tag{3.18}$$

where the arc distance at MSL approximates to S. Also, as shown:

$$\delta/2 = SK/2R = \hat{r} \tag{3.19}$$

Therefore in angular terms:

$$(\hat{c} - \hat{r}) = S(1 - K)/2R \text{ rads} \tag{3.20}$$

Note the difference between equations in linear terms and those in angular.

3.15.3 Reciprocal observations

Reciprocal observations are observations taken from A and B, the arithmetic mean result being accepted. If one assumes a symmetrical line of sight from each end and the observations are taken simultaneously, then the effect of curvature and refraction is cancelled out. For instance, for elevated sights, $(c - r)$ is added to a positive value to increase the height difference. For depressed or downhill sights, $(c - r)$ is added to a negative value and decreases the height difference. Thus the average of the two values is free from the effects of curvature and refraction. This statement is not entirely true as the assumption of symmetrical lines of sight from each end is dependent on uniform ground and atmospheric conditions at each end, at the instant of simultaneous observation.

In practice over short distances, sighting into each other's object lens forms an excellent target, with some form of communication to ensure simultaneous observation.

The following numerical example is taken from an actual survey in which the elevation of A and B had been obtained by precise geodetic levelling and was checked by simultaneous reciprocal trigonometrical levelling.

Worked example

Example 3.8.

Zenith angle at $A = Z_A = 89°\ 59'\ 18.7''$ (VA $0°\ 00'\ 41.3''$)

Zenith angle at $B = Z_B = 90°\ 02'\ 59.9''$ (VA $= -0°\ 02'\ 59.9''$)

Height of instrument at $A = h_A = 1.290$ m

Height of instrument at $B = h_B = 1.300$ m

Slope distance corrected for meteorological conditions = 4279.446 m

Each target was set at the same height as the instrument at its respective station.

As the observations are reciprocal, the corrections for curvature and refraction are ignored:

$$\Delta H_{AB} = S \cos Z_A + h_i - h_t$$

$$= 4279.446 \cos 89^\circ\, 59'\, 18.7'' + 1.290 - 1.300 = 0.846 \text{ m}$$

$$\Delta H_{BA} = 4279.446 \cos 90^\circ\, 02'\, 59.9'' + 1.300 - 1.290 = -3.722 \text{ m}$$

$$\Delta H_{AB} = 3.722 \text{ m}$$

Mean value $\Delta H = 2.284$ m

This value compares favourably with 2.311 m obtained by precise levelling. However, the disparity between the two values 0.846 and −3.722 shows the danger inherent in single observations uncorrected for curvature and refraction. In this case the correction for curvature only is +1.256 m, which, when applied, brings the results to 2.102 m and −2.466 m, producing much closer agreement. To find K simply substitute the mean value $\Delta H = 2.284$ into the equation for a single observation.

From A to B:

$$2.284 = 4279.446 \cos 89^\circ\, 59'\, 18.7'' + 1.290 - 1.300 + (c - r)$$

where $(c - r) = S^2(1 - K)/2R$

and the local value of R for the area of observation = 6 364 700 m

$$2.284 = 0.856 - 0.010 + S^2(1 - K)/2R$$

$$1.438 = 4279.446^2(1 - K)/2 \times 6\,364\,700 \text{ m}$$

$$K = 0.0006$$

From B to A:

$$2.284 = -3.732 + 1.300 - 1.290 + S^2(1 - K)/2R$$

$$K = 0.0006$$

Now this value for K could be used for single ended observations taken within the same area, at the same time, to give improved results.

A variety of formulae are available for finding K directly. For example, using zenith angles:

$$K = 1 - \frac{Z_A + Z_B - 180^\circ}{180^\circ/\pi} \times \frac{R}{S} \tag{3.21}$$

and using vertical angles:

$$K = (\theta + \alpha_0 + \beta_0)/\theta \tag{3.22}$$

where θ = the angle subtended at the center of the Earth by the arc distance $\approx S$ and is calculated using:

$$\theta'' = S\rho/R \quad \text{where } \rho = 206\,265$$

In the above formulae the values used for the angles must be those which would have been observed had $h_i = h_t$ and, in case of vertical angles, entered with their appropriate sign. As shown in *Figure 3.45*,

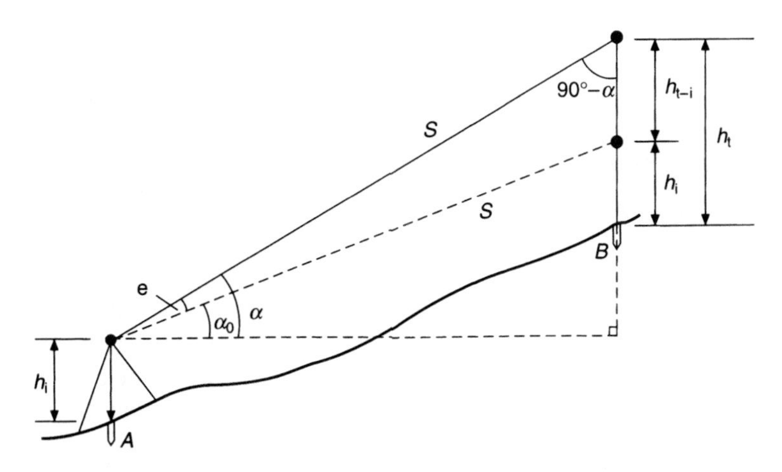

Fig. 3.45 *Correction for unequal instrument and target height*

By sine rule: $\alpha_0 = \alpha - e$ and for an angle of depression it becomes $\beta_0 = \beta + e$.

$$\sin e = \frac{h_{t-i}\ \sin(90° - \alpha)}{S}$$

$$e = \sin^{-1}\left(\frac{h_{t-i}\cos\alpha}{S}\right)$$

$$= \frac{h_{t-1}}{S}\cos\,\alpha + \frac{h^3_{t-i}}{6S^3}\cos^3\,\alpha + \cdots$$

$$\therefore e = (h_{t-i}\cos\,\alpha)/S \qquad (3.23)$$

For zenith angles:

$$e = (h_{t-i}\ \sin\,Z)/S \qquad (3.24)$$

3.15.4 Sources of error

Consider the formula for a single observation, found by substituting *equations 3.8* and *3.17* into *3.13*:

$$\Delta H = S\sin\alpha + h_i - h_t + S^2(1 - K)/2R$$

The obvious sources of error lie in obtaining the slope distance S, the vertical angle α the heights of the instrument and target, the coefficient of refraction K and a value for the local radius of the Earth R. Differentiating gives:

$$\delta(\Delta H) = \delta S\ \sin\,\alpha + S\cos\alpha\cdot\delta\alpha + \delta h_i + \delta h_t + S^2\ \delta K/2R + S^2(1 - K)\ \delta R/2R^2$$

and taking standard errors:

$$\sigma^2_{\Delta H} = (\sigma_s\sin\alpha)^2 + (S\cos\alpha\sigma_\alpha)^2 + \sigma_i^2 + \sigma_t^2 + (S^2\sigma_K/2R)^2 + (S^2(1 - K)\sigma_R/2R^2)^2$$

Taking $S = 2000$ m, $\sigma_S = 0.005$ m, $\alpha = 8°$, $\sigma_\alpha = 7''$ ($= 0.000034$ radians), $\sigma_i = \sigma_t = 2$ mm, $K = 0.15$, $\sigma_K = 1$, $R = 6380$ km, and $\sigma_R = 10$ km, we have

$$\sigma^2_{\Delta H} = (0.7)^2 + (48.0)^2 + 2^2 + 2^2 + (156.7)^2 + (0.4)^2\ \text{mm}^2$$

$$\sigma_{\Delta H} = 164\ \text{mm}$$

Once again it can be seen that the accuracy required to measure S is not critical.

However, the measurement of the vertical angle is critical and the importance of its precision will increase with greater distance. The error in the value of refraction is the most critical component and will increase rapidly as the square of the distance. Thus to achieve reasonable results over long sights, simultaneous reciprocal observations are essential.

3.15.5 Contouring

The ease with which total stations produce horizontal distance, vertical height and horizontal direction makes them ideal instruments for rapid and accurate contouring in virtually any type of terrain. The data recorded may be transformed from direction, distance and elevation of a point, to its position and elevation in terms of three-dimensional coordinates. These points thus comprise a digital terrain or ground model (DTM/DGM) from which the contours are interpolated and plotted.

The total station and a vertical rod that carries a single reflector are used to locate the ground points (*Figure 3.46*). A careful reconnaissance of the area is necessary, in order to plan the survey and define the necessary ground points that are required to represent the characteristic shape of the terrain. Break lines, the tops and bottoms of hills or depressions, the necessary features of water courses, etc., and enough points to permit accurate interpolation of contour lines at the interval required, comprise the field data. As the observation distances are relatively short, curvature and refraction might be ignored. However, in most total stations corrections for curvature and refraction may be applied.

Fig. 3.46 *Contouring with a total station and detail pole*

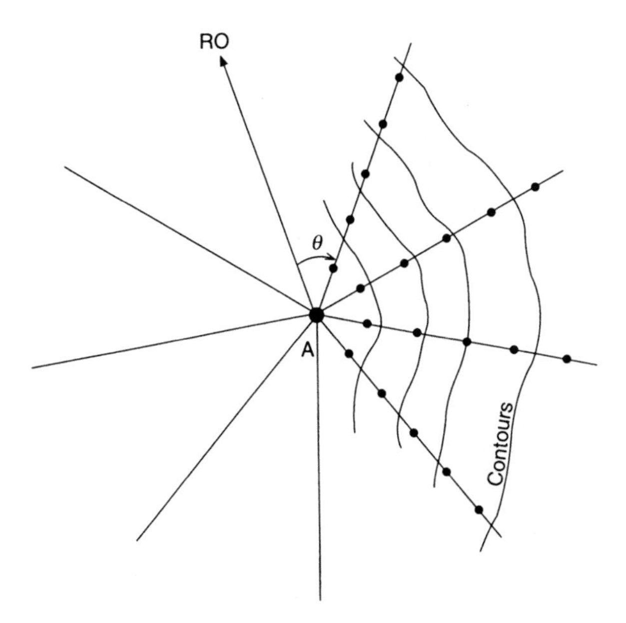

Fig. 3.47 *Radiation method*

From *Figure 3.42*, it can be seen that if the reduced level of point A (RL_A) is known, then the reduced level of ground point B is:

$$RL_B = RL_A + h_i + \Delta h - h_t$$

When contouring, the height of the reflector is set to the same height as the instrument, i.e. $h_t = h_i$, and cancels out in the previous equation. Thus the height displayed by the instrument is the height of the ground point above A:

$$RL_B = RL_A + \Delta h$$

In this way the reduced levels of all the ground points are rapidly acquired and all that is needed are their positions. One method of carrying out the process is by radiation.

As shown in *Figure 3.47*, the instrument is set up on a control point A, whose reduced level is known, and sighted to a second control point (RO). The horizontal circle is set to the direction computed from the coordinates of A and the RO. The instrument is then turned through a chosen horizontal angle (θ) defining the direction of the first ray. Terrain points along this ray are then located by measured horizontal distance and height difference. This process is repeated along further rays until the area is covered. Unless a very experienced person is used to locate the ground points, there will obviously be a greater density of points near the instrument station. The method, however, is quite easy to organize in the field. The angle between successive rays may vary from 20° to 60° depending on the terrain.

Many ground-modelling software packages interpolate and plot contours from strings of linked terrain points. Computer processing is aided if the ground points are located in continuous strings throughout the area, approximately following the line of the contour. They may also follow the line of existing watercourses, roads, hedges, kerbs, etc. (*Figure 3.48*).

Depending on the software package used, the string points may be transformed into a triangular or gridded structure. Heights can then be determined by linear interpolation and the terrain represented by simple planar triangular facets. Alternatively, high-order polynomials may be used to define three-dimensional surfaces fitted to the terrain points. From these data, contours are interpolated and a contour model of the terrain produced.

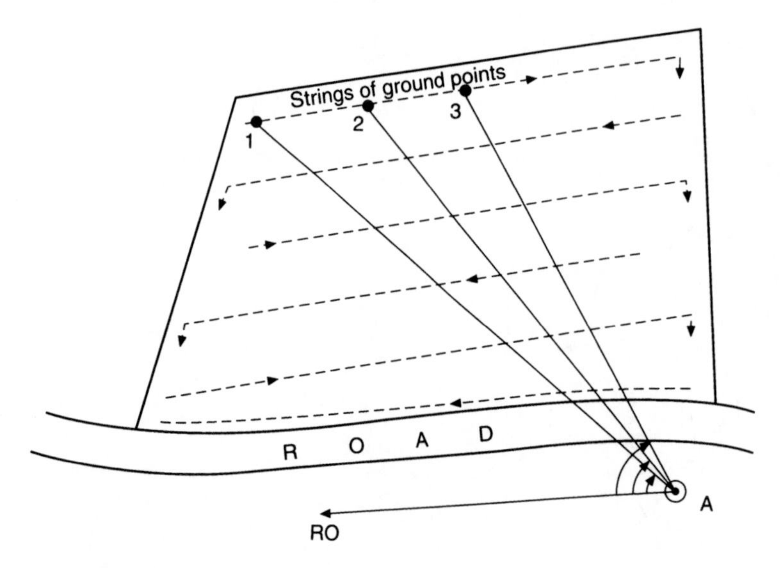

Fig. 3.48 *Plotting*

Worked examples

Example 3.9. (a) Define the coefficient of refraction K, and show how its value may be obtained from simultaneous reciprocal trigonometric levelling observations.

(b) Two triangulation stations A and B are 2856.85 m apart. Observations from A to B gave a mean vertical angle of $+01^\circ\ 35'\ 38''$, the instrument height being 1.41 m and the target height 2.32 m. If the level of station A is 156.86 m OD and the value of K for the area is 0.16, calculate the reduced level of B (radius of Earth = 6372 km). (KU)

(a) Refer to *Section 3.15.2.*

(b) This part will be answered using both the angular and the linear approaches.

Angular method

Difference in height of $AB = \Delta H = D \tan[\alpha + (\hat{c} - \hat{r})]$ where $\hat{c} = \theta/2$ and

$$\theta = \frac{D}{R} = \frac{2856.85}{6\,372\,000} = 0.000\,448 \text{ rad}$$

$$\therefore \hat{c} = 0.000\,224 \text{ rad}$$

$$\hat{r} = K(\theta/2) = 0.16 \times 0.000\,224 = 0.000\,036 \text{ rad}$$

$$\therefore (\hat{c} - \hat{r}) = 0.000\,188 \text{ rad} = 0^\circ\ 00'\ 38.8''$$

$$\therefore \Delta H = 2856.85 \tan(01^\circ\ 35'\ 38'' + 0^\circ\ 00'\ 38.8'') = 80.03 \text{ m}$$

Refer to *Figure 3.42.*

$$\text{RL of } B = \text{RL of } A + h_i + \Delta H - h_t$$

$$= 156.86 + 1.41 + 80.03 - 2.32 = 235.98 \text{ m}$$

Linear method

$$\Delta H = D \tan \alpha + (c - r)$$

where $(c - r) = \left(\frac{D^2}{2R}\right)(1 - K) = \frac{2856.85^2}{2 \times 6\,372\,000} \times 0.084 = 0.54 \text{ m}$

$$D \tan \alpha = 2856.85 \tan(01^\circ\, 35'\, 39'') = 79.49 \text{ m}$$

where $\therefore \Delta H = 79.49 + 0.54 = 80.03$ m

Example 3.10. Two stations *A* and *B* are 1713 m apart. The following observations were recorded: height of instrument at *A* 1.392 m, and at *B* 1.464 m; height of signal at *A* 2.199 m, and at *B* 2 m. Elevation to signal at *B* $1^\circ\, 08'\, 08''$, depression angle to signal at *A* $1^\circ\, 06'\, 16''$. If $1''$ at the Earth's centre subtends 30.393 m at the Earth's surface, calculate the difference of level between *A* and *B* and the refraction correction. (LU)

$$\Delta H = D \tan\left(\frac{\alpha + \beta}{2}\right) + \frac{(h'_t - h'_i) - (h_t - h_i)}{2}$$

where h_i = height of instrument at *A*; h_t = height of target at *B*; h'_i = height of instrument at *B*; h'_t = height of target at *A*.

$$\therefore \Delta H = 1713 \tan\left(\frac{(1^\circ\, 08'\, 08'') + (1^\circ\, 06'\, 15'')}{2}\right) + \frac{(2.199 - 1.464) - (2.000 - 1.392)}{2}$$

$$= 33.490 + 0.064 = 33.55 \text{ m}$$

Using the alternative approach of reducing α and β to their values if $h_i = h_t$.

Correction to angle of elevation:

$$e'' \approx \frac{1.392 - 2.000}{1713.0} \times 206\,265 = -73.2''$$

$$\therefore \alpha = (1^\circ\, 08'\, 08'') - (01'\, 13.2'') = 1^\circ\, 06'\, 54.8''$$

Correction to angle of depression:

$$e'' \approx \frac{(2.199 - 1.464)}{1713.0} \times 206\,265 = 88.5''$$

$$\therefore \beta = (1^\circ\, 06'\, 15'') + (01'\, 28.5'') = 1^\circ\, 07'\, 43.5''$$

$$\therefore \Delta H = 1713 \tan\left(\frac{(1^\circ\, 06'\, 54.8'') + (1^\circ\, 07'\, 43.5'')}{2}\right) = 33.55 \text{ m}$$

Refraction correction $\hat{r} = \frac{1}{2}(\theta + \alpha + \beta)$

where $\theta'' = 1713.0/30.393 = 56.4''$

$$= \frac{1}{2}[56.4'' + (1^\circ\, 06'\, 54.8'') - (1^\circ\, 07'\, 43.5'')] = 3.8''$$

and also $K = \frac{\hat{r}}{\theta/2} = \frac{3.8''}{28.2''} = 0.14$

Example 3.11. Two points A and B are 8 km apart and at levels of 102.50 m and 286.50 m OD, respectively. The height of the target at A is 1.50 m and at B 3.00 m, while the height of the instrument in both cases is 1.50 m on the Earth's surface subtends 1″ of arc at the Earth's centre and the effect of refraction is one seventh that of curvature, predict the observed angles from A to B and B to A.

Difference in level A and $B = \Delta H = 286.50 - 102.50 = 184.00$ m

$$\therefore \text{ by radians } \phi'' = \frac{184}{8000} \times 206\,265 = 4744'' = 1^\circ\, 19'\, 04''$$

$$\text{Angle subtended at the centre of the Earth} = \theta'' = \frac{8000}{31} = 258''$$

$$\therefore \text{ Curvature correction} \quad \hat{c} = \theta/2 = 129'' \quad \text{and} \quad \hat{r} = \hat{c}/7 = 18''$$

Now $\quad \Delta H = D \tan \phi$

where $\quad \phi = \alpha + (\hat{c} + \hat{r})$

$$\therefore \alpha = \phi - (\hat{c} - \hat{r}) = 4744'' - (129'' - 18'') = 4633'' = 1^\circ\, 17'\, 13''$$

Similarly $\quad \phi = \beta - (\hat{c} - \hat{r})$

$$\therefore \beta = \phi + (\hat{c} - \hat{r}) = 4855'' = 1^\circ\, 20'\, 55''$$

The observed angle α must be corrected for variation in instrument and signal heights. Normally the correction is subtracted from the observed angle to give the truly reciprocal angle. In this example, α is the truly reciprocal angle, thus the correction must be added in this reverse situation:

$$e'' \approx [(h_t - h_i)/D] \times 206\,265 = [(3.00 - 1.50)/8000] \times 206\,265 = 39''$$

$$\therefore \alpha = 4633'' + 39'' = 4672'' = 1^\circ\, 17'\, 52''$$

Example 3.12. A gas drilling-rig is set up on the sea bed 48 km from each of two survey stations which are on the coast and several kilometres apart. In order that the exact position of the rig may be obtained, it is necessary to erect a beacon on the rig so that it may be clearly visible from theodolites situated at the survey stations, each at a height of 36 m above the high-water mark.

Neglecting the effects of refraction, and assuming that the minimum distance between the line of sight and calm water is to be 3 m at high water, calculate the least height of the beacon above the high-water mark, at the rig. Prove any equations used.

Calculate the angle of elevation that would be measured by the theodolite when sighted onto this beacon, taking refraction into account and assuming that the error due to refraction is one seventh of the error due to curvature of the Earth. Mean radius of Earth = 6273 km. (ICE)

From *Figure 3.49*:

$$D_1 = (2c_1R)^{\frac{1}{2}} \text{ (equation 3.1)}$$

$$\therefore D_1 = (2 \times 33 \times 6\,273\,000) = 20.35 \text{ km}$$

$$\therefore D_2 = 48 - D_1 = 27.65 \text{ km}$$

$$\therefore \text{ since } D_2 = (2c_2R)^{\frac{1}{2}}$$

$c_2 = 61$ m, and to avoid grazing by 3 m, height of beacon $= 64$ m

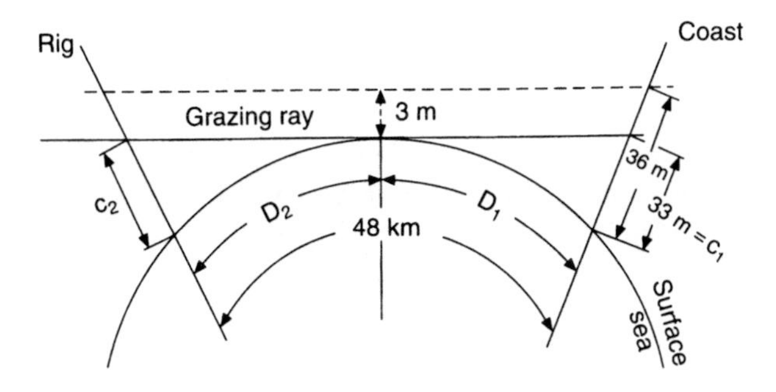

Fig. 3.49 *Lines of sight*

Difference in height of beacon and theodolite = 64 − 36 = 28 m; observed vertical angle $\alpha = \phi - (\hat{c} - \hat{r})$ for angles of elevation, where

$$\phi'' = \frac{28 \times 206\,265}{48000} = 120.3''$$

$$\hat{c} = \theta/2$$

$$\text{where} \quad \theta'' = \left(\frac{48}{6273}\right) \times 206\,265 = 1578.3''$$

$$\therefore \hat{c} = 789.2'' \quad \text{and} \quad \hat{r} = \hat{c}/7 = 112.7''$$

$$\therefore \alpha = 120.3'' - 789.2'' + 112.7'' = -556.2'' = -0°\,09'\,16''$$

The negative value indicates α to be an angle of depression, not elevation, as quoted in the question.

3.16 HEIGHTING WITH GPS

This aim of this short section is to introduce an alternative technology, which can be very useful for heighting, where many points in a given area are required. The subject of satellite positioning is covered in depth in *Chapter 8*. The main advantages of using Global Positioning System (GPS) receivers for heighting are that no line of sight is required between instrument and target and the speed with which heights with their plan positions can be collected. The practical limit is usually the speed at which the GPS receiver can be moved over the ground. Suppose the heights of some open ground are required with a density of not less than a point every 5 m and a vehicle with a GPS antenna mounted on its roof is available. If the GPS records 10 points every second, then the maximum theoretical speed of the vehicle would be 180 km per hour! Although it is unlikely that the surveyor would be travelling at such speeds over rough terrain this example illustrates the potential of the system.

On the negative side, GPS has limits as to the practical precision that can be achieved. Height of one instrument relative to another nearby instrument may be obtained with a precision of about 0.02 m. However, the heights are related to the World Geodetic System 1984 (WGS84) ellipsoidal model of the Earth and to be useful would need to be converted to heights above the local datum. The relationship between WGS84 and the local datum will not be constant and will vary smoothly by up to 0.1 m per kilometre across the area of interest. Therefore external data will be required to apply the appropriate corrections to make the GPS derived heights useful to the surveyor. If the GPS antenna is mounted on a vehicle then the relationship between the antenna and the ground will vary as the vehicle bounces across the terrain adding further random error to the heights of individual points. This may not be so important if the purpose to the heighting is to determine volume of ground to be cut or filled, as the errors in individual heights will tend to cancel each other out.

GPS equipment is rather more expensive than conventional levelling equipment or total stations. GPS equipment works very well in a GPS friendly environment, i.e. where there is an open sky. It becomes much less useful if there are many obstructions such as tall buildings or under tree canopies. In such cases conventional techniques would be more appropriate.

4
Distance measurement

Distance is one of the fundamental measurements in surveying. Although frequently measured as a spatial distance (sloping distance) in three-dimensional space, usually it is the horizontal component which is required.

Distance is required in many instances, e.g. to give scale to a network of control points, to fix the position of topographic detail by offsets or polar coordinates, to set out the position of a point in construction work, etc.

The basic methods of measuring distance are, at the present time, by *taping* or by *electromagnetic* (or *electro-optical*) *distance measurement*, generally designated as EDM. For very rough reconnaissance surveys or approximate estimates pacing may be suitable.

For distances over 5 km, GPS satellite methods, which can measure the vectors between two points accurate to 1 ppm are usually more suitable.

4.1 TAPES

Tapes come in a variety of lengths and materials. For engineering work the lengths are generally 10 m, 30 m, 50 m and 100 m.

Linen or glass fibre tapes may be used for general use, where precision is not a prime consideration. The linen tapes are made from high quality linen, combined with metal fibres to increase their strength. They are sometimes encased in plastic boxes with recessed handles. These tapes are often graduated in 5-mm intervals only.

More precise versions of the above tapes are made of steel and graduated in millimetres.

For high-accuracy work, steel bands mounted in an open frame are used. They are standardized so that they measure their nominal length at a designated temperature usually 20°C and at a designated applied tension usually between 50 N to 80 N. This information is clearly printed on the zero end of the tape. *Figure 4.1* shows a sample of the equipment.

For the most precise work, invar tapes made from 35% nickel and 65% steel are available. The singular advantage of such tapes is that they have a negligible coefficient of expansion compared with steel, and hence temperature variations are not critical. Their disadvantages are that the metal is soft and weak, whilst the price is more than ten times that of steel tapes. An alternative tape, called a Lovar tape, is roughly, midway between steel and invar.

Much ancillary equipment is necessary in the actual taping process, e.g.

(1) Ranging rods are made of wood or steel, 2 m long and 25 mm in diameter, painted alternately red and white and have pointed metal shoes to allow them to be thrust into the ground. They are generally used to align a straight line between two points.
(2) Chaining arrows made from No. 12 steel wire are also used to mark the tape lengths (*Figure 4.2*).

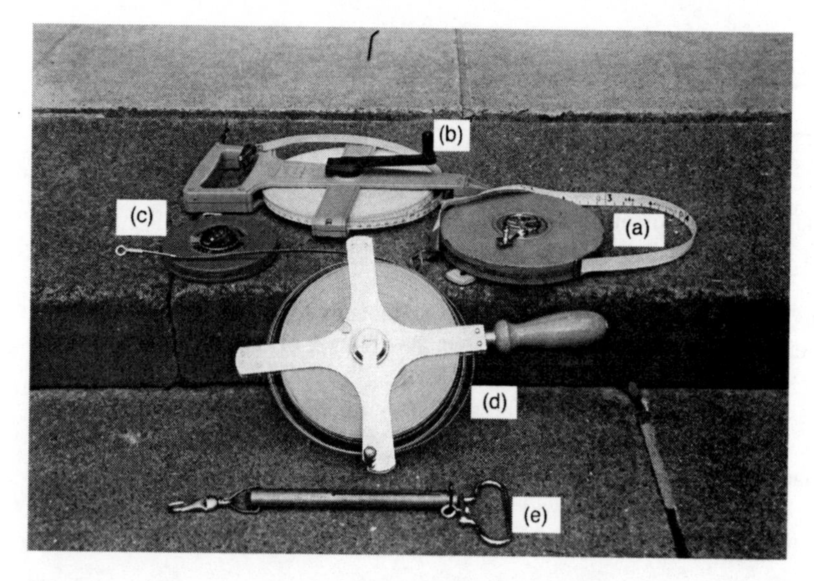

Fig. 4.1 *(a) Linen tape, (b) fibreglass, (c) steel, (d) steel band, (e) spring balance*

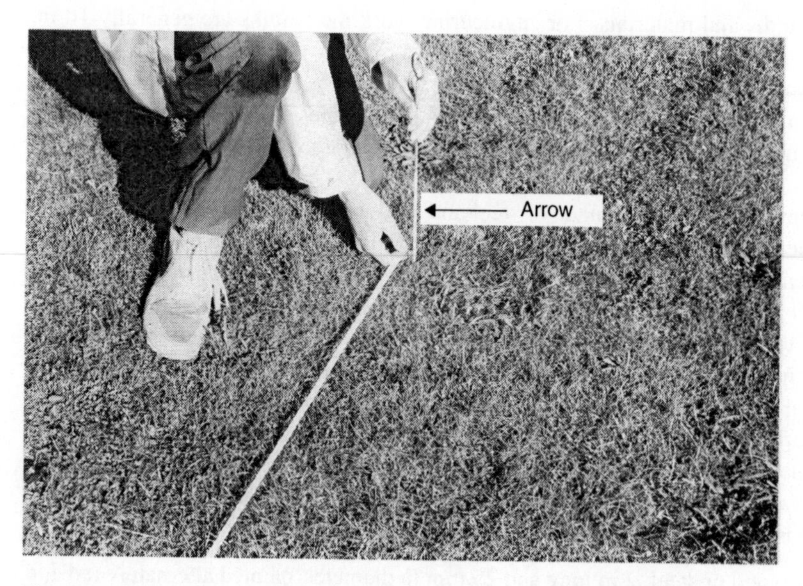

Fig. 4.2 *Using an arrow to mark the position of the end of the tape*

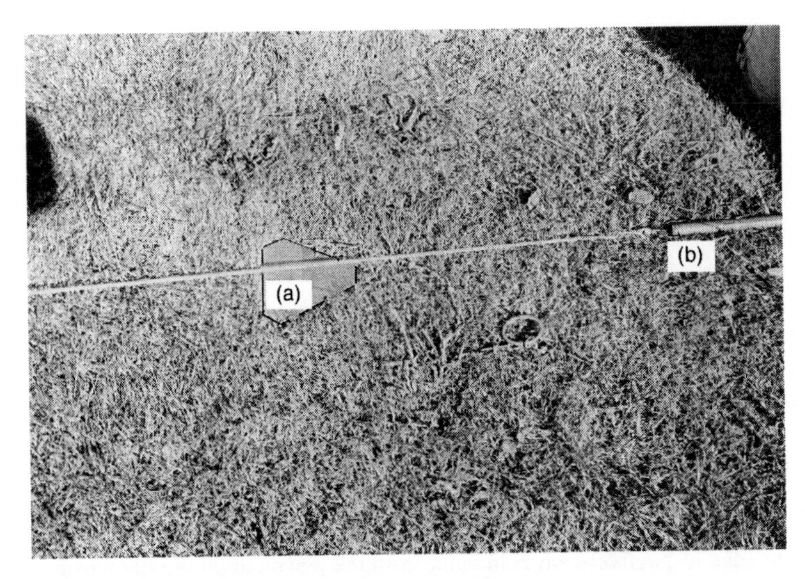

Fig. 4.3 *(a) Measuring plate, (b) spring balance tensioning the tape*

(3) Spring balances are generally used with roller-grips or tapeclamps to grip the tape firmly when the standard tension is applied. As it is quite difficult to maintain the exact tension required with a spring balance, it may be replaced by a tension handle, which ensures the application of correct tension.
(4) Field thermometers are also necessary to record the tape temperature at the time of measurement, thereby permitting the computation of tape corrections when the temperature varies from standard. These thermometers are metal cased and can be clipped onto the tape if necessary, or simply laid on the ground alongside the tape but must be shaded from the direct rays of the sun.
(5) Hand levels may be used to ensure that the tape is horizontal. This is basically a hand-held tube incorporating a spirit bubble to ensure a horizontal line of sight. Alternatively, an Abney level may be used to measure the slope of the ground.
(6) Plumb-bobs may be necessary if stepped taping is used.
(7) Measuring plates are necessary in rough ground, to afford a mark against which the tape may be read. *Figure 4.3* shows the tensioned tape being read against the edge of such a plate. The corners of the triangular plate are turned down to form grips, when the plate is pressed into the earth and thereby prevent its movement.

In addition to the above, light oil and cleaning rags should always be available to clean and oil the tape after use.

4.2 FIELD WORK

4.2.1 Measuring along the ground (Figures 4.3 and 4.4)

The most accurate way to measure distance with a steel band is to measure the distance between pre-set measuring marks, rather than attempt to mark the end of each tape length. The procedure is as follows:

(1) The survey points to be measured are defined by nails in pegs and should be set flush with the ground surface. Ranging rods are then set behind each peg, in the line of measurement.

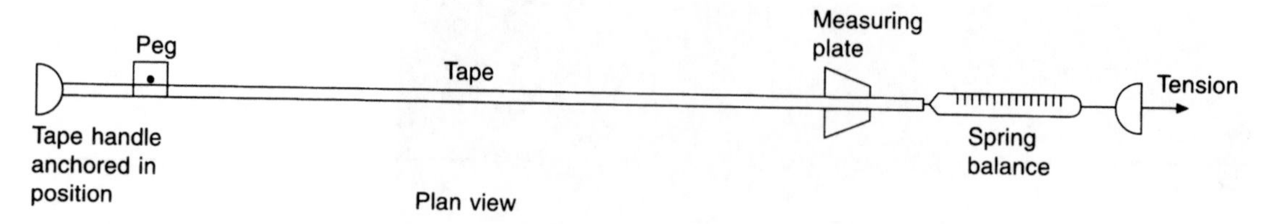

Fig. 4.4 *Plan view*

(2) Using a linen tape, arrows are aligned between the two points at intervals less than a tape length. Measuring plates are then set firmly in the ground at these points, with their measuring edge normal to the direction of taping.
(3) The steel band is then carefully laid out, in a straight line between the survey point and the first plate. One end of the tape is firmly anchored, whilst tension is slowly applied at the other end. At the exact instant of standard tension, both ends of the tape are read simultaneously against the survey station point and the measuring plate edge respectively, on command from the person applying the tension. The tension is eased and the whole process repeated at least four times or until a good set of results is obtained.
(4) When reading the tape, the metres, decimetres and centimetres should be noted as the tension is being applied; thus on the command 'to read', only the millimetres are required.
(5) The readings are noted by the booker and quickly subtracted from each other to give the length of the measured bay.
(6) In addition to 'rear' and 'fore' readings, the tape temperature is recorded, the value of the applied tension, which may in some instances be greater than standard, and the slope or difference in level of the tape ends are also recorded.
(7) This method requires a survey party of five; one to anchor the tape end, one to apply tension, two observers to read the tape and one booker.
(8) The process is repeated for each bay of the line being measured, care being taken not to move the first measuring plate, which is the start of the second bay, and so on.
(9) The data may be booked as follows:

Bay	*Rear*	*Fore*	*Difference*	*Temp.*	*Tension*	*Slope*	*Remarks*
A–1	0.244	29.368	29.124	08°C	70 N	5° 30′	Standard values
	0.271	29.393	29.122				20°C, 70 N
	0.265	29.389	29.124				
	0.259	29.382	29.123				Range 2 mm
		Mean =	29.123				
1–2							2nd bay

The mean result is then corrected for:

(1) Tape standardization.
(2) Temperature.
(3) Tension (if necessary).
(4) Slope.

The final total distance may then be reduced to its equivalent MSL or mean site level.

4.2.2 Measuring in catenary

Although the measurement of base lines in catenary is virtually obsolete, it is still the most accurate method of obtaining relatively short distances over rough terrain and much better than can be obtained with GPS or even a total station. The only difference from the procedures outlined above is that the tape is raised off the ground between two measuring marks and so the tape sags in a curve known as a catenary.

Figure 4.5 shows the basic set-up, with tension applied by levering back on a ranging rod held through the handle of the tape (*Figure 4.6*).

Figure 4.7 shows a typical measuring head with magnifier attached. In addition to the corrections already outlined, a further correction for sag in the tape is necessary.

For extra precision the measuring heads may be aligned in a straight line by theodolite, the difference in height of the heads being obtained by levelling to a staff held directly on the heads.

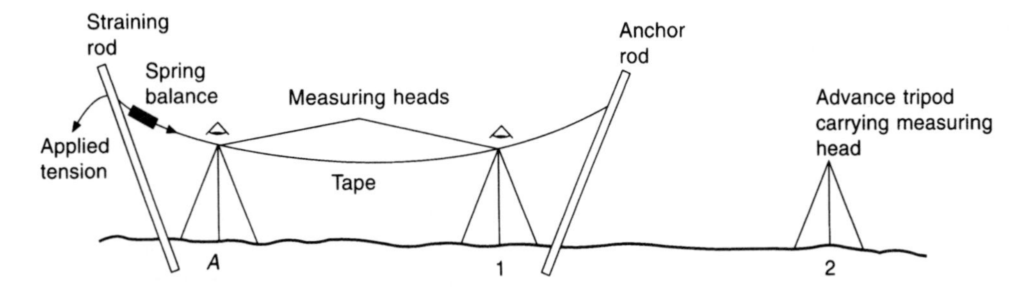

Fig. 4.5 *Suspended tape*

Fig. 4.6 *Ranging rod used to apply tension to a steel band suspended in catenary*

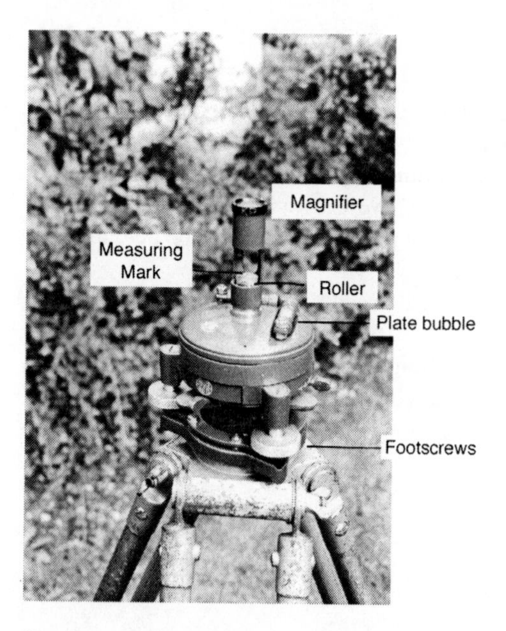

Fig. 4.7 *Measuring head*

4.2.3 Step measurement

Step measurement is the process of breaking the overall distance down into manageable short sections, each much less than a whole tape length. The tape is stretched horizontally and a plumb-bob suspended from the elevated end of the tape. This method of measurement over sloping ground should be avoided if high accuracy is required. The main source of error lies in attempting to accurately locate the suspended end of the tape, as shown in *Figure 4.8*.

Fig. 4.8 *Step measurement*

The steps should be kept short enough to minimize sag in the tape. Thus the sum of the steps equals the horizontal distance required.

4.3 DISTANCE ADJUSTMENT

To eliminate or minimize the systematic errors of taping, it is necessary to adjust each measured bay to its final horizontal equivalent as follows.

4.3.1 Standardization

During a period of use, a tape will gradually alter in length for a variety of reasons. The amount of change can be found by having the tape standardized at either the National Physical Laboratory (NPL) for invar tapes or the Department of Trade and Industry (DTI) for steel tapes, or by comparing it with a reference tape kept purely for this purpose. The tape may then be specified as being 30.003 m at 20°C and 70 N tension, or as 30 m exactly at a temperature other than standard.

Worked examples

Example 4.1. A distance of 220.450 m was measured with a steel band of nominal length 30 m. On standardization the tape was found to be 30.003 m. Calculate the correct measured distance, assuming the error is evenly distributed throughout the tape.

$$\text{Error per 30 m} = 3\,\text{mm}$$

$$\therefore \text{Correction for total length} = \left(\frac{220.450}{30}\right) \times 3\,\text{mm} = 22\,\text{mm}$$

$$\therefore \text{Correct length is } 220.450 + 0.022 = 220.472\ \text{m}$$

Note that:

(1) *Figure 4.9* shows that when the tape is too long, the distance measured appears too short, and the correction is therefore positive. The reverse is the case when the tape is too short.
(2) When setting out a distance with a tape the rules in (1) are reversed.
(3) It is better to compute *Example 4.1* on the basis of the correction (as shown), rather than the total corrected length. In this way fewer significant figures are required.

Example 4.2. A 30-m band standardized at 20°C was found to be 30.003 m. At what temperature is the tape exactly 30 m? Coefficient of expansion of steel = 0.000 011/°C.

$$\text{Expansion per 30 m per °C} = 0.000\,011 \times 30 = 0.000\,33\ \text{m}$$

$$\text{Expansion per 30 m per 9°C} = 0.000\,33 \times 9 = 0.003\ \text{m}$$

$$\text{So the tape is 30 m at } 20°\text{C} - 9°\text{C} = 11°\text{C}$$

Alternatively, using *equation (4.1)* where $\Delta t = (t_s - t_a)$, then

$$t_a = \frac{C_t}{KL} + t_s = -\left(\frac{0.003}{0.000\,011 \times 30}\right) + 20°\text{C} = 11°\text{C}$$

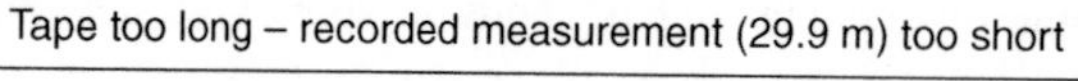

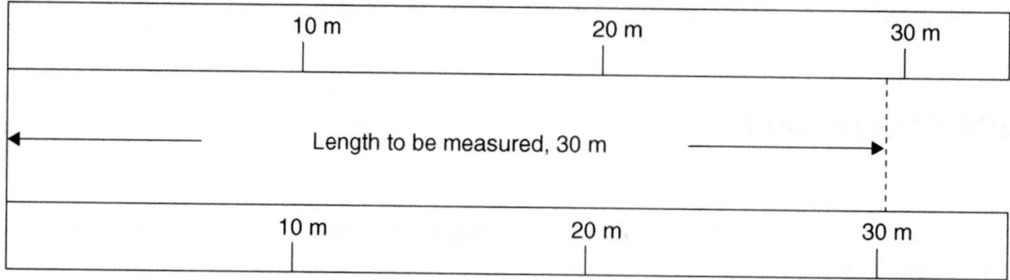

Fig. 4.9 *Measurements with short and stretched tapes*

where C_t = tape correction, K = coefficient of thermal expansion, L = measured length of the tape (m), t_a = actual temperature and t_s = standard temperature (°C). This then becomes the standard temperature for future temperature corrections.

4.3.2 Temperature

Tapes are usually standardized at 20°C. Any variation above or below this value will cause the tape to expand or contract, giving rise to systematic errors. Difficulty of obtaining true temperatures of the tapes lead to the use of invar tapes. Invar is a nickel-steel alloy with a very low coefficient of expansion.

Coefficient of expansion of steel $K = 11.2 \times 10^{-6}$ per °C

Coefficient of expansion of invar $K = 0.5 \times 10^{-6}$ per °C

Temperature correction $$C_t = KL\Delta t \qquad (4.1)$$

where Δt = difference between the standard and field temperatures (°C) $= (t_s - t_a)$.

The sign of the correction is in accordance with the rule specified in note (1) above.

4.3.3 Tension

Generally the tape is used under standard tension, in which case there is no correction. It may, however, be necessary in certain instances to apply a tension greater than standard. From Hooke's law:

stress = strain × a constant

This constant is the same for a given material and is called the *modulus of elasticity* (E). Since strain is a non-dimensional quantity, E has the same dimensions as stress, i.e. N/mm^2:

$$\therefore E = \frac{\text{Direct stress}}{\text{Corresponding strain}} = \frac{\Delta T}{A} \div \frac{C_T}{L}$$

$$\therefore C_T = L \times \frac{\Delta T}{AE} \qquad (4.2)$$

ΔT is normally the total stress acting on the cross-section, but as the tape would be standardized under tension, ΔT in this case is the amount of stress *greater* than standard. Therefore ΔT is the difference between field and standard tension. This value may be measured in the field in kilograms and should be converted to newtons (N) for compatibility with the other units used in the formula, i.e. 1 kgf = 9.806 65 N.

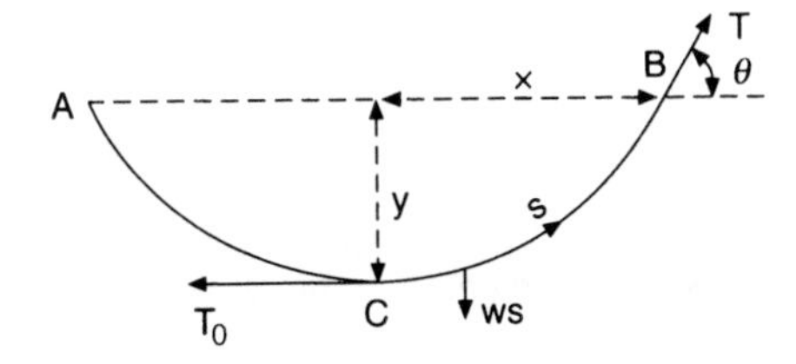

Fig. 4.10 *Catenary*

E is modulus of elasticity in N/mm^2; A is cross-sectional area of the tape in mm^2; L is measured length in m; and C_T is the extension and thus correction to the tape length in m. As the tape is stretched under the extra tension, the correction is positive.

4.3.4 Sag

When a tape is suspended between two measuring heads, A and B, both at the same level, the shape it takes up is a catenary (*Figure 4.10*). If C is the lowest point on the curve, then on length CB there are three forces acting, namely the tension T at B, T_0 at C and the weight of portion CB, where w is the weight per unit length and s is the arc length CB. Thus CB must be in equilibrium under the action of these three forces. Hence

Resolving vertically $\quad T \sin\theta = ws$

Resolving horizontally $\quad T \cos\theta = T_0$

$$\therefore \tan\theta = \frac{ws}{T_0}$$

For a small increment of the tape

$$\frac{dx}{ds} = \cos\theta = (1 + \tan^2\theta)^{-\frac{1}{2}} = \left(1 + \frac{w^2s^2}{T_0^2}\right)^{-\frac{1}{2}} = \left(1 - \frac{w^2s^2}{2T_0^2}\cdots\right)$$

$$\therefore x = \int\left(1 - \frac{w^2s^2}{2T_0^2}\right) ds$$

$$= s - \frac{w^2s^3}{6T_0^2} + K$$

When $x = 0$, $s = 0$, $\quad \therefore K = 0 \quad \therefore x = s - \dfrac{w^2s^3}{6T_0^2}$

The sag correction for the whole span $ACB = C_s = 2(s - x) = 2\left(\dfrac{w^2s^3}{6T_0^2}\right)$

$$\text{but } s = L/2 \quad \therefore C_s = \frac{w^2L^3}{24T_0^2} = \frac{w^2L^3}{24T^2} \text{ for small values of } \theta \tag{4.3}$$

i.e. $T_0 \approx T_0 \cos\theta \approx T$

where w = weight per unit length (N/m)
T = tension applied (N)
L = recorded length (m)
C_s = correction (m)

As $w = W/L$, where W is the *total* weight of the tape, then by substitution in *equation (4.3)*:

$$C_s = \frac{W^2L}{24T^2} \tag{4.4}$$

Although this equation is correct, the sag correction is proportional to the cube of the length as in *equation (4.3)* because by increasing the length of the tape you increase its total weight.

Equations (4.3) and *(4.4)* apply only to tapes standardized on the flat and are always negative. When a tape is standardized in catenary, i.e. it records the horizontal distance when hanging in sag, no correction is necessary provided the applied tension, say T_A, equals the standard tension T_s. Should the tension T_A exceed the standard, then a sag correction is necessary for the excess tension $(T_A - T_S)$ and

$$C_s = \frac{w^2L^3}{24}\left(\frac{1}{T_A^2} - \frac{1}{T_S^2}\right) \tag{4.5}$$

In this case the correction will be positive, in accordance with the basic rule.

4.3.5 Slope

If the difference in height of the two measuring heads is h, the slope distance L and the horizontal equivalent D, then by Pythagoras:

$$D = (L^2 - h^2)^{\frac{1}{2}} \tag{4.6}$$

Alternatively if the vertical angle of the slope of the ground is measured then:

$$D = L\cos\theta \tag{4.7}$$

and the correction $C_\theta = L - D$.

$$C_\theta = L(1 - \cos\theta) \tag{4.8}$$

4.3.6 Altitude

If the surveys are to be connected to the national mapping system of a country, the distances will need to be reduced to the common datum of that system, namely MSL. Alternatively, if the engineering scheme is of a local nature, distances may be reduced to the mean level of the area. This has the advantage that setting-out distances on the ground are, without sensible error, equal to distances computed from coordinates in the mean datum plane.

Consider *Figure 4.11* in which a distance L is measured in a plane situated at a height H above MSL.

By similar triangles $M = \dfrac{R}{R+H} \times L$

$$\therefore \text{Correction} \quad C_M = L - M = L - \frac{RL}{R+H} = L\left(1 - \frac{R}{R+H}\right) = \frac{LH}{R+H}$$

As H is normally negligible compared with R in the denominator

$$C_M = \frac{LH}{R} \tag{4.9}$$

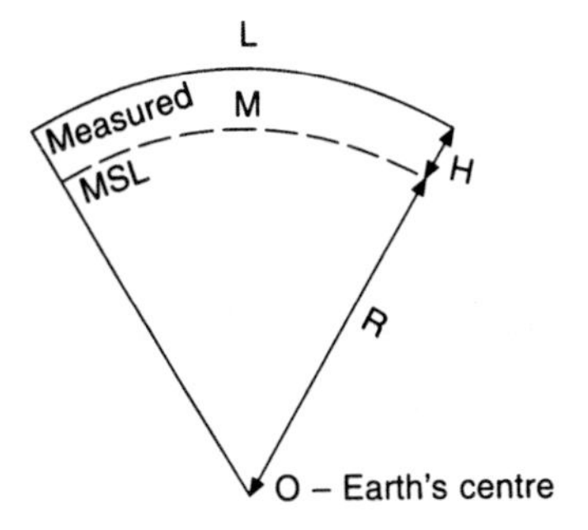

Fig. 4.11 *Distance and altitude*

The correction is negative for surface work above MSL or mean level of the area but may be positive for tunnelling or mining work below MSL or mean level of the area.

4.4 ERRORS IN TAPING

Methods of measuring with a tape have been dealt with, although it must be said that training in the methods is best undertaken in the field. The quality of the end results, however, can only be appreciated by an understanding of the errors involved. Of all the methods of measuring, taping is probably the least automated and therefore most susceptible to personal and natural errors. The majority of errors affecting taping are systematic, not random, and their effect will therefore increase with the number of bays measured.

The errors arise due to defects in the equipment used; natural errors due to weather conditions and human errors resulting in tape-reading errors, etc. They will now be dealt with individually.

4.4.1 Standardization

Taping cannot be more accurate than the accuracy to which the tape is standardized. It should therefore be routine practice to have one tape standardized by the appropriate authority.

This is done on payment of a small fee; the tape is returned with a certificate of standardization quoting the 'true' length of the tape and standard conditions of temperature and tension. This tape is then kept purely as a standard with which to compare working tapes.

Alternatively a base line may be established on site and its length obtained by repeated measurements using, say, an invar tape hired purely for that purpose. The calibration base should be then checked at regular intervals to confirm its stability.

4.4.2 Temperature

When measuring with a steel tape, neglecting temperature effects could be the main source of error. For example, in winter conditions in the UK, with temperatures at 0°C, a 50 m tape, standardized at 20°C, would contract by

$$11.2 \times 10^{-6} \times 50 \times 20 = 11.2 \text{ mm per } 50 \text{ m}$$

Thus even for ordinary precision measurement, the temperature effect cannot be ignored.

Even if the tape temperature is measured there may be an index error in the thermometer used, part of the tape may be in shade and part in the sun, or the thermometer may record ground or air temperature which may not be the same as the tape temperature. Although the use of an invar tape would resolve the problem, this is rarely, if ever, a solution applied on site. This is due to the high cost of such tapes and their fragility. The effect of an error in temperature measurement can be assessed by differentiating *equation (4.1)*, i.e.

$$\delta C_t = KL\,\delta(\Delta t)$$

If $L = 50$ m and the error in temperature is $+2°$C then $\delta C_t = +1.1$ mm. However, if this error remained constant the total error in the measured line would be proportional to the number of tape lengths. Every effort should therefore be made to obtain an accurate value for tape temperature using calibrated thermometers.

4.4.3 Tension

If the tension in the tape is greater or less than standard the tape will stretch or become shorter. Tension applied without the aid of a spring balance or tension handle may vary from length to length, resulting in random error. Tensioning equipment containing error would produce a systematic error proportional to the number of tape lengths. The effect of this error is greater on a light tape having a small cross-sectional area than on a heavy tape.

Consider a 50 m tape with a cross-sectional area of 4 mm^2, a standard tension of 50 N and a value for the modulus of elasticity of $E = 210$ kN/mm^2. Under a pull of 90 N the tape would stretch by

$$C_T = \frac{50\,000 \times 40}{4 \times 210 \times 10^3} = 2.4 \text{ mm}$$

As this value would be multiplied by the number of tape lengths measured it is very necessary to cater for tension in precision measurement, using calibrated tensioning equipment.

4.4.4 Sag

The correction for sag is equal to the difference in length between the arc and its subtended chord and is always negative. As the sag correction is a function of the weight of the tape, it will be greater for heavy tapes than light ones. Correct tension is also very important.

Consider a 50 m heavy tape of $W = 1.7$ kg with a standard tension of 80 N. From *equation (4.4)*:

$$C_s = \frac{(1.7 \times 9.81)^2 \times 50}{24 \times 80^2} = 0.090 \text{ m}$$

and indicates the large corrections necessary.

If the above tape was supported throughout its length to form three equal spans, the correction per span reduces to 0.003 m. This important result shows that the sag correction could be virtually eliminated by the choice of appropriate support.

The effect of an error in tensioning can be found by differentiating *equation (4.4)* with respect to T:

$$\delta C_s = -W^2 L\,\delta T/12T^3$$

In the above case, if the error in tensioning was $+5$ N, then the error in the correction for sag would be -0.01 m. This result indicates the importance of calibrating the tensioning equipment.

The effect of error in the weight (W) of the tape can be found by differentiating *equation (4.4)* with respect to W:

$$\delta C_s = WL\,\delta W/12T^2$$

and shows that an error of $+0.1$ kg in W produces an error of $+0.011$ m in the sag correction.

4.4.5 Slope

Correction for slope is always important.

Consider a 50 m tape measuring on a slope with a difference in height of 5 m between the ends. Upon taking the first term of the binomial expansion of *equation (4.4)* the correction for slope may be approximated as:

$$C_h = -h^2/2L = -25/100 = -0.250\,\text{m}$$

and would constitute a major source of error if ignored. The second-order error resulting from not using the second term $h^4/8L^3$ is less than 1 mm.

Error in the measurement of the difference in height (h) can be assessed using

$$\delta C_h = -h\,\delta h/L$$

Assuming an error of +0.005 m there would be an error of – 0.0005 m (δC_h). Thus error in obtaining the difference in height is negligible and as it is proportional to h, would get smaller on less steep slopes.

By differentiating *equation (4.4)* with respect to θ, we have

$$\delta C_\theta = L \sin\theta\,\delta\theta$$

$$\text{so} \quad \delta\theta'' = \delta C_\theta \times 206\,265/L \sin\theta$$

If $L = 50\,\text{m}$ is required to an accuracy of ± 5 mm on a slope of $5°$ then

$$\delta\theta'' = 0.005 \times 206\,265/50 \sin 5° = 237'' \approx 04'$$

This level of accuracy could easily be achieved using an Abney level to measure slope. As the slopes get less steep the accuracy required is further reduced; however, for the much greater distances obtained using EDM, the measurement of vertical angles is much more critical. Indeed, if the accuracy required above is changed to, say, ± 1 mm, the angular accuracy required changes to $\pm 47''$ and the angle measurement would require the use of a theodolite.

4.4.6 Misalignment

If the tape is not in a straight line between the two points whose distance apart is being measured, then the error in the horizontal plane can be calculated in a similar manner to the error due to slope in the vertical plane. If the amount by which the end of the tape is off line is e, then the resultant error is $e^2/2L$.

A 50 m tape, off line at one end by 0.500 m (an excessive amount), would lead to an error of 2.5 mm in the measured distance. The error is systematic and will obviously result in a recorded distance longer than the actual distance. If we consider a more realistic error in misalignment of, say, 0.05 m, the resulting error is 0.025 mm and completely negligible. Thus for the majority of taping, alignment by eye is quite adequate.

4.4.7 Plumbing

If stepped measurement is used, locating the end of the tape by plumb-bob will inevitably result in error. Plumbing at its best will probably produce a random error of about ± 3 mm. In difficult, windy conditions it would be impossible to use, unless sheltered from the wind by some kind of makeshift wind break, combined with careful steadying of the bob itself.

4.4.8 Incorrect reading of the tape

Reading errors are random and quite prevalent amongst beginners. Care and practice are needed to obviate them.

4.4.9 Booking error

Booking error can be reduced by adopting the process of the booker reading the measurement back to the observer for checking purposes. However, when measuring to millimetres with tensioning equipment, the reading has usually altered by the time it comes to check it. Repeated measurements will generally reveal booking errors, and thus distances should always be measured more than once.

4.5 ACCURACIES

If a great deal of taping measurement is to take place, then it is advisable to construct graphs of all the corrections for slope, temperature, tension and sag, for a variety of different conditions. In this way the corrections can be obtained rapidly and applied more easily. Such an approach rapidly produces in the measurer a 'feel' for the effect of errors in taping.

Random errors increase as the square root of the distance but systematic errors are proportional to distance and reading errors are independent of distance so it is not easy to produce a precise assessment of taping accuracies under variable conditions. Considering taping with a standardized tape corrected only for slope, one could expect an accuracy in the region of 1 in 5000 to 1 in 10 000. With extra care and correcting for all error sources the accuracy would rise to the region of 1 in 30 000. Precise measurement in catenary may be made with accuracies of 1 in 100 000 and better. However, the type of catenary measurement carried out on general site work would probably achieve about 1 in 50 000.

The worked examples should now be carefully studied as they illustrate the methods of applying these corrections both for measurement and setting out.

As previously mentioned, the signs of the corrections for measurement are reversed when setting out. As shown, measuring with a tape which is too long produces a smaller measured distance which requires positive correction. However, a tape that is too long will set out a distance that is too long and will require a negative correction. This can be expressed as follows:

Horizontal distance (D) = measured distance (M) + the algebraic sum of the corrections (C)

i.e. $D = M + C$

When setting out, the horizontal distance to be set out is known and the engineer needs to know its equivalent measured distance on sloping ground to the accuracy required. Therefore $M = D - C$, which has the effect of automatically reversing the signs of the correction. Therefore, compute the corrections in the normal way as if for a measured distance and then substitute the algebraic sum in the above equation.

Worked examples

Example 4.3. A base line was measured in catenary in four lengths giving 30.126, 29.973, 30.066 and 22.536 m. The differences of level were respectively 0.45, 0.60, 0.30 and 0.45 m. The temperature during observation was 10°C and the tension applied 15 kgf. The tape was standardized as 30 m, at 20°C, on the flat with a tension of 5 kg. The coefficient of expansion was 0.000 011 per °C, the weight of the tape 1 kg, the cross-sectional area 3 mm^2, $E = 210 \times 10^3$ N/mm^2 (210 kN/mm^2), gravitational acceleration $g = 9.806\,65$ m/s^2.

(a) Quote each equation used and calculate the length of the base.
(b) What tension should have been applied to eliminate the sag correction? (LU)

(a) As the field tension and temperature are constant throughout, the first three corrections may be applied to the base as a whole, i.e. $L = 112.701$ m, with negligible error.

	+	−
Tension $C_T = \dfrac{L\Delta_T}{AE} = \dfrac{112.701 \times (10 \times 9.806\,65)}{3 \times 210 \times 10^3} =$	+0.0176	
Temperature $C_t = LK\Delta t = 112.701 \times 0.000\,011 \times 10 =$		−0.0124
Sag $C_s = \dfrac{LW^2}{24T^2} = \dfrac{112.701 \times 1^2}{24 \times 15^2} =$		−0.0210
Slope $C_h = \dfrac{h^2}{2L} = \dfrac{1}{2 \times 30}(0.45^2 + 0.60^2 + 0.30^2) + \dfrac{0.45^2}{2 \times 22.536} =$		−0.0154
	+0.0176	−0.0488

$$\text{Horizontal length of base } (D) = \text{measured length } (M) + \text{sum of corrections } (C)$$
$$= 112.701 \text{ m} + (-0.031)$$
$$= 112.670 \text{ m}$$

N.B. In the slope correction the first three bays have been rounded off to 30 m, the resultant second-order error being negligible.

Consider the situation where 112.670 m is the horizontal distance to be set out on site. The equivalent measured distance would be:

$$M = D - C$$
$$= 112.670 - (-0.031) = 112.701 \text{ m}$$

(b) To find the applied tension necessary to eliminate the sag correction, equate the two equations:

$$\frac{\Delta T}{AE} = \frac{W^2}{24T_A^2}$$

where ΔT is the difference between the applied and standard tensions, i.e. $(T_A - T_S)$.

$$\therefore \frac{T_A - T_S}{AE} = \frac{W^2}{24T_A^2}$$

$$\therefore T_A^3 - T_A^2 T_S - \frac{AEW^2}{24} = 0$$

Substituting for T_S, W, A and E, making sure to convert T_S and W to newtons gives

$$T_A^3 - 49T_A^2 - 2\,524\,653 = 0$$

Let $$T_A = (T + x)$$

then $$(T + x)^3 - 49(T + x)^2 - 2\,524\,653 = 0$$

$$T^3\left(1 + \frac{x}{T}\right)^3 - 49T^2\left(1 + \frac{x}{T}\right)^2 - 2\,524\,653 = 0$$

Expanding the brackets binomially gives

$$T^3\left(1 + \frac{3x}{T}\right) - 49T^2\left(1 + \frac{2x}{T}\right) - 2\,524\,653 = 0$$

$$\therefore T^3 + 3T^2x - 49T^2 - 98Tx - 2\,524\,653 = 0$$

$$\therefore x = \frac{2\,524\,653 - T^3 + 49T^2}{3T^2 - 98T}$$

assuming $T = 15$ kgf $= 147$ N, then $x = 75$ N

$\therefore$ at the first approximation $T_A = (T + x) = 222$ N

Example 4.4. A base line was measured in catenary as shown below, with a tape of nominal length 30 m. The tape measured 30.015 m when standardized in catenary at 20°C and 5 kgf tension. If the mean reduced level of the base was 30.50 m OD, calculate its true length at mean sea level.

Given: weight per unit length of tape = 0.03 kg/m (w); density of steel = 7690 kg/m^3 (ρ); coefficient of expansion = 11×10^{-6} per °C; $E = 210 \times 10^3$ N/mm^2; gravitational acceleration $g = 9.806\,65$ m/s^2; radius of the Earth = 6.4×10^6 m (R). (KU)

Bay	*Measured length* (m)	*Temperature* (°C)	*Applied tension* (kgf)	*Difference in level* (m)
1	30.050	21.6	5	0.750
2	30.064	21.6	5	0.345
3	30.095	24.0	5	1.420
4	30.047	24.0	5	0.400
5	30.041	24.0	7	–

	+	−

Standardization:

Error/30 m = 0.015 m
Total length of base = 150.97 m

$$\therefore \text{Correction} = \frac{150.297}{30} \times 0.015 = \qquad +0.0752$$

Temperature:

$$\left.\begin{array}{ll} \text{Bays 1 and 2} & C_t = 60 \times 11 \times 10^{-6} \times 1.6 = 0.0010\,\text{m} \\ \text{Bays 3, 4 and 5} & C_t = 90 \times 11 \times 10^{-6} \times 4 = 0.0040\,\text{m} \end{array}\right\} \qquad +0.0050$$

(Second-order error negligible in rounding off bays to 30 m.)

Tension:

Bay 5 only $C_T = \dfrac{L\Delta T}{AE}$, changing ΔT to newtons

where cross-sectional area $A = \dfrac{w}{\rho}$

$$\therefore A = \frac{0.03}{7690} \times 10^6 = 4\,\text{mm}^2$$

$$\therefore C_T = \frac{30 \times 2 \times 9.81}{4 \times 210 \times 10^3} = \qquad +0.0007$$

Slope:

$$C_h = \frac{h^2}{2L} - \frac{1}{2 \times 30}(0.750^2 + 0.345^2 + 1.420^2 + 0.400^2) = \qquad -0.0476$$

The second-order error in rounding off to 30 m is negligible in this case also. However, care should be taken when many bays are involved, as their accumulative effect may be significant.

Sag:

$$\text{Bay 5 only} \quad C_s = \frac{L^3 w^3}{24}\left(\frac{1}{T_S^2} - \frac{1}{T_A^2}\right)$$

$$= \frac{30^3 \times 0.03^2}{24}\left(\frac{1}{5^2} - \frac{1}{7^2}\right) = \qquad +0.0006$$

Altitude:

$$C_M = \frac{LH}{R} = \frac{150 \times 30.5}{6.4 \times 10^6} = \qquad -0.0007$$

+	−
+0.0815	−0.0483

Therefore Total correction = +0.0332 m
Hence Corrected length = 150.297 + 0.0332 = 150.3302 m

Example 4.5. (a) A standard base was established by accurately measuring with a steel tape the distance between fixed marks on a level bed. The mean distance recorded was 24.984 m at a temperature of 18°C and an applied tension of 155 N. The tape used had recently been standardized in catenary and was 30 m in length at 20°C and 100 N tension. Calculate the true length between the fixed marks given: total weight of the tape = 0.90 kg; coefficient of expansion of steel = 11×10^{-6} per °C; cross-sectional area = 2 mm^2; $E = 210 \times 10^3$ N/mm^2; gravitational acceleration = 9.807 m/s^2.

(b) At a later date the tape was used to measure a 30-m bay in catenary. The difference in level of the measuring heads was 1 m, with an error of 3 mm. Tests carried out on the spring balance indicated that the applied tension of 100 N had an error of 2 N. Ignoring all other sources of error, what is the probable error in the measured bay?

(KU)

(a) If the tape was standardized in catenary, then when laid on the flat it would be too long by an amount equal to the sag correction. This amount, in effect, then becomes the standardization correction:

$$\text{Error per 30 m} = \frac{LW^2}{24T_S^2} = \frac{30 \times (0.90 \times 9.807)^2}{24 \times 100^2} = 0.0097 \text{ m}$$

$$\therefore \text{Correction} = \frac{0.0097 \times 24.984}{30} = 0.0081 \text{ m}$$

$$\text{Tension} = \frac{24.984 \times 55}{2 \times 210 \times 10^3} = 0.0033 \text{ m}$$

$$\text{Temperature} = 24.984 \times 11 \times 10^{-6} \times 2 = -0.0006 \text{ m}$$

$$\therefore \text{Total correction} = 0.0108 \text{ m}$$

$$\therefore \text{Corrected length} = 24.984 + 0.011 = 24.995 \text{ m}$$

(b) *Effect of levelling error*: $C_h = \dfrac{h^2}{2L}$

$$\therefore \delta C_h = \frac{h \times \delta h}{L} = \frac{1 \times 0.003}{30} = 0.0001 \text{ m}$$

Effect of tensioning error: Sag $C_s = \dfrac{LW^2}{24T^2}$

$$\therefore \delta C_s = -\frac{LW^2}{12T^3}\delta T$$

$$\therefore \delta C_s = \frac{30 \times (0.9 \times 9.807)^2 \times 2}{12 \times 100^3} = 0.0004 \text{ m}$$

Tension $C_T = \dfrac{L\Delta T}{AE}$

$$\therefore \delta C_T = \frac{L \times \delta(\Delta T)}{A \times E} = \frac{30 \times 2}{2 \times 210 \times 10^3} = 0.0001 \text{ m}$$

$$\therefore \text{Total error} = 0.0006 \text{ m}$$

Example 4.6. A 30-m invar reference tape was standardized on the flat and found to be 30.0501 m at 20°C and 88 N tension. It was used to measure the first bay of a base line in catenary, the mean recorded length being 30.4500 m.

Using a field tape, the mean length of the same bay was found to be 30.4588 m. The applied tension was 88 N at a constant temperature of 15°C in both cases.

The remaining bays were now measured in catenary, using the field tape only. The mean length of the second bay was 30.5500 m at 13°C and 100 N tension. Calculate its reduced length given: cross-sectional area = 2 mm^2; coefficient of expansion of invar = 6×10^{-7} per °C; mass of tape per unit length = 0.02 kg/m; difference in height of the measuring heads = 0.5 m; mean altitude of the base = 250 m OD; radius of the Earth = 6.4×10^6 m; gravitational acceleration = 9.807 m/s^2; Young's modulus of elasticity = 210 kN/mm^2.

(KU)

To find the corrected length of the first bay using the reference tape:

Standardization:	+	−
Error per 30 m = 0.0501 m		
∴ Correction for 30.4500 m =	+0.0508	
Temperature $= 30 \times 6 \times 10^{-7} \times 5 =$		−0.0001
Sag $= \dfrac{30^3 \times (0.02 \times 9.807)^2}{24 \times 88^2} =$		−0.0056
	+0.0508	−0.0057

Therefore Total correction = +0.0451 m

Hence Corrected length = 30.4500 + 0.0451 = 30.4951 m

(using reference tape). Field tape corrected for sag measures 30.4588 – 0.0056 = 30.4532 m.

Thus the field tape is measuring too short by 0.0419 m (30.4951 – 30.4532) and is therefore too long by this amount. Therefore field tape is 30.0419 m at 15°C and 88 N.

To find length of second bay:

Standardization:	+	−
Error per 30 m = 0.0419 m		
∴ Correction $= \dfrac{30.5500}{30} \times 0.0419 =$	+0.0427	
Temperature $= 30 \times 6 \times 10^{-7} \times 2 =$		−0.000 04
Tension $= \dfrac{30 \times 12}{2 \times 210 \times 10^3} =$	+0.0009	
Sag $= \dfrac{30^3 \times (0.02 \times 9.807)^2}{24 \times 100^2} =$		−0.0043
Slope $= \dfrac{0.500^2}{2 \times 30.5500} =$		−0.0041
Altitude $= \dfrac{30.5500 \times 250}{6.4 \times 10^6} =$		−0.0093
	+0.0436	−0.0177

Therefore Total correction = +0.0259 m

Hence Corrected length of second bay = 30.5500 + 0.0259 = 30.5759 m

N.B. Rounding off the measured length to 30 m is permissible only when the resulting error has a negligible effect on the final distance.

Example 4.7. A copper transmission line of 12 mm diameter is stretched between two points 300 m apart, at the same level with a tension of 5 kN, when the temperature is 32°C. It is necessary to define its limiting positions when the temperature varies. Making use of the corrections for sag, temperature and elasticity normally applied to base-line measurements by a tape in catenary, find the tension at a temperature of −12°C and the sag in the two cases.

Young's modulus for copper is 70 kN/mm^2, its density 9000 kg/m^3 and its coefficient of linear expansion 17.0×10^{-6}/°C.

(LU)

In order first of all to find the amount of sag in the above two cases, one must find (a) the weight per unit length and (b) the sag length of the wire.

(a) $w = \text{area} \times \text{density} = \pi r^2 \rho$
$= 3.142 \times 0.006^2 \times 9000 = 1.02$ kg/m

(b) at 32°C, the sag length of wire $= L_H + \left(\dfrac{L^3 w^2}{24T^2}\right)$

where L is itself the sag length. Thus the first approximation for L of 300 m must be used.

$$\therefore \text{Sag length} = 300 + \left(\frac{300^3 \times (1.02 \times 9.807)^2}{24 \times 5000^2}\right) = 304.5\ \text{m}$$

$$\text{Second approximation} = 300 + \left(\frac{304.5^3 \times (1.02 \times 9.807)^2}{24 \times 5000^2}\right)$$

$$= 304.71\ \text{m} = L_1$$

$$\therefore \text{Sag} = y_1 = \frac{wL_1^2}{8T} = \frac{(1.02 \times 9.807) \times 304.71^2}{8 \times 5000} = 23.22\ \text{m}$$

At -12°C there will be a reduction in L_1 of

$$(L_1 K \Delta t) = 304.71 \times 17.0 \times 10^{-6} \times 44 = 0.23\ \text{m}$$

$$\therefore L_2 = 304.71 - 0.23 = 304.48\ \text{m}$$

$$y_1 \propto L_1^2 \quad \therefore y_2 = y_1 \left(\frac{L_2}{y_1}\right)^2 = 23.22 \frac{(304.48)^2}{(304.71)^2} = 23.18\ \text{m}$$

$$\text{Similarly,} \quad y_1 \propto 1/T_1 \quad \therefore T_2 = T_1 \left(\frac{y_1}{y_2}\right) = 5000 \left(\frac{23.22}{23.18}\right) = 5009\ \text{N or } 5.009\ \text{kN}$$

Exercises

(4.1) A tape of nominal length 30 m was standardized on the flat at the NPL, and found to be 30.0520 m at 20°C and 44 N of tension. It was then used to measure a reference bay in catenary and gave a mean distance of 30.5500 m at 15°C and 88 N tension. As the weight of the tape was unknown, the sag at the mid-point of the tape was measured and found to be 0.170 m.

Given: cross-sectional area of tape = 2 mm^2; Young's modulus of elasticity = 200×10^3 N/mm^2; coefficient of expansion = 11.25×10^{-6} per °C; and difference in height of measuring heads = 0.320 m. Find the horizontal length of the bay. If the error in the measurement of sag was ±0.001 m, what is the resultant error in the sag correction? What does this resultant error indicate about the accuracy to which the sag at the mid-point of the tape was measured?

(KU)

(*Answer*: 30.5995 m and ±0.000 03 m)

(4.2) The three bays of a base line were measured by a steel tape in catenary as 30.084, 29.973 and 25.233 m, under respective pulls of 7, 7 and 5 kg, temperatures of 12°, 13° and 17°C and differences of level of supports of 0.3, 0.7 and 0.7 m. If the tape was standardized on the flat at a temperature of 15°C

under a pull of 4.5 kg, what are the lengths of the bays? 30 m of tape is exactly 1 kg with steel at 8300 kg/m^3, with a coefficient of expansion of 0.000 011 per °C and $E = 210 \times 10^3$ N/mm^2. (LU)

(*Answer*: 30.057 m, 29.940 m and 25.194 m)

(*4.3*) The details given below refer to the measurement of the first 30-m bay of a base line. Determine the correct length of the bay reduced to mean sea level.

With the tape hanging in a catenary at a tension of 10 kg and at a mean temperature of 13°C, the recorded length was 30.0247 m. The difference in height between the ends was 0.456 m and the site was 500 m above MSL. The tape had previously been standardized in catenary at a tension of 7 kg and a temperature of 16°C, and the distance between zeros was 30.0126 m.

$R = 6.4 \times 10^6$ m; weight of tape per m = 0.02 kg; sectional area of tape = 3.6 mm^2; $E = 210 \times 10^3$ N/mm^2; temperature coefficient of expansion of tape = 0.000 011 per °C. (ICE)

(*Answer*: 30.0364 m)

(*4.4*) The following data refer to a section of base line measured by a tape hung in catenary.

Bay	*Observed length* (m)	*Mean temperature* (°C)	*Reduced levels of index marks* (m)	
1	30.034	25.2	293.235	293.610
2	30.109	25.4	293.610	294.030
3	30.198	25.1	294.030	294.498
4	30.075	25.0	294.498	294.000
5	30.121	24.8	294.000	293.355

Length of tape between 0 and 30 m graduations when horizontal at 20°C and under 5 kg tension is 29.9988 m; cross-sectional area of tape = 2.68 mm^2; tension used in the field = 10 kg; temperature coefficient of expansion of tape = 11.16×10^{-6} per °C; elastic modulus for material of tape = 20.4×10^4 N/mm^2; weight of tape per metre length = 0.02 kg; mean radius of the Earth = 6.4×10^6 m. Calculate the corrected length of this section of the line. (LU)

(*Answer*: 150.507 m)

4.6 ELECTROMAGNETIC DISTANCE MEASUREMENT (EDM)

The main instrument for surveyors on site today is the 'total station'. It is an instrument that combines the angle measurements that could be obtained with a traditional theodolite with electronic distance measurements. Taping distance, with all its associated problems, has been rendered obsolete for all base-line measurement. Distance can now be measured easily, quickly and with great accuracy, regardless of terrain conditions. Modern total stations as in *Figure 4.12* and *Figure 4.13* contain algorithms for reducing the slope distance to its horizontal and vertical components. For engineering surveys total stations with automatic data logging are now standard equipment on site. A standard measurement of distance takes between 1.5 and 3 s. Automatic repeated measurements can be used to improve reliability in difficult

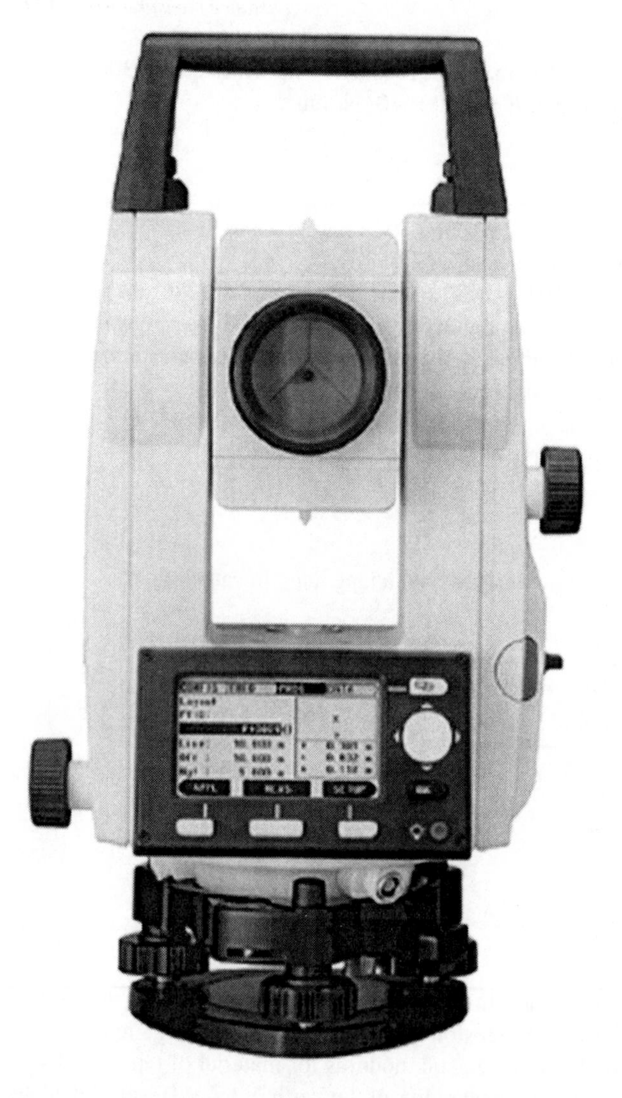

Fig. 4.12 *Total station – courtesy of Leica Geosystems*

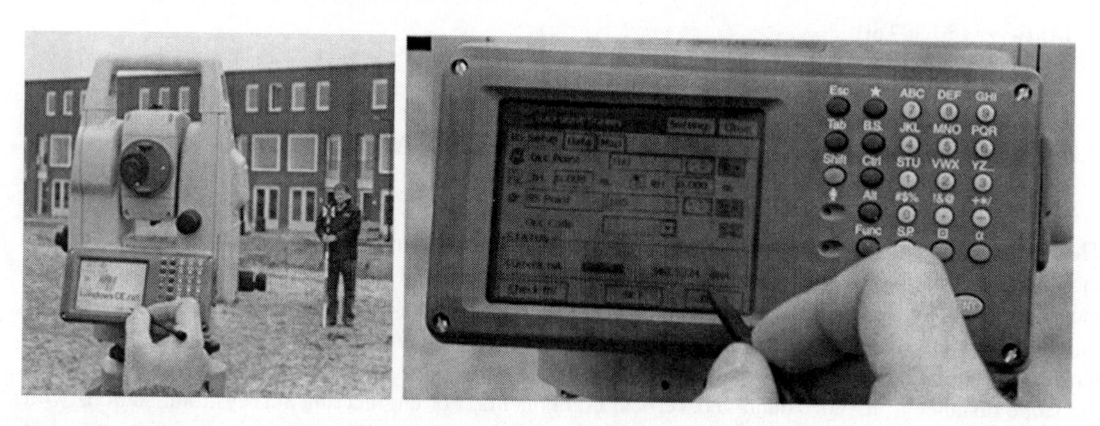

Fig. 4.13 *Total station – courtesy of Topcon*

atmospheric conditions. Tracking modes, for the setting out of distance, repeat the measurement several times a second. Total stations with their inbuilt EDM enable:

(1) Traversing over great distances, with much greater control of swing errors.
(2) The inclusion of many more measured distances into control networks, rendering classical triangulation obsolete. This results in much greater control of scale error.
(3) Setting-out and photogrammetric control, over large areas, by polar coordinates from a single base line.
(4) Deformation monitoring to sub-millimetre accuracies using high-precision EDM, such as the Mekometer ME5000. This instrument has a range of 8 km and an accuracy of ±0.2 mm ±0.2 mm/km of the distance measured ignoring unmodelled refraction effects.

4.6.1 Classification of instruments

Historically EDM instruments have been classified according to the type and wavelength of the electromagnetic energy generated or according to their operational range. Very often one is a function of the other. For survey work most instruments use infra-red radiation (IR). IR has wavelengths of 0.8–0.9 μm transmitted by gallium arsenide (GaAs) luminescent diodes, at a high, constant frequency. The accuracies required in distance measurement are such that the measuring wave cannot be used directly due to its poor propagation characteristics. The measuring wave is therefore superimposed on the high-frequency waves generated, called carrier waves. The superimposition is achieved by amplitude (*Figure 4.14*), frequency (*Figure 4.15*) or impulse modulation (*Figure 4.16*). In the case of IR instruments, amplitude modulation is used. Thus the carrier wave develops the necessary measuring characteristics whilst maintaining the high-frequency propagation characteristics that can be measured with the requisite accuracy.

In addition to IR, visible light, with extremely small wavelengths, can also be used as a carrier. Many of the instruments using visual light waves have a greater range and a much greater accuracy than that required for more general surveying work. Typical of such instruments are the Kern Mekometer ME5000, accurate to ±0.2 mm ±0.2 mm/km, with a range of 8 km, and the Com-Rad Geomensor CR204.

4.7 MEASURING PRINCIPLES

Although there are many EDM instruments available, there are basically only two methods of measurement employed, namely the *pulse method* and the more popular *phase difference method*.

4.7.1 Pulse method

A short, intensive pulse of radiation is transmitted to a reflector target, which immediately transmits it back, along a parallel path, to the receiver. The measured distance is computed from the velocity of the signal multiplied by the time it took to complete its journey, i.e.

$$2D = c \cdot \Delta t$$

$$D = c \cdot \Delta t/2 \qquad (4.10)$$

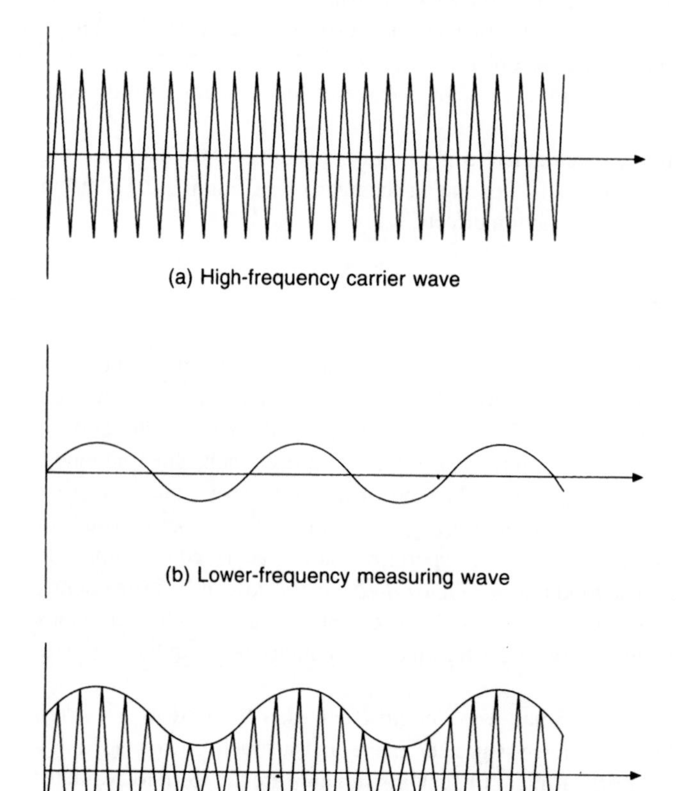

Fig. 4.14 *Amplitude modulation of the carrier wave*

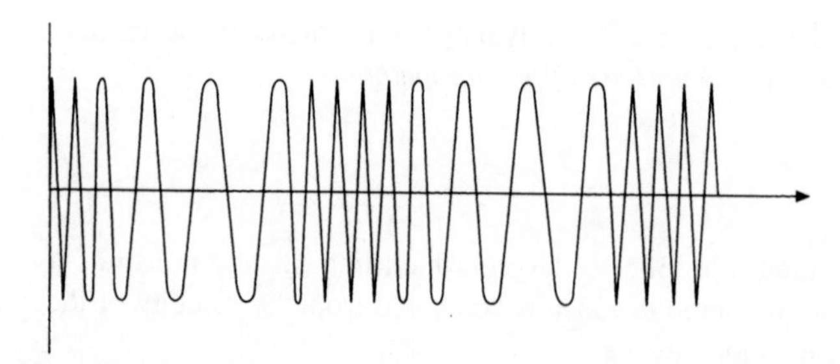

Fig. 4.15 *Frquency modulation of the carrier wave*

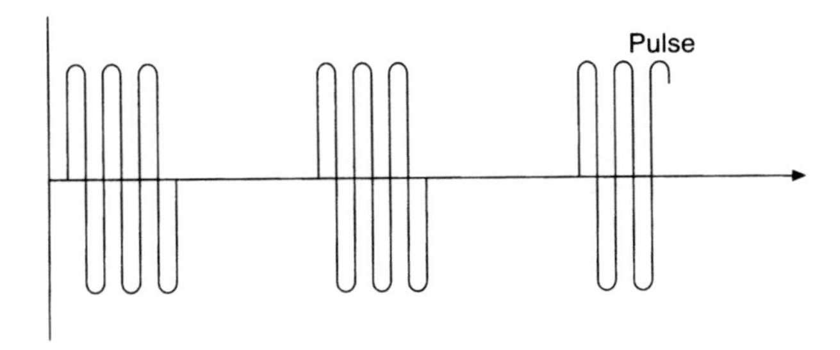

Fig. 4.16 *Impulse modulation of the carrier wave*

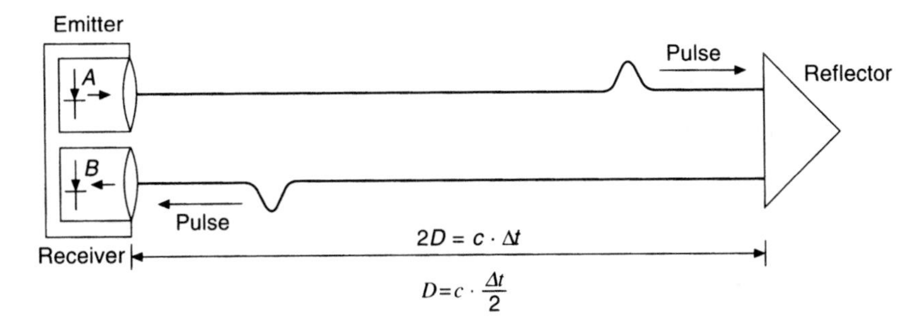

Fig. 4.17 *Principle of pulse distance meter*

If the time of departure of the pulse from gate A is t_A and the time of its reception at gate B is t_B, then $(t_B - t_A) = \Delta t$.

$c =$ the velocity of light in the medium through which it travelled

$D =$ the distance between instrument and target

It can be seen from *equation (4.10)* that the distance is dependent on the velocity of light in the medium and the accuracy of its transit time. Taking an approximate value of 300 000 km/s for the speed of light, 10^{-10} s would be equivalent to 15 mm of measured distance.

The distance that can be measured is largely a function of the power of the pulse. Powerful laser systems can obtain tremendous distances when used with corner-cube prisms and even medium distances when the pulse is 'bounced' off natural or man-made features.

4.7.2 Phase difference method

The majority of EDM instruments, whether infra-red or light, use this form of measurement. Basically the instrument measures the amount ($\delta\lambda$) by which the reflected signal is out of phase with the emitted signal. *Figure 4.18(a)* shows the signals in phase whilst (*b*) shows the amount ($\delta\lambda$) by which they are out of phase. The double distance is equal to the number (M) of full wavelengths (λ) plus the fraction of a wavelength ($\delta\lambda$). The phase difference can be measured by analogue or digital methods. *Figure 4.19* illustrates the digital phase measurement of $\delta\lambda$.

Basically, all the equipment used works on the principle of 'distance "equals" velocity × time'. However, as time is required to such very high accuracies, recourse is made to the measurement of phase difference.

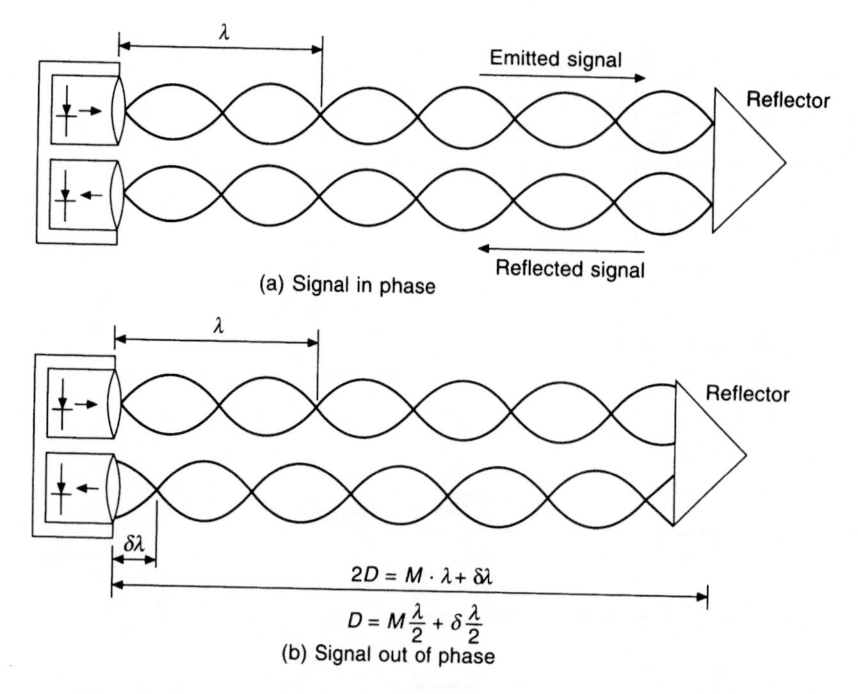

Fig. 4.18 *Principle of phase difference method*

As shown in *Figure 4.19*, as the emitted and reflected signals are in continuous motion, the only constant is the phase difference $\delta\lambda$.

Figure 4.20 shows the path of the emitted radiation from instrument to reflector and back to instrument, and hence it represents twice the distance from instrument to reflector. Any periodic phenomenon which oscillates regularly between maximum and minimum values may be analysed as a simple harmonic motion. Thus, if P moves in a circle with a constant angular velocity ω, the radius vector A makes a phase angle ϕ with the x-axis. A graph of values, computed from

$$y = A\sin(\omega t) \tag{4.11}$$

$$= A\sin\phi \tag{4.12}$$

for various values of ϕ produces the sine wave illustrated and shows

A = amplitude or maximum strength of the signal
ω = angular velocity
t = time
ϕ = phase angle

The value of B is plotted when $\phi = \pi/2 = 90°$, C when $\phi = \pi = 180°$, D when $\phi = 1.5\pi = 270°$ and $A' = 2\pi = 360°$. Thus 2π represents a complete wavelength (λ) and $\phi/2\pi$ a fraction of a wavelength ($\delta\lambda$). The time taken for A to make one complete revolution or cycle is the period of the oscillation and is represented by t s. Hence the phase angle is a function of time. The number of revolutions per second at which the radius vector rotates is called the frequency f and is measured in hertz, where one hertz is one cycle per second.

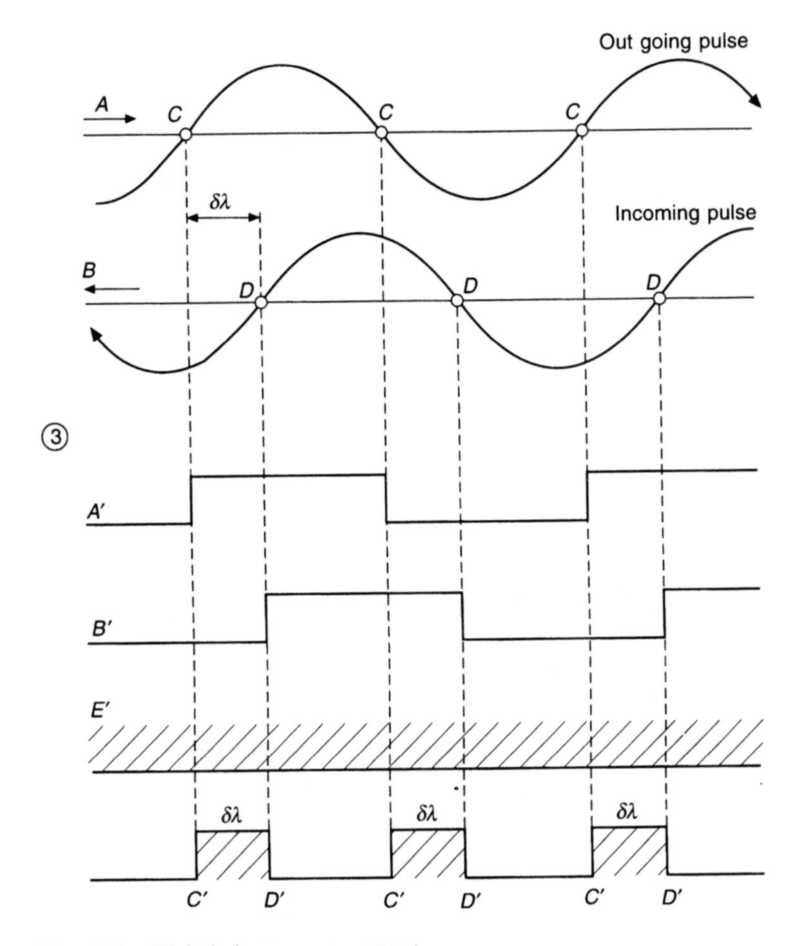

Fig. 4.19 *Digital phase measurement*

With reference to *Figure 4.20*, it can be seen that the double path measurement ($2D$) from instrument to reflector and back to instrument is equal to

$$2D = M\lambda + \delta\lambda \tag{4.13}$$

where

M = the integer number of wavelengths in the medium

$\delta\lambda$ = the fraction of a wavelength $= \dfrac{\phi}{2\pi}\lambda$

As the phase difference method measures only the fraction of wavelength that is out of phase, a second wavelength of different frequency is used to obtain a value for M.

Consider *Figure 4.20*. The double distance is 3.75λ; however, the instrument will only record the phase difference of 0.75. A second frequency is now generated with a wavelength four times greater, producing a phase difference of 0.95. In terms of the basic measuring unit, this is equal to $0.95 \times 4 = 3.80$, and hence the value for M is 3. The smaller wavelength provides a more accurate assessment of the fractional portion, and hence the double path measurement is $3 + 0.75 = 3.75\lambda$. Knowing the value of λ in units of length

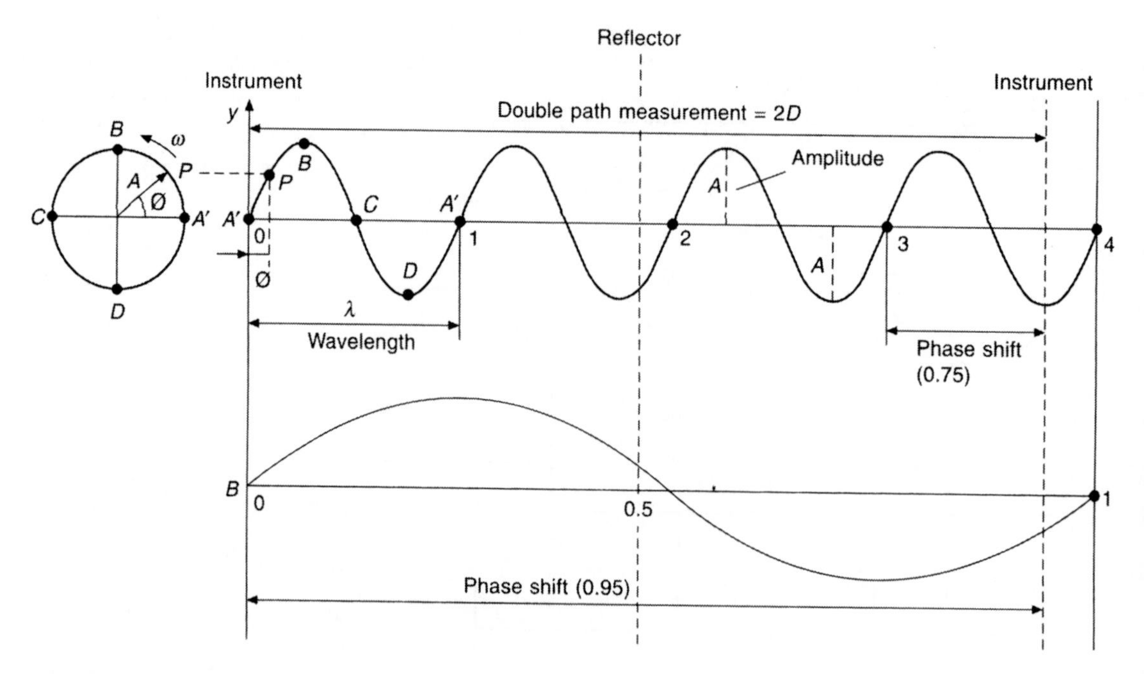

Fig. 4.20 *Principle of electromagnetic distance measurement*

would thus produce the distance. This then is the basic principle of the phase difference method, further illustrated below:

	f	λ	$\lambda/2$	$\delta\lambda$
Fine reading	15 MHz	20 m	10 m	6.325
1st rough reading	1.5 MHz	200 m	100 m	76.33
2nd rough reading	150 kHz	2000 m	1000 m	876.3
		Measured distance		876.325

The first number of each rough reading is added to the initial fine reading to give the total distance. Thus the single distance from instrument to reflector is:

$$D = M(\lambda/2) + \frac{\phi}{2\pi}(\lambda/2) \tag{4.14}$$

where it can be seen that $\lambda/2$ is the main unit length of the instrument. The value chosen for the main unit length has a fundamental effect on the precision of the instrument due to the limited resolution of phase measurement. The majority of EDM instruments use $\lambda/2$ equal 10 m. With a phase resolution of 3×10^{-4} errors of 3 mm would result.

Implicit in the above equation is the assumption that λ is known and constant. However in most EDM equipment this is not so, only the frequency f is known and is related to λ as follows:

$$\lambda = c/f \tag{4.15}$$

where

c = the velocity of electromagnetic waves (light) in a medium

This velocity c can only be calculated if the refractive index n of the medium is known, and also the velocity of light c_o in a vacuum:

$$n = c_o/c \tag{4.16}$$

where n is always greater than unity.

At the 17th General congress on Weights and Measures in 1983 the following exact defining value for c_o was adopted and is still in use today:

$$c_o = 299\,792\,458 \text{ m/s} \tag{4.17}$$

From the standard deviation quoted, it can be seen that this value is accurate to 0.004 mm/km and can therefore be regarded as error free compared with the most accurate EDM measurement.

The value of n can be computed from basic formulae available. However, *Figure 4.14* shows the carrier wave contained within the modulation envelope. The carrier travels at what is termed the phase velocity, whilst the group of frequencies travel at the slower group velocity. The measurement procedure is concerned with the modulation and so it is the group refractive index n_g with which we are concerned.

From *equation (4.16)* $c = c_o/n$ and from *equation (4.15)* $\lambda = c_o/nf$.

Replace n with n_g and substitute in *equation (4.14)*:

$$D = M\frac{(c_o)}{(2n_g f)} + \frac{\phi}{2\pi}\frac{(c_o)}{(2n_g f)} \tag{4.18}$$

Two further considerations are necessary before the final formula can be stated:

(1) The physical centre of the instrument which is plumbed over the survey station does not coincide with the position within the instrument to which the measurements are made. This results in an instrument constant K_1.
(2) Similarly with the reflector. The light wave passing through the atmosphere, whose refractive index is approximately unity, enters the glass prism of refractive index about 1.6 and is accordingly slowed down. This change in velocity combined with the light path through the reflector results in a correction to the measured distance called the reflector constant K_2.

Both these constants are combined and catered for in the instrument; then

$$D = M\frac{(c_o)}{(2n_g f)} + \frac{\phi}{2\pi}\frac{(c_o)}{(2n_g f)} + (K_1 + K_2) \tag{4.19}$$

is the fundamental equation for distances measured with EDM equipment.

An examination of the equation shows the error sources to be:

(1) In the measurement of phase ϕ.
(2) In the measurement of group refractive index n_g.
(3) In the stability of the frequency f.
(4) In the instrument/reflector constants K_1 and K_2.

4.7.3 Corner-cube prism

In contrast to the total station with its EDM which is a complex and sophisticated instrument, the target is simply a glass prism designed to return the signal it receives back to the instrument it came from. Imagine a cube of glass which has been cut across its corner and the corner piece retained. *Figure 4.21(a)* is a picture of a corner-cube prism and *Figure 4.21(b)* represents a corner-cube prism drawn in two dimensions, i.e. with two internal faces 90° apart. The signal enters through the front face of the prism and is refracted.

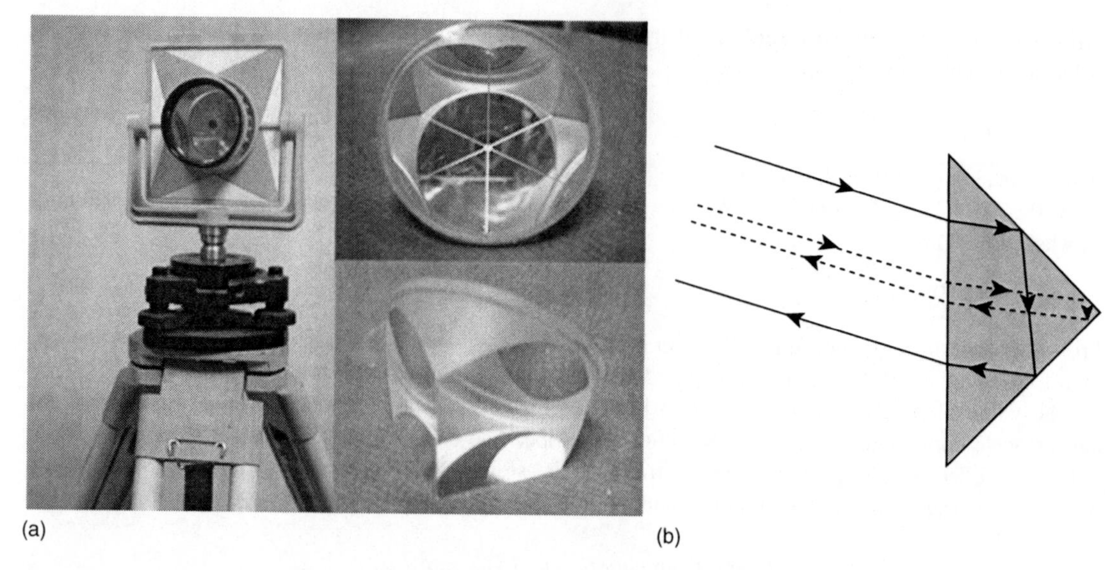

Fig. 4.21 *(a) Corner cube prism. (b) Rays of light through a corner cube prism*

It is then reflected off the two internal faces back to the front face where it is refracted again and returns on the same path by which it entered. For a real corner-cube prism, which is three-dimensional and therefore has three internal faces, the effect is just the same, except that the signal is reflected off the three internal faces but the return signal is returned along the same path as the incoming signal.

This may be verified by holding a corner cube prism, closing one eye, and with the other looking in through the front face at the opposite corner. You will see an image of your eye, with your pupil centred on the opposite corner. Even if you rotate the prism from left to right or up and down your pupil will stay centred on the opposite corner. Therefore if a survey target with a corner-cube prism is set up but not pointed directly at the instrument it will still return a signal to the EDM. Similarly if the total station is moved from one survey point to another and the target is not adjusted to point at the total station in its new position the EDM will still get a returned signal from the target.

In *Figure 4.21*(*b*) the length of the light path into the prism plus the length of the light path out of the prism is the same for both the solid and the dotted rays. Similarly the length of the light path through the prism is the same for both the solid and the dotted rays. Therefore, whatever the refractive index of the glass, the path length of both solid and dotted rays is the same and so a coherent signal entering the prism also emerges as a coherent signal.

4.7.4 Reflective targets

There are a large number of reflective targets available on the market (*Figure 4.22*). Reflective targets are cheap, unobtrusive and expendable. They may be applied to a structure as targets for deformation monitoring where their great advantage is that they do not need to be removed and replaced at successive epochs. However, they are less durable than corner-cube prisms and an EDM will have less range. Often they have a black and grey or silver appearance which can make them difficult to find upon revisiting a site or structure. A photo of a structure from a digital camera with flash will show up where the targets are because the targets, like a corner-cube prism, return the light directly to its source (*Figure 4.23*).

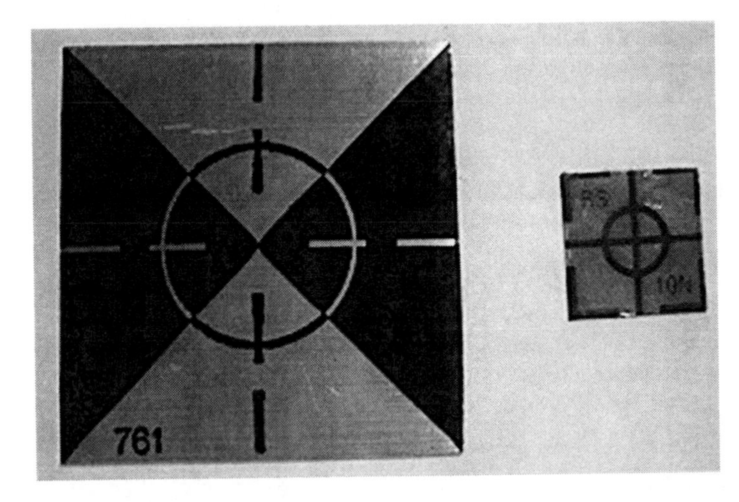

Fig. 4.22 *Reflective targets*

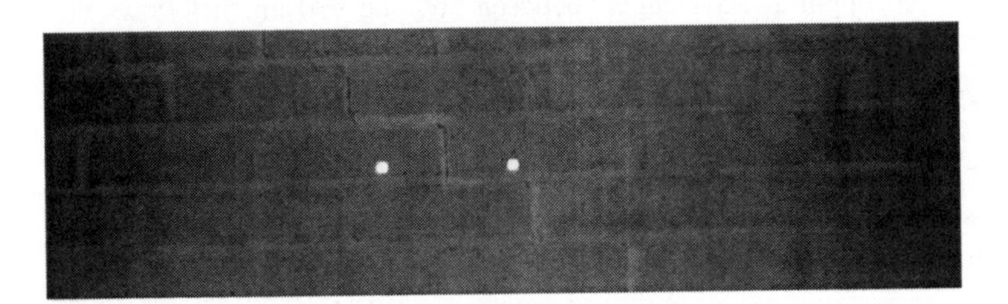

Fig. 4.23 *Flash photo of a wall with reflective targets*

4.8 METEOROLOGICAL CORRECTIONS

Using EDM equipment, the measurement of distance is obtained by measuring the time of propagation of electromagnetic waves through the atmosphere. Whilst the velocity of these waves in a vacuum (c_o) is known, the velocity will be reduced depending on the atmospheric conditions through which the waves travel at the time of measurement. As shown in *equation (4.16)*, knowledge of the refractive index (n) of the prevailing atmosphere is necessary in order to apply a correction for velocity to the measured distance. Thus if D' is the measured distance, the corrected distance D is obtained from

$$D = D'/n \tag{4.20}$$

The value of n is affected by the temperature, pressure and water vapour content of the atmosphere as well as by the wavelength λ of the transmitted electromagnetic waves. It follows from this that measurements of these atmospheric conditions are required at the time and place of measurement.

As already shown, steel tapes are standardized under certain conditions of temperature and tension. In a similar way, EDM equipment is standardized under certain conditions of temperature and pressure. It follows that even on low-order surveys the measurement of temperature and pressure is important.

The refractive index is related to wavelength via the Cauchy equation:

$$n = A + \frac{B}{\lambda^2} + \frac{C}{\lambda^4} \tag{4.21}$$

where A, B and C are constants relative to specific atmospheric conditions. To afford a correction in parts per million (ppm) the refractive number or refractivity (N) is used:

$$N = (n - 1) \times 10^6 \tag{4.22}$$

If $n = 1.000\,300$, then $N = 300$.

A value for n for standard air (0°C temperature, 1013.25 hPa pressure and dry air with 0.03% CO_2) is given by Barrel and Sears as

$$(n - 1) \times 10^6 = \left(287.6155 + \frac{1.628\,87}{\lambda^2} + \frac{0.013\,60}{\lambda^4}\right) \tag{4.23}$$

However, as stated in *Section 4.7.2*, it is the refractive index of the modulated beam, not the carrier, that is required; hence the use of group refractive index where

$$N_g = A + \frac{3B}{\lambda^2} + \frac{5C}{\lambda^4} \tag{4.24}$$

and therefore, for group velocity in standard air with λ in μm:

$$N_g = (n_g - 1) \times 10^6 = 287.6155 + \frac{4.886\,60}{\lambda^2} + \frac{0.068\,00}{\lambda^4} \tag{4.25}$$

The above formula is accurate to ±0.1 ppm at wavelengths between 560 and 900 nm. It follows that different instruments using different wavelengths will have different values for refractivity; for example:

$\lambda = 0.910$ μm gives $N_g = 293.6$

$\lambda = 0.820$ μm gives $N_g = 295.0$

To accommodate the actual atmospheric conditions under which the distances are measured, *equation (4.25)* was modified by Barrel and Sears:

$$N'_g = \frac{N_g \times Q \times P}{T} - \frac{V \times e}{T} \tag{4.26}$$

where T = absolute temperature in degrees Kelvin (°K) = $273.15 + t$
t = dry bulb temperature in °C
P = atmospheric pressure
e = partial water vapour pressure

with P and e in hPa, $Q = 0.2696$ and $V = 11.27$
with P and e in mmHg, $Q = 0.3594$ and $V = 15.02$

The value for e can be calculated from

$$e = e_s - 0.000\,662 \times P \times (t - t_w) \tag{4.27}$$

where t = dry bulb temperature
t_w = wet bulb temperature
e_s = saturation water vapour pressure

The value for e_s can be calculated from

$$\log e_s = \frac{7.5\,t_w}{t_w + 237.3} + 0.7857 \tag{4.28}$$

Using *equation (4.28)*, the following table can be produced:

t_w (°C)	0	10	20	30	40
e_s	6.2	12.3	23.4	42.4	73.7

e_s is therefore quite significant at high temperatures.

Working back through the previous equations it can be shown that at 100% humidity and a temperature of 30°C, a correction of approximately 2 ppm is necessary for the distance measured. In practice, humidity is generally ignored in the velocity corrections for instruments using light waves as it is insignificant compared with other error sources. However, for long lines being measured to very high accuracies in hot, humid conditions, it may be necessary to apply corrections for humidity.

The velocity correction is normally applied by entry of prevailing temperature and pressure into the instrument. In the case of infra-red/light waves, the humidity term is ignored as it is only relevant in conditions of high humidity and temperature. Under normal conditions the error would be in the region of 0.7 ppm.

For maximum accuracy, it may be necessary to compute the velocity correction from first principles, and if producing computer software to reduce the data, this would certainly be the best approach. When using this approach the standard values for t and P are entered into the instrument. The following example will now be computed in detail in order to illustrate the process involved.

Worked example

Example 4.8. An EDM instrument has a carrier wave of 0.91 μm and is standardized at 20°C and 1013.25 hPa. A distance of 1885.864 m was measured with the mean values $P = 1030$ hPa, dry bulb temperature $t = 30°\text{C}$, wet bulb temperature $t_w = 25°\text{C}$. Calculate the velocity correction.

Step 1. Compute the value for partial water vapour pressure e:

$$\log e_s = \frac{7.5t_w}{t_w + 237.3} + 0.7857$$

$$= \frac{7.5 \times 25}{25 + 237.3} + 0.7857$$

$$e_s = 31.66\,\text{hPa}$$

Using *equation (4.27)*:

$$e = e_s - 0.000\,662 \times P \times (t - t_w)$$

$$= 31.66 - 0.000\,662 \times 1030 \times (30 - 25) = 28.25\text{ hPa}$$

Step 2. Compute refractivity (N_g) for standard atmosphere using *equation (4.25)*:

$$N_g = 287.6155 + \frac{4.886\,60}{\lambda^2} + \frac{0.068\,00}{\lambda^4}$$

$$= 287.6155 + \frac{4.886\,60}{0.91^2} + \frac{0.068\,00}{0.91^4}$$

$$= 293.616$$

Step 3. Compute refractivity for the standard conditions of the instrument, i.e. 20°C and 1013.25 hPa, using *equation (4.26)* which for P and e in hPa becomes

$$N'_g = \frac{N_g(0.2696P)}{273.15 + t} - \frac{11.27e}{273.15 + t}$$

$$= \frac{79.159P}{273.15 + t} - \frac{11.27e}{273.15 + t}$$

This in effect is the equation for computing what is called the reference or nominal refractivity for the instrument:

$$N'_g = \frac{79.159 \times 1013.25}{273.15 + 20} - \frac{11.27 \times 28.25}{273.15 + 20}$$

$$= 272.52$$

Step 4. The reference refractivity now becomes the base from which the velocity correction is obtained. Now compute refractivity (N''_g) under the prevailing atmospheric conditions at the time of measurement:

$$N''_g = \frac{79.159 \times 1030}{273.15 + 30} - \frac{11.27 \times 28.25}{273.15 + 30} = 267.90$$

$$\therefore \text{Velocity correction in ppm} = 272.51 - 267.90 = 4.6 \text{ ppm}$$

$$\text{Correction in mm} = 1885.864 \times 4.6 \times 10^{-6} = 8.7 \text{ mm}$$

$$\text{Corrected distance} = 1885.873 \text{ m}$$

Using *equation (4.16)*:

(1) Velocity of light waves under standard conditions:

$$c_s = c_o/n_g = 299\,792\,458/1.000\,2725 = 299\,710\,770 \text{ m/s}$$

(2) Velocity under prevailing conditions:

$$C_p = 299\,792\,458/1.000\,2679 = 299\,712\,150 \text{ m/s}$$

As distance = (velocity × time), the increased velocity of the measuring waves at the time of measuring would produce a positive velocity correction, as shown.

Now considering *equation (4.26)* and differentiating with respect to t, P and e, we have

$$\delta N'_g = \frac{0.2696 N_g P\,\delta t}{(273.15 + t)^2} + \frac{0.2696 N_g\,\delta P}{(273.15 + t)} - \frac{11.27\,\delta e}{(273.15 + t)}$$

At $t = 15°\text{C}$, $P = 1013$ hPa, $e = 10$ hPa and $N_g = 294$, an error of 1 ppm in N'_g and therefore in distance will occur for an error in temperature of ±1°C, in pressure of ±3 hPa and in e of ±39 hPa.

From this it can be seen that the measurement of humidity can ordinarily be ignored for instruments using light. Temperature and pressure should be measured using carefully calibrated, good quality equipment, but not necessarily very expensive, highly accurate instruments. The measurements are usually taken at each end of the line being measured, on the assumption that the mean values of temperature and pressure at the ends of the line are equal to the average values measured along the line. However, tests on a 3-km test line showed the above assumption to be in error by 2°C and –3 hPa. The following measuring procedure is therefore recommended:

(1) Temperature and pressure should be measured at each end of the line.
(2) The above measurements should be taken well above the ground (3 m if possible) to avoid ground radiation effects so that they properly reflect mid-line conditions.
(3) The measurements should be synchronized with the EDM measurements.

(4) If possible, mid-line observations should be included.
(5) Ground-grazing lines should be avoided.

4.9 GEOMETRICAL REDUCTIONS

The measured distance, after the velocity corrections have been applied, is the spatial distance from instrument to target. This distance will almost certainly have to be reduced to the horizontal and then to its equivalent on the ellipsoid of reference.

From *Figure 4.24*, D_1 represents the measured distance (after the velocity corrections have been applied) which is reduced to its chord equivalent D_2; this is in turn reduced to D_3; the chord equivalent of the ellipsoidal distance (at MSL) D_4. Strictly speaking the ellipsoid of reference may be different from the geoid at MSL. A geoid–ellipsoid separation of 6 m would affect the distance by 1 mm/km. In the UK, for instance, maximum separation is in the region of 4.2 m, producing a scale error of 0.7 mm/km. However, if the WGS84 ellipsoid were used in the UK separations of the order of 50 m would be found leading to scale errors of the order of 8 ppm. However, where such information is unavailable, the geoid and ellipsoid are assumed coincident.

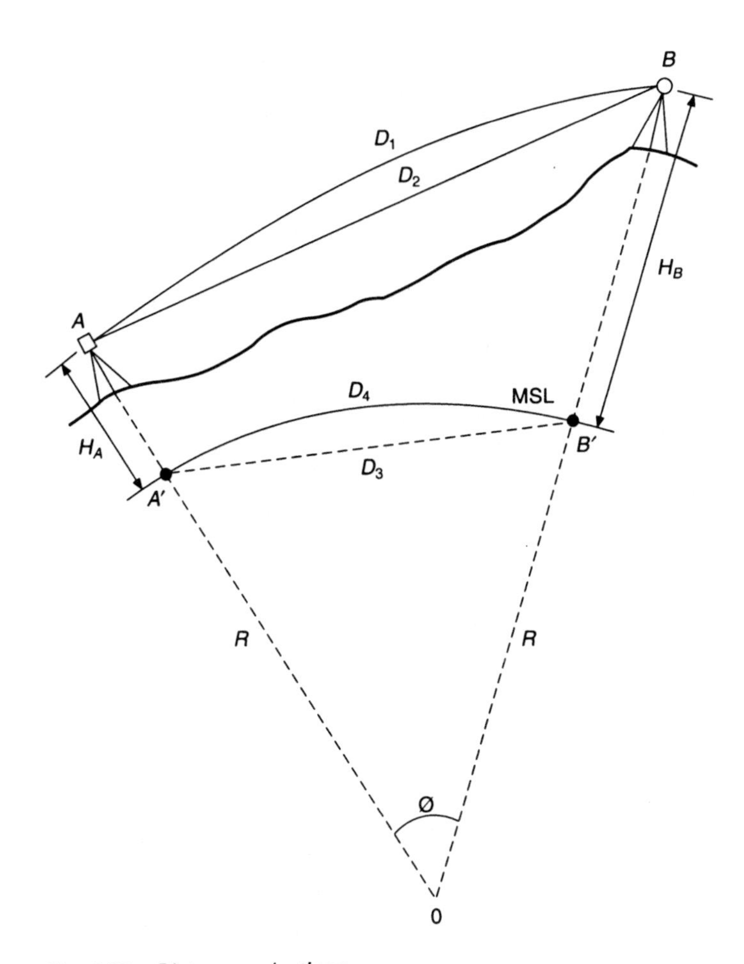

Fig. 4.24 *Distance reductions*

(1) Reduction from D_1 to D_2

$$D_2 = D_1 - C_1$$

where $C_1 = K^2 D_1^3/24R^2$ (4.29)

and K = coefficient of refraction

R, the radius of the ellipsoid in the direction α from A to B, may be calculated from *equations (8.21–8.23)* or with sufficient accuracy for lines less than 10 km from

$$R = (\rho\nu)^{\frac{1}{2}} \tag{4.30}$$

where ρ and ν are defined by *equations (8.21–8.23)*.

For the majority of lines in engineering surveying this first correction may be ignored.

The value for K is best obtained by simultaneous reciprocal observations, although it can be obtained from

$$K = (n_1 - n_2)R/(H_1 - H_2) \tag{4.31}$$

where n_1 and n_2 are the refractive indices at each end of the line and H_1 and H_2 are the respective heights above MSL.

(2) Reduction from D_2 to D_3

In triangle ABO using the cosine rule:

$$\cos\theta = [(R + H_B)^2 + (R + H_A)^2 - D_2^2]/2(R + H_B)(R + H_A)$$

but $\cos\theta = 1 - 2\sin^2(\theta/2)$ and $\sin\theta/2 = D_3/2R$

$$\therefore \cos\theta = 1 - D_3^2/2R^2$$

$$1 - D_3^2/2R^2 = [(R + H_B)^2 + (R + H_A)^2 - D_2^2]/2(R + H_B)(R + H_A)$$

$$D_3 = R\left\{\frac{[D_2 - (H_B - H_A)][D_2 + (H_B - H_A)]}{(R + H_A)(R + H_B)}\right\}^{\frac{1}{2}}$$

$$= \left[\frac{D_2^2 - (H_B - H_A)^2}{(1 + H_A/R)(1 + H_B/R)}\right]^{\frac{1}{2}} \tag{4.32}$$

where H_A = height of the instrument centre above the ellipsoid (or MSL)
H_B = height of the target centre above the ellipsoid (or MSL)

The above rigorous approach may be relaxed for the relatively short lines measured in the majority of engineering surveys.

Pythagoras' theorem may be used as follows:

$$D_M = [D_2^2 - (H_B - H_A)^2]^{\frac{1}{2}} \tag{4.33}$$

where D_M = the horizontal distance at the mean height H_M where $H_M = (H_B + H_A)/2$.

D_M would then be reduced to MSL using the altitude correction:

$$C_A = (D_M H_M)/R \tag{4.34}$$

then $D_3 = D_4 = D_M - C_A$

(3) Reduction of D_3 to D_4

Figure 4.24 shows

$$\theta/2 = D_4/2R = \sin^{-1}(D_3/2R)$$

where

$$\sin^{-1}(D_3/2R) = D_3/2R + D_3^3/8R \times 3! + 9D_3^5/32R^5 \times 5!$$

$$D_4/2R = D_3/2R + D_3^3/48R^3$$

$$D_4 = D_3 + (D_3^3/24R^2) \tag{4.35}$$

For lines less than 10 km, the correction is 1 mm and is generally ignored.

It can be seen from the above that for the majority of lines encountered in engineering (<10 km), the procedure is simply:

(1) Reduce the measured distance D_2 to the mean altitude using *equation (4.33)* $= D_M$.
(2) Reduce the horizontal distance D_M to D_4 at MSL using *equation (4.34)*.

(4) Reduction by vertical angle

Total stations have the facility to reduce the measured slope distance to the horizontal using the vertical angle, i.e.

$$D = S\cos\alpha \tag{4.36}$$

where α = the vertical angle
S = the slope distance
D = the horizontal distance

However, the use of this simple relationship will be limited to short distances if the effect of refraction is ignored.

Whilst S may be measured to an accuracy of, say, ±5 mm, reduction to the horizontal should not result in further degradation of this accuracy. Assume then that the accuracy of reduction must be ±1 mm.

Then $\delta D = -S\sin\alpha\delta\alpha$

and $\delta\alpha'' = \delta D \times 206\,265/S\sin\alpha \quad (4.37)$

where δD = the accuracy of reduction

$\delta\alpha''$ = the accuracy of the vertical angle

Consider $S = 1000$ m, measured on a slope of $\alpha = 5°$; if the accuracy of reduction to the horizontal is to be practically error free, let $\delta D = \pm 0.001$ m and then

$$\delta\alpha'' = 0.001 \times 206\,265/1000 \sin 5° = \pm 2.4''$$

This implies that to achieve error-free reduction the vertical angle must be measured to an accuracy of ±2.4″. If the accuracy of the double face mean of a vertical angle is, say, ±4″, then a further such determination is required to reduce it to $4''/(2)^{\frac{1}{2}} = \pm 2.8''$. However, the effect of refraction assuming an average value of $K = 0.15$ is 2.4″ over 1000 m. Hence the limit has been reached where refraction can be ignored. For $\alpha = 10°$, $\delta\alpha = \pm 1.2''$, and hence refraction would need to be considered over this relatively steep sight. At $\alpha = 2°, \delta\alpha'' = \pm 6''$ and the effect of refraction is negligible. It is necessary to carry out this simple appraisal which depends on the distances and the slopes involved, in order to assess the consequence of ignoring refraction.

In some cases, standard corrections for refraction (and curvature) are built into the total station which may or may not help the reduction accuracy. The true value of refraction can be so variable that it cannot be accounted for unless the effect is made to be self cancelling by using the mean of simultaneous reciprocal vertical angles. If the distances involved are long, then the situation is as shown in *Figure 4.25*. If α and β are the reciprocal angles, corrected for any difference of height between the instrument and the target, as described in *Section 2.15.2*, then

$$\alpha_0 = (\alpha - \beta)/2 \tag{4.38}$$

where β is negative for angle of depression and

$$S \cos \alpha_0 = AC \tag{4.39}$$

However, the distance required is AB' where

$$AB' = AC - B'C \tag{4.40}$$

but $BC = S \sin \alpha_0$

and $B'C = BC \tan \theta/2$

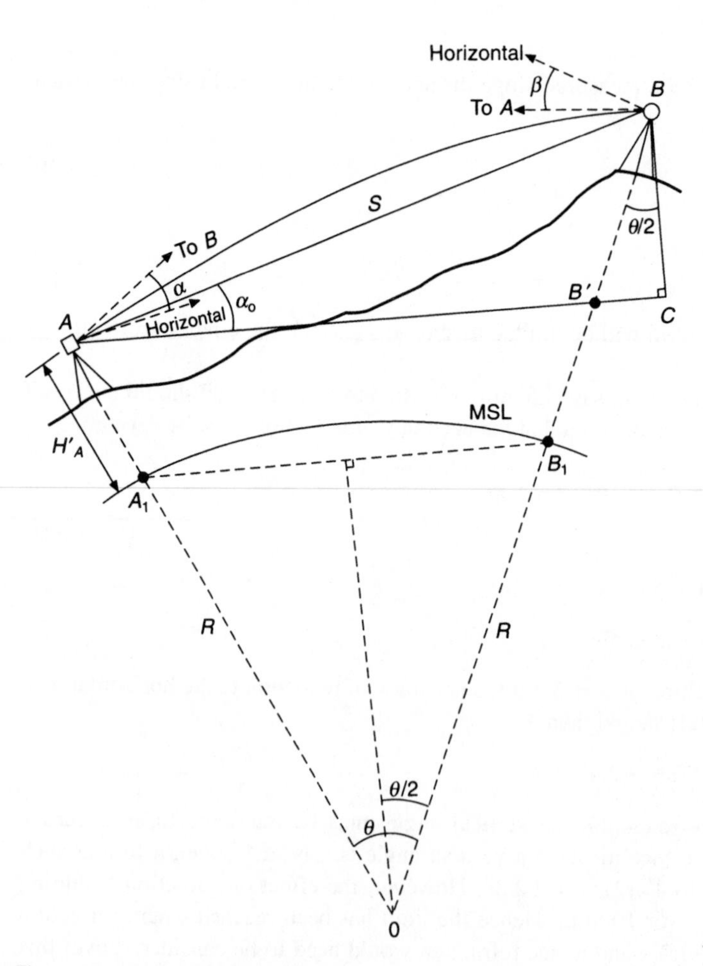

Fig. 4.25 *Reduction by vertical angle*

where

$$\theta'' = \frac{AC \times 206\,265}{R + H'_A} = \frac{S \times 206\,265}{R + H'_A} \quad (4.41)$$

This will give the horizontal distance (AB') at the height of the instrument axis above MSL, i.e. H'_A.

AB' can then be reduced to A_1B_1 at MSL using the altitude correction *(equation (4.34))*.

Alternatively, the above procedure can be rearranged to give

$$AB' = S\cos(\alpha_0 + \theta/2)\sec\theta/2 \quad (4.42)$$

Both procedures will now be demonstrated by an example.

Worked example

Example 4.9. Consider $S = 5643.856$ m and reciprocal vertical angles of $\alpha = 5°\,00'\,22''$, $\beta = 5°\,08'\,35''$. If station A is 42.95 m AOD and the instrument height is 1.54 m, compute the horizontal distance (AB') at instrument height above MSL (*Figure 4.25*) ($R = 6380$ km).

$$\alpha_0 = [5°\,00'\,22'' - (-5°\,08'\,35'')]/2 = 5°\,04'\,29''$$

$$AC = 5643.856 \cos 5°\,04'\,29'' = 6521.732 \text{ m}$$

$$BC = 5643.856 \sin 5°\,04'\,29'' = 499.227 \text{ m}$$

$$\theta' = (5621.732 \times 206\,265)/(6\,380\,000 + 44.49) = 03'\,02''$$

$$B'C = 499.227 \tan 01'\,31'' = 0.220 \text{ m}$$

$$AB' = 5621.732 - 0.220 = 5621.512 \text{ m}$$

Alternatively from *equation (4.42)*:

$$AB' = S\cos(\alpha_0 + \theta/2)\sec\theta/2$$

$$= 5643.856\cos(-5°\,04'\,29'' + 01'\,31'')\sec 01'\,31''$$

$$= 5621.512 \text{ m}$$

AB' is now reduced to A_1B_1 using the correction *(equation (4.34))* where the height above MSL includes the instrument height, i.e. $42.95 + 1.54 = 44.49$ m.

If a value for the angle of refraction ($\hat{r}$) is known and only one vertical angle α is observed then

$$AB' = S\cos(\alpha' + \theta)\sec\theta/2 \quad (4.43)$$

where $\alpha' = \alpha - \hat{r}$.

Adapting the above example and assuming an average value for K of 0.15:

$$r'' = SK\rho/2R$$

$$= \frac{5643.856 \times 0.15 \times 206\,265}{2 \times 6\,380\,000} = 14''$$

$$\alpha' = 5°\,00'\,22'' - 14'' = 5°\,00'\,08''$$

$$AB' = 5643.856\cos(5°\,00'\,08'' + 3'\,02'')\sec 01'\,31''$$

$$= 5621.924 \text{ m}$$

The difference of 0.412 m once again illustrates the dangers of assuming a value for K.

If zenith angles are used then:

$$AB' = S\left[\frac{\sin Z_A - \theta(2-K)/2}{\cos\theta/2}\right] \tag{4.44}$$

where K = the coefficient of refraction.

For most practical purposes *equation (4.44)* is reduced to

$$AB' = S\sin Z_A - \frac{S^2(2-K)\sin 2Z_A}{4R} \tag{4.45}$$

4.10 ERRORS, CHECKING AND CALIBRATION

Although modern EDM equipment is exceptionally well constructed, the effects of age and general wear and tear may alter its performance. It is essential therefore that all field equipment should be regularly checked and calibrated. Checking and calibration are two quite separate activities. Checking is concerned with verifying that the instrument is performing within acceptable tolerances. Calibration is the process of estimating the parameters that need to be applied to correct actual measurements to their true values. For example, in the case of levelling the two-peg test could be used for both checking and calibrating a level. The two-peg test can be used to check whether the level is within ±10 mm at 20 m or whatever is appropriate for the instrument concerned. It could also be used to discover what the actual collimation error is at, say, 20 m, so that a proportional correction, depending on distance, could be applied to each subsequent reading. This latter process is calibration. In the case of levelling both checking and calibration are quite simple procedures. Rather more is involved with checking and calibrating EDM. A feature of calibration is that it should be traceable to superior, usually national, standards. From the point of view of calibration, the errors have been classified under three main headings.

4.10.1 Zero error (independent of distance)

Zero error arises from changes in the instrument/reflector constant due to ageing of the instrument or as a result of repairs. The built-in correction for instrument/reflector constants is usually correct to 1 or 2 mm but may change with different reflectors and so should be assessed for a particular instrument/reflector combination. A variety of other matters may affect the value of the constant and these matters may vary from instrument to instrument. Some instruments have constants which are signal strength dependent, while others are voltage dependent. The signal strength may be affected by the accuracy of the pointing or by prevailing atmospheric conditions. It is very important, therefore, that periodical calibration is carried out.

A simple procedure can be adopted to obtain the zero error for a specific instrument/reflector combination. Consider three points A, B, C set out in a straight line such that AB = 10 m, BC = 20 m and AC = 30 m *(Figure 4.26)*.

Assume a zero error of +0.3 m exists in the instrument; the measured lengths will then be 10.3, 20.3 and 30.3. Now:

$$AB + BC = AC$$

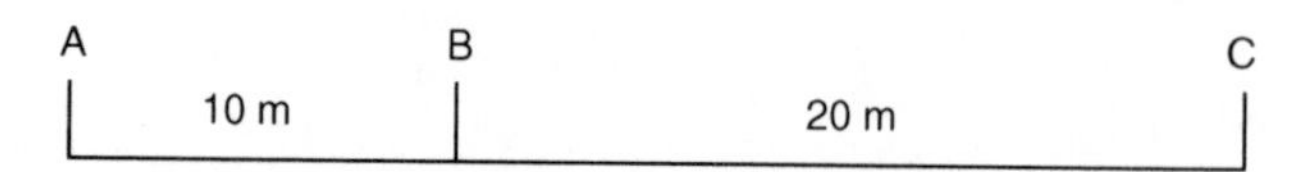

Fig. 4.26 *Simple calibration baseline*

But the measurements $10.3 + 20.3 \neq 30.3$
The error may be found from $10.3 + 20.3 - 30.3 = +0.3$
Now as

Correction = – Error

every measured distance will need a correction of – 0.3 m. Or in more general terms

$$\text{zero error} = k_o = l_{AB} + l_{BC} - l_{AC} \tag{4.46}$$

from which it can be seen that the base-line lengths do not need to be known prior to measurement. If there are more than two bays in the base line of total length L, then

$$k_o = (L - \Sigma l_i)/(n - 1) \tag{4.47}$$

where l_i is the measured length of each of the n sections.

Alternatively the initial approach may be to observe the distances between all possible combinations of points. For example, if the base line comprises three bays AB, BC and CD, we have

$$AB + BC - AC = k_o$$

$$AC + CD - AD = k_o$$

$$AB + BD - AD = k_o$$

$$BC + CD - BD = k_o$$

with the arithmetic mean of all four values being accepted.

The most accurate approach is a least squares solution of the observation equations. Readers unfamiliar with least squares methods may wish to pass over the remainder of this section until they have read *Chapter 7* and Appendix A.

Let the above bays be a, b and c, with measured lengths l and residual errors of measurement r.

Observation equations:

$$a + k_o = l_{AB} + r_1$$

$$a + b + k_o = l_{AC} + r_2$$

$$a + b + c + k_o = l_{AD} + r_3$$

$$b + k_o = l_{BC} + r_4$$

$$b + c + k_o = l_{BD} + r_5$$

$$c + k_o = l_{CD} + r_6$$

which may be written in matrix form as $\mathbf{Ax} = \mathbf{b} + \mathbf{v}$

$$\begin{bmatrix} 1 & 0 & 0 & 1 \\ 1 & 1 & 0 & 1 \\ 1 & 1 & 1 & 1 \\ 0 & 1 & 0 & 1 \\ 0 & 1 & 1 & 1 \\ 0 & 0 & 1 & 1 \end{bmatrix} \begin{bmatrix} a \\ b \\ c \\ k_o \end{bmatrix} = \begin{bmatrix} l_{AB} \\ l_{AC} \\ l_{AD} \\ l_{BC} \\ l_{BD} \\ l_{CD} \end{bmatrix} + \begin{bmatrix} r_1 \\ r_2 \\ r_3 \\ r_4 \\ r_5 \\ r_6 \end{bmatrix}$$

The solution for $\mathbf{x}$ is $\mathbf{x} = (\mathbf{A}^T\mathbf{W}\mathbf{A})^{-1}\mathbf{A}^T\mathbf{W}\mathbf{b}$ and gives a solution for estimates of the length of each bay a, b and c as well as the zero error k_o. Provided that the quality of all the observations of distance is the same then the solutions are simply

$$a = (l_{AB} + l_{AC} + 2l_{AD} - 2l_{BC} - l_{BD} - l_{CD})/4$$

$$b = (-2l_{AB} + l_{AC} + l_{AD} + l_{BC} + l_{BD} - 2l_{CD})/4$$

$$c = (-l_{AB} - l_{AC} + 2l_{AD} - 2l_{BC} + l_{BD} + l_{CD})/4$$

$$k_o = (l_{AB} + l_{BC} + l_{CD} - l_{AD})/2$$

Substituting the values for a, b, c and k_o back into the observation equations gives the residual values r, which can give an indication of the overall magnitude of the other errors.

The distances measured should, of course, be corrected for slope before they are used to find k_o. If possible, the bays should be in multiples of $\lambda/2$ if the effect of cyclic errors is to be cancelled.

4.10.2 Cyclic error (varies with distance)

As already shown, the measurement of the phase difference between the transmitted and received waves enables the fractional part of the wavelength to be determined. Thus, errors in the measurement of phase difference will produce errors in the measured distance. Phase errors are cyclic and not proportional to the distance measured and may be non-instrumental and/or instrumental.

The non-instrumental cause of phase error is spurious signals from reflective objects illuminated by the beam. Normally the signal returned by the reflector will be sufficiently strong to ensure complete dominance over spurious reflections. However, care should be exercised when using vehicle reflectors or reflective material designed for clothing for short-range work.

The main cause of phase error is instrumental and derives from two possible sources. In the first instance, if the phase detector were to deviate from linearity around a particular phase value, the resulting error would repeat each time the distance resulted in that phase. Excluding gross malfunctioning, the phase readout is reliably accurate, so maximum errors from this source should not exceed 2 or 3 mm. The more significant source of phase error arises from electrical cross-talk, or spurious coupling, between the transmit and receive channels. This produces an error which varies sinusoidally with distance and is inversely proportional to the signal strength.

Cyclic errors in phase measurement can be determined by observing to a series of positions distributed over a half wavelength. A bar or rail accurately divided into 10-cm intervals over a distance of 10 m would cover the requirements of most short-range instruments. A micrometer on the bar capable of very accurate displacements of the reflector of +0.1 mm over 20 cm would enable any part of the error curve to be more closely examined.

The error curve plotted as a function of the distance should be done for strong and weakest signal conditions and may then be used to apply corrections to the measured distance. For the majority of short-range instruments the maximum error will not exceed a few millimetres.

Most short-range EDM instruments have values for $\lambda/2$ equal to 10 m. A simple arrangement for the detection of cyclic error which has proved satisfactory is to lay a steel band under standard tension on a horizontal surface. The reflector is placed at the start of a 10-m section and the distance from instrument to reflector obtained. The reflector is displaced precisely 100 mm and the distance is re-measured. The difference between the first and second measurement should be 100 mm; if not, the error is plotted at the 0.100 m value of the graph. The procedure is repeated every 100 mm throughout the 10-m section and an error curve produced. If, in the field, a distance of 836.545 m is measured, the 'cyclic error' correction is abstracted from point 6.545 m on the error curve.

4.10.3 Scale error (proportional to distance)

Scale errors in EDM instruments are largely due to the fact that the oscillator is temperature dependent. The quartz crystal oscillator ensures the frequency (f) remains stable to within ±5 ppm over an operational temperature range of –20°C to 50°C. The modulation frequency can, however, vary from its nominal value due to incorrect factory setting, ageing of the crystal and lack of temperature stabilization. Most modern short-range instruments have temperature-compensated crystal oscillators which have been shown to perform well. However, warm-up effects have been shown to vary from 1 to 5 ppm during the first hour of operation.

Diode errors also cause scale error, as they could result in the emitted wavelength being different from its nominal value.

The magnitude of the resultant errors may be obtained by field or laboratory methods.

The laboratory method involves comparing the actual modulation frequency of the instrument with a reference frequency. The reference frequency may be obtained from off-air radio transmissions such as MSF Rugby (60 kHz) in the UK or from a crystal-generated laboratory standard. The correction for frequency is equal to

$$\left(\frac{\text{Nominal frequency} - \text{Actual frequency}}{\text{Nominal frequency}}\right)$$

A simple field test is to measure a base line whose length is known to an accuracy greater than the measurements under test. The base line should be equal to an integral number of modulation half wavelengths. The base line AB should be measured from a point C in line with AB; then $CB - CA = AB$. This differential form of measurement will eliminate any zero error, whilst the use of an integral number of half wavelengths will minimize the effect of cyclic error. The ratio of the measured length to the known length will provide the scale error.

4.10.4 Multi-pillar base lines

The establishment of multi-pillar base lines for EDM calibration requires careful thought, time and money. Not only must a suitable area be found to permit a base line of, in some cases, over 1 km to be established, but suitable ground conditions must also be present. If possible the bedrock should be near the surface to permit the construction of the measurement pillars on a sound solid foundation. The ground surface should be reasonably horizontal, free from growing trees and vegetation and easily accessible. The construction of the pillars themselves should be carefully considered to provide maximum stability in all conditions of wetting and drying, heat and cold, sun and cloud, etc. The pillar-centring system for instruments and reflectors should be carefully thought out to avoid any hint of centring error. When all these possible error sources have been carefully considered, the pillar separations must then be devised.

The total length of the base line is obviously the first decision, followed by the unit length of the equipment to be calibrated. The inter-pillar distances should be spread over the measuring range of equipment, with their fractional elements evenly distributed over the half wavelength of the basic measuring wave.

Finally, the method of obtaining inter-pillar distances to the accuracy required has to be considered. The accuracy of the distance measurement must obviously be greater than the equipment it is intended to calibrate. For general equipment with accuracies in the range of 3–5 mm, the base line could be measured with equipment of superior accuracy such as those already mentioned.

For even greater accuracy, laser interferometry accurate to 0.1 ppm may be necessary.

When such a base line is established, a system of regular and periodic checking must be instituted and maintained to monitor short- and long-term movement of the pillars. Appropriate computer software must also be written to produce zero, cyclic and scale errors per instrument from the input of the measured field data.

In the past several such base lines have been established throughout the UK, the most recent one by Thames Water at Ashford in Middlesex, in conjunction with the National Physical Laboratory at Teddington, Middlesex. In 2006, it was understood to be no longer operational, along with all the other UK base lines. It is hoped the situation will change. However, this is a good example of a pillar base line and will be described further to illustrate the principles concerned. This is an eight-pillar base line, with a total length of 818.93 m and inter-pillar distances affording a good spread over a 10-m period, as shown below:

	2	3	4	5	6	7	8
1	260.85	301.92	384.10	416.73	480.33	491.88	818.93
2		41.07	123.25	155.88	219.48	231.03	558.08
3			82.18	114.81	178.41	189.96	517.01
4				32.63	96.23	107.78	434.83
5					63.60	75.15	402.20
6						11.55	338.60
7							327.05

With soil conditions comprising about 5 m of gravel over London clay, the pillars were constructed by inserting 8 × 0.410 m steel pipe into a 9-m borehole and filling with reinforced concrete to within 0.6 m of the pillar top. Each pillar top contains two electrolevels and a platinum resistance thermometer to monitor thermal movement. The pillars are surrounded by 3 × 0.510 m PVC pipe, to reduce such movement to a minimum. The pillar tops are all at the same level, with Kern baseplates attached. Measurement of the distances has been carried out using a Kern ME5000 Mekometer and checked by a Terrameter. The Mekometer has in turn been calibrated by laser interferometry. The above brief description serves to illustrate the care and planning needed to produce a base line for commercial calibration of the majority of EDM equipment.

4.11 OTHER ERROR SOURCES

4.11.1 Reduction from slope to horizontal

The reduction process using vertical angles has already been dealt with in *Section 4.10.4*. On steep slopes the accuracy of angle measurement may be impossible to achieve, particularly when refraction effects are considered. An alternative procedure is, of course, to obtain the difference in height (h) of the two measuring sources and correct for slope using Pythagoras. If the correction is C_h, then the first term of a binomial expansion of Pythagoras gives

$$C_L = h^2/2S$$

where S = the slope length measured. Then

$$\delta C_h = h\,\delta h/S \tag{4.48}$$

and for $S = 1000$ m, $h = 100$ m and the accuracy of reduction $\delta C_h = \pm 0.001$ m, substituting in *equation (4.48)* gives $\delta h = \pm 0.010$ m. This implies that the difference in level should be obtained to an accuracy of ± 0.010 m, which is within the accuracy criteria of tertiary levelling. For $h = 10$ m, $\delta h = \pm 0.100$ m, and for $h = 1$ m, $\delta h = \pm 1$ m.

Analysis of this sort will enable the observer to decide on the method of reduction, i.e. vertical angles or differential levelling, in order to achieve the required accuracy.

4.11.2 Reduction to the plane of projection

Many engineering networks are connected to the national grid system of their country; a process which involves reducing the horizontal lengths of the network to mean sea level (MSL) and then to the projection using local scale factors (LSF).

Reduction to MSL is carried out using

$$C_M = \frac{LH}{R} \tag{4.49}$$

where C_M = the altitude correction, H = the mean height of the line above MSL or the height of the measuring station above MSL and R = mean radius of the Earth (6.38×10^6 m).

Differentiating *equation (4.49)* gives

$$\delta C_M = L\,\delta H/R$$

and for $L = 1000$ m, $\delta C_M = \pm 1$ mm, then $\delta H = \pm 6.38$ m. As Ordnance Survey (OS) tertiary bench marks are guaranteed to ± 10 mm, and the levelling process is of more than comparable accuracy, the errors from this source may be ignored.

Reduction of the horizontal distance to MSL theoretically produces the chord distance, not the arc or spheroidal distance. However, the chord/arc correction is negligible at distances of up to 10 km and will not therefore be considered further.

To convert the ellipsoidal distance to grid distance it is necessary to calculate the LSF and multiply the distance by it. The LSF changes from point to point. Considering the OS national grid (NG) system of the UK, the LSF changes at the rate of 0 ppm km^{-1} on the central meridian, 2.5 ppm km^{-1} at 100 km from the central meridian and 7.4 ppm km^{-1} at 300 km from the central meridian.

For details of scale factors, their derivation and application, refer to *Chapter 5*.

The following approximate formula for scale factors will now be used for error analysis, of the UK system.

$$F = F_0[1 + (E_m^2/2R^2)] \tag{4.50}$$

where E_m = the NG difference in easting between the mid-point of the line and the central meridian = 4 000 000 m

F_0 = the scale factor at the central meridian = 0.999 601 27

R = the mean radius of the Earth (6.38×10^6 m)

Then the scale factor correction is $C_F = LF_0[1 + (E_m^2/2R^2)] - L$

and

$$\delta C_F = LF_0(E_m/R^2)\,\delta E_m \tag{4.51}$$

Then for $L = 1000$ m, $\delta C = \pm 1$ mm and $E_m = 120$ km, $\delta E_m = \pm 333$ m; thus the accuracy of assessing one's position on the NG is not critical. Now, differentiating with respect to R

$$\delta C_F = -LF_0E_m^2/R^3\,\delta R \tag{4.52}$$

and for the same parameters as above, $\delta R = \pm 18$ km. The value for $R = 6380$ km is a mean value for the whole Earth and is accurate to about 10 km between latitudes 30° and 60°, while below 30° a more representative value is 6.36×10^6 m.

It can be seen therefore that accuracy of the reduction to MSL and thereafter to NG will usually have a negligible effect on the accuracy of the reduced horizontal distance.

4.11.3 Eccentricity errors

These errors may arise from the manner in which the EDM is mounted with respect to the transit axis of the total station and the type of prism used. Although there are few of (1) and (2) below remaining, they illustrate a problem and its simple solution. In practice these problems will not appear with the majority of total stations where the EDM is located co-axially with the telescope.

(1) Consider telescope-mounted EDM instruments used with a tilting reflector which is offset the same distance, h, above the target as the centre of the EDM equipment is above the line of sight of the telescope (*Figure 4.27*).

In this case the measured distance S is equal to the distance from the centre of the theodolite to the target and the eccentricity e is self-cancelling at instrument and reflector. Hence D and ΔH are obtained in the usual way without further correction.

(2) Consider now a telescope-mounted EDM unit with a non-tilting reflector, as in *Figure 4.28*. The measured slope distance S will be greater than S' by length $AB = h\tan\alpha$. If α is negative, S will be less than S' by $h \tan\alpha$.

Thus if S is used in the reduction to the horizontal D will be too long by $AF = h\sin\alpha$ when α is positive, and too small when α is negative.

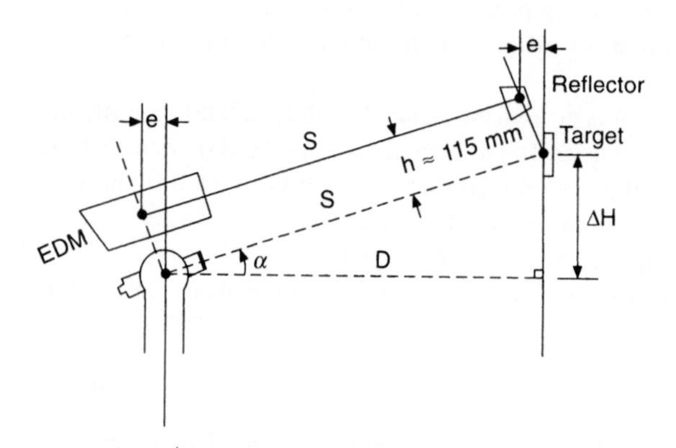

Fig. 4.27 *EDM and prism on tilting reflector*

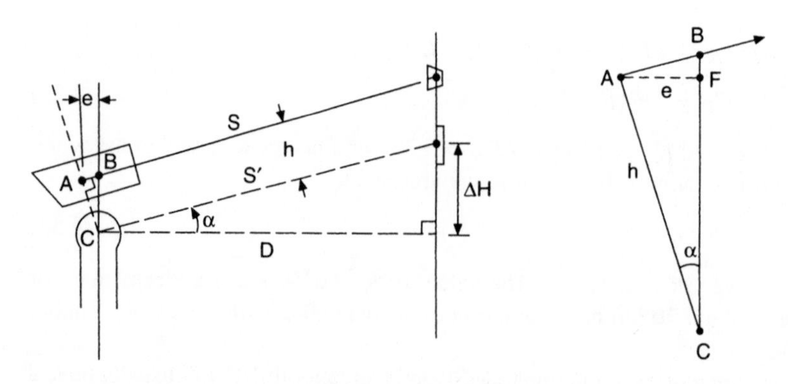

Fig. 4.28 *EDM and prism on non-tilting reflector*

If we assume an approximate value of $h = 115$ mm then the error in D when $\alpha = 5°$ is 10 mm, at 10° it is 20 mm and so on to 30° when it is 58 mm. The errors in ΔH for the above vertical angles are 1 mm, 14 mm and 33 mm, respectively.

(3) When the EDM unit is co-axial with the telescope line of sight and observations are direct to the centre of the reflector, there are no eccentricity corrections.

4.12 INSTRUMENT SPECIFICATIONS

The measuring accuracy of all EDM equipment is specified in manufacturers' literature in different ways but often in the form of $+a$ mm $+b$ ppm or $\pm(a$ mm $+ b$ ppm). This can be a little misleading as the constant uncertainty, a mm, is unrelated to the uncertainty of scale error, b ppm. Therefore these statements should be interpreted as being that the total uncertainty (σ) is given by

$$\sigma = (a^2 + (bL \times 10^{-6})^2)^{1/2} \text{ mm}$$

where L is the length of the line in kilometres.

Using the a typical example of $a = 3$ mm and $b = 5$ ppm the uncertainty of a measured distance if 1.2 km would be

$$\sigma = (3^2 + (5 \times 1.2)^2)^{1/2} \text{ mm} = 7 \text{ mm}$$

In the above specification, a is a result of errors in phase measurement (θ) and zero error (z), i.e.

$$a^2 = \sigma_\theta^2 + \sigma_z^2$$

In the case of b, the resultant error sources are error in the modulation frequency f and in the group refractive index n_g, i.e.

$$b^2 = (\sigma_f/f)^2 + (\sigma_{n_g}/n_g)^2$$

The reason why the specification is expressed in two parts is that θ and z are independent of distance, whilst f and n_g are a function of distance.

For short distances such as a few tens or even hundreds of metres, as frequently encountered in engineering, the a component is more significant and the b component can largely be ignored.

4.13 DEVELOPMENTS IN EDM

Improvements in technology have transformed EDM instruments from large, cumbersome units, which measured distance only, to component parts of total stations.

The average total station is a fully integrated instrument that captures all the spatial data necessary for a three-dimensional position fix. The angles and distances are displayed on a digital readout and can be recorded at the press of a button. Modern total stations have many of the following features:

- *Dual axis compensators* built into the vertical axis of the instrument which constantly monitor the inclination of the vertical axis in two directions. These tilt sensors have a range of a few minutes. The horizontal and vertical angles are automatically corrected, thus permitting single-face observations without significant loss of accuracy.
- A *graphic electronic levelling display* illustrates the levelling situation parallel to a pair of footscrews and at right angles, enables rapid, precise levelling without rotation of the alidade. The problems caused by direct sunlight on plate bubbles are also eradicated.

- On-board *memory* is available in various capacities for logging observations. These devices can usually store at least all the data a surveyor is likely to record in a day. The memory unit can be connected to any external computer or to a special card reader for data transfer. Alternatively, the observations can be downloaded directly into intelligent electronic data loggers. Both systems can be used in reverse to load information into the instruments.

  Some instruments and/or data loggers can be interfaced directly with a computer for immediate processing and plotting of data.
- A *friction clutch and endless drive* eliminates the need for horizontal and vertical circle clamps plus the problem of running out of thread on slow motion screws.
- A *laser plummet* incorporated into the vertical axis, replaces the optical plummet. A clearly visible laser dot is projected onto the ground permitting quick and convenient centring of the instrument.
- *Keyboards* with displays of alphanumeric and graphic data control every function of the instrument. Built-in software with menu and edit facilities, they automatically reduce angular and linear observations to three-dimensional coordinates of the vector observed. This facility can be reversed for setting-out purposes. Keyboards and displays may be on both sides of the instrument allowing observations and readings to be made easily on both faces.
- *Guide light* fitted to the telescope of the instrument enables the target operator to maintain alignment when setting-out points. This light changes colour when the operator moves off-line. With the instrument in the tracking mode, taking measurements every 0.3 s, the guide light speeds up the setting-out process.
- *Automatic target recognition* (ATR) is incorporated in most robotic instruments and is more accurate and consistent than human sighting. The telescope is pointed in the general direction of the target, and the ATR module completes the fine pointing with excellent precision and minimum measuring time as there is no need to focus. It can also be used on a moving reflector. After initial measurement, the reflector is tracked automatically (*Figure 4.29*). A single key touch records all data without interrupting the tracking process. To ensure that the prism is always pointed to the instrument, 360° *prisms* are available from certain manufacturers. ATR recognizes targets up to 1000 m away and maintains lock on prisms moving at a speed of 5 m s^{-1}. A further advantage of ATR is that it can operate in darkness.

Fig. 4.29 *Robotic total station system – courtesy of (a) Leica-Geosystems, (b) Topcon*

- In order to utilize ATR, the instrument must be fitted with *servo motors* to drive both the horizontal and vertical movements of the instrument. It also permits the instrument to automatically turn to a specific bearing (direction) when setting out, calculated from the up-loaded design coordinates of the point.
- *Reflectorless measurement* is also available on many instruments, typically using two different coaxial red laser systems. One laser is invisible and is used to measure long distances, up to a few kilometres to a single reflector, the other is visible, does not require a reflector, and has a limited range of a few hundred metres. The range depends on the reflectivity of the surface being measured to. Concrete and wood give good returns while dark coloured rock has a lesser range. A single key stroke allows one to alternate between the visible or invisible laser.

  Possible uses for this technique include surveying the façades of buildings, tunnel profiling, cooling tower profiling, bridge components, dam faces, overhead cables, points in traffic or other hazardous situations, points where it is impossible to place a target – indeed any situation which is difficult or impossible to access directly. The extremely narrow laser used clearly defines the target point (*Figure 4.30*).
- *Wireless communication* enables communication between instrument and surveyor when the surveyor is remote from the instrument and enables downloading data to a computer without the need for cables.
- *Target identification* enables locking onto only the correct target without the possibility of acquiring false targets.

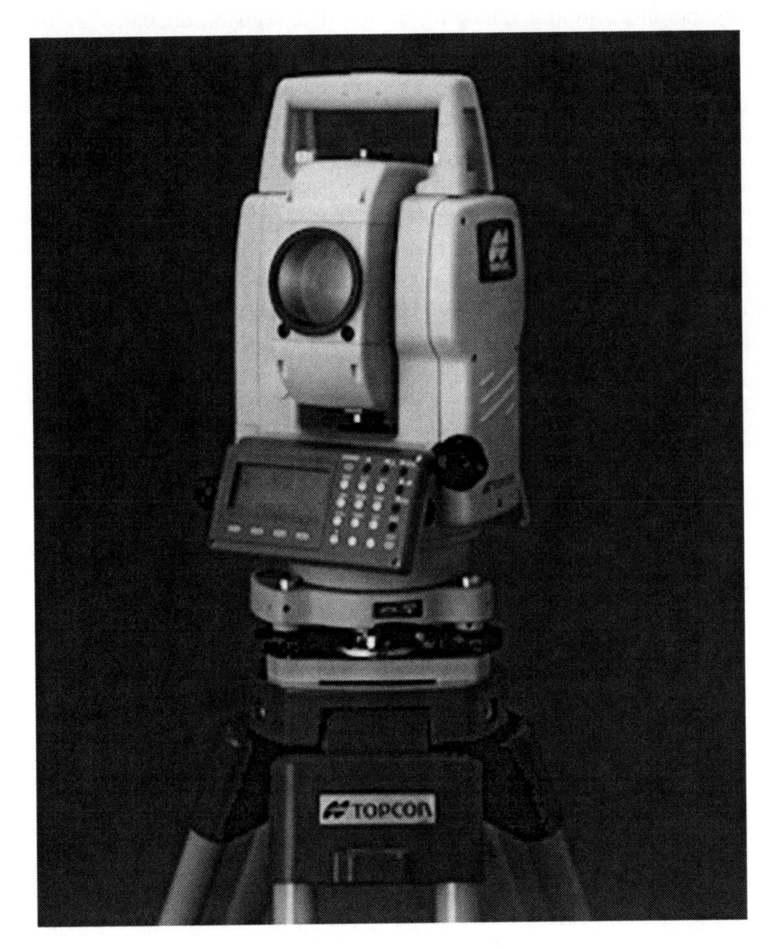

Fig. 4.30 *Reflectorless total station – courtesy of Topcon*

- *Built-in programmes* are available with most total stations. Examples of which are: Traverse coordinate computation with Bowditch adjustment; Missing line measurement in the horizontal and vertical planes to any two points sighted from the instrument; Remote object elevation determines the heights of inaccessible points; Offset measurement gives the distance and bearing to an inaccessible point close to the reflector by obtaining the vector to the reflector and relative direction to the inaccessible point; Resection to a minimum of two known points determines the position of the total station observing those points; Building façade survey allows for the coordination of points on the face of a building or structure using angles only; Three-dimensional coordinate values of points observed; Setting-out data to points whose coordinates have been uploaded into the total station; Coding of the topographic detail with automatic point number incrementing.

The above has detailed many of the developments in total station design and construction which have led to the development of fully automated one-man systems, frequently referred to as robotic surveying systems. Robotic surveying produces high productivity since the fully automatic instrument can be left unmanned and all operations controlled from the target point that is being measured or set out. In this case it is essential that the instrument is in a secure position. The equipment used at the target point would consist of an extendible reflector pole with a circular bubble. It would carry a 360° prism and a control unit incorporating a battery and radio modem. Import and export of data is via the radio modem. Storage capacity would be equal to at least 10 000 surveyed points, plus customized software and the usual facilities to view, edit, code, set-out, etc. The entire measurement procedure is controlled from the reflector pole with facilities for keying the start/stop operation, aiming, changing modes, data registration, calculations and data input.

4.13.1 Machine guidance

Robotic surveying has resulted in the development of several customized systems, not the least of which are those for the control of construction plant on site. These systems, produced by the major companies Leica Geosystems and Trimble, are capable of controlling slip-form pavers, rollers, motorgraders and even road headers in tunnelling. In each case the method is fundamentally the same. See *Figure 4.31*.

The machine is fitted with a customized 360° prism strategically positioned on the machine. The total station is placed some distance outside the working area and continuously monitors the three-dimensional position of the prism. This data is transmitted via the radio link to an industrial PC on board the construction machine. The PC compares the construction project data with the machine's current position and automatically and continuously sends the appropriate control commands to the machine controller to give the necessary construction position required. All the information is clearly displayed on a large screen. Such information comprises actual and required grading profiles; compression factors for each surface area being rolled and the exact location of the roller; tunnel profiles showing the actual position of the cutter head relative to the required position. In slip forming, for example, complex profiles, radii and routes are quickly completed to accuracies of 2 mm and 5 mm in vertical and lateral positions respectively. Not only do these systems provide extraordinary precision, they also afford greater safety, speedier construction and are generally of a higher quality than can be achieved with GPS systems employed to do similar tasks.

Normally, without machine guidance, all these operations are controlled using stringlines, profile boards, batter boards, etc. As these would no longer be required, their installation and maintenance costs are eliminated, they do not interfere with the machines and construction site logistics, and so errors due to displacement of 'wood' and 'string' are precluded.

Worked examples

Example 4.10. The majority of short-range EDM equipment measures the difference in phase of the transmitted and reflected light waves on two different frequencies, in order to obtain distance.

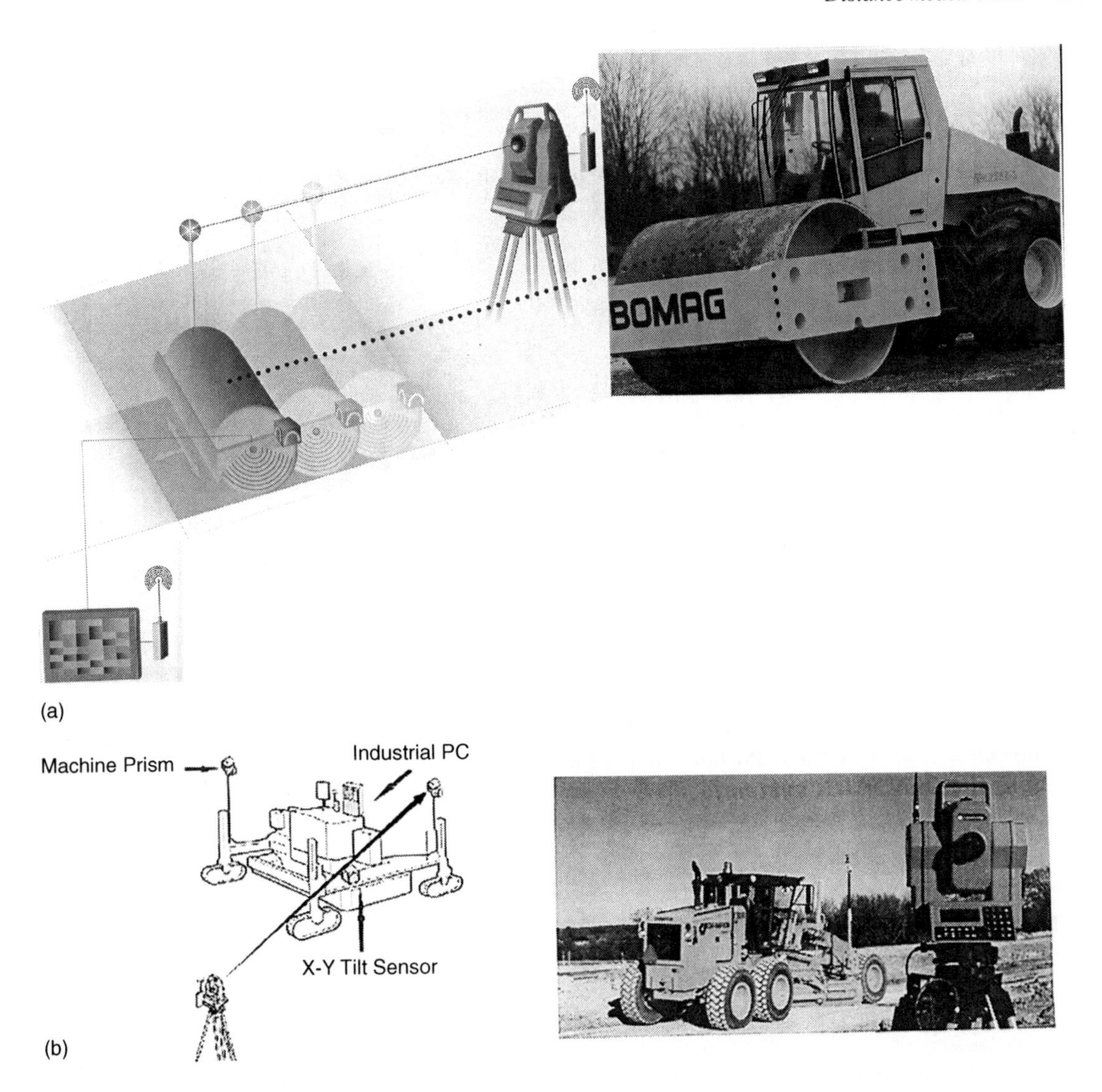

Fig. 4.31 *Machine guidance by robotic surveying system*

The frequencies generally used are 15 MHz and 150 kHz. Taking the velocity of light as 299 793 km/s and a measure distance of 346.73 m, show the computational processes necessary to obtain this distance, clearly illustrating the phase difference technique.

Travel distance $= 2 \times 346.73 = 693.46$ m

Travel time of a single pulse $= t = D/V = 693.46/299\,793\,000$

$= 2.313\ \mu s = 2313$ ns

Standard frequency $= f = 15$ MHz $= 15 \times 10^6$cycles/s

Time duration of a single pulse $= 1/15 \times 10^6$ s $= 66.\bar{6}$ ns $= t_p$.

$\therefore$ No. of pulses in the measured distance $= t/t_p = 2313/66.\bar{6} = 34.673$

i.e. $2D = M\lambda + \delta\lambda = 34\lambda + 0.673\lambda$

However, only the phase difference $\delta\lambda$ is known and not the value of M; hence the use of a second frequency.

A single pulse (λ) takes 66.6 ns, which at 15 MHz = 20 m, and $\lambda/2 = 10$ m.

$$\therefore D = M(\lambda/2) + 0.673(\lambda/2) = 6.73 \text{ m}$$

Now using $f = 150$ kHz $= 150 \times 10^3$ cycles/s, the time duration of a single pulse $= 1/150 \times 10^3 = 6.667$ μs.

$$\text{At 150 kHz, } 6.6674 \ \mu\text{s} = 1000 \text{ m}$$

$$\text{No. of pulses} = t/t_p = 2.313/6.667 = 0.347$$

$$\therefore D = 0.347 \times 1000 \text{ m} = 347 \text{ m}$$

$$\text{Fine measurement using 15 MHz} = 6.73 \text{ m}$$

$$\text{Coarse measurement using 15 kHz} = 347 \text{ m}$$

$$\text{Measured distance} = 346.73 \text{ m}$$

Example 4.11. (a) Using EDM, top-mounted on a theodolite, a distance of 1000 m is measured on an angle of inclination of 09° 00′ 00″. Compute the horizontal distance.

Now, taking $R = 6.37 \times 10^6$ m and the coefficient of refraction $K = 1.10$, correct the vertical angle for refraction effects, and recompute the horizontal distance.

(b) If the EDM equipment used above was accurate to ± 3 mm ± 5 ppm, calculate the required accuracy of the vertical angle, and thereby indicate whether or not it is necessary to correct it for refraction.

(c) Calculate the equivalent error allowable in levelling the two ends of the above measured line.

(KU)

(a) Horizontal distance $= D = S\cos\alpha = 1000\cos 9° = 987.688$ m

$$\hat{r}'' = SK\rho/2R = 987.69 \times 1.10 \times 206\,265/2 \times 6\,370\,000 = 17.6''$$

$$\text{Corrected angle} = 8°\ 59'\ 42''$$

$$\therefore D = 1000\cos 8°\ 59'\ 42'' = 987.702 \text{ m}$$

$$\text{Difference} = 14 \text{ mm}$$

(b) Distance is accurate to $\pm(3^2 + 5^2)^{\frac{1}{2}} = \pm 5.8$ mm

$$D = S\cos\alpha$$
$$\delta D = -S\sin\alpha\,\delta\alpha$$
$$\delta\alpha'' = \delta D\rho/S\sin\alpha = 0.0058 \times 206\,265/10\,000\sin 9° = \pm 7.7''$$

Vertical angle needs to be accurate to ±7.7″, so refraction must be catered for.

(c) To reduce S to D, apply the correction $-h^2/2S = C_h$

$$\delta C_h = h\,\delta h/S$$

$$\text{and} \quad \delta h = \delta C_h S/h$$

$$\text{where} \quad h = S\sin\alpha = 1000\sin 9° = 156.43 \text{ m}$$

$$\delta h = 0.0058 \times 1000/156.43 = \pm 0.037 \text{ m}$$

Example 4.12.

(a) Modern total stations supply horizontal distance (D) and vertical height (ΔH) at the press of a button. What corrections must be applied to the initial field data of slope distance and vertical angle to obtain the best possible values for D and ΔH?

(b) When using EDM equipment of a particular make, why is it inadvisable to use reflectors from other makes of instrument?

(c) To obtain the zero error of a particular EDM instrument, a base line AD is split into three sections AB, BC and CD and measured in the following combinations:

$$AB = 20.512, AC = 63.192, AD = 153.303$$
$$BC = 42.690, BD = 132.803, CD = 90.1201$$

Using all possible combinations, compute the zero error: (KU)

(a) Velocity correction using temperature and pressure measurements at the time of measurement. Vertical angle corrected for refraction to give D and for Earth curvature to give ΔH.

(b) Instrument has a built-in correction for the reflector constant and may have been calibrated for that particular instrument/reflector combination.

(c)

$$20.512 + 42.690 - 63.192 = +0.010$$
$$63.192 + 90.120 - 153.303 = +0.009$$
$$20.512 + 132.803 - 153.303 = +0.012$$
$$42.690 + 90.120 - 132.803 = +0.007$$
$$\text{Mean} = +0.009 \text{ m}$$
$$\text{Correction} = -0.009 \text{ m}$$

Example 4.13. Manufacturers specify the accuracy of EDM equipment as

$\pm a \pm bD$ mm

where b is in ppm of the distance measured, D.

Describe in detail the various errors defined by the variables a and b. Discuss the relative importance of a and b with regard to the majority of measurements taken in engineering surveys.

What calibration procedures are required to minimize the effect of the above errors in EDM measurement. (KU)

For answer, refer to appropriate sections of the text.

5
Angle measurement

As shown in Chapter 1, horizontal and vertical angles are fundamental measurements in surveying.

The vertical angle, as already illustrated, is used in obtaining the elevation of points (trig levelling) and in the reduction of slant distance to the horizontal.

The horizontal angle is used primarily to obtain direction to a survey control point, or to topographic detail points, or to points to be set out.

An instrument used for the measurement of angles is called a theodolite, the horizontal and vertical circles of which can be likened to circular protractors set in horizontal and vertical planes. It follows that, although the points observed are at different elevations, it is always the horizontal angle and not the space angle which is measured. For example, observations to points A and C from B (*Figure 5.1*) will give the horizontal angle $ABC = \theta$. The vertical angle of elevation to A is α and its zenith angle is Z_A.

5.1 THE THEODOLITE

There are basically two types of theodolite, the optical mechanical type or the electronic digital type, both of which may be capable of reading directly to $1'$, $20''$, $1''$ or $0.1''$ of arc, depending upon the precision of the instrument. The selection of an instrument specific to the survey tolerances of the work in hand is usually overridden by the commercial considerations of the company and a $1''$ instrument may be used for all work. When one considers that $1''$ of arc subtends 1 mm in 200 m, it is sufficiently accurate for practically all work carried out in engineering.

Figure 5.2 shows a typical theodolite, whilst *Figure 5.3* shows the main components of a theodolite. This exploded diagram enables the relationships of the various parts to be more clearly understood along with the relationships of the main axes. In a correctly adjusted instrument these axes should all be normal to each other, with their point of intersection being the point about which the angles are measured. Neither figure illustrates the complexity of a modern theodolite or the very high calibre of the process of its production.

The basic features of a typical theodolite are, with reference to *Figure 5.3*, as follows:

(1) The trivet stage forming the base of the instrument connects it to the tripod head.
(2) The tribrach supports the rest of the instrument and with reference to the plate bubble can be levelled using the footscrews which act against the fixed trivet stage.
(3) The lower plate carries the horizontal circle which is made of glass, with graduations from 0° to 360° photographically etched around the perimeter. This process enables lines of only 0.004 mm thickness to be sharply defined on a small-diameter circle (100 mm), thereby resulting in very compact instruments.
(4) The upper plate carries the horizontal circle index and fits concentrically with the lower plate.
(5) The plate bubble is attached to the upper plate and when adjusted, using the footscrews, makes the instrument axis vertical. Some modern digital or electronic theodolites have replaced the spirit bubble with an electronic bubble.

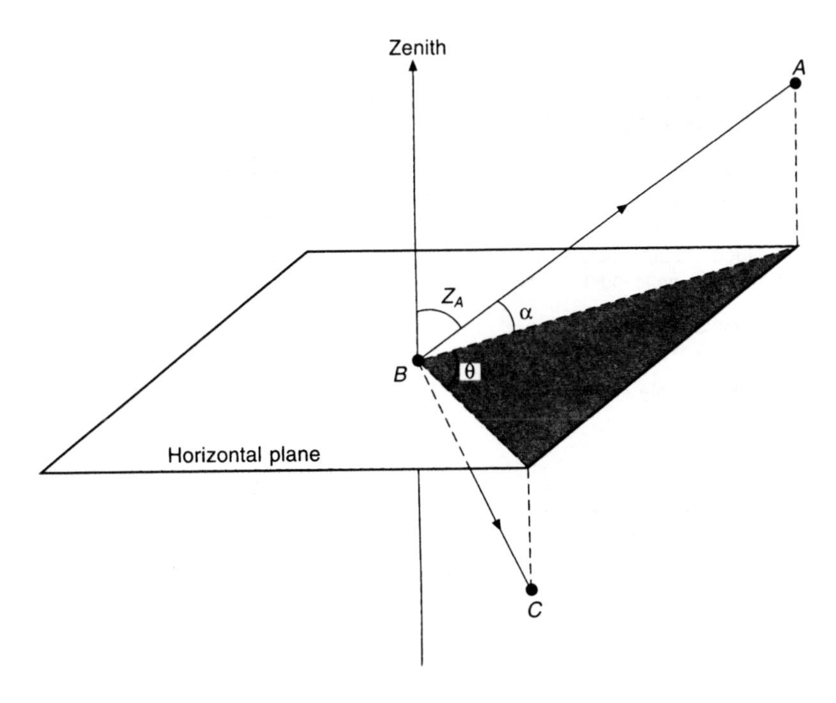

Fig. 5.1 *Horizontal, vertical and zenith angles*

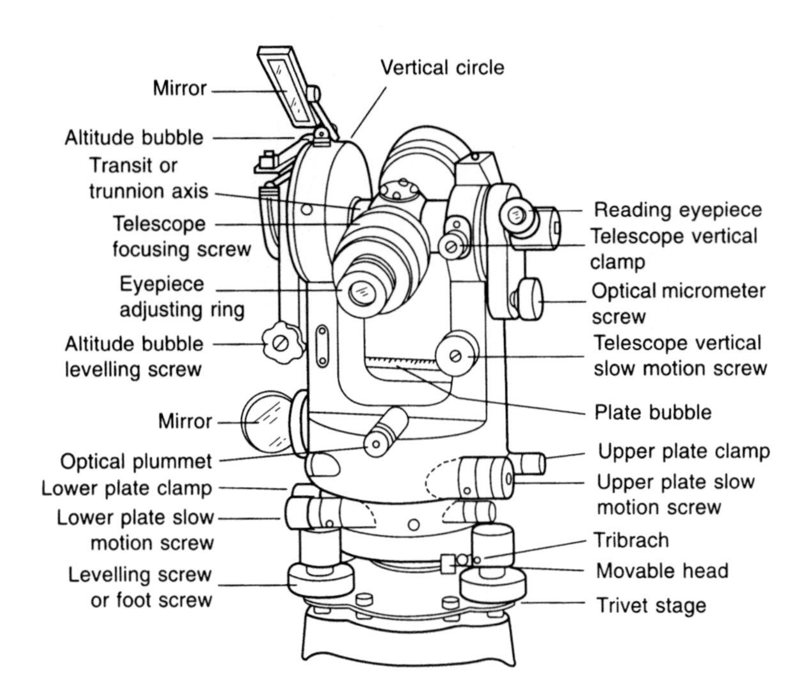

Fig. 5.2 *Typical optical mechanical theodolite*

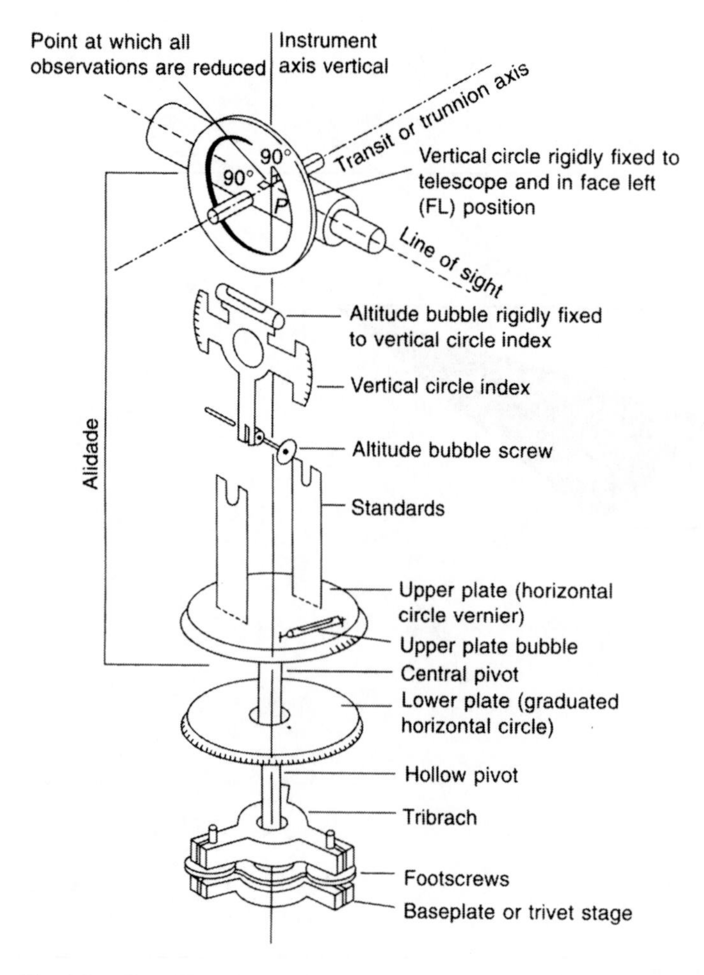

Fig. 5.3 *Simplified theodolite*

(6) The upper plate also carries the standards which support the telescope by means of its transit axis. The standards are tall enough to allow the telescope to be fully rotated about its transit axis.
(7) The vertical circle, similar in construction to the horizontal circle, is fixed to the telescope axis and rotates with the telescope.
(8) The vertical circle index, against which the vertical angles are measured, is set normal to gravity by means of (a) an altitude bubble attached to it, or (b) an automatic compensator. The latter method is now universally employed in modern theodolites.
(9) The lower plate clamp (*Figure 5.2*) enables the horizontal circle to be clamped into a fixed position. The lower plate slow motion screw permits slow movement of the theodolite around its vertical axis, when the lower plate clamp is clamped. Most modern theodolites have replaced the lower plate clamp and slow motion screw with a horizontal circle-setting screw. This single screw rotates the horizontal circle to any reading required.
(10) Similarly, the upper plate clamp and slow motion screw have the same effect on the horizontal circle index.
(11) The telescope clamp and slow motion screw fix and allow fine movement of the telescope in the vertical plane.

(12) The altitude bubble screw centres the altitude bubble, which, as it is attached to the vertical circle index, makes it horizontal prior to reading the vertical circle. As stated in (8), this is now done by means of an automatic compensator.
(13) The optical plummet, built into either the base of the instrument or the tribrach (*Figure 5.12*), enables the instrument to be centred precisely over the survey point. The line of sight through the plummet is coincidental with the vertical axis of the instrument.
(14) The telescopes are similar to those of the optical level but usually shorter in length. They also possess rifle sights or collimators for initial pointing.

5.1.1 Reading systems

The theodolite circles are generally read by means of a small auxiliary reading telescope at the side of the main telescope (*Figure 5.2*). Small circular mirrors reflect light into the complex system of lenses and prisms used to read the circles.

There are basically three types of reading system; optical scale reading, optical micrometer reading and electronic digital display.

(1) The optical scale reading system is generally used on theodolites with a resolution of 20″ or less. Both horizontal and vertical scales are simultaneously displayed and are read directly with the aid of the auxiliary telescope.

The telescope used to give the direct reading may be a 'line microscope' or a 'scale microscope'.

The line microscope uses a fine line etched onto the graticule as an index against which to read the circle.

The scale microscope has a scale in its image plane, whose length corresponds to the line separation of the graduated circle. *Figure 5.4* illustrates this type of reading system and shows the scale from 0′ to 60′ equal in scale of one degree on the circle. This type of instrument is frequently referred to as a direct-reading theodolite and, at best, can be read, by estimation, to 20″.
(2) The optical micrometer system generally uses a line microscope, combined with an optical micrometer using exactly the same principle as the parallel plate micrometer on a precise level.

Figure 5.5 illustrates the principle involved. If the observer's line of sight passes at 90° through the parallel plate glass, the circle reading would be 23° 20′ + S, with the value of S unknown. The parallel plate is rotated using the optical micrometer screw (*Figure 5.2*) until the line of sight is at an exact reading of 23° 20′ on the circle. This is as a result of the line of sight being refracted towards the normal and emerging on a parallel path. The distance S through which the observer's line of sight was displaced is recorded on the micrometer scale as 11′ 40″.

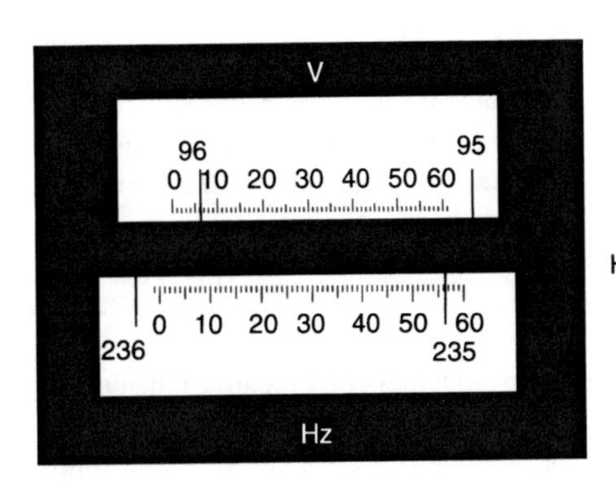

Vertical 96° 06′30″
Horizontal 235° 56′30″

Fig. 5.4 *Wild T16 direct reading theodolite*

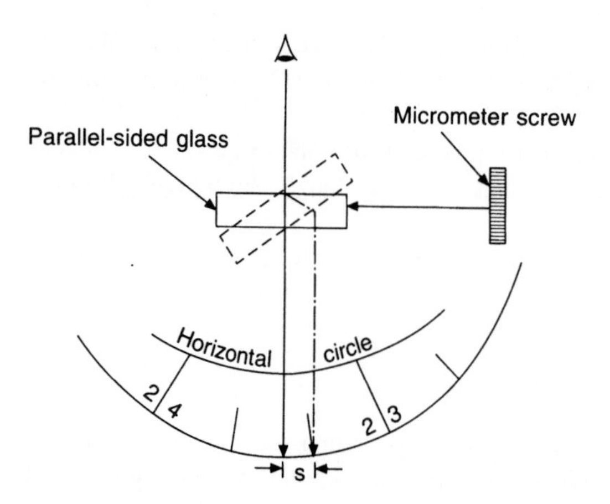

Fig. 5.5 *Micrometer*

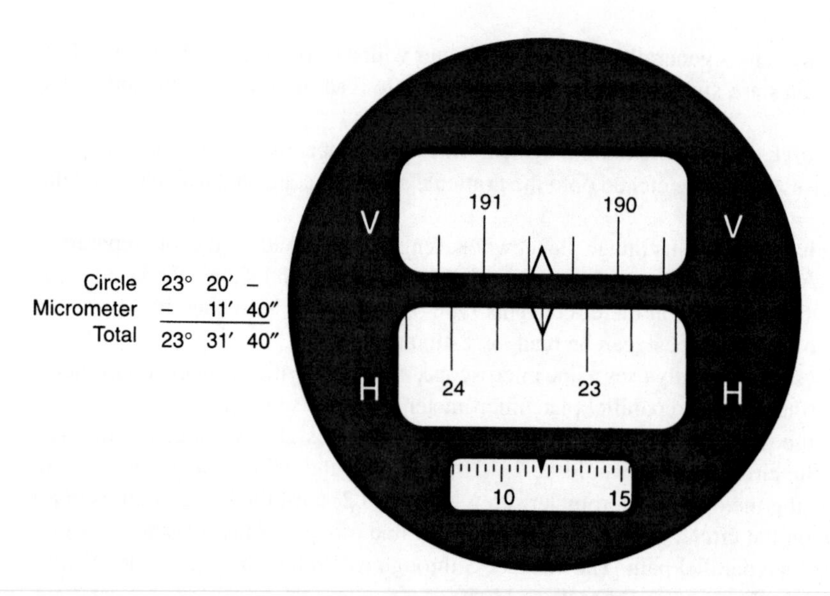

Fig. 5.6 *Watts Microptic No 1 theodolite reading system*

The shift of the image is proportional to the angle of tilt of the parallel plate and is read on the micrometer scale. Before the scale can be read, the micrometer must be set to give an exact reading (23° 20′), as shown on *Figure 5.6*, and the micrometer scale reading (11′ 40″) added on. Thus the total reading is 23° 31′ 40″. In this instance the optical micrometer reads only one side of the horizontal circle, which is common to 20″ instruments.

On more precise theodolites, reading to 1″ of arc and above, a coincidence microscope is used. This enables diametrically opposite sides of the circle to be combined and a single mean reading taken. This mean reading is therefore free from circle eccentricity error.

Figure 5.7 shows the diametrically opposite scales brought into coincidence by means of the optical micrometer screw. The number of divisions on the main scale between 94° and 95° is three; therefore each division represents 20′. The indicator mark can only take up one of two positions, either mid-division or on a full division. In this case it is mid-division and represents a reading of 94° 10′;

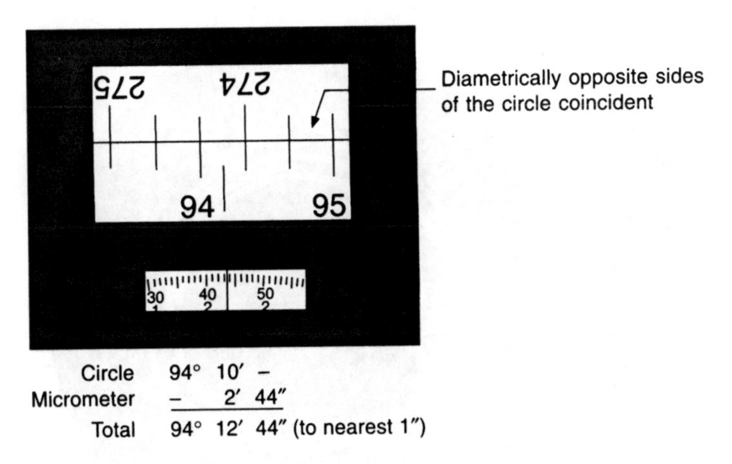

Fig. 5.7 *Wild T2 (old pattern) theodolite reading system*

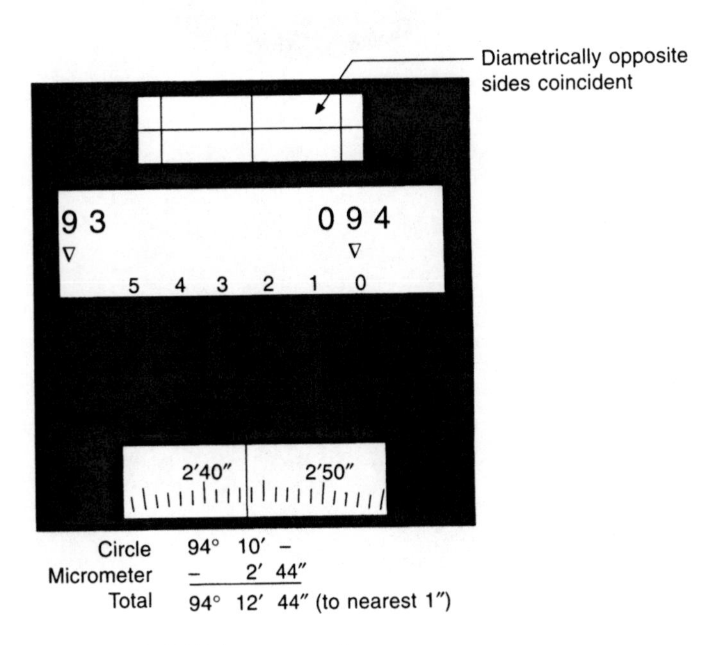

Fig. 5.8 *Wild T2 (new pattern)*

the micrometer scale reads 2′ 44″ to the nearest second, giving a total reading of 94° 12′ 44″. An improved version of this instrument is shown in *Figure 5.8*.

The above process is achieved using two parallel plates rotating in opposite directions, until the diametrically opposite sides of the circle coincide.

(3) There are basically two systems used in the electro-optical scanning process, either the incremental method or the code method (*Figure 5.9*).

The basic concept of the incremental method can be illustrated by considering a glass circle of 70–100 mm diameter, graduated into a series of radial lines. The width of these photographically etched lines is equal to their spacing. The circle is illuminated by a light diode; a photodiode, equal in width to a graduation, forms an index mark. As the alidade of the instrument rotates, the glass circle moves in relation to the diode. The light intensity signal radiated approximates to a sine curve.

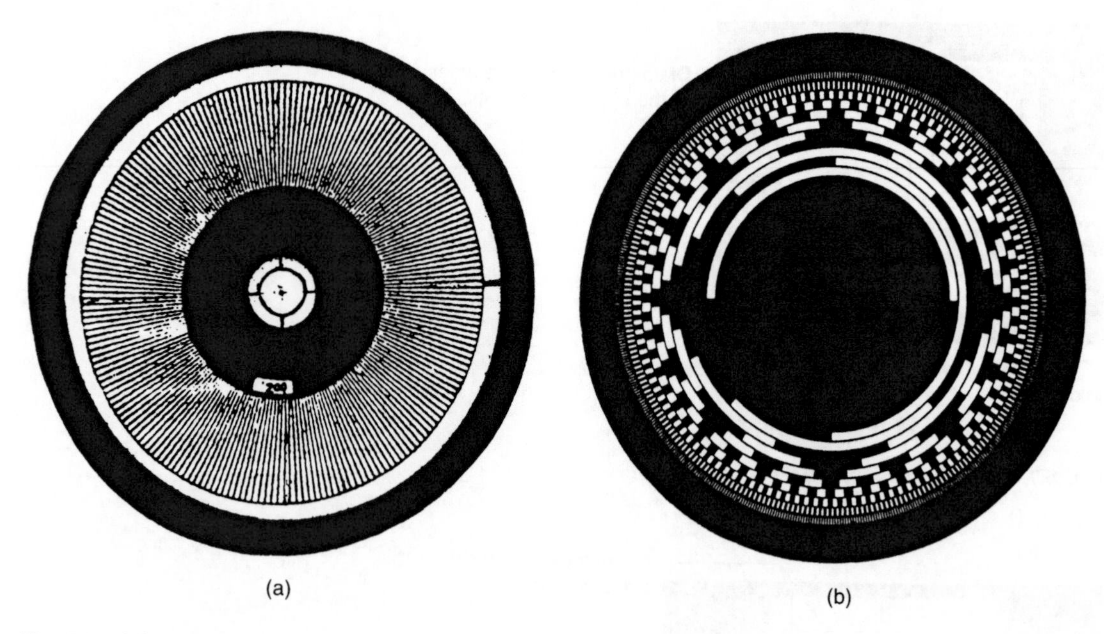

Fig. 5.9 *(a) Incremental disk, (b) binary coded disk*

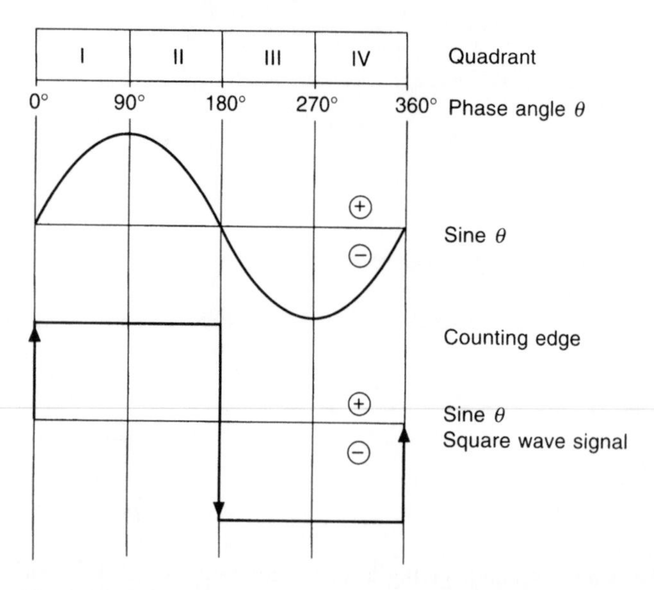

Fig. 5.10 *Sine wave to square wave modulation*

The diode converts this to an electrical signal correspondingly modulated to a square wave signal (*Figure 5.10*). The number of signal periods is counted by means of the leading and trailing edges of the square wave signal and illustrated digitally in degrees, minutes and seconds on the LCD. This simplified arrangement would produce a relatively coarse least count resolution, requiring further refinement.

For example, consider a glass circle that contains 20 000 radial marks, each 5.5 μm thick, with equal width spacing. A section of the circle comprising 200 marks is superimposed on the diametrically

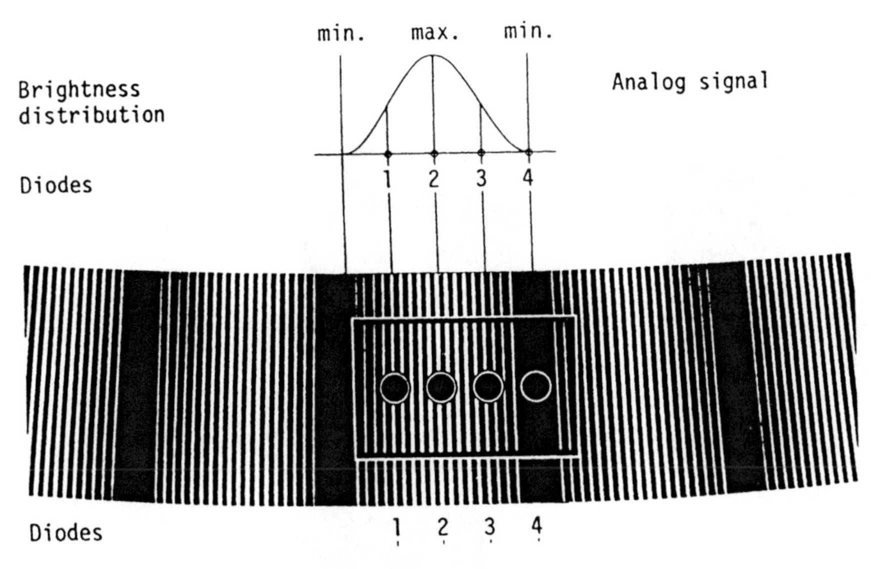

Fig. 5.11 *Fine reading using moiré pattern brightness*

opposite section, forming a moiré pattern. A full period (light–dark variation) corresponds to an angular value of approximately 1 min of arc, with a physical length of 2 mm. A magnification of this period by two provides a length of 4 mm over which the brightness pattern can be electronically scanned. Thus the coarse measurement can be obtained from 40 000 periods per full circle, equivalent to 30″ per period.

The fine reading to 0.3″ is obtained by monitoring the brightness distribution of the moiré pattern using the four diodes shown (*Figure 5.11*). The fine measurement obtains the scanning position location with respect to the leading edge of the square wave form within the last moiré pattern. It is analogous to measuring the fraction of a wavelength using the phase angle in EDM measurement.

The code methods use coded graduated circles (*Figure 5.9(b)*). Luminescent diodes above the glass circle and photodiodes below, one per track, detect the light pattern emitted, depending on whether a transparent track (signal 1) or an opaque track (signal 0) is opposite the diode at that instant. The signal is transferred to the computer for processing into a digital display. If there are n tracks, the full circle is divided into 2^n equal sectors. Thus a 16-track disk has an angular resolution of 2^{16}, which is 65 532 parts of a full circle and is equivalent to a 20″ resolution.

The advantage of the electronic systems over the glass arc scales is that they produce a digital output free from misreading errors and in a form suitable for automatic data recording and processing. *Figure 5.12* illustrates the glass arc and electronic theodolites.

5.2 INSTRUMENTAL ERRORS

In order to achieve reliable measurement of the horizontal and vertical angles, one must use an instrument that has been properly adjusted and adopt the correct field procedure.

In a properly adjusted instrument, the following geometrical relationships should be maintained (*Figure 5.3*):

(1) The plane of the horizontal circle should be normal to the vertical axis of rotation.
(2) The plane of the vertical circle should be normal to the horizontal transit axis.

(a)

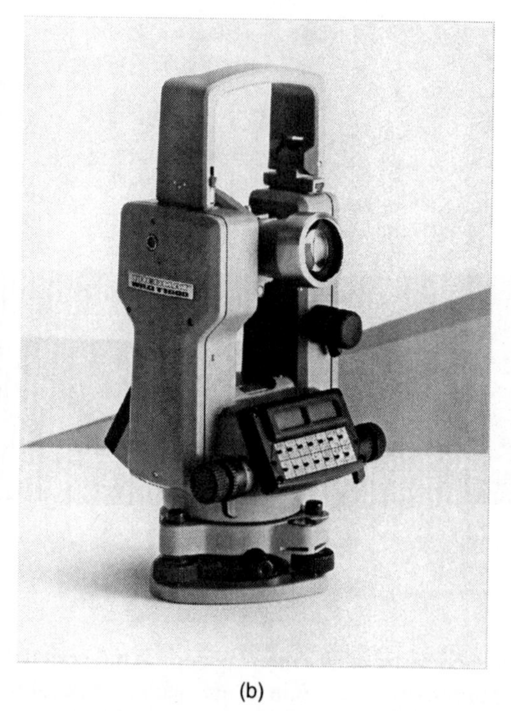

(b)

Fig. 5.12 *(a) Wild T1 glass arc theodolite with optical plummet in the alidade, (b) Wild T1600 electronic theodolite with optical plummet in the tribrach*

(3) The vertical axis of rotation should pass through the point from which the graduations of the horizontal circle radiate.
(4) The transit axis of rotation should pass through the point from which the graduations of the vertical circle radiate.
(5) The principal tangent to the plate bubble should be normal to the main axis of rotation.
(6) The line of sight should be normal to the transit axis.
(7) The transit axis should be normal to the main axis of rotation.
(8) When the telescope is horizontal, the vertical circle indices should be horizontal and reading zero, and the principal tangent of the altitude bubble should, at the same instance, be horizontal.
(9) The main axis of rotation should meet the transit axis at the same point as the line of sight meets this axis.
(10) The line of sight should maintain the same position with change of focus (an important fact when coplaning wires).

Items (1), (2), (3) and (4) above are virtually achieved by the instrument manufacturer and no provision is made for their adjustment. Similarly, (9) and (10) are dealt with, as accurately as possible, in the manufacturing process and in any event are minimized by double face observations. Items (5), (6), (7) and (8) can, of course, be achieved by the usual adjustment procedures carried out by the operator.

The procedure referred to above as 'double face observation' is fundamental to the accurate measurement of angles. An examination of *Figure 5.2* shows that an observer looking through the eyepiece of the telescope would have the vertical circle on the left-hand side of his/her face; this would be termed a 'face left' (FL) observation. If the telescope is now rotated through 180° about its transit axis and then the instrument rotated through 180° about its vertical axis, the vertical circle would be on the right-hand side of the observer's face when looking through the telescope eyepiece. This is called a 'face right' (FR) observation.

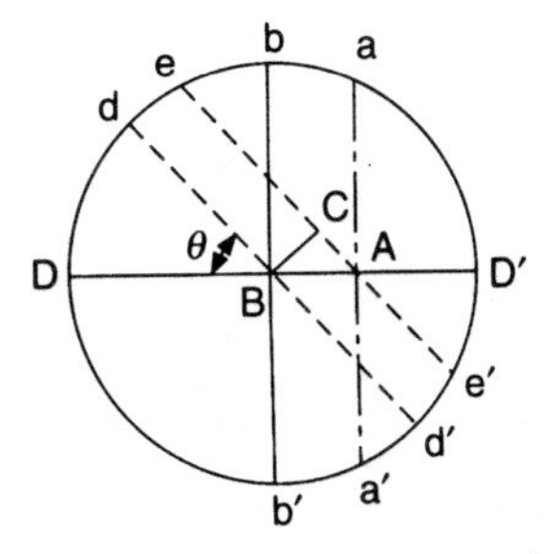

Fig. 5.13 *Eccentricity of centres*

The mean result of a FL and FR observation, called a double face observation, is free from the majority of instrumental errors present in the theodolite.

The main instrumental errors will now be dealt with in more detail and will serve to emphasize the necessity for double face observation.

5.2.1 Eccentricity of centres

This error is due to the centre of the central pivot carrying the alidade (upper part of the instrument) not coinciding with the centre of the hollow pivot carrying the graduated circle (*Figures 5.3* and *5.13*).

The effect of this error on readings is periodic. If B is the centre of the graduated circle and A is the centre about which the alidade revolves, then distance AB is interpreted as an arc ab in seconds on the graduated circle and is called the *error of eccentricity*. If the circle was read at D, on the line of the two centres, the reading would be the same as it would if there were no error. If, at b, it is in error by $ba = E$, the maximum error. In an intermediate position d, the error will be $de = BC = AB \sin\theta = E \sin\theta, \theta$ being the horizontal angle of rotation.

The horizontal circle is graduated clockwise, so, a reading supposedly at b will actually be at a, giving a reading too great by $+E$. On the opposite side, the reading of the point supposedly at b' will actually be at a', thereby reading too small by $-E$. Similarly for the intermediate positions at d and d', the errors will be $+E \sin\theta$ and $-E \sin\theta$. *Thus the mean of the two readings 180° apart, will be free of error.*

Glass-arc instruments in the 20″ class can be read on one side of the graduated circle only, thus producing an error which varies sinusoidally with the angle of rotation. The mean of the readings on both faces of the instrument would be free of error. With 1″ theodolites the readings 180° apart on the circle are automatically averaged and so are free of this error.

Manufacturers claim that this source of error does not arise in the construction of modern glass-arc instruments.

5.2.2 Collimation in azimuth

Collimation in azimuth error refers to the error which occurs in the observed angle due to the line of sight, or more correctly, the line of collimation, not being at 90° to the transit axis (*Figure 5.3*). If the line of sight in *Figure 5.14* is at right angles to the transit axis it will sweep out the vertical plane *VOA* when the telescope is elevated through the vertical angle α.

If the line of sight is not at right angles but in error by an amount e, the line of sight will describe a cone with its axis about the transit axis and an apex angle of just under 90°. So, in *Figure 5.14(b)*, the instrument is at O and the elevated target is at A which is where the line of sight points to. Because of the

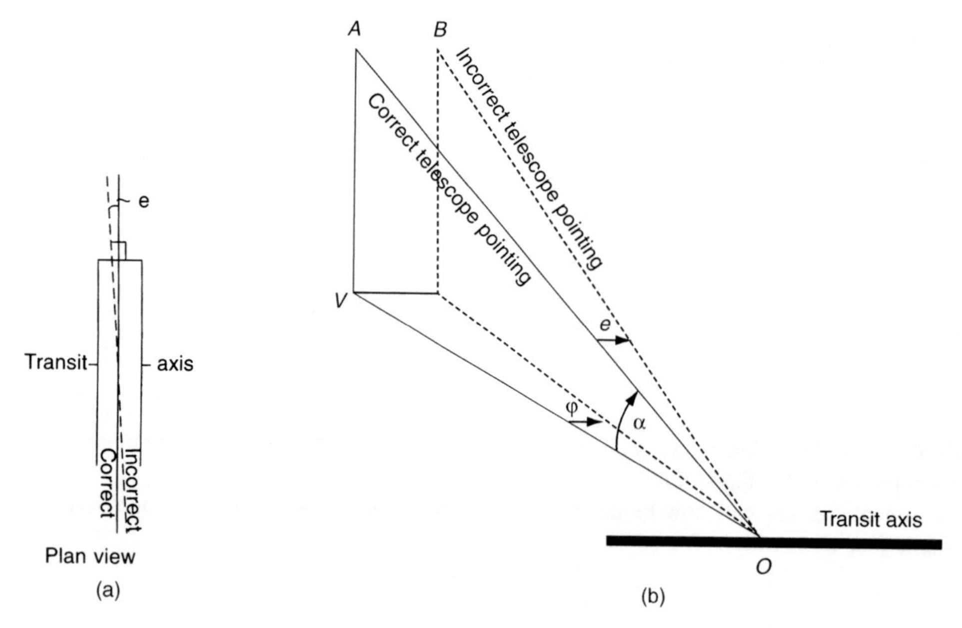

Fig. 5.14 *Collimation in azimuth*

collimation error the axis of the telescope actually points to B so the error in the horizontal pointing is $+\phi$ (positive because the horizontal circle is graduated clockwise).

$$\tan\phi = \frac{AB}{OV} = \frac{OA \tan e}{OV} \quad \text{but} \quad \frac{OA}{OV} = \sec\alpha$$

$$\therefore \tan\phi = \sec\alpha \tan e$$

as ϕ and e are very small, the above may be written

$$\phi = e \sec\alpha \tag{5.1}$$

On changing face VOB will fall to the other side of A and give an equal error of opposite sign, i.e. $-\phi$. *Thus the mean of readings on both faces of the instrument will be free of this error.*

ϕ is the error of one sighting to a target of elevation α. An angle, however, is the difference between two sightings; therefore the error in an angle between two objects of elevation, α_1 and α_2, will be $e(\sec\alpha_1 - \sec\alpha_2)$ and will obviously be zero if $\alpha_1 = \alpha_2$, or if measured in the horizontal plane, $(\alpha = 0)$.

On the opposite face the error in the angle simply changes sign to $-e(\sec\alpha_1 - \sec\alpha_2)$, indicating that the *mean of the two angles taken on each face will be free of error regardless of elevation.*

Similarly it can be illustrated that the true value of a vertical angle is given by $\sin\alpha = \sin\alpha_1 \cos e$ where α is the measured altitude and α_1 the true altitude. However, as e is very small, $\cos e \approx 1$, hence $\alpha_1 \approx \alpha$, proving that the effect of this error on vertical angles is negligible, provided that α is not close to 90°.

5.2.3 Transit axis error

Error will occur in the measurement of the horizontal angle if the transit axis is not at 90° to the instrument axis (*Figure 5.3*). When measuring, the instrument axis should be vertical. If the transit axis is set correctly at right angles to the vertical axis, then when the telescope is elevated it will sweep out the truly vertical

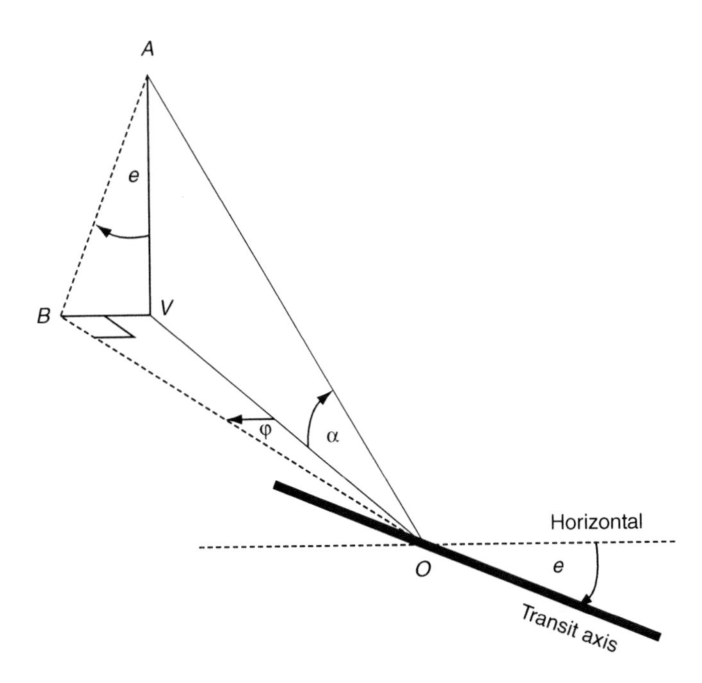

Fig. 5.15 *The effect of transit axis error on horizontal pointing*

plane AOV (*Figure 5.15*). Assuming the transit axis is inclined to the horizontal by e, the telescope will sweep out the plane AOB which is inclined to the vertical by e. This will create an error $-\phi$ in the horizontal reading of the theodolite (negative as the horizontal circle is graduated clockwise).

If α is the angle of elevation then

$$\sin\phi = \frac{BV}{VO} = \frac{AV\tan e}{VO} = \frac{VO\tan\alpha\tan e}{VO} = \tan\alpha\tan e \tag{5.2}$$

Now, as ϕ and e are small, $\phi = e\tan\alpha$.

From *Figure 5.15* it can be seen that the correction, ϕ, to the reading to the elevated target at A, due to a clockwise rotation of the transit axis, is negative because of the clockwise graduations of the horizontal circle. Thus, when looking through the telescope towards an elevated object, if the left-hand end of the transit axis is high, then the correction to the reading is negative, and *vice versa*.

On changing face, AOB will fall to the other side of A and give an equal error of opposite sign. *Thus, the mean of the readings on both faces of the instrument will be free from error.* As previously, the error in the measurement of an angle between two objects of elevations α_1 and α_2 will be

$$e(\tan\alpha_1 - \tan\alpha_2)$$

which on changing face becomes $-e(\tan\alpha_1 - \tan\alpha_2)$ indicating *that the mean of two angles taken one on each face, will be free from error regardless of the elevation of the target*. Also, if $\alpha_1 = \alpha_2$, or the angle is measured in the horizontal plane ($\alpha = 0$), it will be free from error.

Using *Figure 5.16*, if the observed vertical angle is α_1 then the correct vertical angle α can be found from

$$\sin\alpha = \frac{AV}{AO} = \frac{AB\sec e}{AO} = \frac{AO\sin\alpha_1\sec e}{AO} = \sin\alpha_1\sec e \tag{5.3}$$

As e is very small $\sec e \approx 1$, thus $\alpha_1 \approx \alpha$. The effect of this error on vertical angles is negligible provided that the vertical angle is not close to 90°.

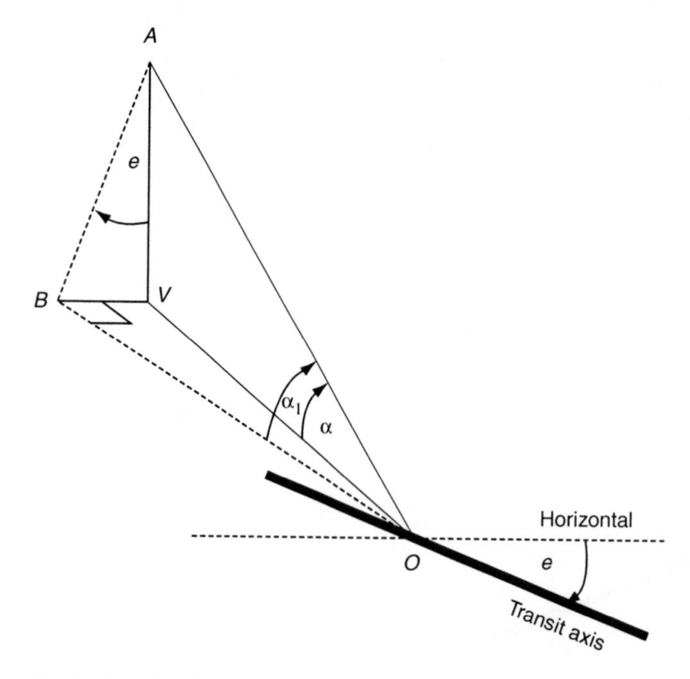

Fig. 5.16 *The effect of transit axis error on vertical angle*

5.2.4 Effect of non-verticality of the instrument axis

If the plate levels of the theodolite are not in adjustment, then the instrument axis will be inclined to the vertical, and hence measured azimuth angles will not be truly horizontal. Assuming the transit axis is in adjustment, i.e. perpendicular to the vertical axis, then error in the vertical axis of e will cause the transit axis to be inclined to the horizontal by e, producing an error in pointing of $\phi = e \tan \alpha$ as derived from *equation (5.2)*. Here, however, the *error is not eliminated by double-face observations* (*Figure 5.17*), but varies with different pointings of the telescope. For example, *Figure 5.18(a)* shows the instrument axis truly vertical and the transit axis truly horizontal. Imagine now that the instrument axis is inclined through e in a plane at 90° to the plane of the paper (*Figure 5.18(b)*). There is no error in the transit axis. If the alidade is now rotated clockwise through 90° into the plane of the paper, it will be as in *Figure 5.18(c)*, and when viewed in the direction of the arrow, will appear as in *Figure 5.18(d)* with the transit axis inclined to the horizontal by the same amount as the vertical axis, e. Thus, the error in the transit axis varies from zero to maximum through 90°. At 180° it will be zero again, and at 270° back to maximum in exactly the same position.

If the horizontal angle between the plane of the transit axis and the plane of dislevelment of the vertical axis is δ, then the transit axis will be inclined to the horizontal by $e \cos \delta$. For example, in *Figure 5.18(b)*, $\delta = 90°$, and therefore as $\cos 90° = 0$, the inclination of the transit axis is zero, as shown.

For an angle between two targets at elevations α_1 and α_2, in directions δ_1 and δ_2, the correction will be $e(\cos \delta_1 \tan \alpha_1 - \cos \delta_2 \tan \alpha_2)$. When $\delta_1 = \delta_2$, the correction is a maximum when α_1 and α_2 have opposite signs. When $\delta_1 = -\delta_2$, that is in opposite directions, the correction is maximum when α_1 and α_2 have the same sign.

If the instrument axis is inclined to the vertical by an amount e and the transit axis further inclined to the horizontal by an amount i, both in the same plane, then the maximum dislevelment of the transit axis on one face will be $(e + i)$, and $(e - i)$ on the reverse face (*Figure 5.19*). Thus, the correction to a

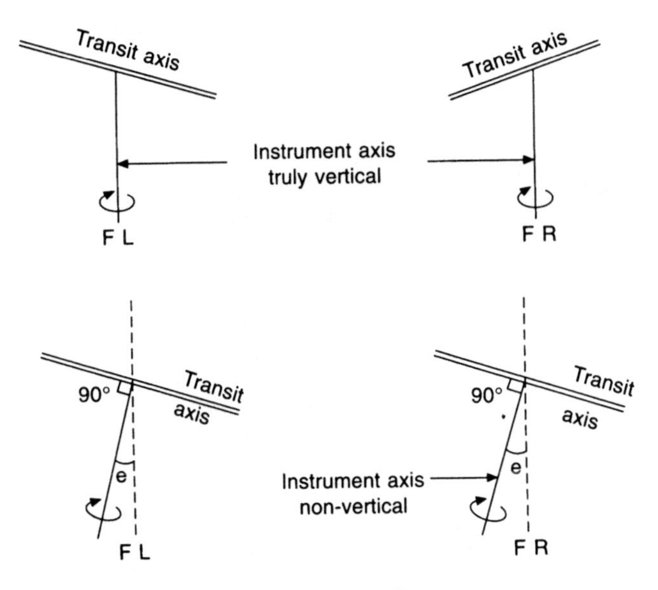

Fig. 5.17 *Non-vertical instrument axis*

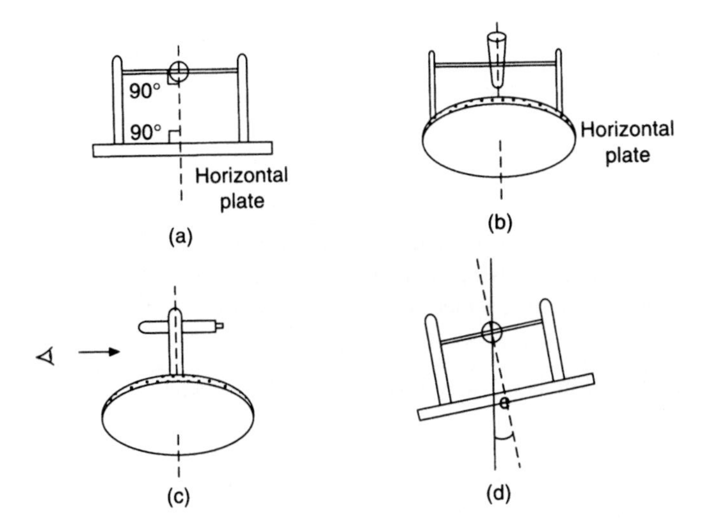

Fig. 5.18 *The effect of non-verticality of the instrument axis*

pointing on one face will be $(e + i)\tan\alpha$ and on the other $(e - i)\tan\alpha$, resulting in a correction of $e\tan\alpha$ to the mean of both face readings.

As shown, the resultant error increases as the angle of elevation α increases and is not eliminated by double face observations. As steep sights frequently occur in mining and civil engineering surveys, it is very important to recognize this source of error and adopt the correct procedures.

Thus, as already illustrated, the correction for a specific direction δ due to non-verticality (e) of the instrument axis is $e\cos\delta\tan\alpha$. The value of $e\cos\delta = E$ can be obtained from

$$E'' = S''\frac{(L - R)}{2} \tag{5.4}$$

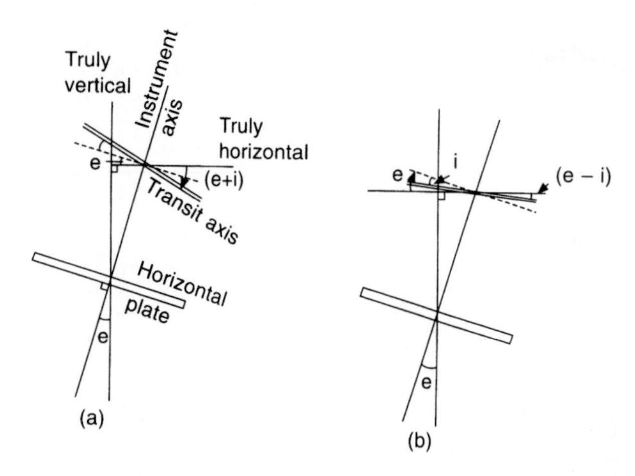

Fig. 5.19 *Non-vertical instrument axis. (a) Face left, (b) face right*

Fig. 5.20 *Reading a bubble*

where S'' = the sensitivity of the plate bubble in seconds of arc per bubble division
L and R = the left- and right-hand readings of the ends of the plate bubble measured out from the centre of the bubble when viewed from the eyepiece end of the telescope. See *Figure 5.20* for an example.

Then the correction to each horizontal circle reading is $C'' = E'' \tan\alpha$, and is positive when $L > R$ and *vice versa*.

For high-accuracy survey work, the accuracy of the correction C will depend upon how accurately E can be assessed. This, in turn, will depend on the sensitivity of the plate bubble and how accurately the ends of the bubble can be read. For very high accuracy involving extremely steep sights, an Electrolevel attached to the theodolite will measure axis tilt directly. This instrument has a sensitivity of 1 scale division equal to 1″ of tilt and can be read to 0.25 div. The average plate bubble has a sensitivity of 20″ per div.

Assuming that one can read each end of the plate bubble to an accuracy of ±0.5 mm, then for a bubble sensitivity of 20″ per (2 mm) div, on a vertical angle of 45°, the error in levelling the instrument (i.e. in the vertical axis) would be $\pm 0.35 \times 20'' \tan 45° = \pm 7''$. It has been shown that the accuracy of reading a bubble through a split-image coincidence system is about ten times greater. Thus, if the altitude bubble, usually viewed through a coincidence system, were used to level the theodolite, error in the axis tilt could be reduced to as little as ±0.7″ provided that the tripod is sufficiently stable.

More recent theodolites have replaced the altitude bubble with automatic vertical circle indexing with stabilization accuracies of ±0.3″. This may therefore be used for high-accuracy levelling of the instrument as follows:

(1) Accurately level the instrument using its plate bubble in the normal way.

(2) Clamp the telescope in any position, place the plane of the vertical circle parallel to two footscrews and note the vertical circle reading.
(3) With telescope remaining clamped, rotate the alidade through 180° and note the vertical circle reading.
(4) Using the two footscrews of (2) above, set the vertical circle to the mean of the two readings obtained in (2) and (3).
(5) Rotate through 90° and by using only the remaining footscrew obtain the same mean vertical circle reading.

The instrument is now precisely levelled to minimize axis tilt and virtually eliminate this source of error on all but the steepest sites.

Vertical angles are not affected significantly by non-verticality of the instrument axis as their horizontal axis of reference is established independently of the plate bubble.

5.2.5 Circle graduation errors

In the construction of the horizontal and vertical circles of a 1″ direct reading theodolite, the graduation lines on a 100-mm-diameter circle have to be set with an accuracy of 0.4 μm. In spite of the sophisticated manufacturing processes available, both regular and irregular errors of graduation occur.

It is possible to calibrate each instrument by producing error curves. However, with a 1″ direct reading theodolite such curves generally show maximum errors in the region of only ±0.3″. In practice, therefore, such errors are dealt with by observing the same angle on different parts of the circle, distributed symmetrically around the circumference. If the angle is to be observed 2, 4 or n times, where a double face measurement is regarded as a single observation, then the alidade is rotated through $180°/n$ prior to each new round of measurements.

5.2.6 Optical micrometer errors

When the optical micrometer is rotated from zero to its maximum position, then the displacement of the circle should equal the least count of the main scale reading. However, due to circle graduation error, plus optical and mechanical defects, this may not be so. The resultant errors are likely to be small but cyclic. Their effects can be minimized by using different micrometer settings.

5.2.7 Vertical circle index error

In the measurement of a vertical angle it is important to note that the vertical circle is attached to and rotates with the telescope. The vertical circle reading is relevant to a fixed vertical circle index which is rendered horizontal by means of its attached altitude bubble (*Figure 5.3*) or by automatic vertical circle indexing.

Vertical circle index error occurs when the index is not horizontal. *Figure 5.21* shows the index inclined at e to the horizontal. The measured vertical angle on FL is M, which requires a correction of $+e$, while on FR the required correction is $-e$. The index error is thus eliminated by taking the mean of the FL and FR readings.

5.3 INSTRUMENT ADJUSTMENT

In order to maintain the primary axes of the theodolite in their correct geometrical relationship (*Figure 5.3*), the instrument should be regularly tested and adjusted. Although the majority of the resultant errors are minimized by double face procedures, this does not apply to plate bubble error. Also, many operations in

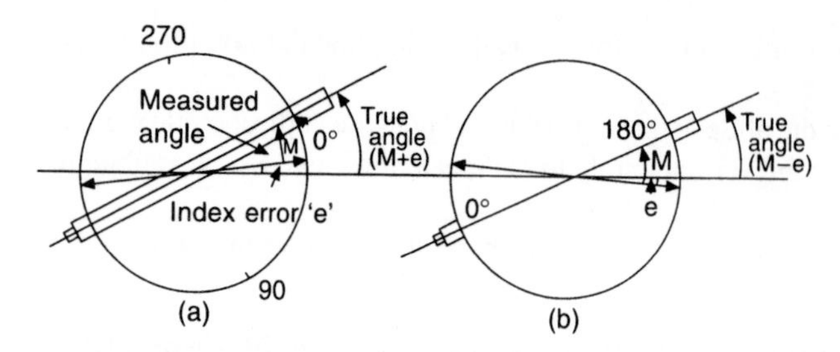

Fig. 5.21 *Vertical circle index error. (a) Face left, (b) face right*

engineering surveying are carried out on a single face of the instrument, and hence regular checking is important.

5.3.1 Tests and adjustments

(1) Plate level test

The instrument axis must be truly vertical when the plate bubble is centralized. The vertical axis of the instrument is perpendicular to the horizontal plate which carries the plate bubble. Thus to ensure that the vertical axis of the instrument is truly vertical, as defined by the bubble, it is necessary to align the bubble axis parallel to the horizontal plate.

Test: Assume the bubble is not parallel to the horizontal plate but is in error by angle e. It is set parallel to a pair of footscrews, levelled approximately, then turned through 90° and levelled again using the third footscrew only. It is now returned to its former position, accurately levelled using the pair of footscrews, and will appear as in *Figure 5.22(a)*. The instrument is now turned through 180° and will appear as in *Figure 5.22(b)*, i.e. the bubble will move off centre by an amount representing twice the error in the instrument ($2e$).

Adjustment: The bubble is brought half-way back to the centre using the pair of footscrews which are turned by a strictly equal and opposite amount. The bubble moves in the direction of the left thumb. See *Figure 5.28*. This will cause the instrument axis to move through e, thereby making it truly vertical and, in the event of there being no adjusting tools available, the instrument may be used at this stage. The

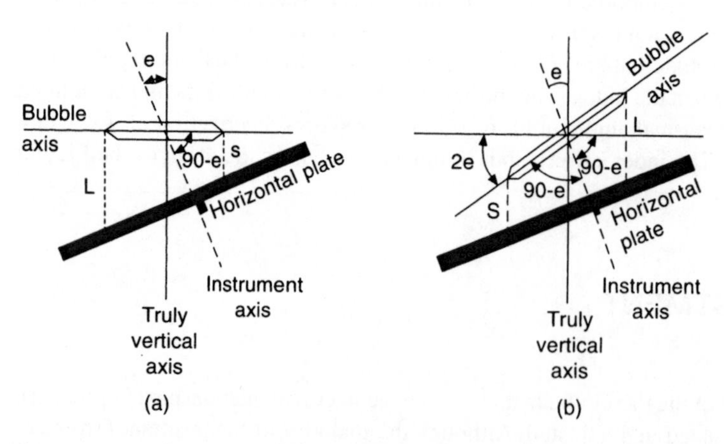

Fig. 5.22 *Misaligned plate bubble. (a) When levelled over two footscrews, (b) when turned through 180°*

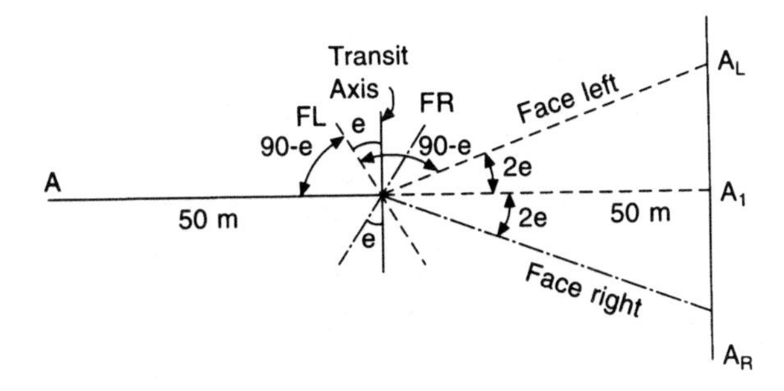

Fig. 5.23 *Collimation in azimuth*

bubble will still be off centre by an amount proportional to e, and should now be centralized by raising or lowering one end of the bubble using its capstan adjusting screws.

(2) Collimation in azimuth

The purpose of this test is to ensure that the line of sight is perpendicular to the transit axis.
Test: The instrument is set up, and levelled, and the telescope directed to bisect a fine mark at A, situated at instrument height about 50 m away (*Figure 5.23*). If the line of sight is perpendicular to the transit axis, then when the telescope is rotated vertically through 180°, it will intersect at A_1. However, assume that the line of sight makes an angle of $(90° - e)$ with the transit axis, as shown dotted in the face left (FL) and face right (FR) positions. Then in the FL position the instrument would establish a fine mark at A_L. Change face, re-bisect point A, transit the telescope and establish a fine mark at A_R. From the sketch it is obvious that distance $A_L A_R$ represents four times the error in the instrument ($4e$).

Adjustment: The cross-hairs are now moved in azimuth using their horizontal capstan adjusting screws, from A_R to a point mid-way between A_R and A_1; this is one-quarter of the distance $A_L A_R$.

This movement of the reticule carrying the cross-hair may cause the position of the vertical hair to be disturbed in relation to the transit axis; i.e. it should be perpendicular to the transit axis. It can be tested by traversing the telescope vertically over a fine dot. If the vertical cross-hair moves off the dot then it is not at right angles to the transit axis and is corrected with the adjusting screws.

This test is frequently referred to as one which ensures the verticality of the vertical hair, which will be true only if the transit axis is truly horizontal. However, it can be carried out when the theodolite is not levelled, and it is for this reason that a dot should be used and not a plumb line as is sometimes advocated.

(3) Spire test (transit axis test)

The spire test ensures that the transit axis is perpendicular to the vertical axis of the instrument.

Test: The instrument is set up and carefully levelled approximately 50 m from a well-defined point of high elevation, preferably greater than 30° (*Figure 5.24*). A well-defined point A, such as a church spire, is bisected and the telescope then lowered to its horizontal position and the vertical cross-hair is used to mark a point on a peg or a wall. If the transit axis is in adjustment the point will appear at A_1 directly below A. If, however, it is in error by the amount e (transit axis shown dotted in FL and FR positions), the mark will be made at A_L. The instrument is now changed to FR, point A bisected again and the telescope lowered to the horizontal, to fix point A_R. The distance $A_L A_R$ is twice the error in the instrument ($2e$).

Adjustment: Length $A_L A_R$ is bisected and a fine mark made at A_1. The instrument is now moved in azimuth, using a plate slow-motion screw until A_1 is bisected. Note that no adjustment of any kind has yet been

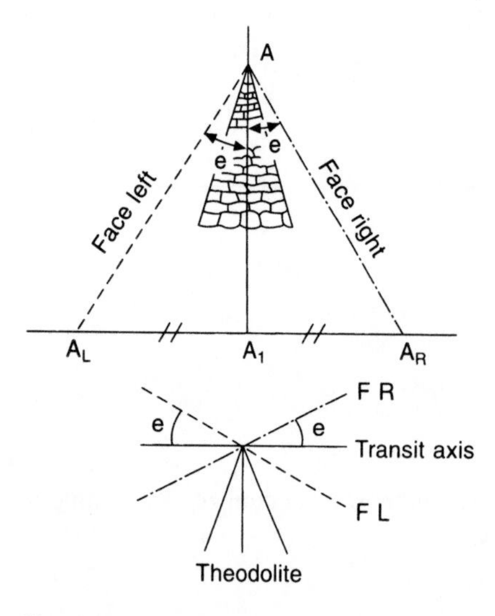

Fig. 5.24 *Spire test (transit axis test)*

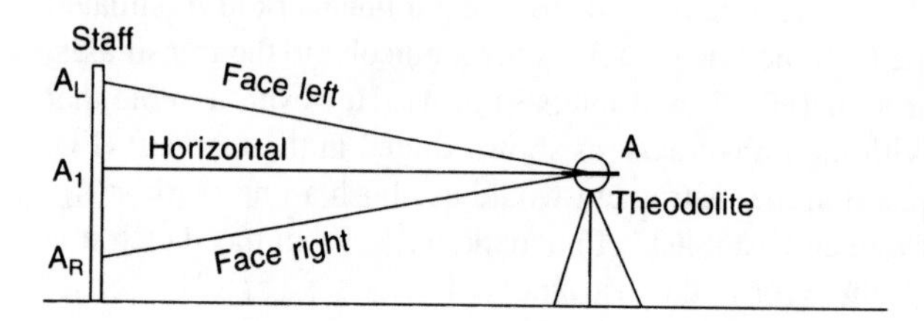

Fig. 5.25 *Vertical circle index test*

made to the instrument. Thus, when the telescope is raised back to A it will be in error by the horizontal distance $A_LA_R/2$. By moving one end of the transit axis using the adjusting screws, the line of sight is made to bisect A. This can only be made to bisect A when the line of sight is elevated. Movement of the transit axis when the telescope is in the horizontal plane A_LA_R, will not move the line of sight to A_1, hence the need to incline steeply the line of sight.

It should be noted that in modern instruments this adjustment cannot be carried out, i.e. there is no facility for moving the transit axis. Manufacturers claim that this error does not occur in modern equipment. Its effect is removed with the mean of FL and FR pointings. Alternatively the effect of the dislevelment of the transit axis can be computed by the same process and formulae described in Section 5.2.4.

(4) Vertical circle index test

This is to ensure that when the telescope is horizontal and the altitude bubble central, the vertical circle reads 0°, 90° or 270° depending on the instrument.

Test: Centralize the altitude bubble using the altitude bubble levelling screw and, by rotating the telescope, set the vertical circle to read zero (or its equivalent for a horizontal sight).

Note the reading on a vertical staff held about 50 m away. Change face and repeat the whole procedure. If error is present, a different reading on each face is obtained, namely A_L and A_R in *Figure 5.25*.

Adjustment: Set the telescope to read the mean of the above two readings, thus making the telescope truly horizontal. The vertical circle will then no longer read zero, and must be brought back to zero without affecting the horizontal pointing of the telescope. This is done by means of the bubble's capstan adjusting screws.

(5) Optical plummet

The line of sight through the optical plummet must coincide with the vertical instrument axis of the theodolite.

If the optical plummet is fitted in the alidade of the theodolite (*Figure 5.12(a)*), rotate the instrument through 360° in 90° intervals and make four marks on the ground. If the plummet is out of adjustment, the four marks will form a square, intersecting diagonals of which will give the correct point. Adjust the plummet cross-hairs to bisect this point.

If the plummet is in the tribrach it cannot be rotated. The instrument is set on its side, firmly on a stable table with the plummet viewing a nearby wall and a mark aligned on the wall. The tribrach is then turned through 180° and the procedure repeated. If the plummet is out of adjustment a second mark will be aligned. The plummet is adjusted to intersect the point midway between the two marks.

5.3.2 Alternative approach

(1) Plate level test

The procedure for this is as already described.

(2) Collimation in azimuth

With the telescope horizontal and the instrument carefully levelled, sight a fine mark and note the reading. Change face and repeat the procedure. If the instrument is in adjustment, the two readings should differ by exactly 180°. If not, the instrument is set to the corrected reading as shown below using the slow-motion screw; the line of sight is brought back on to the fine mark by adjusting the cross-hairs.

e.g. FL reading $01^\circ 30' 20''$
FR reading $180^\circ 31' 40''$

Difference $= 2e = 01' 20''$

$\therefore \quad e = +40''$

Corrected reading $= 181^\circ 31' 00''$ or $01^\circ 31' 00''$

(3) Spire test

First remove collimation in azimuth, then with the instrument carefully levelled, sight a fine point of high elevation and note the horizontal circle reading. Change face and repeat. If error is present, set the horizontal circle to the corrected reading, as above. Adjust the line of sight back on to the mark by raising or lowering the transit axis. (Not all modern instruments are capable of this adjustment.)

(4) Vertical circle index test

Assume the instrument reads 0° on the vertical circle when the telescope is horizontal and in FL position. Carefully level the instrument, make the altitude bubble horizontal and sight a fine point of high elevation. Change face and repeat. The two vertical circle readings should *sum* to 180°, any difference being twice the index error.

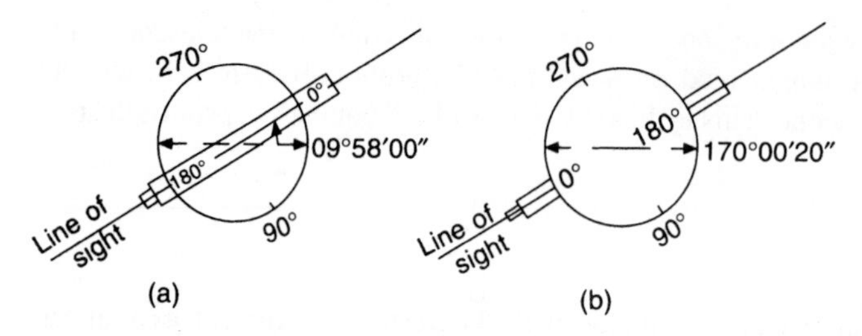

Fig. 5.26 *Vertical circle index test readings: (a) face left, (b) face right*

e.g. FL reading (*Figure 5.26(a)*)	09° 58′ 00″
FR reading (*Figure 5.26(b)*)	170° 00′ 20″
Sum =	179° 58′ 20″
Correct sum =	180° 00′ 00″
$2e =$	−01′ 40″
$e =$	−50″

Thus with the target still bisected, the vertical circle is set to read 170° 00′ 20″ + 50″ = 170° 01′ 10″ by means of the altitude bubble levelling screw. The altitude bubble is then centralized using its capstan adjusting screws. If the vertical circle reads 90° and 270° instead of 0° and 180°, the readings sum to 360°.

These alternative procedures have the great advantage of using the theodolite's own scales rather than external scales, and can therefore be carried out by one person.

5.4 FIELD PROCEDURE

The methods of setting up the theodolite and observing angles will now be dealt with. It should be emphasized, however, that these instructions are no substitute for practical experience.

5.4.1 Setting up using a plumb-bob

Figure 5.27 shows a theodolite set up with the plumb-bob suspended over the survey station. The procedure is as follows:

(1) Extend the tripod legs to the height required to provide comfortable viewing through the theodolite. It is important to leave at least 100 mm of leg extension to facilitate positioning of the plumb-bob.
(2) Attach the plumb-bob to the tripod head, so that it is hanging freely from the centre of the head.
(3) Stand the tripod approximately over the survey station, keeping the head reasonably horizontal.
(4) If the tripod has them, tighten the wing units at the top of the tripod legs and move the whole tripod until the plumb-bob is over the station.
(5) Now tread the tripod feet firmly into the ground.
(6) Unclamp a tripod leg and slide it in or out until the plumb-bob is exactly over the station. If this cannot be achieved in one movement, then use the slide extension to bring the plumb-bob in line with the survey point and another tripod leg. Using this latter leg, slide in or out to bring the plumb-bob onto the survey point.
(7) Remove the theodolite from its case and holding it by its standard, attach it to the tripod head.

Fig. 5.27 *Theodolite with plumb-bob*

(8) The instrument axis is now set truly vertical using the plate bubble as follows:

(a) Set the plate bubble parallel to two footscrews *A* and *B* as shown in (*Figure 5.28*(*a*)) and centre it by equal amounts of simultaneous contra-rotation of both screws. (The bubble follows the direction of the left thumb.)

(b) Rotate alidade through 90° (*Figure 5.28*(*b*)) and centre the bubble using footscrew *C* only.

(c) Repeat (a) and (b) until bubble remains central in both positions. If there is no bubble error this procedure will suffice. If there is slight bubble error present, proceed as follows.

(d) From the initial position at *B* (*Figure 5.28*(*a*)), rotate the alidade through 180°; if the bubble moves off centre bring if half-way back using the footscrews *A* and *B*.

(e) Rotate through a further 90°, placing the bubble 180° different to its position in *Figures 5.28*(*b*). If the bubble moves off centre, bring it half-way back with footscrew *C* only.

(f) Although the bubble is off centre, the instrument axis will be truly vertical and will remain so as long as the bubble remains the same amount off centre (*Section 5.3.2*).

(g) Test that the instrument has been correctly levelled by turning the instrument to any arbitrary direction. If the instrument is correctly levelled the bubble will remain in the same position within its vial no matter where the instrument is pointed.

(9) Check the plumb-bob; if it is off the survey point, slacken off the whole theodolite and shift it laterally across the tripod head, taking care not to allow it to rotate, until the plumb-bob is exactly over the survey point.

(10) Repeat (8) and (9) until the instrument is centred and levelled.

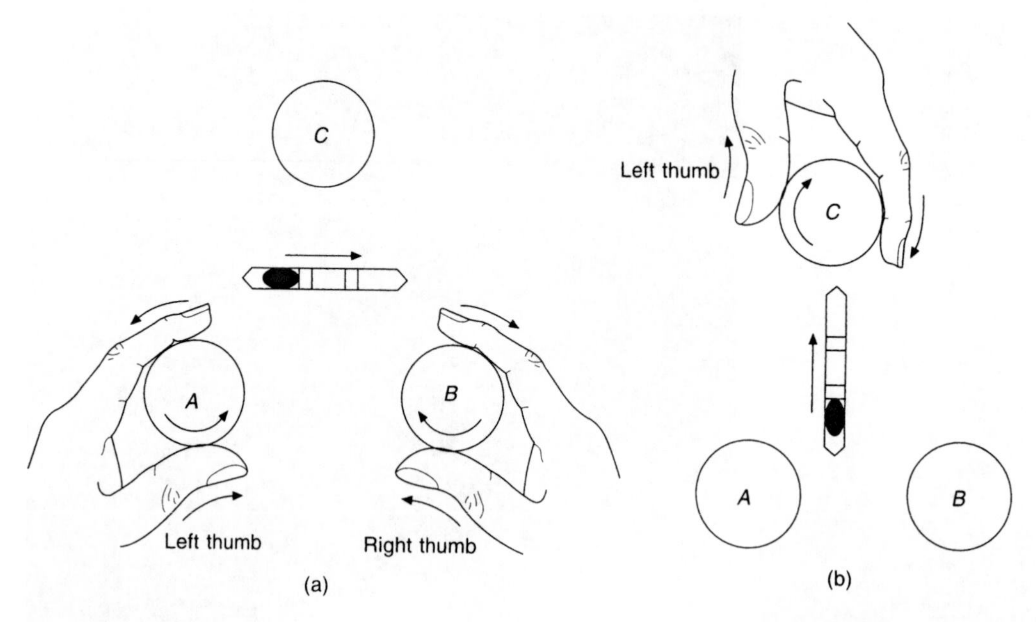

Fig. 5.28 *Footscrews*

5.4.2 Setting up using the optical plumb-bob

It is rare, if ever, that a theodolite is centred over the survey station using only a plumb-bob. All modern instruments have an optical plummet built into the alidade section of the instrument (*Figures 5.2* and *5.12(a)*), or into the tribrach section (*Figures 5.4* and *5.12(b)*). Proceed as follows:

(1) Establish the tripod roughly over the survey point using a plumb-bob as in (1) to (5) of *Section 5.4.1*.
(2) Depending on the situation of the optical plummet, attach the tribrach only, or the theodolite, to the tripod.
(3) Using the footscrews to incline the line of sight through the plummet, centre the plummet exactly on the survey point.
(4) Using the leg extension, slide the legs in or out until the circular bubble of the tribrach/theodolite is exactly centre. Even though the tripod movement may be excessive, the plummet will still be on the survey point. Thus the instrument is now approximately centred and levelled.
(5) Precisely level the instrument using the plate bubble, as described in (8) of *Section 5.4.1*.
(6) Unclamp and move the whole instrument laterally over the tripod until the plummet cross-hair is exactly on the survey point.
(7) Repeat (5) and (6) until the instrument is exactly centred and levelled.

5.4.3 Centring errors

Provided there is no wind, centring with a plumb-bob is accurate to $\pm 3-5$ mm. In windy conditions it is impossible to use unless protected in some way.

The optical plummet is accurate to $\pm 1-0.5$ mm, provided the instrument axis is truly vertical and is not affected by adverse weather conditions.

Forced centring or constrained centring systems are used to control the propagation of centring error in precise traversing. Such systems give accuracies in the region of $\pm 0.1-0.3$ mm. They will be dealt with in *Chapter 6*.

The effect of centring errors on the measured horizontal angle (θ) is shown in *Figure 5.29*.

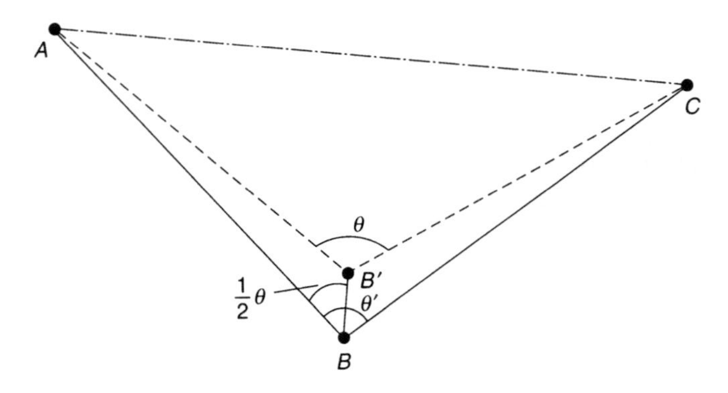

Fig. 5.29 *Centring error*

Due to a miscentring error, the theodolite is established at B', not the actual station B, and the angle θ is observed, not θ'. The maximum angular error (e_θ) occurs when the centring error BB' lies on the bisector of the measured angle and can be shown to be equal to:

$$(\theta - \theta') = e_\theta = 2e_C \tan(½\theta)(L_{AB} + L_{BC})/(L_{AB}L_{BC}) \text{ radians} \tag{5.5}$$

and is positive if B is further from the line AC than B' but negative if B is closer to the line AC than B'

where e_C = the centring error BB'
L_{AB}, L_{BC} = horizontal lengths AB and BC

The effect of target-centring errors on the horizontal angle at B can be obtained as follows. If the target errors are small then at A the target error is $e_{tA} = e_A/L_{AB}$ and similarly the target error at C is $e_{tC} = e_C/L_{BC}$ where e_A and e_C are the component parts of the error in target position at right angles to the lines BA and BC respectively to the right of their lines when viewed from B. The error in the angle would therefore be equal to the difference of these two errors:

$$e_{\theta t} = e_C/L_{BC} - e_A/L_{AB} \text{ radians} \tag{5.6}$$

It can be seen from *equations (5.5)* and *(5.6)* that as the lengths L decrease, the error in the measured angles will increase. Consider the following examples.

Worked examples

Example 5.1. In *Figure 5.29* in the triangle ABC, AB is 700 m and BC is 1000 m. If the error in centring the targets at A and C is 5 mm to the left of line AB as viewed from B and 5 mm to the right of the line BC as viewed from B, what will be the resultant error in the measured angle?

$$e_{\theta t}'' = (0.005/700 + 0.005/1000)\,206\,265 = 2.5''$$

In the scenario leading to *equation (5.6)* the erroneous positions are on the same side of the respective lines from B so the errors are partially self-cancelling, hence the minus sign in the equation. In this worked example the erroneous positions are on opposite sides of their lines and hence the errors add together.

Example 5.2. Consider the same question as in *Example 5.1* with $AB = 70$ m, $BC = 100$ m.

$$e_{\theta t}'' = (0.005/70 + 0.005/100)\,206\,265 = 25''$$

It can be seen that decreasing the lengths by a factor of 10 increases the angular error by the same factor.

Example 5.3. Assuming angle ABC is 90° and the centring error of the theodolite is 3 mm away from the line AC, what is the maximum error in the observed angle due to the centring error?

Then from *equation (5.5)*:

$$e_\theta'' = 206\,265 \times 2 \times 0.003 \times \left(\frac{1}{2}\right)^{\frac{1}{2}} \times (1700)/(700 \times 1000) = 2.1''$$

Example 5.4. If the situations in examples 5.2 and 5.3 both apply what is the total error in the observed angle.

$$e'' = e_\theta'' + e_{\theta t}'' = 2.1'' + 2.5'' = 4.6''$$

5.5 MEASURING ANGLES

Although the theodolite or total station is a very complex instrument the measurement of horizontal and vertical angles is a simple concept. The horizontal and vertical circles of the instrument should be regarded as circular protractors graduated from 0° to 360° in a clockwise manner. Then a simple horizontal angle measurement between three survey points A, B and C in the sense of measuring at A clockwise from B to C would be as shown in *Figure 5.30*.

(1) The instrument is set up and centred and levelled on survey point B. Parallax is removed.
(2) Commencing on, say, 'face left', the target set at survey point A is carefully bisected and the horizontal scale reading noted = 25°.
(3) The instrument is rotated to survey point C which is bisected. The horizontal scale reading is noted = 145°.
(4) The horizontal angle is then the difference of the two directions, i.e. Forward Station (C) minus Back Station (A), (FS − BS) = (145° − 25°) = 120°.
(5) Change face and observe survey point C on 'face right', and note the reading = 325°.
(6) Swing to point A, and note the reading = 205°.
(7) The readings or directions must be subtracted in the same order as in (4), i.e. $C - A$.

Thus (325° − 205°) = 120°

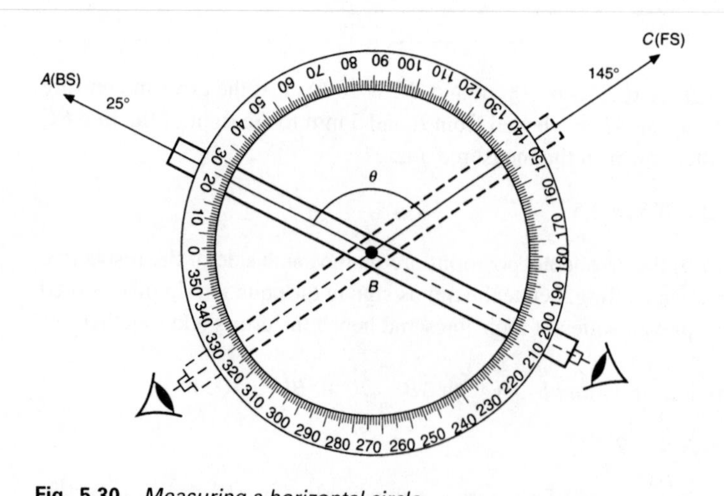

Fig. 5.30 *Measuring a horizontal circle*

(8) Note how changing face changes the readings by 180°, thus affording a check on the observations. The mean of the two values would be accepted if they are in good agreement.
(9) Try to use the same part of the vertical hair when pointing to the target. If the target appears just above the central cross on FL it should appear just below the central cross on FR. This will minimize the effect of any residual rotation of the cross-hairs.

Had the BS to *A* read 350° and the FS to *C* 110°, it can be seen that 10° has been swept out from 350° to 360° and then from 360° or 0° to 110°, would sweep out a further 110°. The total angle is therefore $10° + 110° = 120°$ or $(\text{FS} - \text{BS}) = [(110° + 360°) - 350°] = 120°$.

A further examination of the protractor shows that $(\text{BS} - \text{FS}) = [(25° + 360°) - 145°] = 240°$, producing the external angle. It is thus the manner in which the data are reduced that determines whether or not it is the internal or external angle which is obtained.

A method of booking the data for an angle measured in this manner is shown in *Table 5.1*. This approach constitutes the standard method of measuring single angles in traversing, for instance.

5.5.1 Measurement by directions

The method of directions is generally used when observing a set of angles as in *Figure 5.31*. The angles are observed, commencing from *A* and noting all the readings, as the instrument moves from point to point in a clockwise manner. On completion at *D*, face is changed and the observations repeated moving from *D* in an anticlockwise manner. Finally the mean directions are reduced relative to the starting direction for *PA* by applying the 'orientation correction'. For example, if the mean horizontal circle reading for *PA* is 48° 54′ 36″ and the known bearing for *PA* is 40° 50′ 32″, then the orientation correction applied to all the mean bearings is obviously −8° 04′ 04″.

The observations as above, carried out on both faces of the instrument, constitute a full set. If measuring *n* sets the reading is altered by 180°/*n* each time.

Table 5.1

Sight to	*Face*	*Reading*				*Angle*		
		°	′	″		°	′	″
A	*L*	020	46	28		80	12	06
C	*L*	100	58	34				
C	*R*	280	58	32		80	12	08
A	*R*	200	46	24				
A	*R*	292	10	21		80	12	07
C	*R*	012	22	28				
C	*L*	192	22	23		80	12	04
A	*L*	112	10	19				
					Mean =	80	12	06

Note the built-in checks supplied by changing face, i.e. the reading should change by 180°. Note that to obtain the clockwise angle one always deducts BS (A) reading from the FS (C) reading, regardless of the order in which they are observed.

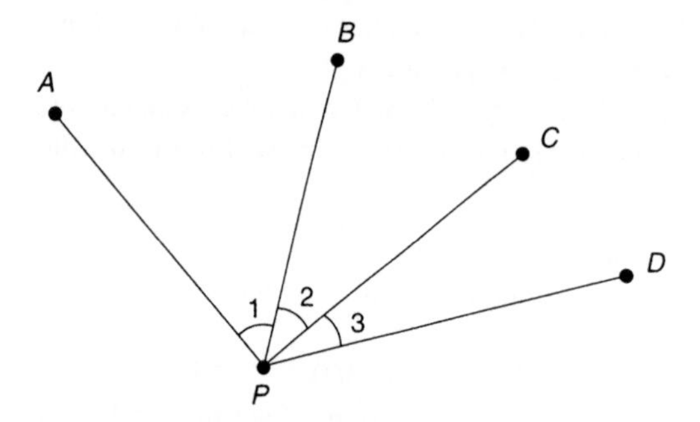

Fig. 5.31 Directions measurements

5.5.2 Further considerations of angular measurement

Considering *Figure 5.31*, the angles may be measured by 'closing the horizon'. This involves observing the points in order from A to D and continuing clockwise back to A, thereby completing the full circle. The difference between the sum of all the angles and 360° is distributed evenly amongst all the angles to bring their sum to 360°. Repeat anticlockwise on the opposite face. *Table 5.2* shows an example.

5.5.3 Vertical angles

In the measurement of horizontal angles the concept is of a measuring index moving around a protractor. In the case of a vertical angle, the situation is reversed and the protractor moves relative to a fixed horizontal index.

Figure 5.32(a) shows the telescope horizontal and reading 90°; changing face would result in a reading of 270°. In *Figure 5.32(b)*, the vertical circle index remains horizontal whilst the protractor rotates with the telescope, as the top of the spire is observed. The vertical circle reading of 65° is the zenith angle, equivalent to a vertical angle of $(90° - 65°) = +25° = \alpha$. This illustrates the basic concept of vertical angle measurement.

Table 5.2

Sight to	*Face*	*Reading*			*Apply misclosure*			*Mean of FL and FR reduced to FL*		
		°	′	″	°	′	″	°	′	″
A	L	20	26	36	20	26	36	20	26	31
B	L	65	37	24	65	37	22	65	37	18
C	L	102	45	56	102	45	52	102	45	54
D	L	135	12	22	135	12	16	135	12	16
A	L	20	26	44	20	26	36	20	26	31
Misclosure				+8			0			
A	R	200	26	26	200	26	26			
D	R	315	12	14	315	12	15			
C	R	282	45	44	282	45	46			
B	R	245	37	12	245	37	15			
A	R	200	26	22	200	26	26			
Misclosure				−4			0			

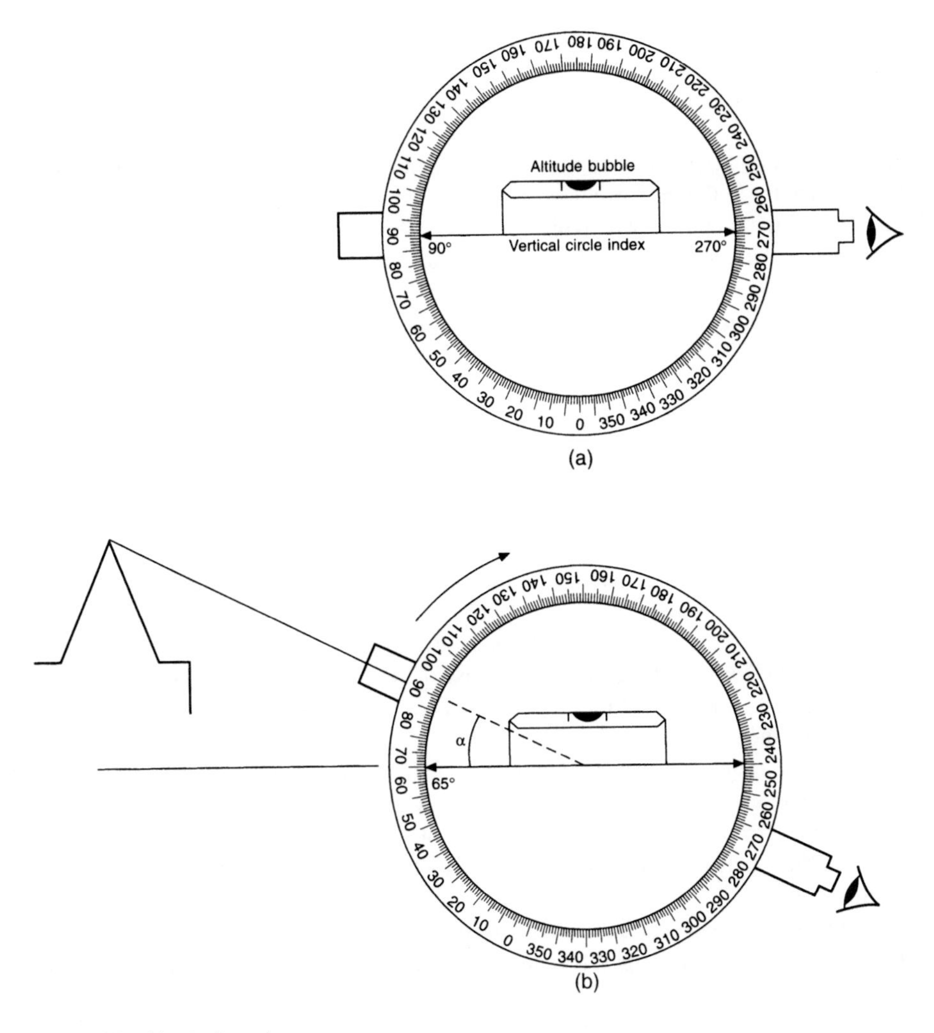

Fig. 5.32 *Vertical angles*

5.6 SOURCES OF ERROR

Error in the measurement of angle occurs because of instrumental, personal or natural factors.

The instrumental errors have been dealt with and, as indicated, can be minimized by taking several measurements of the angle on each face of the theodolite. Regular calibration of the equipment is also of prime importance. The remaining sources will now be dealt with.

5.6.1 Personal error

(1) Careless centring of the instrument over the survey point. Always ensure that the optical plummet is in adjustment. Similarly for the targets.
(2) Lightly clamp the horizontal and vertical movement. Hard clamping can affect the pointing and is unnecessary.

(3) The final movement of the slow motion screws should be clockwise, thus producing a positive movement against the spring. An anticlockwise movement which releases the spring may cause backlash. This was a major source of error in older precise instruments but is less important with modern instruments.
(4) Failure to eliminate parallax and poor focusing on the target can affect accurate pointing. Keep the observed target near the centre of the field of view.
(5) Incorrect levelling of the altitude bubble, where there is one, will produce vertical angle error.
(6) The plate bubble must also be carefully levelled and regularly checked throughout the measuring process, but must not be adjusted during a round of observations.
(7) Make quick, decisive observations. Too much care can be counterproductive.
(8) All movement of the theodolite should be done gently whilst movement around the tripod should be reduced to a minimum.
(9) Do not knock the tripod by tripping. Remove the instrument box, which should be closed, and all other items to a point at least three metres away from the instrument.

5.6.2 Natural errors

(1) Wind vibration may require some form of wind shield to be erected to protect the instrument. Dual axis tilt sensors in modern total stations have greatly minimized this effect.
(2) Vertical and lateral refraction of the line of sight is always a problem. The effect on the vertical angle has already been discussed in *Chapter 3*. Lateral refraction, particularly in tunnels, can cause excessive error in the horizontal angle. A practical solution in tunnels is to use zig-zag traverses with frequent gyro-theodolite azimuths included.
(3) Ensure that the line of sight does not pass near sources of heat such as chimneys or open fires.
(4) Temperature differentials can cause unequal expansion of the various parts of the instrument. Plate bubbles will move off centre towards the hottest part of the bubble tube. Heat shimmer may make accurate pointing impossible. Sheltering the instrument and tripod by means of a large survey umbrella will greatly help in this situation.
(5) Avoid tripod settlement by selecting the site carefully being mindful of ground conditions. If necessary use pegs to pile the ground on which the tripod feet are set or use walk boards to spread the weight of the observer.

All the above procedures should be included in a pre-set survey routine, which should be strictly adhered to. Inexperienced observers should guard against such common mistakes as:

(1) Turning the wrong screw.
(2) Sighting the wrong target.
(3) Using the stadia instead of the cross-hair.
(4) Forgetting to set the micrometer, where there is one.
(5) Misreading the circles.
(6) Transposing figures when booking the data.
(7) Not removing parallax.
(8) Not centring over the survey point.
(9) Failing to level the instrument correctly.

6
Conventional control surveys

A control survey provides a framework of survey points, whose relative positions, in two or three dimensions, are known to specified degrees of accuracy. The areas covered by these points may extend over a whole country and form the basis for the national maps of that country. Alternatively the area may be relatively small, encompassing a construction site for which a large-scale plan is required. Although the areas covered in construction are usually quite small, the accuracy may be required to a very high order. The types of engineering project envisaged are the construction of long tunnels and/or bridges, deformation surveys for dams and reservoirs, three-dimensional tectonic ground movement for landslide prediction, to name just a few. Hence control networks provide a reference framework of points for:

(1) Topographic mapping and large-scale plan production.
(2) Dimensional control of construction work.
(3) Deformation surveys for all manner of structures, both new and old.
(4) The extension and densification of existing control networks.

The methods of establishing the vertical control have already been discussed in *Chapter 3*, so only two-dimensional horizontal control will be dealt with here. Elements of geodetic surveying will be dealt with in *Chapter 8* and so we will concentrate upon plane surveying for engineering control here.

The methods used for control surveys are:

(1) Traversing.
(2) Intersection and resection.
(3) Least squares estimation of survey networks.
(4) Satellite position fixing (see *Chapter 9*).

6.1 PLANE RECTANGULAR COORDINATES

A plane rectangular coordinates system is as defined in *Figure 6.1*.

It is split into four quadrants with the typical mathematical convention of the axis to the north and east being positive and to the south and west, negative.

In pure mathematics, the axes are defined as x and y, with angles measured anticlockwise from the x-axis. In surveying, the x-axis is referred to as the east-axis (E) and the y-axis as the north-axis (N), with angles (α) measured clockwise from the N-axis.

From *Figure 6.1*, it can be seen that to obtain the coordinates of point B, we require the coordinates of point A and the difference in coordinates between the ends of the line AB, i.e.

$$E_B = E_A + \Delta E_{AB} \quad \text{and}$$

$$N_B = N_A + \Delta N_{AB}$$

It can further be seen that to obtain the difference in coordinates between the ends of the line AB we require its horizontal distance and direction.

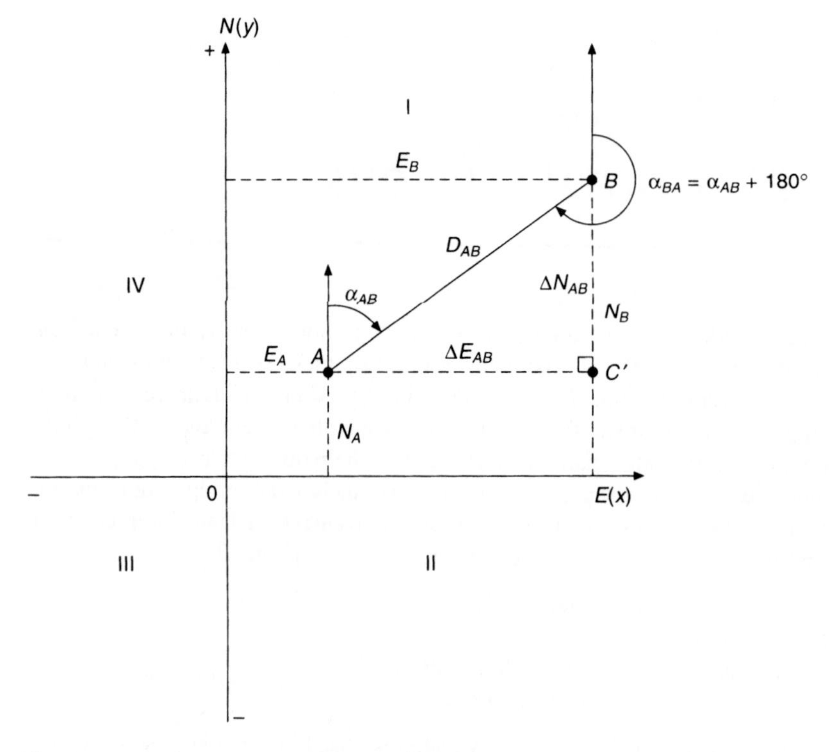

Fig. 6.1 *Plane rectangular coordinate system*

The system used to define a direction is called the *whole circle bearing system* (WCB). A WCB is the direction measured clockwise from 0° full circle to 360°. It is therefore always positive and never greater than 360°.

Figure 6.2 shows the WCB of the lines as follows:

WCB $OA = 40°$

WCB $OB = 120°$

WCB $OC = 195°$

WCB $OD = 330°$

As shown in *Figure 6.2*, the reverse or back bearing is 180° different from the forward bearing, thus:

WCB $AO = 40° + 180° = 220°$

WCB $BO = 120° + 180° = 300°$

WCB $CO = 195° - 180° = 15°$

WCB $DO = 330° - 180° = 150°$

Thus if WCB $< 180°$ it is easier to add 180° to get the reverse bearing, and if $> 180°$ subtract, as shown.

The above statement should not be confused with a similar rule for finding WCBs from the observed angles. For instance (*Figure 6.3*), if the WCB of *AB* is 0° and the observed angle *ABC* is 140°, then the

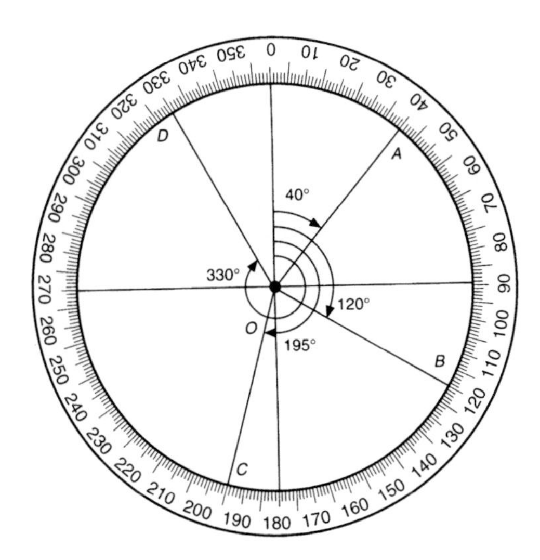

Fig. 6.2 *Whole circle bearings*

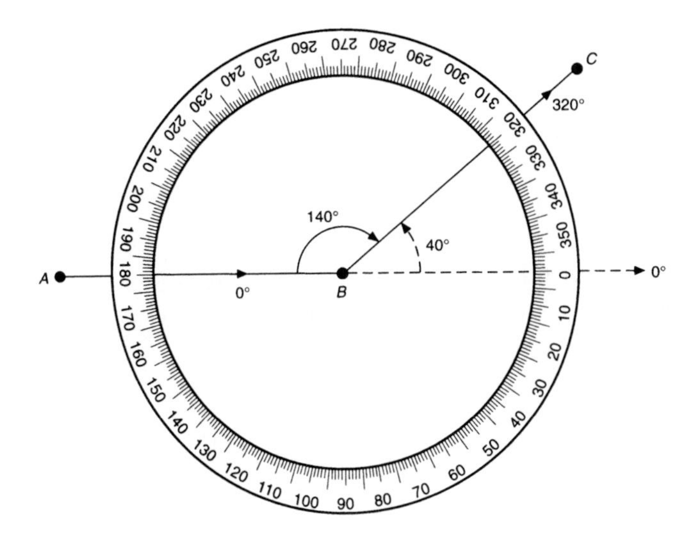

Fig. 6.3 *Whole circle bearing*

relative WCB of BC is 320°, i.e.

$$
\begin{aligned}
\text{WCB of } AB &= 0^\circ \\
\text{Angle } ABC &= 140^\circ \\
\text{Sum} &= 140^\circ \\
&\quad +180^\circ \\
\text{WCB of } BC &= 320^\circ
\end{aligned}
$$

Similarly (*Figure 6.4*), if WCB of AB is 0° and the observed angle ABC is 220°, then the relative WCB of BC is 40°, i.e.

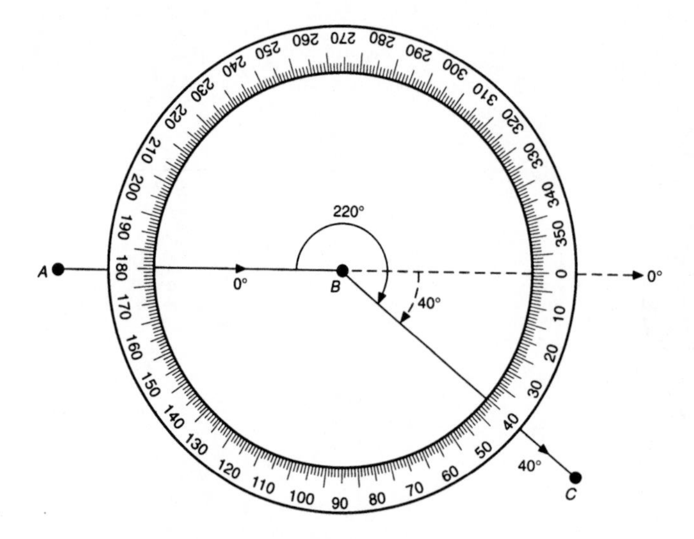

Fig. 6.4 *Whole circle bearing*

$$\begin{aligned} \text{WCB of } AB &= 0^\circ \\ \text{Angle } ABC &= 220^\circ \\ \hline \text{Sum} &= 220^\circ \\ & \;\; -180^\circ \\ \hline \text{WCB of } BC &= 40^\circ \end{aligned}$$

Occasionally, when subtracting 180°, the resulting WCB is still greater than 360°, in this case, one would need to subtract a further 360°. However, this problem is eliminated if the following rule is used.

Add the angle to the previous WCB:

If the *sum* <180°, then *add* 180°
If the *sum* >180°, then *subtract* 180°
If the *sum* >540°, then *subtract* 540°

The application of this rule to traverse networks is shown in *Section 6.2*.

It should be noted that if both bearings are pointing out from B, then

$$\text{WCB } BC = \text{WCB } BA + \text{angle } ABC$$

as shown in *Figure 6.5*, i.e.

$$\text{WCB } BC = \text{WCB } BA\ (30^\circ) + \text{angle } ABC\ (110^\circ) = 140^\circ$$

Having now obtained the WCB of a line and its horizontal distance (polar coordinates), it is possible to transform them to ΔE and ΔN, the difference in rectangular coordinates. From *Figure 6.1*, it can clearly be seen that from the right-angled triangle ABC:

$$\Delta E = D \sin \alpha \tag{6.1a}$$

$$\Delta N = D \cos \alpha \tag{6.1b}$$

where D = horizontal length of the line
α = WCB of the line

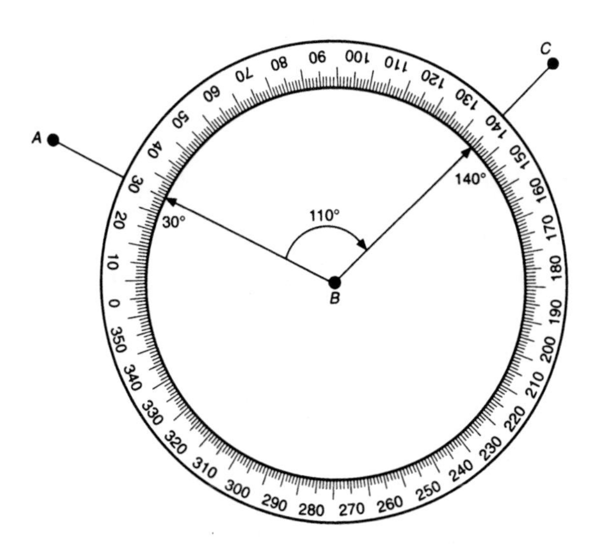

Fig. 6.5 *Whole circle bearing*

ΔE = difference in eastings of the line
ΔN = difference in northings of the line

It is important to appreciate that the difference in coordinates ΔE, ΔN define a *line*, whilst the coordinates E, N define a *point*.

From the above basic equations, the following can be derived:

$$\alpha = \tan^{-1}(\Delta E/\Delta N) = \cot^{-1}(\Delta N/\Delta E) \tag{6.2}$$

$$D = (\Delta E^2 + \Delta N^2)^{\frac{1}{2}} \tag{6.3}$$

$$D = \Delta E/ \sin\alpha = \Delta N/ \cos\alpha \tag{6.4}$$

In *equation (6.2)* it should be noted that the trigonometrical functions of tan and cot can become very unreliable on pocket calculators as α approaches 0° or 180° and 90° or 270° respectively. To deal with this problem:

Use tan when $|\Delta N| > |\Delta E|$, and

use cot when $|\Delta N| < |\Delta E|$

Exactly the same situation occurs with sin and cos in *equation (6.4)*; thus:

Use cos when $|\Delta N| > |\Delta E|$, and

use sin when $|\Delta N| < |\Delta E|$

The two most fundamental calculations in surveying are computing the 'polar' and the 'join'.

6.1.1 Computing the polar

Computing the polar for a line involves calculating ΔE and ΔN given the horizontal distance (D) and WCB (α) of the line.

Worked example

Example 6.1 Given the coordinates of A and the distance and bearing of AB, calculate the coordinates of point B.

$$E_A = 48\,964.38 \text{ m}, \; N_A = 69\,866.75 \text{ m}, \text{WCB } AB = 299°\,58'\,46''$$

$$\text{Horizontal distance} = 1325.64 \text{ m}$$

From the WCB of AB, the line is obviously in the fourth quadrant and signs of ΔE, ΔN are therefore $(-, +)$ respectively. A pocket calculator will automatically provide the correct signs.

$$\Delta E_{AB} = D \sin \alpha = 1325.64 \sin 299°\,58'\,46'' = -1148.28 \text{ m}$$

$$\Delta N_{AB} = D \cos \alpha = 1325.64 \cos 299°\,58'\,46'' = +662.41 \text{ m}$$

$$\therefore E_B = E_A + \Delta E_{AB} = 48\,964.38 - 1148.28 = 47\,816.10 \text{ m}$$

$$N_B = N_A + \Delta N_{AB} = 69\,866.75 + 662.41 = 70\,529.16 \text{ m}$$

This computation is best carried out using the P (Polar) to R (Rectangular) keys of the pocket calculator. However, as these keys work on a pure math basis and not a surveying basis, one must know the order in which the data are input and the order in which the data are output.

The following methods apply to the majority of pocket calculators. However, as new types are being developed all the time, then it may be necessary to adapt to a specific make.

Using P and R keys:

(1) Enter *horizontal distance* (D); press [P → R] or (x ↔ y)
(2) Enter WCB (α); press [=] or (R)
(3) Value displayed is $\pm \Delta N$
(4) Press [x ↔ y] to get $\pm \Delta E$

Operations in brackets are for an alternative type of calculator.

6.1.2 Computing the join

This involves computing the horizontal distance (D) and WCB (α) from the difference in coordinates (ΔE, ΔN) of a line.

Worked example

Example 6.2 Given the following coordinates for two points A and B, compute the length and bearing of AB.

$E_A = 48\,964.38$ m	$N_A = 69\,866.75$ m
$E_B = 48\,988.66$ m	$N_B = 62\,583.18$ m
$\Delta E_{AB} = 24.28$ m	$\Delta N_{AB} = -7283.57$ m

Note:

(1) A rough plot of the E, N of each point will show B to be south-east of A, and line AB is therefore in the second quadrant.
(2) If the direction is from A to B then:

$$\Delta E_{AB} = E_B - E_A$$
$$\Delta N_{AB} = N_B - N_A$$

If the required direction is B to A then:

$$\Delta E_{BA} = E_A - E_B$$

$$\Delta N_{BA} = N_A - N_B$$

(3) As $\Delta N > \Delta E$ use tan:

$$\alpha_{AB} = \tan^{-1}(\Delta E/\Delta N) = \tan^{-1}\{24.28/(-7283.57)\}$$

$$= -0°11'27''$$

It is obvious that as α_{AB} is in the second quadrant and must therefore have a WCB between 90° and 180°, and as we cannot have a negative WCB, $-0°\ 11'27''$ is unacceptable. Depending on the signs of the coordinates entered into the pocket calculator, it will supply the angles as shown in *Figure 6.6*.

If in quadrant I $+\alpha_1 =$ WCB
If in quadrant II $-\alpha_2$, then $(-\alpha_2 + 180) =$ WCB
If in quadrant III $+\alpha_3$, then $(\alpha_3 + 180) =$ WCB
If in quadrant IV $-\alpha_4$, then $(-\alpha_4 + 360) =$ WCB
$\therefore$ WCB $AB = -0°\ 11'27'' + 180° = 179°\ 48'\ 33''$
Horizontal distance $AB = D_{AB} = (\Delta E^2 + \Delta N^2)^{\frac{1}{2}}$
$(24.28^2 + 7283.57^2)^{\frac{1}{2}} = 7283.61$ m
Also as $\Delta N > \Delta E$ use $D_{AB} = \Delta N/\cos\alpha = 7283.57/\cos 179°\ 48'\ 33'' = 7283.61$ m

Note what happens with some pocket calculators when $\Delta E/\sin\alpha$ is used:

$$D_{AB} = \Delta E/\sin\alpha = 24.28/\sin 179°\ 48'\ 33'' = 7289.84 \text{ m}$$

This enormous error of more than 6 m shows that when computing distance it is advisable to use the Pythagoras equation $D = (\Delta E^2 + \Delta N^2)^{\frac{1}{2}}$, at all times. Of the remaining two equations, the appropriate one may be used as a check.

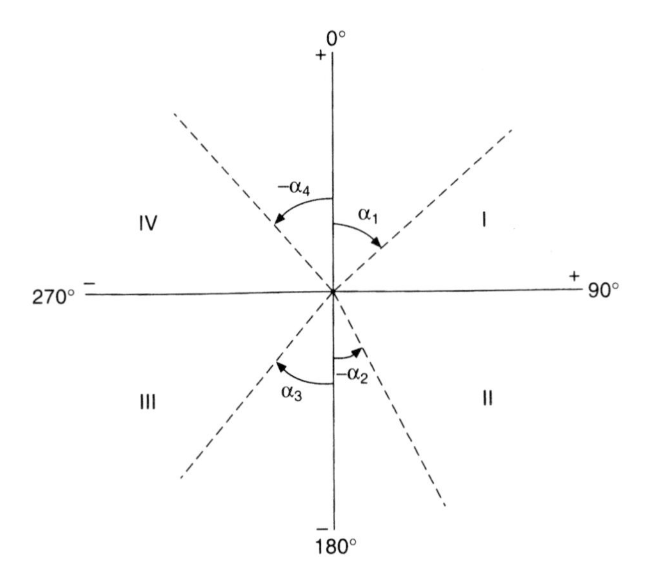

Fig. 6.6 *Calculation of WCB with a calculator*

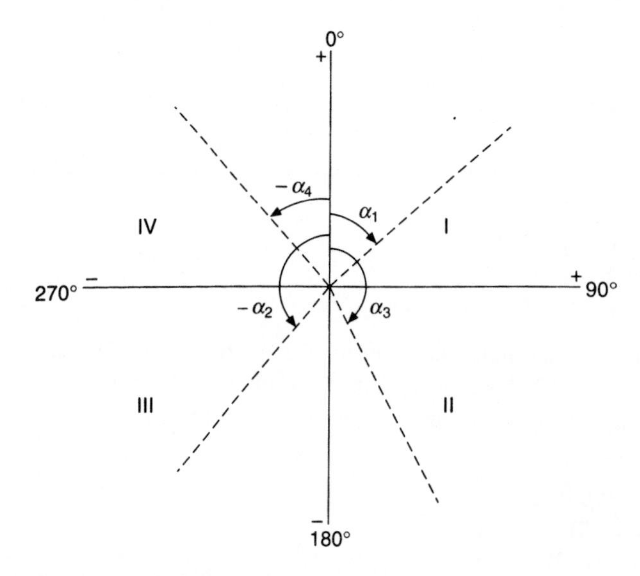

Fig. 6.7 *Calculation of WCB using P and R keys*

Using R and P keys:

(1) Enter $\pm\Delta N$; press [R → P] or (x ↔ y)

(2) Enter $\pm\Delta E$; press [=] or (P)

(3) Value displayed is *horizontal distance* (*D*)

(4) Press [x ↔ y] to obtain WCB in degrees and decimals

(5) If value in '4' is negative, *add* 360°

(6) Change to d.m.s. (°, ′, ″)

When using the P and R keys, the angular values displayed in the four quadrants are as in *Figure 6.7*; thus only a single 'IF' statement is necessary as in (5) above.

6.2 TRAVERSING

Traversing is one of the simplest and most popular methods of establishing control networks in engineering surveying. In underground mining it is the only method of control applicable whilst in civil engineering it lends itself ideally to control surveys where only a few intervisible points surrounding the site are required. Traverse networks have the following advantages:

(1) Little reconnaissance is required compared with that needed for an interconnected network of points.
(2) Observations only involve three stations at a time so planning the task is simple.
(3) Traversing may permit the control to follow the route of a highway, pipeline or tunnel, etc., with the minimum number of stations.

6.2.1 Types of traverse

Using the technique of traversing, the relative position of the control points is fixed by measuring the horizontal angle at each point, between the adjacent stations, and the horizontal distance between consecutive pairs of stations.

The procedures for measuring angles have been dealt with in *Chapter 5* and for measuring distance in *Chapter 4*. For the majority of traverses carried out today the field data would most probably be captured with a total station. Occasionally, steel tapes may be used for distance.

The susceptibility of a traverse to undetected error makes it essential that there should be some external check on its accuracy. To this end the traverse may commence from and connect into known points of greater accuracy than the traverse. In this way the error vector of misclosure can be quantified and distributed throughout the network, to minimize the errors. Such a traverse is called a 'link' traverse.

Alternatively, the error vector can be obtained by completing the traverse back to its starting point. Such a traverse is called a 'polygonal' or 'loop' traverse. Both the 'link' and 'polygonal' traverse are generally referred to as 'closed' traverses.

The third type of traverse is the 'free' or 'open' traverse, which does not close back onto any known point and which therefore has no way of detecting or quantifying the errors.

(1) Link traverse

Figure 6.8 illustrates a typical link traverse commencing from the precisely coordinated point Y and closing onto point W, with terminal orienting bearing to points X and Z. Generally, points X, Y, W and Z would be part of an existing precisely coordinated control network, although this may not always be the case. It may be that when tying surveys into the OS NG, due to the use of very precise EDM equipment the intervening traverse is more precise than the relative positions of the NG stations. This may be just a problem of scale arising from a lack of knowledge, on the behalf of the surveyor, of the positional accuracy of the grid points. In such a case, adjustment of the traverse to the NG could result in distortion of the intervening traverse.

The usual form of an adjustment generally adopted in the case of a link traverse is to hold points Y and W fixed whilst distributing the error throughout the intervening points. This implies that points Y and W are free from error and is tantamount to allocating a weight of infinity to the length and bearing of line YW. It is thus both obvious and important that the control into which the traverse is linked should be of a higher order of precision than the connecting traverse.

The link traverse has certain advantages over the remaining types, in that systematic error in distance measurement and orientation are clearly revealed by the error vector.

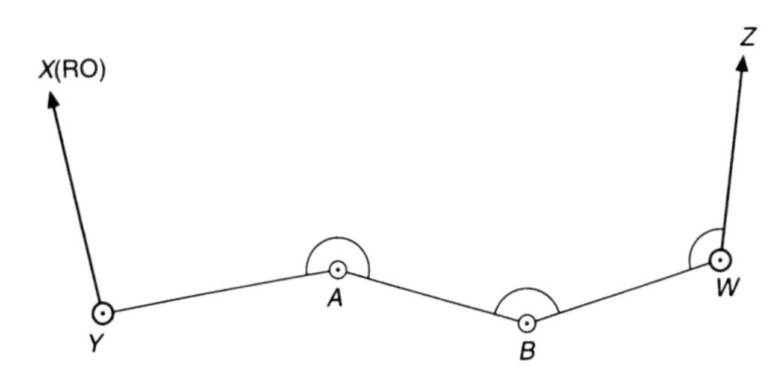

Fig. 6.8 *Link traverse*

(2) Polygonal traverse

Figures 6.9 and *6.10* illustrate the concept of a polygonal traverse. This type of network is quite popular and is used extensively for peripheral control on all types of engineering sites. If no external orientation is available, the control can only be used for independent sites and plans and cannot be directly connected to other survey systems.

In this type of traverse the systematic errors of distance measurement are not eliminated and enter into the result with their full weight. Similarly, orientation error would simply cause the whole network to swing through the amount of error involved and would not be revealed in the angular misclosure.

This is illustrated in *Figures 6.11* and *6.12*. In the first instance a scale error equal to X is introduced into each line of a rectangular-shaped traverse $ABCD$. Then, assuming the angles are error free, the traverse appears to close perfectly back to A, regardless of the totally incorrect coordinates which would give the position of B, C and D as B', C' and D'.

Figure 6.12 shows the displacement of B, C and D to B', C' and D' caused by an orientation error (θ) for AB. This can occur when AB may be part of another network and the incorrect value is taken for its bearing. The traverse will still appear to close correctly, however. Fortunately, in this particular case, the coordinates of B would be known and would obviously indicate some form of mistake when compared with B'.

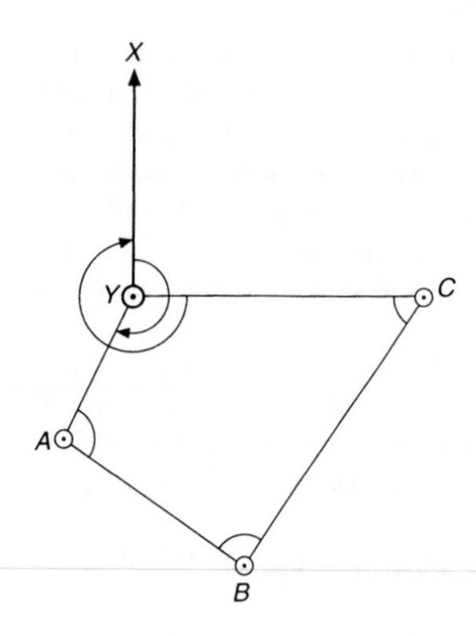

Fig. 6.9 *Loop traverse (oriented)*

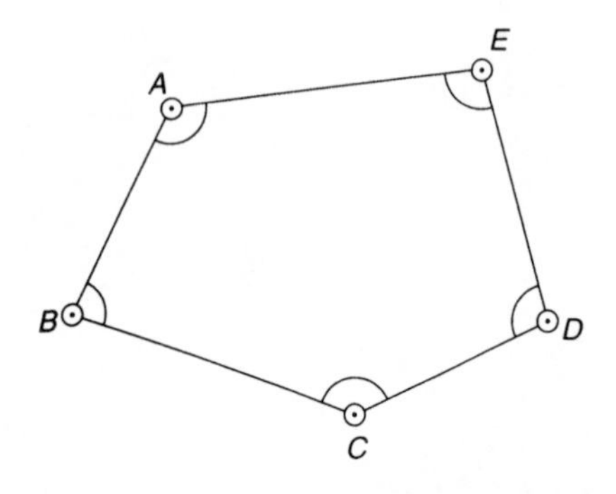

Fig. 6.10 *Loop traverse (independent)*

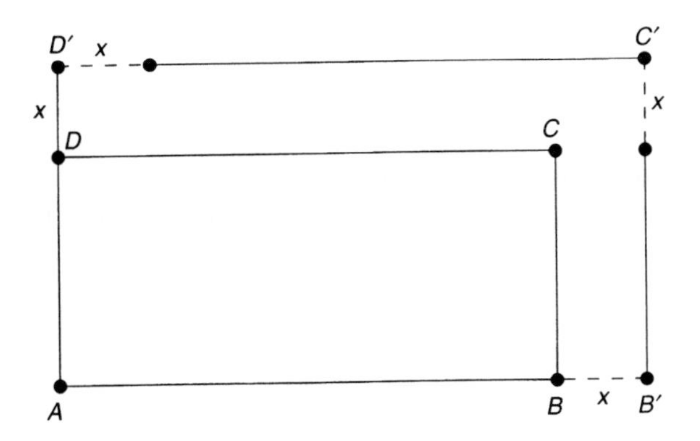

Fig. 6.11 *Scale error in a traverse*

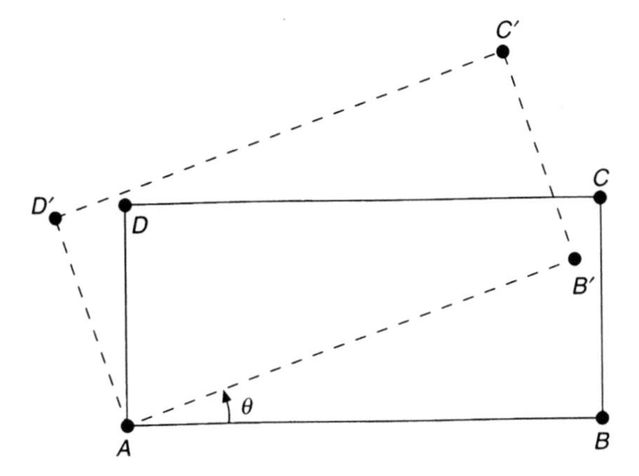

Fig. 6.12 *Orientation error in a traverse*

(3) Open (or free) traverse

Figure 6.13 illustrates the open traverse which does not close into any known point and therefore cannot provide any indication of the magnitude of measuring errors. In all surveying literature, this form of traversing is not recommended due to the lack of checks. Nevertheless, it is frequently utilized in mining and tunnelling work because of the physical restriction on closure.

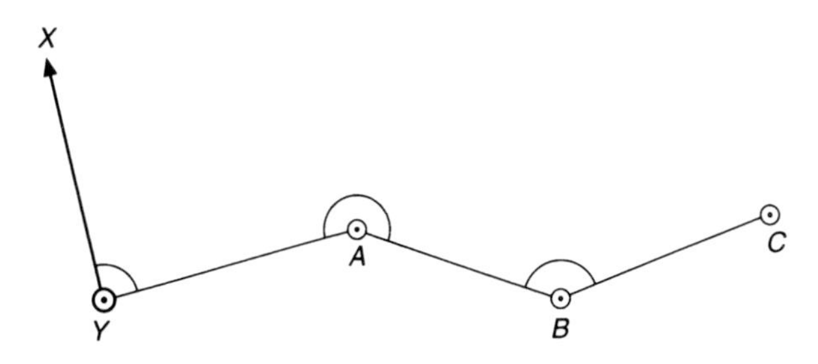

Fig. 6.13 *Open (or free) traverse*

6.2.2 Reconnaissance

Reconnaissance is a vitally important part of any survey project, as emphasized in *Chapter 1*. Its purpose is to decide the best location for the traverse points.

Successive points in the traverse must be intervisible to make observations possible.

If the purpose of the control network is the location of topographic detail only, then the survey points should be positioned to afford the best view of the terrain, thereby ensuring that the maximum amount of detail can be surveyed from each point.

If the traverse is to be used for setting out, say, the centre-line of a road, then the stations should be sited to afford the best positions for setting out the intersection points (IPs) and tangent points (TPs).

The distance between stations should be kept as long as possible to minimize effect of centring errors.

Finally, as cost is always important, the scheme should be one that can be completed in the minimum of time, with the minimum of personnel.

The type of survey station used will also be governed by the purpose of the traverse points. If the survey stations are required as control for a quick, one-off survey of a small area, then wooden pegs about 0.25 m long and driven down to ground level may suffice. A fine point on the top of the peg such as the centre of a nail head may define the control point. Alternatively, longer lasting stations may require construction of some form of commercially manufactured station mark. *Figure 6.14* shows the type of survey station recommended by the Department for Transport (UK) for major road projects. They are recommended to be placed at 250-m intervals and remain stable for at least five years. *Figure 6.15* shows a commercially available earth anchor type of station. Road, masonry or Hilti nails may be used on paved or black-topped surfaces.

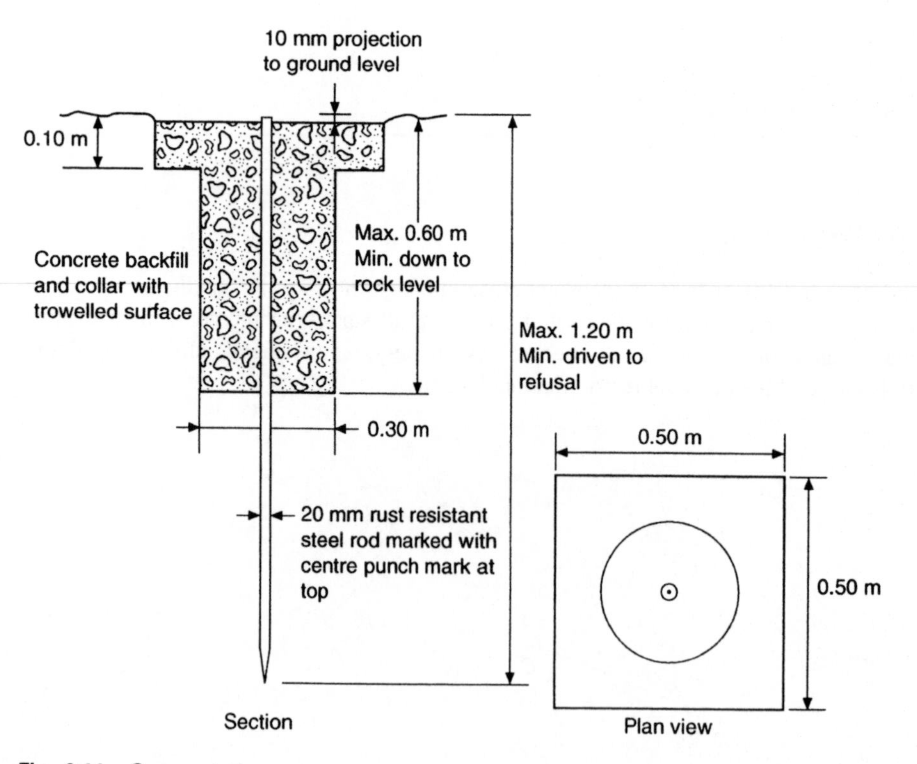

Fig. 6.14 *Survey station*

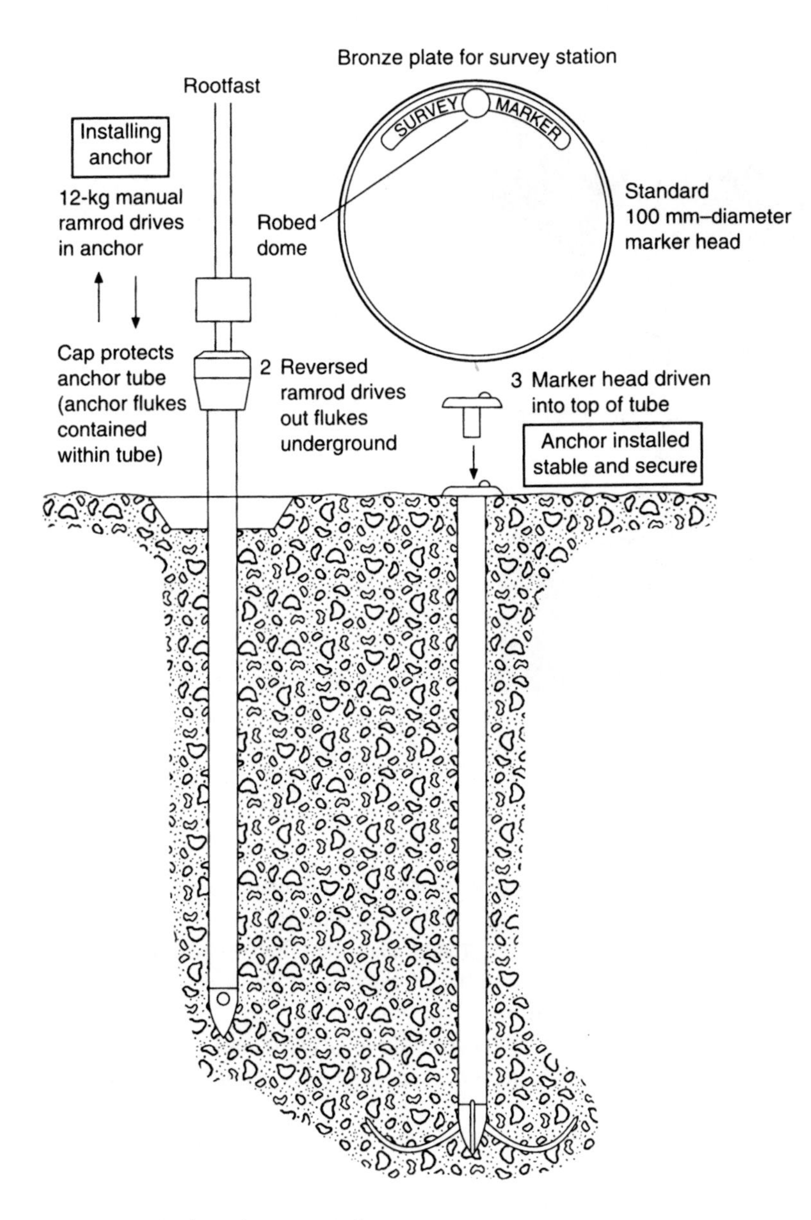

Fig. 6.15 *Anchored survey station*

6.2.3 Sources of error

The sources of error in traversing are:

(1) Errors in the observation of horizontal and vertical angles (angular error).
(2) Errors in the measurement of distance (linear error).
(3) Errors in the accurate centring of the instrument and targets, directly over the survey point (centring error).

Linear and angular errors have been fully dealt with in *Chapters 4* and *5* respectively.

Fig. 6.16 *Interchangeable total station and target with tribrach*

Centring errors were dealt with in *Chapter 5* also, but only insofar as they affected the measurement of a single angle. Their propagation effects through a traverse will now be examined.

In precise traversing the effect of centring errors will be greater than the effect of reading errors if appropriate procedures are not adopted to eliminate them. As has already been illustrated in *Chapter 5*, the shorter the legs, the greater the effect on angular measurements.

The inclusion of short lines cannot be avoided in many engineering surveys, particularly in underground tunnelling work. In order, therefore, to minimize the propagation of centring error, a constrained centring system called the three-tripod system (TTS) is used.

The TTS uses interchangeable levelling heads or tribrachs and targets, and works much more efficiently with a fourth tripod (*Figure 6.16*).

Consider *Figure 6.17*. Tripods are set up at A, B, C and D with the detachable tribrachs carefully levelled and centred over each station. Targets are clamped into the tribrachs at A and C, whilst the theodolite is clamped into the one at B. When the angle ABC has been measured, the target (T_1) is clamped into the tribrach at B, the theodolite into the tribrach at C and the target just moved into the tribrach at D. Whilst the angle BCD is being measured, the tripod and tribrach are removed from A and set up at E in preparation for the next move forward. This technique not only produces maximum speed and efficiency, but also confines the centring error to the station at which it occurred. Indeed, the error in question here is not one of centring in the conventional sense, but one of knowing that the central axes of the targets and theodolite, when moved forward, occupy exactly the same positions as did their previous occupants.

However, such a process does not guarantee that the tribrachs have been set up correctly over the survey station. The coordinates computed by this process will be those of the tribrach, not necessarily those of the survey station.

Figure 6.18 shows how centring errors may be propagated. Consider first the use of the TTS. The target erected at C, 100 m from B, is badly centred, resulting in a displacement of 50 mm to C'. The angle

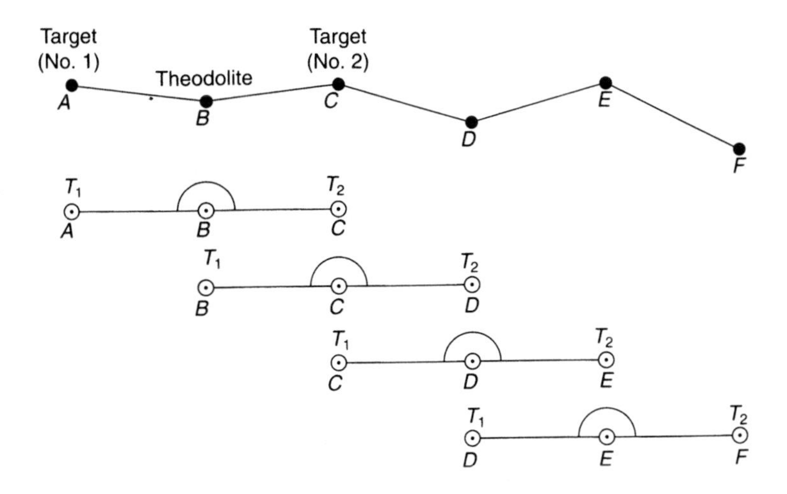

Fig. 6.17 *Conventional three-tripod system*

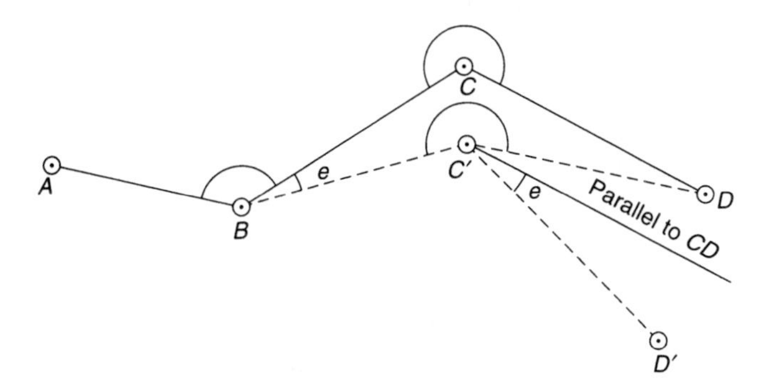

Fig. 6.18 *Propagation of centring error*

measured at B would be ABC' in error by e. The error is 1 in 2000 $\approx$2 min. (*N.B.* If BC was 10 m long, then $e = 20$ min.)

The target is removed from C' and replaced by the theodolite, which measures the angle $BC'D$, thus bringing the survey back onto D. The only error would therefore be a coordinate error at C equal to the centring error and would obviously be much less than the exaggerated 50 mm used here.

Consider now conventional equipment using one tripod and theodolite and sighting to ranging rods. Assume that the rod at C, due to bad centring or tilting, appears to be at C'; the wrong angle, ABC', would be measured. Now, when the *theodolite* is moved it would this time be correctly centred over the station at C and the correct angle BCD measured. However, this correct angle would be added to the previous computed bearing, which would be that of BC', giving the bearing $C'D'$. Thus the error e is propagated from the already incorrect position at C', producing a further error at D' of the traverse. Centring of the instrument and targets precisely over the survey stations is thus of paramount importance; hence the need for constrained centring systems in precise traversing.

It can be shown that if the theodolite and targets are re-centred with an error of ± 0.3 mm, these centring errors alone will produce an error of $\pm 6''$ in bearing after 1500 m of traversing with 100-m sights. If the final bearing is required to $\pm 2''$, the component caused by centring must be limited to about one-third of the total component, which is $\pm 0.6''$, and would therefore require centring errors in the region of ± 0.03 mm. Thus in general a mean error ± 0.1 mm would be compatible with a total mean error of $\pm 6''$ in the

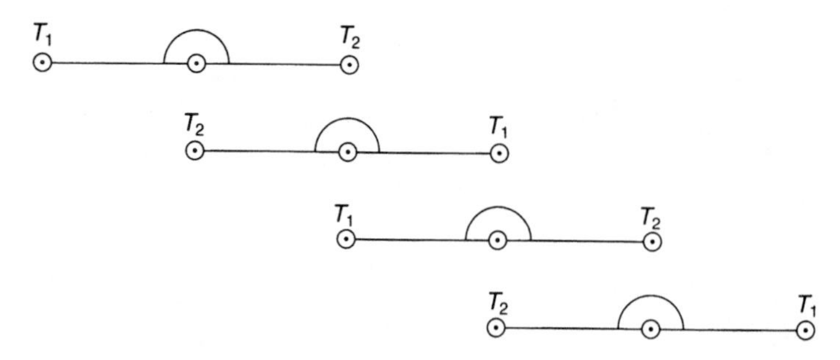

Fig. 6.19 *Suggested three-tripod system*

final bearing of the above traverse. This therefore imposes a very rigorous standard required of constrained centring systems.

When considering the errors of constrained centring, the relevant criterion is the repeatability of the system and not the absolute accuracy of centring over a ground mark. One is concerned with the degree to which the vertical axis of the theodolite placed in a tribrach coincides with the vertical through the centre of the target previously occupying the tribrach.

The error sources are:

(1) The aim mark of the target, eccentric to the vertical axis.
(2) The vertical axis of the total station eccentric to the centring axis as these are separate components.
(3) Variations in clamping pressures.
(4) Tolerance on fits, which is essentially a manufacturing problem.

An alternative arrangement for moving instruments and targets, in which target errors are partly self cancelling, is shown in *Figure 6.19*. If the error in the target T_2 in the first bay is such that the measured angle is too small then the same target in bay 2 will cause the measured angle in bay 2 to be too large.

6.2.4 Traverse computation

The various steps in traverse computation will now be carried out, with reference to the traverse shown in *Figure 6.20*. The observed horizontal angles and distances are shown in columns 2 and 7 of *Table 6.1*.

A common practice is to assume coordinate values for a point in the traverse, usually the first station, and allocate an arbitrary bearing for the first line from that point. For instance, in *Figure 6.20*, point A has been allocated coordinates of E 1000.00, N 2000.00, and line AB a bearing of $0° 00' 00''$. If values of E 0.00, N 0.00 had been chosen there would have been negative values for the coordinates of some of the stations. Negative coordinates can be confusing. This has the effect of establishing a plane rectangular grid and orientating the traverse on it. As shown, AB becomes the direction of the N-axis, with the E-axis at 90° and passing through the grid origin at A.

The computational steps, in the order in which they are carried out, are:

(1) Obtain the angular misclosure W, by comparing the sum of the observed angles (α) with the sum of error-free angles in a geometrically correct figure.
(2) Assess the acceptability or otherwise of W.
(3) If W is acceptable, distribute it throughout the traverse in equal amounts to each angle.
(4) From the corrected angles compute the whole circle bearing of the traverse lines relative to AB.
(5) Compute the coordinates (ΔE, ΔN) of each traverse line.
(6) Assess the coordinate misclosure ($\Delta' E$, $\Delta' N$).

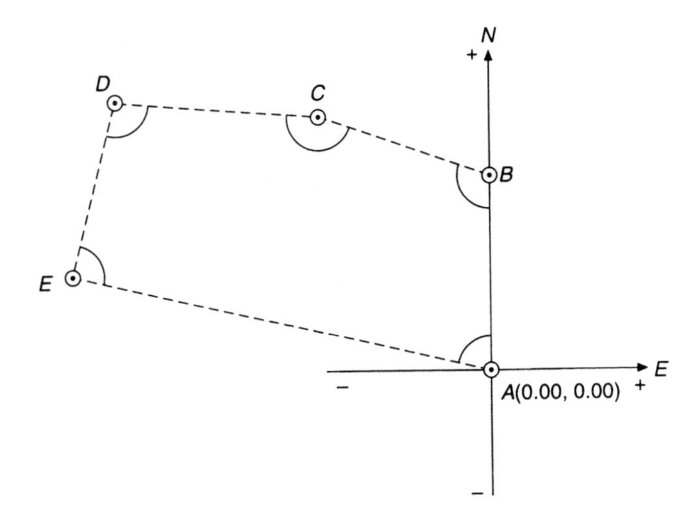

Fig. 6.20 *Polygonal traverse*

(7) Balance the traverse by distributing the coordinate misclosure throughout the traverse lines.
(8) Compute the final coordinates (E, N) of each point in the traverse relative to A, using the balanced values of ΔE, ΔN per line.

The above steps will now be dealt with in detail.

(1) Distribution of angular error

The majority of the systematic errors associated with horizontal angles in a traverse are eliminated by repeated double-face observation. The remaining random errors are distributed equally around the network as follows.

In a polygon the sum of the *internal* angles should equal $(2n - 4)90°$, the sum of the *external* angles should equal $(2n + 4)90°$.

$$\therefore \text{Angular misclosure} = W = \sum_{i=1}^{n} \alpha_i - (2n \pm 4)90° = -50'' \text{ (Table 6.1)}$$

where α = observed angle
n = number of angles in the traverse

The angular misclosure W is now distributed by equal amounts on each angle, thus:

$$\text{Correction per angle} = W/n = +10'' \text{ (Table 6.1)}$$

However, before the angles are corrected, the angular misclosure W must be considered to be acceptable. If W was too great, and therefore indicative of poor observations, the whole traverse may need to be re-measured. A method of assessing the acceptability or otherwise of W is given in the next section.

(2) Acceptable angular misclosure

The following procedure may be adopted provided the variances of the observed angles can be assessed, i.e.

$$\sigma_w^2 = \sigma_{a1}^2 + \sigma_{a2}^2 + \cdots + \sigma_{an}^2$$

Table 6.1 *Bowditch adjustment*

Angle	Observed horizontal angle	Corn.	Corrected horizontal angle	Line	W.C.B.	Horiz. length	Difference in coordinates		δE	δN	Corrected values		Final values		Pt.
	° ′ ″	″	° ′ ″		° ′ ″	m	ΔE	ΔN			ΔE	ΔN	E	N	
(1)	(2)	(3)	(4)	(5)	(6)	(7)	(8)	(9)	(10)	(11)	(12)	(13)	(14) 1000.00	(15) 2000.00	(16) A
ABC	120 25 50	+10	120 26 00	AB	000 00 00 (Assumed)	155.00	0.00	155.00	0.07	0.10	0.07	155.10	1000.07	2155.10	B
BCD	149 33 50	+10	149 34 00	BC	300 26 00	200.00	−172.44	101.31	0.09	0.13	−172.35	101.44	827.72	2256.54	C
CDE	095 41 50	+10	095 42 00	CD	270 00 00	249.00	−249.00	0.00	0.11	0.17	−248.89	0.17	578.83	2256.71	D
DEA	093 05 50	+ 10	093 06 00	DE	185 42 00	190.00	−18.87	−189.06	0.08	0.13	−18.79	−188.93	560.04	2067.78	E
EAB	081 11 50	+10	081 12 00	EA	098 48 00	445.00	439.76	−68.08	0.20	0.30	439.96	−67.78	1000.00	2000.00	A
Sum (2n−4)90°	539 59 10 540 00 00	+50	540 00 00	AB	000 00 00 $\sum L =$	1239.00	−0.55	−0.83	0.55	0.83	0.00 Sum	0.00 Sum			
Error	−50					Correction =	+0.55 Δ′E	+0.83 Δ′N							

Error Vector $= (0.55^2 + 0.83^2)^{\frac{1}{2}} = 0.99(213° 32')$
Accuracy = 1/1252

where σ_{an}^2 = variance of observed angle

σ_w^2 = variance of the sum of the angles of the traverse

Assuming that each angle is measured with equal precision:

$$\sigma_{a1}^2 = \sigma_{a2}^2 = \cdots = \sigma_{an}^2 = \sigma_A^2$$

then $\sigma_w^2 = n \cdot \sigma_A^2$ and

$$\sigma_w = n^{\frac{1}{2}} \cdot \sigma_A \tag{6.5}$$

$$\text{Angular misclosure} = W = \sum_{i=1}^{n} \alpha_i - [(2n \pm 4)90°]$$

where α = mean observed angle

n = number of angles in traverse

then for 95% confidence:

$$P(-1.96\sigma_w < W < +1.96\alpha_w) = 0.95 \tag{6.6}$$

and for 99.73% confidence:

$$P(-3\sigma_w < W < +3\sigma_w) = 0.9973 \tag{6.7}$$

For example, consider a closed traverse of nine angles. Tests prior to the survey showed that an observer with a particular theodolite observes with a standard error (σ_A) of 3″. What would be considered an acceptable angular misclosure for the traverse?

$$\sigma_w = 9^{\frac{1}{2}} \times 3'' = 9''$$

$$P(-1.96 \times 9'' < W < +1.96 \times 9'') = 0.95$$

$$P(-18'' < W < +18'') = 0.95$$

Similarly $P(-27'' < W < +27'') = 0.9973$

Thus, if the angular misclosure W is greater than $\pm 18''$ there is evidence to suggest unacceptable error in the observed angles, provided the estimate for σ_A is reliable. If W exceeds $\pm 27''$ there is definitely angular error present of such proportions as to be quite unacceptable.

Research has shown that a reasonable value for the standard error of the mean of a double face observation is about 2.5 times the least count of the instrument. Thus for a 1-second theodolite:

$$\sigma_A = 2.5''$$

Assuming the theodolite used in the traverse of *Figure 6.20* had a least count of 10″:

$$\sigma_w = 5^{\frac{1}{2}} \times 25'' = \pm 56''$$

Assuming the survey specification requires 95% confidence:

$$P(-110'' < W < 110'') = 0.95$$

Thus as the angular misclosure is within the range of $\pm 110''$, the traverse computation may proceed and after the distribution of the angular error, the WCBs are computed.

(3) Whole circle bearings (WCB)

The concept of WCBs has been dealt with earlier in this chapter and should be referred to for the 'rule' that is adopted. The corrected angles will now be changed to WCBs relative to *AB* using that rule.

	Degree	*Minute*	*Second*	
WCB *AB*	000	00	00	
Angle *ABC*	120	26	00	
Sum	120	26	00	
	+180			
WCB *BC*	300	26	00	
Angle *BCD*	149	34	00	
Sum	450	00	00	
	−180			
WCB *CD*	270	00	00	
Angle *CDE*	95	42	00	
Sum	365	42	00	
	−180			
WCB *DE*	185	42	00	
Angle *DEA*	93	06	00	
Sum	278	48	00	
	−180			
WCB *EA*	98	48	00	
Angle *EAB*	81	12	00	
Sum	180	00	00	
	−180			
WCB *AB*	000	00	00	(Check)

(4) Plane rectangular coordinates

Using the observed distance, reduced to the horizontal, and the bearing of the line, transform this data (polar coordinates) to rectangular coordinates for each line of the traverse. This may be done using *equation (6.1)*.

$$\Delta E = L \sin \text{WCB}$$

$$\Delta N = L \cos \text{WCB}$$

or the P → R keys on a pocket calculator. The results are shown in columns 8 and 9 of *Table 6.1*.

As the traverse is a closed polygon, starting from and ending on point *A*, the respective algebraic sums of the ΔE and ΔN values would equal zero if there were no observational error in the distances present. However, as shown, the error in $\Delta E = -0.55$ m and in $\Delta N = -0.83$ m and is 'the coordinate misclosure'.

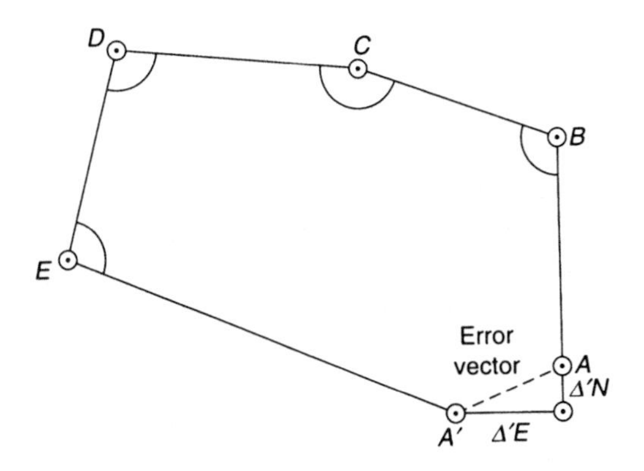

Fig. 6.21 *Coordinate misclosure*

As the correction is always of opposite sign to the error, i.e.

$$\text{Correction} = -\text{Error} \tag{6.8}$$

then the ΔE values must be corrected by $+0.55 = \Delta' E$ and the ΔN values by $+0.83 = \Delta' N$. The situation is as shown in *Figure 6.21*, where the resultant amount of misclosure AA' is called the 'error vector'. This value, when expressed in relation to the total length of the traverse, is used as a measure of the precision of the traverse.

For example:

$$\text{Error vector} = (\Delta' E^2 + \Delta' N^2)^{\frac{1}{2}} = 0.99\,\text{m}$$

$$\text{Accuracy of traverse} = 0.99/1239 = 1/1252$$

(The error vector can be computed using the R → P keys.)

(5) Balancing the traversing

Balancing the traverse, sometimes referred to as 'adjusting' the traverse, involves distributing $\Delta' E$ and $\Delta' N$ throughout the traverse in order to make it geometrically correct.

There is no ideal method of balancing and a large variety of procedures are available, ranging from the very elementary to the much more rigorous. Where a non-rigorous method is used, the most popular procedure is to use the *Bowditch rule*.

The 'Bowditch rule' was devised by Nathaniel Bowditch, surveyor, navigator and mathematician, as a proposed solution to the problem of compass traverse adjustment, which was posed in the American journal *The Analyst* in 1807.

The Bowditch rule is as follows:

$$\delta E_i = \frac{\Delta' E}{\sum_{i=1}^{n} L_i} \times L_i = K_1 \times L_i \tag{6.9}$$

and

$$\delta N_i = \frac{\Delta' N}{\sum_{i=1}^{n} L_i} \times L_i = K_2 \times L_i \tag{6.10}$$

where $\delta E_i, \delta N_i$ = the coordinate corrections

$\Delta' E, \Delta' N$ = the coordinate misclosure (constant)

$\sum_{i=1}^{n} L_i$ = the sum of the lengths of the traverse (constant)

L_i = the horizontal length of the ith traverse leg

K_1, K_2 = the resultant constants

From *equations (6.5)* and *(6.6)*, it can be seen that the corrections made are simply in proportion to the length of the line.

The correction for each length is now computed in order.

For the first line AB:

$$\delta E_1 = (\Delta' E / \Sigma L) L_1 = K_1 \times L_1$$

where $K_1 = +0.55/1239 = 4.4 \times 10^{-4}$

$\therefore \delta E_1 = (4.4 \times 10^{-4})155.00 = +0.07$

Similarly for the second line BC:

$$\delta E_2 = (4.4 \times 10^{-4})200.00 = +0.09$$

and so on:

$$\delta E_3 = (4.4 \times 10^{-4})249.00 = +0.11$$
$$\delta E_4 = (4.4 \times 10^{-4})190.00 = +0.08$$
$$\delta E_5 = (4.4 \times 10^{-4})445.00 = +0.20$$

Sum = +0.55 (Check)

Similarly for the ΔN value of each line:

$$\delta N_1 = (\Delta' N / \Sigma L) L_1 = K_2 L_1$$

where $K_2 = +0.83/1239 = 6.7 \times 10^{-4}$

$\therefore \delta N_1 = (6.7 \times 10^{-4})155.00 = +0.10$

and so on for each line:

$$\delta N_2 = +0.13$$
$$\delta N_3 = +0.17$$
$$\delta N_4 = +0.13$$
$$\delta N_5 = +0.30$$

Sum +0.83 (Check)

These corrections (as shown in columns 10 and 11 of *Table 6.1*) are added algebraically to the values ΔE, ΔN in columns 8 and 9 to produce the balanced values shown in columns 12 and 13.

The final step is to algebraically add the values in columns 12 and 13 to the coordinates in the previous row of columns 14 and 15 respectively to produce the coordinates of each point in turn, as shown in the final three columns of *Table 6.1*.

6.2.5 Link traverse adjustment

A link traverse (*Figure 6.22*) commences from known stations, A and B, and connects to known stations C and D. Stations A, B, C and D are usually fixed to a higher order of accuracy. Their values remain unaltered

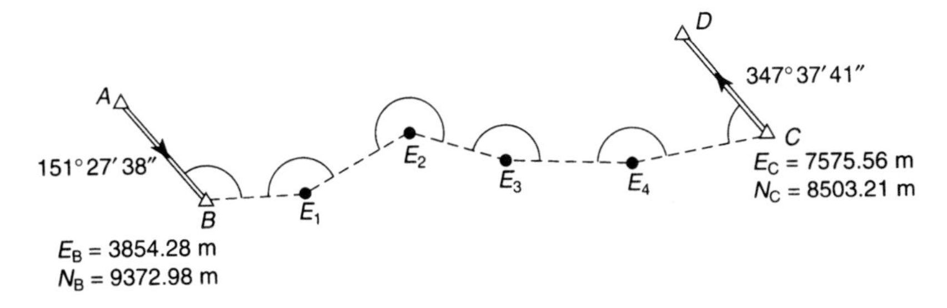

Fig. 6.22 *Link traverse adjustment*

in the subsequent computation. The method of computation and adjustments proceeds as follows:

(1) Angular adjustment

(1) Compute the WCB of CD through the traverse from AB and compare it with the known bearing of CD. The difference (Δ) of the two bearings is the angular misclosure.
(2) As a check on the value of Δ the following rule may be applied. Computed WCB of CD = (sum of observed angles + initial bearing (AB)) $- n \times 180°$ where n is the number of angles. If the result is outside the range 0°–360° add or subtract 360° as appropriate.
(3) The correction per angle would be Δ/n, which is distributed accumulatively over the WCBs as shown in columns 5 and 6 of *Table 6.2*.

(2) Coordinate adjustment

(1) Compute the initial coordinates of C through the traverse from B as origin. Comparison with the known coordinates of C gives the coordinate misclosure $\Delta'E$, and $\Delta'N$.
(2) As the computed coordinates are full, not partial, coordinates, distribute the misclosure accumulatively over stations E_1 to C.

Now study the example given in *Table 6.2*.

6.2.6 The effect of the balancing procedure

The purpose of this section is to show that balancing a traverse does not in any way improve it; it simply makes the figure geometrically correct.

The survey stations set in the ground represent the ‘true’ traverse, which in practice is unknown. Observation of the angles and distances is an attempt to obtain the value of the true traverse. It is never achieved, due to observational error, and hence we have an ‘observed’ traverse, which may approximate very closely to the ‘true’, but is not geometrically correct, i.e. there is coordinate misclosure. Finally, we have the ‘balanced’ traverse after the application of the Bowditch rule. This traverse is now geometrically correct, but in the majority of cases will be significantly different from both the ‘true’ and ‘observed’ network.

As field data are generally captured to the highest accuracy possible, relative to the expertise of the surveyor and the instrumentation used, it could be argued that the best balancing process is that which alters the field data the least.

Basically the Bowditch rule adjusts the positions of the traverse stations, resulting in changes to the observed data. For instance, it can be shown that the changes to the angles will be equal to:

$$\delta\alpha_i = 2\cos\frac{\alpha_i}{2}(\Delta'E\cos\beta + \Delta'N\sin\beta)/\Sigma L \qquad (6.11)$$

Table 6.2 *Bowditch adjustment of a link traverse*

Stns	*Observed angles* ° ′ ″	*Line*	*WCB* ° ′ ″	*Corrn*	*Adjusted WCB* ° ′ ″	*Dist* (m)	*Unadjusted* E	*Unadjusted* N	*Corrn* δE	*Corrn* δN	*Adjusted* E	*Adjusted* N	*Stn*
A		A–B	151 27 38		151 27 38		3854.28	9372.98			3854.28	9372.98	B
B	143 54 47	B–E_1	115 22 25	−4	115 22 21	651.16	4442.63	9093.96	+0.03	−0.05	4442.66	9093.91	E_1
E_1	149 08 11	E_1–E_2	84 30 36	−8	84 30 28	870.92	5309.55	9177.31	+0.08	−0.11	5309.63	9711.20	E_2
E_2	224 07 32	E_2–E_3	128 38 08	−12	128 37 56	522.08	5171.38	8851.36	+0.11	−0.15	5171.49	8851.21	E_3
E_3	157 21 53	E_3–E_4	106 00 01	−16	105 59 45	1107.36	6781.87	8546.23	+0.17	−0.22	6782.04	8546.01	E_4
E_4	167 05 15	E_4–C	93 05 16	−20	93 04 56	794.35	7575.35	8503.49	+0.21	−0.28	7575.56	8503.21	C
C	74 32 48	C–D	347 38 04	−23	347 37 41								
D		C–D	347 37 41		Sum =	3945.87	7575.56	8503.21					
Sum	916 10 26	$\Delta =$	+23		$\Delta' E, \Delta' N$		−0.21	+0.28					
Initial bearing	151 27 38												
Total	1067 38 04												
$-6 \times 180°$	1080 00 00												
	−12 21 56	*Error vector* $= (0.21^2 + 0.28^2)^{\frac{1}{2}} = 0.35$											
	+360 00 00												
CD (comp)	347 38 04	*Proportional error* $= \frac{0.35}{3946} = 1/11\,300$											
CD (known)	347 37 41	Check											
$\Delta =$	+23												

where β is the mean bearing of the lines subtending the angle α. This does not, however, apply to the first and last angle, where the corrections are:

$$\delta\alpha_1 = -(\Delta'E \sin \beta_1 + \Delta'N \cos \beta_1)/\Sigma L \tag{6.12}$$

$$\delta\alpha_n = +(\Delta'E \sin \beta_n + \Delta'N \cos \beta_n)/\Sigma L \tag{6.13}$$

The Bowditch adjustment results in changes to the distances equal to

$$\delta_L = \frac{fL}{t}(\Delta'N \cos \beta + \Delta'E \sin \beta) \tag{6.14}$$

where f = the factor of proportion
t = the error vector.

It can be seen from *equation (6.11)* that in a relatively straight traverse, where the angle (α) approximates to 180°, the corrections to the angles ($\delta\alpha$) will be zero for all but the first and last angles.

6.2.7 Accuracy of traversing

The weak geometry of a traverse means that it generally has only three degrees of freedom (that is three redundant observations), and so it is difficult to estimate its accuracy. Add to that the arbitrary adjustment by the Bowditch method and it becomes virtually impossible. Although there have been many attempts to produce equations defining the accuracy of a traverse, the best approach is to use a least squares adjustment of the traverse network to assess the uncertainty in the derived coordinates. See *Chapter 7*.

6.2.8 Blunders in the observed data

Blunders or mistakes in the measurement of the angles results in gross angular misclosure. Provided there is only a single blunder it can easily be located.

In the case of an angle, the traverse can be computed forward from X (*Figure 6.23*) and then backwards from Y. The point which has the same coordinates in each case, is where the blunder occurred and the angle must be reobserved. This process can be carried out by plotting using a protractor and scale. Alternatively the right angled bisector of the error vector YY' of the plotted traverse, will pass through the required point (*Figure 6.23*). The theory is that BYY' forms an equilateral triangle.

In the case of a blunder in measuring distance, the incorrect leg is the one whose bearing is similar to the bearing of the error vector. If there are several legs with similar bearings the method fails. Again the incorrect leg must be remeasured.

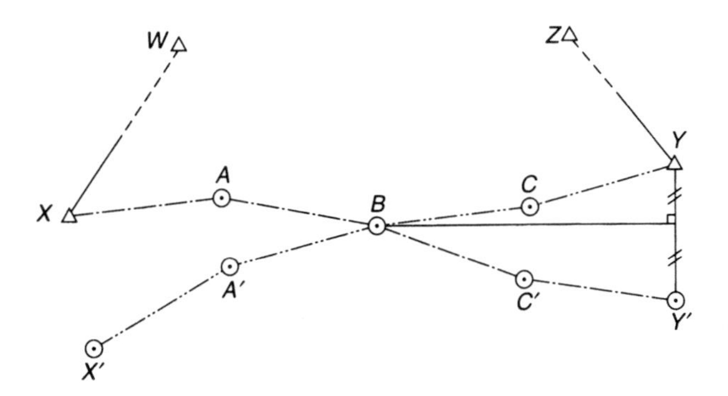

Fig. 6.23 *Detection of angle error in a traverse*

6.3 TRIANGULATION

Because, at one time, it was easier to measure angles than it was distance, triangulation was the preferred method of establishing the position of control points.

Many countries used triangulation as the basis of their national mapping system. The procedure was generally to establish primary triangulation networks, with triangles having sides ranging from 30 to 50 km in length. The primary trig points were fixed at the corners of these triangles and the sum of the measured angles was correct to $\pm 3''$. These points were usually established on the tops of mountains to afford long, uninterrupted sight lines. The primary network was then densified with points at closer intervals connected into the primary triangles. This secondary network had sides of 10–20 km with a reduction in observational accuracy. Finally, a third-order net, adjusted to the secondary control, was established at 3–5-km intervals and fourth-order points fixed by intersection. *Figure 6.24* illustrates such a triangulation system established by the Ordnance Survey of Great Britain and used as control for the production of national maps. The base line and check base line were measured by invar tapes in catenary and connected into the triangulation by angular extension procedures. This approach is classical triangulation, which is

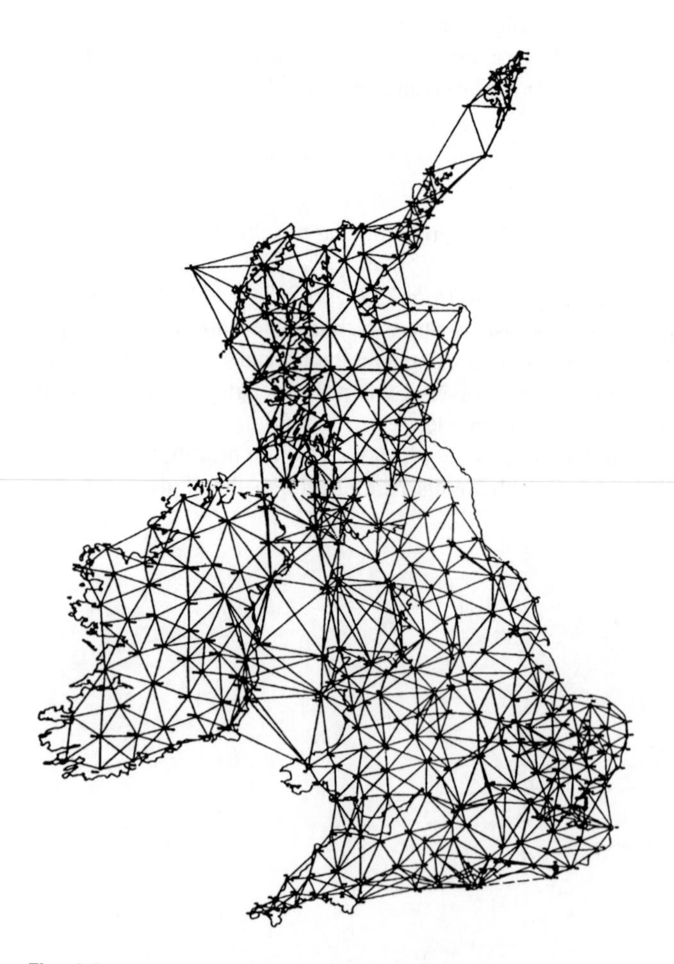

Fig. 6.24 *An example of a triangulation network*

now obsolete. The more modern approach would be to use GPS which would be much easier and would afford greater control of scale error.

Although the areas involved in construction are relatively small compared with national surveys the accuracy required in establishing the control is frequently of a very high order, e.g. for long tunnels or for dam deformation measurements. There are two useful elements from triangulation that still remain applicable. If it is not possible to set a target over the point being observed then a distance cannot be measured. If the inaccessible point is to be coordinated from known points then the process is one of intersection. If the inaccessible point has known coordinates and the instrument station is to be coordinated then the process is one of resection.

6.3.1 Resection and intersection

Using these techniques, one can establish the coordinates of a point P, by observations to or from known points. These techniques are useful for obtaining the position of single points, to provide control for setting out or detail survey in better positions than the existing control may be.

(1) Intersection

This involves sighting in to P from known positions (*Figure 6.25*). If the bearings of the rays are used, then using the rays in combinations of two, the coordinates of P are obtained as follows:

In *Figure 6.26* it is required to find the coordinates of P, using the *bearings* α and β to P from known points A and B whose coordinates are E_A, N_A and E_B, N_B.

$$PL = E_P - E_A \qquad AL = N_P - N_A$$
$$PM = E_P - E_B \qquad MB = N_P - N_B$$

Now as $\quad PL = AL \tan\alpha \qquad (6.15)$

then $\quad E_P - E_A = (N_P - N_A)\tan\alpha$

Similarly $\quad PM = MB \tan\beta$

then $\quad E_P - E_B = (N_P - N_B)\tan\beta \qquad (6.16)$

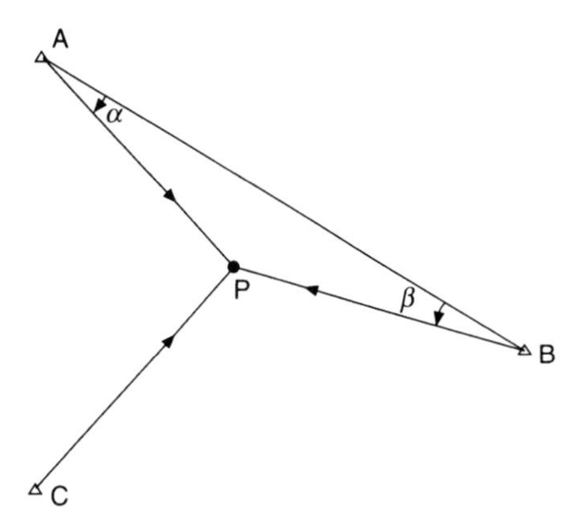

Fig. 6.25 *Intersection by angles*

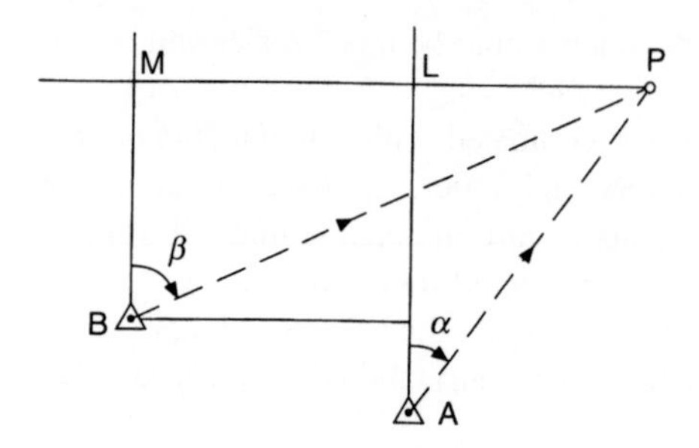

Fig. 6.26 *Intersection by bearings*

Subtracting *equation (6.16)* from *(6.15)* gives

$$E_B - E_A = (N_P - N_A)\tan\alpha - (N_P - N_B)\tan\beta$$

$$= N_P\tan\alpha - N_A\tan\alpha - N_P\tan\beta + N_B\tan\beta$$

$$\therefore N_P(\tan\alpha - \tan\beta) = E_B - E_A + N_A\tan\alpha - N_B\tan\beta$$

Thus $$N_P = \frac{E_B - E_A + N_A\tan\alpha - N_B\tan\beta}{\tan\alpha - \tan\beta} \quad (6.17)$$

Similarly $$N_P - N_A = (E_P - E_A)\cot\alpha$$

$$N_P - N_B = (E_P - E_B)\cot\beta$$

Subtracting $$N_B - N_A = (E_P - E_A)\cot\alpha - (E_P - E_B)\cot\beta$$

Thus $$E_P = \frac{N_B - N_A + E_A\cot\alpha - E_B\cot\beta}{\cot\alpha - \cot\beta} \quad (6.18)$$

Using *equations (6.17)* and *(6.18)* the coordinates of P are computed. It is assumed that P is always to the right of $A \rightarrow B$, in the equations.

If the observed angles α and β, measured at A and B are used (*Figure 6.25*) the equations become

$$E_P = \frac{N_B - N_A + E_A\cot\beta + E_B\cot\alpha}{\cot\alpha + \cot\beta} \quad (6.19)$$

$$N_P = \frac{E_A - E_B + N_A\cot\beta + N_B\cot\alpha}{\cot\alpha + \cot\beta} \quad (6.20)$$

The above equations are also used in the direct solution of triangulation. The inclusion of an additional ray from C, affords a check on the observations and the computation.

(2) Resection

This involves the angular measurement from P out to the known points A, B, C (*Figure 6.27*). It is an extremely useful technique for quickly fixing position where it is best required for setting-out purposes. Where only three known points are used a variety of analytical methods is available for the solution of P.

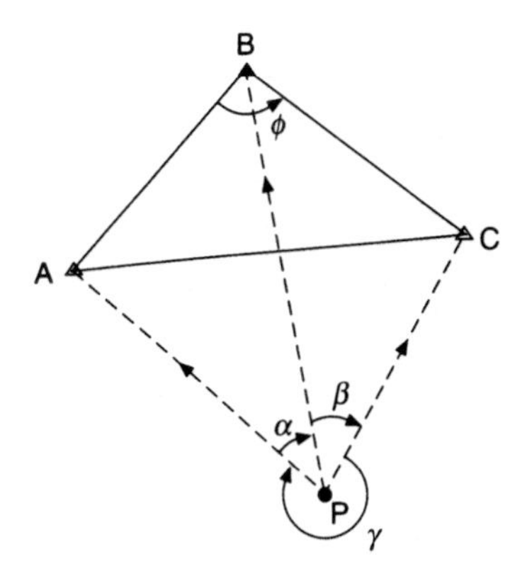

Fig. 6.27 *Resection*

Method 1 From *Figure 6.27*.

Let $BAP = \theta$, then $PCB = (360° - \alpha - \beta - \phi) - \theta = S - \theta$

where ϕ is computed from the coordinates of stations A, B and C; thus S is known.

$$\text{From } \Delta PAB \quad PB = BA \sin\theta / \sin\alpha \tag{6.21}$$

$$\text{From } \Delta PBC \quad PB = BC \sin(S - \theta) / \sin\beta \tag{6.22}$$

Equating *(6.21)* and *(6.22)*:

$$\frac{\sin(S - \theta)}{\sin\theta} = \frac{BA \sin\beta}{BC \sin\alpha} = Q \text{ (known)} \tag{6.23}$$

then $(\sin S \cos\theta - \cos S \sin\theta) / \sin\theta = Q$

$$\sin S \cot\theta - \cos S = Q$$

$$\therefore \cot\theta = (Q + \cos S) / \sin S \tag{6.24}$$

Thus, knowing θ and $(S - \theta)$, the triangles can be solved for lengths and bearings AP, BP and CP, and three values for the coordinates of P obtained if necessary.

The method fails, as do all three-point resections, if P lies on the circumference of a circle passing through A, B and C because it has an infinite number of possible positions which are all on the same circle.

Worked example

Example 6.3 The coordinates of A, B and C (*Figure 6.27*) are:

E_A 1234.96 m	N_A 17 594.48 m
E_B 7994.42 m	N_B 24 343.45 m
E_C 17 913.83 m	N_C 21 364.73 m

Observed angles are:

$$APB = \alpha = 61^\circ 41' 46.6''$$

$$BPC = \beta = 74^\circ 14' 58.1''$$

Find the coordinates of P.

(1) From the coordinates of A and B:

$$\Delta E_{AB} = 6759.46, \ \Delta N_{AB} = 6748.97$$

$$\therefore \text{Horizontal distance } AB = (\Delta E^2 + \Delta N^2)^{\frac{1}{2}} = 9551.91 \text{ m}$$

$$\text{Bearing } AB = \tan^{-1}(\Delta E/\Delta N) = 45^\circ 02' 40.2''$$

(or use the R → P keys on pocket calculator)

(2) Similarly from the coordinates of B and C:

$$\Delta E_{BC} = 9919.41 \text{ m}, \ \Delta N_{BC} = -2978.72 \text{ m}$$

$$\therefore \text{Horizontal distance } BC = 10\,357.00 \text{ m}$$

$$\text{Bearing } BC = 106^\circ 42' 52.6''$$

From the bearings of AB and BC:

$$C\hat{B}A = \phi = 180^\circ 19' 47.6''$$

(3)

$$S = (360^\circ - \alpha - \beta - \phi) = 105.724\,352^\circ$$

$$\text{and } Q = AB \sin \beta / BC \sin \alpha = 1.008\,167$$

$$\therefore \cot \theta = (Q + \cos S)/ \sin S, \text{ from which}$$

$$\theta = 52.554\,505^\circ$$

(4)

$$BP = AB \sin \theta / \sin \alpha = 8613.32 \text{ m}$$

$$BP = BC \sin(S - \theta)/ \sin \beta = 8613.32 \text{ m (Check)}$$

$$\text{Angle } CBP = 180^\circ - [\beta + (S - \theta)] = 52.580\,681^\circ$$

$$\therefore \text{Bearing } BP = \text{Bearing } BC + C\hat{B}P = 159.295\,29^\circ = \delta$$

Now using length and bearing of BP, transform to rectangular coordinate by formulae or P → R keys.

$$\Delta E_{BP} = BP \sin \delta = 3045.25 \text{ m}$$

$$\Delta N_{BP} = BP \cos \delta = -8057.03 \text{ m}$$

$$E_P = E_B + \Delta E_{BP} = 11\,039.67 \text{ m}$$

$$N_P = N_B + \Delta N_{BP} = 16\,286.43 \text{ m}$$

Checks on the observations and the computation can be had by computing the coordinates of P using the length and bearing of AP and CP.

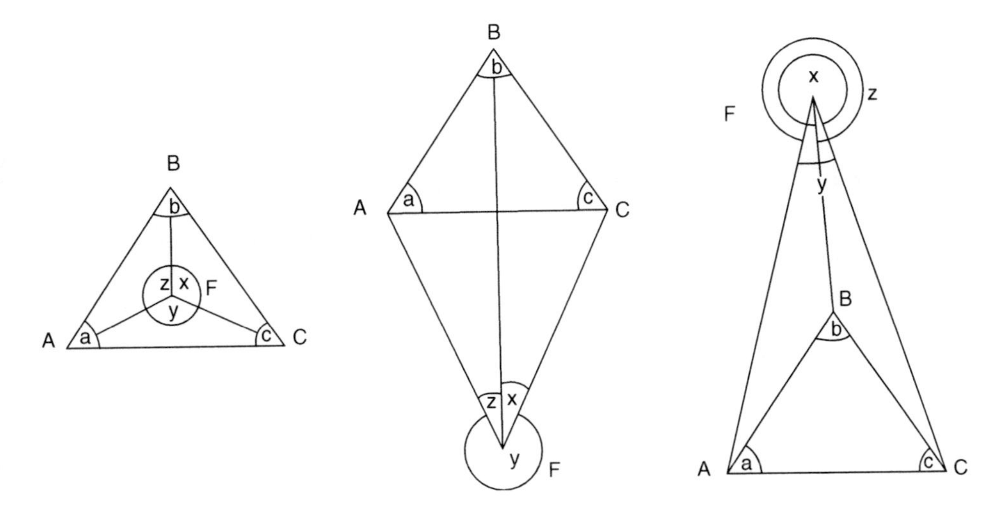

Fig. 6.28 *Resection figures*

Method 2 A, B and C (*Figure 6.27*) are fixed points whose coordinates are known, and the coordinates of the circle centres O_1 and O_2 are:

$$E_1 = \tfrac{1}{2}[E_A + E_B + (N_A - N_B)\cot\alpha]$$

$$N_1 = \tfrac{1}{2}[N_A + N_B - (E_A - E_B)\cot\alpha]$$

$$E_2 = \tfrac{1}{2}[E_B + E_C + (N_B - N_C)\cot\beta]$$

$$N_2 = \tfrac{1}{2}[N_B + N_C - (E_B - E_C)\cot\beta]$$

Thus the bearing δ of $O_1 \to O_2$ is obtained in the usual way, i.e.

$$\delta = \tan^{-1}[(E_2 - E_1)/(N_2 - N_1)]$$

then $$E_P = E_B + 2[(E_B - E_1)\sin\delta - (N_B - N_1)\cos\delta]\sin\delta \qquad (6.25)$$

$$N_P = N_B + 2[(E_B - E_1)\sin\delta - (N_B - N_1)\cos\delta]\cos\delta \qquad (6.26)$$

Method 3 'Tienstra's method' In the three point resection, angles are observed at the unknown station between each of three known stations. Angles at each of the known stations, between the other two known stations, are calculated from coordinates. Three intermediate terms, K_1, K_2 and K_3 are also computed. These are then used in conjunction with the coordinates of the known stations to compute the coordinates of the unknown station, as in the formulae below.

The coordinates of stations A, B and C are known (*Figure 6.28*). The angles x and y are measured. Angle z is calculated from the sum of angles in a circle. The angles a, b and c are computed from coordinates of stations A, B and C. The process is then to compute:

$$K_1 = \frac{1}{\cot a - \cot x} = \frac{\sin a \sin x}{\sin(x - a)}$$

$$K_2 = \frac{1}{\cot b - \cot y} = \frac{\sin b \sin y}{\sin(y - b)}$$

$$K_3 = \frac{1}{\cot c - \cot z} = \frac{\sin c \sin z}{\sin(z - c)}$$

and then compute the coordinates of F from

$$E_F = \frac{K_1E_A + K_2E_B + K_3E_C}{K_1 + K_2 + K_3}$$

$$N_F = \frac{K_1N_A + K_2N_B + K_3N_C}{K_1 + K_2 + K_3}$$

The notation in the diagram is all important in that the observed angles x, y and z, the computed angles a, b and c, and the stations A, B and C must all go in the same direction around the figure, clockwise or anti-clockwise. x must be the angle between the known stations A and B, measured from A to B, clockwise if lettering is clockwise, anti-clockwise if lettering is anti-clockwise. The point F need not lie within the triangle described by the known stations A, B and C but may lie outside, in which case the same rules for the order of the angles apply.

Intersection and resection can also be carried out using observed distances.

Although there are a large number of methods for the solution of a three-point resection, all of them fail if A, B, C and P lie on the circumference of a circle. Many of the methods also give dubious results if A, B and C lie in a straight line. C are should be exercised in the method of computation adopted and independent checks used wherever possible. Field configurations should be used which will clearly eliminate either of the above situations occurring; for example, siting P within a triangle formed by A, B and C, is an obvious solution.

Worked examples

Example 6.4 The following survey was carried out from the bottom of a shaft at A, along an existing tunnel to the bottom of a shaft at E.

Line	*WCB* °	′	″	*Measured distance* (m)	*Remarks*
AB	70	30	00	150.00	Rising 1 in 10
BC	0	00	00	200.50	Level
CD	154	12	00	250.00	Level
DE	90	00	00	400.56	Falling 1 in 30

If the two shafts are to be connected by a straight tunnel, calculate the bearing A to E and the grade.

If a theodolite is set up at A and backsighted to B, what is the value of the clockwise angle to be turned off, to give the line of the new tunnel? (KU)

$$\text{Horizontal distance } AB = \frac{150}{(101)^{\frac{1}{2}}} \times 10 = 149.25 \text{ m}$$

$$\text{Rise from } A \text{ to } B = \frac{150}{(101)^{\frac{1}{2}}} = 14.92 \text{ m}$$

$$\text{Fall from } D \text{ to } E = \frac{400.56}{(901)^{\frac{1}{2}}} = 13.34 \text{ m}$$

$$\text{Horizontal distance } DE = \frac{400.56}{(901)^{\frac{1}{2}}} \times 30 = 400.34 \text{ m}$$

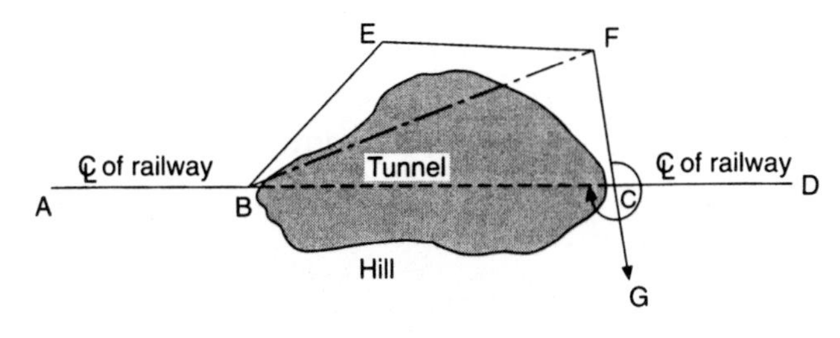

Fig. 6.29 *Railway centre-line*

Coordinates (ΔE, ΔN)	0	0	*A*
$149.25 \begin{smallmatrix}\sin\\\cos\end{smallmatrix} 70^\circ 30' 00''$	140.69	49.82	*B*
200.50 due N	0	200.50	*C*
$200.00 \begin{smallmatrix}\sin\\\cos\end{smallmatrix} 154^\circ 12' 00''$	108.81	−225.08	*D*
400.34 due E	400.34	0	*E*
Total coords of *E*	(*E*) 649.84	(*N*) 25.24	

$\therefore$ Tunnel is rising from *A* to *E* by $(14.92 - 13.34) = 1.58\,\text{m}$

$$\therefore \text{ Bearing } AE = \tan^{-1}\frac{+649.84}{+25.24} = 87^\circ\,47$$

$$\text{Length} = 649.84/\sin 87^\circ 47' = 652.33\,\text{m}$$

$$\text{Grade} = 1.58 \text{ in } 652.33 = 1 \text{ in } 413$$

$$\text{Angle turned off} = BAE = (87^\circ\,47' - 70^\circ\,30') = 17^\circ\,17'\,00''$$

Example 6.5 A level railway is to be constructed from *A* to *D* in a straight line, passing through a large hill situated between *A* and *D*. In order to speed the work, the tunnel is to be driven from both sides of the hill (*Figure 6.29*).

The centre-line has been established from *A* to the foot of the hill at *B* where the tunnel will commence, and it is now required to establish the centre-line on the other side of the hill at *C*, from which the tunnel will be driven back towards *B*.

To provide this data the following traverse was carried out around the hill:

Side	*Bearing* °	′	″	*Horizontal distance* (m)	*Remarks*
AB	88	00	00	–	Centre line of railway
BE	46	30	00	495.8 m	
EF	90	00	00	350.0 m	
FG	174	12	00	–	Long sight past hill

Calculate:

(1) The horizontal distance from F along FG to establish point C.
(2) The clockwise angle turned off from CF to give the line of the reverse tunnel drivage.
(3) The horizontal length of tunnel to be driven. (KU)

Find total coordinates of F relative to B

		ΔE (m)	ΔN (m)	*Station*
$495.8 \frac{\sin}{\cos} 46°\ 30'\ 00''$	→	359.6	341.3	*BE*
$350.0 - 90°\ 00'\ 00''$	→	350.0	–	*EF*
Total coordinates of F		*E* 709.6	*N* 341.3	*F*

$$\text{WCB of } BF = \tan^{-1}\frac{709.60}{341.30} = 64°\ 18'\ 48''$$

$$\text{Distance } BF = 709.60/\sin 64°\ 18'\ 48'' = 787.42\text{ m}$$

Solve triangle BFC for the required data.

The bearings of all three sides of the triangle are known, from which the following values for the angles are obtained:

$$\begin{aligned} \text{FBC} &= 23°\ 41'\ 12'' \\ \text{BCF} &= 86°\ 12'\ 00'' \\ \text{CFB} &= 70°\ 06'\ 48'' \\ \hline & 180°\ 00'\ 00''\ \text{(Check)} \end{aligned}$$

By sine rule:

(a) $$FC = \frac{BF \sin FBC}{\sin BCF} = \frac{787.42 \sin 23°\ 41'\ 12''}{\sin 86°\ 12'\ 00''} = 317.03\text{ m}$$

(b) $360° - BCF = 273°\ 48'\ 00''$

(c) $$BC = \frac{BF \sin CFB}{\sin BCF} = \frac{787.42 \sin 70°\ 06'\ 48''}{\sin 86°\ 12'\ 00''} = 742.10\text{ m}$$

Example 6.6

Table 6.3 *Details of a traverse ABCDEFA*

Line	*Length* (m)	*WCB*	ΔE (m)	ΔN (m)
AB	560.5		0	−560.5
BC	901.5		795.4	−424.3
CD	557.0		−243.0	501.2
DE	639.8		488.7	412.9
EF	679.5	293° 59′		
FA	467.2	244° 42′		

Adjust the traverse by the Bowditch method and determine the coordinates of the stations relative to A(0.0). What are the length and bearing of the line BE? (LU)

Complete the above table of coordinates:

	Line	ΔE (m)	ΔN (m)
$679.5 \begin{smallmatrix}\sin\\ \cos\end{smallmatrix} 293° 59'$	$\rightarrow EF$	−620.8	+276.2
$467.2 \begin{smallmatrix}\sin\\ \cos\end{smallmatrix} 244° 42'$	$\rightarrow FA$	−422.4	−199.7

Table 6.4

Line	*Lengths* (m)	ΔE (m)	ΔN (m)	*Corrected* ΔE	*Corrected* ΔN	*E*	*N*	*Stns*
A						0.0	0.0	*A*
AB	560.5	0	−560.5	0.3	−561.3	0.3	−561.3	*B*
BC	901.5	795.4	−424.3	795.5	−425.7	796.2	−987.0	*C*
CD	557.0	−243.0	501.2	−242.7	500.3	553.5	−486.7	*D*
DE	639.8	488.7	412.9	489.0	411.9	1042.5	−74.8	*E*
EF	679.5	−620.8	276.2	−620.4	275.2	422.1	200.4	*F*
FA	467.2	−422.4	−199.7	−422.1	−200.4	0.0	0.0	*A*
						Check	Check	
Sum	3805.5	−2.1	5.8	0.0	0.0			
Correction to coordinates		2.1	−5.8					

The Bowditch corrections (δE, δN) are computed as follows, and added algebraically to the coordinate differences, as shown in *Table 6.4*.

Line	δE (m)	δN (m)
	$\frac{2.1}{3805.5} \times 560.5$ giving	$\frac{-5.8}{3805.5} \times 560.5$ giving
AB	$K_2 \times 560.5 = 0.3$	$K_1 \times 560.5 = -0.8$
BC	$K_2 \times 901.5 = 0.5$	$K_1 \times 901.5 = -1.4$
CD	$K_2 \times 557.0 = 0.3$	$K_1 \times 557.0 = -0.9$
DE	$K_2 \times 639.8 = 0.3$	$K_1 \times 639.8 = -1.0$
EF	$K_2 \times 679.5 = 0.4$	$K_1 \times 679.5 = -1.0$
FA	$K_2 \times 467.2 = 0.3$	$K_1 \times 467.2 = -0.7$
	Sum = 2.1	Sum = −5.8

To find the length and bearing of *BE*:

$\Delta E = 1042.2, \Delta N = 486.5$

$$\therefore \text{Bearing } BE = \tan^{-1} \frac{1042.2}{486.5} = 64° 59'$$

$$\text{Length } BE = 1042.2/\sin 64° 59' = 1150.1 \text{ m}$$

6.4 NETWORKS

Simple survey figures have their limitations. If, as in intersection or resection, two angles are measured to find the two coordinates, easting and northing, of an unknown point then only the minimum number of observations have been taken and there is no check against error in either observations or the computations. A fully observed traverse is a little better in that there are always three more observations than the strict minimum. If the traverse has many stations then this redundancy of three is spread very thinly in terms of check against error.

A survey is designed for a specific purpose so that a technical or commercial objective can be achieved at a minimum cost. The major questions the planner will ask are: What is the survey for? How extensive must it be? With logistical constraints in mind, where is it? How precise and reliable does it need to be?

The first of these four questions puts the last into context. The second and third require administrative answers which are outside the scope of this book. The last question is entirely technical and is the most difficult to answer.

In this context the terms *precise* and *reliable* have specific meaning. Precision is a measure of the repeatability of the assessment of a parameter under question. Precise observations usually lead to precise coordinates. Accuracy is a measure of truth. Precision is related to accuracy in that it is a practical best estimate of accuracy because true values of survey quantities are never known, they can only be estimated. Measurement is an estimation process. It is therefore quite possible to have a set of measurements that are very precise but wholly inaccurate.

Reliability is an assessment of the fact that what has been found is what it appears to be. A distance measurement made with a tape from one control station to another could be in yards or metres, the difference between them is only about 10%. A way to be assured that the measure is in the units that you believe it to be in, would be to include measurements from other control points in the solution, so that it will be apparent that the suspect measurement does or does not fit at a certain level of statistical confidence.

Consider *Figure 6.30*, where the stations *A*, *B* and *C* are fixed stations with known coordinates. Stations *D* and *E* have yet to be computed. The distance measurements from *D* have been made by pacing and the directions to *A*, *B* and *C* observed with a handheld compass. The angle *ABE* has been observed with a theodolite and the distance *BE* measured with a tape measure. The quality of the observations that will be used to compute the coordinates of *D* is very poor so the precision of the computed coordinates of *D* will also be poor. The observations' reliability will be good because there are many more observations than the strict minimum necessary to estimate the coordinates of *D*.

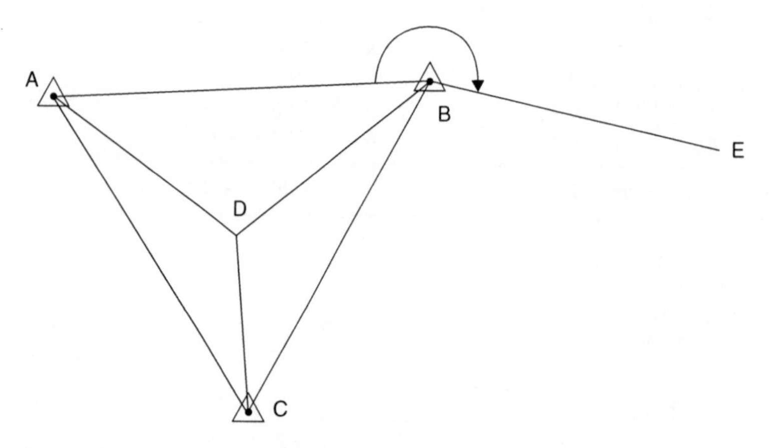

Fig. 6.30 *Precision and reliability*

On the other hand the coordinates of E may be precise because of the quality of the instrumentation that has been used but they will have zero reliability because a gross error in either of the two observations would not be detected. Consider the *independent check* in the context of this situation. If precision is the measure of repeatability, then reliability is the measure of assurance of absence of error.

Statements of precision and reliability may be obtained from the least squares adjustment of a survey that has been carried out. Precision and reliability may also be estimated for a survey that has yet to be undertaken, provided the number and quality of the proposed observations is known. These will then lead to an estimate of the cost of the work. If the estimate is not acceptable, then it may be possible to redesign the survey, by network analysis, or failing that, the requirements of precision and reliability may need to be reconsidered.

Exercises

(*6.1*) In a closed traverse *ABCDEFA* the angles and lengths of sides were measured and, after the angles had been adjusted, the traverse sheet shown below was prepared.

It became apparent on checking through the sheet that it contained mistakes. Rectify the sheet where necessary and then correct the coordinates by Bowditch's method. Hence, determine the coordinates of all the stations. The coordinates of A are E −235.5, N + 1070.0.

Line	*Length* (m)	*WCB*				*Reduced bearing*				ΔE (m)	ΔN (m)
		°	′	″		°	′	″			
AB	355.52	58	30	00	N	58	30	00	E	303.13	185.75
BC	476.65	185	12	30	S	84	47	30	W	−474.70	−43.27
CD	809.08	259	32	40	S	79	32	40	W	−795.68	−146.82
DE	671.18	344	35	40	N	15	24	20	W	−647.08	178.30
EF	502.20	92	30	30	S	87	30	30	E	501.72	−21.83
FA	287.25	131	22	00	S	48	38	00	E	215.58	−189.84

(*Answer*: Mistakes Bearing *BC* to S 5° 12′ 30″ W, hence ΔE and ΔN interchange. ΔE and ΔN of *DE* interchanged. Bearing *EF* to S 87° 29′ 30″ E, giving new ΔN of −21.97 m. Coordinates (*B*) E 67.27, N 1255.18; (*C*) E 23.51, N 781.19; (*D*) E −773.00, N 634.50; (*E*) E −951.99, N 1281.69; (*F*) E −450.78, N 1259.80)

(*6.2*) In a traverse *ABCDEFG*, the line *BA* is taken as the reference meridian. The coordinates of the sides *AB*, *BC*, *CD*, *DE* and *EF* are:

Line	*AB*	*BC*	*CD*	*DE*	*EF*
ΔN	−1190.0	−565.3	590.5	606.9	1017.2
ΔE	0	736.4	796.8	−468.0	370.4

If the bearing of *FG* is 284° 13′ and its length is 896.0 m, find the length and bearing of *GA*. (LU)

(*Answer*: 947.8 m, 216° 45′)

(6.3) The following measurements were obtained when surveying a closed traverse *ABCDEA*:

Line	*EA*	*AB*	*BC*	
Length (m)	793.7	1512.1	863.7	
Included angles	*DEA*	*EAB*	*ABC*	*BCD*
	93° 14′	112° 36′	131° 42′	95° 43′

It was not possible to occupy *D*, but it could be observed from *C* and *E*. Calculate the angle *CDE* and the lengths *CD* and *DE*, taking *DE* as the datum, and assuming all the observations to be correct. (LU)

(*Answer*: $CDE = 96°45'$, $DE = 1847.8$ m, $CD = 1502.0$ m)

(6.4) An open traverse was run from *A* to *E* in order to obtain the length and bearing of the line *AE* which could not be measured direct, with the following results:

Line	*AB*	*BC*	*CD*	*DE*
Length (m)	1025	1087	925	1250
WCB	261° 41′	09° 06′	282° 22′	71° 31′

Find, by calculation, the required information. (LU)

(*Answer*: 1620.0 m, 339° 46′)

(6.5) A traverse *ACDB* was surveyed by theodolite and tape. The lengths and bearings of the lines *AC*, *CD* and *DB* are given below:

Line	*AC*	*CD*	*DB*
Length (m)	480.6	292.0	448.1
Bearing	25° 19′	37° 53′	301° 00′

If the coordinates of *A* are $x = 0$, $y = 0$ and those of *B* are $x = 0$, $y = 897.05$, adjust the traverse and determine the coordinates of *C* and *D*. The coordinates of *A* and *B* must not be altered.

(*Answer*: Coordinate error: $x = 0.71$, $y = 1.41$. (*C*) $x = 205.2$, $y = 434.9$, (*D*) $x = 179.1$, $y = 230.8$).

7
Rigorous methods of control

7.1 INTRODUCTION

The control methods described in the last chapter are limited to specific figures such as intersection, resection and traverse. Only in the case of the traverse is more than the minimum number of observations taken and even then the computational method is arbitrary as far as finding the coordinates of the points from the observations is concerned. A much better, but more complex, process is to use least squares estimation; that is what this chapter is about.

'Least squares' is a powerful statistical technique that may be used for 'adjusting' or estimating the coordinates in survey control networks. The term adjustment is one in popular usage but it does not have any proper statistical meaning. A better term is 'least squares estimation' since nothing, especially observations, are actually adjusted. Rather, coordinates are estimated from the evidence provided by the observations.

The great advantage of least squares over all the methods of estimation, such as traverse adjustments, is that least squares is mathematically and statistically justifiable and, as such, is a fully rigorous method. It can be applied to any overdetermined network, but has the further advantage that it can be used on one-, two- and three-dimensional networks. A by-product of the least squares solution is a set of statistical statements about the quality of the solution. These statistical statements may take the form of standard errors of the computed coordinates, error ellipses or ellipsoids describing the uncertainty of a position in two or three dimensions, standard errors of observations derived from the computed coordinates and other meaningful statistics described later.

The major practical drawback with least squares is that unless the network has only a small number of unknown points, or has very few redundant observations, the amount of arithmetic manipulation makes the method impractical without the aid of a computer and appropriate software.

The examples and exercises in this material use very small networks in order to minimize the computational effort for the reader, while demonstrating the principles. Real survey networks are usually very much larger.

7.2 PRINCIPLE OF LEAST SQUARES

A 'residual' may be thought of as the difference between a computed and an observed value. For example, if in the observation and estimation of a network, a particular angle was observed to be $30° 0' 0''$ and after adjustment of the network the same angle computed from the adjusted coordinates was $30° 0' 20''$, then the residual associated with that observation would be $20''$. In other words:

$$\text{computed value} - \text{observed value} = \text{residual}$$

Any estimation of an overdetermined network is going to involve some change to the observations to make them fit the adjusted coordinates of the control points. The best estimation technique is the one where the observations are in best agreement with the coordinates computed from them. In least squares, at its simplest, the best agreement is achieved by minimizing the sum of the squares of the weighted residuals of all the observations.

7.3 LEAST SQUARES APPLIED TO SURVEYING

In practical survey networks, it is usual to observe more than the strict minimum number of observations required to solve for the coordinates of the unknown points. The extra observations are 'redundant' and can be used to provide an 'independent check' but all the observations can be incorporated into the solution of the network if the solution is by least squares.

All observations have errors so any practical set of observations will not perfectly fit any chosen set of coordinates for the unknown points.

Some observations will be of a better quality than others. For example, an angle observed with a 1″ theodolite should be more precise than one observed with a 20″ instrument. The weight applied to an observation, and hence to its residual, is a function of the previously assessed quality of the observation. In the above example the angle observed with a 1″ theodolite would have a much greater weight than one observed with a 20″ theodolite. How weights are calculated and used will be described later.

If all the observations are to be used, then they will have to be 'adjusted' so that they fit with the computed network. The principle of least squares applied to surveying is that the sum of the squares of the weighted residuals must be a minimum.

7.3.1 A simple illustration

A locus line is the line that a point may lie on and may be defined by a single observation. *Figure 7.1(a)*, (*b*) and (*c*) show the locus lines associated with an angle observed at a known point to an unknown point, a distance measured between a known point and an unknown point and an angle observed at an unknown point between two known points respectively. In each case the locus line is the dotted line. In each case all that can be concluded from the individual observation is that the unknown point lies somewhere on the dotted line, but not where it lies.

In the following, the coordinates of new point P are to be determined from horizontal angles observed at known points A, B, C and D as in *Figure 7.2(a)*. Each observation may be thought of as defining a locus line. For example, if only the horizontal angle at A had been observed then all that could be said about P would be that it lies somewhere on the locus line from A towards P and there could be no solution for the coordinates of P. With the horizontal angles at A and B there are two locus lines, from A towards P and from B towards P. The two lines cross at a unique point and if the observations had been perfect then the

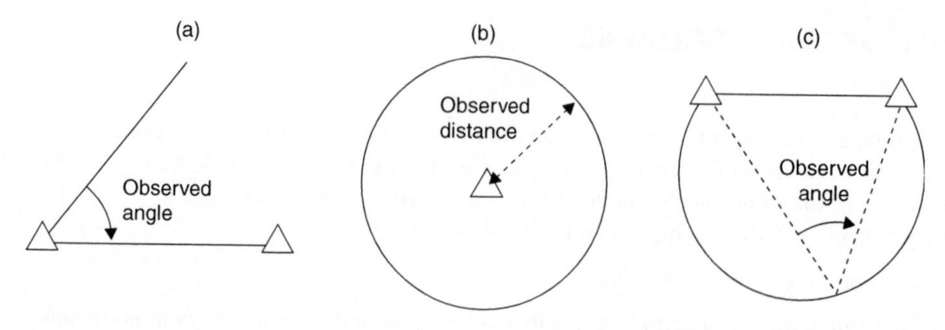

Fig. 7.1 *Locus lines*

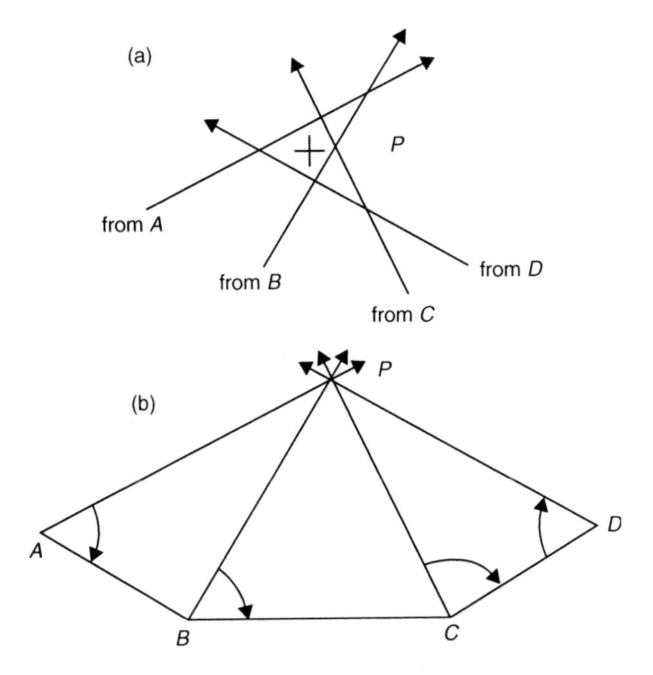

Fig. 7.2 *Intersection of locus lines*

unique point would be exactly at P. But since observations are never perfect when the horizontal angles observed at C and D are added to the solution the four locus lines do not all cross at the same point and the mismatch gives a measure of the overall quality of the observations. *Figure 7.2(b)* shows the detail at point P where the four lines intersect at six different points. The cross is at the unique point where the sum of the squares of the residuals is a minimum.

7.3.2 The mathematical tools

By far the easiest way to handle the enormous amounts of data associated with least squares estimation is to use matrix algebra. In least squares it is necessary to create a system of equations with one equation for each observation and each 'observation equation' contains terms for each of the coordinates of each of the unknown points connected by the observation. So, for example, in a two-dimensional network of 10 points where there are a total of 50 observations there would be a set of 50 simultaneous equations in 20 unknowns. Although this represents only a small network, the mathematical problem it presents would be too difficult to solve by simple algebraic or arithmetic methods. Readers unfamiliar with matrix algebra should first read Appendix A – An introduction to matrix algebra.

7.4 LINEARIZATION

In surveying, equations relating observations with other observations or coordinates are seldom in a linear form. For example, the observation equation for the observation of distance between two points A and B is defined by Pythagoras' equation which relates the differences of the eastings and northings of the points A and B with the distance between the two points.

$$l_{AB} - \{(E_A - E_B)^2 + (N_A - N_B)^2\}^{\frac{1}{2}} = 0$$

So the observation equations are not normally capable of being expressed in terms of a series of unknowns multiplied by their own numerical coefficients. The solution of a series of equations by matrix methods requires that this is so. For example, the following pair of equations:

$$3x + 4y = 24$$
$$4x + 3y = 25$$

can be expressed in matrix terms as:

$$\begin{bmatrix} 3 & 4 \\ 4 & 3 \end{bmatrix} \begin{bmatrix} x \\ y \end{bmatrix} = \begin{bmatrix} 24 \\ 25 \end{bmatrix}$$

and solved as:

$$\begin{bmatrix} x \\ y \end{bmatrix} = \begin{bmatrix} 3 & 4 \\ 4 & 3 \end{bmatrix}^{-1} \begin{bmatrix} 24 \\ 25 \end{bmatrix}$$

The same cannot be done with these non-linear equations!

$$(x - 3)^2 + (2y - 8)^2 - 5 = 0$$
$$(x - 4)^2 + (y - 10)^2 - 49 = 0$$

A route to the solution of these equations, but obviously not the only one, is to make use of the first part of a Taylor expansion of the two functions:

$$f_1(x, y) = 0$$
$$f_2(x, y) = 0$$

The application of Taylor's theorem leads to:

$$f_1(x, y) = 0 = f_1(x_0, y_0) + \left\{ \frac{\partial}{\partial x}(f_1(x, y)) \bigg|_{\substack{x=x_0 \\ y=y_0}} \right\}(x - x_0) + \left\{ \frac{\partial}{\partial y}(f_1(x, y)) \bigg|_{\substack{x=x_0 \\ y=y_0}} \right\}(y - y_0)$$
$$+ \text{ higher order terms} \qquad (7.1)$$

and

$$f_2(x, y) = 0 = f_2(x_0, y_0) + \left\{ \frac{\partial}{\partial x}(f_2(x, y)) \bigg|_{\substack{x=x_0 \\ y=y_0}} \right\}(x - x_0) + \left\{ \frac{\partial}{\partial y}(f_2(x, y)) \bigg|_{\substack{x=x_0 \\ y=y_0}} \right\}(y - y_0)$$
$$+ \text{ higher order terms} \qquad (7.2)$$

where x_0 and y_0 are estimated or provisional values of x and y. $f_1(x_0, y_0)$ is $f_1(x, y)$ where x and y take the values x_0 and y_0 and the notation | means 'under the condition that ...'. Therefore

$$\frac{\partial}{\partial x}(f_1(x, y)) \bigg|_{\substack{x=x_0 \\ y=y_0}}$$

is $f_1(x, y)$ differentiated with respect to x, and with x and y then taking the values x_0 and y_0. $(x - x_0)$ is now the correction to the provisional value of x since

$$x_0 + (x - x_0) = x$$

and so the equations are now linear in $(x - x_0)$ and $(y - y_0)$. A simple one-dimensional example may help to illustrate the mathematical statements above.

Worked example

Example 7.1 Use linearization to solve the equation $f(x) = x^3 - 7x - 6 = 0$ in the region of x is 3.5.

From the graph at *Figure 7.3* it is clear that a better estimate of the value of x may be found from taking the provisional value of $x_0 = 3.5$ and subtracting the x offset, $(x - x_0)$, described by the triangle. The x offset is $f(x)$ when $x = 3.5$, divided by the gradient of the tangent to the curve where x is 3.5, that is:

$$f_1(x) = 0 = f_1(x_0) + \left\{ \frac{\partial}{\partial x}(f_1(x))\bigg|_{x=x_0} \right\}(x - x_0)$$

where:

$$f_1(x) = 0 = x^3 - 7x - 6$$

$$x_0 = 3.5$$

$$f_1(x_0) = 12.375$$

$$\frac{\partial}{\partial x}(f_1(x)) = 3x^2 - 7$$

$$\frac{\partial}{\partial x}(f_1(x))\bigg|_{x=x_0} = 29.75$$

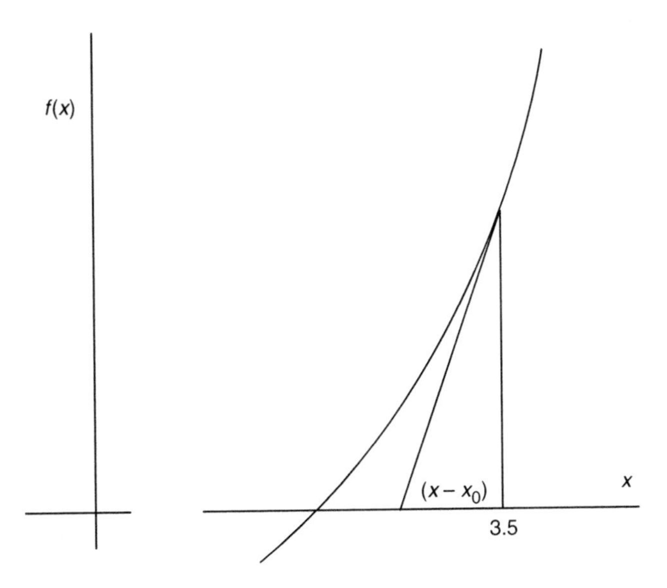

Fig. 7.3 *Linearization*

and on substituting these numbers into the above equation:

$$f_1(x) = 0 = 12.375 + 29.75(x - x_0)$$

$$(x - x_0) = \frac{-12.375}{29.75} = -0.416$$

so $\quad x = 3.5 - 0.416$
$\qquad = 3.084$

The exercise may now be repeated with x_0 taken as 3.084 to find a yet better estimate of x. This is left to the reader, who will find that the solution converges to $x = 3$ exactly.

This is as complex a problem as can be illustrated graphically. The Taylor series is, in effect, a multidimensional application of the idea expressed by the above example where each unknown coordinate represents a dimension. A more complex example follows.

Worked example

Example 7.2 Solve the equations

$$(x - 3)^2 + (2y - 8)^2 - 5 = 0$$

$$(x - 4)^2 + (y - 10)^2 - 49 = 0$$

by firstly linearizing them as the first part of a Taylor expansion. Use provisional values of x and y as:

$$x_0 = 5$$
$$y_0 = 4$$

From these equations:

$$f_1(x_0, y_0) = (x_0 - 3)^2 + (2y_0 - 8)^2 - 5 = -1$$

$$f_2(x_0, y_0) = (x_0 - 4)^2 + (y_0 - 10)^2 - 49 = -12$$

$$\frac{\partial}{\partial x}(f_1(x,y))\Big|_{\substack{x=x_0\\y=y_0}} = (2x - 6)\Big|_{\substack{x=x_0\\y=y_0}} = 4$$

$$\frac{\partial}{\partial x}(f_1(x,y))\Big|_{\substack{x=x_0\\y=y_0}} = (8y - 32)\Big|_{\substack{x=x_0\\y=y_0}} = 0$$

$$\frac{\partial}{\partial x}(f_2(x,y))\Big|_{\substack{x=x_0\\y=y_0}} = (2x - 8)\Big|_{\substack{x=x_0\\y=y_0}} = 2$$

$$\frac{\partial}{\partial x}(f_2(x,y))\Big|_{\substack{x=x_0\\y=y_0}} = (2y - 20)\Big|_{\substack{x=x_0\\y=y_0}} = -12$$

so the equations can be evaluated as:

$$0 = -1 + 4(x - x_0) + 0(y - y_0)$$

$$0 = -12 + 2(x - x_0) - 12(y - y_0)$$

In matrix terms the equation may be expressed as

$$\begin{bmatrix} 4 & 0 \\ 2 & -12 \end{bmatrix} \begin{bmatrix} (x - x_0) \\ (y - y_0) \end{bmatrix} = \begin{bmatrix} 1 \\ 12 \end{bmatrix}$$

and the solution is:

$$(x - x_0) = +0.25$$

$$(y - y_0) = -0.958333$$

which leads to:

$$x = 5.25$$
$$y = 3.041667$$

It will be found that these values do not satisfy the original equations. The reason is that the higher order terms, which were ignored, are significant when the provisional values are not very close to the true values. The closer the provisional values are to the true values, the faster the solution will converge with successive iterations. In this case, if the problem is now re-worked with the derived values of x and y above, then the next and subsequent solutions are:

x	y
4.226630	2.928312
4.017291	2.996559
4.000083	2.999978
...	...
4 exactly	3 exactly

In practical survey networks good provisional coordinates make a successful solution more likely and less iteration will be required.

7.5 DERIVATION OF THE LEAST SQUARES FORMULAE

In surveying, all equations involving observations can be reduced, by linearization where necessary, to:

$$\mathbf{Ax} = \mathbf{b} + \mathbf{v}$$

x is a vector of the terms to be computed which will include the coordinates
b is a vector which relates to the observations
A is a matrix of coefficients
v is a vector of the residuals

When describing the principle of least squares above it was stated that the best agreement between original observations and final coordinates is achieved by minimizing the sum of the squares of the weighted residuals of all the observations. More strictly, the best agreement is defined as being when the 'quadratic form', $\mathbf{v}^T \boldsymbol{\sigma}_{(\mathbf{b})}{}^{-1} \mathbf{v}$ is a minimum, where $\mathbf{v}$ is the vector of the residuals and $\boldsymbol{\sigma}_{(\mathbf{b})}$ is the variance-covariance matrix of the observations. If $\boldsymbol{\sigma}_{(\mathbf{b})}$ is a diagonal matrix, which it often is in survey problems, then the simple definition above will suffice.

The least squares requirement is to minimize the quadratic form $\mathbf{v}^T\boldsymbol{\sigma}_{(\mathbf{b})}{}^{-1}\mathbf{v}$. The term $\boldsymbol{\sigma}_{(\mathbf{b})}{}^{-1}$ is often written as **W** for simplicity and referred to as the weight matrix.

$$\mathbf{v}^T\boldsymbol{\sigma}_{(\mathbf{b})}{}^{-1}\mathbf{v} = \mathbf{v}^T\mathbf{W}\mathbf{v}$$

$$= (\mathbf{Ax} - \mathbf{b})^T\mathbf{W}(\mathbf{Ax} - \mathbf{b})$$

which when multiplied out gives:

$$\mathbf{v}^T\mathbf{W}\mathbf{v} = (\mathbf{Ax})^T\mathbf{WAx} - \mathbf{b}^T\mathbf{WAx} - (\mathbf{Ax})^T\mathbf{Wb} + \mathbf{b}^T\mathbf{Wb}$$

$$= \mathbf{x}^T\mathbf{A}^T\mathbf{WAx} - \mathbf{b}^T\mathbf{WAx} - \mathbf{x}^T\mathbf{A}^T\mathbf{Wb} + \mathbf{b}^T\mathbf{Wb}$$

To find the value of **x** when the quadratic form is a minimum, differentiate the function with respect to **x** and set it equal to 0.

$$\frac{d}{d\mathbf{x}}\left\{\mathbf{x}^T\mathbf{A}^T\mathbf{WAx} - \mathbf{b}^T\mathbf{WAx} - \mathbf{x}^T\mathbf{A}^T\mathbf{Wb} + \mathbf{b}^T\mathbf{Wb}\right\} = 0$$

The first three terms in the brackets { } are, in order, quadratic, bilinear and bilinear in **x**, so the equation becomes:

$$2\mathbf{A}^T\mathbf{WAx} - 2\mathbf{A}^T\mathbf{Wb} = 0$$

So: $\mathbf{A}^T\mathbf{WAx} - \mathbf{A}^T\mathbf{Wb}$

This part of the solution is usually referred to as the normal equations, and is solved for **x** as:

$$\mathbf{x} = (\mathbf{A}^T\mathbf{WA})^{-1}\mathbf{A}^T\mathbf{Wb}$$

The coordinates are only part of the solution. At this stage several sets of useful statistics may be derived and these will be discussed later. An examination of the residuals will show quickly if there are any major discrepancies in the observations.

Since $\mathbf{Ax} = \mathbf{b} + \mathbf{v}$

then $\mathbf{v} = \mathbf{Ax} - \mathbf{b}$

Each row in the **A** matrix represents the coefficients of the terms in the **x** vector that arise from one observation equation. The **b** vector contains all the numerical terms in the observation equation, i.e. those that are not coefficients of terms in the **x** vector. The weight matrix **W** applies the right level of importance to each of the observation equations and **v**, the residuals vector, describes how well the observations fit the computed coordinates. In the following sections of this chapter the terms of the above equation will be more fully described.

7.6 PARAMETER VECTOR

The parameter vector, or **x** vector, lists the parameters to be solved for. Usually these will be coordinates of the unknown points, if the solution is to be a direct one. More often the **x** vector is a vector of corrections to provisional values that have been previously computed or are guessed best estimates of the coordinates of the unknown points. This latter method is known as 'variation of coordinates'.

When formulating the problem the **x** vector is specified first. Once the order of terms in this vector has been decided that dictates the order of terms in each row of the **A** matrix. The parameter vector may also contain a few other terms that need to be solved for at the same time as the coordinates. Two examples might be the local or instrumental scale factor or the orientation of the horizontal circle of the theodolite at each point if directions rather than horizontal angles are observed. Some examples of an **x** vector are:

Height network direct solution	Height network by variation of coordinates	Plan network in 2D by variation of coordinates	Plan network in 3D by variation of coordinates
$\begin{bmatrix} H_W \\ H_X \\ H_Y \\ H_Z \end{bmatrix}$	$\begin{bmatrix} \Delta H_W \\ \Delta H_X \\ \Delta H_Y \\ \Delta H_Z \end{bmatrix}$	$\begin{bmatrix} \Delta E_X \\ \Delta N_X \\ \Delta E_Y \\ \Delta N_Y \\ \Delta E_Z \\ \Delta N_Z \end{bmatrix}$	$\begin{bmatrix} \Delta E_X \\ \Delta N_X \\ \Delta H_X \\ \Delta E_Y \\ \Delta N_Y \\ \Delta H_Y \\ \Delta E_Z \\ \Delta N_Z \\ \Delta H_Z \end{bmatrix}$

The only networks that can be solved directly are those of height, or of difference of position such as may be obtained from the Global Positioning System. All others must be by variation of coordinates because the observation equations cannot be formed as a series of coefficients multiplied by individual parameters. In other words the observation equations must first be linearized and its application will be covered in the section on the design matrix below. In variation of coordinates the **x** vector is a vector of corrections to provisional values. The final adjusted coordinates are then computed from

$$\hat{\mathbf{x}} = \mathbf{x_p} + \delta\mathbf{x}$$

where

$\hat{\mathbf{x}}$ is the final vector of adjusted coordinates
$\mathbf{x_p}$ is the vector of provisional values
$\delta\mathbf{x}$ is the computed vector of corrections to the provisional values.

The dimensions of the **x** vector are $n \times 1$ where n is the number of parameters to be solved for.

7.7 DESIGN MATRIX AND OBSERVATIONS VECTOR

The design matrix and the observations vector will be dealt with together in this section because the elements of both are derived from the same theoretical considerations. The observations vector contains more than just the observations. For each linearized observation equation there is an element in the observations vector that represents the 'observed minus computed' term, described below, and the vector of these terms is often given this name.

7.7.1 Height network

Before dealing with observation equations that need to be linearized, it is worth examining the simple case of a one-dimensional height network. In *Figure 7.4* point A is a datum point and its height is considered fixed. Levelling has been undertaken between the other points in order to find the heights of points X, Y and Z.

The six observations are Δh_{AX}, Δh_{AY}, Δh_{AZ}, Δh_{YX}, Δh_{ZX} and Δh_{ZY}. h_A is the height of point A, etc. The six equations linking the observations of height difference with the parameters, the heights of the

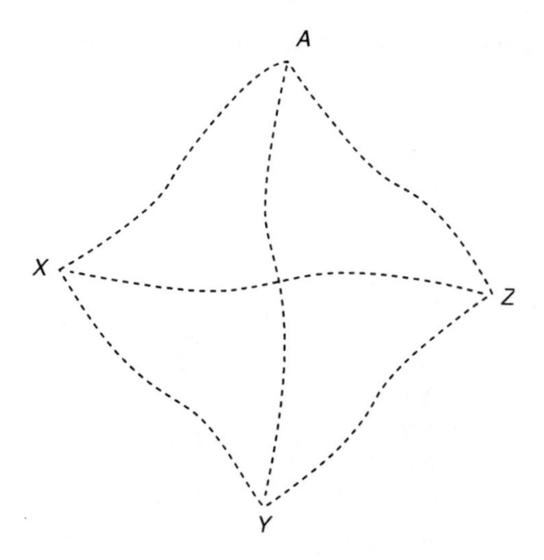

Fig. 7.4 *A levelling network*

points, are:

$$\begin{aligned}
h_X - h_A &= \Delta h_{AX} + v_1 \\
h_Y - h_A &= \Delta h_{AY} + v_2 \\
h_Z - h_A &= \Delta h_{AZ} + v_3 \\
h_X - h_Y &= \Delta h_{YX} + v_4 \\
h_X - h_Z &= \Delta h_{ZX} + v_5 \\
h_Y - h_Z &= \Delta h_{ZY} + v_6
\end{aligned}$$

v_1 is the residual associated with the first observation, etc. Since the height of A is known, these equations can be rewritten to leave only the unknowns on the left-hand side of the equations.

$$\begin{aligned}
h_X &= h_A + \Delta h_{AX} + v_1 \\
h_Y &= h_A + \Delta h_{AY} + v_2 \\
h_Z &= h_A + \Delta h_{AZ} + v_3 \\
h_X - h_Y &= \Delta h_{YX} + v_4 \\
h_X - h_Z &= \Delta h_{ZX} + v_5 \\
h_Y - h_Z &= \Delta h_{ZY} + v_6
\end{aligned}$$

and this may be expressed in matrix terms as:

$$\begin{bmatrix} 1 & 0 & 0 \\ 0 & 1 & 0 \\ 0 & 0 & 1 \\ 1 & -1 & 0 \\ 1 & 0 & -1 \\ 0 & 1 & -1 \end{bmatrix} \begin{bmatrix} h_X \\ h_Y \\ h_Z \end{bmatrix} = \begin{bmatrix} h_A + \Delta h_{AX} \\ h_A + \Delta h_{AY} \\ h_A + \Delta h_{AZ} \\ \Delta h_{YX} \\ \Delta h_{ZX} \\ \Delta h_{ZY} \end{bmatrix} + \begin{bmatrix} v_1 \\ v_2 \\ v_3 \\ v_4 \\ v_5 \\ v_6 \end{bmatrix}$$

This is in the form:

$$\mathbf{Ax} = \mathbf{b} + \mathbf{v}$$

7.7.2 Plan network

Plan networks are solved by variation of coordinates because the observation equations are not linear in the unknowns. In the following pages, the observation equation for each type of observation is developed, linearized, and presented in a standard format. Finally, all the elements that may be in the design matrix are summarized.

The way to set up the linearized equation is as follows.

(1) Form an equation relating the true values of the observations to the true values of the parameters, preferably, but not essentially, in the form

$$\text{observation} = f(\text{parameters}) \quad \{\bar{l} = f(\bar{x})\}$$

(2) Make the equation equate to zero by subtracting one side from the other. The equation becomes:

$$f(\text{observation, parameters}) = 0 \quad \{f(\bar{x}, \bar{l}) = 0\}$$

(3) Linearize as the first part of a Taylor series. The equations, expressed in the section on linearization above, may be rewritten as:

$$f_1(\bar{x}, \bar{l}) = 0 = f_1(x_p, l_o) + \sum \left\{ \frac{\partial}{\partial x}(f_1(x, l)) \bigg|_{\substack{x=x_p \\ l=l_o}} \right\} (x - x_p) + \sum \left\{ \frac{\partial}{\partial y}(f_1(x, l)) \bigg|_{\substack{x=x_p \\ l=l_o}} \right\} (l - l_o)$$

Where $f(\bar{x}, \bar{l})$ is the true value of the function with the true values of the parameters and the true values of the observations. From step 2 above, it must, by definition be zero.

$f(x_p, l_o)$ is the same function but the parameters take their provisional values and the observations their observed values.

$$\left\{ \frac{\partial}{\partial x} f(x, l) \bigg|_{\substack{x=x_p \\ l=l_o}} \right\} (x - x_p)$$

is the function differentiated with respect to a parameter and evaluated using provisional values of the parameters and observed values of the observations. It is then multiplied by the difference between the true value of that parameter and its provisional value. In observation equations, if $f(x, l)$ has been set up in the form described in step 1, then the differentiated function does not contain any observation terms. The term $(x - x_p)$ is what is solved for.

$$\left\{ \frac{\partial}{\partial l} f(x, l) \bigg|_{\substack{x=x_p \\ l=l_o}} \right\} (l - l_o)$$

is a similar function to the one above, except the differentiation is with respect to the observations and $(l - l_o)$ is the residual. These terms appear in observation equations as 1 if the function is formed as in steps 1 and 2 above.

From the above the linearized equation becomes:

$$-\sum \left\{ \frac{\partial}{\partial x}(f_1(x, l)) \bigg|_{\substack{x=x_p \\ l=l_o}} \right\} (x - x_p) = f(x_p, l_o) + (l - l_o)$$

which in matrix terms is one row of the **A** matrix times the **x** vector equals an element of the **b** vector plus the residual. The above will now be applied to the usual survey observation types for a plan, or two-dimensional estimation.

7.7.3 Distance equation

Following the three steps above the distance between two points i and j can be related to their eastings and northings by the equation:

$$l_{ij} - \left\{(E_j - E_i)^2 + (N_j - N_i)^2\right\}^{\frac{1}{2}} = 0$$

Differentiating with respect to the parameters gives:

$$\frac{(E_j - E_i)}{\left\{(E_j - E_i)^2 + (N_j - N_i)^2\right\}^{\frac{1}{2}}}\delta E_j + \frac{-(E_j - E_i)}{\left\{(E_j - E_i)^2 + (N_j - N_i)^2\right\}^{\frac{1}{2}}}\delta E_i + \frac{(N_j - N_i)}{\left\{(E_j - E_i)^2 + (N_j - N_i)^2\right\}^{\frac{1}{2}}}\delta N_j$$
$$+ \frac{-(N_j - N_i)}{\left\{(E_j - E_i)^2 + (N_j - N_i)^2\right\}^{\frac{1}{2}}}\delta N_i = l_{ij} - \left\{(E_j - E_i)^2 + (N_j - N_i)^2\right\}^{\frac{1}{2}}$$

In the four terms on the left-hand side of the equation, the coefficient may be re-expressed as a trigonometrical function of the direction of point j from point i. δE_j is the correction to the provisional value of E_j, etc. On the right-hand side of the equation $\{(E_j - E_i)^2 + (N_j - N_i)^2\}^{\frac{1}{2}}$ is the distance l_{ij} derived from provisional values of the parameters. The right-hand side could now be written as $l_{ij(o-c)}$ where the notation indicates that it is the observed value of the observation minus the computed value of the observation. With a change of sign and re-ordering the terms, the equation may now be more simply expressed in matrix notation as:

$$\begin{bmatrix} -\sin a_{ij} & -\cos a_{ij} & \sin a_{ij} & \cos a_{ij} \end{bmatrix} \begin{bmatrix} \delta E_i \\ \delta N_i \\ \delta E_j \\ \delta N_j \end{bmatrix} = \begin{bmatrix} l_{ij(o-c)} \end{bmatrix}$$

where a_{ij} is the bearing of j from i and is found from provisional values of coordinates of i and j:

$$\tan a_{ij} = \frac{(E_j - E_i)}{(N_j - N_i)}$$

If the units of distance are out of sympathy with the units of the coordinates, and the difference in scale is not known, it can be added to and solved for in the observation equation. This situation might occur when computing on the projection, or when observations have been made with an EDM instrument with a scale error, or even with a stretched tape.

7.7.4 Distance with scale bias equation

The distance with scale bias equation is:

$$l_{ij} - (1 + 10^{-6}s)\left\{(E_j - E_i)^2 + (N_j - N_i)^2\right\}^{\frac{1}{2}} = 0$$

where s is the scale bias in parts per million (ppm).

The final equation above may now be replaced with:

$$\begin{bmatrix} -(1+10^{-6}s)\sin a_{ij} & -(1+10^{-6}s)\cos a_{ij} & (1+10^{-6}s)\sin a_{ij} & (1+10^{-6}s)\cos a_{ij} & 10^{-6}l_{ij} \end{bmatrix} \begin{bmatrix} \delta E_i \\ \delta N_i \\ \delta E_j \\ \delta N_j \\ \delta s \end{bmatrix}$$

$$= \begin{bmatrix} l_{ij(o-c)} \end{bmatrix}$$

7.7.5 Bearing equation

Very seldom would a bearing be observed. It is introduced here as stepping stone towards the angle equation. The bearing equation is:

$$b_{ij} - \tan^{-1}\left\{\frac{E_j - E_i}{N_j - N_i}\right\} = 0$$

Upon differentiating and rearranging as above the final equation in matrix form is:

$$\begin{bmatrix} \frac{-\cos a_{ij}}{l_{ij}\sin 1''} & \frac{\sin a_{ij}}{l_{ij}\sin 1''} & \frac{\cos a_{ij}}{l_{ij}\sin 1''} & \frac{-\sin a_{ij}}{l_{ij}\sin 1''} \end{bmatrix} \begin{bmatrix} \delta E_i \\ \delta N_i \\ \delta E_j \\ \delta N_j \end{bmatrix} = \begin{bmatrix} b_{ij(o-c)} \end{bmatrix}$$

The units of b are arc seconds. Note that the ratio of 1 second of arc to 1 radian is $\sin 1'':1$.

7.7.6 Direction equation

If a round of angles has been observed at a point and it is desired to use the observations in an uncorrelated form to maintain a strictly diagonal weight matrix, then each of the pointings may be used in a separate observation equation. An extra term will be required for each theodolite set up, to account for the unknown amount, z, that the horizontal circle zero direction differs from north. The direction equation is:

$$d_{ij} - \tan^{-1}\left\{\frac{E_j - E_i}{N_j - N_i}\right\} + z = 0$$

This expression is very similar to the bearing equation above, and the derivation of coefficients of the parameters is exactly the same, but with the addition of one extra term. Upon differentiating and rearranging the final equation in matrix form is

$$\begin{bmatrix} \frac{-\cos a_{ij}}{l_{ij}\sin 1''} & \frac{\sin a_{ij}}{l_{ij}\sin 1''} & \frac{\cos a_{ij}}{l_{ij}\sin 1''} & \frac{-\sin a_{ij}}{l_{ij}\sin 1''} & -1 \end{bmatrix} \begin{bmatrix} \delta E_i \\ \delta N_i \\ \delta E_j \\ \delta N_j \\ \delta z'' \end{bmatrix} = \begin{bmatrix} d_{ij(o-c)} \end{bmatrix}$$

7.7.7 Angle equation

An angle is merely the difference of two bearings. In the current notation it is the bearing of point k from point i minus the bearing of point j from point i (*Figure 7.5*).

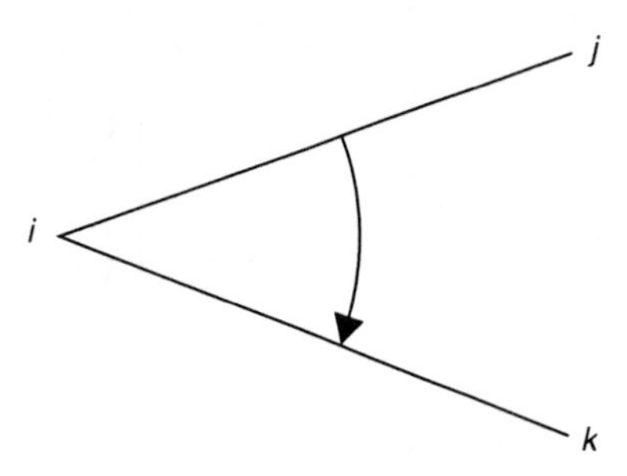

Fig. 7.5 *Angle equation*

The angle equation is:

$$a_{jik} - \tan^{-1}\left\{\frac{E_k - E_i}{N_k - N_i}\right\} + \tan^{-1}\left\{\frac{E_j - E_i}{N_j - N_i}\right\} = 0$$

Again, the terms in this are very similar to those in the bearing equation. Now there are six parameters, i.e. the corrections to the provisional coordinates of points i, j and k. Point i is associated with the directions to point j and k so the coefficients of the corrections to the provisional coordinates of point i are a little more complicated.

$$\left[\left\{\frac{\cos a_{ij}}{l_{ij}\sin 1''} - \frac{\cos a_{ik}}{l_{ik}\sin 1''}\right\}\left\{\frac{\sin a_{ik}}{l_{ik}\sin 1''} - \frac{\sin a_{ij}}{l_{ij}\sin 1''}\right\}\frac{-\cos a_{ij}}{l_{ij}\sin 1''}\ \frac{\sin a_{ij}}{l_{ij}\sin 1''}\ \frac{\cos a_{ik}}{l_{ik}\sin 1''}\ \frac{-\sin a_{ik}}{l_{ik}\sin 1''}\right]\begin{bmatrix}\delta E_i\\ \delta N_i\\ \delta E_j\\ \delta N_j\\ \delta E_k\\ \delta N_k\end{bmatrix} = \left[a_{jik(o-c)}\right]$$

If observations are made to or from points which are to be held fixed, then the corrections to the provisional values of the coordinates of those points, by definition, must be zero. Fixed points therefore do not appear in the **x** vector and therefore there are no coefficients in the **A** matrix.

7.8 WEIGHT MATRIX

Different surveyors may make different observations with different types of instruments. Therefore the quality of the observations will vary and for a least squares solution to be rigorous the solution must take account of this variation of quality. Consider a grossly overdetermined network where an observation has an assumed value for its own standard error, and thus an expected value for the magnitude of its residual. If all the terms in the observation equation are divided by the assumed, or *a priori*, standard error of the observation, then the statistically expected value of the square of the residual will be 1. If all the observation equations are likewise scaled by the assumed standard error of the observation, then the expected values of all the residuals squared will also be 1. In this case the expected value of the mean of the squares of the residuals must also be 1. The square root of this last statistic is commonly known as the 'standard error of an observation of unit weight', although the statistic might be better described as the square root of the 'variance factor' or the square root of the 'reference variance'.

If all the terms in each observation equation are scaled by the inverse of the standard error of the observation then this leads to the solution:

$$\mathbf{x} = (\mathbf{A}^T\mathbf{W}\mathbf{A})^{-1}\mathbf{A}^T\mathbf{W}\mathbf{b}$$

where **W** is a diagonal matrix and the terms on the leading diagonal, w_{ii}, are the inverse of the respective standard errors of the observations squared and all observations are uncorrelated. More strictly, the weight matrix is the inverse of the variance-covariance matrix of the 'estimated observations', that is the observed values of the observations. If:

$$\boldsymbol{\sigma}_{(b)} = \begin{bmatrix} \sigma_{11} & \sigma_{12} & \sigma_{13} & \sigma_{14} & \cdots \\ \sigma_{21} & \sigma_{22} & \sigma_{23} & \sigma_{24} & \cdots \\ \sigma_{31} & \sigma_{32} & \sigma_{33} & \sigma_{34} & \cdots \\ \sigma_{41} & \sigma_{42} & \sigma_{43} & \sigma_{44} & \cdots \\ \cdots & \cdots & \cdots & \cdots & \text{etc.} \end{bmatrix}$$

where $\boldsymbol{\sigma}_{(b)}$ is the variance-covariance matrix of the estimated observations, σ_{11} is the variance of the first observation and σ_{23} is the covariance between the 2nd and 3rd observations, etc., then

$$\mathbf{W} = \boldsymbol{\sigma}_{(b)}{}^{-1}$$

In most practical survey networks it is assumed that all the off-diagonal terms in $\boldsymbol{\sigma}_{(b)}{}^{-1}$ are 0. In other words there are no covariances between observations and so all observations are independent of each other. This will usually be true. An exception is where a round of horizontal angles has been observed at a point. If there is an error in the horizontal pointing to one point and that pointing is used to compute more than one horizontal angle then the error in the pointing will be reflected in both computed horizontal angles in equal, *Figure 7.6(a)*, and possibly opposite, *Figure 7.6(b)*, amounts. In this case the observations will be correlated, positively and negatively, respectively.

The problem is often ignored in practice but a rigorous solution may still be achieved, if observation equations are formed for directions rather than angles.

If there are no covariances, then $\boldsymbol{\sigma}_{(b)}$ becomes:

$$\boldsymbol{\sigma}_{(b)} = \begin{bmatrix} \sigma_{11} & 0 & 0 & 0 & \cdots \\ 0 & \sigma_{22} & 0 & 0 & \cdots \\ 0 & 0 & \sigma_{33} & 0 & \cdots \\ 0 & 0 & 0 & \sigma_{44} & \cdots \\ \cdots & \cdots & \cdots & \cdots & \text{etc.} \end{bmatrix}$$

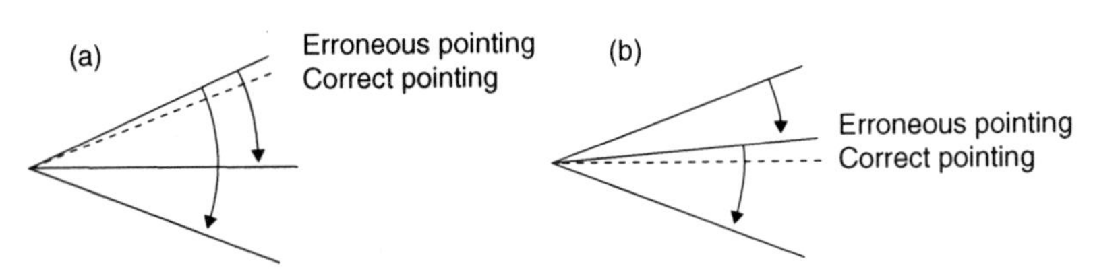

Fig. 7.6 *The effect upon computed angles of an error in a common pointing. (a) Same error in angle; (b) opposite error in angle*

and so since $\boldsymbol{\sigma}_{(b)}$ is now a diagonal matrix its inverse **W** is:

$$\mathbf{W} = \begin{bmatrix} \sigma_1^{-2} & 0 & 0 & 0 & \dots \\ 0 & \sigma_2^{-2} & 0 & 0 & \dots \\ 0 & 0 & \sigma_3^{-2} & 0 & \dots \\ 0 & 0 & 0 & \sigma_4^{-2} & \dots \\ \dots & \dots & \dots & \dots & \text{etc.} \end{bmatrix}$$

where $\sigma_1^2 = \sigma_{11}$, etc.

Generally, the *a priori* standard error of an observation does not need to be known precisely. It is unlikely that it can be estimated to better than two significant figures and even the second significant figure is likely to be no more than guesswork. However, once a weight has been assigned to an observation, it must be held to the same number of significant figures as all the other terms in the least squares computation. Small errors in the assessment of weights are likely to have a negligible effect upon the values of computed coordinates, especially in a well-overdetermined network.

Worked example

Example 7.3 In the levelling network in *Figure 7.4*, point A is a benchmark and has an assumed height of 100.00 m. Levelling has been undertaken along the lines as shown. The observed height differences were:

Line	*Observed height difference*	*Approximate line length*
AX	12.483 m	5 km
AY	48.351 m	10 km
AZ	5.492 m	7 km
XY	35.883 m	7 km
XZ	−7.093 m	12 km
YZ	−42.956 m	9 km

The standard errors of the observed difference heights are believed to be:

$$\sigma_{\Delta h} = 0.017\sqrt{K}\ \text{m}, \text{ where } K \text{ is the line length in km}$$

Find the best estimate of the heights of points X, Y and Z.

The observation equations are:

$$\begin{aligned} h_X &= h_A + \Delta h_{AX} + v_1 \\ h_Y &= h_A + \Delta h_{AY} + v_2 \\ h_Z &= h_A + \Delta h_{AZ} + v_3 \\ h_X - h_Y &= \Delta h_{YX} + v_4 \\ h_X - h_Z &= \Delta h_{ZX} + v_5 \\ h_Y - h_Z &= \Delta h_{ZY} + v_6 \end{aligned}$$

where h_X is the height of point X, etc. This set of equations may be expressed in matrix terms as:

$$\mathbf{Ax} = \mathbf{b} + \mathbf{v}$$

$$\begin{bmatrix} 1 & 0 & 0 \\ 0 & 1 & 0 \\ 0 & 0 & 1 \\ 1 & -1 & 0 \\ 1 & 0 & -1 \\ 0 & 1 & -1 \end{bmatrix} \begin{bmatrix} h_X \\ h_Y \\ h_Z \end{bmatrix} = \begin{bmatrix} h_A + \Delta h_{AX} \\ h_A + \Delta h_{AY} \\ h_A + \Delta h_{AZ} \\ \Delta h_{YX} \\ \Delta h_{ZX} \\ \Delta h_{ZY} \end{bmatrix} + \begin{bmatrix} v_1 \\ v_2 \\ v_3 \\ v_4 \\ v_5 \\ v_6 \end{bmatrix} = \begin{bmatrix} 112.483 \text{ m} \\ 148.351 \text{ m} \\ 105.492 \text{ m} \\ 35.833 \text{ m} \\ -7.093 \text{ m} \\ -42.956 \text{ m} \end{bmatrix} + \begin{bmatrix} v_1 \\ v_2 \\ v_3 \\ v_4 \\ v_5 \\ v_6 \end{bmatrix}$$

Weights are computed using the formula in the question as follows:

Line	*Length* (km)	*Standard error of observation* (m)	*Weight* (m^{-2})
AX	5	0.038	692
AY	10	0.054	348
AZ	7	0.045	496
XY	7	0.045	496
XZ	12	0.059	288
YZ	9	0.051	384

Although the choice of weights is open to interpretation, once they have been chosen, they are fixed and exactly the same values must be used throughout. The weight matrix is:

$$\mathbf{W} = \begin{bmatrix} 692 & 0 & 0 & 0 & 0 & 0 \\ 0 & 348 & 0 & 0 & 0 & 0 \\ 0 & 0 & 496 & 0 & 0 & 0 \\ 0 & 0 & 0 & 496 & 0 & 0 \\ 0 & 0 & 0 & 0 & 288 & 0 \\ 0 & 0 & 0 & 0 & 0 & 384 \end{bmatrix}$$

The necessary matrices and vector are now formed and the computation can proceed to find **x** from:

$$\mathbf{x} = (\mathbf{A}^T\mathbf{WA})^{-1}\mathbf{A}^T\mathbf{Wb}$$

To minimize the computational process it is useful to start by forming the product $\mathbf{A}^T\mathbf{W}$ since this appears twice in the above formula:

$$\mathbf{A}^T\mathbf{W} = \begin{bmatrix} 1 & 0 & 0 & -1 & -1 & 0 \\ 0 & 1 & 0 & 1 & 0 & -1 \\ 0 & 0 & 1 & 0 & 1 & 1 \end{bmatrix} \begin{bmatrix} 692 & 0 & 0 & 0 & 0 & 0 \\ 0 & 348 & 0 & 0 & 0 & 0 \\ 0 & 0 & 496 & 0 & 0 & 0 \\ 0 & 0 & 0 & 496 & 0 & 0 \\ 0 & 0 & 0 & 0 & 288 & 0 \\ 0 & 0 & 0 & 0 & 0 & 384 \end{bmatrix}$$

$$= \begin{bmatrix} 692 & 0 & 0 & -496 & -288 & 0 \\ 0 & 348 & 0 & 496 & 0 & -384 \\ 0 & 0 & 496 & 0 & 288 & 384 \end{bmatrix}$$

Next form the product $\mathbf{A}^T\mathbf{W}\mathbf{A}$:

$$\mathbf{A}^T\mathbf{W}\mathbf{A} = \begin{bmatrix} 692 & 0 & 0 & -496 & -288 & 0 \\ 0 & 348 & 0 & 496 & 0 & -384 \\ 0 & 0 & 496 & 0 & 288 & 384 \end{bmatrix} \begin{bmatrix} 1 & 0 & 0 \\ 0 & 1 & 0 \\ 0 & 0 & 1 \\ -1 & 1 & 0 \\ -1 & 0 & 1 \\ 0 & -1 & 1 \end{bmatrix}$$

$$= \begin{bmatrix} 1476 & -496 & -288 \\ -496 & 1228 & -384 \\ -288 & -384 & 1168 \end{bmatrix}$$

Invert the matrix to get $(\mathbf{A}^T\mathbf{W}\mathbf{A})^{-1}$:

$$(\mathbf{A}^T\mathbf{W}\mathbf{A})^{-1} = \begin{bmatrix} 0.000918866 & 0.000492633 & 0.000388531 \\ 0.000492633 & 0.001171764 & 0.000506708 \\ 0.000388531 & 0.000506708 & 0.001118559 \end{bmatrix}$$

At this stage a check upon the arithmetic used in the inversion would be to confirm that the product:

$(\mathbf{A}^T\mathbf{W}\mathbf{A})(\mathbf{A}^T\mathbf{W}\mathbf{A})^{-1} = \mathbf{I}$, the identity matrix. In this case it is:

$$(\mathbf{A}^T\mathbf{W}\mathbf{A})(\mathbf{A}^T\mathbf{W}\mathbf{A})^{-1} = \begin{bmatrix} 1.0000033 & -0.0000005 & -0.0000004 \\ -0.0000005 & 1.0000044 & -0.0000006 \\ 0.0000004 & -0.0000006 & 1.0000041 \end{bmatrix}$$

which is within acceptable limits of precision. Now form the remaining product in the main equation, $\mathbf{A}^T\mathbf{W}\mathbf{b}$:

$$\mathbf{A}^T\mathbf{W}\mathbf{b} = \begin{bmatrix} 692 & 0 & 0 & -496 & -288 & 0 \\ 0 & 348 & 0 & 496 & 0 & -384 \\ 0 & 0 & 496 & 0 & 288 & 384 \end{bmatrix} \begin{bmatrix} 112.483 \\ 148.351 \\ 105.492 \\ 35.883 \\ -7.093 \\ -42.956 \end{bmatrix} = \begin{bmatrix} 62083.1 \\ 85919.2 \\ 33786.1 \end{bmatrix}$$

and finally find $\mathbf{x}$ from the product of $(\mathbf{A}^T\mathbf{WA})^{-1}$ and $\mathbf{A}^T\mathbf{Wb}$:

$$\mathbf{x} = (\mathbf{A}^T\mathbf{WA})^{-1}\mathbf{A}^T\mathbf{Wb}$$

$$= \begin{bmatrix} 0.000918866 & 0.000492633 & 0.000388531 \\ 0.000492633 & 0.001171764 & 0.000506708 \\ 0.000388531 & 0.000506708 & 0.001118559 \end{bmatrix} \begin{bmatrix} 62083.1 \\ 85919.2 \\ 33786.1 \end{bmatrix}$$

$$= \begin{bmatrix} 112.500\,\text{m} \\ 148.381\,\text{m} \\ 105.449\,\text{m} \end{bmatrix} = \begin{bmatrix} h_X \\ h_Y \\ h_Z \end{bmatrix}$$

The final result is quoted to 6 significant figures (millimetres) and so all the computations leading to the final result were kept to 6 significant figures or better. The computation, so far, only derives the best estimate of the heights of the three points but does not show any error statistics from which an assessment of the quality of the solution may be made. This will be addressed later.

Worked example

Example 7.4 In *Figure 7.7*the coordinates of the points A and C are known and new points B, D and E are to be fixed. All coordinates and distances are in metres.

The known and approximate coordinates of the points are:

A	1000.000	2000.000	known
B	1385.7	1878.2	approximate
C	1734.563	2002.972	known
D	1611.7	2354.7	approximate
E	1238.7	2294.7	approximate

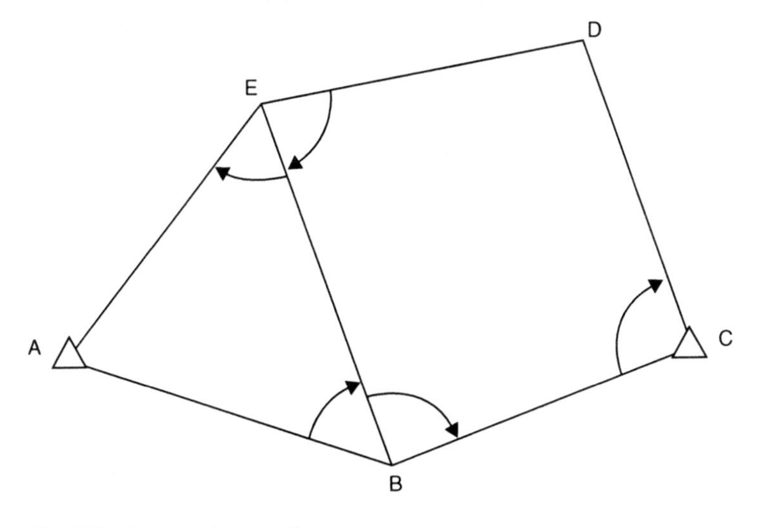

Fig. 7.7 *A control network*

The observed angles are:

ABE	58° 02′ 29″
EBC	89° 45′ 36″
BCD	90° 25′ 48″
DEB	79° 41′ 30″
BEA	58° 26′ 17″

The observed distances are:

AB	404.453 m
AE	379.284 m
BC	370.520 m
BE	441.701 m
CD	372.551 m
DE	377.841 m

The assumed standard errors for all angles is 5″ and for all distances is 0.004 m.

The first step is to define the **x** vector, the vector of parameters to be solved for. As long as all the necessary terms are in the **x** vector it does not matter in which order they appear but the order once defined dictates the order of terms across the **A** matrix. In this case it has been defined to be:

$$\mathbf{x} = \begin{bmatrix} \delta E_B \\ \delta N_B \\ \delta E_D \\ \delta N_D \\ \delta E_E \\ \delta N_E \end{bmatrix}$$

Now identify and compute all the distances and directions that will be needed to form the elements of the **A** matrix. They are formed from the fixed and provisional coordinates of the points.

	Directions	*Distances*
BA	287° 31′ 32.0″	404.4746 m
BE	340° 33′ 35.9″	441.6800 m
BC	70° 19′ 12.9″	370.5043 m
CB	250° 19′ 12.9″	370.5043 m
CD	340° 44′ 42.2″	372.5693 m
EA	219° 00′ 23.9″	379.2437 m
EB	160° 33′ 35.9″	441.6800 m
ED	80° 51′ 42.5″	377.7949 m

Next set up the four elements in each row of the **A** matrix. Use the results of *Sections 7.7.3* and *7.7.7* above as a template. Again the choice of which order to tackle the observations is arbitrary, but once chosen it will dictate the order of the terms in the weight matrix. In this case the order of the angles and distances as stated in the question is used.

Coefficient of	δE_B	δN_B	δE_D	δN_D	δE_E	δN_E
Angles						
ABE	$\frac{\cos a_{BA}}{l_{BA}\sin 1''} - \frac{\cos a_{BE}}{l_{BE}\sin 1''}$	$\frac{\sin a_{BE}}{l_{BE}\sin 1''} - \frac{\sin a_{BA}}{l_{BA}\sin 1''}$	0	0	$\frac{\cos a_{BE}}{l_{BE}\sin 1''}$	$\frac{-\sin a_{BE}}{l_{BE}\sin 1''}$
EBC	$\frac{\cos a_{BE}}{l_{BE}\sin 1''} - \frac{\cos a_{BC}}{l_{BC}\sin 1''}$	$\frac{\sin a_{BC}}{l_{BC}\sin 1''} - \frac{\sin a_{BE}}{l_{BE}\sin 1''}$	0	0	$\frac{-\cos a_{BE}}{l_{BE}\sin 1''}$	$\frac{\sin a_{BE}}{l_{BE}\sin 1''}$
BCD	$\frac{-\cos a_{CB}}{l_{CB}\sin 1''}$	$\frac{\sin a_{CB}}{l_{CB}\sin 1''}$	$\frac{\cos a_{CD}}{l_{CD}\sin 1''}$	$\frac{-\sin a_{CD}}{l_{CD}\sin 1''}$	0	0
DEB	$\frac{\cos a_{EB}}{l_{EB}\sin 1''}$	$\frac{-\sin a_{EB}}{l_{EB}\sin 1''}$	$\frac{-\cos a_{ED}}{l_{ED}\sin 1''}$	$\frac{\sin a_{ED}}{l_{ED}\sin 1''}$	$\frac{\cos a_{ED}}{l_{ED}\sin 1''} - \frac{\cos a_{EB}}{l_{EB}\sin 1''}$	$\frac{\sin a_{EB}}{l_{EB}\sin 1''} - \frac{\sin a_{ED}}{l_{ED}\sin 1''}$
BEA	$\frac{-\cos a_{EB}}{l_{EB}\sin 1''}$	$\frac{\sin a_{EB}}{l_{EB}\sin 1''}$	0	0	$\frac{\cos a_{EB}}{l_{EB}\sin 1''} - \frac{\cos a_{EA}}{l_{EA}\sin 1''}$	$\frac{\sin a_{EA}}{l_{EA}\sin 1''} - \frac{\sin a_{EB}}{l_{EB}\sin 1''}$
Distances						
AB	$\sin a_{AB}$	$\cos a_{AB}$	0	0	0	0
AE	0	0	0	0	$\sin a_{AE}$	$\cos a_{AE}$
BC	$-\sin a_{BC}$	$-\cos a_{BC}$	0	0	0	0
BE	$-\sin a_{BE}$	$-\cos a_{BE}$	0	0	$\sin a_{BE}$	$\cos a_{BE}$
CD	0	0	$\sin a_{CD}$	$\cos a_{CD}$	0	0
DE	0	0	$-\sin a_{DE}$	$-\cos a_{DE}$	$\sin a_{DE}$	$\cos a_{DE}$

Now evaluate all the terms using the directions and distances computed above.

$$\mathbf{A} = \begin{bmatrix} -287 & 331 & 0 & 0 & 440 & 155 \\ 253 & 680 & 0 & 0 & -440 & -155 \\ 187 & -524 & 523 & 183 & 0 & 0 \\ -440 & -155 & -87 & 539 & 527 & -384 \\ 440 & 155 & 0 & 0 & -18 & -498 \\ 0.954 & -0.301 & 0 & 0 & 0 & 0 \\ 0 & 0 & 0 & 0 & 0.629 & 0.777 \\ -0.942 & -0.337 & 0 & 0 & 0 & 0 \\ 0.333 & -0.943 & 0 & 0 & -0.333 & 0.943 \\ 0 & 0 & -0.330 & 0.944 & 0 & 0 \\ 0 & 0 & 0.987 & 0.159 & -0.987 & -0.159 \end{bmatrix}$$

Next compute the weight matrix. Assuming that all the observations are uncorrelated, the weight matrix will be a diagonal matrix. Each of the elements on the leading diagonal will be the inverse of the square of the standard error of the observation taken in the same order as used when setting up the **A** matrix. This will be $0.04''^{-2}$ for angles and 62 500 m^{-2} for distances.

$$\mathbf{W} = \begin{bmatrix} 0.04 & 0 & 0 & 0 & 0 & 0 & 0 & 0 & 0 & 0 & 0 \\ 0 & 0.04 & 0 & 0 & 0 & 0 & 0 & 0 & 0 & 0 & 0 \\ 0 & 0 & 0.04 & 0 & 0 & 0 & 0 & 0 & 0 & 0 & 0 \\ 0 & 0 & 0 & 0.04 & 0 & 0 & 0 & 0 & 0 & 0 & 0 \\ 0 & 0 & 0 & 0 & 0.04 & 0 & 0 & 0 & 0 & 0 & 0 \\ 0 & 0 & 0 & 0 & 0 & 62\,500 & 0 & 0 & 0 & 0 & 0 \\ 0 & 0 & 0 & 0 & 0 & 0 & 62\,500 & 0 & 0 & 0 & 0 \\ 0 & 0 & 0 & 0 & 0 & 0 & 0 & 62\,500 & 0 & 0 & 0 \\ 0 & 0 & 0 & 0 & 0 & 0 & 0 & 0 & 62\,500 & 0 & 0 \\ 0 & 0 & 0 & 0 & 0 & 0 & 0 & 0 & 0 & 62\,500 & 0 \\ 0 & 0 & 0 & 0 & 0 & 0 & 0 & 0 & 0 & 0 & 62\,500 \end{bmatrix}$$

The final part of the vector and matrix preparation is to compute the **b** vector of 'observed minus computed' terms taking the observations in exactly the same order as above.

$$\mathbf{b} = \begin{bmatrix} 53^\circ\,02'\,29'' - 340^\circ\,33'\,35.9'' + 298^\circ\,31'\,32.0'' \\ 89^\circ\,45'\,36'' - 70^\circ\,19'\,12.9'' + 340^\circ\,33'\,35.9'' - 360^\circ \\ 90^\circ\,25'\,48'' - 340^\circ\,44'\,42.2'' + 250^\circ\,19'\,12.9'' \\ 79^\circ\,41'\,30'' - 160^\circ\,33'\,35.9'' + 85^\circ\,51'\,42.5'' \\ 58^\circ\,26'\,17'' - 219^\circ\,00'\,23.9'' + 160^\circ\,33'\,35.9'' \\ 404.453\text{ m} - 404.4746\text{ m} \\ 379.284\text{ m} - 379.2437\text{ m} \\ 370.520\text{ m} - 370.5043\text{ m} \\ 441.701\text{ m} - 441.6800\text{ m} \\ 372.551\text{ m} - 372.5693\text{ m} \\ 377.841\text{ m} - 377.7949\text{ m} \end{bmatrix} = \begin{bmatrix} 25.1'' \\ -1.0'' \\ 18.7'' \\ -23.4'' \\ -31.0'' \\ -0.0216\text{ m} \\ 0.0403\text{ m} \\ 0.0157\text{ m} \\ 0.0210\text{ m} \\ -0.0183\text{ m} \\ 0.0461\text{ m} \end{bmatrix}$$

All the vectors and matrices have now been constructed and the process of matrix manipulation is the same as in the previous example. The objective is to find **x** from

$$\mathbf{x} = (\mathbf{A}^T\mathbf{W}\mathbf{A})^{-1}\mathbf{A}^T\mathbf{W}\mathbf{b}$$

Start by forming the product $\mathbf{A}^T\mathbf{W}$:

$$\mathbf{A}^T\mathbf{W} = \begin{bmatrix} -11.47 & 10.11 & 7.50 & -17.62 & 17.62 & 59599 & 0 & -58849 & 20801 & 0 & 0 \\ 13.23 & 27.18 & -20.97 & -6.22 & 6.22 & -18821 & 0 & -21048 & -58937 & 0 & 0 \\ 0 & 0 & 20.91 & -3.47 & 0 & 0 & 0 & 0 & 0 & -20611 & 61707 \\ 0 & 0 & 7.30 & 21.56 & 0 & 0 & 0 & 0 & 0 & 59004 & 9926 \\ 17.62 & -17.62 & 0 & 21.08 & -0.71 & 0 & 39338 & 0 & -20801 & 0 & -61707 \\ 6.22 & -6.22 & 0 & -15.34 & -19.91 & 0 & 48567 & 0 & 58937 & 0 & -9926 \end{bmatrix}$$

Then $\mathbf{A}^T\mathbf{W}\mathbf{A}$:

$$\mathbf{A}^T\mathbf{W}\mathbf{A} = \begin{bmatrix} 141\,937 & -13\,120 & 5447 & -8126 & -26\,027 & 14\,249 \\ -13\,120 & 104\,111 & -10\,420 & -7179 & 10\,085 & -58\,455 \\ 5447 & -10\,420 & 78\,948 & -7710 & -62\,752 & -8470 \\ -8126 & -7179 & -7710 & 70\,235 & 1565 & -9848 \\ -26\,027 & 10\,085 & -62\,752 & 1565 & 119\,246 & 18\,494 \\ 14\,249 & -58\,455 & -8470 & -9848 & 18\,494 & 112\,623 \end{bmatrix}$$

Invert the matrix to get $(\mathbf{A}^T\mathbf{WA})^{-1}$:

$$(\mathbf{A}^T\mathbf{WA})^{-1} = \begin{bmatrix} 0.00000765 & 0.00000036 & 0.00000158 & 0.00000090 & 0.00000261 & -0.00000101 \\ 0.00000036 & 0.00001487 & 0.00000194 & 0.00000297 & -0.00000149 & 0.00000832 \\ 0.00000158 & 0.00000194 & 0.00002266 & 0.00000271 & 0.00001195 & 0.00000079 \\ 0.00000090 & 0.00000297 & 0.00000271 & 0.00001533 & 0.00000073 & 0.00000285 \\ 0.00000261 & -0.00000149 & 0.00001195 & 0.00000073 & 0.00001578 & -0.00000273 \\ -0.00000101 & 0.00000832 & 0.00000079 & 0.00000285 & -0.00000273 & 0.00001408 \end{bmatrix}$$

Now form the remaining product in the main equation, $\mathbf{A}^T\mathbf{Wb}$:

$$\mathbf{A}^T\mathbf{Wb} = \begin{bmatrix} -2071 \\ -1293 \\ 3693 \\ -992 \\ -1704 \\ 3876 \end{bmatrix}$$

and finally find $\mathbf{x}$ from the product of $(\mathbf{A}^T\mathbf{WA})^{-1}$ and $\mathbf{A}^T\mathbf{Wb}$:

$$\mathbf{x} = (\mathbf{A}^T\mathbf{WA})^{-1}\mathbf{A}^T\mathbf{Wb} = \begin{bmatrix} \delta E_B \\ \delta N_B \\ \delta E_D \\ \delta N_D \\ \delta E_E \\ \delta N_E \end{bmatrix} = \begin{bmatrix} -0.020 \\ 0.019 \\ 0.058 \\ -0.001 \\ 0.002 \\ 0.051 \end{bmatrix}$$

Finally, add these computed corrections to the original provisional coordinates to obtain corrected coordinates.

E_A	1000.000 m		1000.000 m	fixed
N_A	2000.000 m		2000.000 m	fixed
E_B	1385.7 m	−0.020 m	1385.680 m	computed
N_B	1878.2 m	0.019 m	1878.219 m	computed
E_C	1734.563 m		1734.563 m	fixed
N_C	2002.972 m		2002.972 m	fixed
E_D	1611.7 m	0.058 m	1611.758 m	computed
N_D	2354.7 m	−0.001 m	2354.699 m	computed
E_E	1238.7 m	0.002 m	1238.702 m	computed
N_E	2294.7 m	0.051 m	2294.751 m	computed

It must be appreciated that the matrices in the above worked example have been kept to a minimum of size. Practical least squares computations are performed with software because the volumes of data are enormous. The inversion of the $(\mathbf{A}^T\mathbf{WA})$ matrix becomes extremely tedious if the dimensions are large. In practice computing routines that avoid the explicit inversion of this 'normal equations' matrix are used. In the example above, the inversion is of a 6 × 6 matrix only. The size of the normal equations matrix is usually $2n \times 2n$ where n is the number of points to be solved for.

7.9 ERROR ANALYSIS

In control surveying, making measurements and computing coordinates of points from those measurements is only one half of the surveyor's business. The other half is error management, which is concerned with assessing the quality of the work and drawing the appropriate conclusions. In the following sections various quality indicators are considered.

7.9.1 Residuals

The first set of statistics that are available from the least squares computation are the residuals of each of the observations. A residual gives an indication of how well a particular observation fits with the coordinates computed from that, and all the other observations. A quick glance at a set of residuals will give an indication if there are any observations that have a gross error. One gross error will distort the whole network but its worst effect will be in the residual associated with the erroneous observation. The fact that all residuals are large does not necessarily indicate that there is more than one gross error. In this case, however, the observation with the largest residual, ignoring weights, will probably be the one that is in error.

The residuals are computed by putting the final computed values of the parameters, with the observed value of the observation into the original observation equation. In matrix terms:

$$\mathbf{v} = \mathbf{Ax} - \mathbf{b}$$

where **A** and **b** are computed using the final values of **x**. A check can be applied to the computation at this stage. Pre-multiplying both sides of the above equation by $\mathbf{A}^T\mathbf{W}$ gives:

$$\mathbf{A}^T\mathbf{Wv} = \mathbf{A}^T\mathbf{WAx} - \mathbf{A}^T\mathbf{Wb}$$

but in the derivation of the least squares formulae it was shown that:

$$\mathbf{A}^T\mathbf{WAx} = \mathbf{A}^T\mathbf{Wb}$$

which are the normal equations, and therefore,

$$\mathbf{A}^T\mathbf{Wv} = \mathbf{0}, \text{ a null vector}$$

How close $\mathbf{A}^T\mathbf{Wv}$ is to a null vector will give an indication of the arithmetic correctness of the solution and of its completeness, i.e. have there been sufficient iterations.

Worked example

Example 7.5 What are the residuals from the adjustment of the levelling network in *Example 7.3?* The observation equations were:

$$\begin{aligned}
h_X &= h_A + \Delta h_{AX} + v_1 \\
h_Y &= h_A + \Delta h_{AY} + v_2 \\
h_Z &= h_A + \Delta h_{AZ} + v_3 \\
h_X - h_Y &= \Delta h_{YX} + v_4 \\
h_X - h_Z &= \Delta h_{ZX} + v_5 \\
h_Y - h_Z &= \Delta h_{ZY} + v_6
\end{aligned}$$

The known value of h_A was 100.00 m and the computed values of h_X, h_Y and h_Z were 112.500 m, 148.381 m and 105.449 m, respectively. With the original observations also added into the equations above, the residuals are:

$$\begin{aligned}
v_1 &= +0.016 \text{ m} \\
v_2 &= +0.030 \text{ m} \\
v_3 &= -0.044 \text{ m} \\
v_4 &= -0.001 \text{ m} \\
v_5 &= +0.042 \text{ m} \\
v_6 &= +0.023 \text{ m}
\end{aligned}$$

When these values are compared with the *a priori* standard errors of the observations, it does not appear that there are any unexpectedly large residuals.

Worked example

Example 7.6 What are the residuals from the adjustment of the plan network in *Example 7.4*?

The equation $\mathbf{v} = \mathbf{Ax} - \mathbf{b}$ applied to the matrices computed in *Example 7.4* gives:

$$\mathbf{v} = \begin{bmatrix} -4.3'' \\ -0.0'' \\ -2.3'' \\ 5.4'' \\ 0.1'' \\ -0.0029\,\text{m} \\ 0.0006\,\text{m} \\ -0.0035\,\text{m} \\ 0.0015\,\text{m} \\ -0.0018\,\text{m} \\ 0.0005\,\text{m} \end{bmatrix}$$

Again when compared with the *a priori* standard errors none of the residuals appears unexpectedly large.

7.9.2 Standard error of an observation of unit weight

In an excessively overdetermined network the statistically expected value of the square of each of the residuals divided by its *a priori* standard error is 1. This will only be so if all the weights of the observations have been correctly estimated and there are no gross errors. Therefore the expected mean value of this statistic, for all observations, will also be 1. Since networks are not infinitely overdetermined, account must be taken of the 'degrees of freedom' in the network. The variance factor is defined as the sum of the weighted squares of the residuals, divided by the degrees of freedom of the network. The standard error of an observation of unit weight, σ_0, is the positive square root of the variance factor. In notation:

$$\sigma_0^2 = \frac{\sum w_i v_i^2}{m - n}$$

w and v are the weights and residuals of the observations, $m - n$ is the number of degrees of freedom of the network where m and n are, respectively, the number of observations and the number of parameters.

For a small network little can be concluded from σ_0 if it is small, except that there is no evidence of gross error. For large networks, although its expected value is 1, in practice it may lie a little from this value. For a small network where there are 3 degrees of freedom, the range of 0.6 to 1.6 for σ_0 would not be unreasonable. For a network with 30 degrees of freedom a range of 0.8 to 1.2 would be more appropriate. If σ_0 is much greater than this and the residuals do not suggest a gross error, then the probable cause is an incorrect level of weighting of some or all of the observations. If σ_0 is less than the acceptable range, then again, the error is probably in *a priori* assessment of the observation standard errors and hence the weighting.

In the formula above, it is assumed that there are no covariances in $\mathbf{W}^{-1}$. If $\mathbf{W}$ is not a diagonal matrix, then a more complete statement of σ_0 would be

$$\sigma_0^2 = \frac{\mathbf{v}^T\mathbf{W}\mathbf{v}}{m - n}$$

which of course includes the definition, above, as a special case.

The above acceptable values for σ_0 are approximate and derived from experience. A more rigorous assessment of acceptability can be found by using the statistical χ^2 test. If

$$\mathbf{v}^T\mathbf{W}\mathbf{v} \le \chi^2$$

where χ^2 is evaluated at the appropriate probability of error, typically 5%, for $m - n$ degrees of freedom, then it can be stated at the given level of probability of error that the adjustment contains only random errors. This is a one-sided test in that it does not show if the *a priori* standard errors are too large, only if they are too small.

The following table shows the critical values for $\mathbf{v}^T\mathbf{W}\mathbf{v}$ or $\Sigma w_i v_i^2$ against values of probability. In each case if $\mathbf{v}^T\mathbf{W}\mathbf{v}$ or $\Sigma w_i v_i^2$ is less than the value shown in the table then there is no evidence of gross error.

Degrees of freedom (m–n)	*Probability*			
	0.10	0.05	0.02	0.01
1	1.64	1.96	2.33	2.58
2	1.52	1.73	1.98	2.15
3	1.44	1.61	1.81	1.94
4	1.39	1.54	1.71	1.82
5	1.36	1.49	1.64	1.74
6	1.33	1.45	1.58	1.67
8	1.29	1.39	1.51	1.58
10	1.26	1.35	1.45	1.52
12	1.24	1.32	1.42	1.48
15	1.22	1.29	1.37	1.43
20	1.19	1.25	1.32	1.37
40	1.14	1.18	1.23	1.26
100	1.09	1.12	1.15	1.17

Worked example

Example 7.7 The residuals derived in the levelling estimation of *Example 7.5* with their weights were:

$$
\begin{aligned}
v_1 &= +0.016\ \text{m} & w_1 &= 692\ \text{m}^{-2}\\
v_2 &= +0.030\ \text{m} & w_2 &= 348\ \text{m}^{-2}\\
v_3 &= -0.044\ \text{m} & w_3 &= 496\ \text{m}^{-2}\\
v_4 &= -0.001\ \text{m} & w_4 &= 496\ \text{m}^{-2}\\
v_5 &= +0.042\ \text{m} & w_5 &= 288\ \text{m}^{-2}\\
v_6 &= +0.023\ \text{m} & w_6 &= 384\ \text{m}^{-2}
\end{aligned}
$$

m, the number of observations, is 6 and n, the number of parameters to solve for, is 3. The standard error of an observation of unit weight is therefore:

$$\sigma_0^2 = \frac{\sum w_i v_i^2}{m - n} = 0.72 \quad \text{leading to } \sigma_0 = 0.85$$

This appears to confirm that the right weights have been used, but this conclusion can only be tentative as there are so few degrees of freedom in the estimation. Therefore this really only confirms that there have been no gross errors in the observations.

Worked example

Example 7.8 The residuals derived in the plan network estimation of *Example 7.6* with their weights were:

$$
\begin{aligned}
v_1 &= -4.3'' & w_1 &= 0.04''^{-2} \\
v_2 &= -0.0'' & w_2 &= 0.04''^{-2} \\
v_3 &= -2.3'' & w_3 &= 0.04''^{-2} \\
v_4 &= 5.4'' & w_4 &= 0.04''^{-2} \\
v_5 &= 0.1'' & w_5 &= 0.04''^{-2} \\
v_6 &= -0.0029\ \text{m} & w_6 &= 62500\ \text{m}^{-2} \\
v_7 &= 0.0006\ \text{m} & w_7 &= 62500\ \text{m}^{-2} \\
v_8 &= -0.0035\ \text{m} & w_8 &= 62500\ \text{m}^{-2} \\
v_9 &= 0.0015\ \text{m} & w_9 &= 62500\ \text{m}^{-2} \\
v_{10} &= -0.0018\ \text{m} & w_{10} &= 62500\ \text{m}^{-2} \\
v_{11} &= 0.0005\ \text{m} & w_{11} &= 62500\ \text{m}^{-2}
\end{aligned}
$$

m, the number of observations, is 11 and n, the number of parameters, is 6. The standard error of an observation of unit weight is therefore:

$$\sigma_0^2 = \frac{\sum w_i v_i^2}{m - n} = 0.75 \quad \text{leading to } \sigma_0 = 0.87$$

Again this tentatively confirms that the right weights have been used but positively confirms that there have been no gross errors in the observations.

7.10 VARIANCE-COVARIANCE MATRIX OF THE PARAMETERS

Earlier, the weight matrix was defined as the inverse of the variance-covariance matrix of the observations. This latter matrix, when fully populated, contains the variances of the observations, in order on the leading diagonal, and the covariances between them, where they exist, as the off-diagonal terms. Similar variance-covariance matrices can also be set up for all the other vector terms that appear in, or can be derived from, the least squares solution. The derivations of these variance-covariance matrices all make use of the Gauss propagation of error law, which may be interpreted like this.

If two vectors **s** and **t** are related in the equation:

$$\mathbf{s} = \mathbf{Kt}$$

where **K** is a matrix, then their variance-covariance matrices are related by the equation:

$$\sigma_{(\mathbf{s})} = \mathbf{K}\sigma_{(\mathbf{t})}\mathbf{K}^T$$

The least squares solution for the parameters is:

$$\mathbf{x} = (\mathbf{A}^T\mathbf{WA})^{-1}\mathbf{A}^T\mathbf{Wb}$$

where $(\mathbf{A}^T\mathbf{WA})^{-1}\mathbf{A}^T\mathbf{W}$ is the counterpart of the matrix **K** above. To find $\sigma_{(\mathbf{x})}$, apply the Gauss propagation of error law to the least squares solution for **x**.

$$\sigma_{(\mathbf{x})} = (\mathbf{A}^T\mathbf{WA})^{-1}\mathbf{A}^T\mathbf{W}\sigma_{(\mathbf{b})}\{(\mathbf{A}^T\mathbf{WA})^{-1}\mathbf{A}^T\mathbf{W}\}^T$$

But $\sigma_{(\mathbf{b})}$ is $\mathbf{W}^{-1}$ and so can be replaced by it. When the terms in the brackets { } are transposed the expression becomes:

$$\sigma_{(\mathbf{x})} = (\mathbf{A}^T\mathbf{WA})^{-1}\mathbf{A}^T\mathbf{WW}^{-1}\mathbf{WA}(\mathbf{A}^T\mathbf{WA})^{-1}$$

Since $(\mathbf{A}^T\mathbf{W}\mathbf{A})^{-1}$ is a symmetrical matrix, it is the same as its own transpose. By combining terms that are multiplied by their own inverse, this expression reduces to:

$$\boldsymbol{\sigma}_{(\mathbf{x})} = (\mathbf{A}^T\mathbf{W}\mathbf{A})^{-1}$$

The variance-covariance matrix of the parameters is therefore the inverse of the normal equations matrix and this is often written as N^{-1}.

The variance-covariance matrix of the parameters is a symmetrical matrix of the form:

$$\boldsymbol{\sigma}_{(\mathbf{x})} = \begin{bmatrix} \sigma_1^2 & \sigma_{12} & \sigma_{13} & \sigma_{14} & \dots \\ \sigma_{21} & \sigma_2^2 & \sigma_{23} & \sigma_{24} & \dots \\ \sigma_{31} & \sigma_{32} & \sigma_3^2 & \sigma_{34} & \dots \\ \sigma_{41} & \sigma_{42} & \sigma_{43} & \sigma_4^2 & \dots \\ \dots & \dots & \dots & \dots & \dots \end{bmatrix}$$

Matrices, which are symmetrical, such as this one, are often written for convenience as upper triangular matrices, omitting the lower terms as understood, as below:

$$\boldsymbol{\sigma}_{(\mathbf{x})} = \begin{bmatrix} \sigma_1^2 & \sigma_{12} & \sigma_{13} & \sigma_{14} & \dots \\ & {\sigma_2}^2 & \sigma_{23} & \sigma_{24} & \dots \\ & & {\sigma_3}^2 & \sigma_{34} & \dots \\ & & & \sigma_4^2 & \dots \\ \textit{symmetrical} & & & & \end{bmatrix}$$

The terms on the leading diagonal are the variances of the parameters, and the off-diagonal terms are the covariances between them. Covariances are difficult to visualize. A more helpful statistic is the coefficient of correlation r_{12}, which is defined as:

$$r_{12} = \frac{\sigma_{12}}{\sigma_1\sigma_2} \quad \text{and is always between } +1 \text{ and } -1$$

A value of $+1$ indicates that any error in the two parameters will be in the same sense by a proportional amount. -1 indicates that it will be proportional, but in the opposite sense. 0 indicates that there is no relationship between the errors in the two parameters. Matrices of coefficients of correlation are useful for descriptive purposes but do not have any place in these computations.

Worked example

Example 7.9 What are the standard errors of the heights of the points in the levelling network of *Worked example 7.3*? What are the coefficients of correlation between the computed heights?

The inverse of the normal equations matrix, $(\mathbf{A}^T\mathbf{W}\mathbf{A})^{-1}$, was:

$$(\mathbf{A}^T\mathbf{W}\mathbf{A})^{-1} = \begin{bmatrix} 0.000918866 & 0.000492633 & 0.000388531 \\ 0.000492633 & 0.001171764 & 0.000506708 \\ 0.000388531 & 0.000506708 & 0.001118559 \end{bmatrix}$$

Therefore:

$$\begin{aligned} \sigma_{hX} &= 0.0303\,\text{m} \\ \sigma_{hY} &= 0.0342\,\text{m} \\ \sigma_{Hz} &= 0.0334\,\text{m} \\ r_{hXhY} &= 0.475 \\ r_{hXhZ} &= 0.383 \\ r_{hYhZ} &= 0.443 \end{aligned}$$

Worked example

Example 7.10 What are the standard errors of the heights of the eastings and northings in the plan network of *Worked example 7.4*? What are the coefficients of correlation between the eastings and northings of all the points?

The inverse of the normal equations matrix, $(\mathbf{A}^T\mathbf{WA})^{-1}$, was:

$$(\mathbf{A}^T\mathbf{WA})^{-1} = \begin{bmatrix} 0.00000765 & 0.00000036 & 0.00000158 & 0.00000090 & 0.00000261 & -0.00000101 \\ 0.00000036 & 0.00001487 & 0.00000194 & 0.00000297 & -0.00000149 & 0.00000832 \\ 0.00000158 & 0.00000194 & 0.00002266 & 0.00000271 & 0.00001195 & 0.00000079 \\ 0.00000090 & 0.00000297 & 0.00000271 & 0.00001533 & 0.00000073 & 0.00000285 \\ 0.00000261 & -0.00000149 & 0.00001195 & 0.00000073 & 0.00001578 & -0.00000273 \\ -0.00000101 & 0.00000832 & 0.00000079 & 0.00000285 & -0.00000273 & 0.00001408 \end{bmatrix}$$

Therefore:

$$\begin{aligned} \sigma_{EB} &= 0.0028\,\text{m} \quad & \sigma_{NB} &= 0.0039\,\text{m} \\ \sigma_{ED} &= 0.0048\,\text{m} \quad & \sigma_{ND} &= 0.0039\,\text{m} \\ \sigma_{EE} &= 0.0040\,\text{m} \quad & \sigma_{NE} &= 0.0038\,\text{m} \end{aligned}$$

The coefficients of correlation are:

between	N_B	E_D	N_D	E_E	N_E
E_B	0.034	0.120	0.083	0.238	−0.098
N_B		0.106	0.197	−0.097	0.575
E_D			0.146	0.632	0.044
N_D				0.047	0.194
E_E					−0.183

Note the strong positive correlation between N_E and N_B, which is because of the generally north–south distance measurement between points *E* and *B* and also the strong positive correlation between E_D and E_E, which is because of the generally east–west distance measurement between points *D* and *E*. See *Figure 7.7*.

7.11 ERROR ELLIPSES

An error ellipse is a convenient way of expressing the uncertainty of the position of a point in a graphical format. Absolute error ellipses give a measure of the uncertainty of a point relative to the position of the fixed points in a network and relative error ellipses show the uncertainty of one defined point with respect to another defined point in the network.

7.11.1 Absolute error ellipses

An absolute error ellipse is a figure that describes the uncertainty of the computed position of a point. If the eastings and northings of points are successive elements in an $\mathbf{x}$ vector, then their variances will appear as successive elements on the leading diagonal of the variance-covariance matrix of the parameters. The square roots of these variances give the standard errors of the individual northings and eastings of the points. For descriptive purposes it might be imagined that a rectangular box could be drawn about the computed point, with sides of length of $2\sigma_E$ in the east–west direction and $2\sigma_N$ in the north–south direction and the centre of the box at the point. Such a box, it might be supposed, would describe the error in the computed coordinates of the point. Attractive as such a simple description may be, it is inadequate on two counts. Firstly, an ellipse better describes a bivariate distribution and secondly, there is no reason why that ellipse should have its axes pointing north–south and east–west. To achieve the correct orientation, the covariance between the computed eastings and northings is required.

The problem is now to find the orientation of the major axis and the sizes of both semi-major and semi-minor axes of the error ellipse describing the uncertainty of a point. In *Figure 7.8* E and N are the eastings and northings axes. m and n are also a set of orthogonal axes, but rotated by an arbitrary amount a, with respect to E and N. The origin of both axes is at the computed coordinates of the point.

From a consideration of rotation matrices coordinates in the EN system are related to coordinates in the mn system by:

$$\begin{bmatrix} m \\ n \end{bmatrix} = \begin{bmatrix} \cos a & -\sin a \\ \sin a & \cos a \end{bmatrix} \begin{bmatrix} E \\ N \end{bmatrix}$$

From the Gauss propagation of error law, the variance-covariance matrix of the coordinates with respect to the m and n axes is related to the variance-covariance matrix of the coordinates with respect to the E and N axes by:

$$\begin{bmatrix} \sigma_m^2 & \sigma_{mn} \\ \sigma_{mn} & \sigma_n^2 \end{bmatrix} = \begin{bmatrix} \cos a & -\sin a \\ \sin a & \cos a \end{bmatrix} \begin{bmatrix} \sigma_E^2 & \sigma_{EN} \\ \sigma_{EN} & \sigma_N^2 \end{bmatrix} \begin{bmatrix} \cos a & \sin a \\ -\sin a & \cos a \end{bmatrix}$$

When multiplied out this gives:

$$\sigma_m^2 = \sigma_E^2 \cos^2 a - 2\sigma_{EN} \sin a \cos a + \sigma_N^2 \sin^2 a$$

$$\sigma_n^2 = \sigma_E^2 \sin^2 a + 2\sigma_{EN} \sin a \cos a + \sigma_N^2 \cos^2 a$$

To find the direction of the major axis, find the maximum value of σ_n^2 as a changes. This will be when a's rate of change is zero.

$$\frac{d}{da}\sigma_n^2 = 0 = 2\sigma_E^2 \sin a \cos a - 2\sigma_N^2 \sin a \cos a + 2\sigma_{EN}(\cos^2 a - \sin^2 a)$$

$$= (\sigma_E^2 - \sigma_N^2) \sin 2a + 2\sigma_{EN} \cos 2a$$

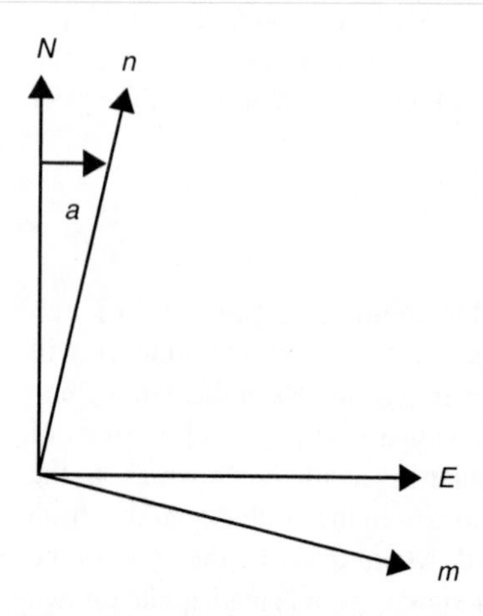

Fig. 7.8 *Coordinate system rotation*

Therefore:

$$\tan 2a = \frac{2\sigma_{EN}}{\sigma_N^2 - \sigma_E^2}$$

This solves for a, the orientation of the major minor axes of the ellipse. $2a$ takes two values in the range 0°–360°, and so a has two values, 90° apart in the range 0°–180°. If $\sigma_{EN} > 0$, the value of a that lies between 0° and 90° gives the direction of the major axis. If $\sigma_{EN} < 0$, the value of a that lies in the same range gives the direction of the minor axis.

Having solved for a, the values of σ_m and σ_n may be found by substituting back into the equations above or may be found directly using terms of the variance-covariance matrix of parameters using the equations below:

$$\sigma_{\max}^2 = \frac{1}{2}\left[\sigma_N^2 + \sigma_E^2 + \left\{\left(\sigma_N^2 - \sigma_E^2\right)^2 + 4\sigma_{EN}^2\right\}^{\frac{1}{2}}\right]$$

$$\sigma_{\min}^2 = \frac{1}{2}\left[\sigma_N^2 + \sigma_E^2 - \left\{\left(\sigma_N^2 - \sigma_E^2\right)^2 + 4\sigma_{EN}^2\right\}^{\frac{1}{2}}\right]$$

Worked example

Example 7.11 What are the terms of the absolute error ellipses at points B, D and E in the plan network of *Worked example 7.4*?

The inverse of the normal equations matrix, $(\mathbf{A}^T\mathbf{WA})^{-1}$, was:

$$(\mathbf{A}^T\mathbf{WA}^{-1}) = \begin{bmatrix} 0.00000765 & 0.00000036 & 0.00000158 & 0.00000090 & 0.00000261 & -0.00000101 \\ & 0.00001487 & 0.00000194 & 0.00000297 & -0.00000149 & 0.00000832 \\ & & 0.00002266 & 0.00000271 & 0.00001195 & 0.00000079 \\ & & & 0.00001533 & 0.00000073 & 0.00000285 \\ & & & & 0.00001578 & -0.00000273 \\ \textit{symmetrical} & & & & & 0.00001408 \end{bmatrix}$$

which is:

$$(\mathbf{A}^T\mathbf{WA})^{-1} = \begin{bmatrix} \sigma_{EB}^2 & \sigma_{EBNB} & \sigma_{EBED} & \sigma_{EBND} & \sigma_{EBEE} & \sigma_{EBNE} \\ & \sigma_{NB}^2 & \sigma_{NBED} & \sigma_{NBND} & \sigma_{NBEE} & \sigma_{NBNE} \\ & & \sigma_{ED}^2 & \sigma_{EDND} & \sigma_{EDEE} & \sigma_{EDNE} \\ & & & \sigma_{ND}^2 & \sigma_{NDEE} & \sigma_{NDNE} \\ & & & & \sigma_{EE}^2 & \sigma_{EENE} \\ \textit{symmetrical} & & & & & \sigma_{NE}^2 \end{bmatrix}$$

Point B	Point D	Point E
$\tan 2a = \dfrac{2\sigma_{EBNB}}{\sigma_{NB}^2 - \sigma_{EB}^2} = 0.1003$	$\tan 2a = \dfrac{2\sigma_{EDND}}{\sigma_{ND}^2 - \sigma_{ED}^2} = -0.7398$	$\tan 2a = \dfrac{2\sigma_{EENE}}{\sigma_{NE}^2 - \sigma_{EE}^2} = 3.2146$
$2a = 5^\circ 44'$ or $185^\circ 44'$	$2a = 143^\circ 30'$ or $323^\circ 30'$	$2a = 72^\circ 43'$ or $252^\circ 43'$
$a = 2^\circ 52'$ or $92^\circ 52'$	$a = 71^\circ 45'$ or $161^\circ 45'$	$a = 36^\circ 22'$ or $126^\circ 22'$
$\sigma_{EBNB} > 0$, so:	$\sigma_{EDND} > 0$, so:	$\sigma_{EENE} < 0$, so:
major axis direction is $2^\circ 52'$	major axis direction is $71^\circ 45'$	major axis direction is $126^\circ 22'$
minor axis direction is $92^\circ 52'$	minor axis direction is $161^\circ 45'$	minor axis direction is $36^\circ 22'$

The magnitudes of the axes for point B are found from

$$\sigma_{\max}^2 = \frac{1}{2}\left[\sigma_{NB}^2 + \sigma_{EB}^2 + \left\{\left(\sigma_{NB}^2 - \sigma_{EB}^2\right)^2 + 4\sigma_{EBNB}{}^2\right\}^{\frac{1}{2}}\right]$$

$$\sigma_{\min}^2 = \frac{1}{2}\left[\sigma_{NB}^2 + \sigma_{EB}^2 - \left\{\left(\sigma_{NB}^2 - \sigma_{EB}^2\right)^2 + 4\sigma_{EBNB}{}^2\right\}^{\frac{1}{2}}\right]$$

with similar formulae for the other two points.

Point B	Point D	Point E
and so $\sigma_{\max} = 0.0039$ m	and so $\sigma_{\max} = 0.0049$ m	and so $\sigma_{\max} = 0.0042$ m
$\sigma_{\min} = 0.0028$ m	$\sigma_{\min} = 0.0038$ m	$\sigma_{\min} = 0.0035$ m

The absolute error ellipses may be plotted on the network diagram, *Figure 7.9*, but with an exaggerated scale.

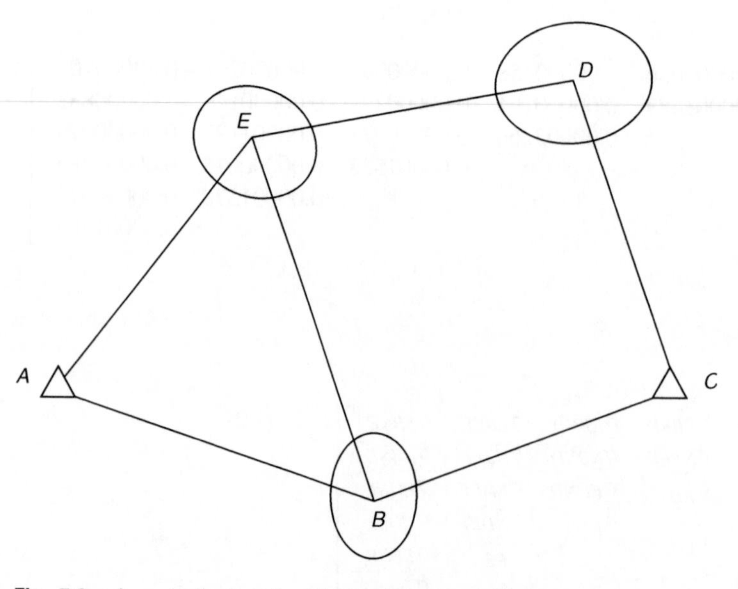

Fig. 7.9 *A control network with absolute error ellipses*

In this case the error ellipses are all small, similar in size and nearly circular. This suggests that the overall geometry of the network is good. How to remedy inconsistencies in a network will be discussed in *Section 7.16* on network design.

Error ellipses are drawn on network diagrams at an appropriate scale. If the semi-major axes of the ellipses are in the order of about 0.5 to 2 centimetres then a scale of 1:1 gives a clear and practical presentation of the uncertainty of the position.

Earlier it was assumed that an ellipse was the correct figure for the description of the error. The equation of the variance in any direction, n

$$\sigma_m^2 = \sigma_E^2 \sin^2 a - 2\sigma_{EN} \sin a \cos a + \sigma_N^2 \cos^2 a$$

does not describe an ellipse for the locus of σ_m^2 as a varies from 0°–360°. The actual locus is a pedal curve which may be shown graphically as the junction of two lines, one in the direction a, and the other at right angles to it and tangential to the ellipse.

It can be seen from *Figure 7.10* that a pedal curve has a greater area that its associated ellipse, especially if the ellipse is highly eccentric. This shows that an ellipse can be an overoptimistic representation of the uncertainty of a point.

The error ellipse, therefore, gives a simplified representation of the error figure. In a random population, one standard error encompasses 68% of the total population. The uncertainty of a position is derived from a bivariate distribution. The chance that the point actually lies within the error ellipse is 39%. If an error ellipse is drawn three times full size then the chance that the point lies inside it, rises to 99%.

If the network has an origin, that is there was a single control point that was held as fixed, then the orientation of the major axes can give an indication of the weaknesses of the network. If all the major axes are pointing roughly on line from the control point to which they refer towards the origin then it suggests that the network is weak in scale. If the major axes are at right angles to the line joining the point to which they refer and the origin, then the network is weak in azimuth.

Absolute error ellipses generally get larger, the further the point is from the origin. If the origin is changed then so are the absolute error ellipses. The absolute error ellipses are therefore 'datum dependent'. In a network which is tied to existing control there is little choice as to which points are held fixed, and overall, the best solutions are obtained when the network is surrounded by fixed points. This, after all, is no more than 'working from the whole to the part'.

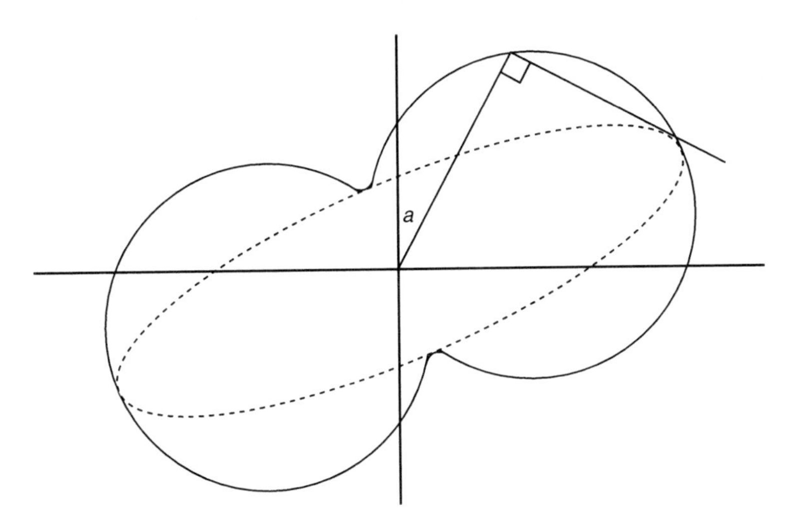

Fig. 7.10 *Pedal curve*

If the standard error of an observation of unit weight does not equal 1 in a large network, and there are no obvious gross errors in the observations, then it indicates that the weights are in error. The situation could be rectified by multiplying all the *a priori* standard errors of the observations by the standard error of an observation of unit weight and then re-computing the adjustment. This would make the standard error of observation of unit weight now equal to 1 but it would leave the computed coordinates unchanged. It would, however, scale all the terms in the variance-covariance matrix of the parameters by the original standard error of an observation of unit weight squared. This is all more easily done by simply multiplying all the terms in the variance-covariance matrix of the parameters rather than by a full readjustment with revised weights of the observations. Alternatively the axes of the error ellipses could be scaled by the standard error of observation of unit weight if these are ultimately all that are required.

7.11.2 Relative error ellipses

A relative error ellipse describes the precision of the coordinates of one point with respect to the coordinates of another point. As such they indicate the internal quality of a network, which is often more important than knowledge of absolute position, especially if the datum has been arbitrarily defined. The relative error ellipse between two points may be constructed from a consideration of the relative position of one station with respect to another and use of the Gauss propagation of error law.

Below is part of an **x** vector and its associated variance-covariance matrix, $\mathbf{N}^{-1}$, where

$$\mathbf{N}^{-1} = (\mathbf{A}^T\mathbf{W}\mathbf{A})^{-1}$$

There is no requirement for the two points concerned to be consecutive in the **x** vector. A relative error ellipse may be found for any pair of points, in this case, points F and J.

$$\mathbf{x} = \begin{bmatrix} \cdots \\ \delta_{EF} \\ \delta_{NF} \\ \cdots \\ \delta_{EJ} \\ \delta_{NJ} \\ \cdots \end{bmatrix} \quad \mathbf{N}^{-1} = \begin{bmatrix} \cdots & \cdots & \cdots & \cdots & \cdots & \cdots & \cdots \\ & \sigma^2_{EF} & \sigma_{EFNF} & \cdots & \sigma_{EFEJ} & \sigma_{EFNJ} & \cdots \\ & & \sigma^2_{NF} & \cdots & \sigma_{NFEJ} & \sigma_{NFNJ} & \cdots \\ & & & \cdots & \cdots & \cdots & \cdots \\ & & & & \sigma^2_{EJ} & \sigma_{EJNJ} & \cdots \\ & \textit{symmetrical} & & & & \sigma^2_{NJ} & \cdots \\ & & & & & & \cdots \end{bmatrix}$$

The relationship between the coordinates of F and J, in terms of their relative position, may be expressed as:

$$\begin{bmatrix} \Delta \mathrm{E} \\ \Delta \mathrm{N} \end{bmatrix} = \begin{bmatrix} E_F - E_J \\ N_F - N_J \end{bmatrix} = \begin{bmatrix} 1 & 0 & -1 & 0 \\ 0 & 1 & 0 & -1 \end{bmatrix} \begin{bmatrix} E_F \\ N_F \\ E_J \\ N_J \end{bmatrix}$$

Using the Gauss propagation of error law to find the variance-covariance matrix of the relative coordinates gives:

$$\begin{bmatrix} \sigma^2_{\Delta E} & \sigma_{\Delta E \Delta N} \\ \sigma_{\Delta E \Delta N} & \sigma^2_{\Delta N} \end{bmatrix} = \begin{bmatrix} 1 & 0 & -1 & 0 \\ 0 & 1 & 0 & -1 \end{bmatrix} \begin{bmatrix} \sigma^2_{EF} & \sigma_{EFNF} & \sigma_{EFEJ} & \sigma_{EFNJ} \\ \sigma_{EFNF} & \sigma^2_{NF} & \sigma_{NFEJ} & \sigma_{NFNJ} \\ \sigma_{EFEJ} & \sigma_{NFEJ} & \sigma^2_{EJ} & \sigma_{EJNJ} \\ \sigma_{EFNJ} & \sigma_{NFNJ} & \sigma_{EJNJ} & \sigma^2_{NJ} \end{bmatrix} \begin{bmatrix} 1 & 0 \\ 0 & 1 \\ -1 & 0 \\ 0 & -1 \end{bmatrix}$$

When selected terms on the right-hand side are multiplied out:

$$\sigma^2_{\Delta E} = \sigma^2_{EF} + \sigma^2_{EJ} - 2\sigma_{EFEJ}$$

$$\sigma^2_{\Delta N} = \sigma^2_{NF} + \sigma^2_{NJ} - 2\sigma_{NFNJ}$$

$$\sigma_{\Delta E \Delta N} = \sigma_{EFNF} + \sigma_{EJNJ} - \sigma_{EFNJ} - \sigma_{NFEJ}$$

These terms may now be used in exactly the same way as their counterparts were used when constructing absolute error ellipses. Absolute error ellipses are drawn at the point to which they refer but relative error ellipses are normally drawn at the mid-point of the line between the two points concerned.

Worked example

Example 7.12 What are the terms of the relative error ellipses between points *B*, *D* and *E* in the plan network of *Worked example 7.4*?

The inverse of the normal equations matrix, $(\mathbf{A}^T\mathbf{WA}^{-1})$, was stated in *Worked example 7.10* as:

$$(\mathbf{A}^T\mathbf{WA})^{-1} = 10^{-8}\begin{bmatrix} 765 & 36 & 158 & 90 & 261 & -101 \\ & 1487 & 194 & 297 & -149 & 832 \\ & & 2266 & 271 & 1195 & 79 \\ & & & 1533 & 73 & 285 \\ & & & & 1578 & -273 \\ & \textit{symmetrical} & & & & 1408 \end{bmatrix}$$

$$= \begin{bmatrix} \sigma^2_{EB} & \sigma_{EBNB} & \sigma_{EBED} & \sigma_{EBND} & \sigma_{EBEE} & \sigma_{EBNE} \\ & \sigma^2_{NB} & \sigma_{NBED} & \sigma_{NBND} & \sigma_{NBEE} & \sigma_{NBNE} \\ & & \sigma^2_{ED} & \sigma_{EDND} & \sigma_{EDEE} & \sigma_{EDNE} \\ & & & \sigma^2_{ND} & \sigma_{NDEE} & \sigma_{NDNE} \\ & & & & \sigma^2_{EE} & \sigma_{EENE} \\ & \textit{symmetrical} & & & & \sigma^2_{NE} \end{bmatrix}$$

The first step is to find the terms needed to construct the relative error ellipse.

Between point *B* and *D*	Between point *B* and *E*	Between point *D* and *E*
$\sigma^2_{\Delta E} = \sigma^2_{EB} + \sigma^2_{ED} - 2\sigma_{EBED}$ $= 10^{-8}(765 + 2266 - 2 \times 158)$ $= 0.00002715$	$\sigma^2_{\Delta E} = \sigma^2_{EB} + \sigma^2_{EE} - 2\sigma_{EBEE}$ $= 10^{-8}(765 + 1578 - 2 \times 261)$ $= 0.00001821$	$\sigma^2_{\Delta E} = \sigma^2_{ED} + \sigma^2_{EE} - 2\sigma_{EDEE}$ $= 10^{-8}(2266 + 1578 - 2 \times 1195)$ $= 0.00001454$
$\sigma^2_{\Delta N} = \sigma^2_{NB} + \sigma^2_{ND} - 2\sigma_{NBND}$ $= 10^{-8}(1487 + 1533 - 2 \times 297)$ $= 0.00002426$	$\sigma^2_{\Delta N} = \sigma^2_{NB} + \sigma^2_{NE} - 2\sigma_{NBNE}$ $= 10^{-8}(1487 + 1408 - 2 \times 832)$ $= 0.00001231$	$\sigma^2_{\Delta N} = \sigma^2_{ND} + \sigma^2_{NE} - 2\sigma_{NDNE}$ $= 10^{-8}(1533 + 1408 - 2 \times 285)$ $= 0.00002371$
$\sigma_{\Delta E \Delta N} = \sigma_{EBNB} + \sigma_{EDND}$ $-\sigma_{EBND} - \sigma_{NBED}$ $= 10^{-8}(36 + 271 - 90 - 194)$ $= 0.00000023$	$\sigma_{\Delta E \Delta N} = \sigma_{EBNB} + \sigma_{EENE}$ $-\sigma_{EBNE} - \sigma_{NBEE}$ $= 10^{-8}(36 - 273 + 101 + 149)$ $= 0.00000013$	$\sigma_{\Delta E \Delta N} = \sigma_{EDND} + \sigma_{EENE}$ $-\sigma_{EDNE} - \sigma_{NDEE}$ $= 10^{-8}(271 - 273 - 79 - 73)$ $= -0.00000154$
$\tan 2a = \dfrac{2\sigma_{\Delta E \Delta N}}{\sigma^2_{\Delta N} - \sigma^2_{\Delta E}} = -0.1660$	$\tan 2a = \dfrac{2\sigma_{\Delta E \Delta N}}{\sigma^2_{\Delta N} - \sigma^2_{\Delta E}} = -0.0440$	$\tan 2a = \dfrac{2\sigma_{\Delta E \Delta N}}{\sigma^2_{\Delta N} - \sigma^2_{\Delta E}} = -0.3351$
$2a = 170° 35'$ or $350° 35'$ $a = 85° 17'$ or $175° 17'$	$2a = 177° 29'$ or $357° 29'$ $a = 88° 44'$ or $178° 44'$	$2a = 161° 28'$ or $341° 28'$ $a = 80° 44'$ or $170° 44'$
$\sigma_{\Delta E \Delta N} > 0$, so: major axis direction is 85° 17′ minor axis direction is 175° 17′	$\sigma_{\Delta E \Delta N} > 0$, so: major axis direction is 88° 44′ minor axis direction is 178° 44′	$\sigma_{\Delta E \Delta N} < 0$, so: major axis direction is 170° 44′ minor axis direction is 80° 44′

The magnitudes of the axes for each relative ellipse between points are found from:

$$\sigma_{max}^2 = \frac{1}{2}\left[\sigma_{\Delta N}^2 + \sigma_{\Delta E}^2 + \left\{\left(\sigma_{\Delta N}^2 - \sigma_{\Delta E}^2\right)^2 + 4\sigma_{\Delta E\Delta N}^2\right\}^{\frac{1}{2}}\right]$$

$$\sigma_{min}^2 = \frac{1}{2}\left[\sigma_{\Delta N}^2 + \sigma_{\Delta E}^2 - \left\{\left(\sigma_{\Delta N}^2 - \sigma_{\Delta E}^2\right)^2 + 4\sigma_{\Delta E\Delta N}^2\right\}^{\frac{1}{2}}\right]$$

and so $\sigma_{max} = 0.0052$ m	and so $\sigma_{max} = 0.0043$ m	and so $\sigma_{max} = 0.0049$ m
$\sigma_{min} = 0.0049$ m	$\sigma_{min} = 0.0035$ m	$\sigma_{min} = 0.0038$ m

The relative error ellipses may be plotted on the network diagram, *Figure 7.11*, but with the same exaggerated scale as that used for the absolute error ellipses.

7.11.3 Eigenvalues, eigenvectors and error ellipses

An alternative approach to finding the parameters of error ellipses is to use eigenvalues and eigenvectors. The eigenvalues of the matrix defined by the relevant parts of the variance-covariance of parameters give the squares of the sizes of the semi-major and semi-minor axes of the error ellipse. The eigenvectors give their directions.

The eigenvalue problem is to find values of λ and $\mathbf{z}$ that satisfy the equation:

$$\mathbf{N}^{-1}\mathbf{z} = \lambda\mathbf{z}$$

$$\text{or} \quad (\mathbf{N}^{-1} - \lambda\mathbf{I})\mathbf{z} = 0$$

where $\mathbf{N}^{-1}$ is defined by the relevant parts of the variance-covariance of parameters.

$$\begin{bmatrix} \sigma_E^2 & \sigma_{EN} \\ \sigma_{EN} & \sigma_N^2 \end{bmatrix}$$

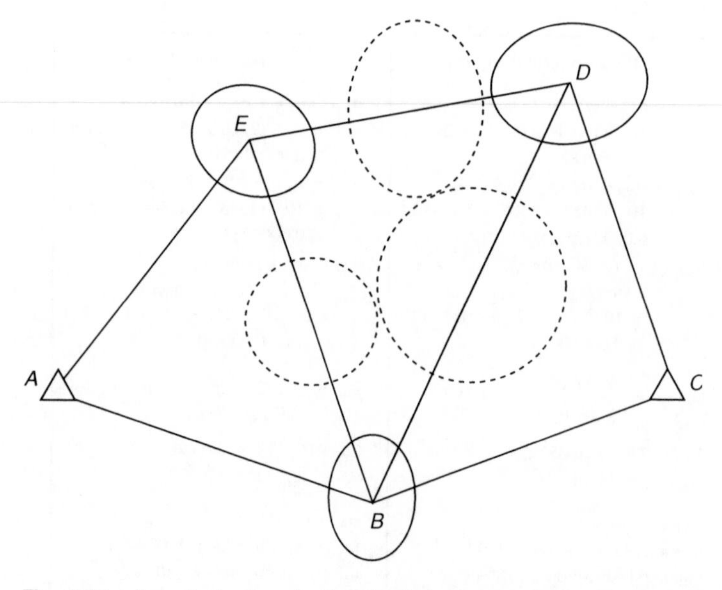

Fig. 7.11 *A control network with absolute and relative error ellipses*

The characteristic polynomial is derived from the determinant of the sub-matrix above:

$$\begin{vmatrix} \sigma_E^2 - \lambda & \sigma_{EN} \\ \sigma_{EN} & \sigma_N^2 - \lambda \end{vmatrix} = 0$$

so that the two solutions for λ are found from the quadratic equation:

$$(\sigma_E^2 - \lambda)(\sigma_N^2 - \lambda) - \sigma_{EN}^2 = 0$$

The solutions for this quadratic equation are:

$$\lambda = \frac{1}{2}\left[\sigma_{\Delta N}^2 + \sigma_{\Delta E}^2 \pm \left\{(\sigma_{\Delta N}^2 - \sigma_{\Delta E}^2)^2 + 4\sigma_{\Delta E \Delta N}^2\right\}^{\frac{1}{2}}\right]$$

The magnitudes of the semi-major and semi-minor axes are the square roots of the two solutions for λ. The directions of the axes are found from the eigenvector associated with each eigenvalue. The eigenvector associated with λ_1 is found from:

$$(\mathbf{N}^{-1} - \lambda\mathbf{I})\mathbf{z} = 0$$

$$= \begin{bmatrix} \sigma_E^2 - \lambda_1 & \sigma_{EN} \\ \sigma_{EN} & \sigma_N^2 - \lambda_1 \end{bmatrix} \begin{bmatrix} z_1 \\ z_2 \end{bmatrix}$$

so that the direction of the axis associated with λ_1 is given by:

$$\tan^{-1}\left\{ - \sigma_{EN}(\sigma_E^2 - \lambda_1)^{-1} \quad \text{or}\right\} \quad \tan^{-1}\{ - (\sigma_N^2 - \lambda_1)\sigma_{EN}^{-1}\}$$

Both formulae give the same answer. λ_2 is found in the same way and will be exactly 90° different from λ_1.

Worked example

Example 7.13 What are the orientations of the axes of the absolute error ellipse at point B in the plan network of *Worked example 7.4*?

The full variance-covariance of parameters is stated in *Worked example 7.11* above. The part that relates to point B is:

$$\begin{bmatrix} \sigma_{EB}^2 & \sigma_{EBNB} \\ \sigma_{EBNB} & \sigma_{NB}^2 \end{bmatrix} = 10^{-8}\begin{bmatrix} 765 & 36 \\ 36 & 1487 \end{bmatrix}$$

The characteristic polynomial is derived from the determinant where:

$$\begin{vmatrix} 0.00000765 - \lambda & 0.00000036 \\ 0.00000036 & 0.00001487 - \lambda \end{vmatrix} = 0$$

When evaluated this leads to:

$$0 = (0.00000765 - \lambda)(0.00001487 - \lambda) - 0.00000036^2$$

$$= 1.1373 \times 10^{-10} - 2.2516\lambda 10^{-5} + \lambda^2 - 1.296 \times 10^{-13}$$

$$= \lambda^2 - 2.525\lambda 10^{-5} + 1.1358 \times 10^{-10}$$

The solutions are:

$$\lambda_1 = 0.000014883\,\text{m}^2$$

$$\lambda_2 = 0.000007633\,\text{m}^2$$

and therefore:

$$\sigma_{\max} = \sqrt{\lambda_1} = 0.0039\,\text{m} \quad \text{and} \quad \sigma_{\min} = \sqrt{\lambda_2} = 0.0028\,\text{m}$$

The eigenvector associated with λ_1 is found from:

$$(\mathbf{N}^{-1} - \lambda\mathbf{I})\mathbf{z} = 0$$

$$= \begin{bmatrix} 0.00000765 - \lambda_1 & 0.00000036 \\ 0.00000036 & 0.00001487 - \lambda_1 \end{bmatrix} \begin{bmatrix} z_1 \\ z_2 \end{bmatrix}$$

$$= \begin{bmatrix} -0.00000723 & 0.00000036 \\ 0.00000036 & -0.00000002 \end{bmatrix} \begin{bmatrix} z_1 \\ z_2 \end{bmatrix}$$

The direction of the axis associated with λ_1 is given by:

$$\tan^{-1}\{-(0.00000036/-0.00000723)\} = 2°\,52'$$

The eigenvector associated with λ_2 is found from:

$$(\mathbf{N}^{-1} - \lambda\mathbf{I})\mathbf{z} = \mathbf{0}$$

$$= \begin{bmatrix} 0.00000765 - \lambda_2 & 0.00000036 \\ 0.00000036 & 0.00001487 - \lambda_2 \end{bmatrix} \begin{bmatrix} z_1 \\ z_2 \end{bmatrix}$$

$$= \begin{bmatrix} 0.00000002 & 0.00000036 \\ 0.00000036 & 0.00000723 \end{bmatrix} \begin{bmatrix} z_1 \\ z_2 \end{bmatrix}$$

and the direction of the axis associated with λ_2 is given by:

$$\tan^{-1}\{-(0.00000036/0.00000002)\} = 92°\,52'$$

7.12 STANDARD ERRORS OF DERIVED QUANTITIES

In engineering surveying it is often necessary to confirm that the distance, and sometimes direction, between two points is within a certain tolerance. It may be that the distance or direction under investigation was observed. If it was, then the same quantity, computed from coordinates, will always have an equal or smaller standard error provided that the computed standard error has not been scaled by the standard error of an observation of unit weight. A measurement relating any two or three points may be computed from the estimated coordinates. This applies to measurements that have been observed and also to any measurements that could have been observed between any of the points. There are two approaches to the problem of computing standard errors of derived quantities. One is to compute the quantities directly from the terms used to describe a relative error ellipse and that is what this section is about. The other is to use the variance-covariance matrix of the computed observations. That will be dealt with in the next section.

In the section above on absolute error ellipses it was shown that, by invoking the Gauss propagation of error law, the variance in any direction could be related to variances and a covariance derived from the variance-covariance matrix of the parameters as:

$$\sigma^2 = \sigma_{\Delta E}^2 \sin^2 a + 2\sigma_{\Delta E \Delta N} \sin a \cos a + \sigma_{\Delta N}^2 \cos^2 a$$

The terms are those of the relative error ellipse. If the distance between the two points is of interest, then a takes the value of the bearing of the line. If direction is of interest then a is at right angles to the line. In the case of direction the quantity computed from the above formula will have units of distance and must be scaled by the length of the line.

Worked example

Example 7.14 Use the data from *Worked example 7.11* to find the standard errors of derived distance and direction between B and D.

Use the computed coordinates of B and D to find the direction and distance BD.

$$\text{Direction } BD = 25^\circ\,22'\,59''$$

$$\text{Distance } BD = 527.394 \text{ m}$$

For the computation of the standard error of the derived distance a takes the value of direction BD in the equation:

$$\sigma^2 = \sigma_{\Delta E}^2 \sin^2 a + 2\sigma_{\Delta E \Delta N} \sin a \cos a + \sigma_{\Delta N}^2 \cos^2 a$$

$$= 0.00002715 \sin^2 a + 2 \times 0.00000023 \sin a \cos a + 0.00002426 \cos^2 a$$

$$= 0.00002497 \text{ m}^2$$

The standard error of the derived distance is the square root:

$$\sigma_{\text{distance } BD} = 0.0050 \text{ m}$$

For the derived direction, a takes the value $a + 90^\circ$:

$$\sigma^2 = \sigma_{\Delta E}^2 \sin^2(a + 90^\circ) + 2\sigma_{\Delta E \Delta N} \sin(a + 90^\circ)\cos(a + 90^\circ) + \sigma_{\Delta N}^2 \cos^2(a + 90^\circ)$$

$$= 0.00002643 \text{ m}^2$$

$$\sigma = 0.0052 \text{ m}$$

The standard error of the derived direction is:

$$\sigma_{\text{direction } BD} = \sigma/\text{Distance } BD$$

$$= 2.0''$$

7.13 BLUNDER DETECTION

If an observation contains a gross error and it is included with a set of otherwise good observations then the least squares process will accommodate it by distorting the network to make it according to the normal least squares criteria. A quick scan down a list of residuals for the largest may identify the erroneous observation but if there are several gross errors in the set of observations then there will be many large residuals. However, by computing the statistic of residual divided by its own standard error it is possible to identify the most significant gross error. The gross error may be dealt with either by correcting the error if it can be traced or by eliminating the rogue observation from the set. Now the next worst gross error may be traced and so on until all gross errors have been removed from the set of observations. The standard error of a residual may be found as the square root of the element of the leading diagonal of the variance-covariance matrix of the residuals.

7.13.1 The variance-covariance matrix of the estimated residuals

The original least squares problem:

$$\mathbf{Ax} = \mathbf{b} + \mathbf{v}$$

was solved for the parameters as:

$$\mathbf{x} = (\mathbf{A}^T\mathbf{WA})^{-1}\mathbf{A}^T\mathbf{Wb}$$

and so, on putting the estimated value of **x** back into the original equation, the estimated residuals become:

$$\mathbf{v} = \left\{\mathbf{A}(\mathbf{A}^T\mathbf{WA})^{-1}\mathbf{A}^T\mathbf{W} - \mathbf{I}\right\}\mathbf{b}$$

Using the Gauss propagation of error law:

$$\sigma_{(\mathbf{v})} = \left\{\mathbf{A}(\mathbf{A}^T\mathbf{WA})^{-1}\mathbf{A}^T\mathbf{W} - \mathbf{I}\right\}\sigma_{(\mathbf{b})}\left\{\mathbf{A}(\mathbf{A}^T\mathbf{WA})^{-1}\mathbf{A}^T\mathbf{W} - \mathbf{I}\right\}^T$$

but $\sigma_{(\mathbf{b})} = \mathbf{W}^{-1}$

$$\text{so}\quad \sigma_{(\mathbf{v})} = \left\{\mathbf{A}(\mathbf{A}^T\mathbf{WA})^{-1}\mathbf{A}^T\mathbf{W} - \mathbf{I}\right\}\mathbf{W}^{-1}\left\{\mathbf{WA}(\mathbf{A}^T\mathbf{WA})^{-1}\mathbf{A}^T - \mathbf{I}\right\}$$

$$= \mathbf{A}(\mathbf{A}^T\mathbf{WA})^{-1}\mathbf{A}^T\mathbf{WW}^{-1}\mathbf{WA}(\mathbf{A}^T\mathbf{WA})^{-1}\mathbf{A}^T - \mathbf{W}^{-1}\mathbf{WA}(\mathbf{A}^T\mathbf{WA})^{-1}\mathbf{A}^T$$

$$- \mathbf{A}(\mathbf{A}^T\mathbf{WA}^{-1})\mathbf{A}^T\mathbf{WW}^{-1} + \mathbf{W}^{-1}$$

which, when terms, preceded by their own inverse, are accounted for, simplifies to:

$$\sigma_{(\mathbf{v})} = \mathbf{W}^{-1} - \mathbf{A}(\mathbf{A}^T\mathbf{WA})^{-1}\mathbf{A}^T$$

The matrix $\mathbf{A}(\mathbf{A}^T\mathbf{WA})^{-1}\mathbf{A}^T$ will be a large one if the number of observations is large. It is unlikely, however, that the covariances between derived observations will be required. If so, then only the leading diagonal of the matrix will be of interest. By considering only non-zero products of the terms of $\mathbf{A}(\mathbf{A}^T\mathbf{WA})^{-1}\mathbf{A}^T$ it can be shown that the variance of an observation may be computed from:

$$\begin{bmatrix} a_1 & a_2 & a_3 & a_4 & a_5 & a_6 \end{bmatrix} \begin{bmatrix} \sigma_1^2 & \sigma_{12} & \sigma_{13} & \sigma_{14} & \sigma_{15} & \sigma_{16} \\ & \sigma_2^2 & \sigma_{23} & \sigma_{24} & \sigma_{25} & \sigma_{26} \\ & & \sigma_3^2 & \sigma_{34} & \sigma_{35} & \sigma_{36} \\ & & & \sigma_4^2 & \sigma_{45} & \sigma_{46} \\ & & & & \sigma_5^2 & \sigma_{56} \\ \textit{symmetrical} & & & & & \sigma_6^2 \end{bmatrix} \begin{bmatrix} a_1 \\ a_2 \\ a_3 \\ a_4 \\ a_5 \\ a_6 \end{bmatrix}$$

where a_1, a_2, a_3, a_4, a_5 and a_6 are the non-zero terms of the row of the **A** matrix relating to the observation concerned and the variance-covariance matrix is a sub-set of the already computed $(\mathbf{A}^T\mathbf{WA})^{-1}$ where all the rows and columns containing variances and covariances of parameters not represented in a_1 to a_6 have been removed.

The above example might be for a derived angle as there are six terms involved. If the variance for a derived distance was required then the matrix product would look like this.

$$\begin{bmatrix} a_1 & a_2 & a_3 & a_4 \end{bmatrix} \begin{bmatrix} \sigma_1^2 & \sigma_{12} & \sigma_{13} & \sigma_{14} \\ & \sigma_2^2 & \sigma_{23} & \sigma_{24} \\ & & \sigma_3^2 & \sigma_{34} \\ \textit{symmetrical} & & & \sigma_4^2 \end{bmatrix} \begin{bmatrix} a_1 \\ a_2 \\ a_3 \\ a_4 \end{bmatrix}$$

The variance of a derived observation that was not observed may also be found in exactly the same way. The terms a_1 to a_4 are computed as if an observation had actually been made and the variance is computed exactly as above.

The test to be applied is to compare the estimated residual with its own standard error. The idea is that if the residual is a significant multiple of its own standard error, then, in a large network, there will be cause for concern that the error in the observation is more than random and so the observation will need to be investigated. Although there are strict statistical tests they do not take account of the fact that the residuals are usually correlated. Therefore the simple rule of thumb often used is that if the residual divided by its own standard error is within the range of plus or minus 4, then the observation will be accepted as containing no significant non-random error, i.e. that:

$$4 > \frac{v}{\sigma_v} > -4$$

Worked example

Example 7.15 Use the data from *Worked example 7.4* and subsequent worked examples to confirm that there are no observations with significant residuals.

Compute $\boldsymbol{\sigma}_{(\mathbf{v})} = \mathbf{W}^{-1} - \mathbf{A}(\mathbf{A}^T\mathbf{W}\mathbf{A})^{-1}\mathbf{A}^T$ as follows from the matrices generated in *Worked example 7.4*.

$$\mathbf{A}(\mathbf{A}^T\mathbf{W}\mathbf{A})^{-1} = 10^{-5}\begin{bmatrix} -108 & 545 & 557 & 149 & 528 & 403 \\ 119 & 956 & -367 & 148 & -688 & 441 \\ 223 & -617 & 1162 & 282 & 765 & -362 \\ -131 & -501 & 450 & 646 & 780 & -622 \\ 388 & -164 & 39 & -58 & 200 & -611 \\ 0.719 & -0.413 & 0.092 & -0.004 & 0.294 & -0.347 \\ 0.086 & 0.553 & 0.813 & 0.268 & 0.781 & 0.922 \\ -0.733 & -0.535 & -0.214 & -0.184 & -0.196 & -0.185 \\ 0.038 & -0.556 & 0.454 & -0.006 & -0.556 & 0.601 \\ 0.033 & 0.217 & -0.491 & 1.358 & -0.325 & 0.244 \\ -0.072 & 0.253 & 1.088 & 0.394 & -0.324 & 0.169 \end{bmatrix}$$

$\mathbf{A}(\mathbf{A}^T\mathbf{W}\mathbf{A})^{-1}\mathbf{A}^T$

$$= 10^{-3}\begin{bmatrix} 5067 & 479 & 124 & 1189 & -1727 & -2.672 & 6.456 & -0.819 & -3.461 & -0.430 & -0.119 \\ 479 & 9137 & -6433 & -6207 & -67 & -1.746 & -0.899 & -4.336 & -2.167 & 2.610 & 2.702 \\ 124 & -6433 & 10240 & 5910 & 1690 & 3.986 & 2.000 & -0.024 & 0.602 & -1.165 & 4.944 \\ 1189 & -6207 & 5910 & 10948 & 1600 & 0.257 & 0.079 & 2.924 & -4.174 & 4.614 & -1.251 \\ -1727 & -67 & 1690 & 1600 & 4462 & 4.198 & -3.492 & -3.102 & -3.588 & -0.675 & -0.708 \\ -2.672 & -1.746 & 3.986 & 0.257 & 4.198 & 0.0081 & -0.0008 & -0.0054 & 0.0020 & -0.0003 & -0.0015 \\ 6.456 & -0.899 & 2.000 & 0.079 & -3.492 & -0.0008 & 0.0121 & -0.0027 & 0.0012 & -0.0002 & -0.0007 \\ -0.819 & -4.336 & -0.024 & 2.924 & -3.102 & -0.0054 & -0.0027 & 0.0087 & 0.0015 & -0.0010 & -0.0002 \\ -3.461 & -2.167 & 0.602 & -4.174 & -3.588 & 0.0020 & 0.0012 & 0.0015 & 0.0129 & 0.0014 & 0.0000 \\ -0.430 & 2.610 & -1.165 & 4.614 & -0.675 & -0.0003 & -0.0002 & -0.0010 & 0.0014 & 0.0144 & 0.0001 \\ -0.119 & 2.702 & 4.944 & -1.251 & -0.708 & -0.0015 & -0.0007 & -0.0002 & 0.0000 & 0.0001 & 0.0143 \end{bmatrix}$$

and the terms on the leading diagonal of $\mathbf{W}^{-1} - \mathbf{A}(\mathbf{A}^T\mathbf{W}\mathbf{A})^{-1}\mathbf{A}^T$ are:

$$\begin{bmatrix} 19.933 & & & & & & & & & & \\ & 15.863 & & & & & & & & & \\ & & 14.760 & & & & & & & & \\ & & & 14.052 & & & & & & & \\ & & & & 20.538 & & & & & & \\ & & & & & 0.000007903 & & & & & \\ & & & & & & 0.000003917 & & & & \\ & & & & & & & 0.000007302 & & & \\ & & & & & & & & 0.000003119 & & \\ & & & & & & & & & 0.000001564 & \\ & & & & & & & & & & 0.000001701 \end{bmatrix}$$

The square roots of these terms are the standard errors of the residuals so, with the residuals, the test statistic can be computed. There is no evidence of a gross error in any of the observations.

Observation		*Residual* v	*Standard error of residual* σ_v	*Test statistic* $\frac{v}{\sigma_v}$	*Is* $4 > \frac{v}{\sigma_v} > -4$?
Angles	*ABE*	−4.3″	4.5″	−1.0	yes
	EBC	−0.0″	4.0″	0.0	yes
	BCD	−2.3″	3.8″	−0.6	yes
	DEB	5.4″	3.7″	1.4	yes
	BEA	0.1″	4.5″	0.0	yes
Distance	*AB*	−0.0029 m	0.0028 m	−1.0	yes
	AE	0.0006 m	0.0020 m	0.3	yes
	BC	−0.0035 m	0.0027 m	−1.3	yes
	BE	0.0015 m	0.0018 m	0.8	yes
	CD	−0.0018 m	0.0013 m	−1.4	yes
	DE	0.0005 m	0.0013 m	0.4	yes

7.13.2 The effect of one gross error

One gross error will distort the whole of a network. Therefore the presence of many test statistics suggesting the presence of gross error does not necessarily mean that there is more than one gross error. In the event that more than one observation's test statistic fails investigate only the observation with the worst test statistic.

Worked example

Example 7.16 In *Worked example 7.15* introduce a 'typo' gross error into angle *ABE* by using **35**° 02′ 29″ instead of the correct value of **53**° 02′ 29″. The above table now appears as:

Observation		*Residual* v	*Standard error of residual* σ_v	*Test statistic* $\frac{v}{\sigma_v}$	*Is* $4 > \frac{v}{\sigma_v} > -4$?
Angles	*ABE*	51662″	4.5″	11571	no
	EBC	−1242″	4.0″	−312	no
	BCD	−322″	3.8″	−84	no
	DEB	−3075″	3.7″	−821	no
	BEA	4477″	4.5″	988	no
Distance	*AB*	6.92 m	0.0028 m	2463	no
	AE	−16.73 m	0.0020 m	−8455	no
	BC	2.12 m	0.0027 m	784	no
	BE	8.97 m	0.0018 m	5080	no
	CD	1.11 m	0.0013 m	889	no
	DE	0.31 m	0.0013 m	236	no

It appears that the test statistic for all observations has failed. However, the first observation, angle *ABE*, has the greatest test statistic but that it is the only observation with a gross error.

7.14 RELIABILITY OF THE OBSERVATIONS

Where there are many more observations than the strict minimum necessary needed to solve for the unknown coordinates then all the observations give a degree of independent check to each other. Where only the minimum number of observations has been made then there is no independent check upon those observations and they will all be unreliable in the sense that if one or more of them are grossly in error there is no way that the error can be detected. Hence any coordinates computed using a grossly erroneous observation will also be grossly in error. *Figure 7.12* illustrates the problem where points A and B are fixed and points C and D are to be found.

In *Figure 7.12* the angles in the triangle ABC and the distances AC and AB have been measured with a low quality total station but the angle DAB and the distance AD have been measured with a high precision instrument. Therefore it would be expected that the uncertainty of the coordinates of C would be greater than those of D. However, the coordinates of C would be very reliable because point C is connected to the fixed points, A and B, by five observations. There are five observations to calculate the two coordinates of point C, so three of the observations are 'redundant'. If one of those observations had a gross error, that error would be easy to detect and deal with. By contrast the coordinates of D would be very unreliable because that point is connected to the fixed points by only two observations. There are only two observations to calculate the two coordinates of point D and therefore a redundancy of zero. If one of those observations had a gross error the error would be undetectable.

7.14.1 The variance-covariance matrix of the estimated observations

The estimated observations are the observations computed from the estimated coordinates. These contrast with the observations actually observed by the surveyor with survey instruments. By a derivation similar to that of the section on the variance-covariance matrix of the residuals above it can be shown that the variance-covariance matrix of the estimated observations, $\sigma_{(l)}$, is:

$$\sigma_{(l)} = \mathbf{A}(\mathbf{A}^T\mathbf{W}\mathbf{A})^{-1}\mathbf{A}^T$$

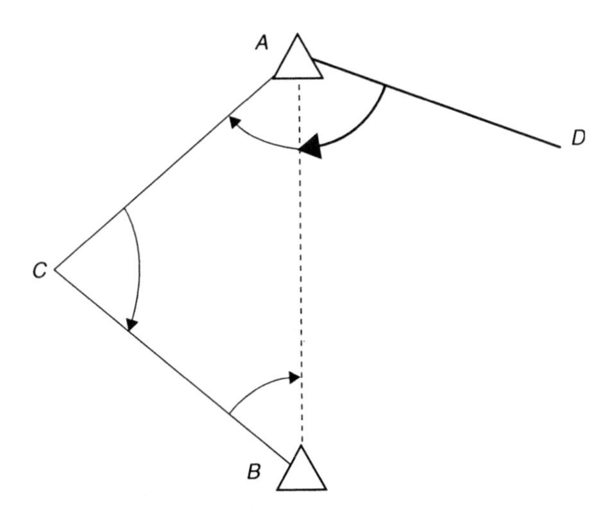

Fig. 7.12 *Precision and reliability*

It will be noted that this matrix has already been computed as part of the process needed to compute the variance-covariance matrix of the residuals and so very little extra work will be required. In this matrix the leading diagonal contains the squares of the standard errors of the estimated observations and the off-diagonal terms are the covariances between them. It is only the leading diagonal that is likely to be needed.

By comparing the standard error of an estimated observation with the standard error of the equivalent observed observation, it is possible to see what the estimation process has done to improve the quality of the observed observation to that of the estimated observation. If there is no improvement then that implies that, whatever the value of the observation, it has been accepted unmodified by the estimation process and therefore the observation is unreliable. Any coordinates that are computed from that observation would also be unreliable. It is likely therefore that if the unreliable observation is removed from the estimation process then the normal equations matrix will become singular and the parameter vector will need to be modified to remove the point that depended only upon the unreliable observation. Although totally unreliable observations may be identified, all observations have different levels of reliability in the network solution. If the ratio of the variances of the computed and observed values of the observation is calculated then this ratio may be tested in an '*F* test' to determine its statistical significance.

This may be tedious for all observations, and a general rule of thumb is that the relationship:

$$\frac{\sigma_{(l)}}{\sigma_{(b)}} < 0.8 \quad \text{should hold for reliable observations}$$

$\sigma_{(l)}$ is the standard error of an estimated observation and $\sigma_{(b)}$ is the standard error of an observed observation. If this test is not satisfied then the observation is considered unreliable and further observations should be undertaken to improve the quality of the network in the suspect area.

Worked example

Example 7.17 Use the data from *Worked example 7.16* to confirm that there are no unreliable observations.

The terms on the leading diagonal of $\boldsymbol{\sigma}_{(\mathbf{l})} = \mathbf{A}(\mathbf{A}^T\mathbf{WA})^{-1}\mathbf{A}^T$ in *Worked example 7.16* were:

$$\begin{bmatrix} 5.067 &&&&&&&&&& \\ & 9.137 &&&&&&&&& \\ && 10.240 &&&&&&&& \\ &&& 10.948 &&&&&&& \\ &&&& 4.462 &&&&&& \\ &&&&& 0.0000081 &&&&& \\ &&&&&& 0.0000121 &&&& \\ &&&&&&& 0.0000087 &&& \\ &&&&&&&& 0.0000129 && \\ &&&&&&&&& 0.0000144 & \\ &&&&&&&&&& 0.0000143 \end{bmatrix}$$

The square roots of these terms are the standard errors of the estimated observations so, with the standard errors of the observed observations, the test statistic can be computed. There is significant evidence of unreliable observations.

Observation		*Standard error of estimated observation* $\sigma_{(l)}$	*Standard error of observed observation* $\sigma_{(b)}$	*Test statistic* $\frac{\sigma_{(l)}}{\sigma_{(b)}}$	*Is* $\frac{\sigma_{(l)}}{\sigma_{(b)}} < 0.8$?
Angles	*ABE*	2.3″	5.0″	0.45	yes
	EBC	3.0″	5.0″	0.60	yes
	BCD	3.2″	5.0″	0.64	yes
	DEB	3.3″	5.0″	0.66	yes
	BEA	2.1″	5.0″	0.42	yes
Distance	*AB*	0.0028 m	0.004 m	0.71	yes
	AE	0.0035 m	0.004 m	0.87	no
	BC	0.0029 m	0.004 m	0.74	yes
	BE	0.0036 m	0.004 m	0.90	no
	CD	0.0038 m	0.004 m	0.95	no
	DE	0.0038 m	0.004 m	0.95	no

7.14.2 Visualization of reliability

Worked example 7.17 has indicated that there are four unreliable distance observations in the network. They have been marked as dotted lines in *Figure 7.13*. There is a weakness in the design of the network in that a small but significant error in any of these observations is at risk of going unnoticed and so causing a small but significant error in one or more of the coordinates. Imagine the network as being made up of slightly flexible struts to represent the measured distances, slightly flexible plates at the points to represent the measured angles and nails at the fixed points attaching them to an immovable board. Now remove one strut or plate, get hold of this structure by any point and try to move it in the plane of the board. This action would cause some minor distortions to the structure depending upon how critical the strut or plate removed was. A critical strut or plate represents unreliable observation.

In this example the distances are all of a similar length of about 400 m and have standard errors of their observations of 0.004 m. An uncertainty of 0.004 m at 400 m is the equivalent of 2″. The observed angles

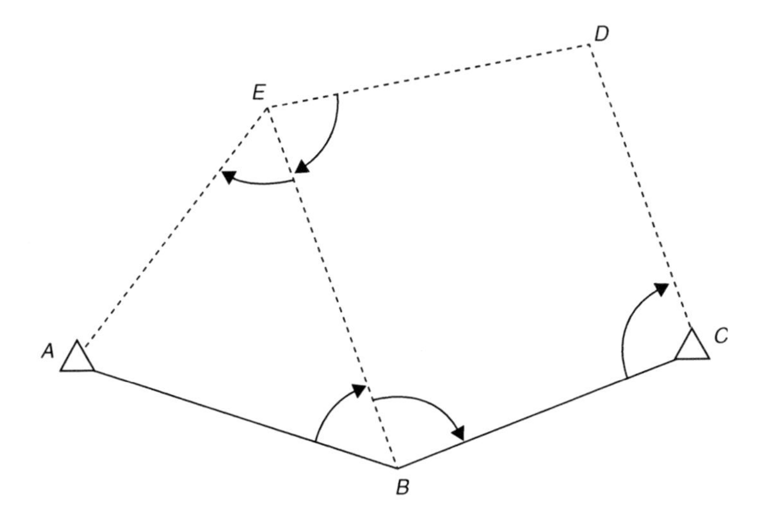

Fig. 7.13 *Unreliable observations*

have standard errors of 5″. Therefore the distances are lending much more strength to the network than the angles. This means that in *Figure 7.13* the angles do little to restrain the distances but the distances do much more to brace the angles. Therefore the distances are less checked, or more unreliable, than the angles.

7.15 SUMMARY OF MATRIX RELATIONSHIPS AND APPLICATIONS OF OUTPUT STATISTICS

In the foregoing the variance-covariance matrices of all the vectors associated with the least squares process were derived. They are summarized below.

Vector	*Variance-covariance matrix of vector*
b 'observed/computed' observations	$\boldsymbol{\sigma}_{(\mathbf{b})} = \mathbf{W}^{-1}$
x parameters	$\boldsymbol{\sigma}_{(\mathbf{x})} = (\mathbf{A}^T\mathbf{W}\mathbf{A})^{-1}$
l estimated observations	$\boldsymbol{\sigma}_{(\mathbf{l})} = \mathbf{A}(\mathbf{A}^T\mathbf{W}\mathbf{A})^{-1}\mathbf{A}^T$
v residuals	$\boldsymbol{\sigma}_{(\mathbf{v})} = \mathbf{W}^{-1} - \mathbf{A}(\mathbf{A}^T\mathbf{W}\mathbf{A})^{-1}\mathbf{A}^T$

Likewise, the applications of the output statistics are also summarized.

Statistics	*Applications*
x	Coordinates or heights of free stations
l	Best estimates of distances or directions between stations
v	First indicator of existence of blunder(s)
σ_0	Detection of the existence of blunder(s). Confirmation of, or scaling factor for, correct weighting. Scaling error ellipses
$\boldsymbol{\sigma}_{(\mathbf{x})}$	Error ellipses. Standard errors of derived distances and directions
$\boldsymbol{\sigma}_{(\mathbf{l})}$	Standard errors of derived quantities
$\boldsymbol{\sigma}_{(\mathbf{l})}$ & $\boldsymbol{\sigma}_{(\mathbf{b})}$	Reliability of observations
$\boldsymbol{\sigma}_{(\mathbf{v})}$ & **v**	Blunder detection

7.16 NETWORK DESIGN

The variance-covariance matrix of the parameters, from which the error ellipses were constructed, was found to be

$$\boldsymbol{\sigma}_{(\mathbf{x})} = (\mathbf{A}^T\mathbf{W}\mathbf{A})^{-1}$$

The terms in the **A** and **W** matrix can all be computed without actual observations being carried out, provided the approximate coordinates of the stations, the proposed observations and the quality of those proposed observations are all known. This enables the surveyor to consider networks for the solution of particular survey problems and test them using all the tests described in preceding sections that do not involve observed observations. This approach allows the surveyor to optimize a network and allows for examining 'what if' scenarios, such as removing observations, moving or removing proposed control stations and the effects of changing the quality of instrumentation. A surveyor can then be assured that an economic, minimal network that will meet some specified criteria is to be observed. Such criteria might be:

- The semi-major axis of all absolute error ellipses must be less than a certain value.
- The standard errors of the computed values of all, or specific, interstation distances must be less than a certain value.
- All observations must be reliable at a certain level.

There are two drawbacks to network analysis. It is reasonably easy to set up a proposed set of observations that meet the survey specification. It is then quite easy to remove some observations and test the new network against the criteria. As the optimum network is approached then much time may be spent finding the remaining few expendable observations, whereas this time might be better spent in the field inadvertently collecting extra but unnecessary observations.

If the network is reduced to the minimum, and for some reason all the planned observations are not made, or are found in the adjustment to be insufficiently precise, or to contain gross errors, then the network will not meet the specification. Although it appears to go against the ideas of network analysis, it is always a good idea to collect extra observations spread over the whole network. Again, a rule of thumb would be to collect 10% extra for large networks rising to 30% extra for small ones. The choice of which extra observations to take would probably be made on the basis of, 'if you can see it – observe it'.

7.17 PRACTICAL CONSIDERATIONS

It is likely that networks of any size will be estimated using software. To do otherwise would be tedious in the extreme. A rigorous solution can only be achieved if all the gross errors in the observations have been removed from the data set first. This can be a tedious process if there are many of them and the geometry of part of the network is weak. The following contains some hints on how to deal with these situations.

First check the data for correct formatting and that there are no 'typos'. When data is manually entered through a keyboard, errors will often be made. It is much easier to check input data independently before processing than to use software outputs to try to identify the source of blunders. Once obvious errors have been removed and the estimation attempted there will one of two outcomes. Either it converges to a solution, or it does not. If it does not converge to a solution then either there are too many gross errors or there is very weak geometry somewhere in the network, or both. Weak geometry occurs when the locus lines from observations associated with a particular point lie substantially parallel to each other in the region of the point. See *Section 7.3.1* for a discussion of locus lines.

It is likely that the software will allow for setting the convergence criteria, either in terms of the largest correction to a provisional coordinate in successive iterations or the maximum number of iterations to be performed. Set the convergence criteria to be rather coarse initially, say to 0.01 m or 0.001 m and the number of iterations to the maximum allowed, or at least the maximum you have patience for. If convergence has not been achieved then it is not possible to identify suspect observations. One possible check for gross errors in the coordinates of fixed points or the provisional coordinates of free points is to use the input coordinates to plot a network diagram to see if there are any obvious errors.

As a very general rule of thumb, points should be within about one third of the shortest distance in the network of their true value to ensure convergence. Imprecise coordinates of free points coupled with weak geometry may cause the estimation to fail. Obviously it would be possible to compute better provisional coordinates by the methods described in *Chapter 5* and of course better coordinates will improve the likelihood of convergence. However, whether it is worth the effort may be questionable. If there are no obvious errors in the plot and convergence has not been achieved, compare the observation values with their dimensions on the plot to identify large gross errors in any of the observations.

If the estimation still fails then it is likely that there are problems associated with the geometry part of the network. These are harder to identify but the following may help. Check to see if there are any points are connected to the rest of the network by only two observations. If there are then check the locus lines at the point associated with the two observations to see the quality of their intersection. If that does not resolve the problem investigate points with three observations and so on.

If this still does not resolve the problem and achieve convergence then more drastic action is called for. There may be problems associated with the coordinates of the fixed points, the coordinates of the free points, or the observations and all problems cannot be resolved simultaneously. Isolate the problems of the coordinates while you resolve the problems with observations by temporarily making all but one point free and adding a realistic fixed bearing to the set of observations. If convergence can now be achieved then there was a problem with one of the fixed coordinates which needs to be resolved.

If convergence was not achieved there is at least one, and probably more than one, gross error in the observations. Convergence needs to be forced. This can be done by removing the fixed bearing and temporarily treating all the coordinates except those of one fixed point as observations of position where the observations have a suitable standard error. Start with a large and therefore weak value for the standard error of a coordinate. If you are confident that you know the coordinates of a point to 10 m, say, use that. If convergence is achieved relax the standard error to 100 m or 1000 m and so on until you have the largest value where convergence is still achieved. If convergence is not achieved reduce the standard error to 1 m, 0.1 m, 0.01 m or whatever is needed to make the estimation converge.

Once convergence has been forced it is possible to investigate the observations in spite of the now highly distorted nature of the network. Examine the estimation output and identify the observation with the statistic of residual divided by its own standard error remembering that most of the coordinates are now also 'observations'. You now have the choice of investigating the observation by reference to the source data, such as a field sheet, and correcting a transcription error. If that was not the problem then you can remove the suspect observation from the data set. If you do, that may remove a suspect observation but it will not solve the problem with it. A better approach is to give that single observation, temporarily, a very large standard error. If the observation was an angle then 100 000″ would be appropriate, if a distance then a standard error of the average distance between stations to one significant figure would suffice. The effect of this is make the observation of no practical significance to the estimation but the estimation process can still compute a residual for the observation.

Continue this process, strictly one observation at a time, until the value of the standard error of an observation of unit weight becomes less than 100. At this stage return to the coordinates and increase their standard errors as much as you can while still achieving convergence. Return to the process of identifying and downweighting suspect observations until the standard error of observation of unit weight is small and relatively stable usually somewhere between 1 and 10. Now all the observations with gross errors have been identified. There are probably still observations with small gross errors but they will not be big enough to cause problems of convergence. Examine the residuals of all the observations that have been downweighted. Some observations will have very large residuals and these are the ones with gross errors. Check these observations against the source data. Some observations will have residuals that are not significantly bigger than good observations. These observations were probably good anyway and the downweighting with a high standard error can be removed so that the observation will subsequently play its proper part in the estimation process. The remaining observations remain suspect.

The problems with the observations have now been substantially, but not completely, resolved. The next step is to investigate the coordinates of the fixed points for error. Make the coordinates of the free points free, i.e. change the 'observations' of coordinates to provisional coordinates of free point. Now examine the residuals of the 'observations' of the coordinates of the fixed points. If there is a gross error in one of the coordinates then that coordinate will have a larger residual than any other coordinates of the same direction, north or east, and that residual will have a sign which is opposite from the sign of other residuals in the same direction.

As the problems of gross errors in the observations and problems with the coordinates of the fixed points have been resolved, now is the time to resolve any remaining problems associated with the observations. Make the coordinates of the fixed points fixed and the coordinates of the free points free. Ensure that the fixed bearing has been removed from the data set. Perform the estimation process in the normal way and downweight suspect observations as above. Correct any errors in the observation data set that you can. Stop when the standard error of an observation does not significantly improve with each downweighting and there is no one residual divided by its standard error that stands out from the rest.

Refine the convergence criteria to be one order of magnitude less than the precision that you require in the computed coordinates. For example, if you require coordinates stated to 0.001 m set the convergence criteria to 0.0001 m. Rerun the estimation process.

Now all the gross errors in the observations have been resolved. If the original weighting for the observations was correct then the standard error of an observation of unit weight should be very close to 1. If this is not so then the overall quality of the observations is not as you supposed and the standard errors of the observations should be scaled by the standard error of an observation of unit weight. For example, if the assumed standard error of the angles was $5''$ and the standard error of an observation of unit weight was 1.6 then the true standard error of the angles was $5'' \times 1.6 = 8''$.

Perform the estimation again and the standard error of an observation of unit weight will now be 1. An alternative approach to debugging the data, especially if there are difficulties with the above, is to start with a small network and build it up by stages. Identify a small group of points which have many observations connecting them and test to see if there is convergence and that the computed standard error of an observation of unit weight is not more than about 5. If it is it indicates strongly the presence of a gross error that needs to be identified and resolved using the ideas above. Once the problems have been resolved add another well-connected point to the network with all the observations that make the connection and repeat the process. Continue adding individual points and later small groups of points until the network is complete. The additional group of points should not exceed 10% of the existing network to ensure stability in the computation.

The same process can be applied to one-, two- and three-dimensional estimations by least squares.

7.18 ESTIMATION IN THREE DIMENSIONS

Most of this chapter, so far, has been concerned with applying least squares principles to estimation in two dimensions. The principles and processes are exactly the same in three dimensions except that everything is 50% bigger. The **x** vector will contain additional terms for the heights of the points and the **A** matrix will contain observations for difference height by levelling, slope distances as opposed to horizontal distances in two dimensions and vertical angles.

7.18.1 The A matrix

The **A** matrix and the observed minus computed vector are constructed in exactly the same way in three dimensions as they are in two dimensions. The only major difference is that in three dimensions there are three parameters, δE_i, δN_i and δH_i, associated with each point. The horizontal angle observation equation remains unchanged except to add 0s as the coefficients of the δHs.

Slope distance equation

The slope distance equation is derived by applying Pythagoras' theorem in three dimensions so the observation equation is

$$l_{ij} - \left\{(E_j - E_i)^2 + (N_j - N_i)^2 + (H_j - H_i)^2\right\}^{\frac{1}{2}} = 0$$

Upon linearizing and putting into matrix form this becomes:

$$\begin{bmatrix} -\sin a_{ij}\cos v_{ij} & -\cos a_{ij}\cos v_{ij} & -\sin v_{ij} & \sin a_{ij}\cos v_{ij} & \cos a_{ij}\cos v_{ij} & \sin v_{ij} \end{bmatrix} \begin{bmatrix} \delta E_i \\ \delta N_i \\ \delta H_i \\ \delta E_j \\ \delta N_j \\ \delta H_j \end{bmatrix} = \begin{bmatrix} l_{ij(o-c)} \end{bmatrix}$$

where a_{ij} is the bearing of j from i and v_{ij} is the vertical angle to j from i and is found from provisional values of coordinates:

$$\tan v_{ij} = \left\{ \frac{H_j - H_i}{\left\{(E_j - E_i)^2 + (N_j - N_i)^2\right\}^{\frac{1}{2}}} \right\}$$

Vertical angle equation

The vertical angle distance equation is:

$$v_{ij} - \tan^{-1}\left[(H_j - H_i)\left\{(E_j - E_i)^2 + (N_j - N_i)^2\right\}^{\frac{1}{2}}\right] = 0$$

Upon linearizing and putting into matrix form this becomes:

$$\begin{bmatrix} \frac{\sin v_{ij}\sin a_{ij}}{l_{ij}\sin 1''} & \frac{\sin v_{ij}\cos a_{ij}}{l_{ij}\sin 1''} & \frac{-\cos v_{ij}}{l_{ij}\sin 1''} & \frac{-\sin v_{ij}\sin a_{ij}}{l_{ij}\sin 1''} & \frac{-\sin v_{ij}\cos a_{ij}}{l_{ij}\sin 1''} & \frac{\cos v_{ij}}{l_{ij}\sin 1''} \end{bmatrix} \begin{bmatrix} \delta E_i \\ \delta N_i \\ \delta H_i \\ \delta E_j \\ \delta N_j \\ \delta H_j \end{bmatrix} = \begin{bmatrix} V_{ij(o-c)} \end{bmatrix}$$

7.18.2 Error ellipsoids

In *Section 7.11* absolute and relative error ellipses in two dimensions were discussed. Two methods for finding the parameters, the sizes of the semi-major and semi-minor axes and the directions of both were presented. The latter method, which involved the use of eigenvalues and eigenvectors, is the most suitable for application in three dimensions. In two dimensions it was only necessary to solve a quadratic equation to find the two eigenvalues. In three dimensions the characteristic polynomial is a cubic and therefore has three roots; therefore the solution is a little more complex. In two dimensions, the direction of each axis was described by a single statistic. In three dimensions, two statistics are required for the direction of an ellipsoid axis, e.g. orientation in the horizontal plane and elevation from the horizontal plane.

In two dimensions, the error ellipses can be adequately described in a graphical format on two-dimensional paper or computer screen. The two-dimensional mediums create practical problems in the presentation of three-dimensional data.

The relevant part of the variance-covariance matrix of parameters for a single point in three dimensions is of the form:

$$\mathbf{N}^{-1} = \begin{bmatrix} \sigma_E^2 & \sigma_{EN} & \sigma_{EH} \\ \sigma_{EN} & \sigma_N^2 & \sigma_{NH} \\ \sigma_{EH} & \sigma_{NH} & \sigma_H^2 \end{bmatrix}$$

and so the eigenvalue problem, $(\mathbf{N}^{-1} - \lambda\mathbf{I})\mathbf{z} = \mathbf{0}$ used to find the magnitudes of the semi axes, is of the form:

$$\begin{vmatrix} \sigma_E^2 - \lambda & \sigma_{EN} & \sigma_{EH} \\ \sigma_{EN} & \sigma_N^2 - \lambda & \sigma_{NH} \\ \sigma_{EH} & \sigma_{NH} & \sigma_H^2 - \lambda \end{vmatrix} = 0$$

Which, when evaluating this determinant leads to the characteristic equation:

$$(\sigma_E^2 - \lambda)\left((\sigma_N^2 - \lambda)(\sigma_H^2 - \lambda) - (\sigma_{NH})^2\right) - \sigma_{EN}\left(\sigma_{EN}(\sigma_H^2 - \lambda) - \sigma_{NH}\sigma_{EH}\right)$$

$$+ \sigma_{EH}\left(\sigma_{EN}\sigma_{NH} - (\sigma_N^2 - \lambda)\sigma_{EH}\right) = 0$$

When multiplied out and terms are gathered together this is:

$$-\lambda^3 + \lambda^2(\sigma_E^2 + \sigma_N^2 + \sigma_H^2) + \lambda\left(-\sigma_E^2\sigma_N^2 - \sigma_E^2\sigma_H^2 - \sigma_N^2\sigma_H^2 + (\sigma_{NH})^2 + (\sigma_{EN})^2 + (\sigma_{EN})^2\right)$$

$$+ \left(\sigma_E^2\sigma_N^2\sigma_H^2 - \sigma_E^2(\sigma_{NH})^2 - \sigma_N^2(\sigma_{EH})^2 - \sigma_H^2(\sigma_{EN})^2 + 2\sigma_{EH}\sigma_{EN}\sigma_{NH}\right) = 0$$

This may be re-expressed as: $f(\lambda) = g_3\lambda^3 + g_2\lambda^2 + g_1\lambda + g_0 = 0$

where: $g_3 = -1$

$$g_2 = \sigma_E^2 + \sigma_N^2 + \sigma_H^2$$

$$g_1 = -\sigma_E^2\sigma_N^2 - \sigma_E^2\sigma_H^2 - \sigma_N^2\sigma_H^2 + (\sigma_{NH})^2 + (\sigma_{EN})^2 + (\sigma_{EH})^2$$

$$g_0 = \sigma_E^2\sigma_N^2\sigma_H^2 - \sigma_E^2(\sigma_{NH})^2 - \sigma_N^2(\sigma_{EH})^2 - \sigma_H^2(\sigma_{EN})^2 + 2\sigma_{EH}\sigma_{EN}\sigma_{NH}$$

A cubic equation has three distinct roots. An ellipsoid must have three axes so all roots will be positive and real. The coefficient of λ^3 is negative so the graph of the function is of the general form of *Figure 7.14*.

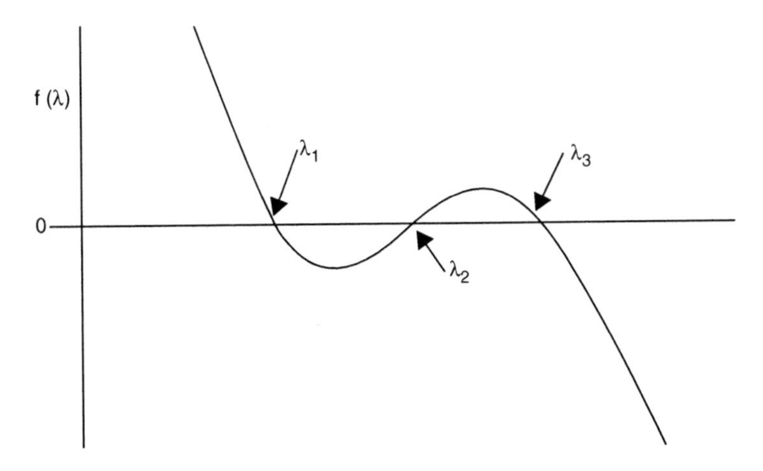

Fig. 7.14 *Cubic equation of* λ

The solution of the characteristic equation may be found by iterative means. Approximate solutions for λ must first be found. In *Figure 7.14* it can be seen that λ_1 lies between 0 and the minimum value of $f(\lambda)$. λ_2 lies between the minimum and maximum valves of $f(\lambda)$. λ_3 is greater than the maximum value of $f(\lambda)$. If the minimum and maximum values of λ are m_1 and m_2 respectively, then initial values for λ_1, λ_2 and λ_3 may be found as follows:

Root	*First approximation*
λ_1	$\lambda_0 = \frac{1}{2}m_1$
λ_2	$\lambda_0 = \frac{1}{2}(m_1 + m_2)$
λ_3	$\lambda_0 = \frac{1}{2}(3m_2 - m_1)$

The values of m_1 and m_2 are found in the normal way, i.e. as solutions to the equation:

$$\frac{df(\lambda)}{d\lambda} = 0$$

Therefore: $3g_3m^2 + 2g_2m + g_1 = 0$

and with $g_3 = -1$, the solutions are: $m_1 = \frac{1}{3}\left\{g_2 - \sqrt{(g_2^2 + 3g_1)}\right\}$ and $m_2 = \frac{1}{3}\left\{g_2 + \sqrt{(g_2^2 + 3g_1)}\right\}$

Better values of λ_1, etc., are found by linearization so:

$$\lambda = \lambda_0 - \frac{f(\lambda_0)}{\left.\frac{d}{dx}(f(\lambda))\right|_{\lambda=\lambda_0}} = \lambda_0 - \frac{\lambda_0^3 - g_2\lambda_0^2 - g_1\lambda_0 - g_0}{3\lambda_0^2 - 2g_2\lambda_0 - g_1}$$

and is iterated to convergence. The magnitudes of λ_1, λ_2 and λ_3, the three semi axes, are now found. The next stage is to find directions of the axes using the eigenvectors, i.e. to find solutions for $\mathbf{z}$ in the equation:

$$(\mathbf{N}^{-1} - \lambda\mathbf{I})\mathbf{z} = \mathbf{0}$$

i.e. $$\begin{bmatrix} \sigma_E^2 - \lambda & \sigma_{EN} & \sigma_{EH} \\ \sigma_{EN} & \sigma_N^2 - \lambda & \sigma_{NH} \\ \sigma_{EH} & \sigma_{NH} & \sigma_H^2 - \lambda \end{bmatrix} \begin{bmatrix} z_1 \\ z_2 \\ z_3 \end{bmatrix} = \mathbf{0}$$

This leads to a solution for z as:

$$\begin{bmatrix} z_1 \\ z_2 \\ z_3 \end{bmatrix} = \begin{bmatrix} (\sigma_N^2 - \lambda)(\sigma_H^2 - \lambda) - (\sigma_{NH})^2 \\ \sigma_{EH}\sigma_{NH} - \sigma_{EN}(\sigma_H^2 - \lambda) \\ (\sigma_N^2 - \lambda)\sigma_{EN} - \sigma_{EN}\sigma_{NH} \end{bmatrix}$$

Finally, the directions of the axis are given by:

the horizontal direction of the axis: $a_\lambda = \tan^{-1}\{z_1 z_2^{-1}\}$

the elevation of the axis: $v_\lambda = \tan^{-1}\{z_3(z_1^2 + z_2^2)^{-\frac{1}{2}}\}$

Worked example

Example 7.18 Find the terms of the error ellipsoid for a point from the following extract of a variance-covariance matrix of parameters. The units are all mm^2.

$$\mathbf{N}^{-1} = \begin{bmatrix} \sigma_E^2 & \sigma_{EN} & \sigma_{EH} \\ \sigma_{EN} & \sigma_N^2 & \sigma_{NH} \\ \sigma_{EH} & \sigma_{NH} & \sigma_H^2 \end{bmatrix} = \begin{bmatrix} 2.678 & 1.284 & 1.585 \\ 1.284 & 3.094 & 2.987 \\ 1.585 & 2.987 & 6.656 \end{bmatrix}$$

$$g_2 = \sigma_E^2 + \sigma_N^2 + \sigma_H^2 = 12.428$$

$$g_1 = -\sigma_E^2\sigma_N^2 - \sigma_E^2\sigma_H^2 - \sigma_N^2\sigma_H^2 + (\sigma_{NH})^2 + (\sigma_{EN})^2 + (\sigma_{EH})^2 = -33.621$$

$$g_0 = \sigma_E^2\sigma_N^2\sigma_H^2 - \sigma_E^2(\sigma_{NH})^2 - \sigma_N^2(\sigma_{EH})^2 - \sigma_H^2(\sigma_{EN})^2 + 2\sigma_{EH}\sigma_{EN}\sigma_{NH} = 24.668$$

$$m_1 = \frac{1}{3}\left\{g_2 - \sqrt{(g_2^2 + 3g_1)}\right\} = 1.7025$$

$$m_2 = \frac{1}{3}\left\{g_2 + \sqrt{(g_2^2 + 3g_1)}\right\} = 6.5829$$

Root	*First approximation*
λ_1	$\lambda_0 = \frac{1}{2}m_1 = 0.8512$
λ_2	$\lambda_0 = \frac{1}{2}(m_1 + m_2) = 4.1427$
λ_3	$\lambda_0 = \frac{1}{2}(3m_2 - m_1) = 9.0231$

$$\lambda = \lambda_0 - \frac{\lambda_0^3 - g_2\lambda_0^2 - g_1\lambda_0 - g_0}{3\lambda_0^2 - 2g_2\lambda_0 - g_1}$$

Subsequent iterations produce the following results for λ_1, λ_2 and λ_3:

	Approximations						
Root	*1st*	*2nd*	*3rd*	*4th*	*5th*	*6th*	√
λ_1	0.8512	1.1544	1.2529	1.2649	1.26507	1.26507	1.12
λ_2	4.1427	2.5989	2.2556	2.1739	2.16777	2.16773	1.47
λ_3	9.0231	8.9954	8.9952	8.9952	8.99519	8.99519	3.00

Find the terms of the eigenvectors:

	z_1	z_2	z_3
λ_1	0.93744	−2.18755	−0.93646
λ_2	−4.76485	−1.02854	−2.36718
λ_3	4.88185	7.73792	−13.18870

Horizontal direction of the axis: $a_\lambda = \tan^{-1}\{z_1 z_2^{-1}\}$

Elevation of the axis: $v_\lambda = \tan^{-1}\{z_3(z_1^2 + z_2^2)^{-\frac{1}{2}}\}$

Axis	*Magnitude*	*Horizontal direction*	*Elevation*
semi-minor	1.12 mm	156° 48′	−21° 29′
semi-middle	1.47 mm	77° 49′	25° 54′
semi-major	3.00 mm	32° 15′	−55° 15′

In *Section 7.11.2* the variances and covariances of the relative positions of two stations were used to find the terms of the two-dimensional relative error ellipse. In three dimensions a very similar process is used to find the terms of the three-dimensional error ellipsoid. By a derivation very similar to that of *Section 7.11.2* it can be shown that:

$$\sigma_{\Delta E}^2 = \sigma_{EF}^2 + \sigma_{EJ}^2 - 2\sigma_{EFEJ}$$

$$\sigma_{\Delta N}^2 = \sigma_{NF}^2 + \sigma_{NJ}^2 - 2\sigma_{NFNJ}$$

$$\sigma_{\Delta H}^2 = \sigma_{HF}^2 + \sigma_{HJ}^2 - 2\sigma_{HFHJ}$$

$$\sigma_{\Delta E \Delta N} = \sigma_{EFNF} + \sigma_{EJNJ} - \sigma_{EFNJ} - \sigma_{NFEJ}$$

$$\sigma_{\Delta E \Delta H} = \sigma_{EFHF} + \sigma_{EJHJ} - \sigma_{EFHJ} - \sigma_{HFEJ}$$

$$\sigma_{\Delta N \Delta H} = \sigma_{NFHF} + \sigma_{NJHJ} - \sigma_{NFHJ} - \sigma_{HFNJ}$$

These terms may now be used in exactly the same way as their counterparts were used when constructing absolute error ellipsoids. Absolute error ellipsoids are constructed at the point to which they refer but relative error ellipses are normally constructed at the mid-point of the line, in three-dimensional space, between the two stations concerned.

Exercises

(*7.1*) Solve the following simultaneous equations by linearizing them as the first part of a Taylor series. Express the formulae in matrix terms, and by substituting with the provisional values of the parameters, solve the equation for a first estimate of the true values of x and y. Iterate the solution until there is no change in the third decimal place. Use $x = 0.5$ radians and $y = 1$ radian as provisional values for the first iteration.

$$\sin^2 x + \cos^3 y - 0.375 = 0$$

$$\sin 3x + \cos 2y - 0.5 = 0$$

(*Answer:* Solution after first iteration $x = 0.526985$ rad, $y = 1.047880$ rad.
Final solution $x = 0.523598775$ radians $= 30°$ exactly, $y = 1.047197551$ radians $= 60°$ exactly)

(7.2) Use linearization to solve the following pair of simultaneous equations.

$$(x - 1)^2 \sin y = 0.00216856$$

$$(\sin 3x - 1)y = -0.748148$$

where x and y are in radians. Use $x = y = 0.9$ as provisional values.

(*Answer:* $x = 0.95000$, $y = 1.05000$)

(*7.3*) A new point, X, is to be added to a control network already containing points A, B and C as in *Figure 7.15*. The coordinates of the points are:

	Easting	*Northing*
A	5346.852 m	4569.416 m
B	6284.384 m	4649.961 m
C	6845.925 m	4469.734 m
X	5280 m	3560 m (approx.)

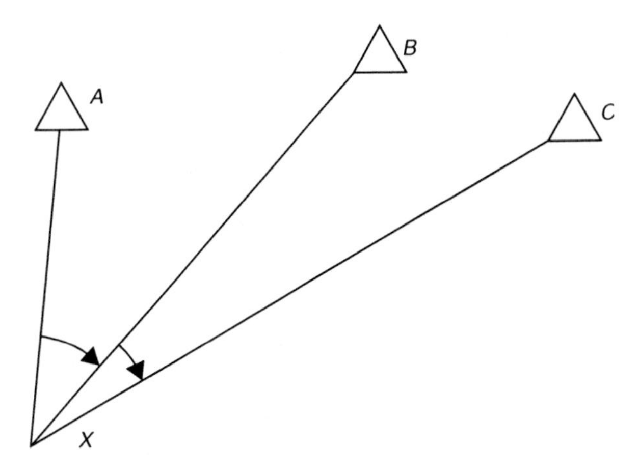

Fig. 7.15 *A control network*

The observations were:

Angles		*Distances*	
AXB	39° 2′ 4″	*AX*	1009.258 m
BXC	17° 14′ 19″	*BX*	1477.819 m
		CX	1806.465 m

The standard errors of the observations are 5″ for angles and 0.02 m for distances. Find the best estimate of the coordinates of point X.

Note that several of the later exercises follow from this one so it is worth retaining all your working for this exercise.

(*Answer:* E_X 5284.061 m N_X 3562.119 m)

(*7.4*) Repeat *Exercise 7.3* using the computed values of the coordinates from that exercise as new provisional values for the coordinates of X. By how much do the computed coordinates of point X change with this further iteration?

(*Answer:* E_X 5284.053 m N_X 3562.127 m; change in E_X − 0.008 m N_X 0.008 m)

(*7.5*) What are the residuals of the observation equations from *Exercise 7.4*?

(*Answer:*	*Observation*	*Residual*
Angle	*AXB*	−5.1″
	BXC	−0.1″
Distance	*AX*	−0.013 m
	BX	0.032 m
	CX	−0.034 m)

(*7.6*) Use the residuals computed in *Exercise 7.5* to find the standard error of an observation of unit weight.

(*Answer:* 1.52)

(7.7) What was the variance-covariance matrix of parameters associated with *Exercise 7.5*? What are the standard errors of the coordinates and what is the coefficient of correlation between them?

$$\left(\textit{Answer:} \begin{bmatrix} 0.000545 & -0.000325 \\ -0.000325 & 0.000404 \end{bmatrix} \quad \sigma_{EX} = 0.023\,\text{m}, \quad \sigma_{NX} = 0.020\,\text{m}, \quad r_{EXNX} = -0.69 \right)$$

(7.8) Using the variance-covariance matrix of parameters derived in *Exercise 7.6* find the terms of the error ellipse at point X.

(*Answer:* $\sigma_{max} = 0.028$ m, $\sigma_{min} = 0.021$ m, orientation of major axis $= 128° 52'$)

(7.9) Using the data from previous exercises confirm or otherwise that there are no observations with significant residuals.

(*Answer:* There are no significant residuals.

Observation		*Residual v*	*Standard error of residual* σ_v	*Test statistic* $\frac{v}{\sigma_v}$	*Is* $4 > \frac{v}{\sigma_v} > -4$?
Angle	*AXB*	−5.1″	4.7″	−1.1	yes
	BXC	−0.1″	4.9″	−0.0	yes
Distance	*AX*	−0.013 m	0.006 m	−2.2	yes
	BX	0.032 m	0.016 m	2.0	yes
	CX	−0.034 m	0.013 m	−2.6	yes)

(7.10) Using the data from previous exercises confirm or otherwise that there are no unreliable observations.

(*Answer:* There is one unreliable observation – distance AX

Observation		*Standard error of estimated observation* $\sigma_{(l)}$	*Standard error of observed observation* $\sigma_{(b)}$	*Test statistic* $\frac{\sigma_{(l)}}{\sigma_{(b)}}$	*Is* $\frac{\sigma_{(l)}}{\sigma_{(b)}} < 0.8$?
Angle	*AXB*	1.7″	5.0″	0.34	yes
	BXC	1.0″	5.0″	0.20	yes
Distance	*AX*	0.019 m	0.020 m	0.95	no
	BX	0.012 m	0.020 m	0.60	yes
	CX	0.015 m	0.020 m	0.75	yes)

(7.11) Find the terms of the error ellipsoid from the following extract of a variance/covariance matrix of parameters. The units are mm^2.

$$\mathbf{N}^{-1} = \begin{bmatrix} rrr\sigma_E^2 & \sigma_{EN} & \sigma_{EH} \\ \sigma_{EN} & \sigma_N^2 & \sigma_{NH} \\ \sigma_{EH} & \sigma_{NH} & \sigma_H^2 \end{bmatrix} = \begin{bmatrix} 3.020 & 0.853 & 4.143 \\ 0.853 & 4.026 & -1.348 \\ 4.143 & -1.348 & 8.563 \end{bmatrix}$$

(*Answer:*

Axis	*Magnitude*	*Horizontal direction*	*Elevation*
semi-minor	0.53 mm	113° 27′	27° 39′
semi-middle	2.11 mm	21° 30′	3° 4′
semi-major	3.29 mm	104° 31′	−62° 4′)

8
Position

8.1 INTRODUCTION

Engineering surveying is concerned essentially with fixing the position of a point in two or three dimensions.

For example, in the production of a plan or map, one is concerned in the first instance with the accurate location of the relative position of survey points forming a framework, from which the position of topographic detail is fixed. Such a framework of points is referred to as a control network.

The same network used to locate topographic detail may also be used to set out points, defining the position, size and shape of the designed elements of the construction project.

Precise control networks are also used in the monitoring of deformation movements on all types of structures.

In all these situations the engineer is concerned with relative position, to varying degrees of accuracy and over areas of varying extent. In order to define position to the high accuracies required in engineering surveying, a suitable homogeneous coordinate system and reference datum must be adopted.

Depending on the accuracies required and the extent of the area of the project it may or may not be necessary to take the shape of the Earth into account. For small projects it may be possible to treat the reference surface of the project area as a plane, in which case the mathematics involved with finding coordinates of points from the observations associated with them relatively simple. If the shape of the reference surface needs to be taken into account then the process is rather more involved.

8.2 REFERENCE ELLIPSOID

Consideration of *Figure 8.1* illustrates that if the area under consideration is of limited extent, the orthogonal projection of *AB* onto a plane surface may result in negligible distortion. Plane surveying techniques could be used to capture field data and plane trigonometry used to compute position. This is the case in the majority of engineering surveys. However, if the area extended from *C* to *D*, the effect of the Earth's curvature is such as to produce unacceptable distortion if treated as a flat surface. It can also be clearly seen that the use of a plane surface as a reference datum for the elevations of points is totally unacceptable.

If *Figure 8.2* is now considered, it can be seen that projecting *CD* onto a surface (*cd*) that was the same shape and parallel to *CD* would be more acceptable. Further, if that surface was brought closer to *CD*, say *c′d′*, the distortion would be even less. This then is the problem of the geodetic surveyor: that of defining a simple mathematical surface that approximates to the shape of the area under consideration and then fitting and orientating it to the Earth's surface. The surface used in surveying is a 'reference ellipsoid'.

Before describing the reference ellipsoid, it is appropriate to review the other surfaces that are related to it.

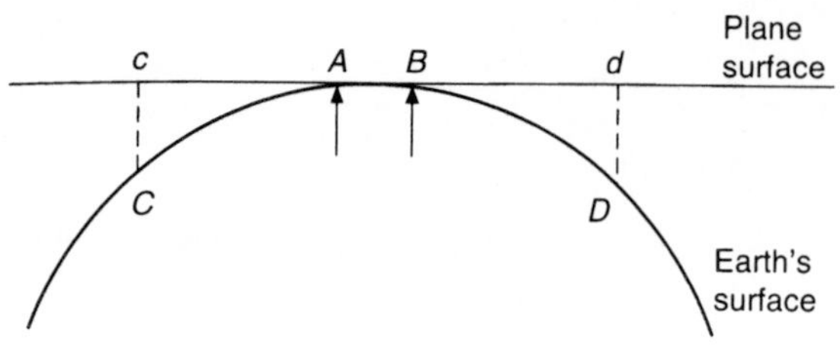

Fig. 8.1 *Projection to a plane surface*

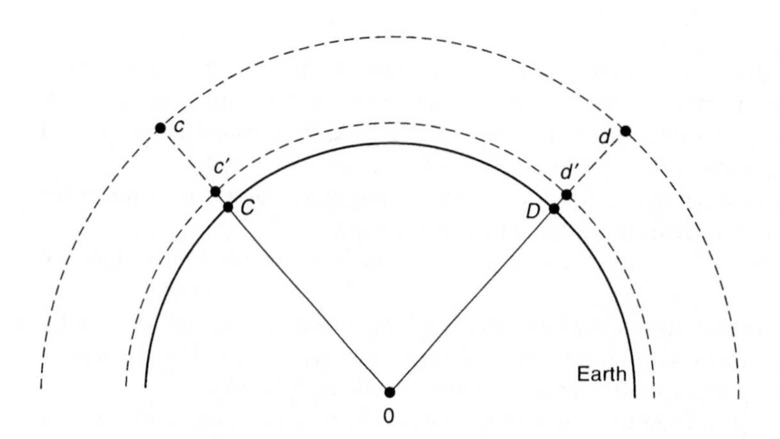

Fig. 8.2 *Projection to a curved surface*

8.2.1 Earth's surface

The Earth's physical surface is where surveying observations are made and points are located. However, due to its irregular surface and overall shape, the physical surface cannot be defined mathematically and so position cannot be easily computed on its surface. It is for this reason that in surveys of limited extent, the Earth may be assumed to be flat, and plane trigonometry used to define position.

An implication of this assumption is that the surface which defines the 'zero of height' is a plane surface and that everywhere, on that plane surface, above it and below it all plumb lines are straight and parallel. In the real world this is not true because the Earth is not flat but round, and being round it is not a sphere or any other regular mathematical figure. Although a defined surface, such as 'Mean Sea Level' (MSL) can be approximated by an ellipsoid there are separations from the best globally fitting ellipsoid of the order of 100 m or 1 part in 60 000. In relative terms these errors appear small, but for topographic surveying they may, and for geodetic surveying they must, be taken into consideration. Gravity below, on, or above any point on the Earth's surface may be described by a vector. The force of gravity, i.e. its magnitude, varies largely with distance from the Earth's mass centre, and to a lesser extent with latitude. It is also affected by the variations in the distribution of the Earth's mass and the changes in its density.

The 'shape' of the Earth is largely defined by gravity. The Earth is mostly a molten mass with a very thin stiff crust some 30 km thick. If the Earth was truly molten, of homogeneous density, and not affected by the gravity field of any external bodies and did not rotate on its own axis (once a day), then through considerations of gravitational attraction, the surface of the Earth would be fully described as a sphere.

The main source of error in this idealized model is that which is due to the Earth's rotation. The centrifugal force at the equator acts in the opposite direction to that of gravitation and so the figure of the Earth is better described by an ellipse of rotation about its minor axis, i.e. an ellipsoid or spheroid. The terms are synonymous.

8.2.2 The geoid

Having rejected the physical surface of the Earth as a computational surface, one is instinctively drawn to a consideration of a mean sea level surface. This is not surprising, as 70% of the Earth's surface is ocean.

If these oceans were imagined to flow in interconnecting channels throughout the land masses, then, ignoring the effects of friction, tides, wind stress, etc., an equipotential surface, approximately at MSL would be formed. An equipotential surface is one on which the gravitational potential is the same at all points. It is a level surface and like contours on a map which never cross, equipotential surfaces never intersect, they lie one within another as in *Figure 8.3*. Such a surface at MSL is called the 'geoid'. It is a physical reality and its shape can be measured. Although the gravity potential is everywhere the same the value of gravity is not. The magnitude of the gravity vector at any point is the rate of change of the gravity potential at that point. The surface of the geoid is smoother than the physical surface of the Earth but it still contains many small irregularities which render it unsuitable for the mathematical location of planimetric position. These irregularities are due to mass anomalies throughout the Earth.

In spite of this, the geoid remains important to the surveyor as it is the surface to which all terrestrial measurements are related.

As the direction of the gravity vector (termed the 'vertical') is everywhere normal to the geoid, it defines the direction of the surveyor's plumb-bob line. Thus any instrument which is made horizontal by means of a spirit bubble will be referenced to the equipotential surface passing through the instrument. Elevations in Great Britain, as described in *Chapter 2*, are related to the equipotential surface passing through MSL, as defined at Newlyn, Cornwall. Such elevations or heights are called orthometric heights (H) and are the linear distances measured along the gravity vector from a point on the surface to the equipotential surface used as a reference datum. As such, the geoid is the equipotential surface that best fits MSL

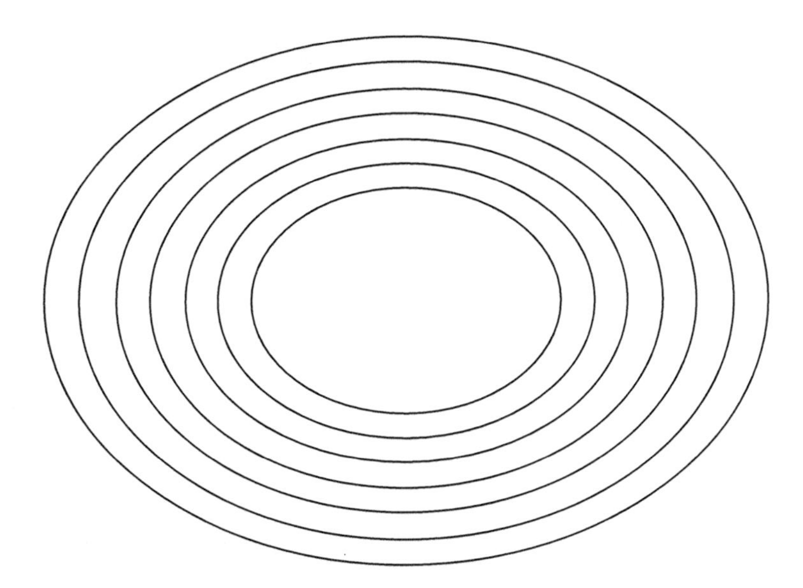

Fig. 8.3 *Equipotential surfaces*

and the heights in question, referred to as heights above or below MSL. It can be seen from this that orthometric heights are datum dependent. Therefore, elevations related to the Newlyn datum cannot be related to elevations that are relative to other datums established for use in other countries. The global MSL varies from the geoid by as much as 3 m at the poles and the equator, largely because the density of sea water changes with its temperature, and hence it is not possible to have all countries on the same datum.

8.2.3 The ellipsoid

An ellipsoid of rotation is the closest mathematically definable shape to the figure of the Earth. Its shape is modelled by an ellipse rotated about its minor axis and the ellipse may be defined by its semi-major axis a (*Figure 8.4*) and the flattening f. Although the ellipsoid is merely a shape and not a physical reality, it represents a smooth surface for which formulae can be developed to compute ellipsoidal distance, azimuth and ellipsoidal coordinates. Due to the variable shape of the geoid, it is not possible to have a single global ellipsoid of reference which is a good fit to the geoid for use by all countries. The best-fitting global geocentric ellipsoid is the Geodetic Reference System 1980 (GRS80), which has the following dimensions:

semi-major axis 6 378 137.0 m

semi-minor axis 6 356 752.314 m

the difference being approximately 21 km.

The most precise global geoid is the Earth Gravitational Model 1996 (EGM96). However, it still remains a complex, undulating figure which varies from the GRS80 ellipsoid by more than 100 m in places. In the UK the geoid–ellipsoid separation is as much as 57 m in the region of the Hebrides. As a 6-m vertical

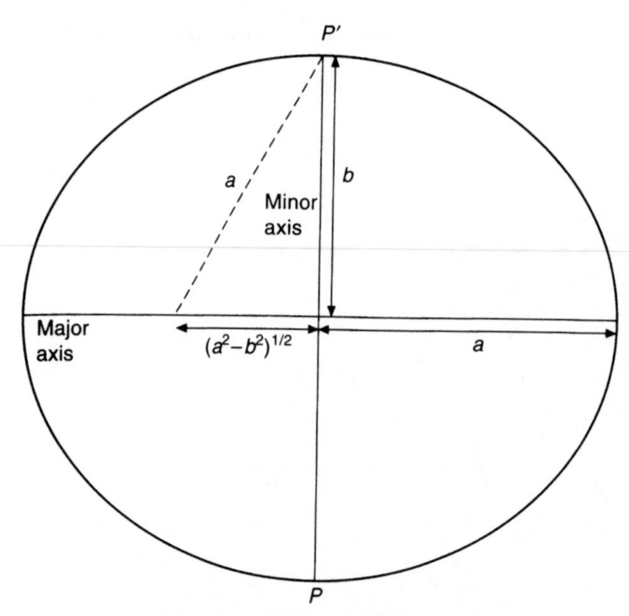

Fig. 8.4 *Elements of an ellipse*

separation between geoid and ellipsoid would result in a scale error of 1 ppm, different countries have adopted local ellipsoids that give the best fit in their particular situation. A small sample of ellipsoids used by different countries is shown below:

Ellipsoid	*a metres*	$1/f$	*Where used*
Airy (1830)	6 377 563	299.3	Great Britain
Everest (1830)	6 377 276	300.8	India, Pakistan
Bessel (1841)	6 377 397	299.2	East Indies, Japan
Clarke (1866)	6 378 206	295.0	North and Central America
Australian National (1965)	6 378 160	298.2	Australia
South American (1969)	6 378 160	298.2	South America

When $f = 0$, the figure described is a sphere. The flattening of an ellipsoid is described by $f = (a - b)/a$. A further parameter used in the definition of an ellipsoid is e, referred to as the first eccentricity of the ellipse, and is equal to $(a^2 - b^2)^{\frac{1}{2}}/a$.

Figure 8.5 illustrates the relationship of all three surfaces. It can be seen that if the geoid and ellipsoid were parallel at A, then the deviation of the vertical would be zero in the plane shown. If the value for geoid–ellipsoid separation (N) was zero, then not only would the surfaces be parallel, they would also be tangential at the point. As the ellipsoid is a smooth surface and the geoid is not, perfect fit can never be achieved. However, the values for deviation of the vertical and geoid–ellipsoid separation can be used as indicators of the closeness of fit.

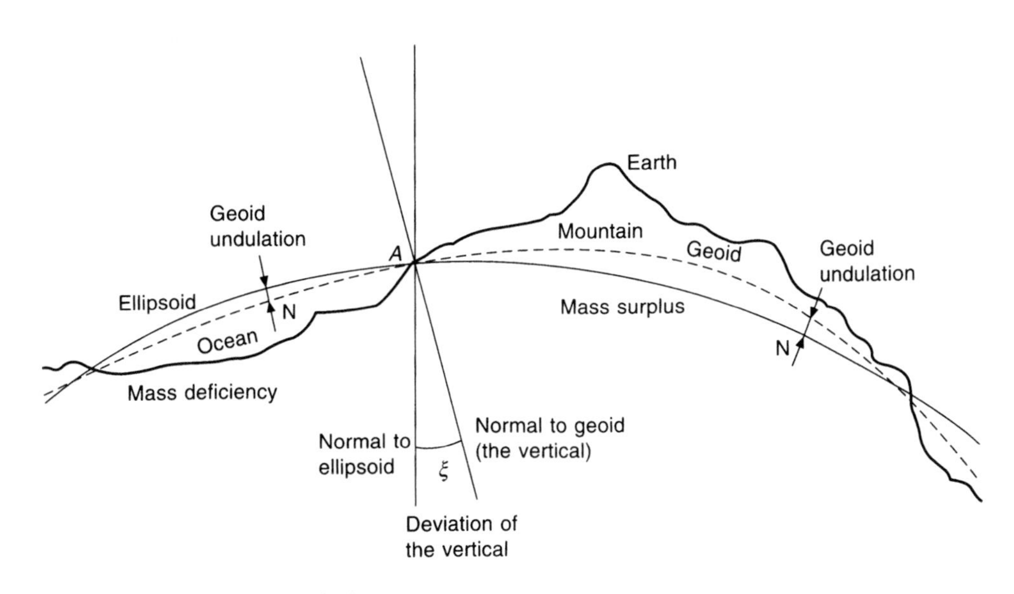

Fig. 8.5 *Deviation of the vertical*

8.3 COORDINATE SYSTEMS

8.3.1 Astronomical coordinates

As shown in *Figure 8.6*, astronomical latitude ϕ_A is defined as the angle that the vertical (gravity vector) through the point in question (P) makes with the plane of the equator, whilst the astronomical longitude λ_A is the angle in the plane of the equator between the zero meridian plane parallel to the vertical at Greenwich and the meridian plane parallel to the vertical through P. Both meridian planes contain the Earth's mean spin axis.

The common concept of a line through the North and South Poles comprising the spin axis of the Earth is not acceptable as part of a coordinate system, as it is constantly moving with respect to the solid body of the Earth. The instantaneous North Pole wanders once around the International Reference Pole (IRP) in a period of about 14 months with a distance from the IRP of about 20 metres. The IRP has been defined, and internationally agreed, by the International Earth Rotation Service (IERS) based in Paris. Similarly, the Greenwich Meridian adopted is not the one passing through the centre of the observatory telescope.

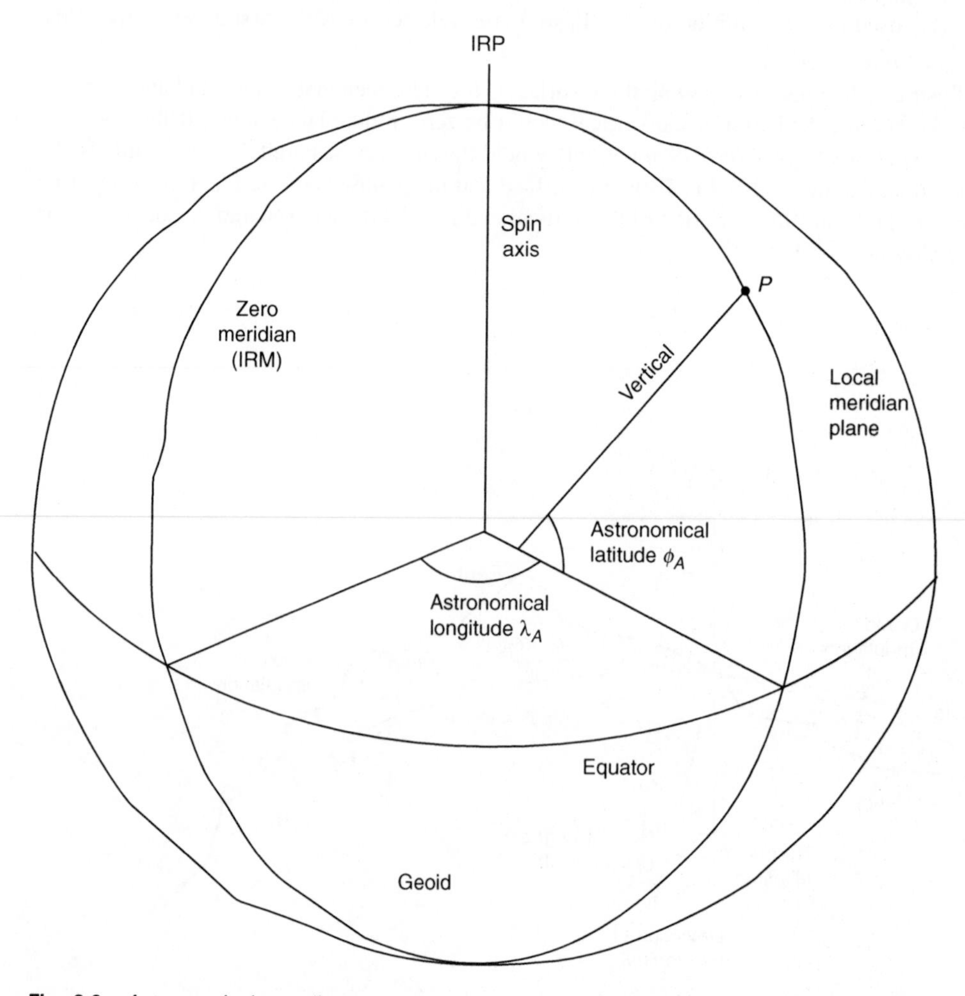

Fig. 8.6 *Astronomical coordinates*

It is one defined as the mean value of the longitudes of a large number of participating observatories throughout the world and is called the IERS Reference Meridian (IRM).

The instantaneous position of the Earth with respect to this axis is constantly monitored by the IERS and published for the benefit of those who need it at http://hpiers.obspm.fr/eop-pc/.

Astronomical position is found from observations to the sun and stars. The methods for doing so are relatively complex, seldom practised today and therefore outside the scope of this book.

Astronomical latitude and longitude do not define position on the Earth's surface but rather the direction and inclination of the vertical through the point in question. Due to the undulation of the equipotential surface deviation of the vertical varies from point to point so latitude and longitude scales are not entirely regular. An astronomical coordinate system is therefore unsatisfactory for precise positioning.

8.3.2 Geodetic coordinates

For a point P at height h, measured along the normal through P, above the ellipsoid, the ellipsoidal latitude and longitude will be ϕ_G and λ_G, as shown in *Figure 8.7*. Thus the ellipsoidal latitude is the angle describing the inclination of the normal to the ellipsoidal equatorial plane. The ellipsoidal longitude is

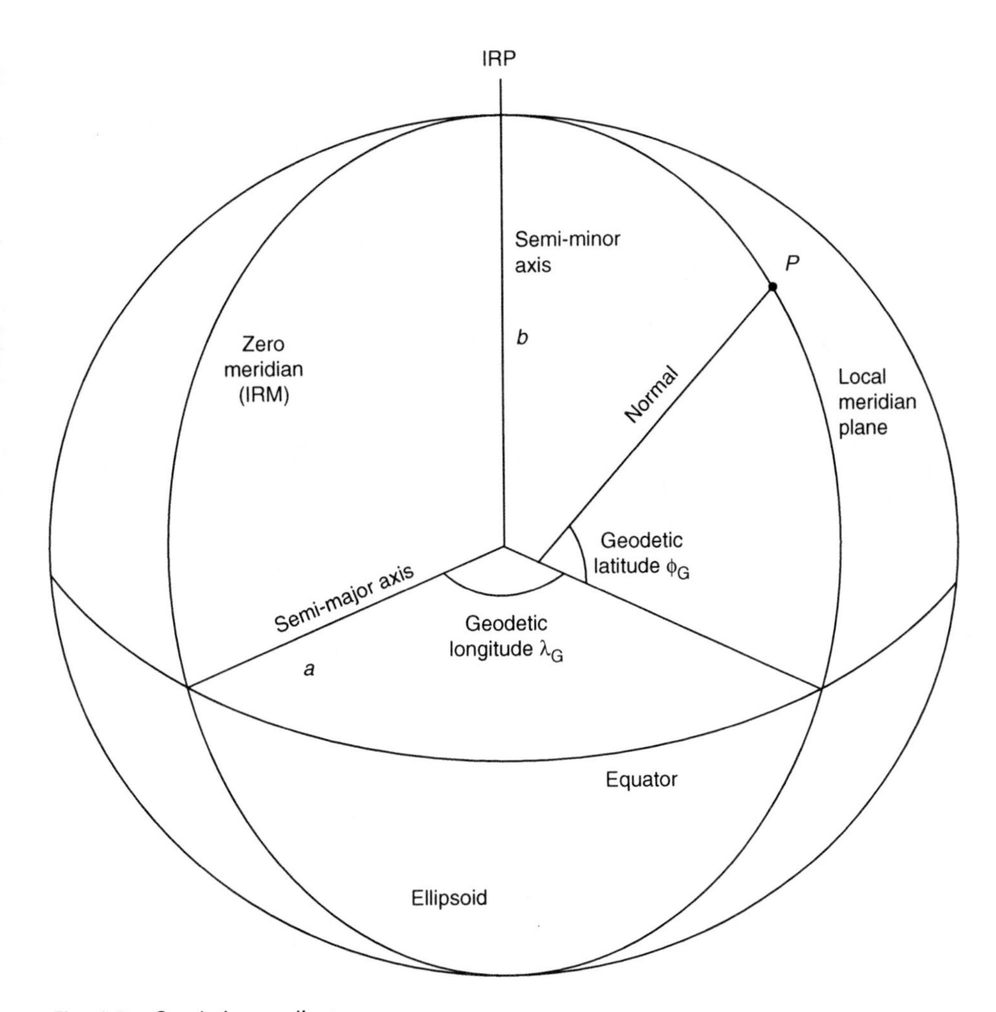

Fig. 8.7 *Geodetic coordinates*

the angle in the equatorial plane between the IRM and the geodetic meridian plane through the point in question P. The height h of P above the ellipsoid is called the ellipsoidal height. Also, the ellipsoidal coordinates can be used to compute azimuth and ellipsoidal distance. These are the coordinates used in classical geodesy to describe position on an ellipsoid of reference.

8.3.3 Cartesian coordinates

As shown in *Figure 8.8*, if the IERS spin axis is regarded as the Z-axis, the X-axis is in the direction of the zero meridian (IRM) and the Y-axis is perpendicular to both, a conventional three-dimensional coordinate system is formed. If we regard the origin of the cartesian system and the ellipsoidal coordinate system as coincident at the mass centre of the Earth then transformation between the two systems may be carried out as follows:

(1) Ellipsoidal to cartesian

$$X = (\nu + h)\cos\phi_G \cos\lambda_G \quad (8.1)$$

$$Y = (\nu + h)\cos\phi_G \sin\lambda_G \quad (8.2)$$

$$Z = [(1 - e^2)\nu + h]\sin\phi_G \quad (8.3)$$

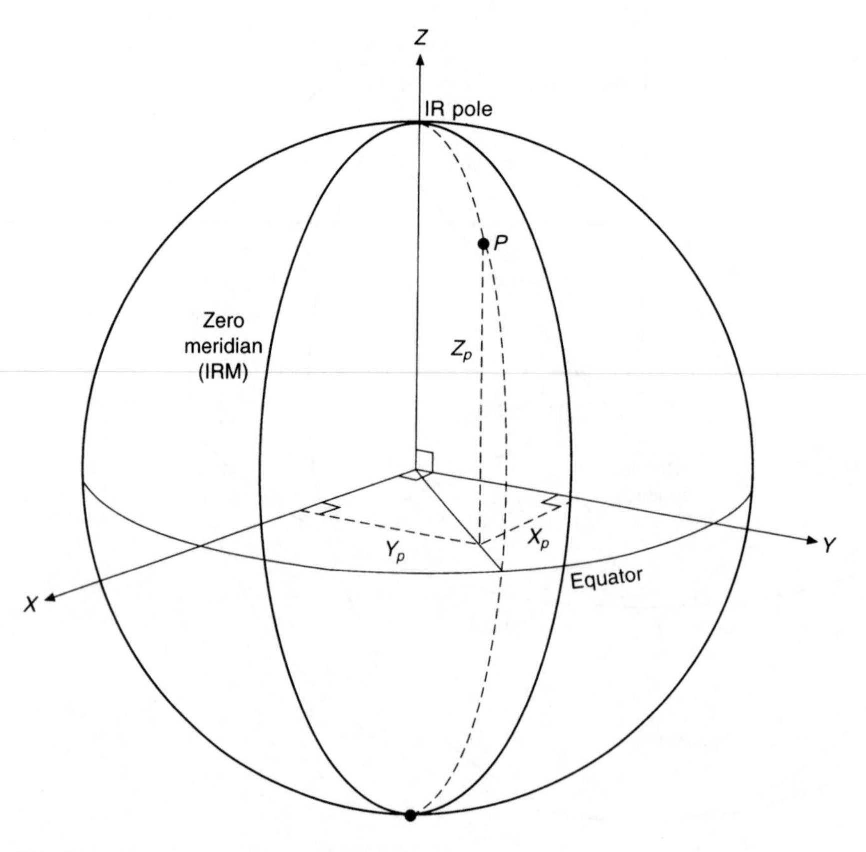

Fig. 8.8 *Geocentric cartesian coordinates*

(2) Cartesian to ellipsoidal

$$\tan \lambda_G = Y/X \tag{8.4}$$

$$\tan \phi_G = (Z + e^2 \nu \sin \phi_G)/(X^2 + Y^2)^{\frac{1}{2}} \tag{8.5}$$

$$h = X \sec \phi_G \sec \lambda_G - \nu \tag{8.6}$$

$$= Y \sec \phi_G \operatorname{cosec} \lambda_G - \nu \tag{8.7}$$

where

$\nu = a/(1 - e^2 \sin^2 \phi_G)^{\frac{1}{2}}$

$e = (a^2 - b^2)^{\frac{1}{2}}/a$

$a =$ semi-major axis

$b =$ semi-minor axis

$h =$ ellipsoidal height

The transformation in *equation (8.5)* is complicated by the fact that ν is dependent on ϕ_G and so an iterative procedure is necessary.

This procedure converges rapidly if an initial value for ϕ_G is obtained from

$$\phi_G = \sin^{-1}(Z/a) \tag{8.8}$$

Alternatively, ϕ_G can be found direct from

$$\tan \phi_G = \frac{Z + \varepsilon b \sin^3 \theta}{(X^2 + Y^2)^{\frac{1}{2}} - e^2 a \cos^3 \theta} \tag{8.9}$$

where $\varepsilon = (a^2/b^2) - 1$

$\tan \theta = a \cdot Z/b(X^2 + Y^2)^{\frac{1}{2}}$

Cartesian coordinates are used in satellite position fixing. Where the various systems have parallel axes but different origins, translation from one to the other will be related by simple translation parameters in X, Y and Z, i.e. ΔX, ΔY and ΔZ.

The increasing use of satellites makes a study of cartesian coordinates and their transformation to ellipsoidal coordinates important.

8.3.4 Plane rectangular coordinates

Geodetic surveys required to establish ellipsoidal or cartesian coordinates of points over a large area require very high precision, not only in the capture of the field data but also in their processing. The mathematical models involved must of necessity be complete and hence are quite involved. To avoid this the area of interest on the ellipsoid of reference, if of limited extent, may be regarded as a plane surface alternatively the curvature may be catered for by the mathematical projection of ellipsoidal position onto a plane surface. These coordinates in the UK are termed eastings (E) and northings (N) and are obtained from:

$E = f_E$ (ϕ_G, λ_G, ellipsoid and projection parameters)

$N = f_N$ (ϕ_G, λ_G, ellipsoid and projection parameters)

How this is done is discussed in later sections of this chapter.

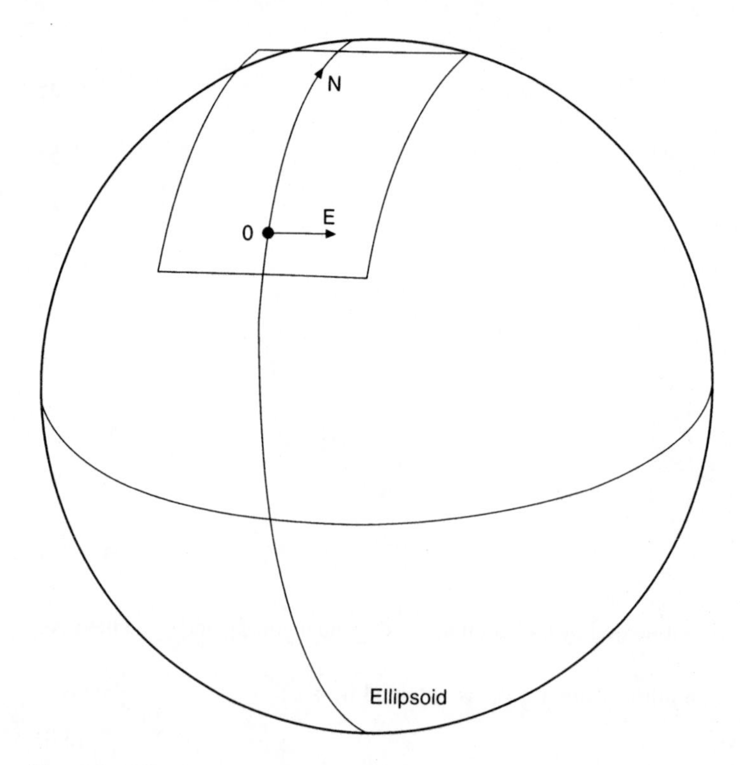

Fig. 8.9 *Plane rectangular coordinates*

The result is the definition of position by plane coordinates (E, N) which can be utilized using plane trigonometry. These positions will contain some distortion compared with their position on the ellipsoid, which is an inevitable result of projecting a curved surface onto a plane. However, the overriding advantage is that only small adjustments need to be made to the observed field data to produce the plane coordinates.

Figure 8.9 illustrates the concept involved and shows the plane tangential to the ellipsoid at the local origin 0. Generally, the projection, used to transform observations in a plane reference system is an orthomorphic projection, which will ensure that at any point in the projection the scale is the same in all directions. The result of this is that, for small areas, shape and direction are preserved and the scale varies but only slightly. Thus when computing engineering surveys in such a reference system, the observed distance, when reduced to its horizontal equivalent at MSL, simply requires multiplication by a local scale factor, and observed horizontal angles generally require no further correction.

8.3.5 Height

In outlining the coordinate systems in general use, the elevation or height of a point has been defined as 'orthometric' or 'ellipsoidal'. With the increasing use of satellites in engineering surveys, it is important to understand the different categories.

Orthometric height (H) is the one most used in engineering surveys and has been defined in *Section 8.2.2*; in general terms, it is referred to as height above MSL.

Ellipsoidal height has been defined in *Section 8.3.2* and is rarely used in engineering surveys for practical purposes. However, satellite systems define position and height in X, Y and Z coordinates, which for use in local systems are first transformed to ϕ_G, λ_G and h using the equations of *Section 8.3.3*. The value of h is the ellipsoidal height, which, as it is not related to gravity, is of no practical use, particularly when

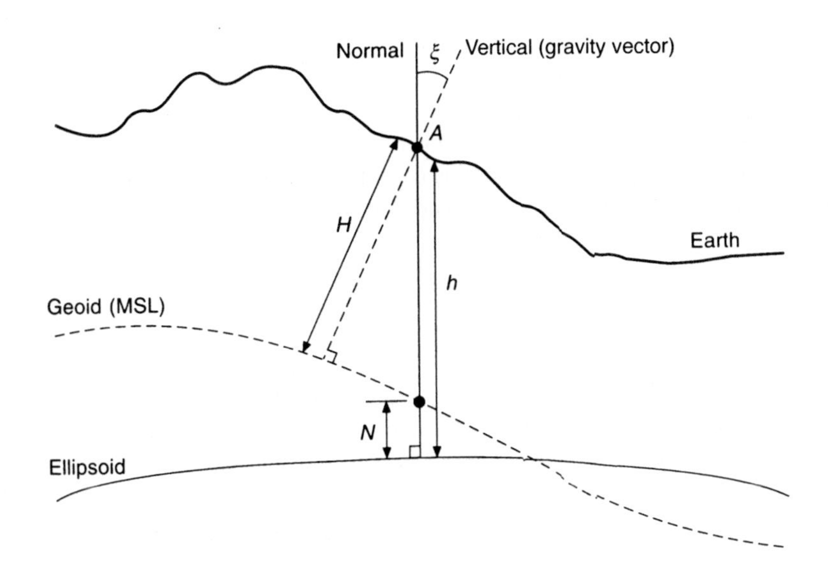

Fig. 8.10 *Ellipsoidal and orthometric heights*

dealing with the direction of water flow. It is therefore necessary to transform h to H, the relationship of which is shown in *Figure 8.10*:

$$h = N + H \cos \xi \tag{8.10}$$

However, as ξ is always less then $60''$, and therefore $0.99999995 < \cos \varepsilon, \leq 1.0$, it can be ignored:

$$\therefore h = N + H \tag{8.11}$$

with an error of less than 0.4 mm at the worst worldwide (0.006 mm in the UK).

The term N is referred to as the 'geoid–ellipsoid separation' or 'geoid height' and to transform ellipsoidal heights to orthometric, must be known to a high degree of accuracy for the particular reference system in use. In global terms N is known (relative to the WGS84 ellipsoid) to an accuracy of 2–6 m. However, for use in local engineering projects N would need to be known to an accuracy greater than h, in order to provide precise orthometric heights from satellite data. To this end, many national mapping organizations, such as the Ordnance Survey in Great Britain, have carried out extensive work to produce an accurate model of the geoid and its relationship to the local ellipsoid.

8.4 LOCAL SYSTEMS

The many systems established by various countries throughout the world for positioning on and mapping of the Earth's surface were established with an astrogeodetic origin to the datum. In the UK the origin of the datum for the Ordnance Survey is at Herstmonceux in Sussex. The origin of the datum for the European system is at Potsdam and the origin of the datum for North America is at Meades Ranch in Kansas. At each of these origins the relationship between the geoid and ellipsoid is defined in terms of the tangential planes between the respective surfaces. Such systems use reference ellipsoids which most closely fit the geoid in their area of coverage and are defined by the following eight parameters:

(1) The size and shape of the ellipsoid, defined by the semi-major axis a, and one other chosen from the semi-minor axis b or the flattening f or the eccentricity e (two parameters).

(2) The minor axis of the ellipsoid is orientated parallel to the mean spin axis of the Earth as defined by IERS (two parameters).
(3) The geodetic, latitude and longitude are defined as being the same as the astronomical, i.e. gravitational latitude and longitude and therefore with respect to the geoid, then the tangential planes must be parallel and the deviations of the vertical at the origin are zero. In this case the vertical to the geoid and the normal to the ellipsoid are coincident. If the tangential planes are further defined as being coincident then the separation between the geoid and ellipsoid surface, N, is also zero. This constitutes the definition of four parameters; two for the deviations of the vertical, East–West and North–South, one for the separation and one for the implicit definition of the zero direction of longitude.

Satellite datums or datums that are required for global use may be defined as follows.

(1) The Earth's mass centre is the origin of the coordinate system, e.g. if a Cartesian coordinate system is being used then at the Earth's mass centre, $X = Y = Z = 0$ (three parameters).
(2) The orientation of one of the coordinate axes is defined in inertial space, in practical terms that means with respect to the 'fixed' stars (two parameters).
(3) The orientation of the zero direction in the plane at right angles to the defined direction, e.g. the direction in which X is always 0 (one parameter).
(4) The shape and size of the reference ellipsoid as above (two parameters).

It follows that all properly defined geodetic systems will have their axes parallel and can be related to each other by simple translations in X, Y and Z.

The goodness of fit can be indicated by an examination of values for the deviation of the vertical (ξ) and geoid–ellipsoid separation (N), as indicated in *Figure 8.5*. Consider a meridian section through the geoid–ellipsoid (*Figure 8.11*), it is obvious that the north–south component of the deviation of the vertical

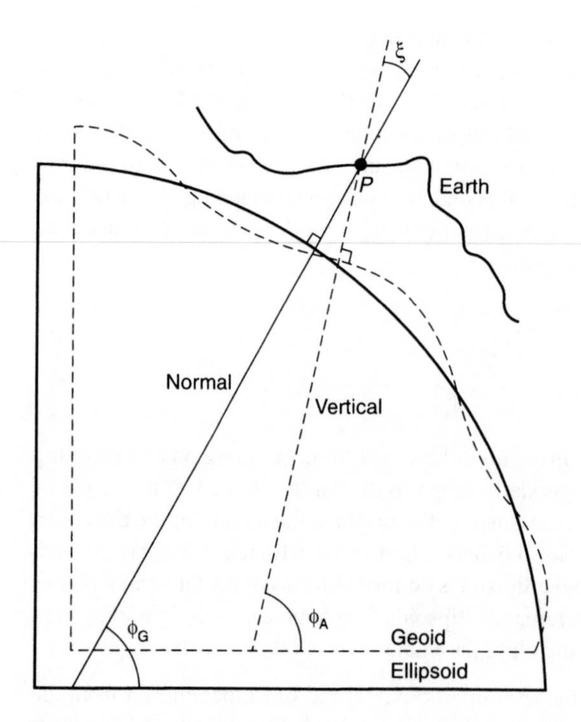

Fig. 8.11 *A meridianal section*

is a function of the ellipsoidal latitude (ϕ_G) and astronomical latitude (ϕ_A), i.e.

$$\xi = \phi_A - \phi_G \tag{8.12}$$

The deviation of the vertical in the east–west direction (prime vertical) is

$$H = (\lambda_A - \lambda_G)\cos\phi \tag{8.13}$$

where ϕ is ϕ_A or ϕ_G, the difference being negligible. It can be shown that the deviation in any azimuth α is given by

$$\psi = -(\xi\cos\alpha + \eta\sin\alpha) \tag{8.14}$$

whilst at 90° to α the deviation is

$$\zeta = (\xi\sin\alpha - \eta\cos\alpha) \tag{8.15}$$

Thus, in very general terms, the process of defining a datum may be as follows. A network of points is established throughout the country to a high degree of observational accuracy. One point in the network is defined as the origin, and its astronomical coordinates, height above the geoid and azimuth to a second point are obtained. The ellipsoidal coordinates of the origin can now be defined as

$$\phi_G = \phi_A - \xi \tag{8.16}$$

$$\lambda_G = \lambda_A - \eta\sec\phi \tag{8.17}$$

$$h = H + N \tag{8.18}$$

However, at this stage of the proceedings there are no values available for ξ, η and N, so they are assumed equal to zero and an established ellipsoid is used as a reference datum, i.e.

$$\phi_G = \phi_A$$

$$\lambda_G = \lambda_A$$

$$h = H$$

plus a and f, comprising five parameters.

As field observations are automatically referenced to gravity (geoid), then directions and angles will be measured about the vertical, with distance observed at ground level. In order to compute ellipsoidal coordinates, the directions and angles must be reduced to their equivalent about the normal on the ellipsoid and distance reduced to the ellipsoid. It follows that as ξ, η and N will be unknown at this stage, an iterative process is involved, commencing with observations reduced to the geoid.

The main corrections involved are briefly outlined as follows:

(1) Deviation of the vertical

As the horizontal axis of the theodolite is perpendicular to the vertical (geoid) and not the normal (ellipsoid), a correction similar to plate bubble error is applied to the directions, i.e.

$$-\zeta\tan\beta \tag{8.19}$$

where ζ is as in *equation (8.15)* and β is the vertical angle. It may be ignored for small values of β.

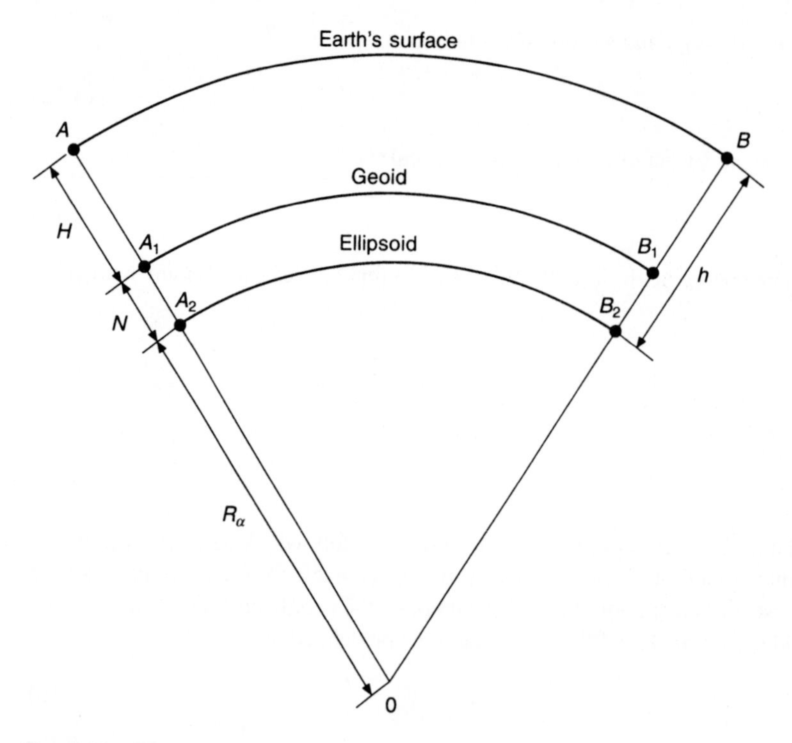

Fig. 8.12 *Distance reduction to the ellipsoid*

(2) Reduction of distance to the ellipsoid

The measured distance is reduced to its horizontal component by applying all the corrections appropriate to the method of measurement as detailed in Chapter 4.

It is then reduced to the ellipsoid by reducing to MSL (geoid), A_1B_1, and from MSL to the ellipsoid, A_2B_2, in one computation. See *Figure 8.12*:

$$A_2B_2 = L - LH/(R_\alpha + H + N) \tag{8.20}$$

where $L = AB$, the mean horizontal distance at ground level

$H =$ mean height above MSL

$N =$ height of the geoid above the ellipsoid

$R_\alpha =$ the radius of curvature of the ellipsoid in the direction α of the line

and

$$R_\alpha = \rho v/(\rho \sin^2 \alpha + v \cos^2 \alpha) \tag{8.21}$$

$$\rho = a(1 - e^2)/(1 - e^2 \sin^2 \phi)^{\frac{3}{2}} = \text{meridional radius of curvature} \tag{8.22}$$

$$v = a/(1 - e^2 \sin^2 \phi)^{\frac{1}{2}} = \text{prime vertical radius of curvature (at 90° to the meridian)} \tag{8.23}$$

As already stated, values of N may not be available and hence the geoidal distance may have to be ignored. It should be remembered that if $N = 6$ m, a scale error of 1 ppm will occur if N is ignored. In the UK, the maximum value for N is about 4.5 m, resulting in a scale error of only 0.7 ppm, and may therefore be ignored for scale purposes. Obviously it cannot be ignored in heighting.

8.5 COMPUTATION ON THE ELLIPSOID

Before proceeding with the computation of ellipsoidal coordinates, it is necessary to consider certain aspects of direction. In plane surveying, for instance, the direction of BA differs from that of AB by exactly 180°. However, as shown in *Figure 8.13*,

$$\text{Azimuth } BA = \alpha_{AB} + 180^\circ + \Delta\alpha = \alpha_{BA} \tag{8.24}$$

where $\Delta\alpha$ is the additional correction due to the convergence of the meridians AP and BP.

Using the corrected ellipsoidal azimuths and distances, the coordinates are now calculated relative to a selected point of origin.

The basic problems are known as the 'direct' and 'reverse' problems and are analogous to computing the 'polar' and 'join' in plane surveying. The simplest computing routine is one that involves 'mid-latitude' formulae.

The mid-latitude formulae are generally expressed as:

$$\Delta\phi'' = \frac{L\cos\alpha_m}{\rho_m \sin 1''}\left(1 + \frac{\Delta\lambda^2}{12} + \frac{\Delta\lambda^2 \sin^2\phi_m}{24}\right) \tag{8.25}$$

$$\Delta\lambda'' = \frac{L\sin\alpha_m \cdot \sec\phi_m}{\nu \sin 1''}\left(1 + \frac{\Delta\lambda^2 \sin^2\phi_m}{24} - \frac{\Delta\phi^2}{24}\right) \tag{8.26}$$

$$\Delta\alpha'' = \Delta\lambda'' \sin\phi_m \left(1 + \frac{\Delta\lambda^2 \sin^2\phi_m}{24} + \frac{\Delta\lambda^2 \cos^2\phi_m}{12} + \frac{\Delta\phi^2}{12}\right) \tag{8.27}$$

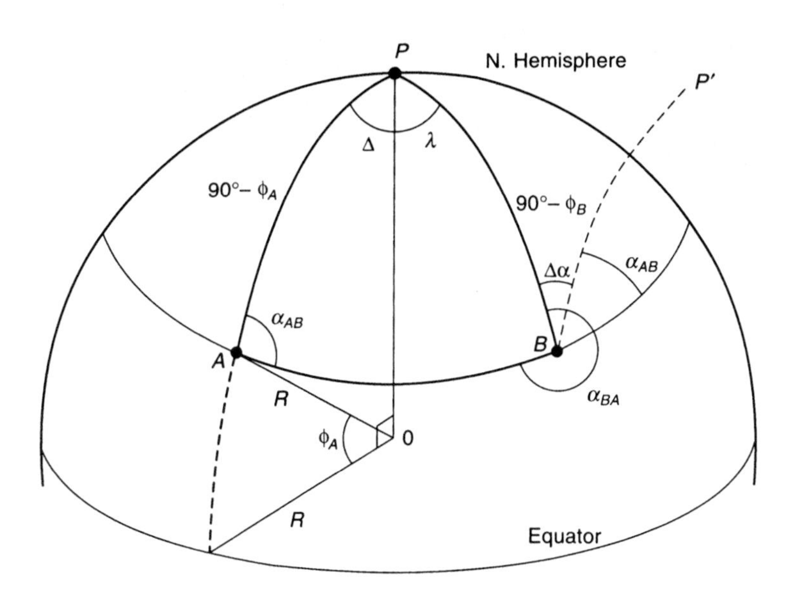

Fig. 8.13 *Geodetic azimuth*

where $\Delta\phi = \phi_A - \phi_B$

$$\Delta\lambda = \lambda_A - \lambda_B$$

$$\Delta\alpha = \alpha_{BA} - \alpha_{AB} \pm 180°$$

$$\alpha_m = \alpha_{AB} + \frac{\Delta\alpha}{2}$$

$$\phi_m = (\phi_A + \phi_B)/2$$

The above formulae are accurate to 1 ppm for lines up to 100 km in length.

(1) The direct problem

Given:

(a) Ellipsoidal coordinates of $A = \phi_A, \lambda_A$
(b) Ellipsoidal distance $AB = L_{AB}$
(c) Ellipsoidal azimuth $AB = \alpha_{AB}$
(d) Ellipsoidal parameters $= a, e^2$

determine:

(a) Ellipsoidal coordinates of $B = \phi_B, \lambda_B$
(b) Ellipsoidal azimuth $BA = \alpha_{BA}$

As the mean values of ρ, v and ϕ are required, the process must be an iterative one and will be outlined using the first term only of the mid-latitude formula:

(a) Determine ρ_A and v_A from *equations (8.22)* and *(8.23)* using ϕ_A
(b) Determine $\Delta\phi'' = L_{AB}\cos\alpha_{AB}/\rho_A \sin 1''$
(c) Determine the first value for $\phi_m = \phi_A + (\Delta\phi/2)$
(d) Determine improved values for ρ and v using ϕ_m and iterate until negligible change in $\Delta\phi$
(e) Determine $\Delta\lambda'' = L_{AB}\sin\alpha_{AB}\sec\phi_m/v_m \sin 1''$
(f) Determine $\Delta\alpha'' = \Delta\lambda''\sin\phi_m$ and so deduce

$$\alpha_m = \alpha_{AB} + (\Delta\alpha/2)$$

(g) Iterate the whole procedure until the differences between successive values of $\Delta\phi$, $\Delta\lambda$ and $\Delta\alpha$ are insignificant. Three iterations normally suffice.
(h) Using the final accepted values, we have:

$$\phi_B = \phi_A + \Delta\phi$$

$$\lambda_B = \lambda_B + \Delta\lambda$$

$$\alpha_{BA} = \alpha_{AB} + \Delta\alpha \pm 180°$$

(2) The reverse problem

Given:

(a) Ellipsoidal coordinates of $A = \phi_A, \lambda_A$
(b) Ellipsoidal coordinates of $B = \phi_B, \lambda_B$
(c) Ellipsoidal parameters $= a, e^2$

determine:

(a) Ellipsoidal azimuths $= \alpha_{AB}$ and α_{BA}
(b) Ellipsoidal distance $= L_{AB}$

This procedure does not require iteration:

(a) Determine $\phi_m = (\phi_A + \phi_B)/2$
(b) Determine $\nu_m = a/(1 - e^2 \sin^2 \phi_m)^{\frac{1}{2}}$ and $\rho_m = a(1 - e^2)/(1 - e^2 \sin^2 \phi_m)^{\frac{3}{2}}$
(c) Determine $\alpha_m = \nu_m \cdot \Delta\lambda \cos \phi_m/(\rho_m \cdot \Delta\phi)$
(d) Determine L_{AB} from $\Delta\phi'' = L_{AB} \cos \alpha_m/\rho_m \sin 1''$ which can be checked using $\Delta\lambda''$
(e) Determine $\Delta\alpha'' = \Delta\lambda'' - \sin \phi_m$, then $\alpha_{AB} = \alpha_m - (\Delta\alpha/2)$ and $\alpha_{BA} = \alpha_{AB} + \Delta\alpha \pm 180°$

Whilst the mid-latitude formula serves to illustrate the procedures involved, computers now permit the use of the more accurate equations.

On completion of all the computation throughout the network, values of ξ, η and N can be obtained at selected stations. The best-fitting ellipsoid is the one for which values of ΣN^2 or $\Sigma(\xi^2 + \eta^2)$ are a minimum. If the fit is not satisfactory, then the values of ξ, η and N as chosen at the origin could be altered or a different ellipsoid selected. Although it would be no problem to change the ellipsoid due to the present use of computers, in the past it was usual to change the values of ξ, η and N to improve the fit.

Although the above is a brief description of the classical geodetic approach, the majority of ellipsoids in use throughout the world were chosen on the basis of their availability at the time. For instance, the Airy ellipsoid adopted by Great Britain was chosen in honour of Professor Airy who was Astronomer Royal at the time and had just announced the parameters of his ellipsoid. In fact, recent tests have shown that the fit is quite good, with maximum values of N equal to 4.5 m and maximum values for deviation of the vertical equal to $10''$.

8.6 DATUM TRANSFORMATIONS

Coordinate transformations are quite common in surveying. They range from simple translations between coordinates and setting-out grids on a construction site, to transformation between global systems.

Whilst the mathematical procedures are well defined for many types of transformation, problems can arise due to varying scale throughout the network used to establish position. Thus in a local system, there may be a variety of parameters, established empirically, to be used in different areas of the system.

8.6.1 Basic concept

From *Figure 8.14* it can be seen that the basic parameters in a conventional transformation between similar *XYZ* systems would be:

(1) Translation of the origin 0, which would involve shifts in X, Y and Z, i.e. ΔX, ΔY, ΔZ.
(2) Rotation about the three axes, θx, θy and θz, in order to render the axes of the systems involved parallel. θx and θy would change the polar axes, and θz the zero meridian.
(3) One scale parameter $(1 + S)$ would account for the difference of scale between different coordinate systems.

In addition to the above, the size (a) of the ellipsoid and its shape (f) may also need to be included. However, not all the parameters are generally used in practice. The most common transformation is the translation in X, Y and Z only (three parameters). Also common is the four-parameter

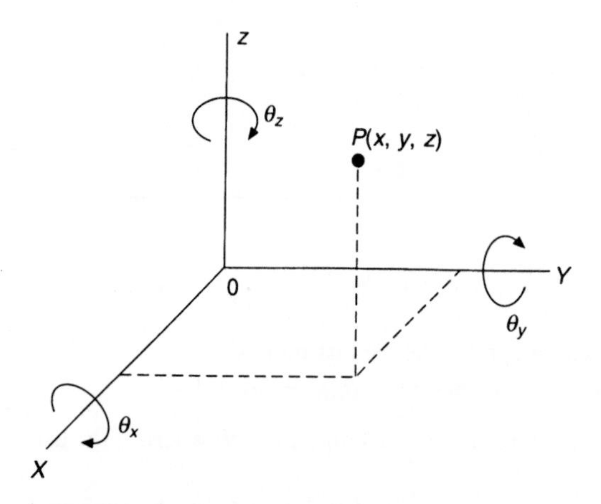

Fig. 8.14 *Transformation parameters*

(ΔX, ΔY, ΔZ + scale) and the five-parameter (ΔX, ΔY, ΔZ + scale + θz). A full transformation would entail seven parameters.

A simple illustration of the process can be made by considering the transformation of the coordinates of $P(X', Y', Z')$ to (X, Y, Z) due to rotation θx about axis OX (*Figure 8.15*):

$$X = X'$$

$$Y = Or - qr = Y' \cos\theta - Z' \sin\theta \tag{8.28}$$

$$Z = mr + Pn = Y' \sin\theta + Z' \cos\theta \tag{8.29}$$

In matrix form:

$$\begin{bmatrix} X \\ Y \\ Z \end{bmatrix} = \begin{bmatrix} 1 & 0 & 0 \\ 0 & \cos\theta & -\sin\theta \\ 0 & \sin\theta & \cos\theta \end{bmatrix} \begin{bmatrix} X' \\ Y' \\ Z' \end{bmatrix} \tag{8.30}$$

$$x = R_\theta \cdot x'$$

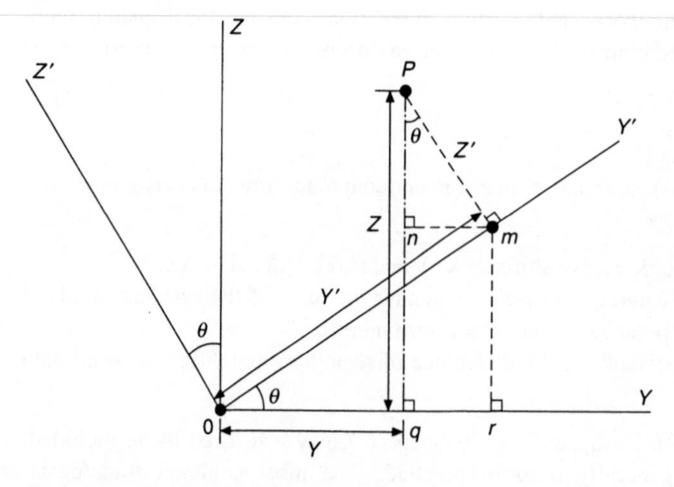

Fig. 8.15 *Coordinate transformation*

where R_θ = rotational matrix for angle θ
x' = the vector of original coordinates

Similar rotation matrices can be produced for rotations about axes $OY(\alpha)$ and $OZ(\beta)$, giving

$$x = R_\theta R_\alpha R_\beta x' \tag{8.31}$$

If a scale change and translation of the origin by ΔX, ΔY, ΔZ is made, the coordinates of P would be

$$\begin{bmatrix} X \\ Y \\ Z \end{bmatrix} = \begin{bmatrix} \Delta X \\ \Delta Y \\ \Delta Z \end{bmatrix} + (1+S)\begin{bmatrix} a_{11} & a_{12} & a_{13} \\ a_{21} & a_{22} & a_{23} \\ a_{31} & a_{32} & a_{33} \end{bmatrix}\begin{bmatrix} X' \\ Y' \\ Z' \end{bmatrix} \tag{8.32}$$

The a coefficients of the rotation matrix would involve the sines and cosines of the angles of rotation, obtained from the matrix multiplication of R_θ, R_α and R_β.

For the small angles of rotation the sines of the angles may be taken as their radian measure ($\sin\theta = \theta$) and the cosines are unity, with sufficient accuracy. In which case the equation simplifies to

$$\begin{bmatrix} X \\ Y \\ Z \end{bmatrix} = \begin{bmatrix} \Delta X \\ \Delta Y \\ \Delta Z \end{bmatrix} + (1+S)\begin{bmatrix} 1 & \beta & -\alpha \\ -\beta & 1 & \theta \\ \alpha & -\theta & 1 \end{bmatrix}\begin{bmatrix} X' \\ Y' \\ Z' \end{bmatrix} \tag{8.33}$$

Equation (8.33) is referred to in surveying as the Helmert transformation and describes the full transformation between the two geodetic datums.

Whilst the X, Y, Z coordinates of three points would be sufficient to determine the seven parameters, in practice as many points as possible are used in a least squares solution. Ellipsoidal coordinates (ϕ, λ, h) would need to be transformed to X, Y and Z for use in the transformations.

As a translation of the origin of the reference system is the most common, a Molodenskii transform permits the transformation of ellipsoidal coordinates from one system to another in a single operation, i.e.

$$\phi = \phi' + \Delta\phi''$$

$$\lambda = \lambda' + \Delta\lambda''$$

$$h = h' + \Delta h$$

where $\Delta\phi'' = (-\Delta X \sin\phi' \cos\lambda' - \Delta Y \sin\phi' \sin\lambda' + \Delta Z \cos\phi'$

$$+ (a'\Delta f + f'\Delta a)\sin 2\phi')/(\rho \sin 1'') \tag{8.34}$$

$$\Delta\lambda'' = (-\Delta X \sin\lambda' + \Delta Y \cos\lambda')/(\nu \sin 1'') \tag{8.35}$$

$$\Delta h = (\Delta X \cos\phi' \cos\lambda' + \Delta Y \cos\phi' \times \sin\lambda' + \Delta Z \sin\phi'$$

$$+ (a'\Delta f + f'\Delta a)\sin^2\phi' - \Delta a) \tag{8.36}$$

In the above formulae:

ϕ', λ', h' = ellipsoidal coordinates in the first system
ϕ, λ, h = ellipsoidal coordinates in the required system
a', f' = ellipsoidal parameters of the first system
$\Delta a, \Delta f$ = difference between the parameters in each system
$\Delta X, \Delta Y, \Delta Z$ = origin translation values
ν = radius of curvature in the prime vertical (*equation (8.23)*)
ρ = radius of curvature in the meridian (*equation (8.22)*)

It must be emphasized once again that whilst the mathematics of transformation are rigorously defined, the practical problems of varying scale, etc., must always be considered.

8.7 ORTHOMORPHIC PROJECTION

The ellipsoidal surface, representing a portion of the Earth's surface, may be represented on a plane using a specific form of projection, i.e.

$$E = f_E(\phi, \lambda) \tag{8.37}$$

$$N = f_N(\phi, \lambda) \tag{8.38}$$

where E and N on the plane of the projection represent ϕ, λ on the reference ellipsoid.

Representation of a curved surface on a plane must result in some form of distortion, and therefore the properties required of the projection must be carefully considered. In surveying, the properties desired are usually:

(1) A line on the projection must contain the same intermediate points as that on the ellipsoid.
(2) The angle between any two lines on the ellipsoid should have a corresponding angle on the projection and scale at a point is the same in all directions. This property is termed orthomorphism and results in small areas retaining their shape.

Using the appropriate projection mathematics the shortest line distance on a curved surface between two points, the geodesic, *AB* in *Figure 8.16* is projected to the curved dotted line *ab*; point *C* on the geodesic will appear at *c* on the projection. The meridian *AP* is represented by the dotted line 'geodetic north', and then:

(1) The angle γ between grid and geodetic north is called the 'grid convergence' resulting from the convergence of meridians.

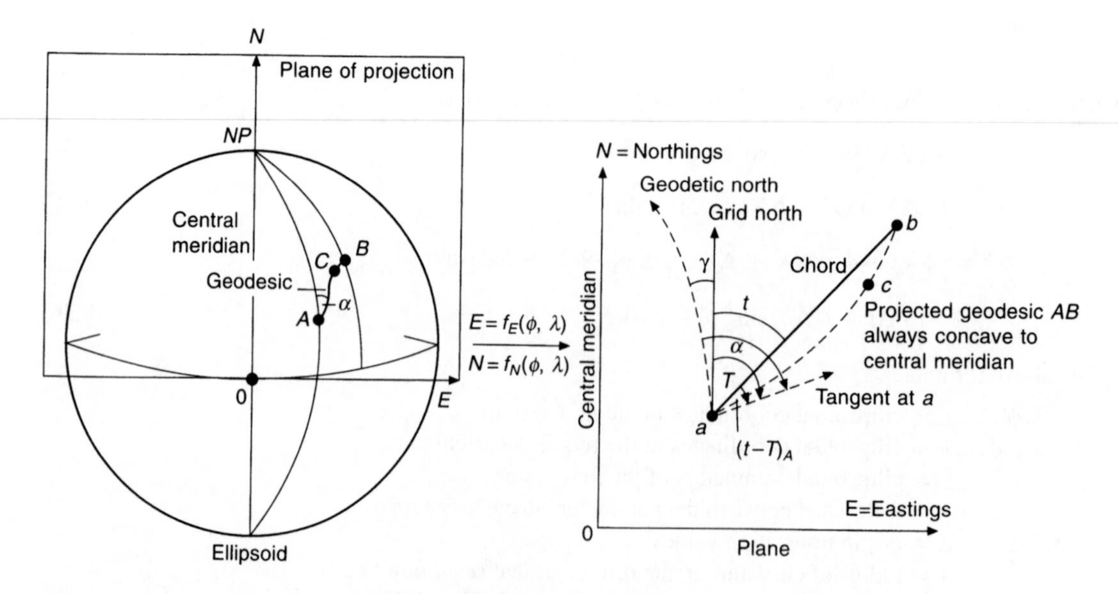

Fig. 8.16 *The geodesic represented on the projection*

(2) The angle α is the azimuth of AB measured clockwise from north.
(3) The angle t is the grid bearing of the chord ab.
(4) The angle T is the angle between grid north and the projected geodesic. From (3) and (4) we have the $(t - T)$ correction.
(5) The line scale factor (F) is the ratio between the length (S) of the geodesic AB as calculated from ellipsoidal coordinates and its grid distance (G) calculated from the plane rectangular coordinates, i.e.

$$F = G/S \tag{8.39}$$

Similarly the point scale factor can be obtained from the ratio between a small element of the geodesic and a corresponding element of the grid distance.
(6) It should be noted that the projected geodesic is always *concave to the central meridian*.

It can be seen from the above that:

(1) The geodetic azimuth can be transformed to grid bearing by the application of 'grid convergence' and the '$t - T$' correction.
(2) The ellipsoidal distance can be transformed to grid distance by multiplying it by the scale factor.

The plane coordinates may now be computed, using this adjusted data, by the application of plane trigonometry. Thus apart from the cartographic aspects of producing a map or plan, the engineering surveyor now has an extremely simple mathematical process for transforming field data to grid data and *vice versa*.

The orthomorphic projection that is now used virtually throughout the world is the transverse Mercator projection, which is ideal for countries having their greatest extent in a north–south direction. This can be envisaged as a cylinder surrounding the ellipsoid (*Figure 8.17*) onto which the ellipsoid positions are projected. The cylinder is in contact with the ellipsoid along a meridian of longitude and the lines of latitude and longitude are mathematically projected onto the cylinder. Orthomorphism is achieved by stretching the scale along the meridians to keep pace with the increasing scale along the parallels. By opening up the cylinder and spreading it out flat, the lines of latitude and longitude form a graticule of complex curves intersecting at right angles, the central meridian being straight.

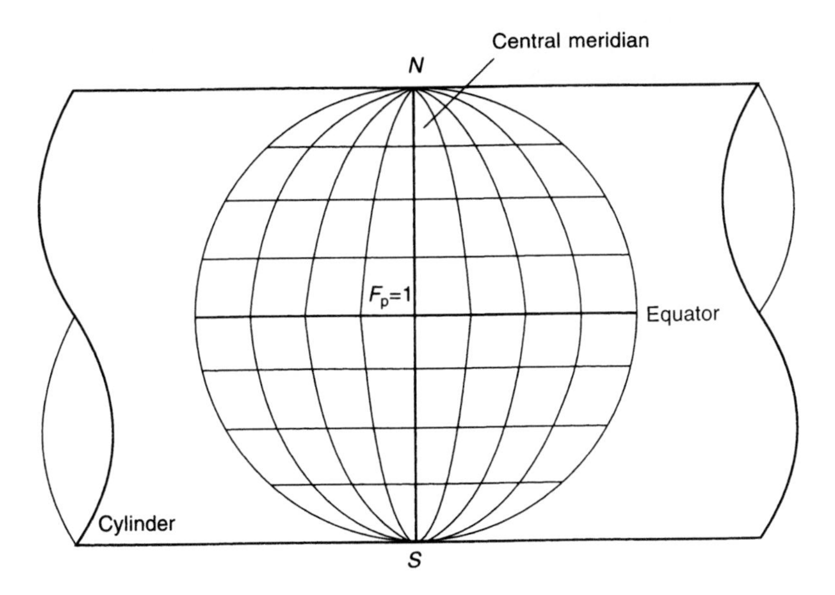

Fig. 8.17 *Cylindrical projection*

It is obvious from *Figure 8.17* that the ratio of distance on the ellipsoid to that on the projection would only be correct along the central meridian, where the cylinder and ellipsoid are in contact, and thus the scale factor would be unity ($F = 1$).

If geographical coordinates of the point are known and projection coordinates are required then:

(1)
$$E = F_0\left[\nu \cdot \Delta\lambda \cos\phi + \nu\frac{\Delta\lambda^3}{6}\cos^3\phi(\psi - t^2) + \nu\frac{\Delta\lambda^5}{120}\cos^5\phi(4\psi^3(1-6t^2) + \psi^2(1+8t^2) - \psi(2t^2) + t^4) + \nu \cdot \frac{\Delta\lambda^7}{5040}\cos^7\phi(61 - 479t^2 + 179t^4 - t^6)\right] \quad (8.40)$$

(2)
$$N = F_0\left[M + \nu\sin\phi\frac{\Delta\lambda^2}{2}\cos\phi + \nu\sin\phi\frac{\Delta\lambda^4}{24}\cos^3\phi(4\psi^2 + \psi - t^2) + \nu\sin\phi\frac{\Delta\lambda^6}{720}\cos^5\phi(8\psi^4(11-24t^2) - 28\psi^3(1-6t^2) + \psi^2(1-32t^2) - \psi(2t^2) + t^4) + \nu\sin\phi\frac{\Delta\lambda^8}{40\,320}\cos^7\phi(1385 - 3111t^2 + 543t^4 - t^6)\right] \quad (8.41)$$

where:

F_0 = scale factor on the central meridian
$\Delta\lambda$ = difference in longitude between the point and the central meridian
$t = \tan\phi$
$\psi = \nu/\rho$
M = meridian distance from the latitude of the origin, and is obtained from:

$$M = a(A_0\phi - A_2\sin 2\phi + A_4\sin 4\phi - A_6\sin 6\phi) \quad (8.42)$$

and $A_0 = 1 - e^2/4 - 3e^4/64 - 5e^6/256$

$A_2 = (3/8)(e^2 + e^4/4 + 15e^6/128)$

$A_4 = (15/256)(e^4 + 3e^6/4)$

$A_6 = 35e^6/3072$

In *equations (8.47)* and *(8.48)* E is the difference in easting from the central meridian and N is the distance north from the projection of the zero northing that passes through the false origin on the central meridian.

The reverse case is where the projection coordinates (E, N) are known and the geographical coordinates (ϕ, $\Delta\lambda$) are required. Find by iteration, ϕ', the latitude for which:

$$N = am_0(A_0\phi' - A_2\sin 2\phi' + A_4\sin 4\phi' - A_6\sin 6\phi')$$

and use it to find ν' and ρ' and hence ϕ and $\Delta\lambda$ from:

$$\phi = \phi' - \frac{t'E^2}{2m_0^2\nu'\rho'} + \frac{t'E^4}{24m_0^4\nu'^3\rho'}\left\{-4\psi'^2 + 9\psi'(1 - t'^2) + 12t'^2\right\}$$

$$- \frac{t'E^6}{720\, m_0^6\nu'^5\rho'}\left\{8\psi'^4(11 - 24t'^2) - 12\psi'^3(21 - 71t'^2)\right.$$

$$\left. +15\psi'^2(15 - 98t'^2 + 15t'^4) + 180\psi'(5t'^2 - 3t'^4) + 360t'^4\right\}$$

$$- \frac{t'E^8}{40\,320\, m_0^8\nu'^7\rho'}(1385 + 3633t'^2 + 4095t'^4 + 1575t'^6) \tag{8.43}$$

$$\Delta\lambda = \frac{E}{m_0\nu'c'} - \frac{E^3}{6\, m_0^3\nu'^3c'}(\psi' + 2t'^2)$$

$$+ \frac{E^5}{120\, m_0^5\nu'^5c'}\left\{-4\psi'^3(1 - 6t'^2) + \psi'^2(9 - 68t'^2) + 72\psi't'^2 + 24t'^4\right\}$$

$$+ \frac{E^7}{5040\, m_0^7\nu'^7c'}(61 + 662t'^2 + 1320t'^4 + 720t'^6) \tag{8.44}$$

where:

$t' = \tan\phi'$

$c' = \cos\phi'$

$\psi' = \nu'/\rho'$

(3) Grid convergence $= \gamma = -\Delta\lambda\sin\phi - \frac{\Delta\lambda^3}{3}\sin\phi\cos^2\phi(2\psi^2 - \psi)$ (8.45)

which is sufficiently accurate for most applications in engineering surveying.

(4) The point scale factor can be computed from:

$$F = F_0\left[1 + (E^2/2R_m^2) + (E^4/24R_m^4)\right] \tag{8.46a}$$

The scale factor for the line AB can be computed from:

$$F = F_0\left[1 + (E_A^2 + E_AE_B + E_B)/6R_m^2\right] \tag{8.46b}$$

where

$R_m = \rho\nu$

E = distance of the point from the central meridian in terms of the difference in Eastings.

In the majority of cases in practice, it is sufficient to take the distance to the easting of the mid-point of the line (E_m), and then:

$$F = F_0(1 + E_m^2/2R^2) \tag{8.47}$$

and $R^2 = \rho\nu$.

(5) The 'arc-to-chord' correction or the $(t - T)$ correction, as it is more commonly called, from A to B, in seconds of arc, is

$$(t - T)''_A = -(N_B - N_A)(2E_A - E_B)/6R^2 \sin 1'' \tag{8.48}$$

with sufficient accuracy for most purposes.

8.8 THE UNIVERSAL TRANSVERSE MERCATOR PROJECTION

The Universal Transverse Mercator Projection (UTM) is a worldwide system of transverse Mercator projections. It comprises 60 zones, each 6° wide in longitude, with central meridians at 3°, 9°, etc. The zones are numbered from 1 to 60, starting with 180° to 174° W as zone 1 and proceeding eastwards to zone 60. Therefore the central meridian (CM) of zone n is given by CM $= 6n° - 183°$. In latitude, the UTM system extends from 84° N to 80° S, with the polar caps covered by a polar stereographic projection.

The scale factor at each central meridian is 0.9996 to counteract the enlargement ratio at the edges of the strips. The false origin of northings is zero at the equator for the northern hemisphere and 10^6 m at the equator for the southern hemisphere. The false origin for eastings is 5×10^5 m west of the zone central meridian.

8.9 ORDNANCE SURVEY NATIONAL GRID

The Ordnance Survey (OS) is the national mapping agency for Great Britain; its maps are based on a transverse Mercator projection of Airy's ellipsoid called the OSGB (36) datum. The current realization of OSGB (36) is the OS's Terrestrial Network 2002 (OSTN02) datum which is a rubber sheet fit of European Terrestrial Reference System 1989 (ETRS89) coordinates, as derived from GPS to the original OSGB (36). For most practical purposes there should be no significant difference between OSGB (36) and OSTN02.

The central meridian selected is 2° W, with the point of origin, called the false origin, at 49° N on this meridian. The scale factor varies as the square of the distance from the central meridian, and therefore in order to reduce scale error at the extreme east and west edges of the country the scale factor on the central meridian was reduced to 0.999 601 27. One can think of this as reducing the radius of the enclosing cylinder as shown in *Figure 8.18*.

The projection cylinder cuts the ellipsoid at two quasi-sub-parallels, approximately 180 km each side of the central meridian, where the scale factor will be unity. Inside these two parallels the scale is too small by less than 0.04%, and outside of them too large by up to 0.05% on the west coast of mainland Scotland.

The central meridian (2° W) which constitutes the N-axis (Y-axis) was assigned a large easting value of E 400 000 m. The E-axis (X-axis) was assigned a value of N −100 000 m at the 49° N parallel of latitude on the CM. Thus a rectangular grid is superimposed on the developed cylinder and is called the OS National Grid (NG) (*Figure 8.19*). The assigned values result in a 'false origin' and positive values only throughout, what is now, a plane rectangular coordinate system. Such a grid thereby establishes the direction of grid north, which differs from geodetic north by γ, a variable amount called the grid convergence. On the central meridian grid north and geodetic north are the same direction.

8.9.1 Scale factors

The concept of scale factors has been fully dealt with and it only remains to deal with their application. It should be clearly understood that scale factors transform distance on the ellipsoid to distance on the

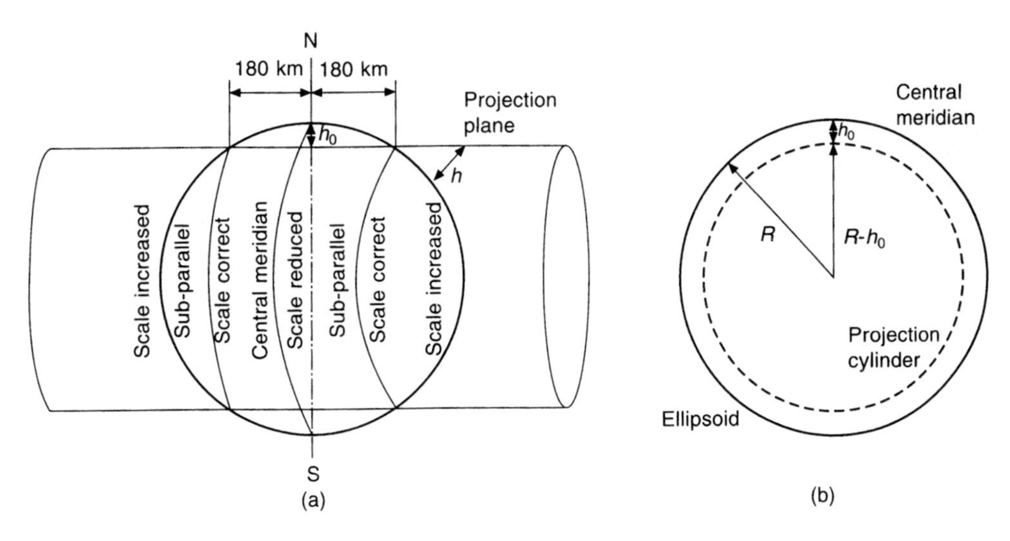

Fig. 8.18 *Scale on the projection*

plane of projection. From *Figure 8.20*, it can be seen that a horizontal distance at ground level AB must first be reduced to its equivalent at MSL (geoid) A_1B_1, using the altitude correction, thence to the ellipsoid $A_1'B_1'$ using the geoid–ellipsoid value (N) and then multiplied by the scale factor to produce the projection distance A_2B_2.

Whilst this is theoretically the correct approach, lack of knowledge of N may result in this step being ignored. In Great Britain, the maximum value is 4.5 m, resulting in a scale error of only 0.7 ppm if ignored. Thus the practical approach is to reduce to MSL and then to the projection plane, i.e. from D to S to G, as in *Figure 8.21*.

The basic equation for scale factor is given in *equation (8.46)*, where the size of the ellipsoid and the value of the scale factor on the central meridian (F_0) are considered. Specific to the OSGB (36) system, the following formula may be developed, which is sufficiently accurate for most purposes.

Scale difference (SD) is the difference between the scale factor at any point (F) and that at the central meridian (F_0) and varies as the square of the distance from the central meridian, i.e.

$$\mathrm{SD} = K(\Delta E)^2$$

where ΔE is the difference in easting between the central meridian and the point in question:

$$F = F_0 + \mathrm{SD} = 0.999\,601\,27 + K(\Delta E)^2$$

Consider a point 180 km east or west of the central meridian where $F = 1$:

$$1 = 0.999\,601\,27 + K(180 \times 10^3)^2$$

$$K = 1.228 \times 10^{-14}$$

and $$F = F_0 + (1.228 \times 10^{-14} \times \Delta E^2) \tag{8.49}$$

where $F_0 = 0.999\,601\,27$

$$\Delta E = E - 400\,000$$

Thus the value of F for a point whose NG coordinates are E 638824, N 309912 is:

$$F = 0.999\,601\,27 + [1.228 \times 10^{-14} \times (638\,824 - 400\,000)^2] = 1.0003016$$

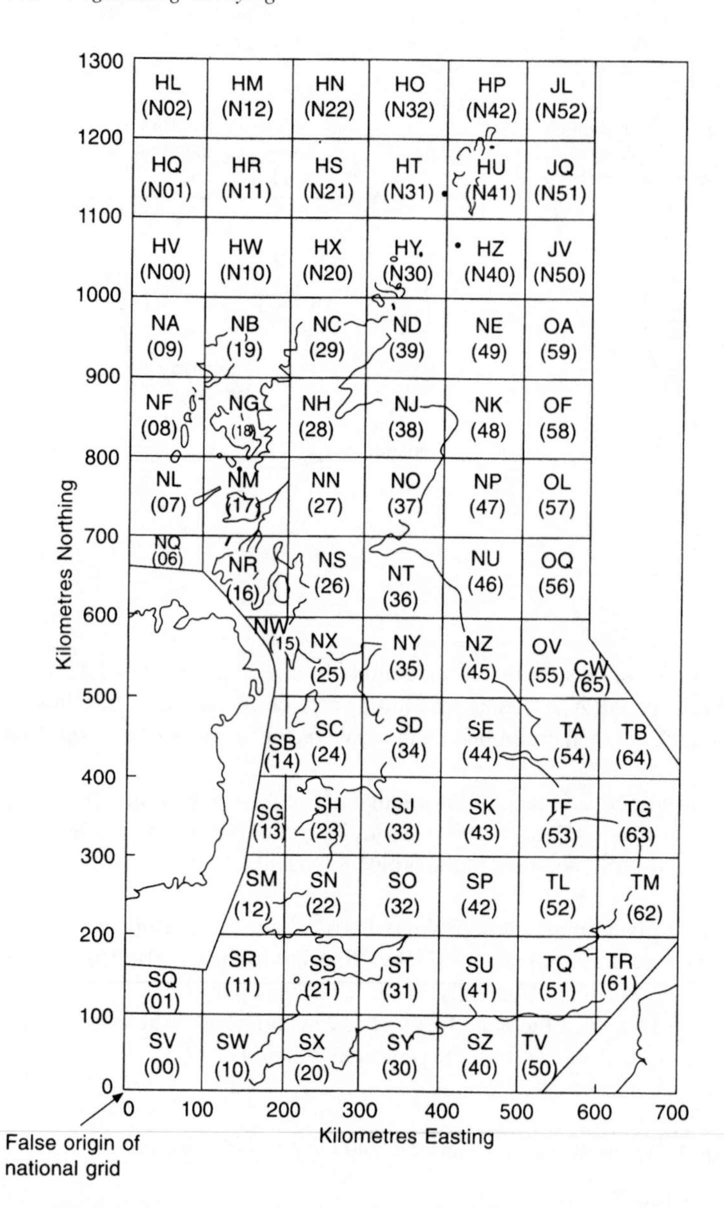

Fig. 8.19 *National reference system of Great Britain showing 100-km squares. The figures used to designate them in the former system, and the letters which have replaced the figures – courtesy Ordnance Survey, Crown Copyright Reserved*

As already intimated in *equation (8.46)*, the treatment for highly accurate work is to compute F for each end of the line and in the middle, and then obtain the mean value from Simpson's rule. However, for most practical purposes on short lines, it is sufficient to compute F at the mid-point of a line. In OSGB (36) the scale factor varies, at the most, by only 6 ppm per km, and hence a single value for F at the centre of a small site can be regarded as constant throughout the area. On long motorway or route projects, however, one would need to use different scale factors for different sections.

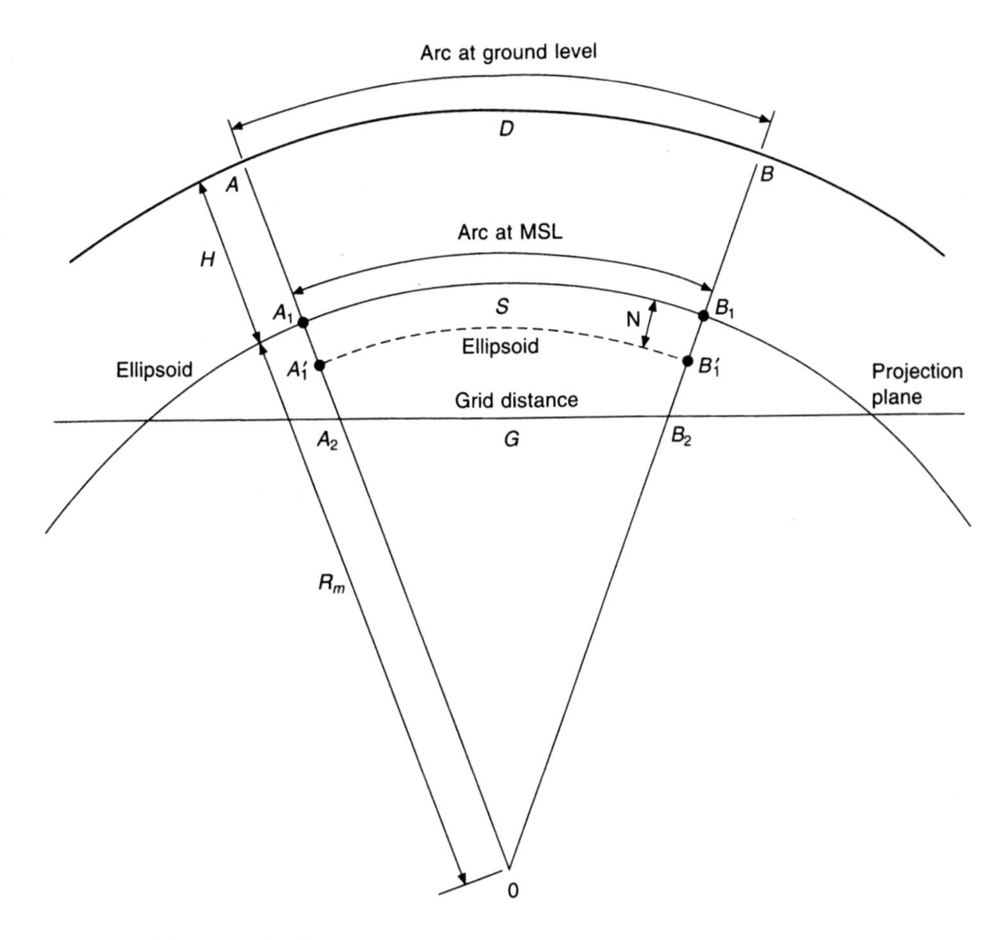

Fig. 8.20 *Distance reduction*

The following examples will serve to illustrate the classical application of scale factors.

Worked examples

Example 8.1 Grid to ground distance Any distance calculated from NG coordinates will be a grid distance. If this distance is to be set out on the ground it must:

(1) Be *divided* by the LSF to give the ellipsoidal distance at MSL, i.e. $S = G/F$.
(2) Have the altitude correction applied in reverse to give the horizontal ground distance.

Consider two points, A and B, whose coordinates are:

A:	E 638 824.076	N 307 911.843
B:	E 644 601.011	N 313 000.421
	$\therefore \Delta E = 5776.935$	$\therefore \Delta N = 5088.578$

$$\text{Grid distance} = (\Delta E^2 + \Delta N^2)^{\frac{1}{2}} = 7698.481 \text{ m} = G$$

$$\text{Mid-easting of } AB = E\ 641\ 712 \text{ m}$$

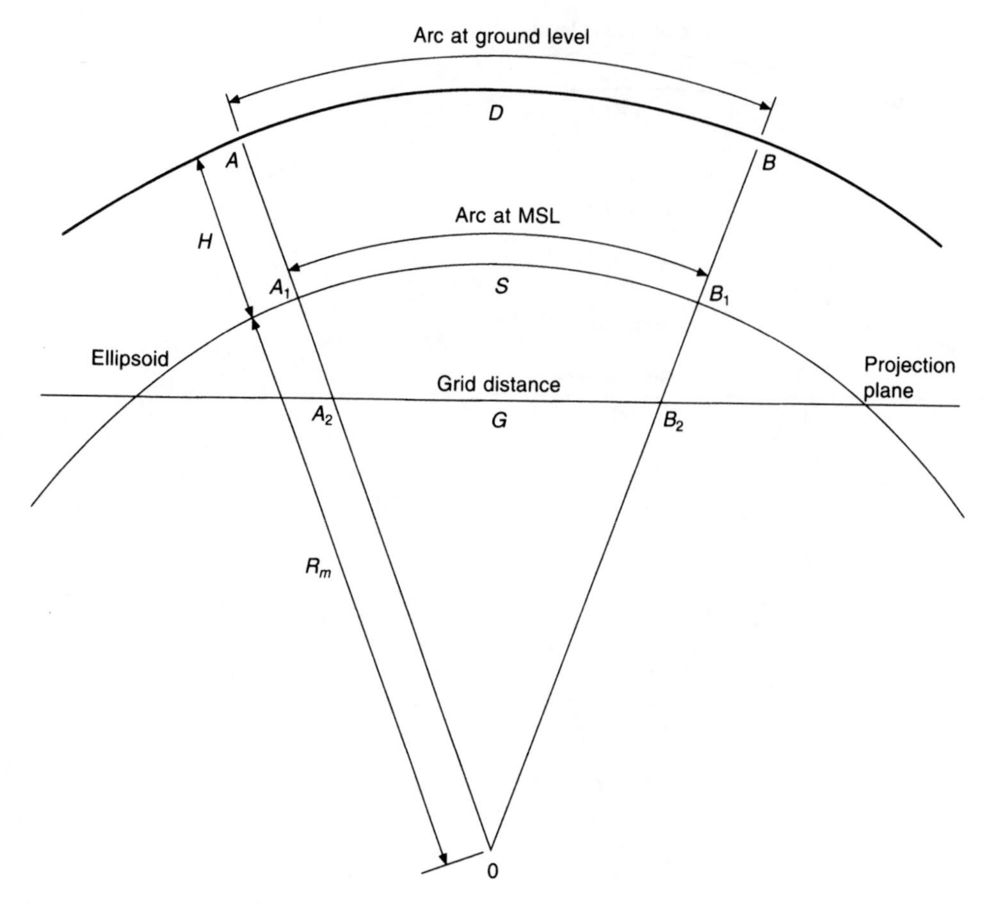

Fig. 8.21 *Distance reduction ignoring separation*

$\therefore F = 1.000\,3188$ (from *equation (8.49)*)

$\therefore$ Ellipsoidal distance at MSL $= S = G/F = 7696.027$ m

Now assuming AB at a mean height (H) of 250 m above MSL, the altitude correction C_m is

$$C_m = \frac{SH}{R} = \frac{7696 \times 250}{6\,384\,100} = +0.301 \text{ m}$$

$\therefore$ Horizontal distance at ground level $= 7696.328$ m

This situation could arise where the survey and design coordinates of a project are in OSGB (36) / OSTN02. Distances calculated from the grid coordinates would need to be transformed to their equivalent on the ground for setting-out purposes.

Example 8.2 Ground to grid distance When connecting surveys to the national grid, horizontal distances measured on the ground must be:

(a) Reduced to their equivalent on the ellipsoid.
(b) *Multiplied* by the LSF to produce the equivalent grid distance, i.e. $G = S \times F$.

Consider now the previous problem worked in reverse:

$$\begin{aligned}
\text{Horizontal ground distance} &= 7696.328\ \text{m}\\
\text{Altitude correction } C_m &= \ \ -0.301\ \text{m}\\
\therefore\ \text{Ellipsoidal distance } S \text{ at MSL} &= 7696.027\ \text{m}\\
F &= 1.0003188\\
\therefore\ \text{Grid distance } G = S \times F &= 7698.481\ \text{m}
\end{aligned}$$

This situation could arise in the case of a link traverse computed in OSTN02. The length of each leg of the traverse would need to be reduced from its horizontal distance at ground level to its equivalent distance on the NG.

There is no application of grid convergence as the traverse commences from a grid bearing and connects into another grid bearing. The application of the $(t - T)$ correction to the angles would generally be negligible, being a maximum of 7″ for a 10-km line and much less than the observational errors of the traverse angles. It would only be necessary to consider both the above effects if the angular error was being controlled by taking gyro-theodolite observations on intermediate lines in the traverse.

The two applications of first reducing to MSL and then to the plane of the projection (NG), illustrated in the examples, can be combined to give:

$$F_a = F(1 - H/R) \tag{8.50}$$

where H is the ground height relative to MSL and is positive when above and negative when below MSL.

Then from *Example 8.1*:

$$F_a = 1.0003188(1 - 250/6\,384\,100) = 1.0002797$$

F_a is then the scale factor adjusted for altitude and can be used directly to transform from ground to grid and *vice versa*.

From *Example 8.2*:

$$7696.328 \times 1.0002797 = 7698.481\ \text{m}$$

8.9.2 Grid convergence

All grid north lines on the NG are parallel to the central meridian (E 400 000 m), whilst the meridians converge to the pole. The difference between these directions is termed the grid convergence γ.

An approximate formula may be derived from the first term of *equation (8.45)*.

$$\gamma = \Delta\lambda \sin\phi$$

but $\Delta\lambda = \Delta E / R\cos\phi_m$

$$\gamma'' = \frac{\Delta E \tan\phi_m \times 206\,265}{R} \tag{8.51}$$

where ΔE = distance from the central meridian

R = mean radius of Airy's ellipsoid $= (\rho\nu)^{\frac{1}{2}}$

ϕ_m = mean latitude of the line

The approximate method of computing γ is acceptable only for lines close to the central meridian, where values correct to a few seconds may be obtained. As the distance from the central meridian increases, so too does the error in the approximate formula and the full *equation (8.45)* is required.

If the NG coordinates of a point are E 626 238 and N 302 646 and the latitude, calculated or scaled from an OS map, is approximately N 52° 34′, then taking $R = 6\,380\,847$ m gives

$$\gamma'' = \frac{226\,238 \tan 52^\circ 34'}{6\,380\,847} \times 206\,265 = 9554'' = 2^\circ 39' 14''$$

8.9.3 (t – T) correction

As already shown, the $(t - T)$ correction occurs because a geodesic on the ellipsoid is a curved line when drawn on the projection and differs in direction, at a point, from the chord. *Figure 8.22* illustrates the angle θ as 'observed' and the angle β as computed from the grid coordinates; then:

$$\beta = \theta - (t - T)_{BA} - (t - T)_{BC}$$

An approximate formula for $(t - T)$ specific to OSGB(36)/OSTN02 is as follows:

$$(t - T)''_A = (2\Delta E_A + \Delta E_B)(N_A - N_B)K \tag{8.52}$$

where ΔE = NG easting −400 000, expressed in km
N = NG northing expressed in km
A = station at which the correction is required
B = station observed to
$K = 845 \times 10^{-6}$

The maximum value for a 10-km line would be about 7″.

The signs of the corrections for $(t - T)$ and grid convergence are best obtained from a diagram similar to that of *Figure 8.23*, where for line AB:

ϕ = grid bearing AB
θ = azimuth AB
then $\theta = \phi - \gamma - (t - T)_A$, or
$\phi = \theta + \gamma + (t - T)_A$

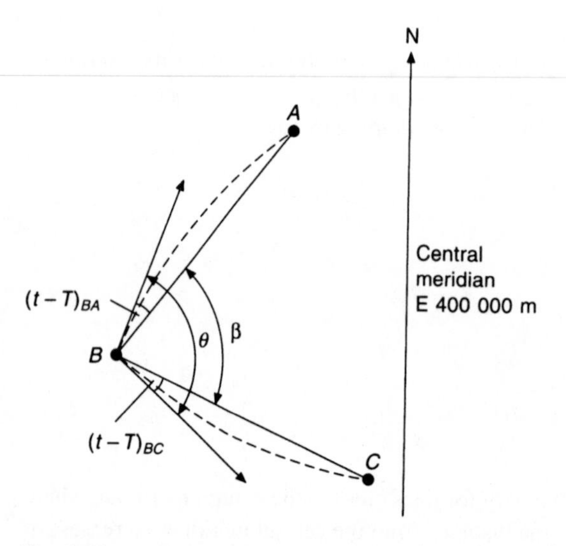

Fig. 8.22 t – T correction

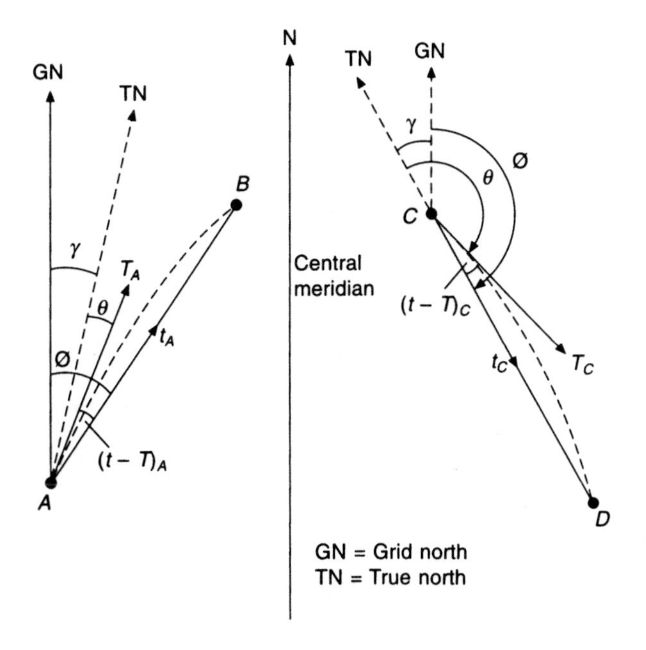

Fig. 8.23 *Sign of $t - T$ correction*

For line *CD*:

$$\theta = \phi + \gamma - (t - T)_C, \text{ or}$$

$$\phi = \theta - \gamma + (t - T)_C$$

8.10 PRACTICAL APPLICATIONS

All surveys connected to the NG should have their measured distances reduced to the horizontal, and then to MSL; and should then be multiplied by the local scale factor to reduce them to grid distance.

Consider *Figure 8.24* in which stations *A*, *B* and *C* are connected into the NG via a link traverse from OS NG stations *W*, *X* and *Y*, *Z*:

(1) The measured distance D_1 to D_4 would be treated as above.
(2) The observed angles should in theory be corrected by appropriate $(t - T)$ corrections as in *Figure 8.22*. These would generally be negligible but could be quickly checked using

$$(t - T)'' = (\Delta N_{AB} \times E/2R^2)206\,265 \tag{8.53}$$

where E = easting of the mid-point of the line
R = an approximate value for the radius of the ellipsoid for the area

(3) There is no correction for grid convergence as the survey has commenced from a grid bearing and has connected into another.
(4) Grid convergence and $(t - T)$ would need to be applied to the bearing of, say, line *BC* if its bearing had been found using a gyro-theodolite and was therefore relative to true north (TN) (see *Figure 8.23*). This procedure is sometimes adopted on long traverses to control the propagation of angular error.

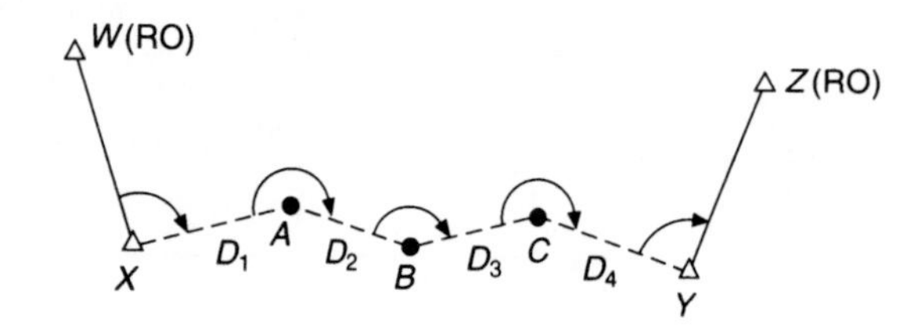

Fig. 8.24 *Link traverse*

When the control survey and design coordinates are on the NG, the setting out by bearing and distance will require the grid distance, as computed from the design coordinates, to be corrected to its equivalent distance on the ground. Thus grid distance must be changed to its MSL value and then divided by the local scale factor to give horizontal ground distance.

The setting-out angle, as computed from the design (grid) coordinates, will require no correction.

Worked examples

Example 8.3 The national grid coordinates of two points, A and B, are A: E_A 238 824.076, N_A 307 911.843; and B: E_B 244 601.011, N_B 313 000.421

Calculate (1) The grid bearing and length of $\overrightarrow{AB}$.

(2) The azimuth of $\overrightarrow{BA}$ and $\overrightarrow{AB}$.

(3) The ground length AB.

Given: (a) Mean latitude of the line = N 54° 00′.

(b) Mean altitude of the line = 250 m AOD.

(c) Local radius of the Earth = 6 384 100 m. (KU)

(1)

$E_A = 238\,824.076$	$N_A = 307\,911.843$
$E_B = 244\,601.011$	$N_B = 313\,000.421$
$\Delta E = 5776.935$	$\Delta N = 5088.578$

$$\text{Grid distance} = (\Delta E^2 + \Delta N^2)^{\frac{1}{2}} = 7698.481 \text{ m}$$

$$\text{Grid bearing } \overrightarrow{AB} = \tan^{-1}\frac{\Delta E}{\Delta N} = 48°\,37'\,30''$$

(2) In order to calculate the azimuth, i.e. the direction relative to true north, one must compute (a) the grid convergence at A and $B(\gamma)$ and (b) the $(t - T)$ correction at A and B (*Figure 8.25*).

(a) Grid convergence at $A = \gamma_A = \dfrac{\Delta E_A \tan \phi_m}{R}$

where ΔE_A = Distance from the central meridian

$= 400\,000 - E_A = 161\,175.924$ m

$$\therefore \gamma_A'' = \frac{161\,176 \tan 54°}{6\,384\,100} \times 206\,265 = 7167'' = 1°\,59'\,27''$$

$$\text{Similarly } \gamma_B'' = \frac{155\,399 \tan 54°}{6\,384\,100} \times 206\,265 = 6911'' = 1°\,55'\,11''$$

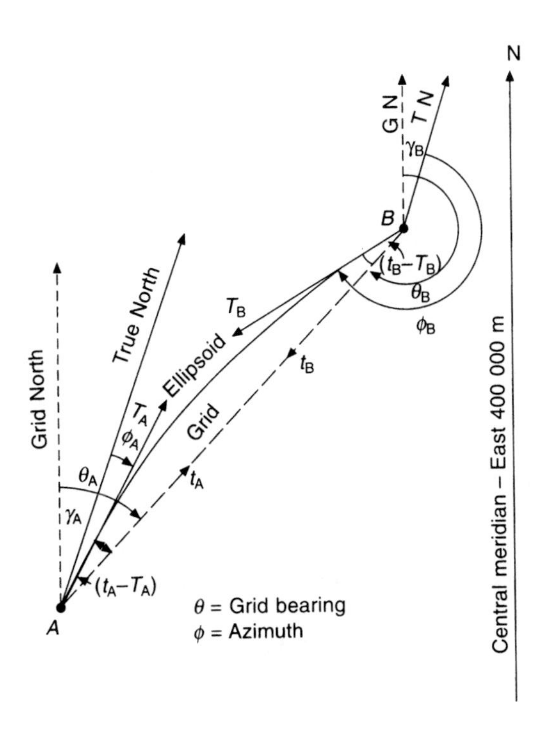

Fig. 8.25 *Grid convergence and $t - T$ corrections*

(b) $(t_A - T_A)'' = (2\Delta E_A + \Delta E_B)(N_A - N_B)K$

$$= 477.751 \times -5.089 \times 845 \times 10^{-6} = -2.05''$$

N.B. The eastings and northings are in km.

$$(t_B - T_B)'' = (2\Delta E_B + \Delta E_A)(N_B - N_A)K$$

$$= 471.974 \times 5.089 \times 845 \times 10^{-6} = +2.03''$$

Although the signs of the $(t - T)$ correction are obtained from the equation you should always draw a sketch of the situation.

Referring to *Figure 8.25*:

$$\text{Azimuth } \vec{AB} = \phi_A = \theta_A - \gamma_A - (t_A - T_A)$$

$$= 48^\circ\,37'\,30'' - 1^\circ\,59'\,27'' - 02'' = 46^\circ\,38'\,01''$$

$$\text{Azimuth } \vec{BA} = \phi_B = \theta_B - \gamma_B - (t_B - T_B)$$

$$= (48^\circ\,37'\,30'' + 180^\circ) - 1^\circ\,55'\,11'' + 2''$$

$$= 226^\circ\,42'\,21''$$

(3) To obtain ground length from grid length one must obtain the LSF adjusted for altitude.

$$\text{Mid-easting of } AB = 241\,712.544 \text{ m} = E$$

$$\text{LSF} = 0.999\,601 + [1.228 \times 10^{-14} \times (E - 400\,000)^2] = F$$

$$\therefore F = 0.999\,908$$

The altitude is 250 m OD, i.e. $H = +250$. LSF F_a adjusted for altitude is:

$$F_a = F\left(1 - \frac{H}{R}\right) = 0.999\,908\left(1 - \frac{250}{6\,384\,100}\right) = 0.999\,869$$

$$\therefore \text{Ground length } AB = \text{grid length}/F_a$$

$$\therefore AB = 7698.481/0.999\,869 = 7699.483 \text{ km}$$

Example 8.4 As part of the surveys required for the extension of a large underground transport system, a baseline was established in an existing tunnel and connected to the national grid via a wire correlation in the shaft and precise traversing therefrom.

Thereafter, the azimuth of the base was checked by gyro-theodolite using the reversal point method of observation as follows:

Reversal points	*Horizontal circle readings*			*Remarks*
	°	′	″	
r_1	330	20	40	Left reversal
r_2	338	42	50	Right reversal
r_3	330	27	18	Left reversal
r_4	338	22	20	Right reversal

$$\text{Horizontal circle reading of the baseline} = 28°\,32'\,46''$$

$$\text{Grid convergence} = 0°\,20'\,18''$$

$$(t - T) \text{ correction} = 0°\,00'\,04''$$

$$\text{NG easting of baseline} = 500\,000 \text{ m}$$

Prior to the above observations, the gyro-theodolite was checked on a surface baseline of known azimuth. The following mean data were obtained.

$$\text{Known azimuth of surface base} = 140°\,25'\,54''$$

$$\text{Gyro azimuth of surface base} = 141°\,30'\,58''$$

Determine the national grid bearing of the underground baseline. (KU)

Refer to *Chapter 13* for information on the gyro-theodolite.

Using Schuler's mean

$$N_1 = \frac{1}{4}(r_1 + 2r_2 + r_3) = 334^\circ\,33'\,24''$$

$$N_2 = \frac{1}{4}(r_2 + 2r_3 + r_4) = 334^\circ\,29'\,54''$$

$$\therefore N = (N_1 + N_2)/2 = 334^\circ\,31'\,39''$$

Horizontal circle reading of the base $= 28^\circ\,32'\,46''$

$$\therefore \text{ Gyro azimuth of the baseline} = 28^\circ\,32'\,46'' - 334^\circ\,31'\,39''$$

$$= 54^\circ\,01'\,07''$$

However, observations on the surface base show the gyro-theodolite to be 'over-reading' by $(141^\circ\,30'\,58'' - 140^\circ\,25'\,54'') = 1^\circ\,05'\,04''$.

$$\therefore \text{ True azimuth of baseline } \phi = \text{gyro azimuth} - \text{instrument constant}$$

$$= 54^\circ\,01'\,07'' - 1^\circ\,05'\,04''$$

$$= 52^\circ\,56'\,03''$$

Now by reference to *Figure 8.26*, the sign of the correction to give the NG bearing can be seen, i.e.

$$\begin{aligned} \text{Azimuth } \phi &= 52^\circ\,56'\,03'' \\ \text{Grid convergence } \gamma &= -0^\circ\,20'\,18'' \\ (t - T) &= -0^\circ\,00'\,04'' \\ \hline \therefore \text{NG bearing } \theta &= 52^\circ\,35'\,41'' \end{aligned}$$

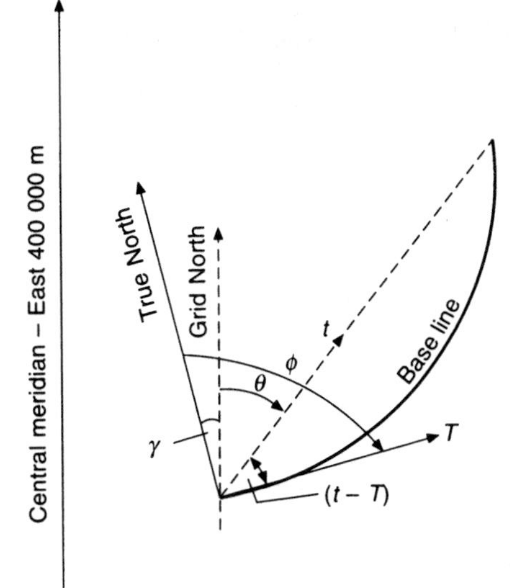

Fig. 8.26 *Signs of grid convergence and t – T corrections*

Exercises

(*8.1*) Explain the meaning of the term 'grid convergence'. Show how this factor has to be taken into account when running long survey lines by theodolite.

From a point *A* in latitude 53°N, longitude 2°W, a line is run at right angles to the initial meridian for a distance of 31 680 m in a westernly direction to point *B*.

Calculate the true bearing of the line at *B*, and the longitude of that point. Calculate also the bearing and distance from *B* of a point on the meridian of *B* at the same latitude as the starting point *A*. The radius of the Earth may be taken as 6273 km. (LU)

(*Answer*: 269° 37′ 00″; 2° 28′ 51″ W; 106.5 m)

(*8.2*) Two points, *A* and *B*, have the following coordinates:

	Latitude				*Longitude*			
	°	′	″		°	′	″	
A	52	21	14	N	93	48	50	E
B	52	24	18	N	93	42	30	E

Given the following values:

Latitude	1″ *of latitude*	1″ *of longitude*
52° 20′	30.423 45 m	18.638 16 m
52° 25′	30.423 87 m	18.603 12 m

find the azimuths of *B* from *A* and of *A* from *B*, also the distance *AB*. (LU)

(*Answer*: 308° 23′ 36″, 128° 18′ 35″, 9021.9 m)

(*8.3*) At a terminal station *A* in latitude N 47°22′40″, longitude E 0° 41′ 10″, the azimuth of a line *AB* of length 29 623 m was 23° 44′ 00″.

Calculate the latitude and longitude of station *B* and the reverse azimuth of the line from station *B* to the nearest second. (LU)

Latitude	1″ *of longitude*	1″ *of latitude*
47° 30′	20.601 m	30.399
47° 35′	20.568 m	30.399

(*Answer*: N 47° 37′ 32″; E 0° 50′ 50″; 203° 51′ 08″)

9 Satellite positioning

9.1 INTRODUCTION

Before commencing this chapter, the reader should have studied *Chapter 8* and acquired a knowledge of geoid models, ellipsoids, transformations and heights, i.e. *Sections 8.1 to 8.8*. The subject of satellite positioning is changing fast. Throughout this chapter a number of websites are referred to for further information and data. The websites are mostly government or academic and are considered to be likely to be maintained during the life of this edition of this book, although of course that cannot be guaranteed.

The concept of satellite position fixing commenced with the launch of the first Sputnik satellite by the USSR in October 1957. This was rapidly followed by the development of the Navy Navigation Satellite System (NNSS) by the US Navy. This system, commonly referred to as the Transit system, was created to provide a worldwide navigation capability for the US Polaris submarine fleet. The Transit system was made available for civilian use in 1967 but ceased operation in 1996. However, as the determination of position required very long observation periods and relative positions determined over short distances were of low accuracy, its application was limited to geodetic and low dynamic navigation uses.

In 1973, the US Department of Defense (DoD) commenced the development of NAVSTAR (Navigation System with Time and Ranging) Global Positioning System (GPS), and the first satellites were launched in 1978.

The system is funded and controlled by the DoD but is partially available for civilian and foreign users. The accuracies that may be obtained from the system depend on the degree of access available to the user, the sophistication of his/her receiver hardware and data processing software, and degree of mobility during signal reception.

In very broad terms, the geodetic user in a static location may obtain 'absolute' accuracy (with respect to the mass centre of the Earth within the satellite datum) to better than ± 1 metre and position relative to another known point, to a few centimetres over a range of tens of kilometres, with data post-processing. At the other end of the scale, a technically unsophisticated, low dynamic (ship or land vehicle) user, with limited access to the system, might achieve real time 'absolute' accuracy of 10–20 metres.

The GPS navigation system relies on satellites that continuously broadcast their own position in space and in this the satellites may be thought of as no more than control stations in space. Theoretically, a user who has a clock, perfectly synchronized to the GPS time system, is able to observe the time delay of a GPS signal from its own time of transmission at the satellite, to its time of detection at the user's equipment. The time delay, multiplied by the mean speed of light, along the path of the transmission from the satellite to the user equipment, will give the range from the satellite at its known position, to the user. If three such ranges are observed simultaneously, there is sufficient information to compute the user's position in three-dimensional space, rather in the manner of a three-dimensional trilateration. The false assumption in all this is that the user's receiver clock is perfectly synchronized with the satellite clocks.

In practice, although the satellite clocks are almost perfectly synchronized to the GPS time system, the user clock will have an error or offset. So the user is not directly able to measure the range to a particular satellite, but only the 'pseudo-range', i.e. the actual range with an unknown, but instantaneously fixed offset. This is the clock error times the speed of light. There are four unknown parameters to be solved for in the navigation solution, the three coordinates of user position and the receiver clock offset. A four-parameter solution therefore requires simultaneous observations to four satellites. At least four satellites must be visible at all times, to any observer, wherever he/she may be on or above the surface of the Earth. Not only must at least four satellites be visible but also they, or the best four if there are more, must be in a good geometric arrangement with respect to the user.

Now that GPS is fully operational, relative positioning to several millimetres, with short observation periods of a few minutes, have been achieved. For distances in excess of 5 km GPS is generally more accurate than EDM traversing. Therefore GPS has a wide application in engineering surveying. The introduction of GPS has had an even greater impact on practice in engineering surveying than that of EDM. Apart from the high accuracies attainable, GPS offers the following significant advantages:

(1) The results from the measurement of a single line, usually referred to as a baseline, will yield not only the distance between the stations at the end of the line but their component parts in the *X*/*Y*/*Z* or Eastings/Northings/Height or latitude/longitude/height directions.
(2) No line of sight is required. Unlike all other conventional surveying systems a line of sight between the stations in the survey is not required. Each station, however, must have a clear view of the sky so that it can 'see' the relevant satellites. The advantage here, apart from losing the requirement for intervisibility, is that control no longer needs to be placed on high ground and can be in the same location as the engineering works concerned.
(3) Most satellite surveying equipment is suitably weatherproof and so observations, with current systems, may be taken in any weather, by day or by night. A thick fog will not hamper survey operations.
(4) Satellite surveying can be a one-person operation with significant savings in time and labour.
(5) Operators do not need high levels of skill.
(6) Position may be fixed on land, at sea or in the air.
(7) Base lines of hundreds of kilometres may be observed thereby removing the need for extensive geodetic networks of conventional observations.
(8) Continuous measurement may be carried out resulting in greatly improved deformation monitoring.

However, GPS is not the answer to every survey problem. The following difficulties may arise:

(1) A good electronic view of the sky is required so that the satellites may be 'seen' and 'tracked'. There should not be obstructions that block the 'line of sight' from the receiver to the satellite. This is usually not a problem for the land surveyor but may become one for the engineering surveyor as a construction rises from the ground. Satellite surveying cannot take place indoors, nor can it take place underground.
(2) The equipment concerned is expensive. A pair of GPS receivers costs about the same as three or four total stations, though this will vary from manufacturer to manufacturer. Like total stations, however, prices are falling while capabilities are increasing.
(3) Because satellites orbit the whole Earth, the coordinate systems that describe the positions of satellites are global rather than local. Thus, if coordinates are required in a local datum or on a projection, then the relationship between the local projection and datum, and the coordinate system of the satellite, must also be known.
(4) The value of height determined by satellite is not that which the engineering surveyor would immediately recognize. Since the coordinate system of GPS is Earth mass centred, then any height of a point on the Earth's surface will be relative to some arbitrarily defined datum, such as the surface of an ellipsoid. If height above the geoid (or mean sea level) is required, then the separation between the geoid and the chosen ellipsoid will also be required. Some GPS receivers may have a geoid model in their software to solve this problem; however, the model may be coarse.

9.2 GPS SEGMENTS

The GPS system can be broadly divided into three segments: the *space* segment, the *control* segment and the *user* segment.

The *space* segment is composed of satellites (*Figure 9.1*). The constellation consists at the time of writing of 29 satellites including spares. The satellites are in almost circular orbits, at a height of 20 200 km above the Earth or about three times the radius of the Earth and with orbit times of just under 12 hours. The six orbital planes are equally spaced (*Figure 9.2*), and are inclined at 55° to the equator. Individual satellites may appear for up to five hours above the horizon. The system has been designed so that at least four satellites will always be in view at least 15° above the horizon.

The GPS satellites weigh, when in final orbit, approximately 850 kg. The design life of the satellites is 7.5 years but they carry 10 years' worth of propulsion consumables. Two sun-seeking single degree of

Fig. 9.1 *GPS satellite (courtesy of Air Force Link)*

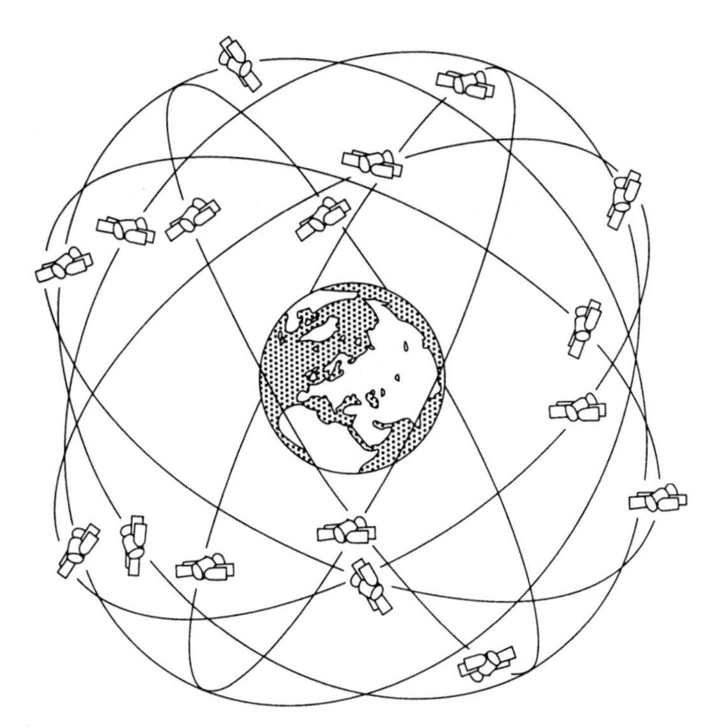

Fig. 9.2 *The original planned GPS constellation: 24 satellites in 6 orbital planes, at 55° inclination and 20 200 km altitude with 12-hour orbits (courtesy Leica Geosystems)*

freedom solar arrays, which together cover over 7 m^2 provide the electrical power. Power is retained during eclipse periods by three nickel-cadmium batteries. Reaction wheels control the orientation and position of the satellite in space. Thermal control louvers, layered insulation and thermostatically controlled heaters control the temperature of this large satellite. The satellite is built with a rigid body of aluminium boarded honeycomb panels. The satellite may be 'navigated' to a very limited extent in space with small hydrazine jets. There are two small trim thrusters and 20 even smaller attitude control thrusters. Antennae transmit the satellite's signals to the user. Each satellite carries two rubidium and two caesium atomic clocks to ensure precise timing.

As far as the user is concerned, each GPS satellite broadcasts on two L Band carrier frequencies. L1 = 1575.42 MHz (10.23 × 154) and L2 = 1227.6 MHz (10.23 × 120). The carriers are phase modulated to carry two codes, known as the P code or Precise code or PPS (Precise Positioning Service) and the C/A code or Course/Acquisition code or SPS (Standard Positioning Service) (*Figure 9.3*). The C/A code has a 'chipping rate', which is a rate of phase modulation, of 1.023×10^6 bits/sec and the code repeats every millisecond. This means that the sequence that makes up the C/A code is only 1023 bits long. Multiplied by the speed of light, each bit is then 293 m long and the whole code about 300 km. By contrast, the P code chips at 10.23×10^6 bits/sec and repeats every 267 days although each satellite only uses a seven-day segment of the whole code. The P code is thus about 2.4×10^{14} bits long. Without prior knowledge of its structure, the P code will appear as Pseudo Random Noise (PRN). This means that it is relatively easy for the user's equipment to obtain lock onto the C/A code, since it is short, simple and repeats 1000 times a second. Without knowledge of the P code, it is impossible in practice to obtain lock because the P code is so long and complex. This is the key to selective access to the GPS system. Only those users approved by the US DoD will be able to use the P code. A 50 Hz data stream that contains the following information further modulates each code:

- The satellite ephemeris, i.e. its position in space with respect to time
- Parameters for computing corrections to the satellite clock

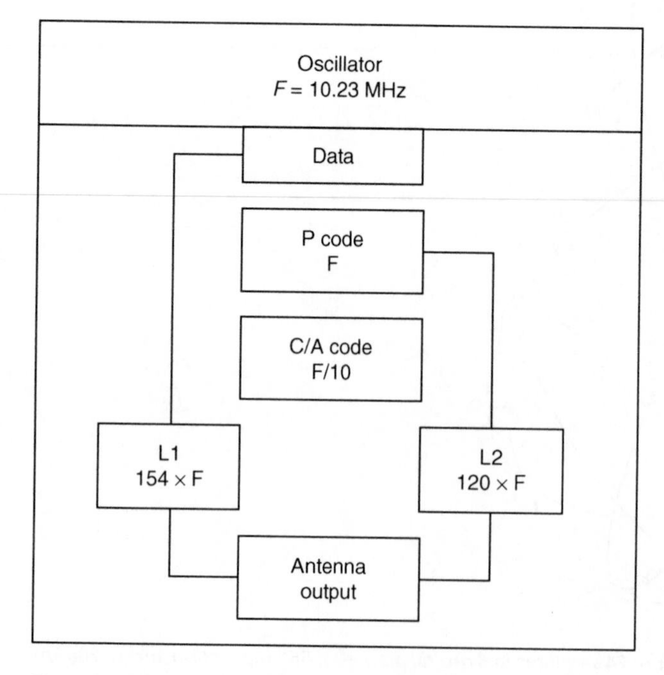

Fig. 9.3 *GPS signal generation*

- The Hand Over Word (HOW) for time synchronization that allows the user with access to the P code to transfer from the C/A to P code
- Information on other satellites of the constellation, including status and ephemerides

The satellite navigation message, which describes the satellite positions, is uploaded to the satellites by the Operational Control Segment (OCS). The OCS operates as three elements:

- Monitor stations at Ascension Island, Diego Garcia, Kwajalein and Hawaii
- A master control station at Colorado Springs, USA
- An upload station at Vandenberg Air Force Base, USA

The monitor stations are remote, unmanned stations, each with a GPS receiver, a clock, meteorological sensors, data processor and communications. Their functions are to observe the broadcast satellite navigation message and the satellite clock errors and drifts.

The data is automatically gathered and processed by each monitor station and is transmitted to the master control station. By comparing the data from the various monitor stations the master control station can compute the errors in the current navigation messages and satellite clocks, and so can compute updated navigation messages for future satellite transmission. These navigation messages are passed to the upload station and are in turn processed for transmission to the satellites by the ground antenna. The monitor stations then receive the updated navigation messages from the satellites and so the data transmission and processing circle is complete.

The master control station is also connected to the time standard of the US Naval Observatory in Washington, DC. In this way, satellite time can be synchronized and data relating it to universal time transmitted. Other data regularly updated are the parameters defining the ionosphere, to facilitate the computation of refraction corrections to the distances measured. The user segment consists essentially of a portable receiver/processor with power supply and an omnidirectional antenna (*Figure 9.4*). The processor is basically a microcomputer containing all the software for processing the field data.

9.3 GPS RECEIVERS

Basically, a receiver obtains pseudo-range or carrier phase data to at least four satellites. As GPS receiver technology is developing so rapidly, it is only possible to deal with some of the basic operational characteristics. The type of receiver used will depend largely upon the requirements of the user. For instance, if GPS is to be used for absolute as well as relative positioning, then it is necessary to use pseudo-ranges. If high-accuracy relative positioning were the requirement, then the carrier phase would be the observable involved. For real-time pseudo-range positioning, the user's receiver needs access to the navigation message (Broadcast Ephemerides). If carrier phase observations are to be used, the data may be post-processed and an external precise ephemeris may also be used.

Most modern receivers are 'all in view', that is they have enough channels to track all visible satellites simultaneously. A channel consists of the hardware and software necessary to track a satellite's code and/or carrier phase measurement continuously.

When using the carrier phase observable, it is necessary to remove the modulations. Modern geodetic receivers may work in a code correlation or codeless way. Code correlation uses a delay lock loop to maintain alignment with the incoming, satellite-generated signal. The incoming signal is multiplied by its equivalent part of the generated signal, which has the effect of removing the codes. It does still retain the navigation message and can therefore utilize the Broadcast Ephemeris.

In the codeless mode a receiver uses signal squaring to multiply the received signal by itself, thereby doubling the frequency and removing the code modulation. This process, whilst reducing the signal-to-noise ratio, loses the navigation message.

Fig. 9.4 *Handheld GPS receiver*

Code correlation needs access to the P code if tracking on L2 frequency. As the P code may be changed to the Y code and made unavailable to civilian users, L2 tracking would be impossible. However, with code correlation receivers are able to track satellites at lower elevations.

Some receivers used for navigation purposes generally track all available satellites obtaining L1 pseudo-range data and for entering the majority of harbours need to be able to accept differential corrections (DGPS) from an on-shore reference receiver.

Geodetic receivers used in engineering surveying may be single or dual frequency, with from 12 to 24 channels in order to track all the satellites available. Some geodetic receivers also have channels available for GLONASS, the Russian system equivalent to GPS.

All modern receivers can acquire the L1 pseudo-range observable using a code correlation process illustrated later. When the pseudo-range is computed using the C/A code it can be removed from the signal in order to access the L1 carrier phase and the navigation message. These two observations could be classified as civilian data. Dual frequency receivers also use code correlation to access the P code pseudo-range data and the L2 carrier phase. However, this is only possible with the 'permission' of the US military who can prevent access to the P code. This process is called Anti-Spoofing (AS). When AS is operative a signal squaring technique may be used to access the L2 carrier.

9.4 SATELLITE ORBITS

The German astronomer Johannes Kepler (1571–1630) established three laws defining the movement of planets around the sun, which also apply to the movement of satellites around the Earth.

(1) Satellites move around the Earth in elliptical orbits, with the centre of mass of the Earth situated at one of the focal points G (*Figure 9.5*). The other focus G' is unused. The implications of this law are that a satellite will at times be closer to or further away from the Earth's surface depending upon which part of its orbit it is in. GPS satellite orbits are nearly circular and so have very small eccentricity.
(2) The radius vector from the Earth's centre to the satellite sweeps out equal areas at equal time intervals (*Figure 9.6*). Therefore a satellite's speed is not a constant. The speed will be a minimum when the satellite is at apogee, at its furthest from the centre of the Earth and a maximum when it is at perigee, the point of closest approach.
(3) The square of the orbital period is proportional to the cube of the semi-major axis a, i.e. $T^2 = a^3 \times$ constant. The value of the constant was later shown by Newton to be $\mu/4\pi^2$ where μ is the Earth's gravitational constant and is equal to 398 601 km^3 s^{-2}. Therefore $T^2 = a^3\mu/4\pi^2$. So, whatever the satellite's orbital eccentricity, providing the semi-major axis is the same, then so will be the period.

Therefore these laws define the geometry of the orbit, the velocity variation of the satellite along its orbital path, and the time taken to complete an orbit.

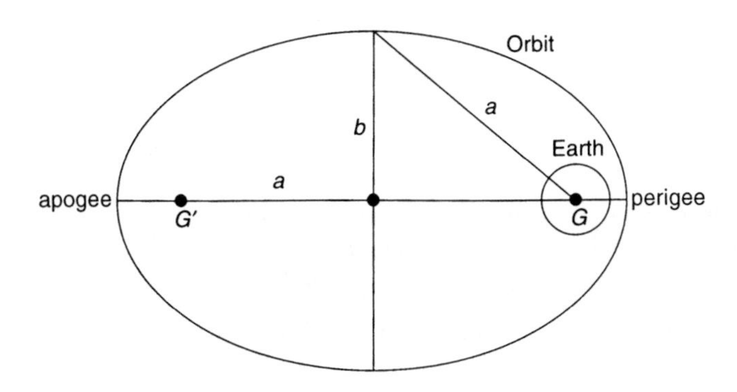

Fig. 9.5 *An elliptical orbit*

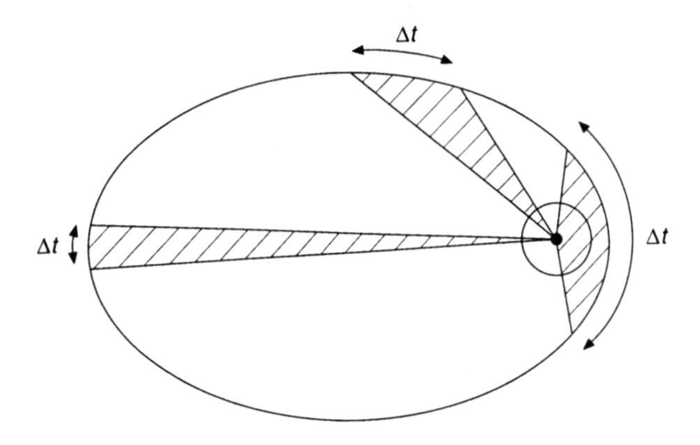

Fig. 9.6 *Kepler's second law*

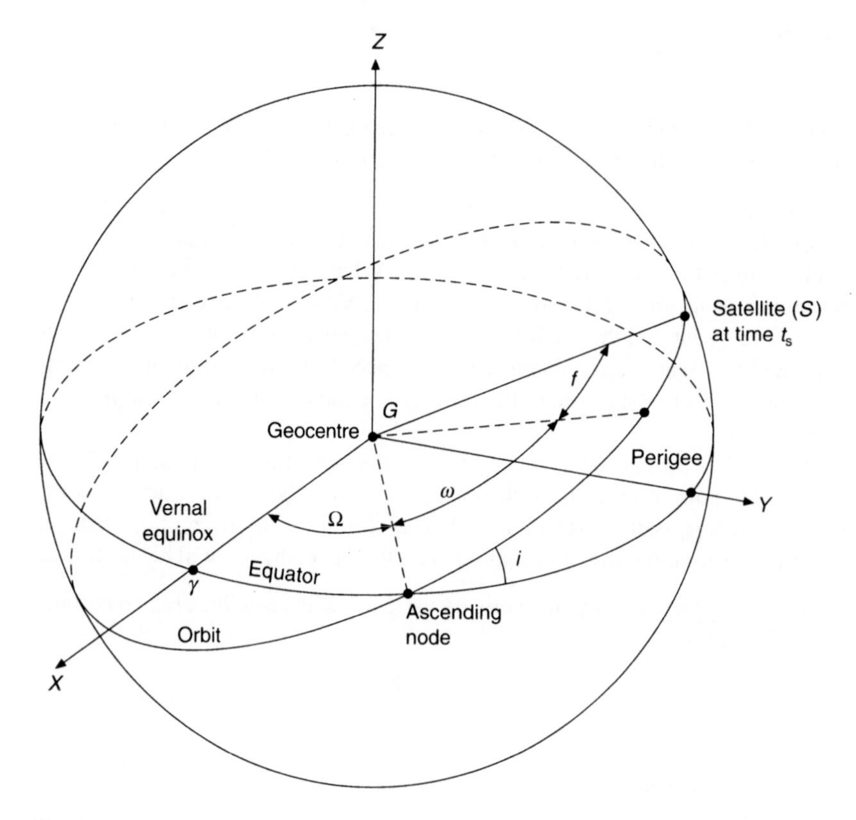

Fig. 9.7 *The orbit in space*

Whilst a and e, the semi-major axis and the eccentricity, define the shape of the ellipse (see *Chapter 8*), its orientation in space must be specified by three angles defined with respect to a space-fixed reference coordinate system. The spatial orientation of the orbital ellipse is shown in *Figure 9.7* where:

(1) Angle Ω is the right ascension (RA) of the ascending node of the orbital path, measured on the equator, eastward from the vernal equinox (γ).
(2) i is the inclination of the orbital plane to the equatorial plane.
(3) ω is the argument of perigee, measured in the plane of the orbit from the ascending node.

Having defined the orbit in space, the satellite is located relative to the point of perigee using the angle f, called the 'true anomaly' at the time of interest.

The line joining perigee and apogee is called the 'line of apsides' and is the X-axis of the orbital space coordinate system. The Y-axis is in the mean orbital plane at right angles to the X-axis. The Z-axis is normal to the orbital plane and will be used to represent small perturbations from the mean orbit. The XYZ space coordinate system has its origin at G. It can be seen from *Figure 9.8* that the space coordinates of the satellite at time t are:

$$X_0 = r\cos f$$

$$Y_0 = r\sin f$$

$$Z_0 = 0 \text{ (in a pure Keplerian or } \textit{normal} \text{ orbit)}$$

where r = the distance from the Earth's centre to the satellite.

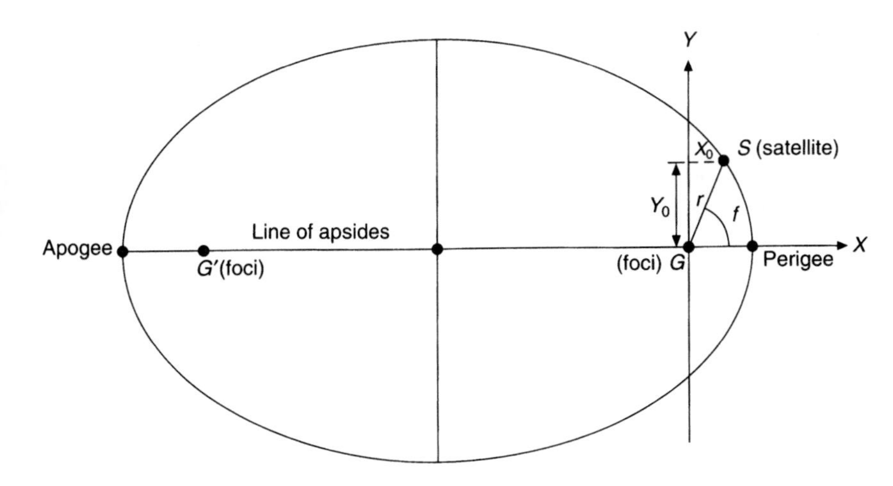

Fig. 9.8 *Orbital coordinate system*

The space coordinates can easily be computed using the information contained in the broadcast ephemeris. The procedure is as follows:

(1) Compute T, which is the orbital period of the satellite, i.e. the time it takes to complete its orbit. Using Kepler's third law:

$$T = 2\pi a(a/\mu)^{\frac{1}{2}} \tag{9.1}$$

μ is the Earth's gravitational constant and is equal to 3.986005×10^{14} m^{3}s^{-2}.

(2) Compute the 'mean anomaly' M, which is the angle swept out by the satellite in the time interval $(t_s - t_p)$ from

$$M = 2\pi(t_s - t_p)/T \tag{9.2}$$

where t_s = the time of the satellite signal transmission and
t_p = the time of the satellite's passage through perigee (obtained from the broadcast ephemeris).

M defines the position of the satellite in orbit but only for ellipses with $e = 0$, i.e. circles. To correct for this it is necessary to obtain the 'eccentric anomaly' E and hence the 'true anomaly' f (*Figure 9.9*) for the near-circular GPS orbits.

(3) From Kepler's equation: $E - e \sin E = M$ (9.3)

where E and M are in radians. The equation is solved iteratively for E. e is the eccentricity of an ellipse, calculated from

$$e = (1 - b^2/a^2)^{\frac{1}{2}}$$

Now the 'true anomaly' f is computed from

$$\cos f = (\cos E - e)/(1 - e \cos E) \tag{9.4}$$

or

$$\tan(f/2) = [(1 + e)/(1 - e)]^{\frac{1}{2}} \tan(E/2) \tag{9.5}$$

Use the formula (9.4) or (9.5) which is most sensitive to change in f.

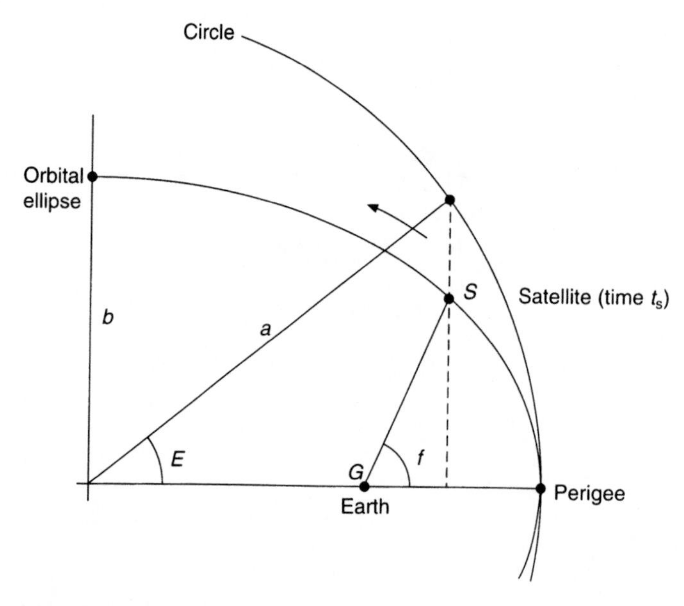

Fig. 9.9 *The orbital ellipse*

(4) Finally, the distance from the centre of the Earth to the satellite (GS), equal to r, is calculated from

$$r = a(1 - e\cos E) \tag{9.6}$$

and, as first indicated,

$$\begin{aligned} X_0 &= r\cos f \\ Y_0 &= r\sin f \\ Z_0 &= 0 \end{aligned} \tag{9.7}$$

It is assumed here that the position of the satellite may be determined from the terms that define a normal or Keplerian orbit at the time (t_s) of observation. The actual orbit of the satellite departs from the Keplerian orbit due to the effects of:

(1) the non-uniformity of the Earth's gravity field
(2) the attraction of the moon and sun
(3) atmospheric drag
(4) direct and reflected solar radiation pressure
(5) earth tides
(6) ocean tides

These forces produce orbital perturbations, the total effect of which must be mathematically modelled to produce a precise position for the satellite at the time of observation. As already illustrated, the pure, smooth Keplerian orbit is obtained from the elements:

a – semi-major axis
e – eccentricity
which give the size and shape of the orbit

i – inclination
Ω – right ascension of the ascending node
which orient the orbital plane in space with respect to the Earth
ω – argument of perigee
t_p – ephemeris reference time
which fixes the position of the satellite within its orbit

Additional parameters given in the broadcast ephemeris describe the deviations of the satellite motion from the pure Keplerian form. There are two ephemerides available: the broadcast, shown below, and the precise.

M_0 = mean anomaly
Δn = mean motion difference
e = eccentricity
$\sqrt{a}$ = square root of semi-major axis
Ω = right ascension
i_0 = inclination
ω = argument of perigee
$\dot{\Omega}$ = rate of right ascension
$\dot{i}$ = rate of inclination
C_{uc}, C_{us} = correction terms to argument of latitude
C_{rc}, C_{rs} = correction terms to orbital radius
C_{ic}, C_{is} = correction terms to inclination
t_p = ephemeris reference time

Using the broadcast ephemeris, plus two additional values from the WGS84 model, namely:

ω_e – the angular velocity of the Earth ($7.2921151467 \times 10^{-5}$ rad s^{-1})
μ – the gravitational/mass constant of the Earth (3.986005×10^{14} m^3 s^{-2})

The Cartesian coordinates in a perturbed satellite orbit can be computed using:

u – the argument of latitude (the angle in the orbital plane, from the ascending node to the satellite)
r – the geocentric radius, as follows

$$\begin{aligned} X_0 &= r\cos u \\ Y_0 &= r\sin u \\ Z_0 &= 0 \end{aligned} \tag{9.8}$$

where $r = a(1 - e\cos E) + C_{rc}\cos 2(\omega + f) + C_{rs}\sin 2(\omega + f)$

$$u = \omega + f + C_{uc}\cos 2(\omega + f) + C_{us}\sin 2(\omega + f)$$

where $a(1-e\cos E)$ is the elliptical radius, C_{rc} and C_{rs} the *c*osine and *s*ine correction terms of the geocentric radius and C_{uc}, C_{us} the correction terms for u.

It is now necessary to rotate the orbital plane about the X_0-axis, through the inclination i, to make the orbital plane coincide with the equatorial plane and the Z_0-axis coincide with the Z-axis of the Earth fixed system (IRP). Thus:

$$\begin{aligned} X_E &= X_0 \\ Y_E &= Y_0\cos i \\ Z_E &= Y_0\sin i \end{aligned} \tag{9.9}$$

where

$$i = i_0 + \dot{i}(t_s - t_p) + C_{ic}\cos 2(\omega + f) + C_{is}\sin 2(\omega + f)$$

and i_0 is the inclination of the orbit plane at reference time t_p
$\dot{i}$ is the linear rate of change of inclination
C_{ic}, C_{is} are the amplitude of the cosine and sine correction terms of the inclination of the orbital plane.

Finally, although the Z_E-axis is now correct, the X_E-axis aligns with the Ascending Node and requires a rotation about Z towards the Zero Meridian (IRM) usually referred to as the Greenwich Meridian. The required angle of rotation is the Right Ascension of the Ascending Node minus the Greenwich Apparent Sidereal Time (GAST) and is in effect the longitude of the ascending node of the orbital plane (λ_0) at the time of observation t_s.

To compute λ_0 we use the right ascension parameter Ω_0, the change in GAST using the Earth's rotation rate ω_e during the time interval $(t_s - t_p)$ and change in longitude since the reference time, thus:

$$\lambda_0 = \Omega_0 + (\dot{\Omega} - \omega_e)(t_s - t_p) - \omega_e t_p$$

and

$$X = X_E \cos\lambda_0 - Y_E \sin\lambda_0$$

$$Y = X_E \sin\lambda_0 + Y_E \cos\lambda_0 \qquad (9.10)$$

$$Z = Z_E$$

Full details of the computing process, summarized above, may be found in the GPS Interface Control Document at http://www.navcen.uscg.gov/pubs/gps/icd200/default.htm. Alternatively try keywords *GPS* and *ICD* in a search engine.

The accuracy of the orbit deduced from the Broadcast Ephemeris is about 10 m at best and is directly reflected in the *absolute* position of points. Whilst this may be adequate for some applications, such as navigation, it would not be acceptable for most engineering surveying purposes. Fortunately, differential procedures and the fact that engineering generally requires *relative* positioning using carrier phase, substantially eliminates the effect of orbital error. However, relative positioning accuracies better than 0.1 ppm of the length of the base line can only be achieved using a Precise Ephemeris. Several different Precise Ephemerides may be downloaded from http://www.ngs.noaa.gov/GPS/GPS.html.

The GPS satellite coordinates are defined with respect to an ellipsoid of reference called the World Global System 1984 (WGS84). The system has its centre coinciding with the centre of mass of the Earth and orientated to coincide with the IERS axes, as described in *Chapter 8*. Its size and shape is the one that best fits the geoid over the whole Earth, and is identical to the Ellipsoid GRS80 with $a = 6\,378\,137.0$ m and $1/f = 298.257223563$.

In addition to being a coordinate system, other values, such as a description of the Earth's gravity field, the velocity of light and the Earth's angular velocity, are also supplied. Consequently, the velocity of light as quoted for the WGS84 model must be used to compute ranges from observer to satellite, and the subsequent position, based on all the relevant parameters supplied.

The final stage of the positioning process is the transformation of the WGS84 coordinates to local geodetic or plane rectangular coordinates and height. This is usually done using the Helmert transformation outlined in *Chapter 8*. The translation, scale and rotational parameters between GPS and national mapping coordinate systems have been published. The practical problems involved have already been mentioned in *Chapter 8*. If the parameters are unavailable, they can be obtained by obtaining the WGS84 coordinates of points whose local coordinates are known. A least squares solution will produce the parameters required. (The transformation processes are dealt with later in the chapter.)

It must be remembered that the height obtained from satellites is the ellipsoidal height and will require accurate knowledge of the geoid–ellipsoid separation (N) to change it to orthometric height.

9.5 BASIC PRINCIPLE OF POSITION FIXING

Position fixing in three dimensions may involve the measurement of distance (or range) to at least three satellites whose X, Y and Z position is known, in order to define the user's X_p, Y_p and Z_p position.

In its simplest form, the satellite transmits a signal on which the time of its departure (t_D) from the satellite is modulated. The receiver in turn notes the time of arrival (t_A) of this time mark. Then the time which it took the signal to go from satellite to receiver is $(t_A - t_D) = \Delta t$ called the delay time. The measured range R is obtained from

$$R_1 = (t_A - t_D)c = \Delta tc \tag{9.11}$$

where c = the velocity of light.

Whilst the above describes the basic principle of range measurement, to achieve it one would require the receiver to have a clock as accurate as the satellite's and perfectly synchronized with it. As this would render the receiver impossibly expensive, a correlation procedure, using the pseudo-random binary codes (P or C/A), usually 'C/A', is adopted. The signal from the satellite arrives at the receiver and triggers the receiver to commence generating its own internal copy of the C/A code. The receiver-generated code is cross-correlated with the satellite code (*Figure 9.10*). The ground receiver is then able to determine the time delay (Δt) since it generated the same portion of the code received from the satellite. However, whilst this eliminates the problem of the need for an expensive receiver clock, it does not eliminate the problem of exact synchronization of the two clocks. Thus, the time difference between the two clocks, termed clock bias, results in an incorrect assessment of Δt. The distances computed are therefore called 'pseudo-ranges'. The use of four satellites rather than three, however, can eliminate the effect of clock bias.

A line in space is defined by its difference in coordinates in an X, Y and Z system:

$$R = (\Delta X^2 + \Delta Y^2 + \Delta Z^2)^{\frac{1}{2}}$$

If the error in R, due to clock bias, is δR and is constant throughout, then:

$$\begin{aligned}
R_1 + \delta R &= [(X_1 - X_p)^2 + (Y_1 - Y_p)^2 + (Z_1 - Z_p)^2]^{\frac{1}{2}} \\
R_2 + \delta R &= [(X_2 - X_p)^2 + (Y_2 - Y_p)^2 + (Z_2 - Z_p)^2]^{\frac{1}{2}} \\
R_3 + \delta R &= [(X_3 - X_p)^2 + (Y_3 - Y_p)^2 + (Z_3 - Z_p)^2]^{\frac{1}{2}} \\
R_4 + \delta R &= [(X_4 - X_p)^2 + (Y_4 - Y_p)^2 + (Z_4 - Z_p)^2]^{\frac{1}{2}}
\end{aligned} \tag{9.12}$$

where X_n, Y_n, Z_n = the coordinates of satellites 1, 2, 3 and 4 (n = 1 to 4)
X_p, Y_p, Z_p = the coordinates required for point P
R_n = the measured ranges to the satellites

Solving the four equations for the four unknowns X_p, Y_p, Z_p and δR also solves for the error due to clock bias.

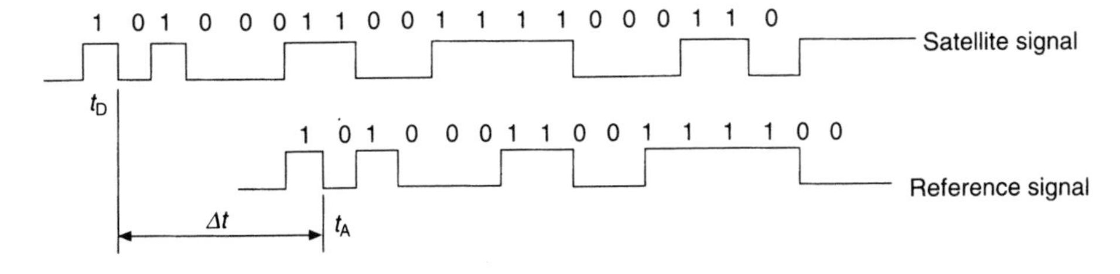

Fig. 9.10 *Correlation of the pseudo-binary codes*

Whilst the use of pseudo-range is sufficient for navigational purposes and constitutes the fundamental approach for which the system was designed, a much more accurate measurement of range is required for positioning in engineering surveying. Measuring phase difference by means of the carrier wave in a manner analogous to EDM measurement does this. As observational resolution is about 1% of the signal wavelength λ, the following table shows the reason for using the carrier waves; this is referred to as the carrier phase observable.

GPS signal	*Approximate wavelength* λ	*1% of* λ
C/A code	300 m	3 m
P code	30 m	0.3 m
Carrier	200 mm	2 mm

Carrier phase is the difference between the incoming satellite carrier signal and the phase of the constant-frequency signal generated by the receiver. It should be noted that the satellite carrier signal when it arrives at the receiver is different from that initially transmitted, because of the relative velocity between transmitter and receiver; this is the well-known Doppler effect. The carrier phase therefore changes according to the continuously integrated Doppler shift of the incoming signal. This observable is biased by the unknown offset between the satellite and receiver clocks and represents the difference in range to the satellite at different times or epochs. The carrier phase movement, although analogous to EDM measurement, is a one-way measuring system, and thus the number of whole wavelengths (N) at lock-on is missing; this is referred to as the integer or phase ambiguity. The value of N can be obtained from GPS network adjustment or from double differencing or eliminated by triple differencing.

9.6 DIFFERENCING DATA

Whilst the system was essentially designed to use pseudo-range for navigation purposes, it is the carrier phase observable which is used in engineering surveying to produce high accuracy relative positioning. Carrier phase measurement is similar to the measuring process used in EDM. However, it is not a two-way process, as with EDM. The observations are ambiguous because of the unknown integer number of cycles between the satellite and receiver at lock-on. Once the receiver has acquired the satellite signals, the number of cycles can be tracked and counted (carrier phase) with the initial integer number of cycles, known as the integer ambiguity, still unknown (*Figure 9.11*).

By a process of differencing between the phase received at one station from two satellites, from one satellite at two stations, and between each of these over a period of time it is possible to solve for the relative position of one station with respect to the other.

There are two major problems here. One is that, although the position on a single radio sine wave from the satellite may be determined to about 1% of the wavelength, it is not possible to differentiate one sine wave from another. Therefore, although the distance from a satellite may be computed to a few millimetres (the wavelength on the L2 frequency is approximately 19 cm) there is an 'ambiguity' in that distance of a whole number (an integer) of 19 cm wavelengths. A major part of the solution is to solve for this integer ambiguity. This is done by a series of 'difference' solutions that are described below.

The other major problem is that of 'cycle slips'. In the 'triple difference solution', which is differencing over time, the cycles are counted from the start to the end of the observing period. If one or more of the cycles are missed then the measure will be in error by the number of cycle slips multiplied by the wavelength of the signal. Cycle slips may be 'repaired' if more than the minimum number of observations

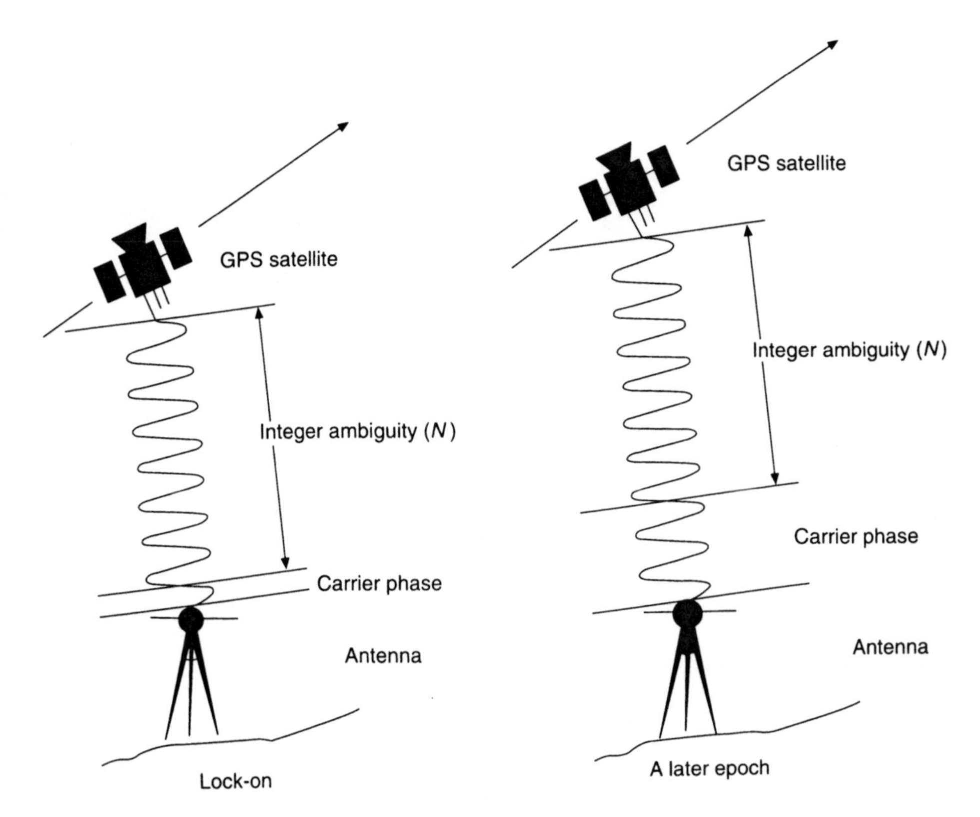

Fig. 9.11 *Integer ambiguity*

is taken and observations are recorded on a continuous basis. More flexible methods of using the carrier phase observable have been developed. Kinematic surveying involves taking observations at two stations for a period of time, say 10 minutes. One instrument is then moved from station to station remaining for only a short period at each station. The post-processing again is all on a differencing basis. Semi-kinematic surveying uses the whole number nature of the integer ambiguity and even allows for switching off the instrument between repeated visits to the stations.

What follows is a demonstration of the concepts involved in the differencing processes and shows how a solution may be achieved. It does not represent how the software of any specific instrument works.

In *Figure 9.12* and in the text that follows, elements associated with satellites have superscripts and elements associated with ground stations have subscripts.

R_j = position vector of ground station j
r^a = position vector of satellite a
p_j^a = actual range from satellite a to ground station j
$= \left\{(x^a - x_j)^2 + (y^a - y_j)^2 + (z^a - z_j)^2\right\}^{\frac{1}{2}}$
where x^a, y^a, z^a are the coordinates of satellite a
and x_j, y_j, z_j are the coordinates of station j
P_j^a = pseudo-range
c = speed of radio waves in vacuo
T_j = receiver clock time of receipt of signal at station j from satellite a
t^a = satellite clock time of transmission of signal at satellite a
g_j = GPS time of receipt of signal at station j

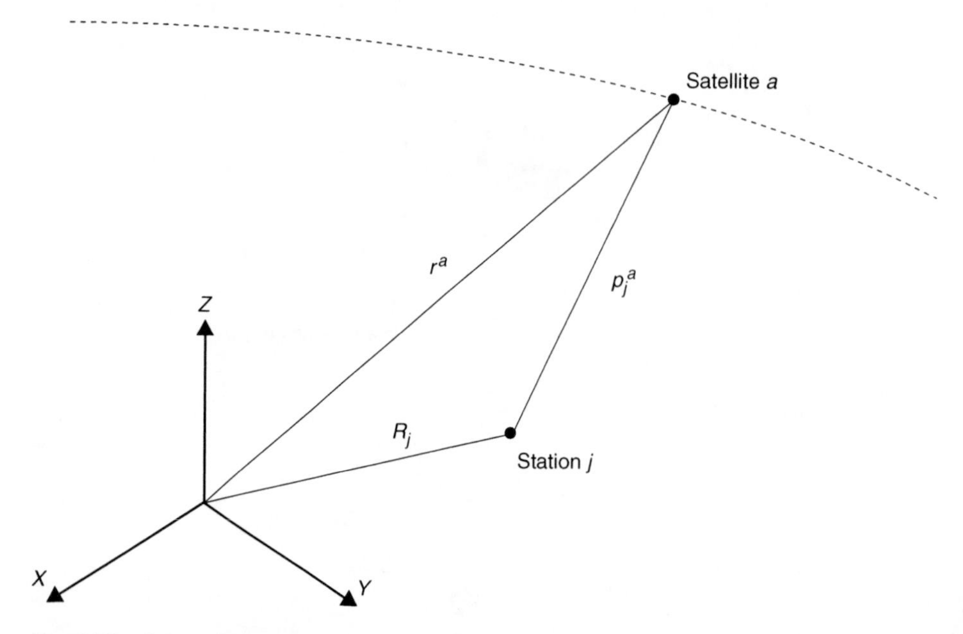

Fig. 9.12 *Ranges between and positions of satellite and ground station*

g^a = GPS time of transmission of signal at satellite a
$\mathrm{d}T_j$ = receiver clock offset at time of receipt of signal at station j from satellite a
$\mathrm{d}t^a$ = satellite clock offset at time of transmission of signal at satellite a
F_j^a = frequency of signal received from satellite a as measured at station j

The instantaneous phase ϕ_j^a of the signal from the ath satellite measured at the jth station receiver is

$$\phi_j^a = F_j^a P_j^a / c = F_j^a \{T_j - t^a\}$$

Using the common GPS time scale yields:

$$\begin{aligned}\phi_j^a &= F_j^a \{(g_j - \mathrm{d}T_j) - (g^a - \mathrm{d}t^a)\} \\ &= F_j^a \{g_j - g^a\} + F_j^a \{-\mathrm{d}T_j + \mathrm{d}t^a\} \\ &= F_j^a p_j^a / c + F_j^a \{-\mathrm{d}T_j + \mathrm{d}t^a\} \\ &= F_j^a |r^a - R_j| / c + F_j^a \{-\mathrm{d}T_j + \mathrm{d}t^a\}\end{aligned} \qquad (9.13)$$

The measurement of the phase ϕ_j^a needs the signal from the satellite to be a pure sine wave. The signal that actually comes from the satellite has been phase modulated by the P code and the data, and if it is on the L1 frequency, by the C/A code as well. If the data has been decoded and the code(s) are known, then they can be applied in reverse to form the 'reconstructed carrier wave'. If these are not known, then squaring the received carrier wave may form a sine wave. Although the advantage is that data and codes do not need to be known the squared wave signal is very noisy compared with the original signal, it has half the amplitude and twice the frequency. Reconstructed carrier is an unmodulated wave that therefore does not contain time information and so it is not possible to determine the time of a measurement. In other words, the initial phase (or range) is not known.

The equation of a single phase measurement, *equation (9.13)*, implicitly contains as unknowns the coordinates of the satellite and ground station positions and the satellite and receiver clock offsets. Also in

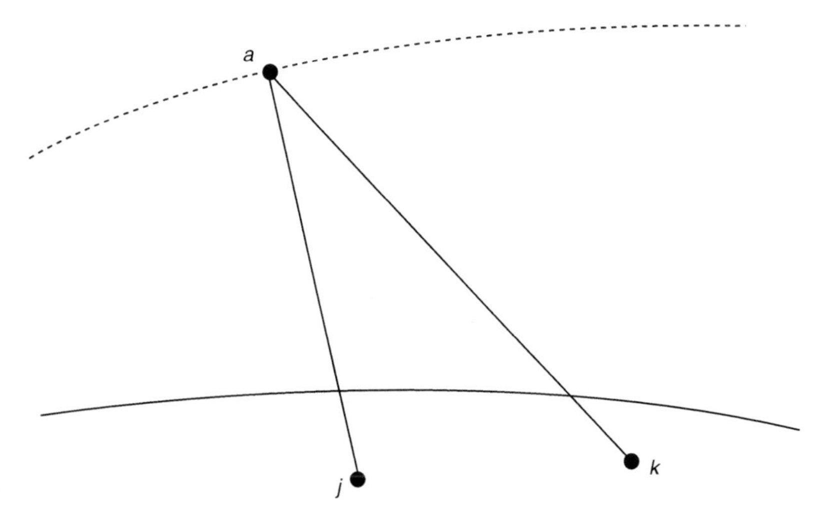

Fig. 9.13 *Single difference*

the measurement of the phase ϕ_j^a are the unknown number of whole cycles between the station and the satellite.

The basic observable of GPS is the phase measurement, but it is necessary to use a linear combination of the observables for further processing. These combinations are in the form of single differences, double differences and triple differences and are used to solve for the clock errors and the integer ambiguities.

A *single difference* is most commonly understood as the difference in phase of 'simultaneous' measurement between one satellite position and two observing stations (*Figure 9.13*). The processing eliminates the effects of errors in the satellite clock. The single difference equation is:

$$\mathrm{d}\phi_{jk}^a = \phi_k^a - \phi_j^a - 2\pi n_{jk}^a$$

which becomes

$$\mathrm{d}\phi_{jk}^a = F_k^a|r^a - R_k|/c - F_j^a|r^a - R_j|/c - F_k^a\,\mathrm{d}T_k + F_j^a\,\mathrm{d}T_j + (F_k^a - F_j^a)\,\mathrm{d}t^a - 2\pi n_{jk}^a$$

where $2\pi n_{jk}^a$ is the difference between the integer ambiguities. If GPS time is taken as the time system of the satellite then $\mathrm{d}t^a = 0$ so the equation becomes

$$\mathrm{d}\phi_{jk}^a = F_k^a|r^a - R_k|/c - F_j^a|r^a - R_j|/c - F_k^a\,\mathrm{d}T_k + F_j^a\,\mathrm{d}T_j - 2\pi n_{jk}^a$$

and so errors in the satellite clock are eliminated. If the equation is now expanded then:

$$-\mathrm{d}\phi_{jk}^a + F_k^a\{(x^a - x_k)^2 + (y^a - y_k)^2 + (z^a - z_k)^2\}^{\frac{1}{2}}/c - F_j^a\{(x^a - x_j)^2 + (y^a - y_j)^2$$
$$+ (z^a - z_j)^2\}^{\frac{1}{2}}/c - F_k^a\,\mathrm{d}T_k + F_j^a\,\mathrm{d}T_j - 2\pi n_{jk}^a = 0$$

The unknowns are now those of the positions of each ground station and the satellite and also the two receiver clock offsets. There is also the unknown integer number of difference cycles between each ground station and the satellite. To solve for this, it is necessary to know n_{jk}^a by some means and also to know the satellite position since the x, y and z components of the satellite position will not be repeated in any subsequent observation equations. The equation contains the integer ambiguity element $2\pi n_{jk}^a$.

Orbital and atmospheric errors are virtually eliminated in relative positioning, as the errors may be assumed identical. Baselines up to 50 km in length would be regarded as short compared with the height of the satellites (20 200 km). Thus it could be argued that the signals to each end of the baseline would pass through the same column of atmosphere, resulting in equal errors cancelling each other out. The above

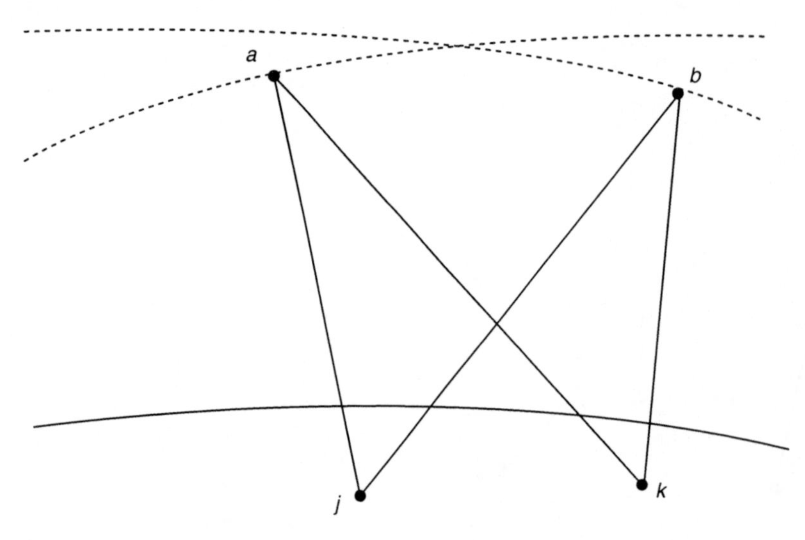

Fig. 9.14 *Double difference*

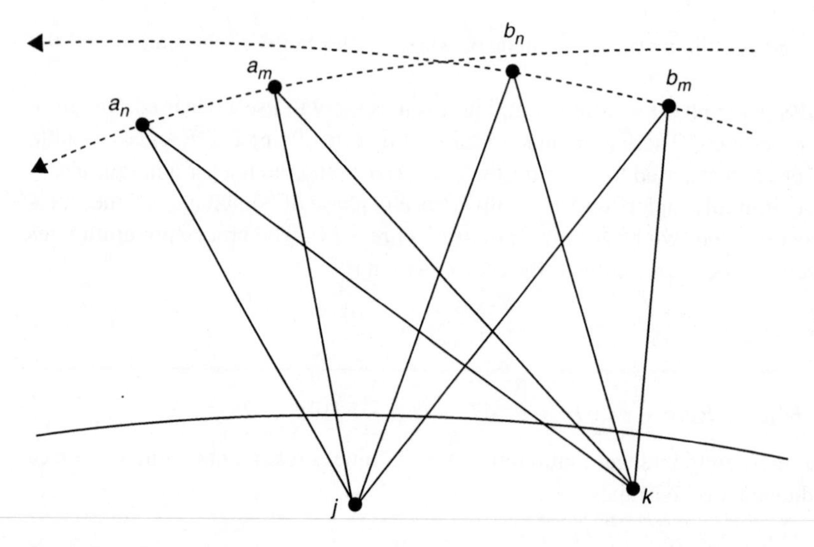

Fig. 9.15 *Triple difference*

differencing procedure is sometimes referred to as the 'between-station' difference and is the basis of differential GPS.

A *double difference* is the difference of two single differences observed at the same ground stations but with respect to two different satellites at the same time (*Figure 9.14*). The double difference formula cancels out the errors in both the satellite and the ground station clocks. This is the observable that is used in most GPS surveying software. The equation thus becomes:

$$\mathrm{d}\phi_{jk}^{ab} = \mathrm{d}\phi_{jk}^{a} - \mathrm{d}\phi_{jk}^{b} - 2\pi n_{jk}^{ab}$$

$$= F_k^a|r^a - R_k|/c - F_j^a|r^a - R_j|/c - F_k^b|r^b - R_k|/c + F_j^b|r^b - R_j|/c - 2\pi n_{jk}^{ab}$$

and contains coordinates of satellites and stations but is also still an integer ambiguity term.

A solution to this problem is to observe a *triple difference* phase difference. This is the difference of two double differences related to the same constellation of receivers and satellites but at different epochs or events (*Figure 9.15*). In addition to all the errors removed by double differencing, the integer ambiguity is also removed. The triple difference equation becomes

$$\mathrm{d}\phi_{mn} = \{\mathrm{d}\phi_{jk}^{ab}\}_n - \{\mathrm{d}\phi_{jk}^{ab}\}_m$$

and the $2\pi n_{jk}^{ab}$ is eliminated provided that the observations between epochs m and n are continuous. By using triple differences, the phase ambiguity is eliminated. However, by using the triple difference solution in the double difference equations to solve for the integer ambiguity, which of course must be an integer, and by using that integer value back in the double difference solution, a more precise estimate of the baseline difference in coordinates is obtained, than by using triple differences alone.

Triple differencing reduces the number of observations and creates a high noise level. It can, however, be useful in the first state of data editing, particularly the location of cycle slips and their subsequent correction. The magnitude of a cycle slip is the difference between the initial integer ambiguity and the subsequent one, after signal loss. It generally shows up as a 'jump' or 'gap' in the residual output from a least squares adjustment. Graphical output of the residuals in single, double and triple differencing clearly illustrates the cycle slip and its magnitude (*Figure 9.16*).

Much of the above processing is transparent to the user in modern GPS processing software.

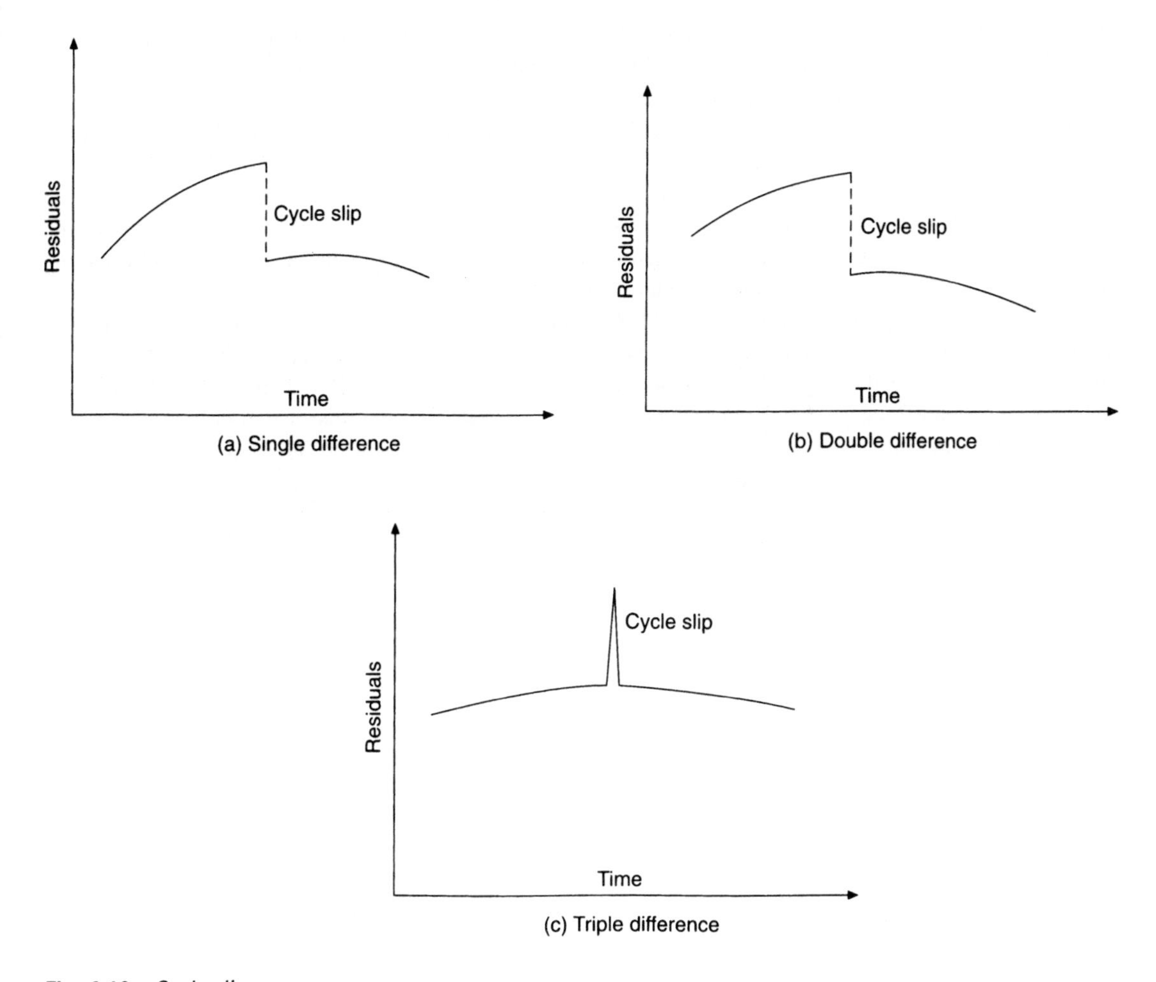

Fig. 9.16 *Cycle slips*

9.7 GPS OBSERVING METHODS

The use of GPS for positioning to varying degrees of accuracy, in situations ranging from dynamic (navigation) to static (control networks), has resulted in a wide variety of different field procedures using one or other of the basic observables. Generally pseudo-range measurements are used for navigation, whilst the higher precision necessary in engineering surveys requires carrier frequency phase measurements.

The basic point positioning method used in navigation gives the X, Y, Z position to an accuracy of better than 20 m by observation to four satellites. However, the introduction of Selective Availability (SA), see below, degraded this accuracy to 100 m or more and so led to the development of the more accurate differential technique. In this technique the vector between two receivers (baseline) is obtained, i.e. the difference in coordinates (ΔX, ΔY, ΔZ). If one of the receivers is set up over a fixed station whose coordinates are known, then comparison with the observed coordinates enables the differences to be transmitted as corrections to the second receiver (rover). In this way, all the various GPS errors are lumped together in a single correction. At its simplest the corrections transmitted could be in a simple coordinate format, i.e. δX, δY, δZ, which are easy to apply. Alternatively, the difference in coordinate position of the fixed station may be used to derive corrections to the ranges to the various satellites used. The rover then applies those corrections to its own observations before computing its position.

The fundamental assumption in Differential GPS (DGPS) is that the errors within the area of survey would be identical. This assumption is acceptable for most engineering surveying where the areas involved are small compared with the distance to the satellites.

Where the area of survey becomes extensive this argument may not hold and a slightly different approach is used called Wide Area Differential GPS.

It can now be seen that, using DGPS, the position of a roving receiver can be found relative to a fixed master or base station without significant errors from satellite and receiver clocks, ionospheric and tropospheric refraction and even ephemeris error. This idea has been expanded to the concept of having permanent base stations established throughout a wide area or even a whole country.

As GPS is essentially a military product, the US Department of Defense has retained the facility to reduce the accuracy of the system by interfering with the satellite clocks and the ephemeris of the satellite. This is known as Selective Availability (SA) of the Standard Positioning Service (SPS). This form of degradation has been switched off since May 2000 and it is unlikely, though possible, that it will be reintroduced as there are other ways that access to the system can be denied to a hostile power. The P can also be altered to a Y code, to prevent imitation of the PPS by hostile forces, and made unavailable to civilian users. This is known as Anti-Spoofing (AS). However, the carrier wave is not affected and differential methods should correct for most SA effects.

Using the carrier phase observable in the differential mode produces accuracies of 1 ppm of the baseline length. Post-processing is needed to resolve for the integer ambiguity if the highest quality results are to be achieved. Whilst this, depending on the software, can result in even greater accuracies than 1 ppm (up to 0.01 ppm), it precludes real-time positioning. However, the development of Kinematic GPS and 'on-the-fly' ambiguity resolution makes real-time positioning possible and greatly reduces the observing times.

The following methods are based on the use of carrier phase measurement for relative positioning using two receivers.

9.7.1 Static positioning

This method is used to give high precision over long baselines such as are used in geodetic control surveys. At its simplest, one receiver is set up over a station of known X, Y, Z coordinates, preferably in the

WGS84 reference system, whilst a second receiver occupies the station whose coordinates are required. Observation times may vary from 45 min to several hours. This long observational time is necessary to allow a change in the relative receiver/satellite geometry in order to calculate the initial integer ambiguity terms.

More usually baselines are observed when the precise coordinates of neither station are known. The approximate coordinates of one station can be found by averaging the pseudo-range solution at that station. Provided that those station coordinates are known to within 10 m it will not significantly affect the computed difference in coordinates between the two stations. The coordinates of a collection of baselines, provided they are interconnected, can then be estimated by a least squares free network adjustment. Provided that at least one, and preferably more, stations are known in WGS84 or the local datum then the coordinates of all the stations can be found in WGS84 or the local datum.

Accuracies in the order of 5 mm ± 1 ppm of the baseline are achievable as the majority of errors in GPS, such as clock, orbital and atmospheric errors, are eliminated or substantially reduced by the differential process. The use of permanent active GPS networks established by a government agency or private company results in a further increase in accuracy for static positioning.

Apart from establishing high precision control networks, it is used in control densification, measuring plate movement in crustal dynamics and oil rig monitoring.

9.7.2 Rapid static

Rapid static surveying is ideal for many engineering surveys and is halfway between static and kinematic procedures. The ‘master’ receiver is set up on a reference point and continuously tracks all visible satellites throughout the duration of the survey. The ‘roving’ receiver visits each of the remaining points to be surveyed, but stays for just a few minutes, typically 2–10 min.

Using difference algorithms, the integer ambiguity terms are quickly resolved and position, relative to the reference point, obtained to sub-centimetre accuracy. Each point is treated independently and as it is not necessary to maintain lock on the satellites, the roving receiver may be switched off whilst travelling between stations. Apart from a saving in power, the necessity to maintain lock, which is very onerous in urban surveys, is removed.

This method is accurate and economic where there are many points to be surveyed. It is ideally suited for short baselines where systematic errors such as atmospheric, orbital, etc., may be regarded as equal at all points and so differenced out. It can be used on large lines (> 10 km) but may require longer observing periods due to the erratic behaviour of the ionosphere. If the observations are carried out at night when the ionosphere is more stable observing times may be reduced.

9.7.3 Reoccupation

This technique is regarded as a third form of static surveying or as a pseudo-kinematic procedure. It is based on repeating the survey after a time gap of one or two hours in order to make use of the change in receiver/satellite geometry to resolve the integer ambiguities.

The master receiver is once again positioned over a known point, whilst the roving receiver visits the unknown points for a few minutes only. After one or two hours, the roving receiver returns to the first unknown point and repeats the survey. There is no need to track the satellites whilst moving from point to point. This technique therefore makes use of the first few epochs of data and the last few epochs that reflect the relative change in receiver/satellite geometry and so permit the ambiguities and coordinate differences to be resolved.

Using dual frequency data gives values comparable with the rapid static technique. Due to the method of changing the receiver/satellite geometry, it can be used with cheaper single-frequency receivers (although extended measuring times are recommended) and a poorer satellite constellation.

9.7.4 Kinematic positioning

The major problem with static GPS is the time required for an appreciable change in the satellite/receiver geometry so that the initial integer ambiguities can be resolved. However, if the integer ambiguities could be resolved (and constrained in a least squares solution) prior to the survey, then a single epoch of data would be sufficient to obtain relative positioning to sub-centimetre accuracy. This concept is the basis of kinematic surveying. It can be seen from this that, if the integer ambiguities are resolved initially and quickly, it will be necessary to keep lock on these satellites whilst moving the antenna.

9.7.4.1 Resolving the integer ambiguities

The process of resolving the integer ambiguities is called initialization and may be done by setting up both receivers at each end of a baseline whose coordinates are accurately known. In subsequent data processing, the coordinates are held fixed and the integers determined using only a single epoch of data. These values are now held fixed throughout the duration of the survey and coordinates estimated every epoch, provided there are no cycle slips.

The initial baseline may comprise points of known coordinates fixed from previous surveys, by static GPS just prior to the survey, or by transformation of points in a local coordinate system to WGS84.

An alternative approach is called the 'antenna swap' method. An antenna is placed at each end of a short base (5–10 m) and observations taken over a short period of time. The antennae are interchanged, lock maintained, and observations continued. This results in a big change in the relative receiver/satellite geometry and, consequently, rapid determination of the integers. The antennae are returned to their original position prior to the surveys.

It should be realized that the whole survey will be invalidated if a cycle slip occurs. Thus, reconnaissance of the area is still of vital importance, otherwise reinitialization will be necessary. A further help in this matter is to observe to many more satellites than the minimum four required.

9.7.4.2 Traditional kinematic surveying

Assuming the ambiguities have been resolved, a master receiver is positioned over a reference point of known coordinates and the roving receiver commences its movement along the route required. As the movement is continuous, the observations take place at pre-set time intervals, often less than 1 s. Lock must be maintained to at least four satellites, or re-established when lost. In this technique it is the trajectory of the rover that is surveyed and points are surveyed by time rather than position, hence linear detail such as roads, rivers, railways, etc., can be rapidly surveyed. Antennae can be fitted to fast moving vehicles, or even bicycles, which can be driven along a road or path to obtain a three-dimensional profile.

9.7.4.3 Stop and go surveying

As the name implies, this kinematic technique is practically identical to the previous one, only in this case the rover stops at the point of detail or position required (*Figure 9.17*). The accent is therefore on individual points rather than a trajectory route, so data is collected only at those points. Lock must be maintained, though the data observed when moving is not necessarily recorded. This method is ideal for engineering and topographic surveys.

9.7.4.4 Real-time kinematic (RTK)

The previous methods that have been described all require post-processing of the results. However, RTK provides the relative position to be determined instantaneously as the roving receiver occupies a position.

Fig. 9.17 *The roving receiver*

The essential difference is in the use of mobile data communication to transmit information from the reference point to the rover. Indeed, it is this procedure that imposes limitation due to the range over which the communication system can operate.

The system requires two receivers with only one positioned over a known point. A static period of initialization will be required before work can commence. If lock to the minimum number of satellites is lost then a further period of initialization will be required. Therefore the surveyor should try to avoid working close to major obstructions to line of sight to the satellites. The base station transmits code and carrier phase data to the rover. On-board data processing resolves the ambiguities and solves for a change in coordinate differences between roving and reference receivers. This technique can use single or dual frequency receivers. Loss of lock can be regained by remaining static for a short time over a point of known position.

The great advantage of this method for the engineering surveyor is that GPS can be used for setting-out on site. The setting-out coordinates can be entered into the roving receiver, and a graphical output indicates the direction and distance through which the pole-antenna must be moved. The positions of the point to be set-out and the antenna are shown. When the two coincide, the centre of the antenna is over the setting-out position.

9.7.4.5 Real-time kinematic on the fly

Throughout all the procedures described above, it can be seen that initialization or reinitialization can only be done with the receiver static. This may be impossible in high accuracy hydrographic surveys or road profiling in a moving vehicle. Ambiguity Resolution On the Fly (AROF) enables ambiguity resolution whilst the receiver is moving. The techniques require L1 and L2 observations from at least five satellites with a good geometry between the observer and the satellites. There are also restrictions on the minimum periods of data collection and the presence of cycle slips. Both these limitations restrict this method of surveying to GPS friendly environments. Depending on the level of ionospheric disturbances, the maximum range from the reference receiver to the rover for resolving ambiguities whilst the rover is in motion is about 10 km, with an achievable accuracy of 10–20 mm.

For both RTK and AROF the quality of data link between the reference and roving receiver is important. Usually this is by radio but it may also be by mobile phone. When using a radio the following issues should be considered:

- In many countries the maximum power of the radio is legally restricted and/or a radio licence may be required. This in turn restricts the practical range between the receivers.
- The radio will work best where there is a direct line of sight between the receivers. This may not always be possible to achieve so for best performance the reference receiver should always be sited with the radio antenna as high as possible.
- Cable lengths should be kept as short as possible to reduce signal losses.

9.8 ERROR SOURCES

The final position of the survey station is influenced by:

(1) The error in the range measurement.
(2) The satellite–receiver geometry.
(3) The accuracy of the satellite ephemerides.
(4) The effect of atmospheric refraction.
(5) The multipath environment at the receivers.
(6) The quantity and quality of satellite data collected.
(7) The connections between the observed GPS network and the existing control.
(8) The processing software used.

It is necessary, therefore, to consider the various errors involved, some of which have already been mentioned.

The majority of the error sources are eliminated or substantially reduced if relative positioning is used, rather than single-point positioning. This fact is common to many aspects of surveying. For instance, in simple levelling it is generally the difference in elevation between points that is required. Therefore, if we consider two points A and B whose heights H_A and H_B were obtained by measurements from the same point which had an error δH in its assumed height then:

$$\Delta H_{AB} = (H_A + \delta H) - (H_B + \delta H)$$

with the result that δH is differenced out and difference in height is much more accurate than the individual heights. Thus, if the absolute position of point A fixed by GPS was 10 m in error, the same would apply to point B but their relative position would be comparatively error free. Then knowing the actual coordinates of A and applying the computed difference in position between A and B would bring B to its correct relative position. This should be borne in mind when examining the error sources in GPS.

9.8.1 Receiver clock error

This error is a result of the receiver clock not being compatible and in the same time system as the satellite clock. Range measurement (pseudo-range) is thus contaminated. As the speed of light is approximately 300 000 km s^{-1}, then an error of 0.01 s results in a range error of about 3000 km. As already shown this error can be evaluated using four satellites or cancelled using differencing software.

9.8.2 Satellite clock error

Excessive temperature variations in the satellite may result in variation of the satellite clock from GPS time. Careful monitoring allows the amount of drift to be assessed and included in the broadcast message and therefore substantially eliminated if the user is using the same data. Differential procedures eliminate this error.

9.8.3 Satellite ephemeris error

Orbital data has already been discussed in detail with reference to Broadcast and Precise Ephemeris. Errors are still present and influence baseline measurement in the ratio:

$$\delta b/b = \delta S/R_s$$

where

δb = error in baseline b

δS = error in satellite orbit

R_s = satellite range

The specification for GPS is that orbital errors should not exceed 3.7 m, but this is not always possible. Error in the range of 10–20 m may occur using the Broadcast Ephemeris. Thus, for an orbital error of 10 m on a 10 km baseline with a range of 20 000 km, the error in the baseline would be 5 mm. This error is substantially eliminated over moderate length baselines using differential techniques.

9.8.4 Atmospheric refraction

Atmospheric refraction error is usually dealt with in two parts, namely ionospheric and tropospheric. The effects are substantially reduced by DGPS compared with single-point positioning. Comparable figures are:

	Single point	*Differential*
Ionosphere	15–20 m	2–3 m
Troposphere	3–4 m	1 m

If refraction were identical at each end of a small baseline, then the total effect would be cancelled when using DGPS.

The ionosphere is the region of the atmosphere from 50 to 1000 km in altitude in which ultraviolet radiation has ionized a fraction of the gas molecules, thereby releasing free electrons. GPS signals are slowed down and refracted from their true path when passing through this medium. The effect on range measurement can vary from 5 to 150 m. As the ionospheric effect is frequency dependent, carrier wave measurement using the different L-band frequencies, i.e. L1 and L2, can be processed to eliminate substantially the ionospheric error. If the ionosphere were of constant thickness and electron density, then DGPS, as already mentioned, would eliminate its effect. This, unfortunately, is not so and residual effects remain. Positional and temporal variation in the electron density makes complete elimination over longer baselines impossible and may require complex software modelling.

Ionospheric refraction effects vary with the solar cycle. The cycle lasts about 11 years and was at a minimum in 2005 and will be at a maximum in 2011 therefore users will find increasing difficulty with GPS work during that period. After 2011 of course the situation will improve until about 2016 when the cycle will start again. Observations by night, when the observer is not facing the sun, are less affected.

The troposphere is even more variable than the ionosphere and is not frequency dependent. However, being closer to the ground, the temperature, pressure and humidity at the receiver can be easily measured and the integrated effect along the line of sight through the troposphere to the satellite modelled. If conditions are identical at each end of a baseline, then its effect is completely eliminated. Over longer baselines measurements can be taken and used in an appropriate model to reduce the error by as much as 95%.

9.8.5 Multipath error

This is caused by the satellite signals being reflected off local surfaces such as buildings or the ground, resulting in a time delay and consequently a greater apparent range to the satellite (*Figure 9.18*). At the frequencies used in GPS they can be of considerable amplitude, due to the fact that the antenna must be designed to track several satellites and cannot therefore be more directional.

Manufacturers have developed a number of solutions to minimize the effects of multipath. A ground plane fitted to the antenna will shield the antenna from signals reflected from the ground but may induce surface waves reflected from buildings onto the ground plane which are then fed to the antenna. This problem can be solved if a choke ring antenna is used. A series of concentric circular troughs each at a depth of a quarter of the GPS signal wavelength are placed on the ground plane. These troughs have high impedance to surface waves. However, choke ring antennae are much larger, heavier and more costly than ordinary antenna and they cannot attenuate signals arriving from above. Phased array antennae are designed to have high gain from the direct path to the satellite but attenuate signals from other directions. However, different satellites have different multipath geometry so there will be several simultaneous direction patterns required and of course the antenna must respond to changing satellite positions. Such an antenna will be very expensive. There are a number of software solutions called multipath mitigation approaches. These include:

- Those associated with narrow pre-correlation of the signal (narrow correlator).
- Using only the first part of the signal arriving at the antenna before it is corrupted by the reflected part (leading edge).
- Modelling the shape of the total signal including the direct signal and its reflected components (correlation function shape).
- Modifying the C/A code waveform (modified waveforms).

However, these methods only mitigate against the effects of multipath, they do not remove it entirely. Whatever hardware and software the surveyor is using it is prudent to take care with the placing survey stations to ensure that they are clear of any reflecting surfaces. In built-up areas, multipath may present insurmountable problems. Multipath errors cannot be eliminated by differential techniques because the multipath environment is specific to each site.

9.8.6 Dilution of Precision

The quality of an instantaneous positioning solution is largely a function of two parameters; the quality of an individual pseudo-range to a satellite and the geometry of the relative position of observer and satellites. This situation has its parallels in trilateration in terms of the quality of the EDM in use and the positions of the known and unknown control stations.

Dilution of Precision (DOP) is the concept whereby the problem of geometry is analysed and a numerical parameter is derived to describe the quality of the geometric relationship between the user's equipment and the chosen satellites.

Depending upon the user's application, there are several interrelated DOP statistics; Geometrical, Position, Horizontal, Vertical and Time Dilutions of Precision. In each case the DOP statistics are the amplification factor, which, when multiplied by the pseudo-range measurement error, give the error of the computed position or time, etc. These statistics are only a function of the effect of satellite and user geometry. The reader is probably a surveyor and therefore most interested in three-dimensional position and hence the PDOP of the satellites. A sailor navigating on the relatively flat sea would be more interested in HDOP, since they will already have sufficient information about their height. If a particular user were using the GPS for time transfer, their interest would only be in TDOP, since they will not need to know where they are.

We will consider the surveyor's PDOP, but similar arguments can apply for any of the DOPs. PDOP is a dimensionless number which will vary from about 1.6 in the best possible geometrical configurations

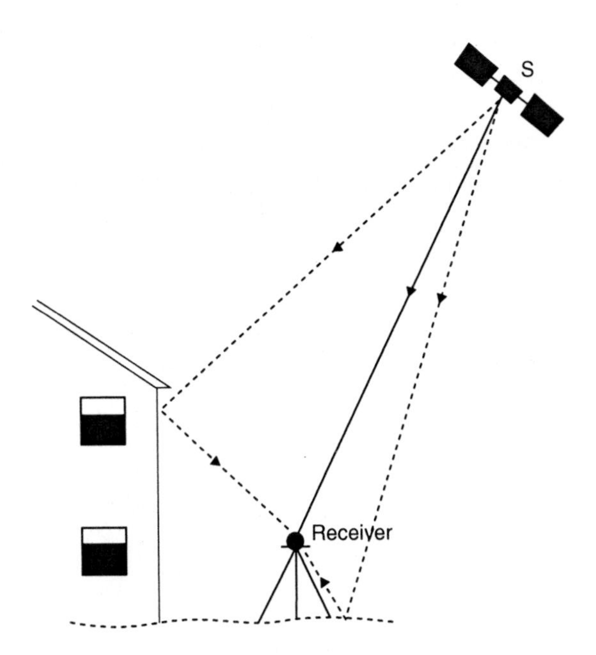

Fig. 9.18 *Multipath effect*

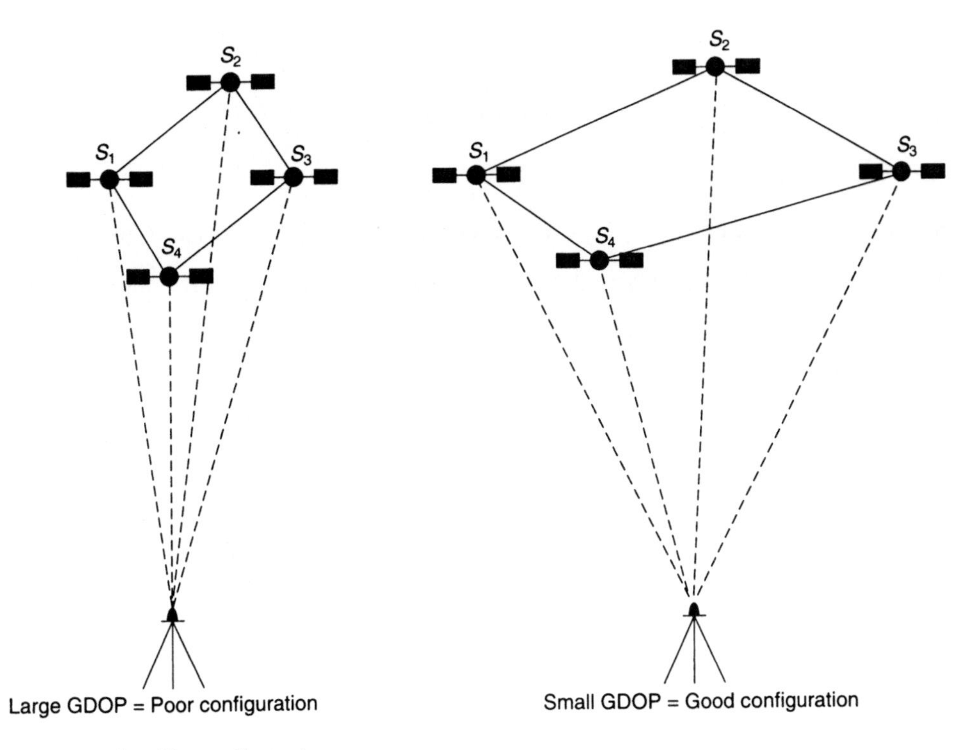

Fig. 9.19 *Satellite configuration*

with four satellites, to much larger numbers when satellites are badly positioned for a particular user. For example, if the user was able to measure pseudo-range from a user set to a satellite to 5 metres and at a particular instant the satellite and user geometry was such that the PDOP was 2.2, then the user error of position would be $2.2 \times 5 = 11$ m.

PDOP and all the other DOPs are:

- Independent of the coordinate system employed, both in terms of scale (unit of distance) and orientation.
- A means of user selection of the best satellites from those that are visible.
- The amplification factor of pseudo-range measurement error into the user error due to the effect of satellite geometry.

The basic equations that relate the position of the user and the satellites are derived from considerations of Pythagoras' theorem applied in three dimensions as in *equation (9.12)* and rearranged here.

$$[(X_1 - X_p)^2 + (Y_1 - Y_p)^2 + (Z_1 - Z_p)^2]^{\frac{1}{2}} + T = R_1$$

$$[(X_2 - X_p)^2 + (Y_2 - Y_p)^2 + (Z_2 - Z_p)^2]^{\frac{1}{2}} + T = R_2$$

$$[(X_3 - X_p)^2 + (Y_3 - Y_p)^2 + (Z_3 - Z_p)^2]^{\frac{1}{2}} + T = R_3$$

$$[(X_4 - X_p)^2 + (Y_4 - Y_p)^2 + (Z_4 - Z_p)^2]^{\frac{1}{2}} + T = R_4$$

where X_i, Y_i, Z_i = the coordinates of satellites ($i = 1$ to 4) (known)
X_p, Y_p, Z_p = the coordinates required for point P (unknown)
R_i = the measured ranges to the satellites (measured)
T = the clock bias times the speed of light (unknown)

These are four non-linear equations in four unknowns.

$$[(X_i - X_p)^2 + (Y_i - Y_p)^2 + (Z_i - Z_p)^2]^{\frac{1}{2}} + T = R_i \quad (i = 1 \text{ to } 4)$$

$$\text{or} \quad R_i - T - [(X_i - X_p)^2 + (Y_i - Y_p)^2 + (Z_i - Z_p)^2]^{\frac{1}{2}} = 0 \qquad (9.14)$$

Linearizing *equation (9.14)* as the first part of a Taylor series and evaluating the terms leads to:

$$(X_o - X_i)/(R_{ic} - T_o)\,\delta X + (Y_o - Y_i)/(R_{ic} - T_o)\,\delta Y + (Z_o - Z_i)/(R_{ic} - T_o)\,\delta Z + \delta T = \delta R_i \qquad (9.15)$$

where X_i, Y_i, Z_i = the coordinates of satellites from the broadcast ephemeris (known)
X_o, Y_o, Z_o = the provisional values of the coordinates of point P (defined)
T_o = the provisional value the clock bias times the speed of light (defined)
R_{ic} = the pseudo-range to the ith satellite computed from the known satellite coordinates, the provisional ground coordinates of the user's position and provisional value of the clock offset times the speed of light
$\delta X, \delta Y, \delta Z, \delta T$ = the corrections to the provisional values of X, Y, Z and T to get their true values
δR_i = the difference between the observed and the computed measurements of pseudo-range. $\delta R_i = R_i - R_{ic}$, i.e. observed – computed pseudo-range.

In matrix form, *equation (9.15)* applied to the measurements from four satellites may be expressed as:

$$\begin{bmatrix} a_{11} & a_{12} & a_{13} & 1 \\ a_{21} & a_{22} & a_{23} & 1 \\ a_{31} & a_{32} & a_{33} & 1 \\ a_{41} & a_{42} & a_{43} & 1 \end{bmatrix} \begin{bmatrix} \delta x \\ \delta y \\ \delta z \\ \delta T \end{bmatrix} = \begin{bmatrix} \delta R_1 \\ \delta R_2 \\ \delta R_3 \\ \delta R_4 \end{bmatrix}$$

where a_{ij} is the direction cosine of the angle between the range to the ith satellite and jth coordinate axis, since, for example:

$$a_{23} = \frac{z_o - z_2}{R_{2c} - T_o} = \frac{z_o - z_2}{\{(x_o - x_2)^2 + (y_o - y_2)^2 + (z_o - z_2)^2\}^{\frac{1}{2}}} = \frac{z \text{ component of distance to satellite}}{\text{distance to satellite}}$$

This matrix equation can be expressed in the form:

$$\mathbf{Ax} = \mathbf{r}$$

where $\mathbf{A}$ is the direction cosine matrix
$\mathbf{x}$ is the vector of unknown corrections
$\mathbf{r}$ is the vector of 'observed-computed' pseudo-ranges
and so $\mathbf{x} = \mathbf{A}^{-1}\mathbf{r}$

Since the equation is now linear, we can use it to express the relationship between the errors in the pseudo-range measurements and the user quantities. Then using the Gauss propagation of error equation:

$$\boldsymbol{\sigma}_{(\mathbf{x})} = \mathbf{A}^{-1}\boldsymbol{\sigma}_{(\mathbf{r})}\mathbf{A}^{-T}$$

where $\boldsymbol{\sigma}_{(\mathbf{x})}$ is the variance-covariance matrix of the parameters and
$\boldsymbol{\sigma}_{(\mathbf{r})}$ is the variance-covariance matrix of the observations

If there is no correlation between measurements to different satellites and if the standard errors of measurements to each satellite are the same, then $\boldsymbol{\sigma}_{(\mathbf{r})}$ may be written as:

$$\boldsymbol{\sigma}_{(\mathbf{r})} = \sigma_u \begin{bmatrix} 1 & 0 & 0 & 0 \\ 0 & 1 & 0 & 0 \\ 0 & 0 & 1 & 0 \\ 0 & 0 & 0 & 1 \end{bmatrix}$$

where σ_u is the standard error of a measured pseudo-range. Therefore

$$\boldsymbol{\sigma}_{(\mathbf{x})} = \mathbf{A}^{-1}\mathbf{A}^{-T}\sigma_u = (\mathbf{A}^T\mathbf{A})^{-1}\sigma_u$$

A number of useful statistics may be derived from the $(\mathbf{A}^T\mathbf{A})^{-1}$ term, such as:

$$\text{GDOP} = (\text{trace}(\mathbf{A}^T\mathbf{A})^{-1})^{\frac{1}{2}}$$

where the trace of a matrix is the sum of the terms on the leading diagonal. If σ_x^2, σ_y^2, σ_z^2 and σ_T^2 are the variances of user position and time then:

$$\text{GDOP} = (\sigma_x^2 + \sigma_y^2 + \sigma_z^2 + \sigma_T^2)^{\frac{1}{2}}$$

$$\text{PDOP} = (\sigma_x^2 + \sigma_y^2 + \sigma_z^2)^{\frac{1}{2}}$$

$$\text{TDOP} = \sigma_T$$

So, for example, the error in three-dimensional position is $\sigma_u \times$ PDOP.

If the ends of the unit vector from the user's equipment to each satellite all lie in the same plane, for example the plane at right angles to the z axis, *Figure 9.20*, then $a_{13} = a_{23} = a_{33} = a_{43} = a_3$ and the direction cosine matrix becomes:

$$\mathbf{A} = \begin{bmatrix} a_{11} & a_{21} & a_3 & 1 \\ a_{21} & a_{22} & a_3 & 1 \\ a_{31} & a_{23} & a_3 & 1 \\ a_{41} & a_{24} & a_3 & 1 \end{bmatrix}$$

The matrix is singular since one column is a multiple of another, the third and fourth columns, and so a solution is not possible. The matrix will also be singular if there is any direction such that the angle

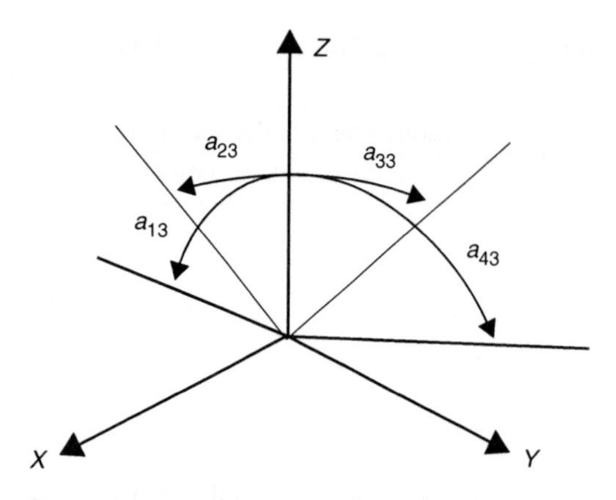

Fig. 9.20 *Satellite directions*

between it and the directions to the satellites are all the same. This can be difficult to spot in practice but may explain why, even though there are many satellites in view and they are well spread about the sky, that the DOP statistics are very large. This situation can be visualized if the observer can 'draw' a circle in the sky connecting all the satellites. In practice, if the satellite and user geometry approaches this situation the DOP statistics will become very large and the quality of the solution will degrade rapidly.

9.9 GPS SURVEY PLANNING

There are many factors that need to be considered when planning survey work. Some will be familiar from conventional surveying, e.g. with total stations, but there are some special factors that apply only to GPS surveying. This section covers the factors to be considered in planning and executing the fieldwork. It does not cover the computing processes that will vary with different manufacturers' software.

For positioning with a single receiver, i.e. absolute positioning, there is little to consider other than to ensure that there is a clear view of the sky, there are no nearby buildings that may cause multipath effects and no nearby radio/microwave transmitters that may interfere with the signal. The instantaneous solution is good to about 10–20 m. If the solution is averaged over $\frac{1}{2}$–2 hours then the solution may be as good as 5 m in plan and 10 m in height.

For relative positioning, i.e. baseline measurements, then there are several extra factors to be taken into account. The shorter the baseline the more precisely it will be determined. In relative positioning it is assumed that the effects of the ionosphere and troposphere are the same for both ends of the line. The shorter the line the truer this will be. It is usually better to measure two baselines of 5 km and add the results than to measure a single 10 km baseline. This idea can be applied to the observation of networks. It is better to observe short lines from a small number of reference stations than all lines from one central reference station.

The *independent check* is one of the fundamental principles of survey and it must also be applied to GPS surveying. This can be achieved by independently measuring baselines between stations. Alternatively treat the survey as a traverse and close all figures, then adjust as a network. If the observer re-observes a baseline at different times this will ensure that satellite geometry, multipath effects and tropospheric and ionospheric refraction effects are different. The ionospheric effect is less by night than during the day so it will be possible to resolve the integer ambiguities for longer lines at night.

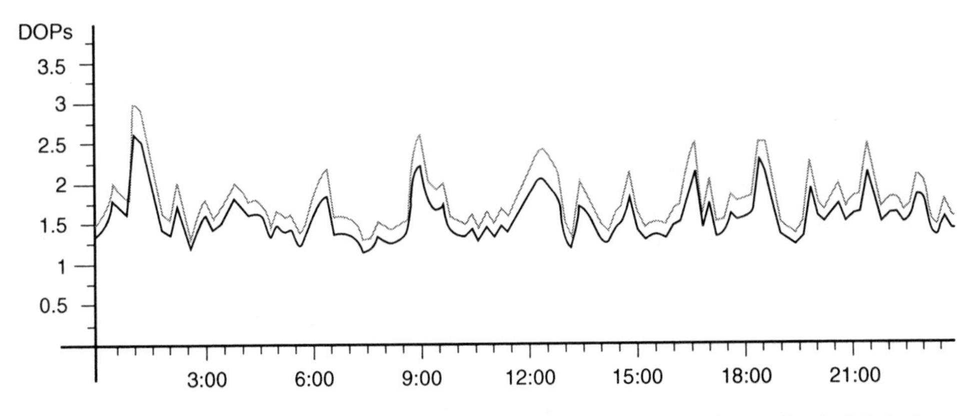

Fig. 9.21 *Predicted DOP values at Nottingham, UK, for 10 April 2007. Upper line is GDOP, lower line is PDOP*

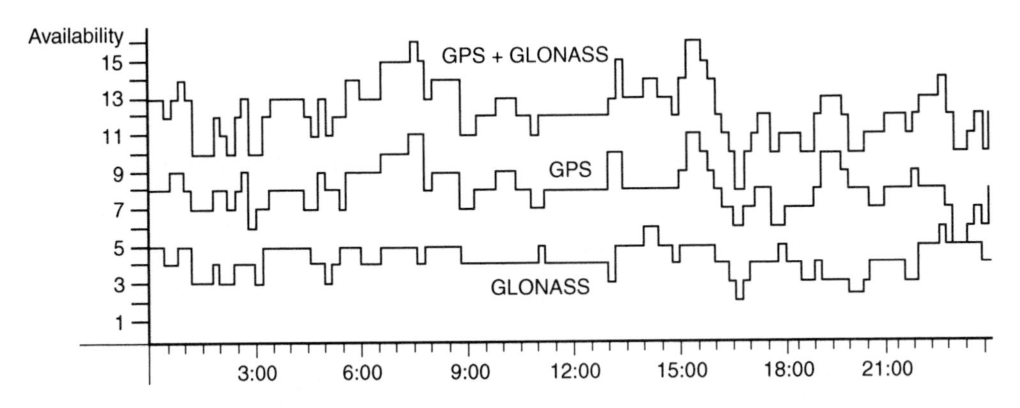

Fig. 9.22 *Predicted satellite availability at Nottingham, UK, for 10 April 2007*

Satellite coverage during the day varies in terms of the number of satellites available and the GDOP. Use mission planning software to select the best times when to observe: *Figure 9.21* shows Predicted DOP values at Nottingham, UK, for 10 April 2007 with a 10° horizon cut-off angle. Ensure that the there are at least four satellites available when measuring baselines and preferably five or six when operating in any form of kinematic mode. *Figure 9.22* shows, for the same place and time, the availability of all satellites (top line), GPS satellites (middle line), and GLONASS satellites (bottom line).

Figure 9.23 shows a sky plot for a 24-hour period. The centre is in the observer's zenith. Satellites marked 'G' are GPS and 'R' indicates GLONASS. Note the large *hole* in the sky surrounding the north celestial pole, shown as a small circle. This exists because the satellite orbits' inclinations are considerably less than 90° so no satellite's ground track goes anywhere near the poles. When planning kinematic surveys it is usually better to approach an obstruction from the south, as the obstruction will then block fewer satellites. Note the manufacturer's recommendations for DOP limitations for baselines and kinematic work. *Figure 9.24* shows a sky plot for a 1-hour period. The satellite number appears at the beginning of the selected 1-hour period.

Satellite coverage must be common to both receivers. To ensure that the rover receiver is observing the same satellites as the reference receiver, place the reference receiver where there are ***no*** obstructions. It does not matter that the reference receiver is not at a survey point that is needed; its function is merely to act as a reference receiver. Ensure also that the reference receiver is at a secure location and will not be interfered with or stolen.

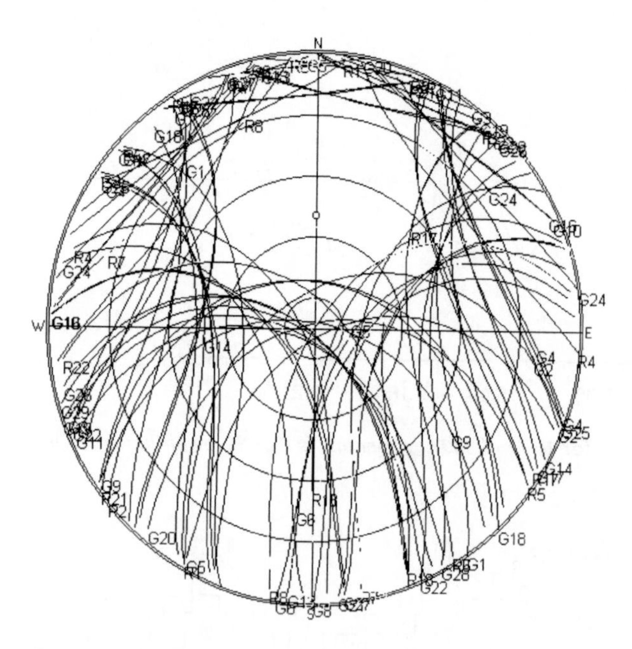

Fig. 9.23 *Predicted 24-hour sky plot at Nottingham, UK, for 10 April 2007 (Topcon Pinnacle software)*

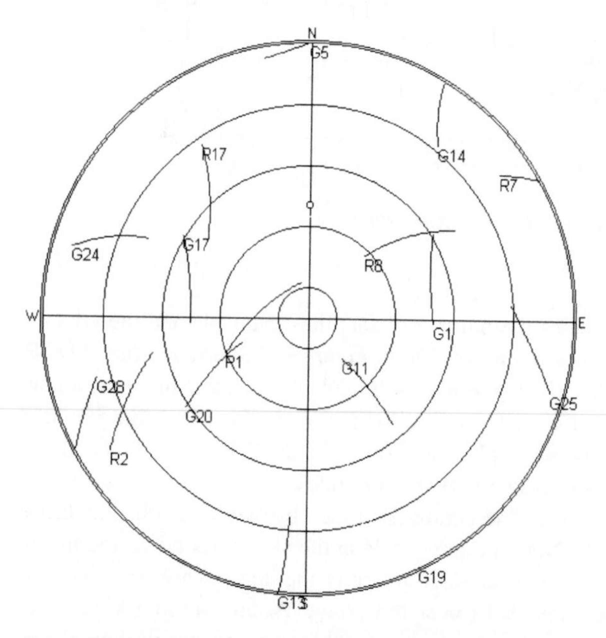

Fig. 9.24 *Predicted 1-hour sky plot at Nottingham, UK, for 11:00–12:00 on 10 April 2007 (Topcon Pinnacle software)*

The GPS results will be in a unique WGS84 style of datum. It is likely that the surveyor will require coordinates in the local datum. At least three, and preferably more, stations in the local datum must be observed so that the transformation parameters can be computed and applied. The stations with known local coordinates should surround the project area. If there are more than three stations one could be in the middle of the project area. When establishing new control stations ensure that they have a good view of the sky. The control should be well monumented as further work will be related to it.

From a business point of view time is money. Consider the number of points, accuracy required, time at the station and travelling times between stations in assessing the time required for the whole project. Do not try to cut observing times short to increase the number points. It is better to remain 20% longer at each station and thereby ensure that the ambiguities *will* be resolved to avoid having to repeat observations of some of the baselines. Be cautious when working in an urban environment because multipath is likely to corrupt the results. GPS is not always the best surveying technique, conventional observations with a total station may be more accurate and more reliable. For control surveys consider using the precise ephemeris rather than the broadcast ephemeris especially if the project area is large. Make sure when observing a network that all baselines are connected at both ends to other baselines in the network. This will ensure that there is reliability for all the observations and hence all the computed points. Check all the instrumentation before starting a major project by undertaking a simple local network and fully computing it. In selecting the GPS equipment for the task it is usually better to use all equipment from one manufacturer. If that is not possible avoid mixing antenna types. Use a daily routine that includes downloading data at the end of the day and putting all batteries on charge overnight.

When collecting detail points by any non-static methods ensure that a suitable precision value is set for individual detail points. Even when working with RTK it is worth recording all the raw data so that if problems are discovered later it may be possible to recompute the survey without the need to revisit the field. If positions are successfully recorded using RTK then it will not be necessary to post-process the data. If there is a loss of lock then pick up at least one previously observed point to ensure that the receiver has correctly reinitialized. Visit extra points of detail that are already known if GPS derived positions are to be added to an existing map or point list for an area.

The better the DOP, the better the final solution is likely to be. Avoid periods when there is rapidly changing DOP, especially when it is getting large. If there are local obstructions that may affect satellite signals use prediction software to check the DOP values with the visibility restrictions taken into account.

Sites for GPS reference stations should be visited before work starts, to ensure that there are no microwave, telecommunications, radio or TV transmitters in the area. There should be no buildings or other reflecting surfaces that could cause multipath and there must be no significant obstructions at the reference receiver. The reference receiver site must be secure if the instrument is to be left unguarded. Also ensure that there is an adequate power supply and sufficient memory in the data recorder and that the antenna is oriented according to the manufacture's recommendations.

GPS equipment is expensive. Consider whether it is better to buy the equipment outright, hire the equipment when it is required and operate it yourself or call upon the services of a specialist GPS survey company. Which you choose will depend upon the current level of GPS expertise in your company, how often you undertake tasks suitable for GPS and whether you already own your GPS equipment.

Although the field procedures for GPS surveying are not complex it is easy to make mistakes that invalidate the recorded data. The most common problems associated with GPS work are battery failure during operation, damage to cables and connections, failure to record antenna height and misnaming stations. Field operators should receive adequate training and supervised experience. They should also be knowledgeable about survey techniques so that they understand what they are doing. Processing GPS data with the manufacturer's software is more complex, especially if there are problems with the data. Knowledge of datums, projections and adjustments is essential and therefore a qualified surveyor should undertake the task. Manufacturers may provide training as part of their aftersales service.

An excellent publication *Guidelines for the use of GPS in Surveying and Mapping*, which is an RICS guidance note and covers all aspects of practical GPS work, is available from RICS Books.

9.10 TRANSFORMATION BETWEEN REFERENCE SYSTEMS

As all geodetic systems are theoretically parallel, it would appear that the transformation of the Cartesian coordinates of a point in one system (WGS84) to that in another system (OSGB36), for instance, would simply involve a three-dimensional translation of the origin of one coordinate system to the other, i.e. ΔX, ΔY, ΔZ (*Figure 9.25*). However, due to observational errors, the orientation of the coordinate axis of both systems may not be parallel and must therefore be made so by rotations θ_x, θ_y, θ_z about the X-, Y-, Z-axis. The size and shape of the reference ellipsoid is not relevant when working in three-dimensional Cartesian coordinates, hence six parameters should provide the transformation necessary.

However, it is usual to include a seventh parameter S which allows the scale of the axes to vary between the two coordinate systems.

A clockwise rotation about the X-axis (θ_x) has been shown in *Chapter 8, Section 8.6* to be:

$$\begin{bmatrix} X \\ Y \\ Z \end{bmatrix}_{\theta_x} = \begin{bmatrix} 1 & 0 & 0 \\ 0 & \cos\theta_x & -\sin\theta_x \\ 0 & \sin\theta_x & \cos\theta_x \end{bmatrix} \begin{bmatrix} X \\ Y \\ Z \end{bmatrix}_{\text{WGS84}} = R_{\theta_x} x_{\text{WGS84}} \tag{9.16}$$

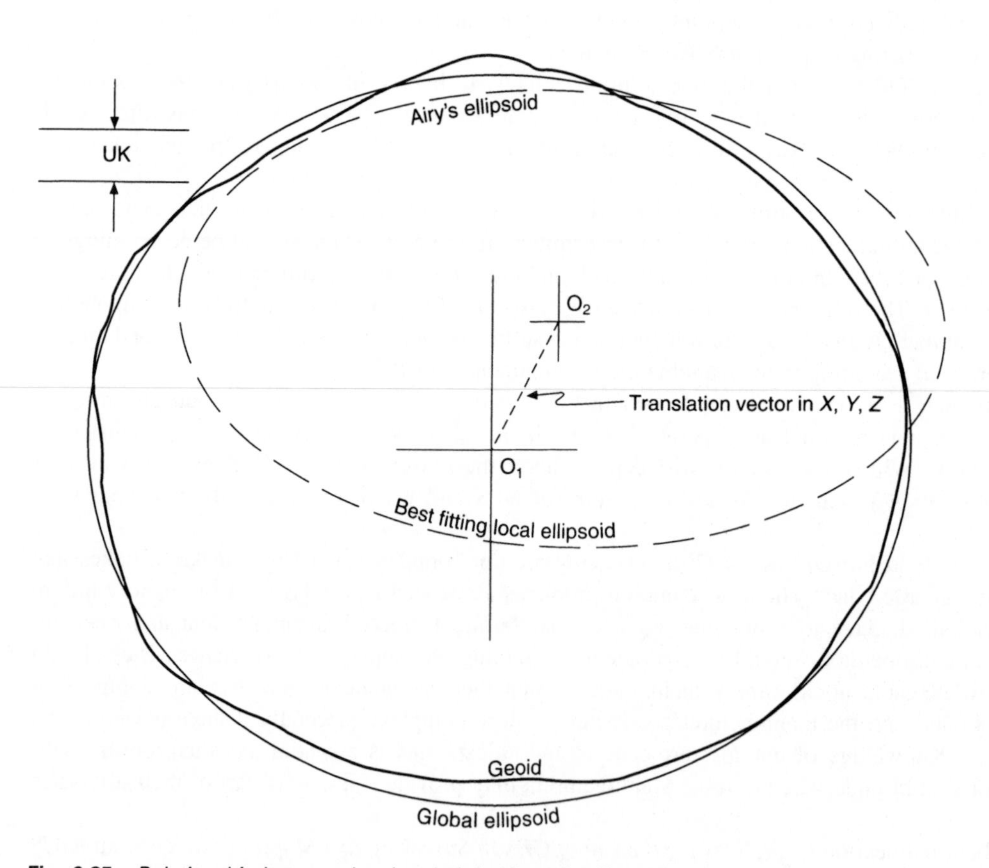

Fig. 9.25 *Relationship between local and global best-fit ellipsoids*

Similarly, rotation about the Y-axis (θ_y) will give:

$$\begin{bmatrix} X \\ Y \\ Z \end{bmatrix}_{\theta_{xy}} = \begin{bmatrix} \cos\theta_y & 0 & -\sin\theta_y \\ 0 & 1 & 0 \\ \sin\theta_y & 0 & \cos\theta_y \end{bmatrix} \begin{bmatrix} X \\ Y \\ Z \end{bmatrix}_{\theta_x} = R_{\theta_x} R_{\theta_y} x_{\text{WGS84}} \tag{9.17}$$

Finally, rotation about the Z-axis (θ_z) gives:

$$\begin{bmatrix} X \\ Y \\ Z \end{bmatrix}_{\theta_{xyz}} = \begin{bmatrix} \cos\theta_z & \sin\theta_z & 0 \\ -\sin\theta_z & \cos\theta_z & 0 \\ 0 & 0 & 1 \end{bmatrix} \begin{bmatrix} X \\ Y \\ Z \end{bmatrix}_{\theta_{xy}} = R_{\theta_x} R_{\theta_y} R_{\theta_z} x_{\text{WGS84}} \tag{9.18}$$

where R are the rotation matrices. Combining the rotations, including the translation to the origin and applying the scale factor S, gives:

$$\begin{bmatrix} X \\ Y \\ Z \end{bmatrix} = \begin{bmatrix} \Delta X \\ \Delta Y \\ \Delta Z \end{bmatrix} + (1+S) \begin{bmatrix} r_{11} & r_{12} & r_{13} \\ r_{21} & r_{22} & r_{23} \\ r_{31} & r_{32} & r_{33} \end{bmatrix} \begin{bmatrix} X \\ Y \\ Z \end{bmatrix}_{\text{WGS84}} \tag{9.19}$$

where $r_{11} = \cos\theta_y \cos\theta_z$

$r_{12} = \cos\theta_x \sin\theta_z + \sin\theta_x \sin\theta_y \cos\theta_z$

$r_{13} = \sin\theta_x \sin\theta_z - \cos\theta_x \sin\theta_y \cos\theta_z$

$r_{21} = -\cos\theta_y \sin\theta_z$

$r_{22} = \cos\theta_x \cos\theta_z - \sin\theta_x \sin\theta_y \cos\theta_z$

$r_{23} = \sin\theta_x \cos\theta_z + \cos\theta_x \sin\theta_y \sin\theta_z$

$r_{31} = \sin\theta_y$

$r_{32} = -\sin\theta_x \cos\theta_y$

$r_{33} = \cos\theta_x \cos\theta_y$

In matrix form:

$$x = \Delta + S \cdot R \cdot \overline{x} \tag{9.20}$$

where x = vector of 3-D Cartesian coordinates in a local system

Δ = the 3-D shift vector of the origins (Δx, Δy, Δz)

S = scale factor

R = the orthogonal matrix of the three successive rotation matrices, θ_x, θ_y, θ_z

$\overline{x}$ = vector of 3-D Cartesian coordinates in the GPS satellite system, WGS84

This seven-parameter transformation is called the Helmert transformation and, whilst mathematically rigorous, is entirely dependent on the rigour of the parameters used. In practice, these parameters are computed from the inclusion of at least three *known* points in the networks. However, the coordinates of the known points will contain observational error which, in turn, will affect the transformation parameters. Thus, the output coordinates will contain error. It follows that any transformation in the 'real' world can only be a 'best estimate' and should contain a statistical measure of its quality.

As all geodetic systems are theoretically aligned with the International Reference Pole (IRP) and International Reference Meridian (IRM), which is approximately Greenwich, the rotation parameters are usually less than 5 seconds of arc. In which case, $\cos\theta \approx 1$ and $\sin\theta \approx \theta$ rads making the Helmert transformation linear, as follows:

$$\begin{bmatrix} X \\ Y \\ Z \end{bmatrix} = \begin{bmatrix} \Delta X \\ \Delta Y \\ \Delta Z \end{bmatrix} + \begin{bmatrix} 1+S & -\theta_z & \theta_y \\ \theta_z & 1+S & -\theta_x \\ -\theta_y & \theta_x & 1+S \end{bmatrix} \begin{bmatrix} X \\ Y \\ Z \end{bmatrix}_{\text{WGS84}} \tag{9.21}$$

The rotations θ are in radians, the scale factor S is unitless and, as it is usually expressed in ppm, must be divided by a million.

When solving for the transformation parameters from a minimum of three known points, the XYZ_{LOCAL} and XYZ_{WGS84} are known for each point. The difference in their values would give ΔX, ΔY and ΔZ, which would probably vary slightly for each point. Thus a least squares estimate is taken, the three points give nine observation equations from which the seven transformation parameters are obtained.

It is not always necessary to use a seven-parameter transformation. Five-parameter transformations are quite common, comprising three translations, a longitude rotation and scale change. For small areas involved in construction, small rotations can be described by translations (3), and including scale factors gives a four-parameter transformation. The linear equation *(9.21)* can still be used by simply setting the unused parameters to zero.

It is important to realize that the Helmert transformation is designed to transform between two datums and cannot consider the scale errors and distortions that exist throughout the Terrestrial Reference Framework of points that exist in most countries. For example, in Great Britain a single set of transformation parameters to relate WGS84 to OSGB36 would give errors in some parts of the country as high as 4 m. It should be noted that the Molodensky datum transformation (*Chapter 8, Section 8.6*) deals only with ellipsoidal coordinates (ϕ, λ, h), their translation of origin and changes in reference ellipsoid size and shape. Orientation of the ellipsoid axes is not catered for. However, the advantage of Molodensky is that it provides a single-stage procedure between data.

The next step in the transformation process is to convert the X, Y, Z Cartesian coordinates in a local system to corresponding ellipsoidal coordinates of latitude (ϕ), longitude (λ) and height above the local ellipsoid of reference (h) whose size and shape are defined by its semi-major axis a and eccentricity e. The coordinate axes of both systems are coincident and so the Cartesian to ellipsoidal conversion formulae can be used as given in *Chapter 8, equations (8.4)* to *(8.7)*, i.e.

$$\tan\lambda = Y/X \tag{9.22}$$

$$\tan\phi = (Z + e^2 v \sin\phi)/(X^2 + Y^2)^{\frac{1}{2}} \tag{9.23}$$

$$h = [X/(\cos\phi\cos\lambda)] - v \tag{9.24}$$

where $v = a/(1 - e^2\sin^2\phi)^{\frac{1}{2}}$

$$e = (a^2 - b^2)^{\frac{1}{2}}/a$$

An iterative solution is required in *equation (9.23)*, although a direct formula exists as shown in *Chapter 8, equation (8.9)*.

The final stage is the transformation from the ellipsoidal coordinates ϕ, λ and h to plane projection coordinates and height above mean sea level (MSL). In Great Britain this would constitute grid eastings and northings on the Transverse Mercator projection of Airy's Ellipsoid and height above MSL, as defined by continuous tidal observations from 1915 to 1921 at Newlyn in Cornwall, i.e. E, N and H.

An example of the basic transformation formula is shown in *Chapter 8, Section 8.7, equations (8.40)* to *(8.42)*. The Ordnance Survey offer the following approach:

N_0 = northing of true origin (−100 000 m)
E_0 = easting of true origin (400 000 m)
F_0 = scale factor of central meridian (0.999 601 2717)
ϕ_0 = latitude of true origin (49°N)
λ_0 = longitude of true origin (2°W)
a = semi-major axis (6 377 563.396 m)
b = semi-minor axis (6 356 256.910 m)
e^2 = eccentricity squared = $(a^2 - b^2)/a^2$

$$n = (a - b)/(a + b)$$
$$\nu = aF_0(1 - e^2 \sin^2 \phi)^{-\frac{1}{2}}$$
$$\rho = aF_0(1 - e^2)(1 - e^2 \sin^2 \phi)^{-\frac{3}{2}}$$
$$\eta^2 = \frac{\nu}{\rho} - 1$$

$$M = bF_0 \left[\begin{array}{l} \left(1 + n + \frac{5}{4}n^2 + \frac{5}{4}n^3\right)(\phi - \phi_0) - \left(3n + 3n^2 + \frac{21}{8}n^3\right)\sin(\phi - \phi_0)\cos(\phi + \phi_0) \\ + \left(\frac{15}{8}n^2 + \frac{15}{8}n^3\right)\sin(2(\phi - \phi_0))\cos(2(\phi - \phi_0)) \\ - \frac{35}{24}n^2 \sin(3(\phi - \phi_0))\cos(3(\phi - \phi_0)) \end{array} \right] \quad (9.25)$$

$$\mathrm{I} = M + N_0$$

$$\mathrm{II} = \frac{\nu}{2}\sin\phi\cos\phi$$

$$\mathrm{III} = \frac{\nu}{24}\sin\phi\cos^3\phi(5 - \tan^2\phi + 9\eta^2)$$

$$\mathrm{IIIA} = \frac{\nu}{720}\sin\phi\cos^5\phi(61 - 58\tan^2\phi + \tan^4\phi)$$

$$\mathrm{IV} = \nu\cos\phi$$

$$\mathrm{V} = \frac{\nu}{6}\cos^3\phi\left(\frac{\nu}{\rho} - \tan^2\phi\right)$$

$$\mathrm{VI} = \frac{\nu}{120}\cos^5\phi(5 - 18\tan^2\phi + \tan^4\phi + 14\eta^2 - 58(\tan^2\phi)\eta^2)$$

then: $N = \mathrm{I} + \mathrm{II}(\lambda - \lambda_0)^2 + \mathrm{III}(\lambda - \lambda_0)^4 + \mathrm{IIIA}(\lambda - \lambda_0)^6$

$E = E_0 + \mathrm{IV}(\lambda - \lambda_0) + \mathrm{V}(\lambda - \lambda_0)^3 + \mathrm{VI}(\lambda - \lambda_0)^5$

The computation must be done with sufficient precision with the angles in radians.

As shown in *Chapter 8, Section 8.3.5, Figure 8.10*, it can be seen that the ellipsoidal height h is the linear distance, measured along the normal, from the ellipsoid to a point above or below the ellipsoid. These heights are not relative to gravity and so cannot indicate flow in water, for instance.

The orthometric height H of a point is the linear distance from that point, measured along the gravity vector, to the equipotential surface of the Earth that approximates to MSL. The difference between the two heights is called the geoid-ellipsoid separation, or the geoid height and is denoted by N, thus:

$$h = N + H \quad (9.26)$$

In relatively small areas, generally encountered in construction, GPS heights can be obtained on several benchmarks surrounding and within the area, provided the benchmarks are known to be stable. The difference between the two sets of values gives the value of N at each benchmark. The geoid can be regarded as a plane between these points or a contouring program could be used, thus providing corrections for further GPS heighting within the area. Accuracies relative to tertiary levelling are achievable.

It is worth noting that if, within a small area, *height differences* are required and the geoid–ellipsoid separations are constant, then the value of N can be ignored and ellipsoidal heights only used.

The importance of orthometric heights (H) relative to ellipsoidal heights (h) cannot be over-emphasized. As *Figure 9.26* clearly illustrates, the orthometric heights, relative to the geoid, indicate that the lake is

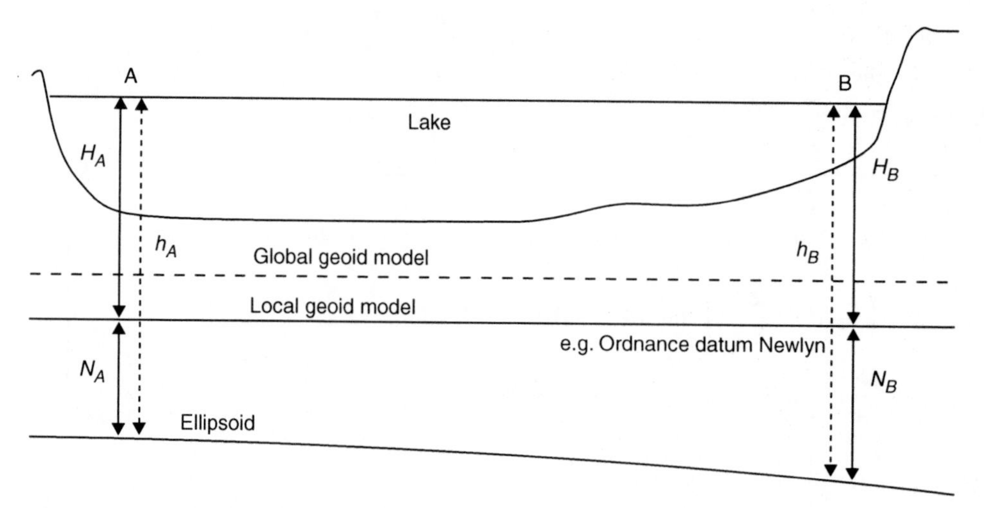

Fig. 9.26 *Orthometric (H) and ellipsoidal (h) heights. The local geoid model through Newlyn lies approximately 800 mm below the global geoid model*

level, i.e. $H_A = H_B$. However, the ellipsoidal heights would indicate water flowing from B to A, i.e. $h_B > h_A$. As the engineer generally requires difference in height (ΔH), then from GPS ellipsoidal heights the following would be needed:

$$\Delta H_{AB} = \Delta h_{AB} - \Delta N_{AB} \tag{9.27}$$

An approximate method of obtaining N on a small site has already been mentioned. On a national basis an accurate national geoid model is required.

In Great Britain the complex, irregular surface of the geoid was established by a combination of astro-geodetic, gravimetric and satellite observations (OSGM02) to such an accuracy that precise GPS heights can be transformed to orthometric with the same accuracy achievable as with precise spirit levelling. However, over distances greater than 5 km, standard GPS heights (accurate to 20–50 mm), when transformed using OSGM02 (accurate to 3 mm$(k)^{1/2}$, where k is the distance in kilometres between points), will produce relative orthometric values as good as those achieved by standard (tertiary) levelling. This means, in effect, that the National GPS Network of points can also be treated as benchmarks.

To summarize, the transformation process when using GPS is:

$(X, Y, Z)_{GPS}$	Using Keplerian elements and time parameters
$\downarrow$	
$(X, Y, Z)_{LOCAL}$	Using transformation parameters
$\downarrow$	
$(\phi, \lambda, h)_{LOCAL}$	Using elliposidal conversion formule
$\downarrow$	
$(E, N, H)_{LOCAL}$	Using projection parameters and geoid-ellipsoid separation

9.11 DATUMS

As already mentioned, the Broadcast Ephemeris is sufficiently accurate for relative positioning over a limited area. For instance, an error of 20 m in the satellite position would produce an error of only 10 mm in a 10 km baseline. However, to achieve an accuracy of 1 mm would require satellite positioning accurate to 2 m and so require the use of a Precise Ephemeris.

Whilst the Broadcast Ephemeris is computed from only five monitoring stations, a Precise is computed from many stations situated throughout the world. The IGS RAPID orbits (IGR) of the International GPS Geodynamic Service (IGS) are available just over one day after the event it describes. Precise Ephemeredes are available from a variety of government, commercial and academic sources including http://www.ngs.noaa.gov/GPS/GPS.html.

9.11.1 Global datums

Modern engineering surveying uses GPS in an increasing number of situations. Indeed it is often the primary method of survey. Global datums are established by assigning Cartesian coordinates to various positions throughout the world. Observational errors in these positions will obviously be reflected in the datum.

The WGS84 was established from the coordinate position of about 1600 points around the globe, fixed largely by TRANSIT satellite observations. At the present time its origin is geocentric (i.e. the centre of mass of the whole Earth) and its axes virtually coincide with the International Reference Pole and International Reference Meridian. Designed to best fit the global geoid as a whole means it does not fit many of the local ellipsoids in use by many countries. In Great Britain, for instance, it lies about 50 m below the geoid and slopes from east to west, resulting in the geoid–ellipsoid separation being 10 m greater in the west than in the east.

It is also worth noting that the axes are stationary with respect to the *average* motions of this dynamically changing Earth. For instance, tectonic plate movement causes continents to move relative to each other by about 10 cm per year. Local movements caused by tides, pressure weather systems, etc., can result in movement of several centimetres. The result is that the WGS84 datum appears to move relative to the various countries. In Great Britain, the latitudes and longitudes are changing at a rate of 2.5 cm per year in a north-easterly direction. In time, this effect will be noticeable in large-scale mapping.

It can be seen from the above statements that constant monitoring of the WGS84 system is necessary to maintain its validity. In 1997, 13 tracking stations situated throughout the globe had their positional accuracies redefined to an accuracy of better than 5 cm, thereby bringing the origin, orientation and scale of the system to within the accuracy of its theoretical specification. Another global datum almost identical to the WGS84 Reference System is the International Terrestrial Reference Frame (ITRF) produced by the International Earth Rotation Service (IERS) in Paris, France. The system was produced from the positional coordinates of over 500 stations throughout the world, fixed by a variety of geodetic space positioning techniques such as Satellite Laser Ranging (SLR), Very Long Baseline Interferometry (VLBI), Lunar Laser Ranging (LLR), Doppler Ranging Integrated on Satellite (DORIS) and GPS. Combined with the constant monitoring of Earth rotation, crustal plate movement and polar motion, the IERS have established a very precise terrestrial reference frame, the latest version of which is the ITRF2000. This TRF has been established by the civil GPS community, not the US military. It comprises a list of Cartesian coordinates (X, Y, Z), with the change in position (dX, dY, dZ) in metres per year for each station. The ITRF2000 is available as a SINEX format text file from the IERS website. Details are at http://www.iers.org/iers/publications/tn/tn31/. The ITRF is the most accurate global TRF and for all purposes is identical to the WGS84 TRF. A new ITRF2005 is in preparation which will be based on the time series of station positions and earth orientation parameters using observations from at least 1999–2005.

9.11.2 Local datums

Historically, the majority of local datums were made accessible to the user by means of a TRF of points coordinated by triangulation. These points gave horizontal position only and the triangulation point, monumented by pillars in the UK, were situated on hilltops. A vertical TRF of benchmarks, established by spirit levelling, required low-lying, easily traversed routes. Hence, there were two different but loosely connected systems.

A TRF established by GPS gives a single three-dimensional system of easily accessible points that can be transformed to give more accurate position in the local system. As WGS84 is continually changing position due to tectonic movement, the local system must be based on WGS84 at a certain time. Thus, we have local datums like the North American Datum 1983 and the European Terrestrial Reference System 1989 (ETRS89). In 1989 a high precision European Three-Dimensional Reference Frame (EUREF) was established by GPS observations on 93 stations throughout Europe (ETRF89). The datum used (ETRS89) was consistent with WGS84/1TRF2000 and extends into Great Britain, where it forms the datum for the Ordnance Survey National GPS Network.

The OS system will be briefly described here as it illustrates a representative model that will be of benefit to the everyday user of GPS. The National GPS network TRF comprises two types of GPS station consisting of:

- Active network: this is a primary network of about 60 continuously observing, permanent GPS receivers whose precise coordinates are known. Using a single dual frequency receiver and data downloaded from these stations, which are located within 100 km of any point in Britain, precise positioning can be achieved to accuracies of 10 mm. The data is available in RINEX format for the previous 30 days from http://gps.ordnancesurvey.co.uk/active.asp.
- Passive network: this is a secondary network of about 900 easily accessible stations at a density of 20–35 km. These stations can be used for control densification or in kinematic form using two receivers to obtain real-time positioning to accuracies of 50–100 mm. Their details can be found at http://gps.ordnancesurvey.co.uk/passive.asp.

The coordinates obtained by the user are ETRS89 and can be transformed to and from WGS84/1TRF2000 by a six-parameter transformation published by IERS on their internet site. Of more interest to local users would be the transformation from ETRS89 to OSGB36, which is the basic mapping system for the country. The establishment of the OSGB36 TRF by triangulation has resulted in variable scale changes throughout the framework, which renders the use of a single Helmert transformation unacceptable. The OS have therefore developed a 'rubber-sheet' transformation called OSTN02, which copes not only with a change of datum but also with the scale distortions and removes the need to compute local transformation parameters by the inclusion of at least three known points within the survey. The transformation may be done using the Ordnance Survey's 'Coordinate Converter' online at http://gps.ordnancesurvey.co.uk/convert.asp. The same transformation software is included in some GPS manufacturers' processing software. Using the National GPS Network in a static, post-processing mode, can produce horizontal accuracy of several millimetres. The same mode of operation using 'active' stations and one hour of data would give accuracies of about 20 mm. Heighting accuracies would typically be twice these values. However, it must be understood that these accuracies apply to position computed in ETRS89 datum. Transformation to the National Grid using OSTN02 would degrade the above accuracy, and positional errors in the region of 200 mm have been quoted. The final accuracy would, nevertheless, be greater than the OS base-map, where errors of 500 mm in the position of detail at the 1/1250 scale are the norm. Thus, transformation to the National Grid should only be used when integration to the OS base-map is required.

The evolution of GPS technology and future trends have been illustrated by a description of the OS National GPS Network and its application. Other countries are also establishing continuously operating GPS reference station networks. For instance, in Japan a network of about 1000 stations is deployed throughout the country with a spacing of about 20–30 km (GEONET). In Germany the Satellite Positioning Service of the German State Survey (SAPOS) has established a network of about 250 permanently observing reference stations. Three products are available, at a cost, to the user. A DGPS service, for a one-off fee, allows the user with a single receiver to obtain 0.5–3 m accuracy using only a single GPS frequency. This is suitable for GIS, vehicle navigation and most maritime applications. A 'High Precision Real Time' positioning service using carrier phase measurements can give accuracies of 10–20 mm in plan and 20–60 mm in height in real time. The fee for use is charged by the minute and the signals are transmitted by VHF radio and GSM phone. The applications for this service include engineering surveying.

The 'High Precision' positioning service using dual frequency carrier phase observations delivers RINEX files to the user via email or internet for post-processing for the highest accuracy solutions; 1–10 mm is claimed. The fee is also by the minute. Further details are at http://www.sapos.de. Other countries are developing similar systems.

Many innovative techniques have been used in GPS surveying to resolve the problem of ambiguity resolution (AR), reduce observation time and increase accuracy. They involve the development of sophisticated AR algorithms to reduce the time for static surveying, or carry out 'on-the-fly' carrier phase AR whilst the antenna is continually moving. Other techniques such as 'stop and go' and 'rapid static' techniques were also developed. They all, however, require the use of at least two, multi-channel, dual frequency receivers, and there are limitations on the length of baseline observed (<15 km). However, downloading the data from only one of the 'active stations' in the above network results in the following benefit:

(1) For many engineering surveying operations the use of only one single-frequency receiver, thereby reducing costs. However, this may result in antenna phase centre variation due to the use of different antennae.
(2) Baseline lengths greatly extended for rapid static and kinematic procedures.
(3) 'On-the-fly' techniques can be used with ambiguity resolution algorithms from a single epoch.

Active network data is not available until two or three hours after it has been recorded.

9.12 VIRTUAL REFERENCE STATIONS AND NETWORKED GPS

Precise relative positioning requires simultaneous observations with two receivers. One of those receivers can be at a permanent installation such as with the SAPOS system or the OS's active network. One drawback of the OS's active network is that, other than by coincidence, the user will be at a considerable distance from the nearest active station and so unmodelled differences in ionospheric and tropospheric effects will limit possible accuracy. The idea of the virtual reference station (VRS) is to create a reference station by computer at about a metre from the user. A network of several continuously operated reference stations which transmit raw data to a central server which in turn creates a database is established. The surveyor who wants to do RTK survey connects to the central server by phone or modem. The position of the rover is sent to the central server which creates corrections for ionospheric, tropospheric and ephemeris errors for the virtual reference station and transmits them to the rover. This facility is available in the SAPOS system described above and on a number of other networks around the world, some of which are mentioned at http://www.trimble.com/vrsinstallations.shtml. Position accuracy of 1–2 cm and height accuracy of 2–4 cm is claimed.

In the UK an alternative approach is being launched in conjunction with the Ordnance Survey. Leica's SmartNet is a subscription service in which RTK and DGPS corrections are transmitted to the user through GPRS or GSM phones. The user receives data from an augmented OS active network and through a reference station network technique, refraction and orbit errors are significantly reduced.

9.13 GLONASS

Just as GPS was designed to replace the TRANSIT system, so the Russian GLObal NAvigation Satellite System (GLONASS) replaced their earlier system, TSIKADA. The space segment of GLONASS was originally planned to be 24 satellites, including three spares, in circular orbits. There were to be eight in each of three orbital planes which were to be inclined at 64.8° and with relative right ascensions of ascending nodes of 120°. The system was designed so that there would always be at least six satellites in view.

The semi-major axis of the orbits is approximately 25 400 km so that the orbital period is 11 h 16 m. This means that the satellites repeat their ground tracks every 17th orbit, which is every eight days. Therefore, unlike GPS, where the orbital period is 12 sidereal hours and the constellation repeats its position with respect to a stationary observer at the same time every sidereal day, GLONASS does not repeat daily for a stationary observer. The different pattern for GLONASS is presumed to avoid any small Earth gravitational resonance effect that might affect GPS.

The functions of the Control Segment are similar to those of GPS. One major difference between GPS and GLONASS is in the signal structure of GLONASS. Whereas all GPS satellites transmit on the same two frequencies but have different C/A and P codes, the GLONASS satellites transmit the same codes but on different frequencies for each satellite. The pair of frequencies for each satellite is 1246.0 MHz + 7/16 n MHz (L2) and 1602.0 MHz + 9/16 n MHz (L1) where n, the number of the satellite, has an integer value from 1 to 24. A precise code with a bandwidth of 5.11 MHz is on both L1 and L2, whilst the coarser code of 0.511 MHz is on L1 only. Thus, the satellites provide the signals for pseudo-range and carrier phase measurement. Whereas the GPS uses the WGS84 datum and UTC time frame GLONASS uses the PZ90 datum and UTC(Russia) time frame. The Broadcast Ephemeris for GPS is in terms of Keplerian elements but for GLONASS it is in terms of position, velocity and acceleration in Earth-centred, Earth-fixed coordinates. One advantage that GLONASS does have is that of much better protection against cross-correlation interference between satellite signals.

The overall performance of GLONASS has been variable. At one stage the full constellation was in orbit but because of technical failures only a handful of satellites were still working by the millennium. There is now a programme of development with 15 satellites operational and the likelihood of further improvements in the future. Receivers that can use the code and carrier signals from both GPS and GLONASS are therefore able to work in more restricted sites because they have a greater choice of satellites above the horizon. The most recent information can be found at http://www.glonass-ianc.rsa.ru.

9.14 GPS SYSTEM FUTURE

The GPS Block I satellites were only launched for system testing. The 24 operational or Block II satellites were due for launch from October 1986 to December 1988 but the space shuttle Challenger accident seriously delayed the programme. However, the launch programme came back on a new schedule and the full constellation of 24 satellites was operational in 1996. As the Block II satellites reach the end of their design life they are being replaced by Block IIR (replenishment) satellites. The Block IIR satellites have the same signals in space as the Block II satellites. However, they autonomously navigate, that is they create their own navigation message and maintain full accuracy for at least 180 days. They have improved reliability and integrity of broadcast signal. There is additional radiation hardening of the satellite and cross-link ranging between satellites. This allows much greater flexibility in control of the system and some of the Control Segment's functions have been transferred to the satellites themselves. There are two atomic clocks on at all times. The satellite has a larger fuel capacity.

In 1996, a major policy statement from the White House was issued. In effect, it stated that at some date between 2000 and 2006, Selective Availability (SA) would be removed and therefore there would be full access to both frequencies. In May 2000, SA was removed.

In January 1999, the White House through the Office of the Vice President announced that a third civilian signal would be located at 1176.45 MHz. This is the L5 signal proposed for future Block IIF satellites. It was also announced that a second civilian signal would be located at 1227.6 MHz with the current military signal and that would be implemented on satellites scheduled for launch from 2003. This is the L2C signal on future IIR-M satellites.

In May/June 2001, the GPS Joint Programme Office (JPO) announced details of the proposed developments. Details are at http://www.navcen.uscg.gov/gps/modernization/default.htm.

In March 1997 Dr James Schlesinger, a former Secretary of Defense, summed up the situation with respect to GPS. 'The nation's reliance on GPS has become an issue of national security in its broadest sense that goes beyond merely national defense.' Originally GPS was conceived as a Military Support System for War, but President Reagan, in 1983, decided that there should be civil access to the system. Therefore, it is now a critical dual-use US national asset. On the other hand, it has become more essential to military forces than had ever been previously imagined but it has also become indispensable to civil and commercial users as well. However, it is still funded, managed, and operated by the US Department of Defense.

The civil and military communities have different needs for the twenty-first century's upgraded GPS system. The military require greater security of the system, a quicker fix from a cold start, more powerful signals, their own signals for fast acquisition and better security codes. These will be achieved with a new M code on the L1 and L2 frequencies. The civil market requires that the GPS signal is accurate, fully available, has full coverage, measurable integrity, redundant signals, more power and that SA remains switched off. A second civil signal is required for simpler ionospheric corrections and a third civil signal for high accuracy, real-time applications. Those signals which are required for 'safety of life' applications require spectrum protection. This will be achieved with the new L2C and L5 signals and codes.

The L2C code on the L2 signal and the M code on both the L1 and L2 frequencies will be implemented on the Block IIR-M satellites and the L5 signal will be implemented on the Block IIF satellites.

The L2C code is not just to be a replica of the existing L1 C/A code. In 2002 there were estimated to be about 50 000 dual-frequency receivers in use and to be worth about $1bn not counting spares, software and associated communications. Unlike leisure industry receivers, these receivers are used to *add value to society*, for example to monitor earthquakes, volcanoes, continental drift, and weather. They add value for cadastral and land survey purposes, for guidance and control of mining, construction and agricultural operations and are used in land and offshore oil and mineral exploration and marine surveys.

A primary objective of the introduction of the new L2C code is to remove the need for (semi-) codeless tracking now used for L2 measurements. A 'C/A' type code on the L2 signal would achieve this objective. Nevertheless, the L2C code has no data on one of its two codes and gives a 3 dB better performance. Therefore, there can be carrier phase measurements without the need for ambiguity resolution.

For single-frequency users, the objective is to make L2 valuable for single-frequency GPS applications. On the L1 C/A code, a strong GPS signal interferes with weak GPS signals but on the L2C, the cross-correlation will be more than 250 times better. Therefore it will be possible to read the message with a signal that is barely detectable and so it is probable that this will be the signal of choice for emergency phone applications, called E911 in the US, and for positioning inside buildings.

Receiver technology has improved since the 1970s when the simple L1 C/A code was developed. Today's technology can accept a more complex and therefore useful signal. The first Block IIR-M satellite was launched in 2005 and it is expected that Block IIF satellite launches will start in 2008. By 2011–12 there should be 28 satellites broadcasting the L2C signal and 18 broadcasting the L5 signal.

It is expected that the benefits available from the civil signals will be many. The L1 frequency will still have the lowest ionospheric refraction error. The L5 will have the highest power and will receive some protection because it is in the Aeronautical Radio Navigation Service band. The L2C will give the best cross-correlation, threshold tracking and data recovery performance. It will have a better message structure than the L1 C/A code and will be fully available years before L5. With its lower code clock rate it will be better than L5 for many consumer applications because power consumption relates to code clock rate and so could present less power drain problems for wristwatch and cell telephone navigation applications. As chip size relates to thermal dissipation a slower clock aids miniaturization.

GPS III is a GPS modernization programme that is designed to take GPS forward for the first third of the twenty-first century. The programme is still very much at the definition stage. The Statement of Objectives at http://www.navcen.uscg.gov/gps/modernization/SOO.doc states that the objective is to create a new architecture for enhanced position, velocity, and timing signals, and related services to meet the needs of the next generation of GPS users. The system is to be a best value solution and the security infrastructure is to provide user access to and protection of the entire system.

Additional mission capabilities may include the military application of 'Blue Force Tracking', i.e. friendly force tracking and search and rescue missions.

9.15 WIDE AREA AUGMENTATION SYSTEM (WAAS)

The US Federal Aviation Authority (FAA) is developing the Wide Area Augmentation System (WAAS). The WAAS is 'safety-critical' navigation system that improves the accuracy, integrity and availability of GPS so that GPS can be a primary means of navigation for aircraft en route travel and non-precision approaches. The WAAS improves real time civil accuracy of GPS to about 7 metres in three dimensions and improves system availability and provides integrity information for the GPS constellation.

GPS signals are received by about 25 Wide area ground Reference Stations (WRSs). They determine the errors in the GPS signals and relay the data to the Wide area Master Station (WMS). The WMS computes correction information and assesses the integrity of the system. Correction messages are then uplinked to the geostationary communication satellites. The aircraft listens to the GPS signals and the message broadcast on the GPS L1 frequency by geostationary communication satellites. Benefits claimed for the WAAS are that it enables GPS to be the primary means of navigation including take-off, approach and landing. Aircraft routes are more direct and there will not be the need for aircraft to be routed according to the constraints of existing ground-based navigation systems. There will be increased capacity in airspace without increased risk and there will be reduced and simplified navigation equipment on aircraft.

As far as the engineering surveyor is concerned WAAS satellites may be useful as additional satellites for pseudo-range measurements.

9.16 EUROPEAN GEOSTATIONARY NAVIGATION OVERLAY SERVICE (EGNOS)

EGNOS is the European counterpart of the US WAAS. It is a joint project of the European Space Agency (ESA), the European Commission (EC) and Eurocontrol, the European Organization for the Safety of Air Navigation. EGNOS will augment the GPS and GLONASS systems to make them suitable for safety critical applications.

The EGNOS space segment is the set of navigation transponders on three geostationary satellites as well as the signals from the GPS and GLONASS constellation. There are transponders on the satellites but they have no signal generators on board. Instead, each transponder uses a 'bent pipe' system where the satellite transmits signals uplinked from the ground, which is where all signal processing takes place.

On the ground there are about 34 Ranging and Integrity Monitoring Stations (RIMS), four master control centres, including three reserves, and six uplink stations also including three reserves. The RIMS measure the positions of the EGNOS satellites and compare their own positions from positions calculated from the GPS and GLONASS satellite signals. The RIMS then send this data to the master control centres. The master control centres determine the accuracy of the GPS and GLONASS signals and determine the position inaccuracies due to the ionosphere. They send signals to the uplink stations that send them to the EGNOS satellites. The EGNOS satellites then transmit the correction signals to GPS/GLONASS users who are equipped with an EGNOS enabled receiver. To enable a receiver the EGNOS PRN numbers are PRN 120, 124 and 126. Operational stability for the system is expected in late 2006. The latest position may be checked at http://esamultimedia.esa.int/docs/egnos/estb/newsletter.htm.

In Japan, they are developing a similar Multi-Functional Transport Satellite (MTSAT) system and China is understood to be testing its own two satellites, Satellite Navigation Augmentation System (SNAS). India too has proposals for their own system.

9.17 GALILEO

Galileo is a joint initiative of the European Commission (EC) and the European Space Agency (ESA). When fully operational the system will have 30 satellites, in three planes, placed equally around the equator with semi-major axes of 30 000 km and an inclination of 56°. The satellites are to be for 'search and rescue' as well as navigation. The reason for this European initiative is that satellite navigation users in Europe can only get positions from US GPS or Russian GLONASS satellites. The military operators of both systems will give no guarantee that they will maintain an uninterrupted service.

The first experimental Galileo satellite has been launched and the remaining satellites should be in orbit by 2008.

With GPS, GLONASS and Galileo it will be possible to get high accuracy positions in the urban canyons of high-rise cities because there could be as many as 90 satellites available. But the strongest argument, the one most likely to get the system 'off the ground', is the business case for Galileo namely the European aerospace, manufacturing and service provider business opportunities. The system cost is currently estimated to be 3200 million euros with a total expected return to Europe over 20 years of 90 000 million euros.

The European strategy appears to be that we recognize our reliance on satellite systems for safety-critical and commercial applications and that EGNOS will compensate for the shortcomings of GPS and GLONASS for civil access to the necessary accuracy and integrity but Europe's civil satellite infrastructure is not under Europe's control.

There will be two Galileo Control Centres (GCC) in Europe controlling the satellites and carrying out the navigation mission management. Twenty Galileo Sensor Stations (GSS) in a global network will send data to the GCCs through a communications network where integrity information and time signal synchronization of all satellite and ground station clocks will be computed. Data exchange between the GCCs and satellites will be through 15 uplink stations.

The services to be delivered have been identified as Open Access for most non-specialist users, a Safety-of-Life capability with appropriate accuracy and integrity, Commercial services and a more secure Public Regulated Navigation service. Galileo will be interoperable with GPS, GLONASS and EGNOS and this has influenced the design of the satellite signals and reference standards. The Open, Commercial and Safety-of-Life services will have two open navigation signals with separate frequencies for ionospheric correction. Data is to be added to one ranging code and the other ranging code is to be used for more precise and robust navigation measurements. In principle, no ranging code is to be encrypted so that the signals may support Open and Safety-of-Life services. Service providers may encrypt one of the data-less ranging codes for commercial applications, such as for a Universal Mobile Telecommunications System–GSM. However, whether this will be compatible with a Safety-of-Life service is questionable.

There will be a third navigation signal that will enable three carrier phase ambiguity resolution which might be an encrypted ranging code for commercial service exploitation. There will be integrity data for safety of life applications and there may be encrypted data for commercial service exploitation. Commercial data might include GPS signal corrections or location maps.

The Public Regulated Navigation service will also have two navigation signals with encrypted codes and data. The signals will be on separate frequencies for ionospheric correction and on separate frequencies from Open services so that there may be local jamming of Open signals without affecting the Public Regulated Navigation Services.

The applications for the Public Regulated Navigation Services are seen as those for national security, police, emergency services, critical energy, transport and telecommunications applications and economic and industrial activities of European strategic interest such as banking services. It is likely there will be a restricted distribution of receivers managed by national governments.

A 'Local component', driven by user and market needs, public regulation, and finance may have differential corrections for single frequency users leading to an accuracy of better than one metre. Three carrier phase ambiguity resolution will improve that to better than 0.1 metre. The Local component may report integrity with a time to alarm of one second. Such a system would enhance mobile phone networks, e.g. E911 in urban canyons and have indoor applications. Pseudolites, local ground-based transmitters that broadcast a Galileo type signal, could be included for increasing the availability of the Galileo service in a defined local area.

The search and rescue transponder on each Galileo satellite will be coordinated by the Cospas-Sarsat International Satellite System for search and rescue operations. It will have to be compatible with the Global Maritime Distress and Safety System (GMDSS) and be able to respond to signals from Emergency Position Indicating Radio Beacons (EPIRB). An added facility may be for an acknowledgement signal to be received by the vessel or aircraft in distress through the broadcast Galileo signals.

The table below shows the planned signals and frequencies.

Signal	*Central frequency* (MHz)	*Service*
E5a/*GPS L5*	1176.45	Open, Safety-of-Life
E5b	1196.91–1207.14	Open, Safety-of-Life, Commercial
GPS L2	*1227.60*	
E6	1278.75	Commercial, Public Regulated
E2	1561	Open, Safety-of-Life, Public Regulated
GPS L1	*1575.42*	
E1	1590	Open, Safety-of-Life, Public Regulated

The satellite launch programme has similar time scales to that of EGNOS. Further information and updates may be found at http://www.esa.int/esaNA/galileo.html.

9.18 THE FUTURE FOR SURVEYORS AND ENGINEERS

The future for GPS, WAAS and EGNOS and Galileo looks promising. The future for GLONASS is harder to assess but could also match expectations. In the worst case, if only the GPS modernization programme and Galileo go ahead, then there will still be many more Galileo and GPS satellites and all of those GPS satellites launched after 2005 will have significantly enhanced capabilities. However, it will take some time before there are sufficient satellites with enhanced capabilities for the new signals and codes to be universally useful. How the equipment manufacturers will exploit the new resources is also hard to assess, but it is certain that they will and we should expect our speed of survey production and capability in difficult environments to improve significantly. It is less likely that prices for geodetic survey receivers will fall significantly because of the overall limited volume of the market.

When Galileo is operational then the possibilities for combined GPS, GLONASS and Galileo receivers mean that the enhanced coverage will permit satellite survey in some previously very difficult environments. However, geodetic survey receivers using GPS, Galileo and GLONASS will most likely command a high premium.

Undoubtedly there will be many different and varied applications of satellite technology and the survey market will only have a very small share of the total applications market. It seems likely that equipment manufacturers select signals based on:

- Purpose of product
- Chipset availability
- Satellites and signals currently available and their limitations
- Satellites and signals expected to be available in the next 2–4 years
- Manufacturing cost
- Expected market
- Competitor activity

However, it is likely that manufacturers will also be making their current products more attractive by future proofing them, i.e. making existing products capable of using new signals and codes when they become available.

9.19 APPLICATIONS

The previous pages have already indicated the basic application of GPS in engineering surveying such as in control surveys, topographic surveys and setting-out on site. Indeed, much three-dimensional spatial data normally captured using conventional surveying techniques with a total station can be captured by GPS, even during the night, provided sufficient satellites are visible. However, GPS is not the solution for all survey data capture problems. Where GPS works, i.e. in an open skies environment, it works very well. However, once the skies become obstructed as they often are on the construction site, then conventional survey techniques may be more appropriate. GPS does not, at present, work well or at all close to buildings, indoors, underground or underwater.

On a national scale, horizontal and, to a certain extent, vertical control, used for mapping purposes and previously established by classical triangulation with all its built-in scale error, are being replaced by three-dimensional GPS networks. In relation to Great Britain and the Ordnance Survey, this has been dealt with in previous pages. The great advantage of this to the engineering surveyor is that, when using GPS at the local level, there is no requirement for coordinate transformations. Kinematic methods can be used for rapid detailing, and Real-Time Kinematic (RTK) for setting-out.

Whilst the above constitutes the main area of interest for the engineering surveyor, other applications will be briefly mentioned to illustrate the power and versatility of GPS.

9.19.1 Machine guidance and control

Earth moving is required to shape the ground prior to construction taking place. In this context machine guidance is concerned with guiding the operator to move the blade or dozer, scraper, excavator or grader using GPS as the reference. On the other hand, machine control is when this information is used directly to control the machine's hydraulics to automate the blade.

Initially a GPS base station is established on site. The base station collects the code and phase data and this is broadcast by radio throughout the whole site. Rover units working in RTK mode get corrections from the base station. Shortly after, position at the machine is determined with 20–30 mm precision at a data rate of at least 0.1 second with an even smaller latency. Usually there are two GPS antennae attached to each end of the blade so that the grade profile may be determined precisely.

The machine operator needs to be guided in terms of what to do with his/her machine, therefore the GPS derived coordinates and site design components need to be transparent to the operator in the cab. The site DTM and project design are loaded into the control unit aboard the machine. As the task progresses and changes are made to the plan, these can be added either with updated physical media or by transferring the data by radio link. The operator may be guided on the task by light bars telling him/her whether to raise, lower or angle the machine's blade or there may be a visual display of the task so that the ground ahead can be compared with a design visualization.

With machine guidance there is no need for conventional setting-out because no profile or batter boards need to be established. The role of the surveyor is now more concerned with setting up the GPS system and ensuring that the correct coordinate systems are in use, that the GPS is functioning correctly as well as ensuring the design data is correctly formatted for the task in hand. As well as using the system to make the process of cut, fill and grading more efficient the GPS can be used to do a survey of the work achieved so far on a daily basis. This can be used to calculate payment for the task and to determine whether the specification for the work has been achieved.

The same design file can be used for the initial rough cuts and moves of mass of dirt as well as the finished grades. Where there is automatic control of the blade this helps with more accurate grading with fewer passes. When the GPS is interfaced to the machines' hydraulic valves there can be automatic control of elevation and cross-slope.

For large sites where parts are beyond the range of the base station radio repeater radios may extend operations up to about 20 km. GPS has its limitations in terms of the precision that can be achieved in real time. If the system is augmented by lasers then final profiles can be achieved down to a few millimetres.

Cut and fill to planes defined by conventional profiles and batter boards can easily be achieved. What is more difficult is to create a smooth surface defined by vertical profiles in more than one direction such as in landscaping, especially golf courses. With a DTM and GPS machine control this is more easily achieved.

GPS antennae mounted on masts attached to plant blades are subject to harsh shocks and vibrations so the antennae need to be firmly attached and robust. Antennae are expensive and therefore vulnerable to theft and damage when the site is not active or guarded. Using GPS means there will be fewer personnel such as surveyors on site and therefore the site becomes safer. Where there are areas of danger on the site which are invisible to the operator, such as pipelines or contaminated ground, they can be included in the data for the control unit and an audible warning given to the operator if the machine gets too close. See *Figures 9.27* and *9.28*.

9.19.2 Plate tectonics

Plate tectonics is centred on the theories of continental drift and is the most widely accepted model describing crustal movement. GPS is being used on a local and regional basis to measure three-dimensional movement. Locally, inter-station vectors across faults are being continually monitored to millimetre accuracy, whilst regionally, GPS networks have been established across continental plate boundaries. The information obtained adds greatly to the study of earthquake prediction, volcanoes and plate motion.

9.19.3 Precision farming

The concepts are similar to those used in machine guidance and control; however, DGPS levels of accuracy are often sufficient. The unit steers the agricultural machine along parallel, curved, or circular evenly spaced swaths taking account of the swath width.

Automated steering may be used to free the operator from steering the equipment until he/she comes to the end or the corner of the field. Yield monitoring may also be incorporated into the system. The system may also be used to vary the levels of seed or fertilizer distribution across the field to ensure there is a minimum of waste.

Fig. 9.27 *Site vision GPS' machine guidance by Trimble, showing two GPS antennae and the in-cab control system*

9.19.4 Geographical information systems (GIS)

GPS is the ideal tool for the collection of spatially related data because the speed at which the position element of the data may be captured and because the precision of the position element is likely to be acceptable for almost all applications that the engineering surveyor will be involved with. GPS, in any mode from stop-and-go to RTK, lends itself particularly well to the process of recording vector data such as pipeline routes, property boundaries and pavement edges. Because the data is in electronic form it is easily compiled into files for use within a GIS.

9.19.5 Navigation

GPS is now used in all aspects of navigation. GPS voice navigation systems are standard with many models of car. Simply typing the required destination into the unit results in a graphical display of the route, along with voiced directions. Similar systems are used by private boats as digital charts and by aeroplanes, whilst handheld receivers are now standard equipment for many walkers, and cyclists. GPS has many applications for sports monitoring and provides the spatial and time element of the data for sports performance analysis. GPS enabled wristwatches and mobile phones have been available for several years now.

Fig. 9.28 *GPS guided plant – courtesy of AGTEK Development*

GPS is used in fleet management for determining the vehicles' positions and status so that they can be transmitted to a central control, thereby permitting better management of the assets. At the individual vehicle level the driver can use GPS as an aid to destination and route location.

GPS is used by surveying ships for major offshore hydrographic surveys. Ocean-going liners use it for navigation purposes, whilst most harbours have a DGPS system to enable precise docking. Even recreational craft rely on it almost to the exclusion of traditional methods of navigation and pilotage.

At the present time, aircraft landing and navigation are controlled by a variety of disparate systems. GPS is well used and may eventually provide a single system for all aircraft operations. The uses to which GPS can be put are limited only by the imagination of the user. They can range from the complexities of measuring tectonic movement to the simplicity of spreading fertilizer in precision farming, and include many areas of scientific study, such as meteorology, oceanography, and geophysics. As satellite systems continue to develop the applications will continue to grow.

9.19.6 Deformation monitoring

The advantage of using GPS is that it enables near real-time monitoring for deformation if data is recorded and processed online. There are a number of such projects currently operational and are described at their respective websites. Examples include monitoring by Nottingham University of the Humber Bridge and a bridge in Nottingham for movement, http://www.gmat.unsw.edu.au/wang/jgps/v3n12/v3n12p28.pdf.

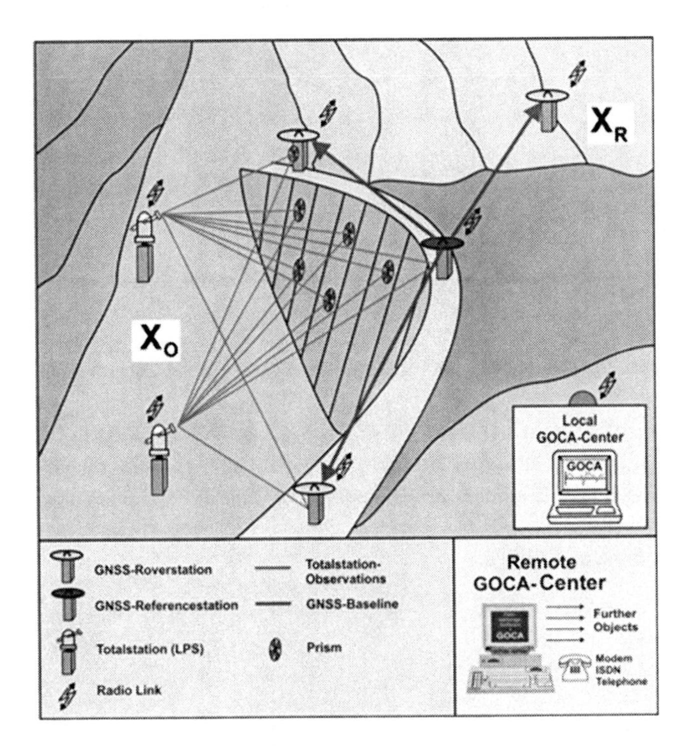

Fig. 9.29 *GOCA – GPS-based online Control and Alarm Systems*

Although GPS can give three-dimensional time-tagged positions at the rate of 10–20 Hz this means that oscillations of more than half this value cannot be easily detected. Since information about actual vibration is of use to engineers the output of other technologies such as accelerometers may be incorporated with that of GPS through mathematical techniques like Kalman filtering. Kalman filtering is a 'predictor–corrector' process that enables solutions of position and velocity of greater precision than those obtainable with either GPS or the accelerometers alone.

The GPS-based online Control and Alarm System (GOCA) developed at Fachhochschule Karlsruhe-University of Applied Sciences, Germany, can be used for the online monitoring of movements associated with, and preceding, natural disasters such as landslides, volcanoes and earthquakes. The system may be entirely GPS based or use total stations as well, as in *Figure 9.29*. The system can also be used to monitor safety-critical movement of buildings or dams. The online modelling of a classical deformation network leads to the computation and visualization of time series data of individual receivers. With filtering and analysis of the data it is possible to detect movements that are significant, but much smaller than the precision of individual GPS measurements. These movements may trigger an automatic alarm and give warning of an impending disaster. Details are at http://www.goca.info/.

10
Curves

In the geometric design of motorways, railways, pipelines, etc., the design and setting out of curves is an important aspect of the engineer's work.

The initial design is usually based on a series of straight sections whose positions are defined largely by the topography of the area. The intersections of pairs of straights are then connected by horizontal curves (see *Section 10.2*). In the vertical design, intersecting gradients are connected by curves in the vertical plane.

Curves can be listed under three main headings, as follows:

(1) Circular curves of constant radius.
(2) Transition curves of varying radius (spirals).
(3) Vertical curves of parabolic form.

10.1 CIRCULAR CURVES

Two straights, D_1T_1 and D_2T_2 in *Figure 10.1*, are connected by a circular curve of radius R:

(1) The straights when projected forward, meet at I: the *intersection point.*
(2) The angle Δ at I is called the *angle of intersection* or *the deflection angle*, and equals the angle T_10T_2 subtended at the centre of the curve 0.
(3) The angle ϕ at I is called the *apex angle*, but is little used in curve computations.
(4) The curve commences from T_1 and ends at T_2; these points are called the *tangent points*.
(5) Distances T_1I and T_2I are the tangent lengths and are equal to $R\tan\Delta/2$.
(6) The length of curve T_1AT_2 is obtained from:

$$\text{Curve length} = R\Delta \text{ where } \Delta \text{ is expressed in radians, or}$$

$$\text{Curve length} = \frac{\Delta^\circ \cdot 100}{D^\circ} \text{ where } \textit{degree of curve } (D) \text{ is used (see } \textit{Section 10.1.1})$$

(7) Distance T_1T_2 is called the *main chord* (C), and from *Figure 10.1*:

$$\sin\frac{\Delta}{2} = \frac{T_1B}{T_1O} = \frac{\frac{1}{2}\,\text{chord}(C)}{R} \qquad \therefore C = 2R\sin\frac{\Delta}{2}$$

(8) IA is called the *apex distance* and equals

$$IO - R = R\sec\Delta/2 - R = R(\sec\Delta/2 - 1)$$

(9) AB is the *rise* and equals $R - OB = R - R\cos\Delta/2$

$$\therefore AB = R(1 - \cos\Delta/2)$$

These equations should be deduced using a curve diagram (*Figure 10.1*).

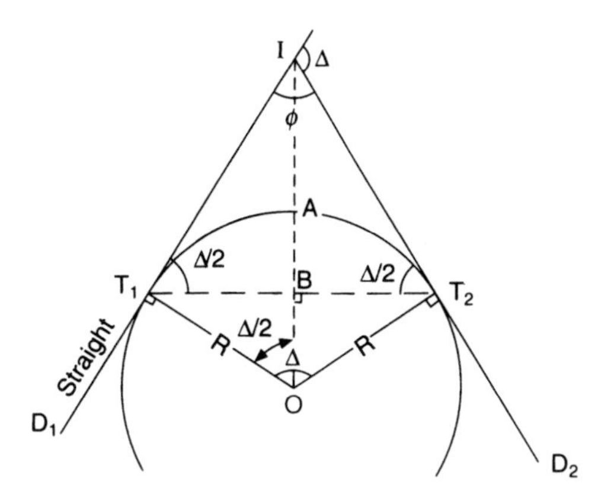

Fig. 10.1 *Circular curve*

10.1.1 Curve designation

Curves are designated either by their *radius* (R) or their *degree of curvature* (D°). The degree of curvature is defined as the angle subtended at the centre of a circle by an *arc* of 100 m (*Figure 10.2*).

$$\text{Thus} \quad R = \frac{100 \text{ m}}{D \text{ rad}} = \frac{100 \times 180^\circ}{D^\circ \times \pi}$$

$$\therefore R = \frac{5729.578}{D^\circ} \text{ m} \tag{10.1}$$

Thus a 10° curve has a radius of 572.9578 m.

10.1.2 Through chainage

Through chainage is the horizontal distance from the start of a scheme for route construction.

Consider *Figure 10.3*. If the distance from the start of the route (Chn 0.00 m) to the tangent point T_1 is 2115.50 m, then it is said that the chainage of T_1 is 2115.50 m, written as (Chn 2115.50 m).

If the route centre-line is being staked out at 20-m chord intervals, then the peg immediately prior to T_1 must have a chainage of 2100 m (an integer number of 20 m intervals). The next peg on the centre-line must therefore have a chainage of 2120 m. It follows that the length of the first sub-chord on the curve from T_1 must be (2120 − 2115.50) = 4.50 m.

Similarly, if the chord interval had been 30 m, the peg chainage prior to T_1 must be 2100 m and the next peg (on the curve) 2130 m, thus the first sub-chord will be (2130 − 2115.50) = 14.50 m.

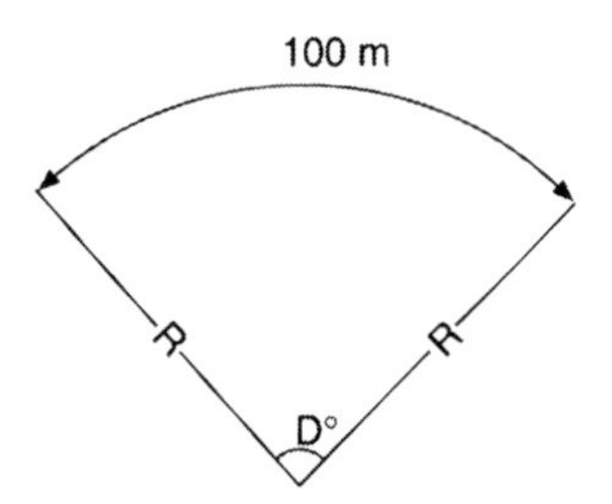

Fig. 10.2 *Radius and chainage*

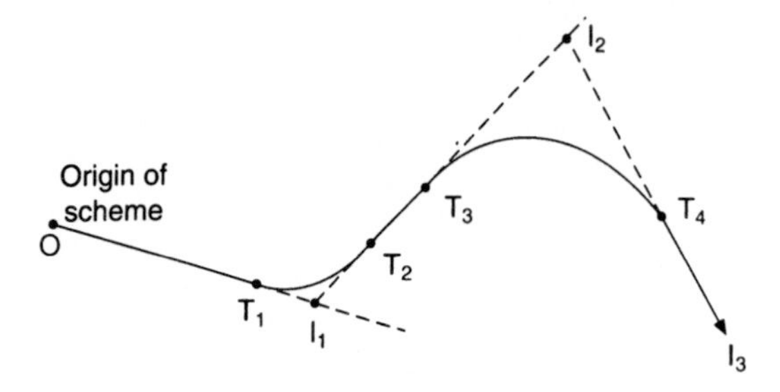

Fig. 10.3 *Through chainage*

A further point to note in regard to chainage is that if the chainage at I_1 is known, then the chainage at $T_1 =$ Chn $I_1 -$ distance I_1T_1, the tangent length. However the chainage at $T_2 =$ Chn $T_1 +$ curve length, as chainage is measured along the route under construction.

10.2 SETTING OUT CURVES

This is the process of establishing the centre-line of the curve on the ground by means of pegs at 10 m to 30 m intervals. In order to do this the tangent and intersection points must first be fixed in the ground, in their correct positions.

Consider *Figure 10.3*. The straights OI_1, I_1I_2, I_2I_3, etc., will have been designed on the plan in the first instance. Using railway curves, appropriate curves will now be designed to connect the straights. The tangent points of these curves will then be fixed, making sure that the tangent lengths are equal, i.e. $T_1I_1 = T_2I_1$ and $T_3I_2 = T_4I_2$. The coordinates of the origin, point O, and all the intersection points only will now be carefully scaled from the plan. Using these coordinates, the bearings of the straights are computed and, using the tangent lengths on these bearings, the coordinates of the tangent points are also computed. The difference of the bearings of the straights provides the deflection angles (Δ) of the curves which, combined with the tangent length, enables computation of the curve radius, through chainage and all setting-out data. Now the tangent and intersection points are set out from existing control survey stations and the curves ranged between them using the methods detailed below.

10.2.1 Setting out with theodolite and tape

The following method of setting out curves is the most popular and it is called *Rankine's deflection or tangential angle method*, the latter term being more definitive.

In *Figure 10.4* the curve is established by a series of chords T_1X, XY, etc. Thus, peg 1 at X is fixed by sighting to I with the theodolite reading zero, turning off the angle δ_1 and measuring out the chord length T_1X along this line. Setting the instrument to read the second deflection angle gives the direction T_1Y, and peg 2 is fixed by measuring the chord length XY from X until it intersects at Y. The procedure is now continued, the angles being set out from T_1I, and the chords measured from the previous station.

It is thus necessary to be able to calculate the setting-out angles δ as follows:

Assume $0A$ bisects the chord T_1X at right angles; then

$$A\hat{T}_1O = 90° - \delta_1, \quad \text{but} \quad I\hat{T}_1O = 90°$$

$$\therefore I\hat{T}_1A = \delta_1$$

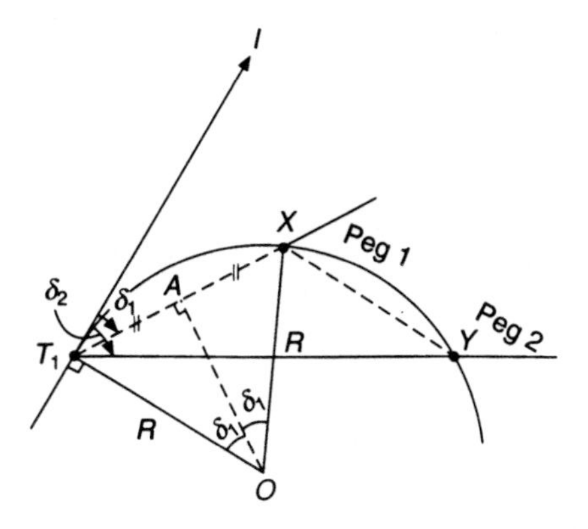

Fig. 10.4 *Setting out with thedolite and tape*

By radians, arc length $T_1X = R2\delta_1$

$$\therefore \delta_1 \text{ rad} = \frac{\text{arc } T_1X}{2R} \approx \frac{\text{chord } T_1X}{2R}$$

$$\therefore \delta_1^\circ = \frac{\text{chord } T_1X \times 180^\circ}{2R \cdot \pi} = 28.6479\frac{\text{chord}}{R} = 28.6479\frac{C}{R} \tag{10.2a}$$

$$\text{or} \quad \delta^\circ = \frac{D^\circ \times \text{chord}}{200} \text{ where } \textit{degree of curve} \text{ is used} \tag{10.2b}$$

(Using *equation* (*10.2a*) the angle is obtained in degree and decimals of a degree; a single key operation converts it to degrees, minutes, seconds.)

An example will now be worked to illustrate these principles.

The centre-line of two straights is projected forward to meet at I, the deflection angle being 30°. If the straights are to be connected by a circular curve of radius 200 m, tabulate all the setting-out data, assuming 20-m chords on a through chainage basis, the chainage of I being 2259.59 m.

$$\text{Tangent length} = R\tan\Delta/2 = 200\tan 15^\circ = 53.59 \text{ m}$$

$$\therefore \text{Chainage of } T_1 = 2259.59 - 53.59 = 2206 \text{ m}$$

$$\therefore \text{1st sub-chord} = 14\text{m}$$

Length of circular arc $= R\Delta = 200(30^\circ \cdot \pi/180)$ m $= 104.72$ m
From which the number of chords may now be deduced i.e.,

1st sub-chord = 14 m

2nd, 3rd, 4th, 5th chords = 20 m each

Final sub-chord = 10.72 m

Total = 104.72 m (*Check*)

$$\therefore \text{Chainage of } T_2 = 2206 \text{ m} + 104.72 \text{ m} = 2310.72 \text{ m}$$

Deflection angles:

$$\text{For 1st sub-chord} = 28.6479 \cdot \frac{14}{200} = 2° \, 00' \, 19''$$

$$\text{Standard chord} = 28.6479 \cdot \frac{20}{200} = 2° \, 51' \, 53''$$

$$\text{Final sub-chord} = 28.6479 \cdot \frac{10.72}{200} = 1° \, 32' \, 08''$$

Check: The sum of the deflection angles $= \Delta/2 = 14° \, 59' \, 59'' \approx 15°$

Chord number	*Chord length* (m)	*Chainage* (m)	*Deflection angle* °	′	″	*Setting-out angle* °	′	″	*Remarks*
1	14	2220.00	2	00	19	2	00	19	peg 1
2	20	2240.00	2	51	53	4	52	12	peg 2
3	20	2260.00	2	51	53	7	44	05	peg 3
4	20	2280.00	2	51	53	10	35	58	peg 4
5	20	2300.00	2	51	53	13	27	51	peg 5
6	10.72	2310.72	1	32	08	14	59	59	peg 6

The error of 1″ is, in this case, due to the rounding-off of the angles to the nearest second and is negligible.

10.2.2 Setting out with two theodolites

Where chord taping is impossible, the curve may be set out using two theodolites at T_1 and T_2 respectively, the intersection of the lines of sight giving the position of the curve pegs.

The method is explained by reference to *Figure 10.5*. Set out the deflection angles from T_1I in the usual way. From T_2, set out the same angles from the main chord T_2T_1. The intersection of the corresponding angles gives the peg position.

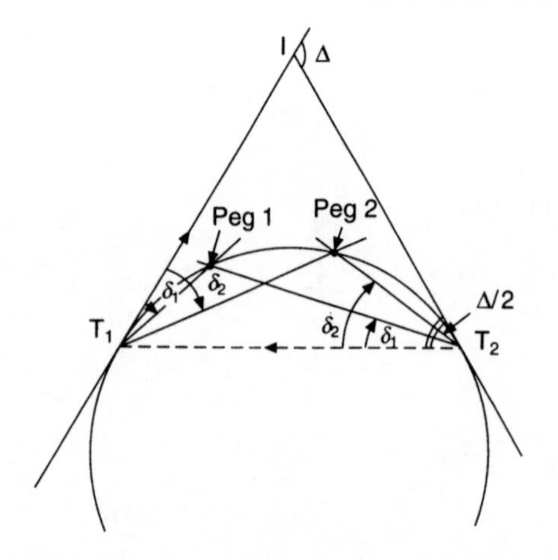

Fig. 10.5 *Setting out with two thedolites*

If T_1 cannot be seen from T_2, sight to I and turn off the corresponding angles $\Delta/2 - \delta_1$, $\Delta/2 - \delta_2$, etc.

10.2.3 Setting-out using EDM

When setting-out by EDM, the total distance from T_1 to the peg is set out, i.e. distances T_1A, T_1B and T_1C, etc., in *Figure 10.6*. However, the chord and sub-chord distances are still required in the usual way, plus the setting-out angles for those chords. Thus all the data and setting-out computation as shown the table above must first be carried out prior to computing the distances to the pegs direct from T_1. These distances are computed using the equation point 7, in *Section 10.1*, i.e.

$$T_1A = 2R\sin\delta_1 = 2R\sin 2°\,00'\,19'' = 14.00\text{ m.}$$

$$T_1B = 2R\sin\delta_2 = 2R\sin 4°\,52'\,12'' = 33.96\text{ m.}$$

$$T_1C = 2R\sin\delta_3 = 2R\sin 7°\,44'\,05'' = 53.83\text{ m.}$$

$$T_1T_2 = 2R\sin(\Delta/2) = 2R\sin 15°\,00'\,00'' = 103.53\text{ m.}$$

In this way the curve is set-out by measuring the distances directly from T_1 and turning off the necessary direction in the manner already described.

10.2.4 Setting-out using coordinates

In this procedure the coordinates along the centre-line of the curve are computed relative to the existing control points. Consider *Figure 10.7*:

(1) From the design process, the coordinates of the tangent and intersection point are obtained.
(2) The chord intervals are decided in the usual way and the setting-out angles $\delta_1, \delta_2, \ldots, \delta_n$, computed in the usual way (*Section 10.2.1*).
(3) From the known coordinates of T_1 and I, the bearing T_1I is computed.

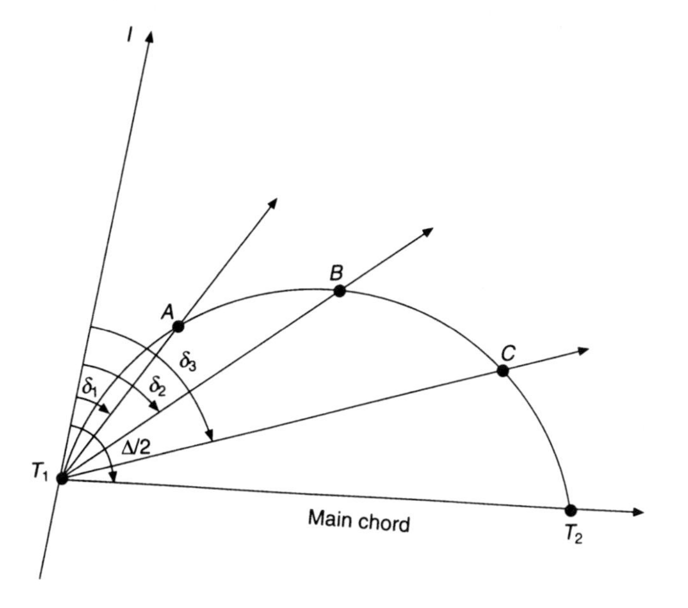

Fig. 10.6 *Setting out by EDM*

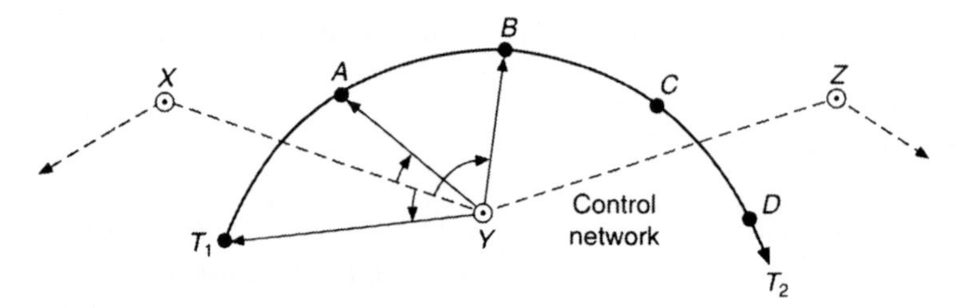

Fig. 10.7 *Setting out using coordinates*

(4) Using the setting-out angles, the bearings of the rays T_1A, T_1B, T_1C, etc., are computed relative to T_1I. The distances are obtained as in *Section 10.2.3*.
(5) Using the bearings and distances in (4) the coordinates of the curve points *A*, *B*, *C*, etc., are obtained.
(6) These points can now be set out from the nearest control points either by 'polars' or by 'intersection', as follows:
(7) Using the coordinates, compute the bearing and distance from, say, station *Y* to T_1, *A* and *B*.
(8) Set up theodolite at *Y* and backsight to *X*; set the horizontal circle to the known bearing *YX*.
(9) Now turn the instrument until it reads the computed bearing YT_1 and set out the computed distance in that direction to fix the position of T_1. Repeat the process for *A* and *B*. The ideal instrument for this is a total station, many of which will have onboard software to carry out the computation in real time. However, provided that the ground conditions are suitable and the distances within, say, a 50 m tape length, a theodolite and steel tape would suffice.

Other points around the curve are set out in the same way from appropriate control points.

Intersection may be used, thereby precluding distance measurement, by computing the bearings to the curve points from *two* control stations. For instance, the theodolites are set up at *Y* and *Z* respectively. Instrument *Y* is orientated to *Z* and the bearing *YZ* set on the horizontal circle. Repeat from *Z* to *Y*. The instruments are set to bearings *YB* and *ZB* respectively, intersecting at peg *B*. The process is repeated around the curve.

Using coordinates eliminates many of the problems encountered in curve ranging and does not require the initial establishment of tangent and intersection points.

10.2.5 Setting out with two tapes (method of offsets)

Theoretically this method is exact, but in practice errors of measurement propagate round the curve. It is therefore generally used for minor curves.

In *Figure 10.8*, line *OE* bisects chord T_1A at right-angles, then $ET_1O = 90° - \delta$, $\therefore\ CT_1A = \delta$, and triangles CT_1A and ET_1O are similar, thus:

$$\frac{CA}{T_1A} = \frac{T_1E}{T_1O} \quad \therefore CA = \frac{T_1E}{T_1O} \times T_1A$$

$$\text{i.e. offset} \quad CA = \frac{\frac{1}{2}\,\text{chord} \times \text{chord}}{\text{radius}} = \frac{\text{chord}^2}{2R} \tag{10.3}$$

From *Figure 10.8*, assuming lengths $T_1A = AB = AD$

$$\text{then angle } DAB = 2\delta, \quad \text{and so offset} \quad DB = 2CA = \frac{\text{chord}^2}{R} \tag{10.4}$$

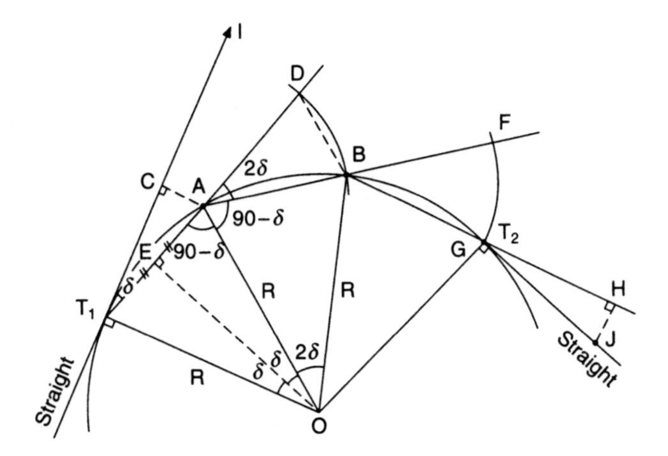

Fig. 10.8 *Setting out with two tapes*

The remaining offsets round the curve to T_2 are all equal to DB whilst, if required, the offset HJ to fix the line of the straight from T_2, equals CA.

The method of setting out is as follows:

It is sufficient to approximate distance T_1C to the chord length T_1A and measure this distance along the tangent to fix C. From C a right-angled offset CA fixes the first peg at A. Extend T_1A to D so that AD equals a chord length; peg B is then fixed by pulling out offset length from D and chord length from A, and where they meet is the position B. This process is continued to T_2.

The above assumes equal chords. When the first or last chords are sub-chords, the following (*Section 10.2.6*) should be noted.

10.2.6 Setting out by offsets with sub-chords

In *Figure 10.9* assume T_1A is a sub-chord of length x; from *equation* (*10.3*) the offset $CA = O_1 = x^2/2R$.

As the normal chord AB differs in length from T_1A, the angle subtended at the centre will be 2θ not 2δ. Thus, as shown in *Figure 10.8*, the offset DB will not in this case equal $2CA$.

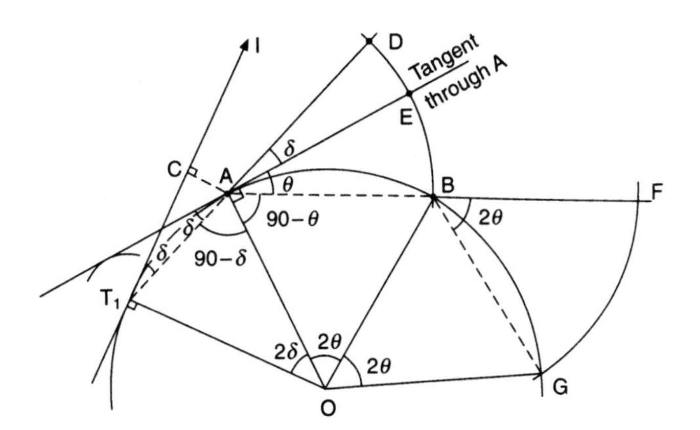

Fig. 10.9 *Setting out by offsets with sub-chords*

Construct a tangent through point A, then from the figure it is obvious that angle $EAB = \theta$, and if chord $AB = y$, then offset $EB = y^2/2R$.

Angle $DAE = \delta$, therefore offset DE will be directly proportional to the chord length, thus:

$$DE = \frac{O_1}{x}y = \frac{x^2}{2R}\frac{y}{x} = \frac{xy}{2R}$$

Thus the total offset $DB = DE + EB$

$$= \frac{y}{2R}(x + y) \tag{10.5}$$

$$\text{i.e.} = \frac{\text{chord}}{2R}(\text{sub-chord} + \text{chord})$$

Thus having fixed B, the remaining offsets to T_2 are calculated as y^2/R and set out in the usual way.

If the final chord is a sub-chord of length x_1, however, then the offset will be:

$$\frac{x_1}{2R}(x_1 + y) \tag{10.6}$$

Note the difference between *equations (10.5)* and *(10.6)*.

A more practical approach to this problem is actually to establish the tangent through A in the field. This is done by swinging an arc of radius equal to CA, i.e. $x^2/2R$ from T_1. A line tangential to the arc and passing through peg A will then be the required tangent from which offset EB, i.e. $y^2/2R$, may be set off.

10.2.7 Setting out with inaccessible intersection point

In *Figure 10.10* it is required to fix T_1 and T_2, and obtain the angle Δ, when I is inaccessible.

Project the straights forward as far as possible and establish two points A and B on them. Measure distance AB and angles BAC and DBA then:

angle $IAB = 180° - B\hat{A}C$ and angle $IBA = 180° - D\hat{B}A$, from which angle BIA is deduced and angle Δ. The triangle AIB can now be solved for lengths IA and IB. These lengths, when subtracted from the computed tangent lengths ($R \tan \Delta/2$), give AT_1 and BT_2, which are set off along the straight to give positions T_1 and T_2 respectively.

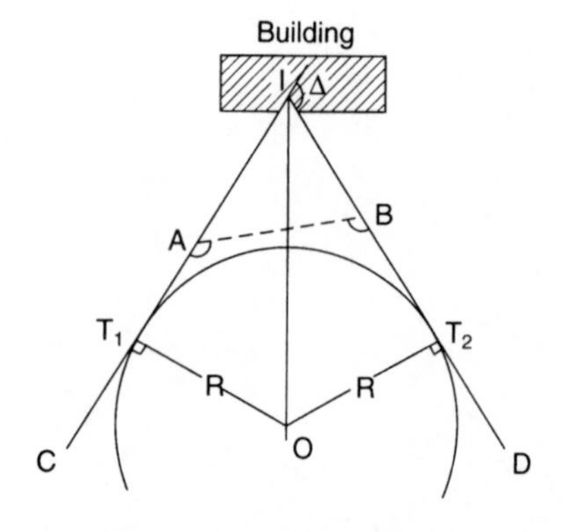

Fig. 10.10 *Setting out with inaccessible intersection point*

10.2.8 Setting out with theodolite at an intermediate point on the curve

Due to an obstruction on the line of sight (*Figure 10.11*) or difficult communications and visibility on long curves, it may be necessary to continue the curve by ranging from a point on the curve. Assume that the setting-out angle to fix peg 4 is obstructed. The theodolite is moved to peg 3, backsighted to T_1 with the instrument reading 180°, and then turned to read 0°, thus giving the direction $3 - T$. The setting-out angle for peg 4, δ_4, is turned off and the chord distance measured from 3. The remainder of the curve is now set off in the usual way, that is, δ_5 is set on the theodolite and the chord distance measured from 4 to 5.

The proof of this method is easily seen by constructing a tangent through peg 3, then angle $A3T_1 = AT_13 = \delta_3 = T3B$. If peg 4 was fixed by turning off δ from this tangent, then the required angle from $3T$ would be $\delta_3 + \delta = \delta_4$.

10.2.9 Setting out with an obstruction on the curve

In this case (*Figure 10.12*) an obstruction on the curve prevents the chaining of the chord from 3 to 4. One may either:

(1) Set out the curve from T_2 to the obstacle.
(2) Set out the chord length $T_14 = 2R \sin \delta_4$ (EDM).
(3) Set out using intersection from theodolites at T_1 and T_2.
(4) Use coordinate method.

10.2.10 Passing a curve through a given point

In *Figure 10.13*, it is required to find the radius of a curve which will pass through a point P, the position of which is defined by the distance IP at an angle of ϕ to the tangent.

Consider triangle IPO:

$$\text{angle } \beta = 90^\circ - \Delta/2 - \phi \text{ (right-angled triangle } IT_2O)$$

$$\text{by sine rule: } \sin \alpha = \frac{IO}{PO} \sin \beta \quad \text{but} \quad IO = R \sec \frac{\Delta}{2}$$

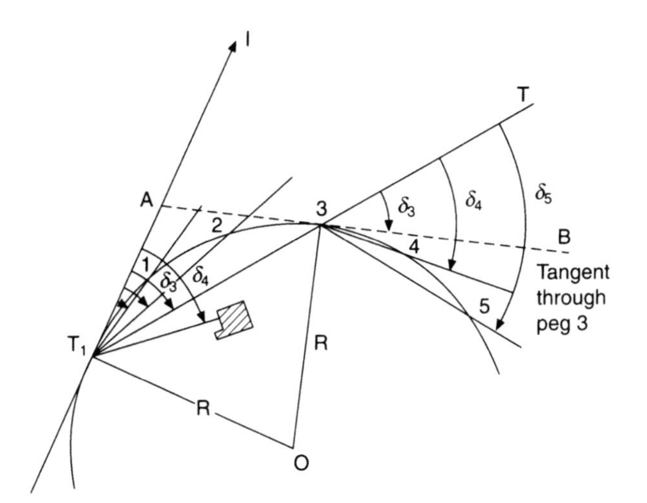

Fig. 10.11 *Setting out from an intermediate point*

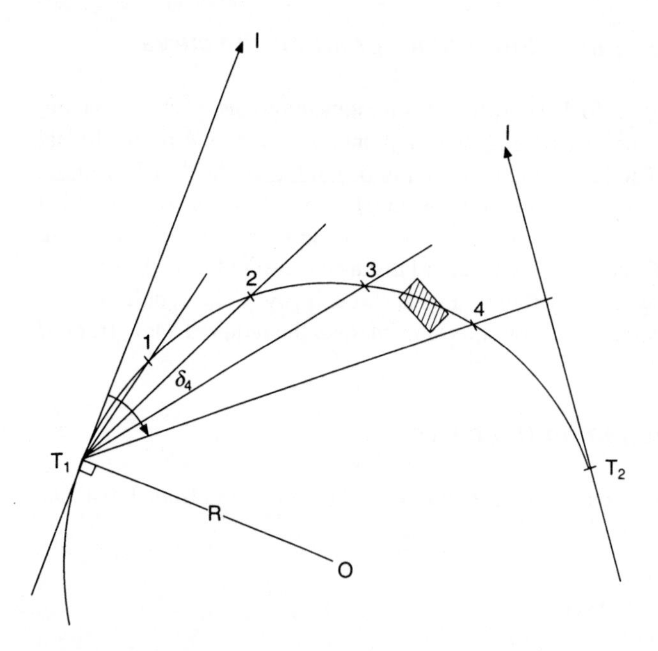

Fig. 10.12 *Obstruction on the curve*

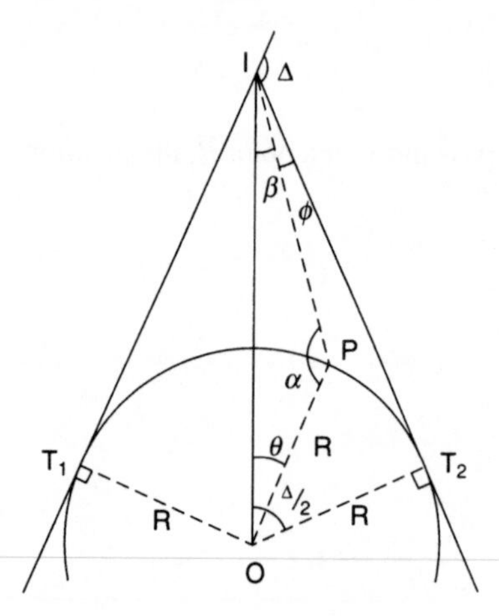

Fig. 10.13 *Curve passing through a given point*

$$\therefore \sin\alpha - \sin\beta \frac{R \sec \Delta/2}{R} = \sin\beta \sec\frac{\Delta}{2}$$

then $\theta = 180° - \alpha - \beta$, and by the sine rule: $R = IP\dfrac{\sin\beta}{\sin\theta}$

10.3 COMPOUND AND REVERSE CURVES

Although equations are available which solve compound curves (*Figure 10.14*) and reverse curves (*Figure 10.15*), they are difficult to remember so it is best to treat the problem as two simple curves with a common tangent point t.

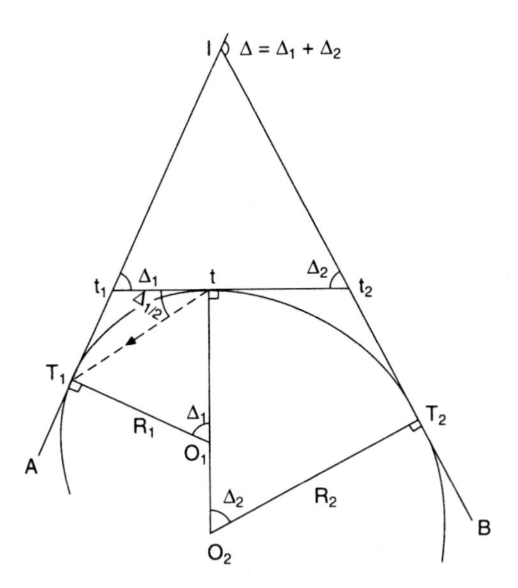

Fig. 10.14 *Compound curve*

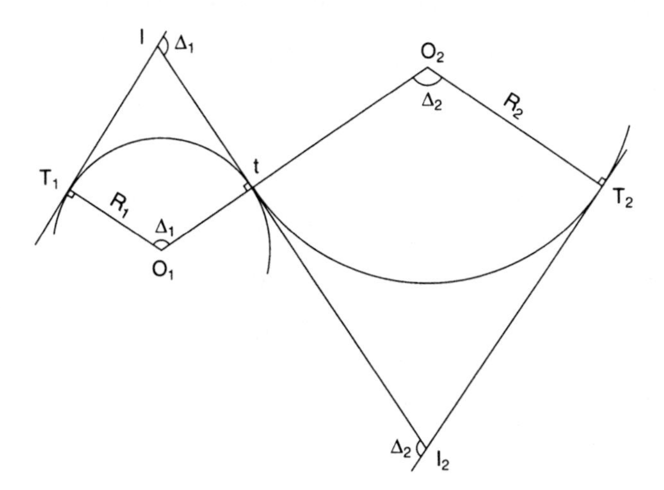

Fig. 10.15 *Reverse curve*

In the case of the compound curve, the total tangent lengths T_1I and T_2I are found as follows:

$$R_1 \tan \Delta_1/2 = T_1t_1 = t_1t \quad \text{and} \quad R_2 \tan \Delta_2/2 = T_2t_2 = t_2t, \text{as } t_1t_2 = t_1t + t_2t$$

then triangle t_1It_2 may be solved for lengths t_1I and t_2I which, if added to the known lengths T_1t_1 and T_2t_2 respectively, give the total tangent lengths.

In setting out this curve, the first curve R_1 is set out in the usual way to point t. The theodolite is moved to t and backsighted to T_1, with the horizontal circle reading $(180° - \Delta_1/2)$. Set the instrument to read zero and it will then be pointing to t_2. Thus the instrument is now oriented and reading zero, prior to setting out curve R_2.

In the case of the reverse curve, both arcs can be set out from the common point t.

10.4 SHORT AND/OR SMALL-RADIUS CURVES

Short and/or small-radius curves such as for kerb lines, bay windows or for the construction of large templates may be set out by the following methods.

10.4.1 Offsets from the tangent

The position of the curve (in *Figure 10.16*) is located by right-angled offsets Y set out from distances X, measured along each tangent, thereby fixing half the curve from each side.

The offsets may be calculated as follows for a given distance X. Consider offset Y_3, for example.

$$\text{In} \quad \Delta ABO,\ AO^2 = OB^2 - AB^2 \ \therefore (R - Y_3)^2 = R^2 - X_3^2 \quad \text{and} \quad Y_3 = R - (R^2 - X_3^2)^{\frac{1}{2}}$$

Thus for any offset Y_i at distance X_i along the tangent:

$$Y_i = R - (R^2 - X_i^2)^{\frac{1}{2}} \tag{10.7}$$

10.4.2 Offsets from the long chord

In this case (*Figure 10.17*) the right-angled offsets Y are set off from the long chord C, at distances X to each side of the centre offset Y_0.

An examination of *Figure 10.17*, shows the central offset Y_0 equivalent to the distance T_1A on *Figure 10.16*; thus:

$$Y_0 - R - [R^2 - (C/2)^2]^{\frac{1}{2}}$$

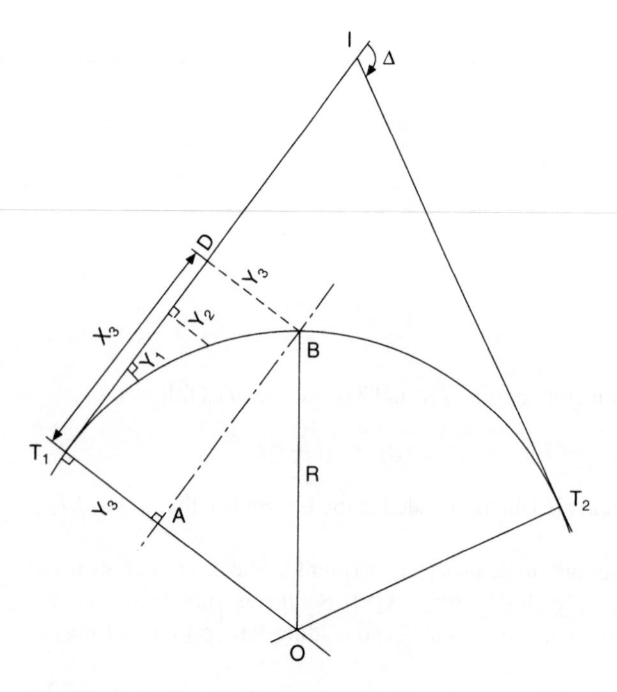

Fig. 10.16 *Offsets from the tangent*

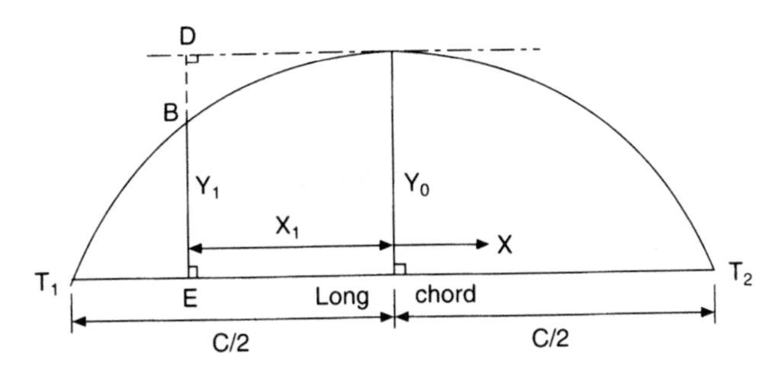

Fig. 10.17 *Offsets from the long chord*

Similarly, DB is equivalent to DB on *Figure 10.16*, thus: $DB = R - (R^2 - X_1^2)^{\frac{1}{2}}$ and offset $Y_1 = Y_0 - DB \therefore Y_1 = Y_0 - [R - (R^2 - X_1^2)^{\frac{1}{2}}]$ and for any offset Y_i at distance X_i each side of the mid-point of T_1T_2:

$$\text{mid-point of } T_1T_2: \quad Y_i = Y_0 - [R - (R^2 - X^2)^{\frac{1}{2}}] \tag{10.8}$$

Therefore, after computation of the central offset, further offsets at distances X_i, each side of Y_0, can be found.

10.4.3 Halving and quartering

Referring to *Figure 10.18*:

(1) Join T_1 and T_2 to form the long chord. Compute and set out the central offset Y_0 to A from B (assume $Y_0 = 20$ m), as in *Section 10.4.2*.
(2) Join T_1 and A, and now halve this chord and quarter the offset. That is, from mid-point E set out offset $Y_1 = 20/4 = 5$ m to D.
(3) Repeat to give chords T_1D and DA; the mid-offsets FG will be equal to $Y_1/4 = 1.25$ m.

Repeat as often as necessary on both sides of the long chord.

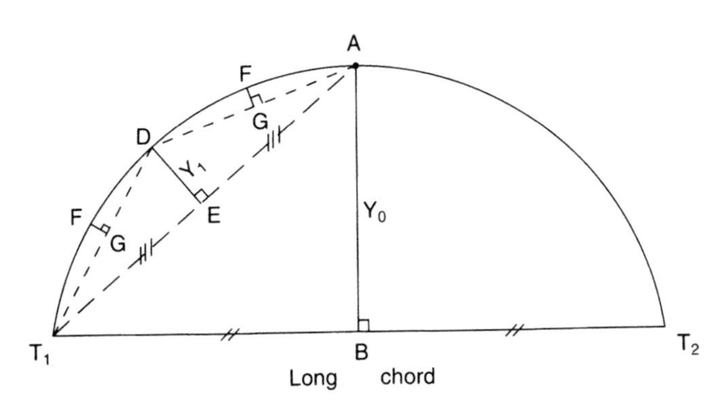

Fig. 10.18 *Halving and quartering*

Worked examples

Example 10.1 The tangent length of a simple curve was 202.12 m and the deflection angle for a 30-m chord 2° 18′.

Calculate the radius, the total deflection angle, the length of curve and the final deflection angle. (LU)

$$2^\circ\,18' = 2.3^\circ = 28.6479 \cdot \frac{30}{R} \qquad \therefore R = 373.67\text{ m}$$

$$202.12 = R\tan\Delta/2 = 373.67\tan\Delta/2 \qquad \therefore \Delta = 56^\circ\,49'\,06''$$

Length of curve = $R\Delta$ rad = 373.67 × 0.991 667 rad = 370.56 m
Using 30-m chords, the final sub-chord = 10.56 m

$$\therefore \text{final deflection angle} = \frac{138' \times 10.56}{30} = 48.58' = 0^\circ\,48'\,35''$$

Example 10.2 The straight lines *ABI* and *CDI* are tangents to a proposed circular curve of radius 1600 m. The lengths *AB* and *CD* are each 1200 m. The intersection point is inaccessible so that it is not possible directly to measure the deflection angle; but the angles at *B* and *D* are measured as:

$$A\hat{B}D = 123^\circ\,48', \quad B\hat{D}C = 126^\circ\,12' \text{ and the length } BD \text{ is } 1485\text{ m}$$

Calculate the distances from *A* and *C* of the tangent points on their respective straights and calculate the deflection angles for setting out 30-m chords from one of the tangent points. (LU)

Referring to *Figure 10.19*:

$$\Delta_1 = 180^\circ - 123^\circ\,48' = 56^\circ\,12', \quad \Delta_2 = 180^\circ - 126^\circ\,12' = 53^\circ\,48'$$

$$\therefore \Delta = \Delta_1 + \Delta_2 = 110^\circ$$

$$\phi = 180^\circ - \Delta = 70^\circ$$

Tangent lengths IT_1 and $IT_2 = R\tan\Delta/2 = 1600\tan 55^\circ = 2285$ m

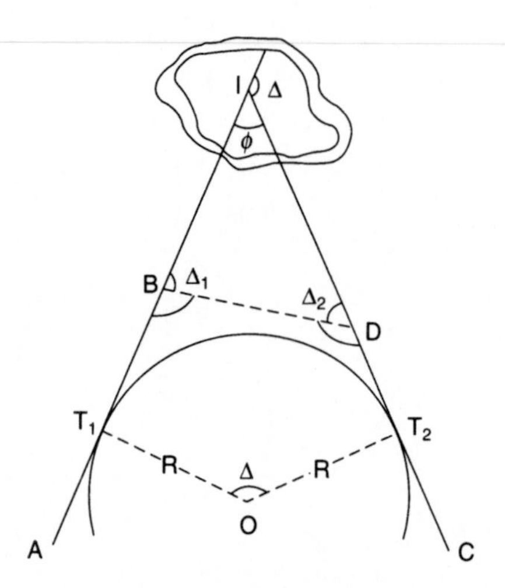

Fig. 10.19 *Inaccessible intersection point*

By sine rule in triangle *BID*:

$$BI = \frac{BD \sin \Delta_2}{\sin \phi} = \frac{1484 \sin 53^\circ 48'}{\sin 70^\circ} = 1275.2 \text{ m}$$

$$ID = \frac{BD \sin \Delta_1}{\sin \phi} = \frac{1485 \sin 56^\circ 15'}{\sin 70^\circ} = 1314 \text{ m}$$

Thus: $AI = AB + BI = 1200 + 1275.2 = 2475.2$ m

$CI = CD + ID = 1200 + 1314 = 2514$ m

$\therefore AT_1 = AI - IT_1 = 2475.2 - 2285 = 190.2$ m

$CT_2 = CI - IT_2 = 2514 - 2285 = 229$ m

Deflection angle for 30-m chord $= 28.6479 \times 30/1600 = 0.537148^\circ$

$= 0^\circ 32' 14''$

Example 10.3 A circular curve of 800 m radius has been set out connecting two straights with a deflection angle of 42°. It is decided, for construction reasons, that the mid-point of the curve must be moved 4 m towards the centre, i.e. away from the intersection point. The alignment of the straights is to remain unaltered.

Calculate:

(1) The radius of the new curve.
(2) The distances from the intersection point to the new tangent points.
(3) The deflection angles required for setting out 30-m chords of the new curve.
(4) The length of the final sub-chord. (LU)

Referring to *Figure 10.20*:

$$IA = R_1(\sec \Delta/2 - 1) = 800(\sec 21^\circ - 1) = 56.92 \text{ m}$$

$$\therefore IB = IA + 4 \text{ m} = 60.92 \text{ m}$$

(1) Thus, $60.92 = R_2(\sec 21^\circ - 1)$, from which $R_2 = 856$ m
(2) Tangent length $= IT_1 = R_2 \tan \Delta/2 = 856 \tan 21^\circ = 328.6$ m

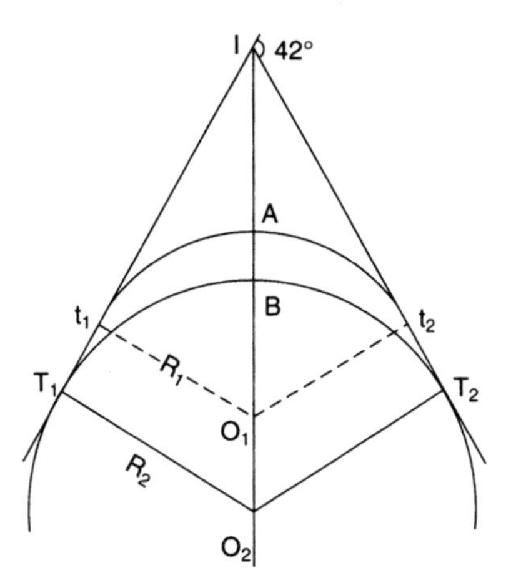

Fig. 10.20 *A realigned road*

(3) Deflection angle for 30-m chord $= 28.6479 \cdot C/R = 28.6479 \cdot \dfrac{30}{856} = 1°\,00'\,14''$

(4) Curve length $= R\Delta$ rad $= \dfrac{856 \times 42° \times 3600}{206\,265} = 627.5$ m

$\therefore$ Length of final sub-chord = 27.5 m

Example 10.4 The centre-line of a new railway is to be set out along a valley. The first straight AI bears 75°, whilst the connecting straight IB bears 120°. Due to site conditions it has been decided to join the straights with a compound curve.

The first curve of 500 m radius commences at T_1, situated 300 m from I on straight AI, and deflects through an angle of 25° before joining the second curve.

Calculate the radius of the second curve and the distance of the tangent point T_2 from I on the straight IB. (KU)

Referring to *Figure 10.14*:

$\Delta = 45°, \quad \Delta_1 = 25° \quad \therefore \Delta_2 = 20°$

Tangent length $T_1t_1 = R_1 \tan \Delta_1/2 = 500 \tan 12°\,30' = 110.8$ m. In triangle t_1It_2:

Angle $t_2It_1 = 180° - \Delta = 135°$

Length $It_1 = T_1I - T_1t_1 = 300 - 110.8 = 189.2$ m

By sine rule:

$$t_1t_2 = \frac{It_1 \sin t_2It_1}{\sin \Delta_2} = \frac{189.2 \sin 135°}{\sin 20°} = 391.2 \text{ m}$$

$$It_2 = \frac{It_1 \sin \Delta_1}{\sin \Delta_2} = \frac{189.2 \sin 25°}{\sin 20°} = 233.8 \text{ m}$$

$$\therefore tt_2 = t_1t_2 - T_1t_1 = 391.2 - 110.8 = 280.4 \text{ m}$$

$$\therefore 280.4 = R_2 \tan \Delta_2/2 = R_2 \tan 10°; \quad \therefore R_2 = 1590 \text{ m}$$

$$\text{Distance} \quad IT_2 = It_2 = t_2T_2 = 233.8 + 280.4$$

$$= 514.2 \text{ m}$$

Example 10.5 Two straights intersecting at a point B have the following bearings, BA 270°, BC 110°. They are to be joined by a circular curve which must pass through a point D which is 150 m from B and the bearing of BD is 260°.

Find the required radius, tangent lengths, length of curve and setting-out angle for a 30-m chord. (LU)

Referring to *Figure 10.21*:
From the bearings, the apex angle $= (270° - 110°) = 160°$

$\therefore \Delta = 20°$

and angle $DBA = 10°$ (from bearings)

$\therefore OBD = \beta = 70°$

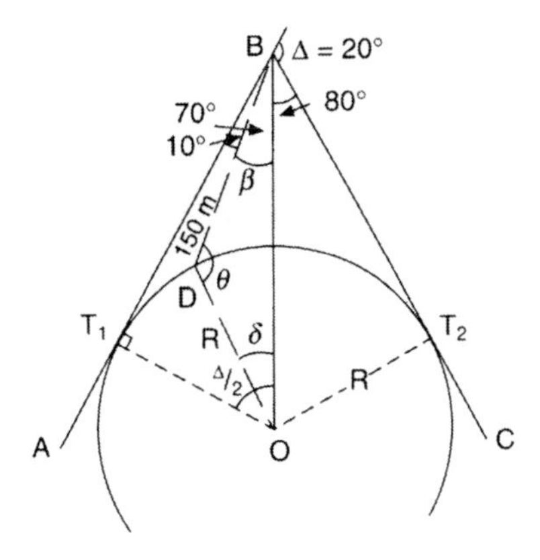

Fig. 10.21 *Curve passing through a given point*

In triangle BDO by sine rule:

$$\sin\theta = \frac{OB}{OD}\sin\beta = \frac{R\sec\Delta/2}{R}\sin\beta = \sec\frac{\Delta}{2}\sin\beta$$

$$\therefore \sin\theta = \sec 10^\circ \times \sin 70^\circ$$

$$\theta = \sin^{-1} 0.954\,190 = 72^\circ\,35'\,25'' \text{ or}$$

$$(180^\circ - 72^\circ\,35'\,25'') = 107^\circ\,24'\,35''$$

An examination of the figure shows that δ must be less than 10°,

$$\therefore \theta = 170^\circ\,24'\,35''$$

$$\delta = 180^\circ - (\theta + \beta) = 2^\circ\,35'\,25''$$

$$\text{By sine rule } DO = R = \frac{DB\sin\beta}{\sin\delta} = \frac{150\sin 70^\circ}{\sin 2^\circ\,35'\,25''}$$

$$\therefore R = 3119 \text{ m}$$

Tangent length $= R\tan\Delta/2 = 3119\tan 10^\circ = 550$ m

$$\text{Length of curve} = R\Delta \text{ rad} = \frac{3119 \times 20^\circ \times 3600}{206\,265} = 1089 \text{ m}$$

$$\text{Deflection angle for 30-m chord} = 28.6479 \times \frac{30}{3119} = 0^\circ\,16'\,32''$$

Example 10.6 Two straights AEI and CFI, whose bearings are respectively 35° and 335°, are connected by a straight from E to F. The coordinates of E and F in metres are:

E E 600.36 N 341.45

F E 850.06 N 466.85

Calculate the radius of a connecting curve which shall be tangential to each of the lines AE, EF and CF. Determine also the coordinates of I, T_1, and T_2, the intersection and tangent points respectively. (KU)

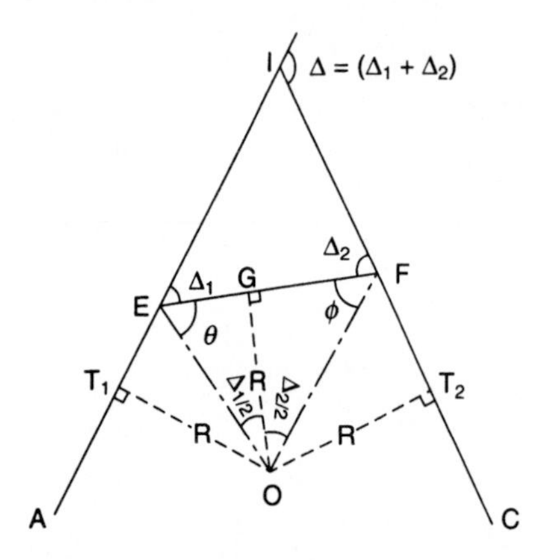

Fig. 10.22 *Curve fit to two straight lines defined by coordinates and bearings*

Referring to *Figure 10.22*:

$$\text{Bearing } AI = 35°, \text{bearing } IC = (335° - 180°) = 155°$$
$$\therefore \Delta = 155° - 35° = 120°$$

By coordinates

$$\text{Bearing } EF = \tan^{-1}\frac{+249.70\text{ N}}{+125.40\text{ N}} = 63°\,20'$$

$$\text{Length } EF = 249.7/\sin 63°\,20' = 279.42\text{ m}$$

From bearings AI and EF, angle $IEF = \Delta_1 = (63°\,20' - 35°)$

$$= 28°\,20'$$

From bearings CI and EF, angle $IFE = \Delta_2 = (155° - 63°\,20')$

$$= 91°\,40', \text{check } (\Delta_1 + \Delta_2) = \Delta = 120°$$

In triangle EFO

$$\text{Angle } FEO = (90° - \Delta_1/2) = \theta = 75°\,50'$$
$$\text{Angle } EFO = (90° - \Delta_2/2) = \phi = 44°\,10'$$
$$EG = GO\cot\theta = R\cot\theta$$
$$GF = GO\cot\phi = R\cot\phi$$
$$\therefore EG + GF = EF = R(\cos\theta + \cos\phi)$$
$$\therefore R = \frac{EF}{(\cot\theta + \cot\phi)} = \frac{279.42}{\cot 75°\,50' + \cot 44°\,10'}$$
$$= 217.97\text{ m}$$
$$ET_1 = R\tan\Delta_1/2 = 217.97\tan 14°\,10' = 55.02\text{ m}$$
$$FT_2 = R\tan\Delta_2/2 = 217.97\tan 45°\,50' = 224.4\text{ m}$$

bearing $ET_1 = 215°$

bearing $FT_2 = 155°$

$\therefore$ Coordinates of $T_1 = 55.02 \sin 215° = -31.56$ E

$55.02 \cos 215° = -45.07$ N

$\therefore$ Total coordinates of $T_1 = $ E $600.36 - 31.56 = $ E 568.80 m

$= $ N $341.45 - 45.07 = $ N 296.38 m

Similarly:

Coordinates of $T_2 = 224.4 \sin 155° = +98.84$ E

$224.4 \cos 155° = -203.38$ N

$\therefore$ Total coordinates of $T_2 = $ E $850.06 + 94.84 = $ E 944.90 m

$= $ N $466.85 - 203.38 = $ N 263.47 m

$T_1I = R \tan \Delta/2 = 217.97 \tan 60° = 377.54$ m

Bearing of $T_1I = 35°$

$\therefore$ Coordinates of $I = 377.54 \sin 35° = +216.55$ E

$377.54 \cos 35° = +309.26$ N

$\therefore$ Total coordinates of $I = $ E $586.20 + 216.55 = $ E 802.75

$= $ N $321.23 + 309.26 = $ N 630.49

The coordinates of I can be checked via T_2I.

Example 10.7 The coordinates in metres of two points B and C with respect to A are:

B	470 E	500 N
C	770 E	550 N

Calculate the radius of a circular curve passing through the three points, and the coordinates of the intersection point I, assuming that A and C are tangent points. (KU)

Referring to *Figure 10.23*:

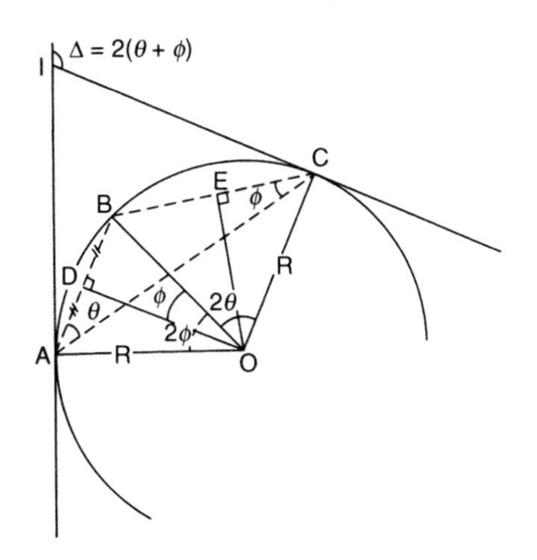

Fig. 10.23 *Curve defined by three points*

By coordinates

$$\text{Bearing}\quad AB = \tan^{-1}\frac{+470\,\text{E}}{+500\,\text{N}} = 43°\,14'$$

$$\text{Bearing}\quad AC = \tan^{-1}\frac{+770\,\text{E}}{550\,\text{N}} = 54°\,28'$$

$$\text{Bearing}\quad BC = \tan^{-1}\frac{+330\,\text{E}}{+50\,\text{N}} = 80°\,32'$$

$$\text{Distance}\quad AB = 500/\cos 43°\,14' = 686\ \text{m}$$

From bearings of AB and AC, angle $BAC = \theta = 11°\,14'$

From bearings of CA and CB, angle $BCA = \phi = 26°\,04'$

As a check, the remaining angle, calculated from the bearings of BA and BC = $142°\,42'$, summing to 180°.

In right-angled triangle DOB:

$$OB = R = \frac{DB}{\sin\phi} = \frac{343}{\sin 26°\,04'} = 781\ \text{m}$$

This result could now be checked through triangle OEC.

$$\Delta = 2(\phi + \theta) = 74°\,36'$$

$$\therefore AI = R\tan\Delta/2 = 781\tan 37°\,18' = 595\ \text{m}$$

$$\text{Bearing}\ AI = \text{bearing}\ AC - \Delta/2 = 54°\,82' - 37°\,18'$$
$$= 17°\,10'$$

$\therefore$ Coordinates of I equal:

$$595\sin 17°\,10' = +176\ \text{E}$$
$$595\cos 17°\,10' = +569\ \text{N}$$

Exercises

(*10.1*) In a town planning scheme, a road 9 m wide is to intersect another road 12 m wide at 60°, both being straight. The kerbs forming the acute angle are to be joined by a circular curve of 30 m radius and those forming the obtuse angle by one of 120 m radius.

Calculate the distances required for setting out the four tangent points.

Describe how to set out the larger curve by the deflection angle method and tabulate the angles for 15-m chords. (LU)

(*Answer*: 75, 62, 72, 62 m. $\delta = 3°\,35'$)

(*10.2*) A straight BC deflects 24° right from a straight AB. These are to be joined by a circular curve which passes through a point P, 200 m from B and 50 m from AB.

Calculate the tangent length, length of curve and deflection angle for a 30-m chord. (LU)

(*Answer*: $R = 3754$ m, $IT = 798$ m, curve length = 1572 m, $0°\,14'$)

(*10.3*) A reverse curve is to start at a point A and end at C with a change of curvature at B. The chord lengths AB and BC are respectively 661.54 m and 725.76 m and the radii likewise 1200 and 1500 m.

Due to irregular ground the curves are to be set out using two theodolites and no tape or chain.
Calculate the data for setting out and describe the procedure in the field. (LU)

(*Answer*: Tangent lengths: 344.09, 373.99; curve length: 670.2, 733, per 30-m chords: $\delta_1 = 0° 42' 54''$, $\delta_2 = 0° 34'' 30''$)

(*10.4*) Two straights intersect making a deflection angle of 59° 24′, the chainage at the intersection point being 880 m. The straights are to be joined by a simple curve commencing from chainage 708 m.

If the curve is to be set out using 30-m chords on a through chainage basis, by the method of offsets from the chord produced, determine the first three offsets.

Find also the chainage of the second tangent point, and with the aid of sketches, describe the method of setting out. (KU)

(*Answer*: 0.066, 1.806, 2.985 m, 864.3 m)

(*10.5*) A circular curve of radius 250 m is to connect two straights, but in the initial setting out it soon becomes apparent that the intersection point is in an inaccessible position. Describe how it is possible in this case to determine by what angle one straight deflects from the other, and how the two tangent points may be accurately located and their through chainages calculated.

On the assumption that the chainages of the two tangent points are 502.2 m and 728.4 m, describe the procedure to be adopted in setting out the first three pegs on the curve by a theodolite (reading to 20″) and a steel tape from the first tangent point at 30-m intervals of through chainage, and show the necessary calculations.

If it is found to be impossible to set out any more pegs on the curve from the first tangent point because of an obstruction between it and the pegs, describe a procedure (without using the second tangent point) for accurately locating the fourth and succeeding pegs. No further calculations are required. (ICE)

(*Answer*: 03° 11′ 10″, 06° 37′ 20″, 10° 03′ 40″)

10.5 TRANSITION CURVES

The *transition curve* is a curve of constantly changing radius. If used to connect a straight to a curve of radius R, then the commencing radius of the transition will be the same as the straight (∞), and the final radius will be that of the curve R (see *Figure 10.28*).

Consider a vehicle travelling at speed (V) along a straight. The forces acting on the vehicle will be its weight W, acting vertically down, and an equal and opposite force acting vertically up through the wheels. When the vehicle enters the curve of radius R at tangent point T_1, an additional centrifugal force (P) acts on the vehicle, as shown in *Figures 10.24* and *10.25*. If P is large the vehicle will be forced to the outside of the curve and may skid or overturn. In *Figure 10.25* the resultant of the two forces is shown as N, and if the road is super-elevated normal to this force, there will be no tendency for the vehicle to skid. It should be noted that as

$$P = WV^2/Rg \tag{10.9}$$

super-elevation will be maximum at minimum radius R.

It therefore requires a length of spiral curve to permit the gradual introduction of super-elevation, from zero at the start of the transition to maximum at the end, where the radius is the minimum safe radius R.

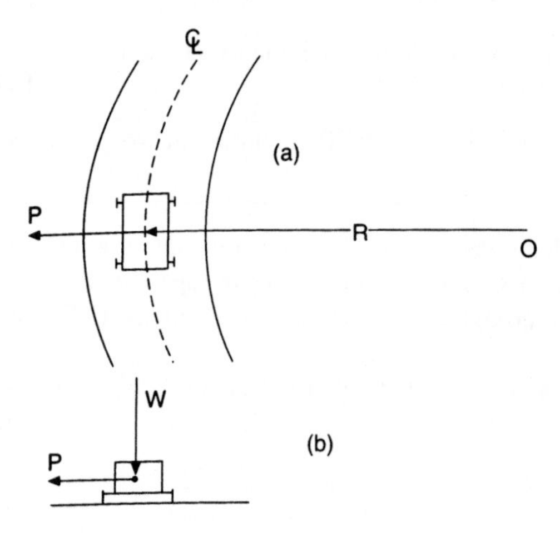

Fig. 10.24 *Centrifugal force*

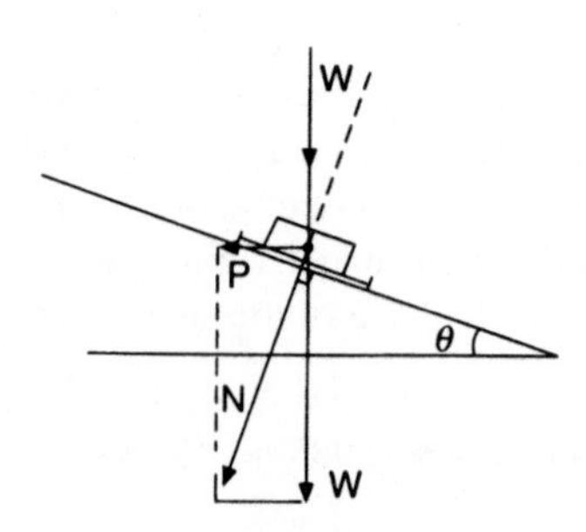

Fig. 10.25 *Super-elevation*

10.5.1 Principle of the transition

The purpose then of a transition curve is to:

(1) Achieve a gradual change of direction from the straight (radius ∞) to the curve (radius R).
(2) Permit the gradual application of super-elevation to counteract centrifugal force and minimize passenger discomfort.

Since P cannot be eliminated, it is allowed for by permitting it to increase uniformly along the curve. From *equation (10.9)*, as P is inversely proportional to R, the basic requirement of the ideal transition curve is that its radius should decrease uniformly with distance along it. This requirement also permits the uniform application of super-elevation; thus at distance l along the transition the radius is r and $rl = c$ (constant):

$$\therefore l/c = l/r$$

From *Figure 10.26*, tt_1 is an infinitely small portion of a transition δl of radius r; thus:

$$\delta l = r\delta\phi$$

$$\therefore l/r = \delta\phi/\delta l \quad \text{which on substitution above gives}$$

$$l/c = \delta\phi/\delta l$$

integrating: $\phi = l^2/2c \quad \therefore l = (2c\phi)^{\frac{1}{2}}$

putting $\quad a = (2c)^{\frac{1}{2}}$

$$l = a(\phi)^{\frac{1}{2}} \tag{10.10}$$

when $\quad c = RL, a = (2RI)^{\frac{1}{2}}$ and *equation (10.10)* may be written:

$$l = (2RL\phi)^{\frac{1}{2}} \tag{10.11}$$

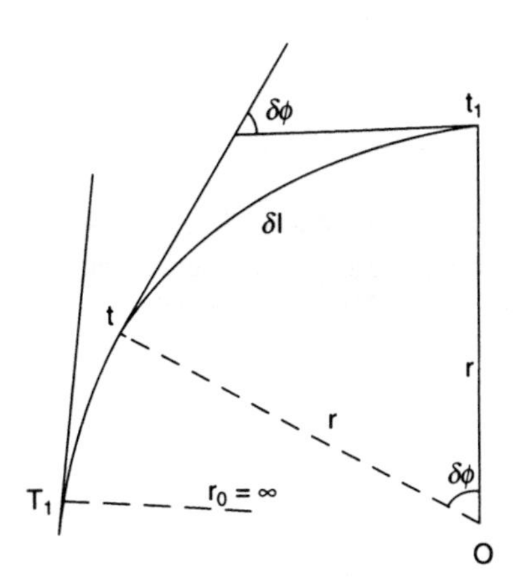

Fig. 10.26 *The transition curve*

The above expressions are for the *clothoid curve*, sometimes called the *Euler spiral*, which is the one most used in road design.

10.5.2 Curve design

The basic requirements in the design of transition curves are:

(1) The value of the minimum safe radius (R), and
(2) The length (L) of the transition curve. *Sections 10.5.5* or *10.5.6*

The value R may be found using either of the approaches *Sections 10.5.3* or *10.5.4*.

10.5.3 Centrifugal ratio

Centrifugal force is defined as $P = WV^2/Rg$; however, this 'overturning force' is counteracted by the mass (W) of the vehicle, and may be expressed as P/W, termed the *centrifugal ratio*. Thus, centrifugal ratio:

$$P/W = V^2/Rg \quad (10.12)$$

where V is the design speed in m/s, g is acceleration due to gravity in m/s^2 and R is the minimum safe radius in metres.

When V is expressed in km/h, the expression becomes:

$$P/W = V^2/127R \quad (10.13)$$

Commonly used values for centrifugal ratio are:

0.21 to 0.25 on roads, 0.125 on railways

Thus, if a value of $P/W = 0.25$ is adopted for a design speed of $V = 50$ km/h, then

$$R = \frac{50^2}{127 \times 0.25} = 79 \text{ m}$$

The minimum safe radius R may be set either equal to or greater than this value.

10.5.4 Coefficient of friction

The alternative approach to find R is based on Transport Research Laboratory (TRL) values for the coefficient of friction between the car tyres and the road surface.

Figure 10.27(a) illustrates a vehicle passing around a correctly super-elevated curve. The resultant of the two forces is N. The force F acting towards the centre of the curve is the friction applied by the car tyres to the road surface. These forces are shown in greater detail in *Figure 10.27(b)* from which it can be seen that:

$$F_2 = \frac{WV^2}{Rg}\cos\theta \quad \text{and} \quad F_1 = W\cos(90-\theta) = W\sin\theta$$

$$\therefore F = F_2 - F_1 = \frac{WV^2}{Rg}\cos\theta - W\sin\theta$$

$$\text{Similarly } N_2 = \frac{WV^2}{Rg}\sin\theta \quad \text{and} \quad N_1 = W\cos\theta$$

$$\therefore N = N_2 + N_1 = \frac{WV^2}{Rg}\sin\theta + W\cos\theta$$

$$\text{Then } \frac{F}{N} = \frac{\dfrac{WV^2}{Rg}\cos\theta - W\sin\theta}{\dfrac{WV^2}{Rg}\sin\theta + W\cos\theta} = \frac{\dfrac{V^2}{Rg} - \tan\theta}{\dfrac{V^2}{Rg}\tan\theta + 1}$$

For Highways Agency requirements, the maximum value for $\tan\theta = 0.07 = 7\%$, and as V^2/Rg cannot exceed 0.25 the term in the denominator can be ignored and

$$\frac{F}{N} = \frac{V^2}{Rg} - \tan\theta = \frac{V^2}{127R} - \tan\theta \tag{10.14}$$

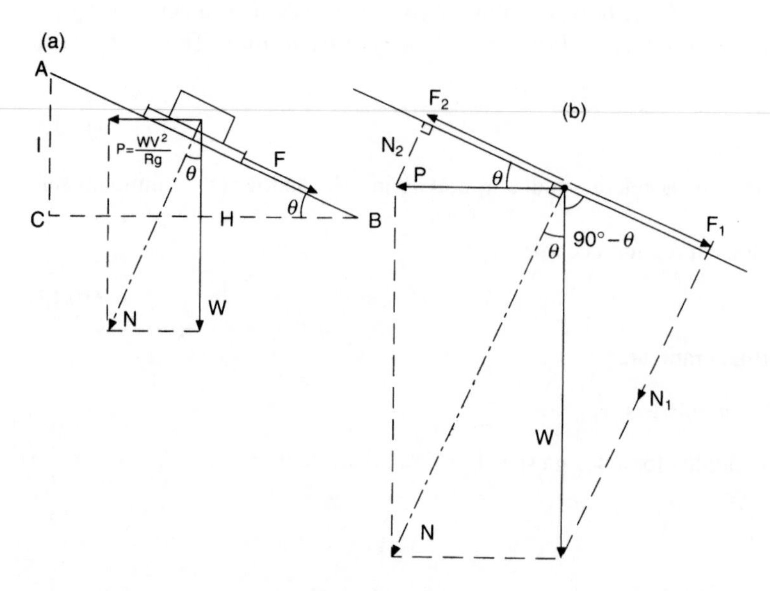

Fig. 10.27 *Forces on a super-elevated road*

To prevent vehicles slipping sideways, F/N must be greater than the coefficient of friction μ between tyre and road. The TRL quote value for μ of 0.15, whilst 0.18 may be used up to 50 km/h, thus:

$$V^2/127R \ngtr \tan\theta + \mu \qquad (10.15)$$

For example, if the design speed is to be 100 km/h, super-elevation limited to 7% and $\mu = 0.15$, then:

$$\frac{100^2}{127R} = 0.07 + 0.15$$

$$\therefore R = 360 \text{ m}$$

In the UK, the geometric parameters used in design are normally related to design speed. *Table 10.3* shows typical desirable and absolute minimum values for horizontal and vertical curvature; there is also an additional lower level designated 'limiting radius', specific of horizontal curvature.

Designs for new roads should aim to achieve the desirable values for each design parameter. However, absolute minimum values can be used wherever substantial saving in construction or environmental costs can be achieved.

The Highways Agency Technical Standard TD9/93 advises that in the design of new roads, the use of radii tighter than the limiting value is undesirable and not recommended.

10.5.5 Rate of application of super-elevation

It is recommended that on motorways super-elevation should be applied at a rate of 0.5%, on all-purpose roads at 1%, and on railways at 0.2%. Thus, if on a motorway the super-elevation were computed as 0.5 m, then 100 m of transition curve would be required to accommodate 0.5 m at the required rate of 0.5%, i.e. 0.5 m in 100 m = 0.5%. In this way the length L of the transition is found.

The amount of super-elevation is obtained as follows:

From the triangle of forces in *Figure 10.27(a)*

$$\tan\theta = V^2/Rg = 1/H = 1 \text{ in } H$$

$$\text{thus } H = Rg/V^2 = 127R/V^2, \quad \text{where } V \text{ is in km/h}$$

The percentage super-elevation (or crossfall), S may be found from:

$$S = V^2/2.828R \qquad (10.16)$$

The Highways Agency recommend the crossfall should never be greater than 7%, or less than 2.5%, to allow rainwater to run off the road surface.

It is further recommended that adverse camber should be replaced by a favourable crossfall of 2.5% when the value of V^2/R is greater than 5 and less than 7 (see *Table 10.3*).

Driver studies have shown that whilst super-elevation is instrumental to driver comfort and safety, it need not be applied too rigidly. Thus for sharp curves in urban areas with at-grade junctions and side access, super-elevation should be limited to 5%.

The rate of crossfall, combined with the road width, allows the amount of super-elevation to be calculated. Its application at the given rate produces the length L of transition required.

10.5.6 Rate of change of radial acceleration

An alternative approach to finding the length of the transition is to use values for 'rate of change of radial acceleration' which would be unnoticeable to passengers when travelling by train. The appropriate values were obtained empirically by W.H. Shortt, an engineer working for the railways; hence it is usually referred to as *Shortt's Factor.*

$$\text{Radial acceleration} = V^2/R$$

Thus, as radial acceleration is inversely proportional to R it will change at a rate proportional to the rate of change in R. The transition curve must therefore be long enough to ensure that the rate of change of radius, and hence radial acceleration, is unnoticeable to passengers.

Acceptable values for rate of change of radial acceleration (q) are 0.3 m/s^3, 0.45 m/s^3 and 0.6 m/s^3.

Now, as radial acceleration is V^2/R and the time taken to travel the length L of the transition curve is L/V, then:

$$\text{Rate of change of radial acceleration} = q = \frac{V^2}{R} \div \frac{L}{V} = \frac{V^3}{RL}$$

$$\therefore L = \frac{V^3}{Rq} = \frac{V^3}{3 \cdot 6^3 \cdot R \cdot q} \tag{10.17}$$

where the design speed (V) is expressed in km/h.

Although this method was originally devised for railway practice, it is also applied to road design. q should normally not be less than 0.3 m/s^3 for unrestricted design, although in urban areas it may be necessary to increase it to 0.6 m/s^3 or even higher, for sharp curves in tight locations.

10.6 SETTING-OUT DATA

Figure 10.28 indicates the usual situation of two straights projected forward to intersect at I with a clothoid transition curve commencing from tangent point T_1 and joining the circular arc at t_1. The second equal transition commences at t_2 and joins at T_2. Thus the composite curve from T_1 to T_2 consists of a circular arc with transitions at entry and exit.

(1) Fixing the tangent points T_1 and T_2

In order to fix T_1 and T_2 the tangent lengths T_1I and T_2I are measured from I back down the straights, or they are set out direct by coordinates.

$$T_1I = T_2I = (R + S)\tan \Delta/2 + C \tag{10.18}$$

where $S = \text{shift} = L^2/24R - L^4/(3! \times 7 \times 8 \times 2^3R^3) + L^6/(5! \times 11 \times 12 \times 2^5R^5)$

$$- L^8/(7! \times 15 \times 16 \times 2^7R^7)\ldots$$

and $C = L/2 - L^3/(2! \times 5 \times 6 \times 2^2R^2) + L^5/(4! \times 9 \times 10 \times 2^4R^4)$

$$- L^7/(6! \times 13 \times 14 \times 2^6R^6)\ldots$$

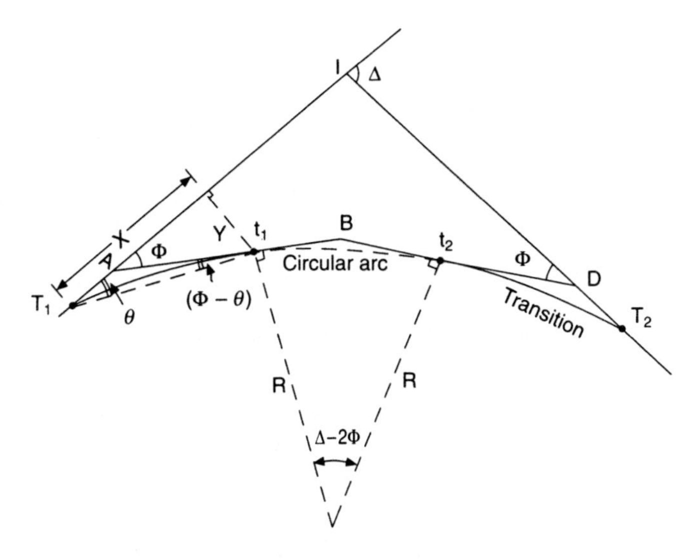

Fig. 10.28 *Transition and circular curves*

The values of S and C are abstracted from the *Highway Transition Curve Tables* (*Metric*) (see *Table 10.2*).

(2) Setting out the transitions

Referring to Figure 10.29:

The theodolite is set at T_1 and oriented to I with the horizontal circle reading zero. The transition is then pegged out using deflection angles (θ) and chords (Rankine's method) in exactly the same way as for a simple curve.

The data are calculated as follows:

(a) The length of transition L is calculated (see design factors in *Section 10.5.5* and *10.5.6*), assume $L = 100$ m.

(b) It is then split into, say, 10 arcs, each 10 m in length (ignoring through chainage), the equivalent chord lengths being obtained from:

$$A - \frac{A^3}{24R^2} + \frac{A^5}{1920R^4}, \quad \text{where } A \text{ is the arc length}$$

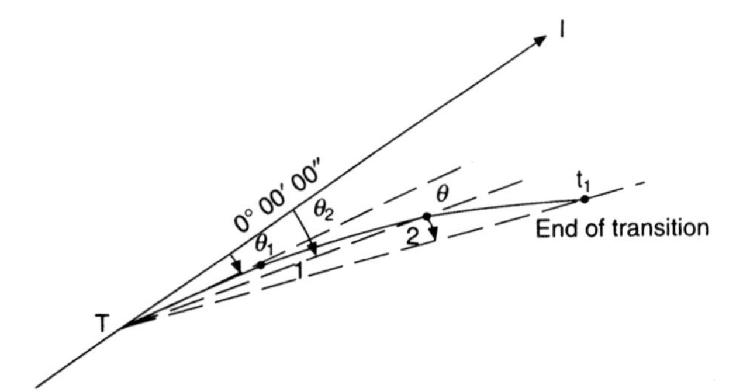

Fig. 10.29 *Setting out the transitions*

(c) The setting-out angles $\theta_1, \theta_2, \ldots, \theta_n$ are obtained as follows:

Basic formula for clothoid: $l = (2RL\phi)^{\frac{1}{2}}$

$$\therefore \Phi = \frac{l^2}{2RL} = \frac{L}{2R} \quad \text{when } l = L \tag{10.19}$$

(l is any distance along the transition other than total distance L)

then $$\theta = \Phi/3 - 8\Phi^3/2835 - 32\Phi^5/467\,775\ldots \tag{10.20}$$

$= \Phi/3 - N$, where N may be taken from tables and ranges in value from $0.1''$ when $\Phi = 3°$, to $34' \, 41.3''$ when $\Phi = 86°$ (see *Table 10.1*).

Now $$\frac{\phi_1}{\Phi} = \frac{l_1^2}{L^2} \tag{10.21}$$

$\therefore \phi_1 = \Phi \dfrac{l_1^2}{L^2}$ where l_1 = chord length = 10 m, say

and $\theta_1 = \phi_1/3 - N_1$ (where N_1 is the value relative to ϕ_1 in *Table 10.1*)

Similarly $\phi_2 = \Phi \dfrac{l_2^2}{L^2}$ where l_2 = 20 m

and $\theta_2 = \phi_2/3 - N_2$ and so on.

Table 10.1 *Interpolated deflection angles*

For any point on a spiral where angle consumed = ϕ, the true deflection is $\phi/3$ minus the correction tabled below. The back angle is $2\phi/3$ plus the same correction.

Angle consumed ϕ	$\phi/3$		*Deduct N*		*Deflection angle*			*Angle consumed* ϕ	$\phi/3$		*Deduct N*		*Deflection angle*		
°	°	′	′	″	°	′	″	°	°	′	′	″	°	′	″
			NIL												
2	0	40		0.0		40	0	45	15	00	4	46.2	14	55	13.8
3	1	00		0.1		59	59.9	46	15	20	5	6.0	15	14	54.0
4	1	20		0.2	1	19	59.8	47	15	40	5	26.6	15	34	33.4
5	1	40		0.4	1	39	59.6	48	16	00	15	48.1	15	54	11.9
6	2	00		0.7	1	59	59.3	49	16	20	15	10.6	15	13	49.4
7	2	20		1.0	2	19	59.0	50	16	40	16	34.1	16	33	25.9
Continued at 1° intervals of ϕ															
41	13	40	3	35.9	13	36	24.1	84	28	00	32	14.4	27	27	45.6
42	14	00	3	52.1	13	56	7.7	85	28	20	33	26.9	27	46	33.1
43	14	20	4	9.4	14	15	50.6	86	28	40	34	41.3	28	5	18.7
44	14	40	4	27.4	14	35	32.6								

Note that:

(a) The values for l_1, l_2, etc., are accumulative.
(b) Thus the values obtained for θ_1, θ_2, etc., are the final setting-out angles and are obviously not to be summed.
(c) Although the chord length used is accumulative, the method of setting out is still the same as for the simple curve.

(3) Setting-out circular arc t_1t_2

In order to set out the circular arc it is first necessary to establish the direction of the tangent t_1B (*Figure 10.28*). The theodolite is set at t_1 and backsighted to T_1 with the horizontal circle reading $[180° - (\Phi - \theta)]$, setting the instrument to zero will now orient it in the direction t_1B with the circle reading zero, prior to setting-out the simple circular arc. The angle $(\Phi - \theta)$ is called the *back-angle to the origin* and may be expressed as follows:

$$\theta = \Phi/3 - N$$

$$\therefore (\Phi - \theta) = \Phi - (\Phi/3 - N) = 2/3\Phi + N \quad (10.22)$$

and is calculated using *equation* (*10.22*) or obtained from *Table 10.1*.

The remaining setting-out data are obtained as follows:

(a) As each transition absorbs an angle Φ, then the angle subtending the circular arc $= (\Delta - 2\Phi)$.
(b) Length of circular arc $= R(\Delta - 2\Phi)$, which is then split into the required chord lengths C.
(c) The deflection angles $\delta° = 28.6479 \cdot C/R$ are then set out from the tangent t_1B in the usual way.

The second transition is best set out from T_2 to t_2. Setting-out from t_2 to T_2 involves the 'osculating-circle' technique (see *Section 10.9*).

The preceding formulae for clothoid transitions are specified in accordance with the *Highway Transition Curve Tables* (*Metric*) compiled by the County Surveyors' Society. As the equations involved in the setting-out data are complex, the information may be taken straight from tables. However, approximation of the formulae produces two further transition curves, the *cubic spiral* and the *cubic parabola* (see *Section 10.7*).

In the case of the clothoid, *Figure 10.28* indicates an offset Y at the end of the transition, distance X along the straight, where

$$X = L - L^3/(5 \times 4 \times 2!R^2) + L^5/(9 \times 4^2 \times 4!R^4) - L^7/(13 \times 4^3 \times 6!R^6) + \ldots \quad (10.23)$$

$$Y = L^2/(3 \times 2R) - L^4/(7 \times 3! \times 2^3R^3) + L^6/(11 \times 5! \times 2^5R^5) - L^8/(15 \times 7! \times 2^7R^7) + \ldots \quad (10.24)$$

The clothoid is always set out by deflection angles, but the values for X and Y are useful in the large-scale plotting of such curves, and are taken from tables.

Refer to the end of the chapter for derivation of clothoid formulae.

10.6.1 Highway transition curve tables (metric)

Highway transition curve tables have now largely been superseded by the use of software for design purposes. However, this section and its associated table, *Table 10.2*, have been retained from previous editions of this book because the numerical data give a useful insight into the values involved in the design of transition curves.

An examination of the complex equations defining the clothoid transition spiral indicates the obvious need for tables of prepared data to facilitate the design and setting out of such curves. These tables have

Table 10.2 *Highway transition curves*

Gain of Accn. m/s^3	0.30	0.45	0.60	Increase in Degree of Curve per Metre = $D/L = 0° 8' 0.0''$
Speed Value km/h	84.4	96.6	106.3	RL Constant = 42971.835

Degree of Curvature Based on 100 m. Standard Arc

Radius R	Degree of Curve D			Spiral Length L	Φ Angle Consumed			Shift S	R + S R	C	Long Chord	Coordinates X	Coordinates Y	Θ Deflection Angle from Origin			Back Angle to Origin		
Metres	°	′	″	Metres	°	′	″	Metres	Metres	Metres	Metres	Metres	Metres	°	′	″	°	′	″
8594.3669	0	40	0.0	5.00	0	1	0.0	0.0001	8594.3670	2.5000	5.0000	5.0000	0.0005	0	0	20.0	0	0	40.0
4297.1835	1	20	0.0	10.00	0	4	0.0	0.0010	4297.1844	5.0000	10.0000	10.0000	0.0039	0	1	20.0	0	2	40.0
2864.7890	2	0	0.0	15.00	0	9	0.0	0.0033	2864.7922	7.5000	15.0000	15.0000	0.0131	0	3	0.0	0	6	0.0
2148.5917	2	40	0.0	20.00	0	16	0.0	0.0078	2148.5995	10.0000	20.0000	20.0000	0.0310	0	5	20.0	0	10	40.0
1718.8734	3	20	0.0	25.00	0	25	0.0	0.0152	1718.8885	12.5000	24.9999	24.9999	0.0606	0	8	20.0	0	16	40.0
1432.3945	4	0	0.0	30.00	0	36	0.0	0.0262	1432.4207	14.9999	29.9999	29.9997	0.1047	0	12	0.0	0	24	0.0
1227.7667	4	40	0.0	35.00	0	49	0.0	0.0416	1227.8083	17.4999	34.9997	34.9993	0.1663	0	16	20.0	0	32	40.0
1074.2959	5	20	0.0	40.00	1	4	0.0	0.0621	1074.3579	19.9998	39.9994	39.9986	0.2482	0	21	20.0	0	42	40.0
954.9297	6	0	0.0	45.00	1	21	0.0	0.0884	955.0180	22.4996	44.9989	44.9975	0.3534	0	27	0.0	0	54	0.0
859.4367	6	40	0.0	50.00	1	40	0.0	0.1212	859.5579	24.9993	49.9981	49.9958	0.4848	0	33	20.0	1	6	40.0
781.3061	7	20	0.0	55.00	2	1	0.0	0.1613	781.4674	27.4989	54.9970	54.9932	0.6452	0	40	20.0	1	20	40.0
716.1972	8	0	0.0	60.00	2	24	0.0	0.2094	716.4067	29.9982	59.9953	59.9895	0.8377	0	48	0.0	1	36	0.0
661.1051	8	40	0.0	65.00	2	49	0.0	0.2663	661.3714	32.4974	64.9930	64.9843	1.0650	0	56	19.9	1	52	40.1
613.8834	9	20	0.0	70.00	3	16	0.0	0.3325	614.2159	34.9962	69.9899	69.9772	1.3300	1	5	19.9	2	10	40.1
572.9578	10	0	0.0	75.00	3	45	0.0	0.4090	573.3668	37.4946	74.9857	74.9679	1.6357	1	14	59.8	2	30	0.2
537.1479	10	40	0.0	80.00	4	16	0.0	0.4964	537.6443	39.9926	79.9803	79.9556	1.9850	1	25	19.8	2	50	40.2
505.5510	11	20	0.0	85.00	4	49	0.0	0.5953	506.1463	42.4900	84.9733	84.9399	2.3807	1	36	19.7	3	12	40.3
477.4648	12	0	0.0	90.00	5	24	0.0	0.7066	478.1715	44.9867	89.9645	89.9201	2.8256	1	47	59.5	3	36	0.5
452.3351	12	40	0.0	95.00	6	1	0.0	0.8310	453.1661	47.4825	94.9534	94.8953	3.3227	2	0	19.3	4	0	40.7
429.7183	13	20	0.0	100.00	6	40	0.0	0.9692	430.6875	49.9774	99.9398	99.8647	3.8748	2	13	19.1	4	26	40.9
409.2556	14	0	0.0	105.00	7	21	0.0	1.1218	410.3774	52.4712	104.9232	104.8273	4.4846	2	26	58.8	4	54	1.2
390.6530	14	40	0.0	110.00	8	4	0.0	1.2897	391.9427	54.9637	109.9031	109.7822	5.1550	2	41	18.4	5	22	41.6
373.6681	15	20	0.0	115.00	8	49	0.0	1.4734	375.1416	57.4546	114.8790	114.7280	5.8888	2	56	17.9	5	52	42.1
358.0986	16	0	0.0	120.00	9	36	0.0	1.6738	359.7725	59.9439	119.8503	119.6636	6.6886	3	11	57.3	6	24	2.7

been produced by the County Surveyors' Society under the title *Highway Transition Curve Tables (Metric)*, and contain a great deal of valuable information relating to the geometric design of highways. A very brief sample of the tables is given here simply to convey some idea of the format and information contained therein (*Table 10.2*).

As shown in *Section 10.6(3)*, $\theta = \phi/3 - N$ and the 'back-angle' is $2\phi/3 + N$, all this information for various values of ϕ is supplied in *Table 10.1* and clearly shows that for large values of ϕ, N cannot be ignored.

Part only of *Table 10.2* is shown and it is the many tables like this that provide the bulk of the design data. Much of the information and its application to setting out should be easily understood by the reader, so only a brief description of its use will be given here.

Use of tables

(1) Check the angle of intersection of the straights (Δ) by direct measurement in the field.
(2) Compare Δ with 2Φ, if $\Delta \leq 2\Phi$, then the curve is wholly transitional.
(3) Abstract $(R + S)$ and C in order to calculate the tangent lengths $= (R + S)\tan\Delta/2 + C$.
(4) Take Φ from tables and calculate length of circular arc using $R(\Delta - 2\Phi)$, or, if working in 'degree of curvature' D, use

$$\frac{100(\Delta - 2\Phi)}{D}$$

(5) Derive chainages at the beginning and end of both transitions.
(6) Compute the setting-out angles for the transition $\theta_1 \ldots \theta$ from $\phi_1/\Phi = l_1^2/L^2$ from which $\theta_1 = \phi_1/3 - N_1$, and so on, for accumulative values of l.
(7) As control for the setting out, the end point of the transition can be fixed first by turning off from T_1 (the start of the transition) the 'deflection angle from the origin' θ and laying out the 'long chord' as given in the tables. Alternatively, the right-angled offset Y distance X along the tangent may be used.
(8) When the first transition is set out, set up the theodolite at the end point and with the theodolite reading $(180° - (\frac{2}{3}\Phi + N))$, backsight to T_1. Turn the theodolite to read 0° when it will be pointing in a direction tangential to the start of the circular curve prior to its setting out. This process has already been described.
(9) As a check on the setting out of the circular curve take $(R + S)$ and S from the tables to calculate the apex distance $= (R + S)(\sec\Delta/2 - 1) + S$, from the intersection point I of the straights to the centre of the circular curve.
(10) The constants RL and D/L are given at the head of the tables and can be used as follows:

(a) Radius at any point P on the transition $= r_p = RL/l_p$
(b) Degree of curve at $P = D_p = (D/L) \times l_p$, where l_p is the distance to P measured along the curve from T_1
Similarly:
(c) Angle consumed at $P = \phi_p = \dfrac{l_p^2}{2RL}$ or $\dfrac{l_p^2}{200} \times \dfrac{D}{L}$

(d) Setting-out angle from T_1 to $P = \theta_p = \dfrac{\phi_p}{3} - N_p$ or

$$\frac{l^2}{600} \times \frac{D}{L} - N_p$$

Worked examples

Example 10.8 It is required to connect two straights ($\Delta = 50°$) with a composite curve, comprising entry transition, circular arc and exit transition. The initial design parameters are $V = 105$ km/h, $q = 0.6$ m/s^3, $\mu = 0.15$ and crossfall 1 in 14.5 ($\tan\theta = 0.07$).

(1) $$V^2/127R = \tan\theta + \mu$$

$$105^2/127R = 0.07 + 0.15 = 0.22$$

$$R = 105^2/127 \times 0.22 = 394.596 \text{ m}$$

The nearest greater value is 409.2556 m for a value of $V = 106.3$ km/h (*Table 10.2*).

(2) From *Table 10.2*, the length L of transition is given as 105 m. Purely to illustrate the application of formulae:

$$(L = V^3/3.6^3 \times R \times q = 106.3^3/3.6^3 \times 409.2556 \times 0.6 = 104.8 \text{ m})$$

(3) From *Table 10.2*, $\Phi = 7°\,21'\,00''$

Check $\Phi = L/2R = 105/2 \times 409.2556 = 0.12828169$ rads $= 7°\,21'\,00''$

Now as $2\Phi < \Delta$, there is obviously a portion of circular arc subtended by angle of $(\Delta - 2\Phi)$.

(4) From *Table 10.2*, find the tangent length:

$$(R + S)\tan\Delta/2 + C$$

where $(R + S) = 410.3374$, $C = 52.4721$ } see *Table 10.2*

$$T_1I = T_2I = 410.3774 \tan 25° + 52.4712 = 243.8333 \text{ m}$$

(5) The chainage of the intersection point $I = 5468.39$ m (see *Figure 10.28*)

Chn $T_1 = 5468.399 - 243.833 = 5224.566$ m

Chn $t_1 = 5224.566 + 105 = 5329.566$ m

Length of circular arc $= R(\Delta - 2\Phi) = 252.1429$ m

or using 'degree of curve' $(D) = 100(\Delta - 2\Phi)/D = 252.1429$ m (check)

Chn $t_2 = 5329.566 + 252.143 = 5581.709$ m

Chn $T_2 = 5581.709 + 105 = 5686.709$ m

(6) As a check, the ends of the transition could be established using the offset $Y = 4.485$ m at distance $X = 104.827$ m along the tangent (see *Table 10.2*) or by the 'long chord' $= 104.923$ m at an angle θ to the tangent of $2°\,26'\,58.8''$. This procedure will establish t_1 and t_2 relative to T_1 and T_2 respectively. Now the setting-out angles $(\theta_1, \theta_2, \ldots, \theta)$ are computed. If the curve was to be set out by 10-m chords and there was no through chainage, their values could simply be abstracted from *Table 10.2* as shown:

10 m $= 0°\,01'\,20.0''$

20 m $= 0°\,05'\,20.0''$, and so on

These values would be subtracted from 360°, for use in setting-out the 2nd transition from T_2 back to t_2.

However, the usual way is to set out on a through chainage basis.

Hence using 10-m standard chords, as the chainage at $T_1 = 5224.566$, the first sub-chord $= 10 - 4.566 = 5.434$ m. The setting-out angles are computed from *equations (10.20)* and *(10.21)*, i.e.

(a) $\Phi = 7^\circ\, 21'\, 00''$ from *Table 10.2*

and $\theta = 2^\circ\, 26'\, 58.8''$

as $\theta_1/\theta = (l_1/L)^2$

then $\theta_1 = 2^\circ\, 26'\, 58.8''\, (5.434/105)^2 = 0^\circ\, 00'\, 23.6'' - N_1$

(b) Alternatively using 'degree of curve' D:

$$\theta_1 = \frac{l_1^2}{600} \cdot \frac{D}{L} - N_1$$

D/L is given in *Table 10.2* $= 0^\circ\, 08'\, 00''$

$$\therefore \frac{D}{600 \cdot L} = 0.8'' \text{ (a constant)}$$

and $\theta_1 = 0.8''\, (5.434)^2 = 0^\circ\, 00'\, 23.6'' - N_1$

(It can be seen from *Table 10.1* that the deduction N_1 is negligible and will have a maximum value for the formal setting-out angle θ of just over 1″.)

Using either of the above approaches and remembering that the chord lengths are accumulative in the computation, the setting-out angles are:

Entry Transition

Chord m	*Chainage* m	*Setting-out Angle* (θ)	*Deduction N*	*Remarks*
0	5224.566	–	–	T_1 – Start of spiral
5.434	5230	0° 00′ 23.6″	–	Peg. 1
10	5240	0° 03′ 10.6″	–	Peg. 2
10	5250	0° 08′ 37.6″	–	Peg. 3
		etc.		

When computing the setting-out angles for the exit transition from T_2 back to t_2 one starts with the length of the final sub-chord at T_2.

(7) When the entry transition is set out to t_1, the theodolite is moved to t_1 and backsighted to T_1, with the horizontal circle reading:

180° – (back-angle to the origin)

$= 180^\circ - (\Phi - \theta)$

$= 180^\circ - 4^\circ\, 54'\, 01.2''$ (taken from final column of *Table 10.2*)

$= 175^\circ\, 05'\, 58.8$

Rotating the upper circle to read $0^\circ\, 00'\, 00''$ will establish the line of sight tangential to the route of the circular arc, ready for setting out.

(8) As the chainage of t_1 = 5329.566, the first sub-chord = 0.434 m, thereby putting Peg 1 on chainage 5330 m.

The setting out angles are now calculated in the usual way as shown in *Section 10.2.1*.

If it is decided to set out the exit transition from the end of the circular arc, then the theory of the osculating circle is used (see *Section 10.9.2*).

10.7 CUBIC SPIRAL AND CUBIC PARABOLA

Approximation of the clothoid formula produces the cubic spiral and cubic parabola, the latter being used on railway and tunnelling work because of the ease in setting out by offsets. The cubic spiral can be used for minor roads, as guide for excavation prior to the clothoid being set out, or as check on clothoid computation.

$Y = L^2/6R$, which when $L = l$, $Y = y$, becomes

$$y = l^3/6RL \tag{10.25}$$

Approximating *equation* (*10.23*) gives

$X = L$, thus $x = l$

$$\therefore y = x^3/6RL \text{ (the equation for a cubic parabola)} \tag{10.26}$$

In both cases:

Tangent length $T_1I = (R + S)\tan \Delta/2 + C$

$$\text{where} \quad S = L^2/24R \tag{10.27}$$

$$\text{and} \quad C = L/2 \tag{10.28}$$

$$\Phi = L/2R = l^2/2RL \tag{10.29}$$

$$\text{and} \quad \theta = \Phi/3 \tag{10.30}$$

The deflection angles for these curves may be obtained as follows (the value of N being ignored):

$$\theta_1/\theta = l_1^2/L^2, \text{ where } l \text{ is the chord/arc length} \tag{10.31}$$

When the value of $\Phi \approx 24°$, the radius of these curves starts to increase again, which makes them useless as transitions.

Refer to *worked examples* for application of the above equations, and note 'back-angle to origin' would be $(\Phi - \theta)$ or $\frac{2}{3}\Phi$.

10.8 CURVE TRANSITIONAL THROUGHOUT

A curve transitional throughout (*Figure 10.30*) comprises two transitions meeting at a common tangent point t.

$$\text{Tangent length } T_1I = X + Y \tan \Phi \tag{10.32}$$

where X and Y are obtained from *equations* (*10.23*) and (*10.24*) and $\Phi = \Delta/2 = L/2R$.

$$\therefore \Delta = L/R \tag{10.33}$$

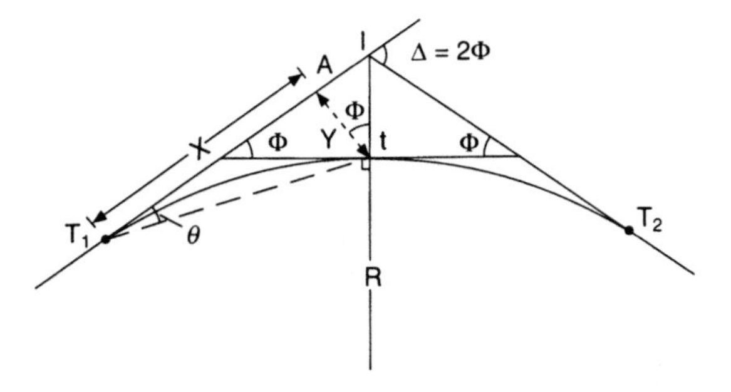

Fig. 10.30 *Transitional curve throughout*

10.9 THE OSCULATING CIRCLE

Figure 10.31 illustrates a transition curve T_1PE. Through P; where the transition radius is r, a simple curve of the same radius is drawn and called the *osculating circle.*

At T_1 the transition has the same radius as the straight T_1I, that is, ∞, but diverges from it at a constant rate. Exactly the same condition exists at P with the osculating circle, that is, the transition has the same radius as the osculating circle, r, but diverges from it at a constant rate. Thus if chords $T_1t = Pa = Pb = l$, then:

$$\text{angle} \quad IT_1t = aPb = \theta_1$$

This is the theory of the osculating circle, and its application is described in the following sections.

10.9.1 *Setting out with the theodolite at an intermediate point along the transition curve*

Figure 10.32 illustrates the situation where the transition has been set out from T_1 to P_3 in the normal way. The sight T_1P_4 is obstructed and the theodolite must be moved to P_3 where the remainder of the transition will be set out. The direction of the tangent P_3E is first required from the back-angle $(\phi_3 - \theta_3)$.

From the figure it can be seen that the angle from the tangent to the chord P_3P_4' on the osculating circle is $\delta_1^\circ = 28.6479 \times l/r_3$. The angle between the chord on the osculating circle and that on the transition

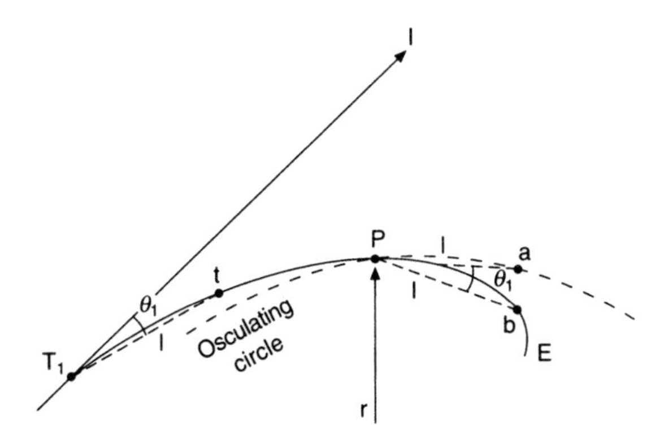

Fig. 10.31 *Osculating circle*

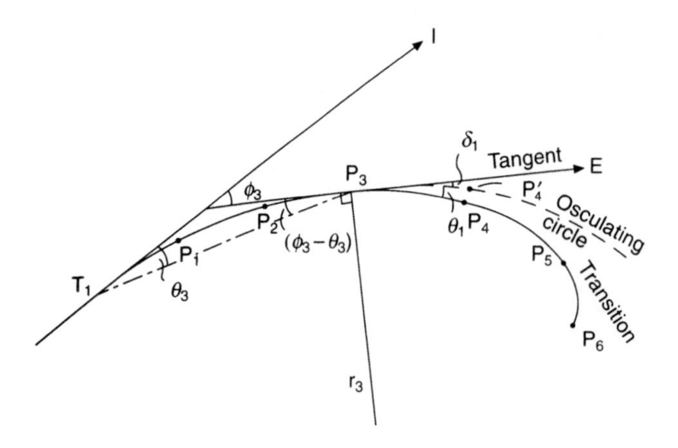

Fig. 10.32 *Setting out the transition from an intermediate point*

is $P_4'P_3P_4 = \theta_1$, thus the setting-out angle from the tangent to $P_4 = (\delta_1 + \theta_1)$, to $P_5 = (\delta_2 + \theta_2)$ and to $P_6 = (\delta_3 + \theta_3)$, etc.

For example, assuming $\Delta = 60°$, $L = 60$ m, l = chord = 10 m, $R = 100$ m and $T_1P_3 = 30$ m, calculate the setting-out angles for the remainder of the transition from P_3.

From basic formula:

$$\phi_3 = \frac{l_3^2}{2RL} = \frac{30^2}{2 \times 100 \times 60} = 4°\,17'\,50''\ (-N_3,\text{ if clothoid})$$

or, if curve is defined by its 'degree of curvature' = D, then

$$\phi_3 = \frac{l_3^2}{200} \cdot \frac{D}{L}\ (-N_3,\text{ if clothoid})$$

Thus the back-angle to the origin is found $\frac{2}{3}\,\phi_3$, and the tangent established as already shown.

Now from $\Phi = L/2R$, and $\theta = \Phi/3$, the angles θ_1, θ_2, and θ_3 are found as normal. In practice these angles would already be available, having been used to set out the first 30 m of the transition.

Before the angles to the osculating circle can be found, the value of r_3 must be known, thus, from $rl = RL$:

$$r_3 = RL/l_3 = 100 \times 60/30 = 200\text{ m}$$

or 'degree of curvature' at P_3, 30 m from $T_1 = \dfrac{D}{L} \times l_3$

$$\therefore \delta_1^\circ = 28.6479\left(\frac{10}{200}\right) = 1°\,25'\,57''$$

$$\delta_2 = 2\delta_1 \text{ and } \delta_3 = 3\delta_1 \text{ as for a simple curve}$$

The setting-out angles are then $(\delta_1 + \theta_1)$, $(\delta_2 + \theta_2)$, $(\delta_3 + \theta_3)$.

10.9.2 Setting out transition from the circular arc

Figure 10.33 indicates the second transition in *Figure 10.28* to be set out from t_2 to T_2. The tangent t_2D would be established by backsighting to t_1 with the instrument reading $[180° - (\Delta - 2\Phi)/2]$, setting to zero to fix direction t_2D. It can now be seen that the setting-out angles here would be $(\delta_1 - \theta_1)$, $(\delta_2 - \theta_2)$, etc., computed in the usual way.

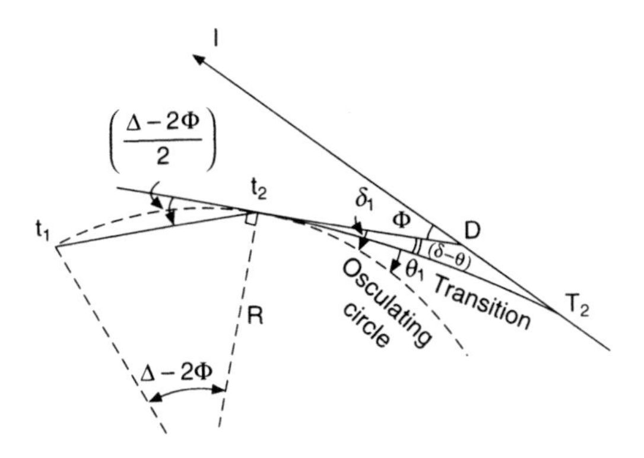

Fig. 10.33 *Setting out the transition from the end of the circular curve*

Consider the previous example, where the chainage at the end of the circular arc (t_2) was 5581.709 m, then the first sub-chord = 8.291 m.

The setting-out angles for the transition are computed in the usual way. Use, say, $D/600L = 0.8''$, then:

$$\theta_1 = 0.8''(8.291)^2 = 0° \, 00' \, 55.0''$$

$$\theta_2 = 0.8''(18.291)^2 = 0° \, 04' \, 27.6''$$

$$\theta_3 = 0.8''(28.291)^2 = 0° \, 10' \, 40.3''$$

and so on to the end of the transition.

Now compute the setting-out angles for a circular curve in the usual way:

$$\delta° = 28.6479 \times (C/R)$$

$$= 28.6479 \times (8.291/409.2556) = 0° \, 34' \, 49.3''$$

Also $\delta°_{10} = 28.6479(10/409.2556) = 0° \, 42' \, 00.0''$

Remembering that the angles (δ) for a circular arc are accumulative, the setting-out angles for the transition are:

Chord (m)	*Chainage* (m)	δ			θ			*Setting-out angle* $\delta - \theta$			*Remarks*
		°	′	″	°	′	″	°	′	″	
8.291	5990	0	34	49.3	0	00	55.0	0	33	54.3	Peg 1
10	5600	01	16	49.3	0	04	27.6	01	12	22.3	Peg 2
10	5610	01	58	49.3	0	10	40.3	01	48	09.0	Peg 3
etc.	etc.	etc.			etc.			etc.			etc.

10.9.3 Transitions joining arcs of different radii (compound curves)

Figure 10.14 indicates a compound curve requiring transitions at T_1, t and T_2. To permit the entry of the transitions the circular arcs must be shifted forward as indicated in *Figure 10.34* where

$$S_1 = L_1^2/24R_1 \quad \text{and} \quad S_2 = L_2^2/24R_2$$

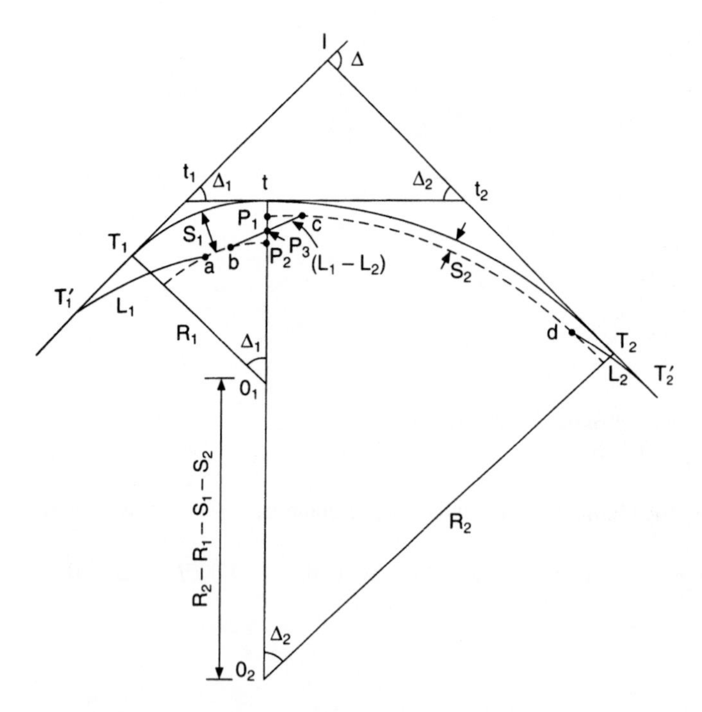

Fig. 10.34 *Compound curve construction*

The lengths of transition at entry (L_1) and exit (L_2) are found in the normal way, whilst the transition connecting the compound arcs is:

$$bc = L = (L_1 - L_2)$$

The distance $P_1P_2 = (S_1 - S_2)$ is bisected by the transition curve at P_3. The curve itself is bisected and length $bP_3 = P_3c$. As the curves at entry and exit are set out in the normal way, only the fixing of their tangent points T_1' and T_2' will be considered. In triangle t_1It_2:

$$t_1t_2 = t_1t + tt_2 = (R_1 + S_1)\tan\Delta_1/2 + (R_2 + S_2)\tan\Delta_2/2$$

from which the triangle may be solved for t_1I and t_2I.

$$\text{Tangent length} \quad T_1'I = T_1't_1 + t_1I = (R_1 + S_1)\tan\Delta_1/2 + L_1/2 + t_1I$$

$$\text{and} \quad T_2'I = T_2't_2 + t_2I = (R_2 + S_2)\tan\Delta_2/2 + L_2/2 + t_2I$$

The curve bc is drawn enlarged in *Figure 10.35* from which the method of setting out, using the osculating circle, may be seen.

Setting-out from b, the tangent is established from which the setting-out angles would be $(\delta_1 - \theta_1)$, $(\delta_2 - \theta_2)$, etc., as before, where δ_1, the angle to the osculating circle, is calculated using R_1.

If setting out from C, the angles are obviously $(\delta_1 + \theta_1)$, etc., where δ_1 is calculated using R_2.

Alternatively, the curve may be established by right-angled offsets from chords on the osculating circle, using the following equation:

$$y = \frac{x^3}{6RL} = \frac{x^3}{L^3}\frac{L^2}{6R} \quad \text{where} \quad \frac{L^2}{6R} = 4S$$

$$\therefore\ y = \frac{4x^3}{L^3}(S_1 - S_2) \tag{10.34}$$

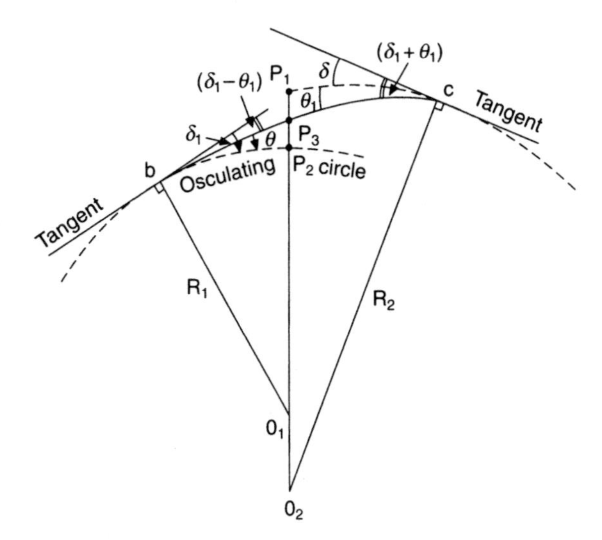

Fig. 10.35 *Transition between arcs of different radii*

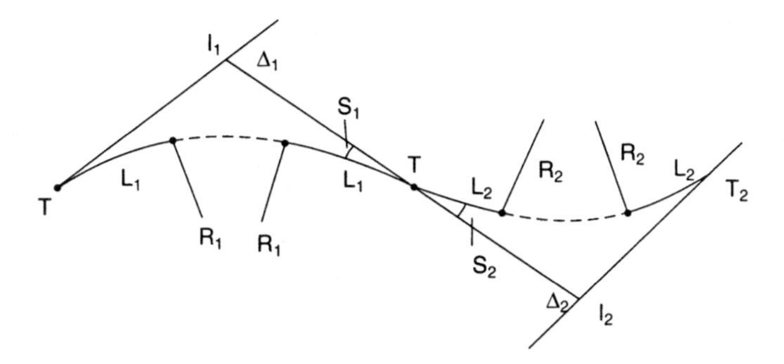

Fig. 10.36 *Reverse compound curve*

It should be noted that the osculating circle provides only an approximate solution, but as the transition is usually short, it may be satisfactory in practice. In the case of a reverse compound curve (*Figure 10.36*):

$$S = (S_1 + S_2), \quad L = (L_1 + L_2) \quad \text{and} \quad y = \frac{4x^3}{L^3}(S_1 + S_2) \tag{10.35}$$

otherwise it may be regarded as two separate curves.

10.9.4 Coordinates on the transition spiral (*Figure 10.37*)

The setting-out of curves by traditional methods of angles and chords has been dealt with. However, these methods are frequently superseded by the use of coordinates (as indicated in *Section 10.2.4*) of circular curves. The method of calculating coordinates along the centre-line of a transition curve is probably best illustrated with a worked example.

Consider *Worked example 10.8* dealing with the traditional computation of a clothoid spiral using Highway Transition Curve Tables. As with circular curves, it is necessary to calculate the traditional setting-out data first, as an aid to calculating coordinates.

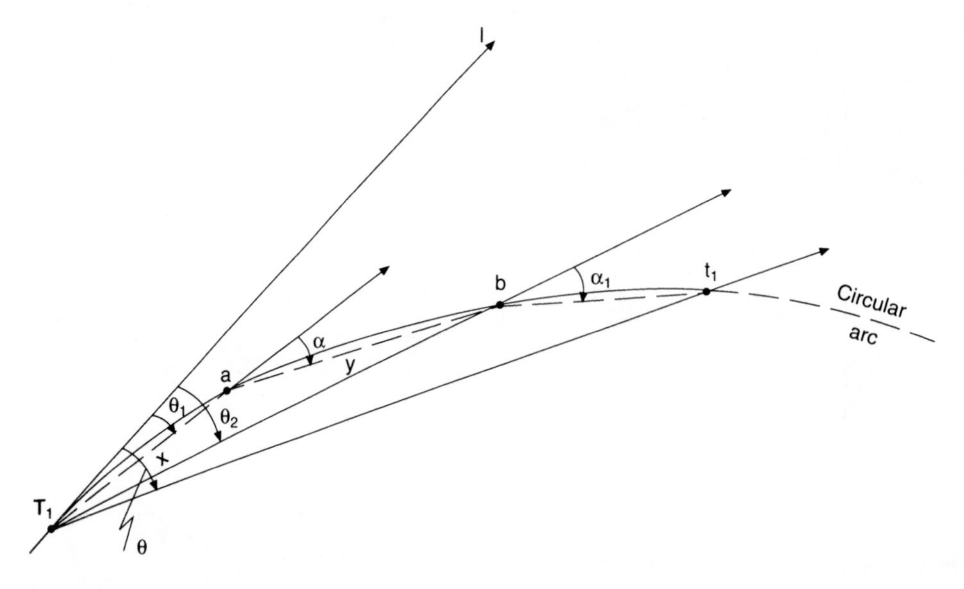

Fig. 10.37 *Coordinates along a transition curve*

In *Example 10.8* the first three setting-out angles are:

$$\theta_1 = 0^\circ\, 00'\, 23.6'', \text{ sub-chord} = 5.434 \text{ m } (x)$$

$$\theta_2 = 0^\circ\, 03'\, 10.6'', \text{ standard chord} = 10 \text{ m } (y)$$

$$\theta_3 = 0^\circ\, 08'\, 37.6'', \text{ standard chord} = 10 \text{ m } (y)$$

If working in coordinates, the coordinates of T_1, the tangent point and I, the intersection point, would be known, say.

$$T_1 = \text{E } 500.000, \text{ N } 800.000$$

And calculated using the coordinates of I, the WCB of $T_1I - 10^\circ\, 25'\, 35.0''$.

$$\text{Now} \quad \text{WCB} T_1 a = \text{WCB } T_1 I + \theta_1 = 10^\circ\, 25'\, 58.6''$$

and a sub-chord length $T_1 a = x = 5.434$ m.

Using the P and R keys, or the traditional formula the coordinates of the line $T_1 a$ are calculated, thus:

$$\Delta\text{E} = 0.9840 \text{ m}, \Delta\text{N} = 5.3442 \text{ m}$$

$$\text{E}_a = 500.000 + 0.9840 = 500.9840 \text{ m}$$

$$\text{N}_a = 800.000 + 5.3442 = 805.3442 \text{ m}$$

In triangle $T_1 ab$:

$$aT_1 b = (\theta_2 - \theta_1) = 0^\circ\, 02'\, 47.0''$$

Using the sine rule, the angle at b can be calculated:

$$\frac{T_1a}{\sin T_1ba} = \frac{ab}{\sin aT_1b}$$

$$\sin T_1ba = \frac{T_1a \sin aT_1b}{ab} = \frac{5.434 \sin 0^\circ\, 02'\, 47.0''}{10}$$

$$T_1ba = 00^\circ\, 01'\, 30.8'', \text{ and}$$

$$\alpha = aT_1b + T_1ba = 00^\circ\, 04'\, 17.8''$$

and $\quad \text{WCB } ab = \text{WCB } T_1a + \alpha = 10^\circ\, 30'\, 16.4''$

$$\text{dist. } ab = 10 \text{ m}$$

$$\therefore \Delta E_{ab} = 1.8231, \quad \Delta N_{ab} = 9.8324$$

and

$$E_b = E_a + \Delta E_{ab} = 500.9840 + 1.8231 = 502.8071 \text{ m}$$

$$N_b = N_a + \Delta N_{ab} = 805.3442 + 9.8324 = 815.1766 \text{ m}$$

This procedure is now repeated to the end of the spiral, as follows:

In triangle T_1bt_1,

(1) calculate distance T_1b from the coordinates of T_1 and b
(2) calculate angle T_1t_1b by sine rule, i.e.

$$\frac{T_1b}{\sin T_1t_1b} = \frac{bt_1}{\sin bT_1t_1}$$

where bt_1 is a known chord or sub-chord length and angle $bT_1t_1 = (\theta - \theta_2)$ in this instance
(3) calculate $\alpha_1 = bT_1t_1 + T_1t_1b$
(4) WCB T_1b = WCB $T_1l + \theta_2$
(5) WCB bt_1 = WCB $T_1b + \alpha_1$ and length is known
(6) computed ΔE, ΔN of the line bt_1 and add then algebraically to the E_b and N_b respectively, to give E_{t_1}, N_{t_1}

Using the 'back angle to the origin' $(\Phi - \theta)$ at the end of the spiral (t_1), the WCB of the tangent to the circular arc can be found and the coordinates along the centre-line calculated as shown in *Section 10.2.4*.

Worked examples

Example 10.9 Consider *Figure 10.35* in which $R_1 = 700$ m, $R_2 = 1500$ m, $q = 0.3\,\text{m/s}^3$ and $V = 100\,\text{km/h}$:

$$L_1 = V^3/3.6^3 R_1 q = 102 \text{ m}$$

$$L_2 = V^3/3.6^3 R_2 q = 48 \text{ m}$$

$$L = (L_1 - L_2) = 54 \text{ m}$$

Setting-out from b:

$$\Phi = L/2R_1 = 54/2 \times 700 = 2° 12' 36''$$

$$\text{Check:} \quad \Phi = R_1 L/2R_1^2 = 2° 12' 36''$$

$$\text{as} \quad \theta = \Phi/3 = 0° 44' 12''$$

Then, assuming 10 m chords and no through chainage:

$$\theta_1 = 0° 44' 12'' (10/54)^2 = 0° 01' 31''$$

$$\theta_2 = 0° 44' 12'' (20/54)^2 = 0° 06' 04'', \text{etc.}$$

For the osculating circle (R_1):

$$\delta_{10}^{\circ} = 28.6479(10/700) = 0° 24' 33''$$

Thus the setting-out angles for the 'suspended' transition are

Chord	δ			θ			$\delta - \theta$			*Remarks*
(m)	°	′	″	°	′	″	°	′	″	
10	0	24	33	0	01	31	0	23	02	Peg 1
10	0	49	06	0	06	04	0	43	02	Peg 2
etc.	etc.			etc.			etc.			etc.

Example 10.10 Part of a motorway scheme involves the design and setting out of a simple curve with cubic spiral transitions at each end. The transitions are to be designed such that the centrifugal ratio is 0.197, whilst the rate of change of centripetal acceleration is 0.45 m/s^3 at a design speed of 100 km/h.

If the chainage of the intersection to the straights is 2154.22 m and the angle of deflection 50°, calculate:

(a) The length of transition to the nearest 10 m.
(b) The chainage at the beginning and the end of the total composite curve.
(c) The setting-out angles for the first three 10-m chords on a through chainage basis.

Briefly state where and how you would orient the theodolite in order to set out the circular arc. (KU)

Referring to *Figure 10.28*:

$$\text{Centrifugal ratio} \quad P/W = V^2/127R$$

$$\therefore R = \frac{100^2}{127 \times 0.197} = 400 \text{ m}$$

$$\text{Rate of change of centripetal acceleration} = q = \frac{V^3}{3.6^3 RL}$$

$$\text{(a)} \ \therefore L = \frac{100^3}{3.6^3 \times 400 \times 0.45} = 120 \text{ m}$$

(b) To calculate chainage:

$$S = \frac{L^2}{24R} = \frac{120^2}{24 \times 400} = 1.5 \text{ m}$$

$$\text{Tangent length} = (R + S)\tan \Delta/2 + L/2$$

$$= (400 + 1.5)\tan 25° + 60 = 247.22 \text{ m}$$

$$\therefore \text{Chainage at } T_1 = 2154.22 - 247.22 = 1907 \text{ m}$$

To find length of circular arc:

$$\text{length of circular arc} = R(\Delta - 2\Phi) \quad \text{where } \Phi = L/2R$$

$$\text{thus} \quad 2\Phi = \frac{L}{R} = \frac{120}{400} = 0.3 \text{ rad}$$

$$\text{and} \quad \Delta = 50° = 0.872\,665 \text{ rad}$$

$$\therefore \; R(\Delta - 2\Phi) = 400(0.872\,655 - 0.3) = 229.07 \text{ m}$$

$$\text{Chainage at} \quad T_2 = 1907.00 + 2 \times 120 + 229.07 = 2376.07 \text{ m}$$

(c) To find setting-out angles from *equation (10.31)*:

$$\theta_1/\theta = l_1^2/L^2$$

$$\theta = \frac{\Phi}{3} = \frac{L}{6R} = \frac{120}{6 \times 400} \text{ rad}$$

$$\theta'' = \frac{120 \times 206\,265}{6 \times 400} = 10\,313''$$

As the chainage of $T_1 = 1907$, then the first chord will be 3 m long to give a round chainage of 1910 m.

$$\therefore \; \theta_1 = \theta\frac{l_1^2}{L^2} = 10\,313'' \times \frac{3^2}{120^2} = 0° \, 00' \, 06.5''$$

$$\theta_2 = 10\,313'' \times \frac{13^2}{120^2} = 0° \, 02' \, 01''$$

$$\theta_3 = 10\,313'' \times \frac{23^2}{120^2} = 0° \, 06' \, 19''$$

For final part of answer refer to *Sections 10.6(3)* and *10.7*.

Example 10.11 A transition curve of the cubic parabola type is to be set out from a straight centre-line. It must pass through a point which is 6 m away from the straight, measured at right angles from a point on the straight produced 60 m from the start of the curve.

Tabulate the data for setting out a 120-m length of curve at 15-m intervals.

Calculate the rate of change of radial acceleration for a speed of 50 km/h. (LU)

The above question may be read to assume that 120 m is only a part of the total transition length and thus L is unknown.

From expression for a cubic parabola:

$$y = x^3/6RL = cx^3 \quad y = 6 \text{ m}, \quad x = 60 \text{ m} \;\; \therefore c = \frac{1}{36\,000} = \frac{1}{6\,RL}$$

The offsets are now calculated using this constant:

$$y_1 = \frac{15^3}{3600} = 0.094 \text{ m}$$

$$y_2 = \frac{30^3}{36\,000} = 0.750 \text{ m}$$

$$y_3 = \frac{45^3}{36\,000} = 2.531 \text{ m and so on}$$

Rate of change of radial acceleration $= q = V^3/3.6^3RL$:

$$\text{now} \quad \frac{1}{6RL} = \frac{1}{36\,000} \quad \therefore \frac{1}{RL} = \frac{1}{6000}$$

$$\therefore q = \frac{50^3}{3.6^3 \times 6000} \times 0.45 \text{ m/s}^3$$

Example 10.12 Two straights of a railway track of gauge 1.435 m have a deflection angle of 24° to the right. The straights are to be joined by a circular curve having cubic parabola transition spirals at entry and exit. The ratio of super-elevation to track gauge is not to exceed 1 in 12 on the combined curve, and the rate of increase/decrease of super-elevation on the spirals is not to exceed 1 cm in 6 m. If the through chainage of the intersection point of the two straights is 1488.8 m and the maximum allowable speed on the combined curve is to be 80 km/h, determine:

(a) The chainages of the four tangent points.
(b) The necessary deflection angles (to the nearest 20″) for setting out the first four pegs past the first tangent point, given that pegs are to be set out at the 30-m points of the through chainage.
(c) The rate of change of radial acceleration on the curve when trains are travelling at the maximum permissible speed. (ICE)

(a) Referring to *Figure 10.28*, the four tangent points are T_1, t_1, t_2, T_2.
Referring to *Figure 10.27(a)*, as the super-elevation on railways is limited to 0.152 m, then $AB = 1.435 \text{ m} \approx CB$.

$$\therefore \text{ Super-elevation} = AC = \frac{1.435}{12} = 0.12 \text{ m} = 12 \text{ cm}$$

$$\text{Rate of application} = 1 \text{ cm in } 6 \text{ m}$$

$$\therefore \text{ Length of transition} = L = 6 \times 12 = 72 \text{ m}$$

From *Section 10.5.5*, $\tan\theta = \dfrac{V^2}{127R} = \dfrac{1}{12}$

$$\therefore \frac{80^2}{127R} = \frac{1}{12} \quad \therefore R = 604.72 \text{ m}$$

$$\text{Shift} = S = \frac{L^2}{24R} = \frac{72^2}{24 \times 604.72} = 0.357 \text{ m}$$

$$\text{Tangent length} = (R + S)\tan\Delta/2 + L/2 = 605.077 \tan 12° + 36$$
$$= 164.6 \text{ m}$$

$$\therefore \text{ Chainage } T_1 = 1488.8 - 164.6 = 1324.2 \text{ m}$$

$$\text{Chainage } t_1 = 1324.2 + 72 = 1396.2 \text{ m (end of transition)}$$

To find length of circular curve:

$$2\Phi = \frac{L}{R} = \frac{72}{604.72} = 0.119\,063\,\text{rad}$$

$$\Delta = 24° = 0.418\,879\,\text{rad}$$

$$\therefore \text{ Length of curve} = R(\Delta - 2\Phi) = 604.72(0.418\,879 - 0.119\,063)$$

$$= 181.3 \text{ m}$$

$$\therefore \text{ Chainage } t_2 = 1396.2 + 181.3 = 1577.5 \text{ m}$$

$$\text{Chainage } T_2 = 1577.5 + 72 = 1649.5 \text{ m}$$

(b) From chainage of T_1 the first chord = 5.8 m

$$\theta = \frac{L}{6R} \times 206\,265 = \frac{72 \times 206\,265}{6 \times 604.72} = 4093''$$

$$\therefore \theta_1 = \theta \frac{l_1^2}{L^2} = 4093'' \times \frac{5.8^2}{72^2} = 27'' = 0°\,00'\,27'' \quad \text{peg 1}$$

$$\theta_2 = 4093 \times \frac{35.8^2}{72^2} = 1012'' = 0°\,16'\,52'' \quad \text{peg 2}$$

$$\theta_3 = 4093 \times \frac{65.8^2}{72^2} = 3418'' = 0°\,56'\,58'' \quad \text{peg 3}$$

$$\theta_4 = 4093'' = \theta \text{ (end of transition)} = 1°\,08'\,10'' \quad \text{peg 4}$$

(c) $$q = \frac{V^3}{3.6^3 RL} = \frac{80^3}{3.6^3 \times 604.72 \times 72} = 0.25 \text{ m/s}^3$$

Example 10.13 A compound curve *AB*, *BC* is to be replaced by a single arc with transition curves 100 m long at each end. The chord lengths *AB* and *BC* are respectively 661.54 and 725.76 m and radii 1200 m and 1500 m. Calculate the single arc radius:

(a) If *A* is used as the first tangent point.
(b) If *C* is used as the last tangent point. (LU)

Referring to *Figure 10.14* assume $T_1 = A$, $t = B$, $T_2 = C$, $R_1 = 1200$ m and $R_2 = 1500$ m. The requirements in this question are the tangent lengths *AI* and *CI*.

$$\text{Chord } AB = 2R_1 \sin \frac{\Delta_1}{2}$$

$$\therefore \sin \frac{\Delta_1}{2} = \frac{661.54}{2 \times 1200}$$

$$\therefore \Delta_1 = 32°$$

Similarly,

$$\sin \frac{\Delta_2}{2} = \frac{725.76}{3000}$$

$$\therefore \Delta_2 = 28°$$

Distance $At_1 = t_1B = R_1 \tan \frac{\Delta_1}{2} = 1200 \tan 16° = 344$ m

and $Bt_2 = t_2C = R_2 \tan \frac{\Delta_2}{2} = 1500 \tan 14° = 374$ m

$\therefore t_1t_2 = 718$ m

By sine rule in triangle t_1It_2:

$$t_1I = \frac{718 \sin 28°}{\sin 120°} = 389 \text{ m}$$

and $$t_2I = \frac{718 \sin 32°}{\sin 120°} = 439 \text{ m}$$

$$\therefore AI = At_1 + t_1I = 733 \text{ m}$$

$$CI = Ct_2 + t_2I = 813 \text{ m}$$

To find single arc radius:

(a) From tangent point A:

$$AI = (R + S) \tan \Delta/2 + L/2$$

where $S = L^2/24R$ and $\Delta = \Delta_1 + \Delta_2 = 60°, L = 100$ m

then $733 = \left(R + \frac{L^2}{24R}\right) \tan 30° + 50$ from which

$$R = 1182 \text{ m}$$

(b) From tangent point C:

$$CI = (R + S) \tan \Delta/2 + L/2$$

$$813 = (R + L^2/24R) \tan 30° + 50 \text{ from which}$$

$$R = 1321 \text{ m}$$

Example 10.14 Two straights with a deviation angle of 32° are to be joined by two transition curves of the form $\lambda = a(\Phi)^{\frac{1}{2}}$ where λ is the distance along the curve, Φ the angle made by the tangent with the original straight and a is a constant.

The curves are to allow for a final 150 mm cant on a 1.435 m track, the straights being horizontal and the gradient from straight to full cant being 1 in 500.

Tabulate the data for setting out the curve at 15-m intervals if the ratio of chord to curve for 16° is 0.9872. Find the design for this curve. (LU)

Referring to *Figure 10.30*:

Cant = 0.15 m, rate of application = 1 in 500

$\therefore L = 500 \times 0.15 = 75$ m

As the curve is wholly transitional $\Phi = \Delta/2 = 16°$

$\therefore$ from $\Phi = \frac{L}{2R}$, $R = 134.3$ m

From ratio of chord to curve:

Chord $\quad T_1t = 75 \times 0.9872 = 74$ m

$$\therefore X = T_1t\cos\theta = 73.7\text{ m } (\theta = \Phi/3)$$

$$Y = T_1t\sin\theta = 6.9\text{ m}$$

$$\therefore \text{ Tangent length} = X + Y\tan\Phi = 73.7 + 6.9\tan 16° = 75.7\text{ m}$$

Setting-out angles:

$$\theta_1 = (5°\,20')\,\frac{15^2}{75^2} = 12'\,48''$$

$$\theta_2 = (5°\,20')\,\frac{30^2}{75^2} = 51'\,12''$$

and so on to θ_5.

Design speed:

From *Figure 10.27(a)* $\quad \tan\theta \approx \dfrac{AC}{CB} = \dfrac{0.15}{1.435} = \dfrac{V^2}{Rg}$

from which $\quad V = 11.8\text{ m/s} = 42\text{ km/h}$

Exercises

(*10.6*) The centre-line of a new road is being set out through a built-up area. The two straights of the road T_1I and T_2I meet giving a deflection angle of 45°, and are to be joined by a circular arc with spiral transitions 100 m long at each end. The spiral from T_1 must pass between two buildings, the position of the pass point being 70 m along the spiral from T_1 and 1 m from the straight measured at right angles.

Calculate all the necessary data for setting out the first spiral at 30-m intervals; thereafter find:

(a) The first three angles for setting out the circular arc, if it is to be set out by 10 equal chords.
(b) The design speed and rate of change of centripetal acceleration, given a centrifugal ratio of 0.1.
(c) The maximum super-elevation for a road width of 10 m. (KU)

(*Answer*: data $R = 572$ m, $T_1I = 237.23$ m, $\theta_1 = 9'\,01''$, $\theta_2 = 36'\,37''$, $\theta_3 = 1°\,40'\,10''$)

(*Answer*: (a) $1°\,44'\,53''$, $3°\,29'\,46''$, $5°\,14'\,39''$, (b) 85 km/h, 0.23 m/s^3, (c) 1 m)

(*10.7*) A circular curve of 1800 m radius leaves a straight at through chainage 2468 m, joins a second circular curve of 1500 m radius at chainage 3976.5 m, and terminates on a second straight at chainage 4553 m. The compound curve is to be replaced by one of 2200 m radius with transition curves 100 m long at each end.

Calculate the chainages of the two new tangent points and the quarter point offsets of the transition curves. (LU)

(*Answer*: 2114.3 m, 4803.54 m; 0.012, 0.095, 0.32, 0.758 m)

(*10.8*) A circular curve must pass through a point P which is 70.23 m from I, the intersection point and on the bisector of the internal angle of the two straights AI, IB. Transition curves 200 m long are to be applied at each end and one of these must pass through a point whose coordinates are 167 m from the first tangent point along AI and 3.2 m at right angles from this straight. IB deflects 37° 54′ right from AI produced.

Calculate the radius and tabulate the data for setting out a complete curve. (LU)

(*Answer*: $R = 1200$ m, $AI = IB = 512.5$ m, setting-out angles or offsets calculated in usual way)

(*10.9*) The limiting speed around a circular curve of 667 m radius calls for a super-elevation of 1/24 across the 10-m carriageway. Adopting the Department of Transport recommendations of a rate of 1 in 200 for the application of super-elevation along the transition curve leading from the straight to the circular curve, calculate the tangential angles for setting out the transition curve with pegs at 15-m intervals from the tangent point with the straight. (ICE)

(*Answer*: $L = 83$ m, $2' 20''$, $9' 19''$, $20' 58''$, $37' 16''$, $58' 13''$, $1° 11' 18''$)

(*10.10*) A circular curve of 610 m radius deflects through an angle of $40° 30'$. This curve is to be replaced by one of smaller radius so as to admit transitions 107 m long at each end. The deviation of the new curve from the old at their mid-points is 0.46 m towards the intersection point.

Determine the amended radius assuming that the shift can be calculated with sufficient accuracy on the old radius. Calculate the lengths of track to be lifted and of new track to be laid. (LU)

(*Answer*: $R = 590$ m, new track $= 521$ m, old track $= 524$ m)

(*10.11*) The curve connecting two straights is to be wholly transitional without intermediate circular arc, and the junction of the two transitions is to be 5 m from the intersection point of the straights which deflects through an angle of 18°.

Calculate the tangent distances and the minimum radius of curvature. If the super-elevation is limited to 1 vertical to 16 horizontal, determine the correct velocity for the curve and the rate of gain of radial acceleration. (LU)

(*Answer*: 95 m, 602 m, 68 km/h, 0.06 m/s^3)

10.10 VERTICAL CURVES

Vertical curves (VC) are used to connect intersecting gradients in the vertical plane. Thus, in route design they are provided at all changes of gradient. They should be of sufficiently large curvature to provide comfort to the driver, that is, they should have a low 'rate of change of grade'. In addition, they should afford adequate 'sight distances' for safe stopping at a given design speed.

The type of curve generally used to connect the intersecting gradients g_1 and g_2 is the simple parabola. Its use as a *sag* or *crest* curve is illustrated in *Figure 10.38*.

10.10.1 Gradients

In vertical curve design the gradients are expressed as percentages, with a negative for a downgrade and a positive for an upgrade,

e.g. A downgrade of 1 in 20 = 5 in 100 = $-5\% = -g_1\%$

An upgrade of 1 in 25 = 4 in 100 = $+4\% = +g_2\%$

The angle of deflection of the two intersecting gradients is called the *grade angle* and equals A in *Figure 10.38*. The grade angle simply represents the change of grade through which the vertical curve deflects and is the algebraic difference of the two gradients:

$$A\% = (g_1\% - g_2\%)$$

In the above example $A\% = (-5\% - 4\%) = -9\%$ (negative indicates a sag curve).

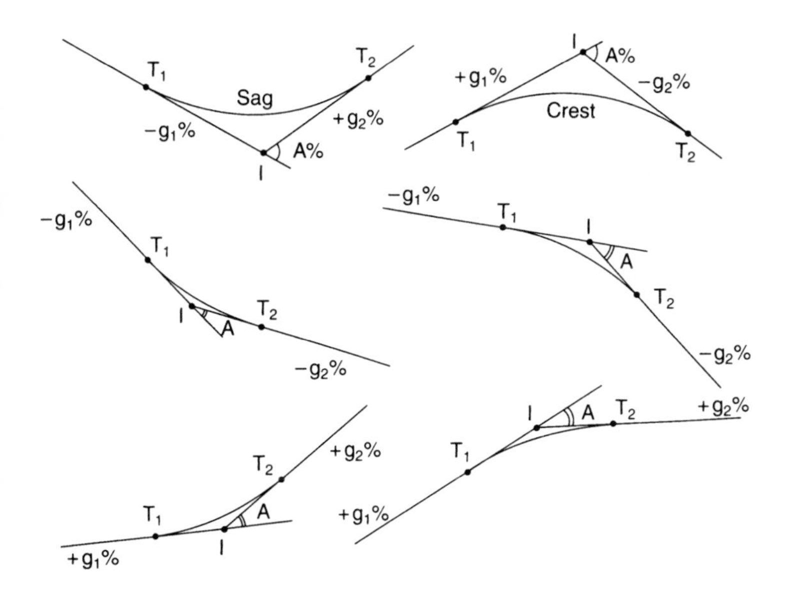

Fig. 10.38 *Sag and crest curves*

10.10.2 Permissible approximations in vertical curve computation

In the UK, civil engineering road design is carried out in a accordance with the Highways Agency's *Design Manual for Roads and Bridges*, Volume 6.

However, practically all the geometric design is in the Highways Agency's TD 9/93, hereafter referred to simply as TD 9/93 and can be found at http://www.standardsforhighways.co.uk/dmrb/index.htm.

In TD 9/93 the desirable maximum gradients for vertical curve design are:

Motorways	3%
Dual carriageways	4%
Single carriageways	6%

Due to the shallowness of these gradients, the following VC approximations are permissible, thereby resulting in simplified computation (*Figure 10.39*).

(1) Distance $T_1D = T_1BT_2 = T_1CT_2 = (T_1I + IT_2)$, without sensible error. This is very important and means that all distances may be regarded as horizontal in both the computation and setting out of vertical curves.
(2) The curve is of equal length each side of I. Thus $T_1C = CT_2 = T_1I = IT_2 = L/2$, without sensible error.
(3) The curve bisects BI at C, thus $BC = CI = Y$ (the mid-offset).
(4) From similar triangles T_1BI and T_1T_2J, if $BI = 2Y$, then $T_2J = 4Y$. $4Y$ represents the vertical divergence of the two gradients over half the curve length ($L/2$) and therefore equals $AL/200$.
(5) The basic equation for a simple parabola is

$$y = C \cdot l^2$$

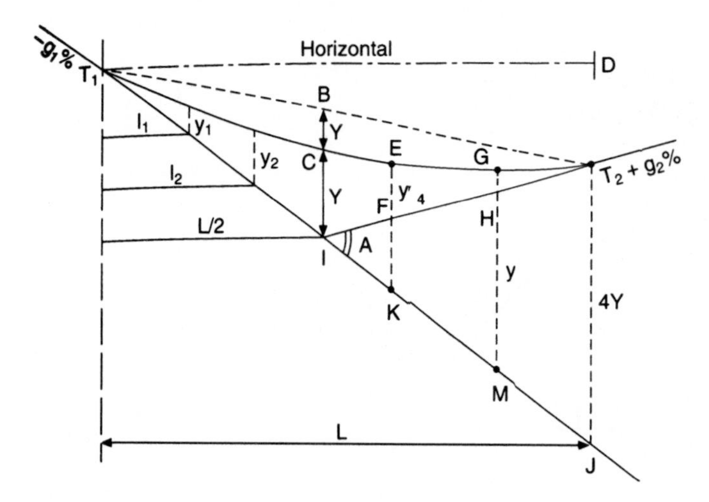

Fig. 10.39 *Vertical curve approximations*

where y is the *vertical* offset from gradient to curve, distance l from the start of the curve, and C is a constant. Thus, as the offsets are proportional to distance squared, the following equation is used to compute them:

$$\frac{y_1}{Y} = \frac{l_1^2}{(L/2)^2} \tag{10.36}$$

where Y = the mid-offset = $AL/800$ (see *Section 10.10.7*).

10.10.3 Vertical curve design

In order to set out a vertical curve in the field, one requires levels along the curve at given chainage intervals. Before the levels can be computed, one must know the length L of the curve. The value of L is obtained from parameters supplied in *Table 3* of TD 9/93 (reproduced below as *Table 10.3*) and the appropriate parameters are K-values for specific design speeds and sight distances; then

$$L = KA \tag{10.37}$$

where A = the difference between the two gradients (grade angle)
K = the design speed related coefficient (*Table 10.3*)

e.g. A + 4% gradient is linked to a − 3% gradient by a crest curve. What length of curve is required for a design speed of 100 km/h?

$$A = (4\% - (-3\%)) = +7\% \text{ (positive for crest)}$$

From *Table 10.3*:

C1 Desirable minimum crest K-value = 100
C2 One step below desirable minimum crest K-value = 55
∴ from $L = KA$
Desirable minimum length = $L = 100 \times 7 = 700$ m
One step below desirable minimum length = $L = 55 \times 7 = 385$ m

Wherever possible the vertical and horizontal curves in the design process should be coordinated so that the sight distances are correlated and a more efficient overtaking provision is ensured.

The various design factors will now be dealt with in more detail.

Table 10.3

Design speed (kph)	*120*	*100*	*85*	*70*	*60*	*50*	V^2/R
A *Stopping sight distance*, m							
Desirable minimum	295	215	160	120	90	70	
One step below desirable minimum	215	160	120	90	70	50	
B *Horizontal curvature*, m							
Minimum R^* without elimination of adverse camber and transitions	2880	2040	1440	1020	720	510	5
Minimum R^* with super-elevation of 2.5%	2040	1440	1020	720	510	360	7.07
Minimum R^* with super-elevation of 3.5%	1440	1020	720	510	360	255	10
Desirable minimum R with super-elevation of 5%	1020	720	510	360	255	180	14.14
One step below desirable minimum R with super-elevation of 7%	720	510	360	255	180	127	20
Two steps below desirable minimum radius with super-elevation of 7%	510	360	255	180	127	90	28.28
C *Vertical curvature*							
Desirable minimum* crest K-value	182	100	55	30	17	10	
One step below desirable minimum crest K-value	100	55	30	17	10	6.5	
Absolute minimum sag K-value	37	26	20	20	13	9	
Overtaking sight distances							
Full overtaking sight distance FOSD, m	*	580	490	410	345	290	
FOSD overtaking crest K-value	*	400	285	200	142	100	

*Not recommended for use in the design of single carriageways.

The V^2/R values shown above simply represent a convenient means of identifying the relative levels of design parameters, irrespective of design speed.

(Reproduced with permission of the Controller of Her Majesty's Stationery Office)

10.10.3.1 K-value

Rate of change of gradient (r) is the rate at which the curve passes from one gradient (g_1%) to the next (g_2%) and is similar in concept to rate of change of radial acceleration in horizontal transitions. When linked to design speed it is termed *rate of vertical acceleration* and should never exceed 0.3 m/s^2.

A typical example of a badly designed vertical curve with a high rate of change of grade is a hump-backed bridge where usually the two approaching gradients are quite steep and connected by a very short length of vertical curve. Thus one passes through a large grade angle A in a very short time, with the result that often a vehicle will leave the ground and/or cause great discomfort to its passengers. Fortunately, in the UK, few of these still exist.

Commonly-used design values for r are:

3%/100 m on crest curves

1.5%/100 m on sag curves

thereby affording much larger curves to prevent rapid change of grade and provide adequate sight distances.

Working from first principles if $g_1 = -2\%$ and $g_2 = +4\%$ (sag curve), then the change of grade from -2% to $+4\% = 6\%$ (A), the grade angle. Thus, to provide for a rate of change of grade of 1.5%, one would require 400 m (L) of curve. If the curve was a crest curve, then using 3% gives 200 m (L) of curve:

$$\therefore L = 100A/r \tag{10.38}$$

Now, expressing rate of change of grade as a single number we have

$$K = 100/r \tag{10.39}$$

and as shown previously, $L = KA$.

10.10.3.2 Sight distances

Sight distance is a safety design factor which is intrinsically linked to rate of change of grade, and hence to K-values.

Consider once again the hump-backed bridge. Drivers approaching from each side of this particular vertical curve cannot see each other until they arrive, simultaneously, almost on the crest; by which time it may be too late to prevent an accident. Had the curve been longer and flatter, thus resulting in a low rate of change of grade, the drivers would have had a longer sight distance and consequently more time in which to take avoiding action.

Thus, sight distance, i.e. the length of road ahead that is visible to the driver, is a safety factor, and it is obvious that the sight distance must be greater than the stopping distance in which the vehicle can be brought to rest.

Stopping distance is dependent upon:

(1) Speed of the vehicle.
(2) Braking efficiency.
(3) Gradient.
(4) Coefficient of friction between tyre and road.
(5) Road conditions.
(6) Driver's reaction time.

In order to cater for all the above variables, the height of the driver's eye above the road surface is taken as being only 1.05 m; a height applicable to sports cars whose braking efficiency is usually very high. Thus, other vehicles, such as lorries, with a much greater eye height, would have a much longer sight distance in which to stop.

10.10.3.3 Sight distances on crests

Sight distances are defined as follows:

(1) Stopping sight distance (SSD) (*Figure 10.40*)

The SSD is the sight distance required by a driver to stop a vehicle when faced with an unexpected obstruction on the carriageway. It comprises two elements:

(a) The *perception-reaction distance*, which is the distance travelled from the time the driver sees the obstruction to the time it is realized that the vehicle must stop; and
(b) The *braking distance*, which is the distance travelled before the vehicle halts short of the obstruction.

The above are a function of driver age and fatigue, road conditions, etc., and thus the design parameters are based on average driver behaviour in wet conditions. *Table 10.3* provides values for desirable and absolute minimum SSD.

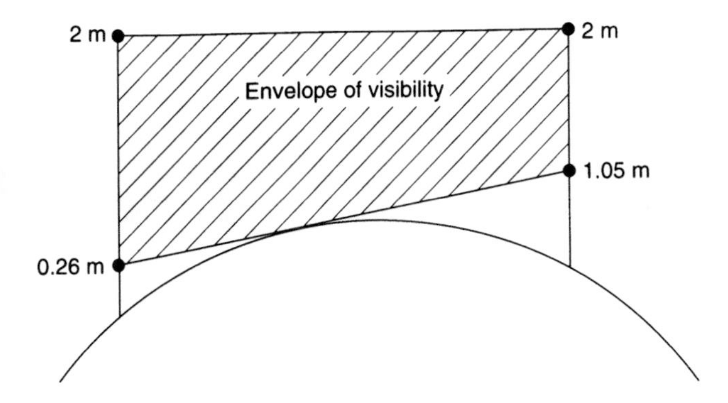

Fig. 10.40 *Visibility on a vertical curve*

It has been shown that 95% of drivers' eye height is 1.05 m or above; the upper limit of 2 m represents large vehicles.

The height of the obstruction is between 0.26 m and 2.0 m. Forward visibility should be provided in both horizontal and vertical planes between points in the centre of the lane nearest the inside of the curve.

(2) Full overtaking sight distance (FOSD) (Figure 10.41)

On single carriageways, overtaking in the lane of the opposing traffic occurs. To do so in safety requires an adequate sight distance which will permit the driver to complete the normal overtaking procedure.

The FOSD consists of four elements:

(a) The *perception/reaction distance* travelled by the vehicle whilst the decision to overtake or not is made.
(b) The *overtaking distance* travelled by the vehicle to complete the overtaking manoeuvre.
(c) The *closing distance* travelled by the oncoming vehicle whilst overtaking is occurring.
(d) The *safety distance* required for clearance between the overtaking and oncoming vehicles at the instant the overtaking vehicle has returned to its own lane.

It has been shown that 85% of overtaking takes place in 10 seconds and *Table 10.3* gives appropriate FOSD values relative to design speed.

It should be obvious from the concept of FOSD that it is used in the design of single carriageways only, where safety when overtaking is the prime consideration.

For instance, consider the design of a crest curve on a dual carriageway with a design speed of 100 km/h.

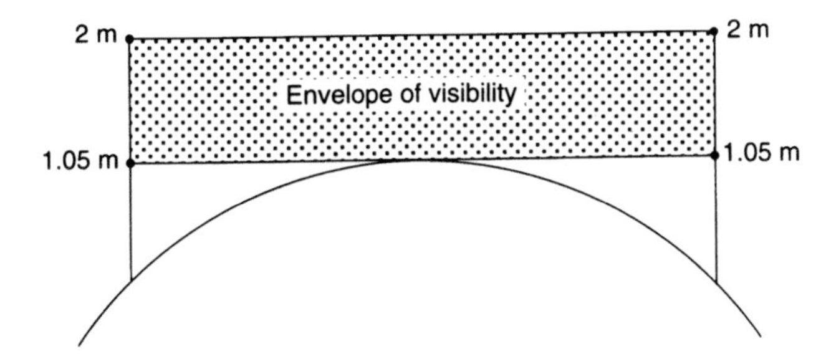

Fig. 10.41 *Overtaking visibility*

From *Table 10.3*:

Desirable minimum K-value = 100
One step below desirable minimum K-value = 55
FOSD K-value = 400

As overtaking is not a safety hazard on a dual carriageway, FOSD is not necessary and one would use:

$$L = 100\,\text{A (desirable minimum)}$$
or $$L = 55\,\text{A (one step below desirable minimum)}$$

Had the above road been a single carriageway then FOSD would be required and:

$$L = 400\,\text{A}$$

If this resulted in too long a curve, with excessive earthworks, then it might be decided to prohibit overtaking entirely, in which case:

$$L = 55\,\text{A}$$

would be used.

Although equations are unnecessary when using design tables, they can be developed to calculate curve lengths L for given sight distances S, as follows:

(a) When $S < L$ (*Figure 10.42*)

From basic equation $y = Cl^2$

$$Y = C(L/2)^2, h_1 = C(l_1)^2 \text{ and } h_2 = C(l_2)^2$$

then $$\frac{h_1}{Y} = \frac{l_1^2}{(L/2)^2} = \frac{4l_1^2}{L^2} \quad \text{and} \quad \frac{h_2}{Y} = \frac{4l_2^2}{L^2}$$

thus $$l_1^2 = \frac{h_1 L^2}{4Y} \text{ but since } 4Y = \frac{AL}{200}$$

$$l_1^2 = \frac{200 h_1 L}{A}$$

and $$l_1 = (l_1)^{\frac{1}{2}} \left(\frac{200L}{A}\right)^{\frac{1}{2}}$$

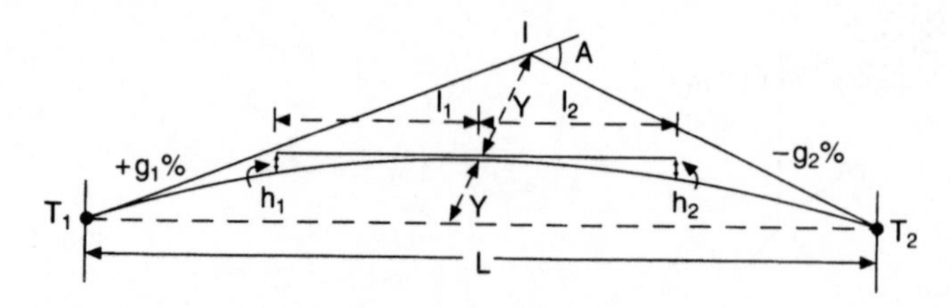

Fig. 10.42 *Calculation of curve lengths*

Similarly $l_2 = (h_2)^{\frac{1}{2}} \left(\frac{200L}{A}\right)^{\frac{1}{2}}$

$$\therefore S = (l_1 + l_2) = [(h_1)^{\frac{1}{2}} + (h_2)^{\frac{1}{2}}] \left(\frac{200L}{A}\right)^{\frac{1}{2}} \quad (10.40)$$

and
$$L = \frac{S^2 A}{200[(h_1)^{\frac{1}{2}} + (h_2)^{\frac{1}{2}}]^2} \quad (10.41)$$

when $h_1 = h_2 = h$

$$L = \frac{S^2 A}{800h} \quad (10.42)$$

(b) When $S > L$ it can similarly be shown that

$$L = 2S - \frac{200}{A}[(h_1)^{\frac{1}{2}} + (h_2)^{\frac{1}{2}}]^2 \quad (10.43)$$

and when $h_1 = h_2 = h$

$$L = 2S - \frac{800h}{A} \quad (10.44)$$

When $S = L$, substituting in either of *equations* (*10.42*) or (*10.44*) will give the correct solution,

e.g. (10.42) $L = \frac{S^2 A}{800h} = \frac{L^2 A}{800h} = \frac{800h}{A}$

and (10.44) $L = 2S - \frac{800h}{A} = 2L - \frac{800h}{A} = \frac{800h}{A}$

N.B. If the relationship of S to L is not known then both cases must be considered; one of them will not fulfil the appropriate argument $S < L$ or $S > L$ and is therefore wrong.

10.10.3.4 Sight distances on sags

Visibility on sag curves is not obstructed as it is in the case of crests; thus sag curves are designed for at least absolute minimum comfort criteria of 0.3 m/sec^2. However, for design speeds of 70 km/h and below in unlit areas, sag curves are designed to ensure that headlamps illuminate the road surface for at least absolute minimum SSD. The relevant K values are given in 'Absolute minimum saq K-value', *Table 10.3*.

The headlight is generally considered as being 0.6 m above the road surface with its beam tilted up at 1° to the horizontal. As in the case of crests, equations can be developed if required.

Consider *Figure 10.43* where L is greater than S. From the equation for offsets:

$$\frac{BC}{T_2D} = \frac{S^2}{L^2} \quad \therefore BC = \frac{S^2(T_2D)}{L^2}$$

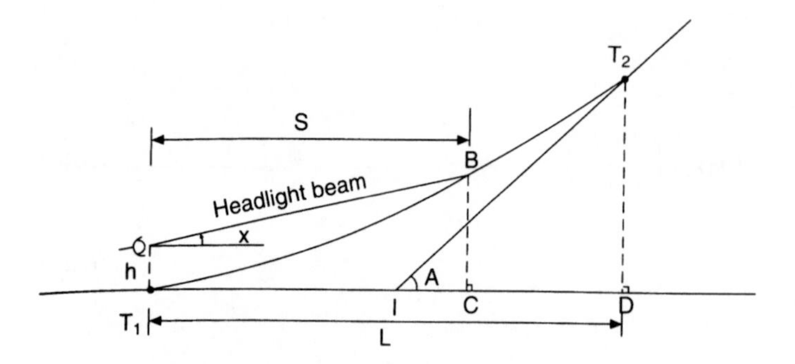

Fig. 10.43 *Clearance on a vertical curve where S < L*

but T_2D is the vertical divergence of the gradients and equals

$$\frac{A \cdot L}{100 \cdot 2} \quad \therefore BC = \frac{A \cdot S^2}{200L} \tag{a}$$

also $\quad BC = h + S \tan x \quad$ (b)

Equating (a) and (b): $L = S^2A(200h + 200S \tan x)^{-1}$ (10.45)

putting $x = 1°$ and $h = 0.6$ m

$$L = \frac{S^2A}{120 + 3.5S} \tag{10.46}$$

Similarly, when S is greater than L (*Figure 10.44*):

$$BC = \frac{A}{100}\left(S - \frac{L}{2}\right) = h + S \tan x$$

equating: $L = 2S - (200h + 200S \tan x)/A$ (10.47)

when $x = 1°$ and $h = 0.6$ m

$$L = 2S - (120 + 3.5S)/A \tag{10.48}$$

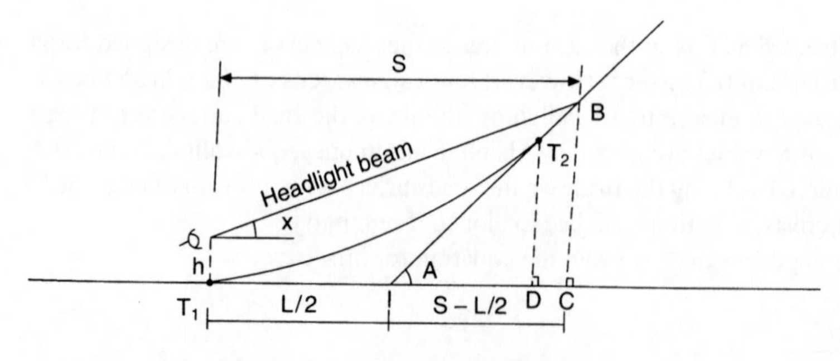

Fig. 10.44 *Clearance on a vertical curve where S > L*

10.10.4 Passing a curve through a point of known level

In order to ensure sufficient clearance at a specific point along the curve it may be necessary to pass the curve through a point of known level. For example, if a bridge parapet or road furniture were likely to intrude into the envelope of visibility, it would be necessary to design the curve to prevent this.

This technique will be illustrated by the following example. A downgrade of 4% meets a rising grade of 5% in a sag curve. At the start of the curve the level is 123.06 m at chainage 3420 m, whilst at chainage 3620 m there is an overpass with an underside level of 127.06 m. If the designed curve is to afford a clearance of 5 m at this point, calculate the required length (*Figure 10.45*).

To find the offset distance CE:
From chainage horizontal distance $T_1E = 200$ m at -4%

$$\therefore \text{ Level at } E = 123.06 - 8 = 115.06 \text{ m}$$
$$\text{Level at } C = 127.06 - 5 = 122.06 \text{ m}$$
$$\therefore \text{ Offset } CE = 7 \text{ m}$$

From offset equation $\dfrac{CE}{T_2B} = \dfrac{(T_1E)^2}{(T_1B)^2}$

but T_2B = the vertical divergence $= \dfrac{A}{100}\dfrac{L}{2}$, where $A = 9$

$$\therefore CE = \frac{AL}{200}\frac{200^2}{L^2} = \frac{1800}{L}$$

$$L = 257 \text{ m}$$

10.10.5 To find the chainage of highest or lowest point on the curve

The position and level of the highest or lowest point on the curve is frequently required for drainage design.

With reference to *Figure 10.45*, if one considers the curve as a series of straight lines, then at T_1 the grade of the line is -4% gradually changing throughout the length of the curve until at T_2 it is $+5\%$. There has thus been a change of grade of 9% in distance L. At the lowest point the grade will be horizontal, having just passed through -4% from T_1. Therefore, the chainage of the lowest point from the start of the curve is, by simple proportion,

$$D = \frac{L}{9\%} \times 4\% = \frac{L}{A} \times g_1 \tag{10.49}$$

which in the previous example is $\dfrac{257}{9\%} \times 4\% = 114.24$ m from T_1.

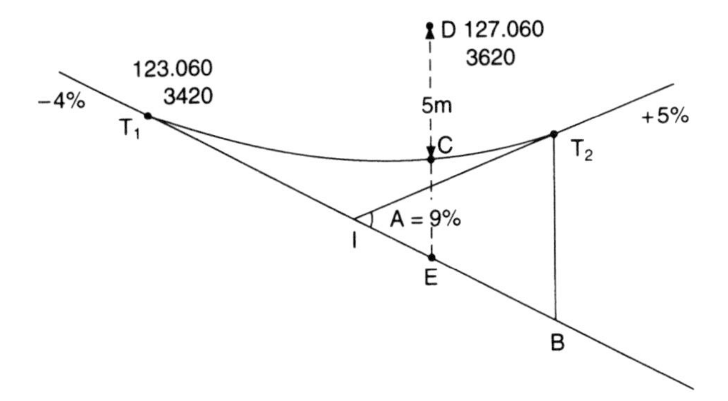

Fig. 10.45 *Clearance at a given point*

Knowing the chainage, the offset and the curve level at that point may be found.

This simple approach suffices as the rate of change of grade is constant for a parabola, i.e. $y = Cl^2$, $\therefore d^2y/dl^2 = 2C$.

10.10.6 Vertical curve radius

Due to the very shallow gradients involved in vertical curve (VC) design, the parabola may be approximated to a circular curve. In this way vertical accelerations (V^2/R) may be easily assessed.

In circular curves (*Section 10.1*) the main chord from T_1 to $T_2 = 2R \sin \Delta/2$, where Δ is the deflection angle of the two straights. In vertical curves, the main chord may be approximated to the length (L) of the VC and the angle Δ to the grade angle A, i.e.

$$\Delta \approx A\%$$

$$\therefore \sin \Delta/2 \approx \Delta/2 \text{ rads} \approx A/200$$

$$\therefore L \approx 2RA/200 = AR/100 \quad (10.50)$$

and as

$$K = L/A = R/100, \text{ then:}$$

$$R = 100L/A = 100K \quad (10.51)$$

It is important to note that the reduced levels of vertical curve must always be computed. Scaling levels from a longitudinal section, usually having a vertical scale different from the horizontal, will produce a curve that is neither parabolic nor circular.

10.10.7 Vertical curve computation

The computation of a vertical curve will now be demonstrated using an example.

A '2nd difference' ($\delta^2y/\delta l^2$) arithmetical check on the offset computation should automatically be applied. The check works on the principle that the change of grade of a parabola ($y = C \cdot l^2$) is constant, i.e. $\delta^2y/\delta l^2 = 2C$. Thus, if the first and last chords are sub-chords of lengths different from the remaining standard chords, then the change of grade will be constant only for the equal-length chords.

For example, a 100-m curve is to connect a downgrade of 0.75% to an upgrade of 0.25%. If the level of the intersection point of the two grades is 150 m, calculate:

(1) Curve levels at 20-m intervals, showing the second difference (d^2y/dl^2), check on the computations.
(2) The position and level of the lowest point on the curve.

Method

(a) Find the value of the central offset Y.
(b) Calculate offsets.
(c) Calculate levels along the gradients.
(d) Add/subtract (b) from (c) to get curve levels.

(a) *Referring to Figure 10.39*:
Grade angle $A = (-0.75 - 0.25) = -1\%$ (this is seen automatically).

$L/2 = 50$ m, thus as the grades IT_2, and IJ are diverging at the rate of 1% (1 m per 100 m) in 50 m, then

$$T_2J = 0.5 \text{ m} = 4Y \quad \text{and} \quad Y = 0.125 \text{ m}$$

The computation can be quickly worked mentally by the student. Putting the above thinking into equation form gives:

$$4Y = \frac{A}{100}\frac{L}{2} \quad \therefore Y = \frac{AL}{800} = \frac{1 \times 100}{800} = 0.125 \tag{10.52}$$

(b) *Offsets from equation* (10.36):

There are two methods of approach.

(1) The offsets may be calculated from one gradient throughout; i.e. y_1, y_2, EK, GM, T_2J, from the grade T_1J.
(2) Calculate the offsets from one grade, say, T_1I, the offsets being equal on the other side from the other grade IT_2.

Method (1) is preferred due to the smaller risk of error when calculating curve levels at a constant interval and grade down T_1J.

From *equation (10.36)*: $y_1 = Y \times \dfrac{l_1^2}{(L/2)^2}$

	1st diff.	*2nd diff.*
$T_1 = 0\,\text{m}$		
	0.02	
$y_1 = 0.125\dfrac{20^2}{50^2} = 0.02\,\text{m}$		0.04
	0.06	
$y_2 = 0.125\dfrac{40^2}{50^2} = 0.08\,\text{m}$		0.04
	0.10	
$y_3 = 0.125\dfrac{60^2}{50^2} = 0.18\,\text{m}$		0.04
	0.14	
$y_4 = 0.125\dfrac{80^2}{50^2} = 0.32\,\text{m}$		0.04
	0.18	
$y_2 = T_2J = 4Y = 0.50\,\text{m}$		

The second difference arithmetical check, which works only for equal chords, should be applied before any further computation.

(c) First find level at T_1 from known level at I:

Distance from I to T_1 = 50 m, grade = 0.75% (0.75 m per 100 m)

$$\therefore \text{Rise in level from } I \text{ to } T_1 = \frac{0.75}{2} = 0.375\,\text{m}$$

Level at T_1 = 150.000 + 0.375 = 150.375 m

Levels are now calculated at 20-m intervals along T_1J, the fall being 0.15 m in 20 m. Thus, the following table may be made.

Chainage (m)	*Gradient levels*	*Offsets*	*Curve levels*	*Remarks*
0	150.375	0	150.375	Start of curve T_1
20	150.225	0.02	150.245	
40	150.075	0.08	150.155	
60	149.925	0.18	150.105	
80	149.775	0.32	150.095	
100	149.625	0.50	150.125	End of curve T_2

$$\text{Position of lowest point on curve} = \frac{100 \text{ m}}{1\%} \times 0.75\% = 75 \text{ m from } T_1$$

$$\therefore \text{ Offset at this point} = y_2 = 0.125 \times 75^2/50^2 = 0.281 \text{ m}$$

$$\text{Tangent level 75 m from } T_1 = 150.375 - 0.563 = 149.812 \text{ m}$$

$$\therefore \text{ Curve level} = 149.812 + 0.281 = 150.093 \text{ m}$$

10.10.8 Computer-aided drawing and design (CADD)

The majority of survey packages now available contain a road design module and many examples can be found on the internet.

Basically all the systems work from a digital ground model (DGM), established by ground survey methods or aerial photogrammetry. Thus not only is the road designed, but earthwork volumes, setting out data and costs are generated.

In addition to road design using straights and standard curves, polynomial alignment procedures are available if required. The engineer is not eliminated from the CAD process; he/she must still specify such parameters as minimum permissible radius of curvature, maximum slope, minimum sight distances, coordination of horizontal and vertical alignment, along with any political, economic or aesthetic decisions. A series of gradients and curves can be input, until earthwork is minimized and balanced out. This is clearly illustrated by the generation of resultant mass-haul diagrams.

When a satisfactory road design has been arrived at, plans, longitudinal sections, cross-sections and mass-haul diagrams can be quickly produced. Three-dimensional views with colour shading are also available for environmental impact studies. Bills of quantity, total costs and all setting-out data is provided as necessary.

Thus the computer provides a fast, flexible and highly economic method of road design, capable of generating contract drawings and schedules on request.

Worked examples

Example 10.15 An existing length of road consists of a rising gradient of 1 in 20, followed by a vertical parabolic crest curve 100 m long, and then a falling gradient of 1 in 40. The curve joins both gradients tangentially and the reduced level of the highest point on the curve is 173.07 m above datum.

Visibility is to be improved over this stretch of road by replacing this curve with another parabolic curve 200 m long.

Find the depth of excavation required at the mid-point of the curve. Tabulate the reduced levels of points at 30-m intervals on the new curve.

What will be the minimum visibility on the new curve for a driver whose eyes are 1.05 m above the road surface? (ICE)

The first step here is to find the level of the start of the new curve; this can only be done from the information on the highest point P (*Figure 10.45*).

Old curve $\quad A = 7.5\%, L = 100$ m

Chainage of highest point P from $T_1 = \dfrac{100}{7.5\%} \times 5\% = 67$ m

Distance T_2C is the divergence of the grades (7.5 m per 100 m) over half the length of curve (50) = $7.5 \times 0.5 = 3.75$ m $= 4Y$.

$\therefore$ Central offset $\quad Y = 3.75/4 = 0.938$ m

Thus offset $PB = 0.938 \cdot \dfrac{67^2}{50^2} = 1.684$ m

Therefore the level of B on the tangent $= 173.07 + 1.684 = 174.754$ m. This point is 17 m from I, and as the new curve is 200 m in length, it will be 177 m from the start of the new curve T_3.

$\therefore$ Fall from B to T_3 of new curve $= 5 \times 1.17 = 5.85$ m
$\therefore$ Level of $T_3 = 174.754 - 5.85 = 168.904$ m

It can be seen that as the value of A is constant, when L is doubled, the value of Y, the central offset to the new curve, is doubled, giving 1.876 m.

$\therefore$ Amount of excavation at mid-point $= 0.938$ m

New curve offsets	*1st diff.*	*2nd diff.*
	0.169	
$y_1 = 1.876 \times \dfrac{30^2}{100^2} = 0.169$		0.337
	0.506	
$y_2 = 1.876 \times \dfrac{60^2}{100^2} = 0.675$		0.339
	0.845	
$y_3 = 1.876 \times \dfrac{90^2}{100^2} = 1.520$		0.336
	1.181	
$y_4 = 1.876 \times \dfrac{120^2}{100^2} = 2.701$		0.339
	1.520	
$y_5 = 1.876 \times \dfrac{150^2}{100^2} = 4.221$		0.337
	1.857	
$y_6 = 1.876 \times \dfrac{180^2}{100^2} = 6.078$		0.431*
	1.426	
$y_7 = 4y \qquad = 7.504$		

*Note change due to change in chord length from 30 m to 20 m.

Levels along the tangent T_3C are now obtained at 30-m intervals.

Chainage (m)	*Tangent levels*	*Offsets*	*Curve levels*	*Remarks*
0	168.904	0	168.904	T_3 of new curve
30	170.404	0.169	170.235	
60	171.904	0.675	171.229	
90	173.404	1.520	171.884	
120	174.904	2.701	172.203	
150	176.404	4.221	172.183	
180	177.904	6.078	171.826	
200	178.904	7.504	171.400	T_4 of new curve

From *Figure 10.46* it can be seen that the minimum visibility is half the sight distance and could thus be calculated from the necessary equation. However, if the driver's eye height of $h = 1.05$ m is taken as an offset then

$$\frac{h}{Y} = \frac{D^2}{(L/2)^2}, \quad \text{thus } \frac{1.05}{1.876} = \frac{D^2}{100^2}$$

$$\therefore D = 75 \text{ m}$$

Example 10.16 A rising gradient g_1 is followed by another rising gradient g_2 (g_2 less than g_1). The gradients are connected by a vertical curve having a constant rate of change of gradient. Show that at any point on the curve the height y above the first tangent point A is given by

$$y = g_1 x = -\frac{(g_1 - g_2)x^2}{2L}$$

where x is the horizontal distance of the point from A, and L is the horizontal distance between the two tangent points.

Draw up a table of heights above A for 100-m pegs from A when $g_1 = +5\%$, $g_2 = +2\%$ and $L = 1000$ m. At what horizontal distance from A is the gradient $+3\%$? (ICE)

Figure 10.47 from equation for offsets.

$$\frac{BC}{Y} = \frac{x^2}{(L/2)^2}$$

$$\therefore BC = Y \cdot \frac{4x^2}{L^2} \quad \text{but} \quad Y = \frac{AL}{8} \text{ stations} = \frac{(g_1 - g_2)L}{8}$$

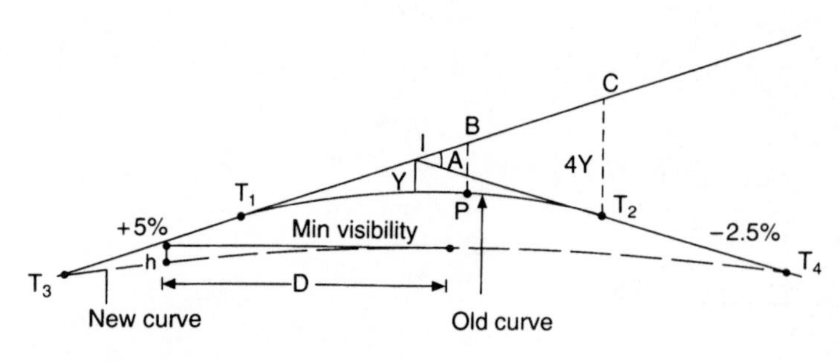

Fig. 10.46 *Minimum visibility*

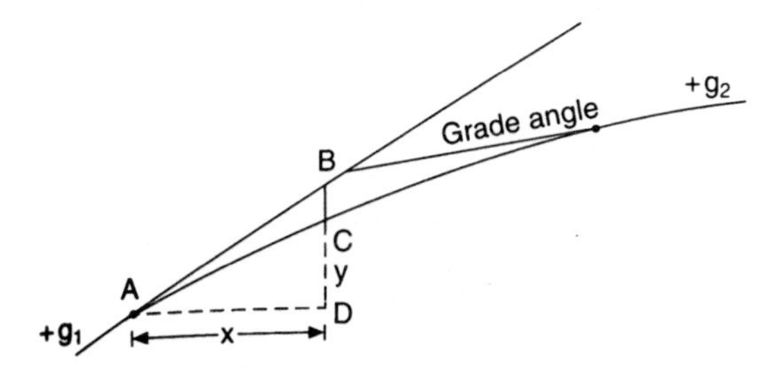

Fig. 10.47 *Curve of decreasing gradient*

$$\therefore BC = \frac{(g_1 - g_2)L4x^2}{8L^2} = \frac{(g_1 - g_2)x^2}{2L}$$

Now $BC = g_1 x$

Thus, as $y = BD - BC = g_1 x - \dfrac{(g_1 - g_2)x^2}{2L}$

Using the above formula (which is correct only if horizontal distances x and L are expressed in stations, i.e. a station = 100 m)

$$y_1 = 5 - \frac{3 \times 1^2}{20} = 4.85 \text{ m}$$

$$y_2 = 10 - \frac{3 \times 2^2}{20} = 9.4 \text{ m}$$

$$y_3 = 15 - \frac{3 \times 3^2}{20} = 13.65 \text{ m and so on}$$

Grade angle = 3% in 1000 m

Change of grade from 5% to 3% = 2%

$$\therefore \text{Distance} = \frac{1000}{3\%} \times 2\% = 667 \text{ m}$$

Example 10.17 A falling gradient of 4% meets a rising gradient of 5% at chainage 2450 m and level 216.42 m. At chainage 2350 m the underside of a bridge has a level of 235.54 m. The two grades are to be joined by a vertical parabolic curve giving 14 m clearance under the bridge. List the levels at 50-m intervals along the curve. (KU)

To find the offset to the curve at the bridge (Figure 10.48)

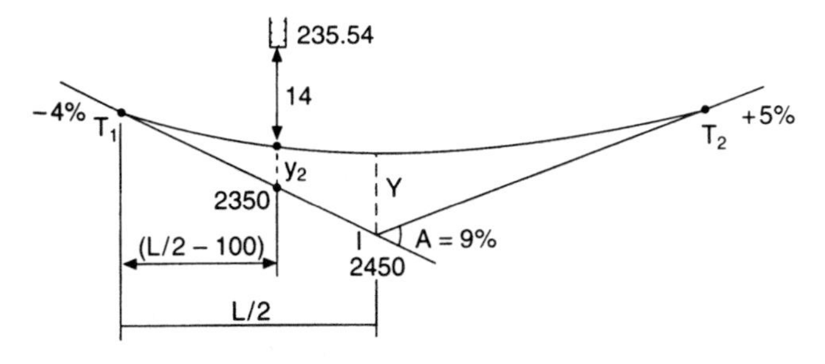

Fig. 10.48 *Offset on a curve, at a given point*

$$\text{Level on gradient at chainage } 2350 = 216.42 + 4 = 220.42\text{ m}$$
$$\text{Level on curve at chainage } 2350 = 235.54 - 14 = 221.54\text{ m}$$
$$\therefore \text{Offset at chainage } 2350 = y_2 = 1.12\text{ m}$$

From equation for offsets $\dfrac{y_2}{Y} = \dfrac{(L/2 - 100)^2}{(L/2)^2}$

where $Y = \dfrac{AL}{800}$ and $A = 9\%$

$$\frac{1.12 \times 800}{9 \times L} = \left(1 - \frac{200}{L}\right)^2, \quad \text{and putting } x = \frac{200}{L}$$

$$1.12 \times 4x = 9(1 - x)^2$$

from which $x^2 - 2.5x + 1 = 0$, giving

$x = 2$ or 0.5 $\therefore L = 400$ m (as $x = 2$ is not possible)

Now $Y = \dfrac{9 \times 400}{800} = 4.5$ m from which the remaining offsets are found as follows:

at chainage 50 m offset $y_1 = 4.5\dfrac{50^2}{200^2} = 0.28$ m

at chainage 100 m offset $y_2 = 4.5\dfrac{100^2}{200^2} = 1.12$ m

at chainage 150 m offset $y_3 = 4.5\dfrac{150^2}{200^2} = 2.52$ m

at chainage 200 m offset $Y = 4.50$ m

To illustrate the alternative method, these offsets may be repeated on the other gradient at 250 m $= y_3$, 300 m $= y_2$, 350 m $= y_1$. The levels are now computed along each gradient from I to T_1 and T_2 respectively.

Chainage (m)	*Gradient levels*	*Offsets*	*Curve levels*	*Remarks*
0	224.42		224.42	Start of curve T
50	222.42	0.28	222.70	
100	220.42	1.12	221.54	
150	218.42	2.52	220.94	
200	216.42	4.50	220.92	Centre of Curve I
250	218.92	2.52	221.44	
300	221.42	1.12	222.54	
350	223.92	0.28	224.20	
400	226.42		226.42	End of curve T_2

Example 10.18 A vertical parabolic curve 150 m in length connects an upward gradient of 1 in 100 to a downward gradient of 1 in 50. If the tangent point T_1 between the first gradient and the curve is taken as datum, calculate the levels of points at intervals of 25 m along the curve until it meets the second gradient at T_2. Calculate also the level of the summit giving the horizontal distance of this point from T_1.

If an object 75 mm high is lying on the road between T_1 and T_2 at 3 m from T_2, and a car is approaching from the direction of T_1, calculate the position of the car when the driver first sees the object if his eye is 1.05 m above the road surface. (LU)

To find offsets:

$$A = 3\% \quad \therefore 4Y = \frac{L}{200} \times 3\% = 2.25 \text{ m}$$

and $\quad Y = 0.562$ m

$$\therefore y_1 = 0.562 \times \frac{25^2}{75^2} = 0.062 \quad y_4 = 0.562 \times \frac{100^2}{75^2} = 1.000$$

$$y_2 = 0.562 \times \frac{50^2}{75^2} = 0.250 \quad y_5 = 0.562 \times \frac{125^2}{75^2} = 1.562$$

$$y_3 = 0.562 \times \frac{75^2}{75^2} = 0.562 \quad y_6 = 4y = 2.250$$

Second difference checks will verify these values.

With T_1 at datum, levels are now calculated at 25-m intervals for 150 m along the 1 in 100 (1%) gradient.

Chainage (m)	*Gradient levels*	*Offsets*	*Curve levels*	*Remarks*
0	100.0	0	100.000	Start of curve T_1
25	100.25	0.062	100.188	
50	100.75	0.250	100.250	
75	101.00	0.562	100.188	
100	101.25	1.000	100.000	
125	101.50	1.562	99.688	
150		2.250	99.250	End of curve T_2

Distance to highest point from $T_1 = \dfrac{150}{3\%} \times 1\% = 50$ m

Sight distance ($S < L$)

From *equation (10.40)*

$$S = \left[(h_1)^{\frac{1}{2}} + (h_2)^{\frac{1}{2}}\right]\left(\frac{200L}{A}\right)^{\frac{1}{2}} \quad \text{when} \quad h_1 = 1.05 \text{ m}, h_2 = 0.075 \text{ m}$$

$\therefore S = 130$ m, and the car is 17 m and T_1 and between T_1 and T_2

Example 10.19 A road gradient of 1 in 60 down is followed by an up-gradient of 1 in 30, the valley thus formed being smoothed by a circular curve of radius 1000 m in the vertical plane. The grades, if produced, would intersect at a point having a reduced level of 299.65 m and a chainage of 4020 m.

It is proposed to improve the road by introducing a longer curve, parabolic in form, and in order to limit the amount of filling it is decided that the level of the new road at chainage 4020 m shall be 3 m above the existing surface.

Determine:

(a) The length of new curve.
(b) The levels of the tangent points.

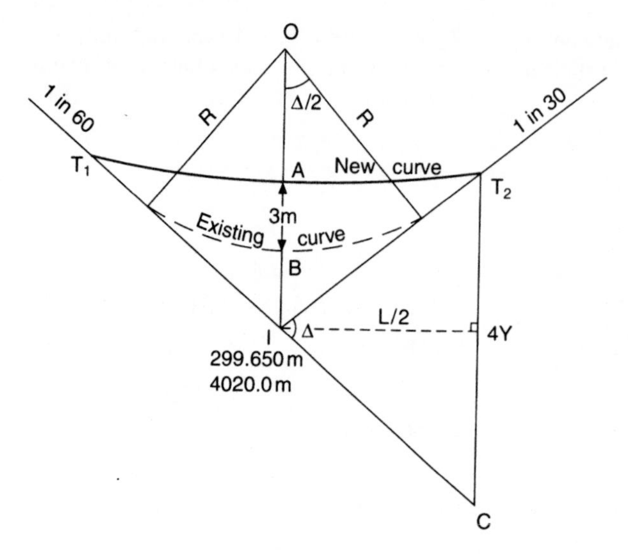

Fig. 10.49 *Central offset to a curve*

(c) The levels of the quarter points.
(d) The chainage of the lowest point on the new curve. (LU)

To find central offset Y to new curve (Figure 10.49):
From simple curve data $\Delta = \cot 60 + \cot 30 = 2° 51' 51''$
Now $BI = R(\sec \Delta/2 - 1) = 0.312$ m

$\therefore$ Central offset $AI = Y = 3.312$ m and $T_2C = 4Y = 13.248$ m

To find length of new curve:
Grade 1 in 60 = 1.67%, 1 in 30 = 3.33%

$\therefore$ Grade angle $\Delta = 5\%$

(1) Then from T_2IC, $L/2 = \dfrac{13.248}{5} \times 100$

$\therefore L = 530$ m

(2) Rise from I to $T_1 = 1.67 \times 2.65 = 4.426$ m

$\therefore$ Level at $T_1 = 299.65 + 4.426 = 304.076$ m

Rise from I to $T_2 = 3.33 \times 2.65 = 8.824$ m

$\therefore$ Level at $T_2 = 299.65 + 8.824 = 308.474$ m

(3) Levels at quarter points:
1st quarter point is 132.5 m from T_1

$\therefore$ Level on gradient $= 304.076 - (1.67 \times 1.325) = 301.863$ m

$$\text{Offset} = 3.312 \times \frac{1^2}{2^2} = 0.828 \text{ m}$$

$\therefore$ Curve level $= 301.863 + 0.828 = 302.691$ m

2nd quarter point is 397.5 m

$\therefore$ Level on gradient $= 304.076 - (167 \times 3.975) = 310.714$ m

$$\text{Offset} = 3.312 \times \frac{3^2}{2^2} = 7.452 \text{ m}$$

$\therefore$ Curve level $= 310.714 + 7.452 = 318.166$ m

(4) Position of lowest point on curve from $T_1 = \dfrac{530}{5\%} \times 1.67\% = 177$ m

Chainage at $T_1 = 4020 - 265 = 3755$ m
Chainage of lowest point $3755 + 177 = 3932$ m

Exercises

(*10.12*) A vertical curve 120 m long of the parabola type is to join a falling gradient of 1 in 200 to a rising gradient of 1 in 300. If the level of the intersection of the two gradients is 30.36 m give the levels at 15-m intervals along the curve.

If the headlamp of a car was 0.375 m above the road surface, at what distance will the beam strike the road surface when the car is at the start of the curve? Assume the beam is horizontal when the car is on a level surface. (LU)

(*Answer*: 30.660, 30.594, 30.541, 30.504, 30.486, 30.477, 30.489, 30.516, 30.588; 103.8 m)

(*10.13*) A road having an up-gradient of 1 in 15 is connected to a down-gradient of 1 in 20 by a vertical parabolic curve 120 m in length. Determine the visibility distance afforded by this curve for two approaching drivers whose eyes are 1.05 m above the road surface.

As part of a road improvement scheme a new vertical parabolic curve is to be set out to replace the original one so that the visibility distance is increased to 210 m for the same height of driver's eye.

Determine:

(a) The length of new curve.
(b) The horizontal distance between the old and new tangent points on the 1 in 5 gradient.
(c) The horizontal distance between the summits of the two curves. (ICE)

(*Answer*: 92.94 m, (a) 612 m, (b) 246 m, (c) 35.7 m)

(*10.14*) A vertical parabolic sag curve is to be designed to connect a down-gradient of 1 in 20 with an up-gradient of 1 in 15, the chainage and reduced level of the intersection point of the two gradients being 797.7 m and 83.544 m respectively.

In order to allow for necessary headroom, the reduced level of the curve at chainage 788.7 m on the down-gradient side of the intersection point is to be 85.044 m.

Calculate:

(a) The reduced levels and chainages of the tangent points and the lowest point on the curve.
(b) The reduced levels of the first two pegs on the curve, the pegs being set at the 30-m points of through chainage. (ICE)

(*Answer*: (a)T_1 = 745.24 m, 86.166 m, T_2 = 850.16 m, 87.042 m, lowest pt = 790.21 m, 85.041 m, (b) 85.941 m, 85.104 m)

(*10.15*) The surface of a length of a proposed road consists of a rising gradient of 2% followed by a falling gradient of 4% with the two gradients joined by a vertical parabolic summit curve 120 m in length. The two gradients produced meet a reduced level of 28.5 m OD.

Compute the reduced levels of the curve at the ends, at 30-m intervals and at the highest point.

What is the minimum distance at which a driver, whose eyes are 1.125 m above the road surface, would be unable to see an obstruction 100 mm high? (ICE)

(*Answer*: 27.300, 27.675, 27.600, 27.075, 26.100 m; highest pt, 27.699 m, 87 m)

Derivation of clothoid spiral formulae

AB is an infinitely small portion (δl) of the transition curve T_1t_1. (*Figure 10.50*)

$$\delta x/\delta l = \cos\phi = (1 - \phi^2/2! + \phi^4/4! - \phi^6/6! \ldots)$$

The basic equation for a clothoid curve is: $\phi = l^2/2RL$

$$\therefore \delta x/\delta l = \left(1 - \frac{(l^2/2RL)^2}{2!} + \frac{(l^2/2RL)^4}{4!} - \frac{(l^2/2RL)^6}{6!} \cdots\right)$$

$$\text{Integrating: } x = l\left(1 - \frac{l^4}{40(RL)^2} + \frac{l^8}{3456(RL)^4} - \frac{l^{12}}{599\,040(RL)^6}\right)$$

$$= l - \frac{l^5}{40(RL)^2} + \frac{l^9}{3456(RL)^4} - \frac{l^{13}}{599\,040(RL)^6} + \cdots$$

when $x = X$, $l = L$ and:

$$X = L - \frac{L^3}{5 \cdot 4 \cdot 2!R^2} + \frac{L^5}{9 \cdot 4^2 \cdot 4!R^4} - \frac{L^7}{13 \cdot 4^3 \cdot 6!R^6} + \cdots \quad (10.53)$$

Similarly, $\delta y/\delta l = \sin\phi = (\phi - \phi^3/3! + \phi^5/5! \ldots)$.
Substituting for ϕ as previously:

$$\delta y/\delta l = \frac{l^2}{2RL} - \frac{(l^2/2RL)^3}{6} + \frac{(l^2/2RL)^5}{120} \cdots$$

$$\text{Integrating: } y = l\left(\frac{l^2}{6RL} - \frac{l^6}{336(RL)^3} + \frac{l^{10}}{42\,240(RL)^5} \cdots\right)$$

$$= \frac{l^3}{6RL} - \frac{l^7}{336(RL)^3} + \frac{l^{11}}{42\,240(RL)^5} \cdots$$

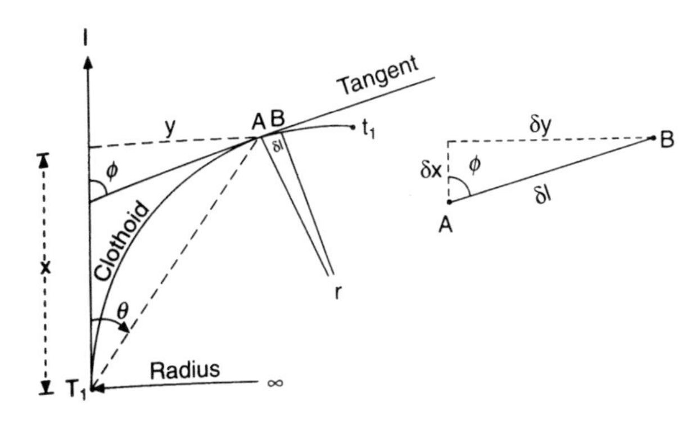

Fig. 10.50 *Clothoid spiral*

when $y = Y, l = L$, and

$$Y = \frac{L^2}{3 \cdot 2R} - \frac{L^4}{7 \cdot 3!2^3R^3} + \frac{L^6}{11 \cdot 5!2^5R^5} \cdots \tag{10.54}$$

The basic equation for a clothoid is $l = a(\phi)^{\frac{1}{2}}$, where $l = L$, $\phi = \Phi$ and $L = a(\Phi)^{\frac{1}{2}}$, then squaring and dividing gives:

$$L^2/l^2 = \Phi/\phi \tag{10.55}$$

From *Figure 10.50*:

$$\tan\theta = \frac{y}{x} = \frac{l(l^2/6RL - l^6/336(RL)^3 + l^{10}/42\,240\,(RL)^5)}{l(1 - l^4/40(RL)^2 + l^8/3456(RL)^4)}$$

However, $l = (2RL\phi)^{\frac{1}{2}}$:

$$\therefore \tan\theta = \frac{(2RL\phi)^{\frac{1}{2}}(\phi/3 - \phi^3/42 + \phi^5/1320)}{(2RL\phi)^{\frac{1}{2}}(1 - \phi^2/10 + \phi^4/216)}$$

$$= \left(\frac{\phi}{3} - \frac{\phi^3}{42} + \frac{\phi^5}{1320}\right)\left(1 - \frac{\phi^2}{10} + \frac{\phi^4}{216}\right)^{-1}$$

Let $x = -(\phi^2/10 - \phi^4/216)$ and expanding the second bracket binomially,

i.e. $(1 + x)^{-1} = 1 - nx + \dfrac{n(n-1)x^2}{2!} \cdots$

$$= 1 + \frac{\phi^2}{10} - \frac{\phi^4}{216} + \frac{2}{2!}\left(\frac{\phi^2}{10} - \frac{\phi^4}{216}\right)^2$$

$$= 1 + \frac{\phi^2}{10} - \frac{\phi^4}{216} + \frac{\phi^4}{100} + \frac{\phi^8}{81 \cdot 4!4!} - \frac{2\phi^6}{45 \cdot 2!4!}$$

$$= 1 + \frac{\phi^2}{10} - \frac{\phi^4}{216} + \frac{\phi^4}{100}$$

$$\therefore \tan\theta = \left(\frac{\phi}{3} - \frac{\phi^3}{42} + \frac{\phi^5}{11.5!}\right)\left(1 + \frac{\phi^2}{10} - \frac{\phi^4}{216} + \frac{\phi^4}{100}\right)$$

$$= \frac{\phi}{3} + \frac{\phi^3}{105} + \frac{26\phi^5}{155\,925}$$

However, as $\theta = \tan\theta - \dfrac{1}{3}\tan^3\theta + \dfrac{1}{5}\tan^5\theta$

then

$$\theta = \frac{\phi}{3} - \frac{8\phi^3}{2835} - \frac{32\phi^5}{467\,775} \cdots \tag{10.56}$$

$$= \frac{\phi}{3} - N$$

When $\theta = \text{maximum}, \phi = \Phi$ and

$$\theta = \frac{\Phi}{3} - N$$

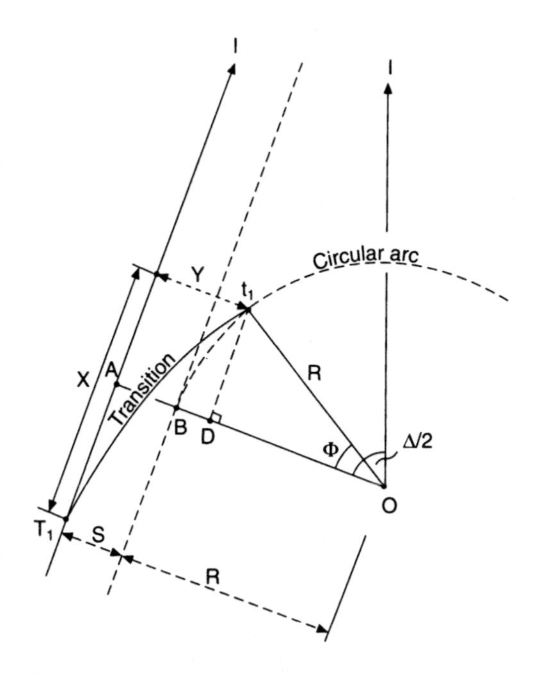

Fig. 10.51 *Transition and circular arcs*

From *Figure 10.51*:

$$BD = BO - DO = R - R\cos\Phi = R(1 - \cos\Phi)$$

$$= R\left(\frac{\Phi^2}{2!} + \frac{\Phi^4}{4!} + \frac{\Phi^6}{6!}\cdots\right)$$

but $\Phi = L/2R$,

$$\therefore BD = \frac{L^2}{2!2^2R} + \frac{L^4}{4!2^4R^3} + \frac{L^6}{6!2^6R^5}$$

and $$Y = \frac{L^2}{3 \cdot 2R} - \frac{L^4}{7 \cdot 3!2^3R^3} + \frac{L^6}{11 \cdot 5!2^5R^5}$$

$$\therefore \text{Shift} = S = (Y - BD) = \frac{L^2}{24R} - \frac{L^4}{3!7 \cdot 8 \cdot 2^3R^3} + \frac{L^6}{5!11 \cdot 12 \cdot 2^5R^5}$$

From *Figure 10.51*:

$$Dt_1 = R\sin\phi = R\left(\Phi - \frac{\Phi^3}{3!} + \frac{\Phi^5}{5!}\cdots\right)$$

but $\Phi = L/2R$

$$\therefore Dt_1 = \frac{L}{2} - \frac{L^3}{3!2^3R^2} + \frac{L^5}{5!2^5R^4}\cdots$$

$$\text{also } X = L - \frac{L^3}{5 \cdot 4 \cdot 2!R^2} + \frac{L^5}{9 \cdot 4^2 \cdot 4!R^4} \cdots$$

$$\text{Tangent length} = T_1 I = (R + S)\tan\frac{\Delta}{2} + AT_1 = (R + S)\tan\frac{\Delta}{2} + (X - Dt_1)$$

$$= (R + S)\tan\frac{\Delta}{2} + \left(\frac{L}{2} - \frac{L^3}{2!5 \cdot 6 \cdot 2^2 R^2} + \frac{L^5}{4!9 \cdot 10 \cdot 2^4 R^4}\right)$$

$$= (R + S)\tan\frac{\Delta}{2} + C \quad (10.57)$$

11
Earthworks

Estimation of areas and volumes is basic to most engineering schemes such as route alignment, reservoirs, tunnels, etc. The excavation and hauling of material on such schemes is the most significant and costly aspect of the work, on which profit or loss may depend.

Areas may be required in connection with the purchase or sale of land, with the subdivision of land or with the grading of land.

Earthwork volumes must be estimated to enable route alignment to be located at such lines and levels that cut and fill are balanced as far as practicable; and to enable contract estimates of time and cost to be made for proposed work; and to form the basis of payment for work carried out.

The tedium of earthwork computation has now been removed by the use of computers. Digital ground models (DGM), in which the ground surface is defined mathematically in terms of x, y and z coordinates, are stored in the computer memory. This data bank may now be used with several alternative design schemes to produce the optimum route in both the horizontal and vertical planes. In addition to all the setting-out data, cross-sections are produced, earthwork volumes supplied and mass-haul diagrams drawn. Quantities may be readily produced for tender calculations and project planning. The data banks may be updated with new survey information at any time and further facilitate the planning and management not only of the existing project but of future ones.

To understand how software does each stage of the earthwork computations, one requires a knowledge of the fundamentals of areas and volumes, not only to produce the software necessary, but to understand the input data required and to be able to interpret and utilize the resultant output properly.

11.1 AREAS

The computation of areas may be based on data scaled from plans or drawings, or data gained directly from survey field data.

11.1.1 Plotted areas

(1) It may be possible to sub-divide the plotted area into a series of triangles, measures the sides a, b, c, and compute the areas using:

$$\text{Area} = [s(s-a)(s-b)(s-c)]^{\frac{1}{2}} \qquad \text{where } s = (a+b+c)/2$$

The accuracy achieved will be dependent upon the scale error of the plan and the accuracy to which the sides are measured.

(2) Where the area is irregular, a sheet of gridded tracing material may be superimposed over it and the number of squares counted. Knowing the scale of the plan and the size of the squares, an estimate of the area can be obtained. Portions of squares cut by the irregular boundaries can be estimated.

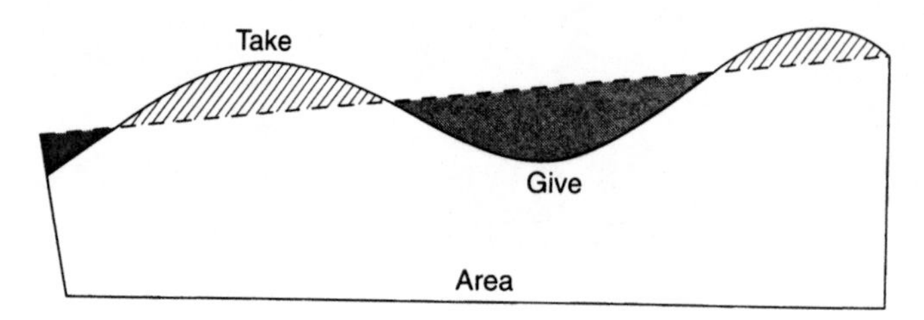

Fig. 11.1 *Areas of give and take*

(3) Alternatively, irregular boundaries may be reduced to straight lines using *give-and-take lines*, in which the areas 'taken' from the total area balance out with extra areas 'given' (*Figure 11.1*).

(4) If the area is a polygon with straight sides it may be reduced to a triangle of equal area. Consider the polygon *ABCDE* shown in *Figure 11.2*

Take *AE* as the base and extend it as shown, Join *CE* and from *D* draw a line parallel to *CE* on to the base at *F*. Similarly, join *CA* and draw a line parallel from *B* on to the base at *G*. Triangle *GCF* has the same area as the polygon *ABCDE*.

(5) The most common mechanical method of measuring areas from paper plans is to use an instrument called a *planimeter* (*Figure 11.3(a)*). This comprises two arms, *JF* and *JP*, which are free to move relative to each other through the hinged point at *J* but fixed to the plan by a weighted needle at *F*. *M* is the graduated measuring wheel and *P* the tracing point. As *P* is moved around the perimeter of the area, the measuring wheel partly rotates and partly slides over the plan with the varying movement of the tracing point (*Figure 11.3(b)*). The measuring wheel is graduated around the circumference into 10 divisions, each of which is further sub-divided by 10 into one-hundredths of a revolution, whilst a vernier enables readings to one thousandths of a revolution. The wheel is connected to a dial that records the numbered revolutions up to 10. On a *fixed-arm planimeter* one revolution of the wheel may represent 100 mm^2 on a 1:1 basis; thus, knowing the number of revolutions and the scale of the plan, the area is easily computed. In the case of a *sliding-arm planimeter* the sliding arm *JP* may be set to the scale of the plan, thereby facilitating more direct measurement of the area.

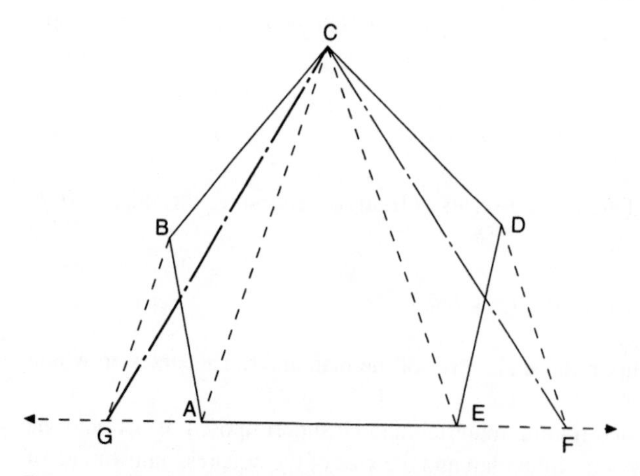

Fig. 11.2 *Reduction of a polygon to a triangle*

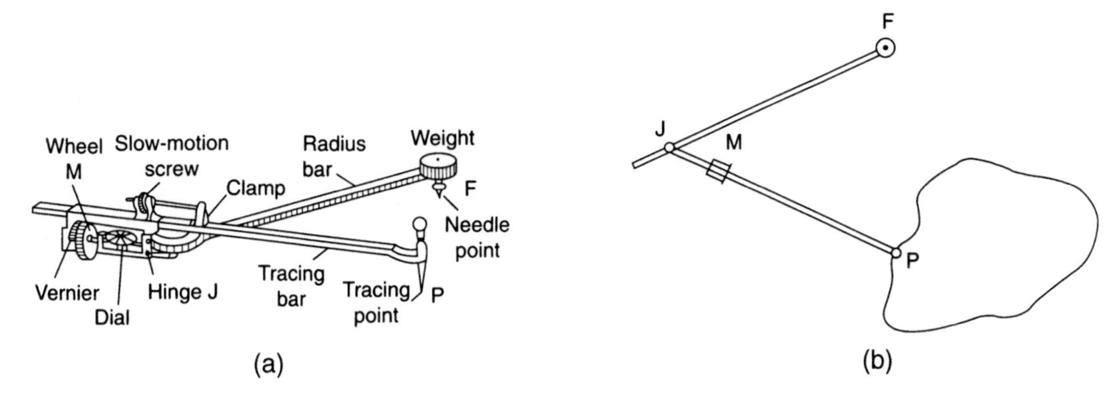

Fig. 11.3 *A polar planimeter*

In the normal way, needle point F is fixed *outside* the area to be measured, the initial reading noted, the tracing point traversed around the area and the final reading noted. The difference of the two readings gives the number of revolutions of the measuring wheel, which is a direct measure of the area. If the area is too large to enable the whole of its boundary to be traversed by the tracing point P when the needle point F is outside the area, then the area may be sub-divided into smaller more manageable areas, or the needle point can be transposed *inside* the area.

As the latter procedure requires the application of the *zero circle* of the instrument, the former approach is preferred.

The zero circle of a planimeter is that circle described by the tracing point P, when the needle point F is at the centre of the circle, and the two arms JF and JP are at right angles to each other. In this situation the measuring wheel is normal to its path of movement and so slides without rotation, thus producing a zero change in reading. The value of the zero circle is supplied with the instrument.

If the area to be measured is greater than the zero circle (*Figure 11.4*(*a*)) then only the tinted area is measured, and the zero circle value must be added to the difference between the initial and final wheel readings. In such a case the final reading will always be *greater* than the initial reading. If the final reading is *smaller* than the initial reading, then the situation is as shown in *Figure 11.4*(*b*) and the measured area, shown tinted, must be subtracted from the zero circle value.

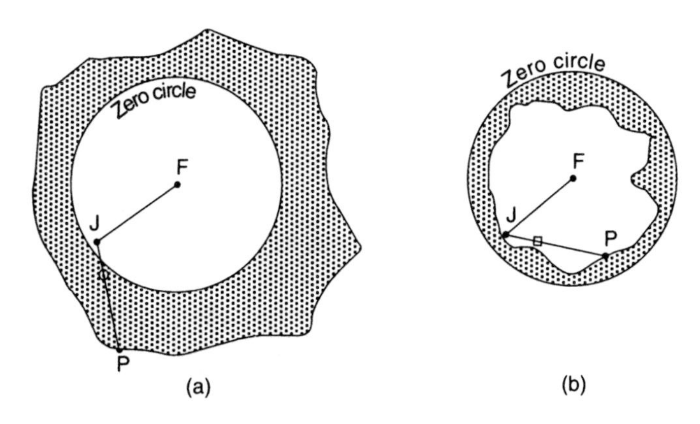

Fig. 11.4 *Measured areas and the zero circle*

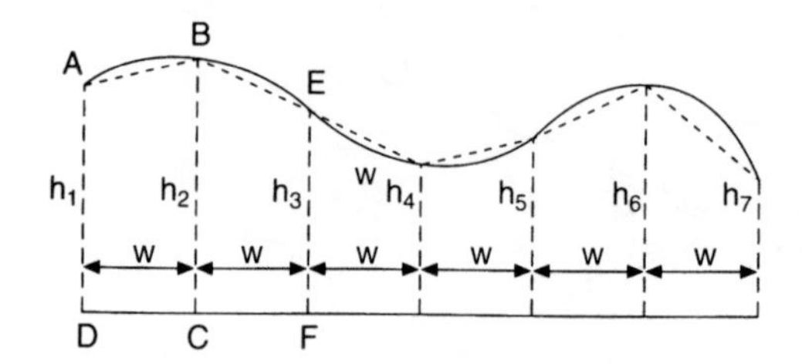

Fig. 11.5 *Trapezoidal and Simpson's rule*

Worked example

Example 11.1

(a) *Forward movement*

Initial reading	2.497
Final reading	6.282
Difference	3.785 revs
Add zero circle	18.546
Area =	22.331 revs

(b) *Backward movement*

Initial reading	2.886
Final reading	1.224
Difference	1.662 revs
Subtract from zero circle	18.546
Area =	16.884 revs

If one revolution corresponds to an area of (A), then on a plan of scale 1 in M, the actual area in (a) above is $22.331 \times A \times M^2$. If the area can be divided into strips then the area can be found using either the trapezoidal rule or Simpson's rule, as follows (*Figure 11.5*).

11.1.2 Calculated areas

(a) *Trapezoidal rule*
In *Figure 11.5*:

$$\text{Area of 1st trapezoid } ABCD = \frac{h_1 + h_2}{2} \times w$$

$$\text{Area of 2nd trapezoid } BEFC = \frac{h_2 + h_3}{2} \times w \quad \text{and so on.}$$

Total area = sum of trapezoids

$$= A = w\left(\frac{h_1 + h_7}{2} + h_2 + h_3 + h_4 + h_5 + h_6\right) \tag{11.1}$$

N.B. (i) If the first or last ordinate is zero, it must still be included in the equation.
(ii) The formula represents the area bounded by the broken line under the curving boundary; thus, if the boundary curves outside then the computed area is too small, and *vice versa*.

(b) *Simpson's rule*

$$A = w[(h_1 + h_7) + 4(h_2 + h_4 + h_6) + 2(h_3 + h_5)]/3 \tag{11.2}$$

i.e. one-third the distance between ordinates, multiplied by the sum of the *first* and *last* ordinates, plus four times the sum of the *even* ordinates, plus twice the sum of the *odd* ordinates.

N.B. (i) This rule assumes a boundary modelled as a parabola across pairs of areas and is therefore more accurate than the trapezoidal rule. If the boundary were a parabola the formula would be exact.

(ii) The equation requires an *odd* number of ordinates and consequently an even number of areas.

11.1.3 Areas by coordinates

Using appropriate field data it may be possible to define the area by its rectangular coordinates. For example:

The area enclosed by the traverse *ABCDA* in *Figure 11.6* can be found by taking the area of the rectangle *a′cDd* and subtracting the surrounding triangles, etc., as follows:

$$
\begin{aligned}
\text{Area of rectangle } a'cDd &= a'c \times a'd \\
&= 263 \times 173 = 45\,499\ \text{m}^2 \\
\text{Area of rectangle } a'bBa &= 77 \times 71 = 5\,467\ \text{m}^2 \\
\text{Area of triangle } AaB &= 71 \times 35.5 = 2\,520.5\ \text{m}^2 \\
\text{Area of triangle } BBC &= 77 \times 46 = 3\,542\ \text{m}^2 \\
\text{Area of triangle } Ccd &= 173 \times 50 = 8\,650\ \text{m}^2 \\
\text{Area of triangle } DdA &= 263 \times 12.5 = 3\,287.5\ \text{m}^2 \\
\text{Total} &= 23\,467\ \text{m}^2
\end{aligned}
$$

$$\therefore \text{Area } ABCDA = 45\,499 - 23\,467 = 22\,032\ \text{m}^2 \approx 22\,000\ \text{m}^2$$

The following rule may be used when the *total coordinates* only are given. Multiply the algebraic *sum* of the northing of each station and the one following by the algebraic *difference* of the easting of each station and the one following. The area is half the algebraic sum of the products. Thus, from *Table 11.1* and *Figure 11.6*

$$\text{Area } ABCDA \approx 22\,032\ \text{m}^2 \approx 22\,000\ \text{m}^2$$

The value of $22\,000\ \text{m}^2$ is more realistic considering the number of significant figures involved in the computations.

Table 11.1

Stns	*E*	*N*	*Difference of E*	*Sum of N*	*Double area* +	*Double area* −
A	0.0	0.0	−71	71		5 041
B	71	71	−92	219		20 148
C	163	148	−100	123		12 300
D	263	−25	263	−25		6 575
A	0.0	0.0				—
					Σ	44 064

Area *ABCDA* = $22\,032\ \text{m}^2 \approx 22\,000\ \text{m}^2$

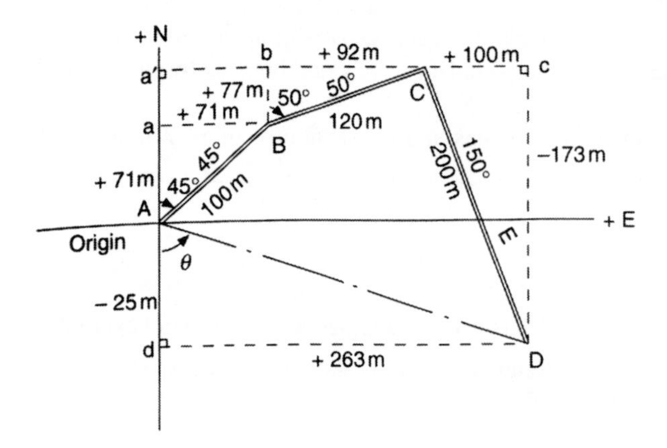

Fig. 11.6 *Area by coordinates*

This latter rule is the one most commonly used and is easily remembered if written as follows:

$$\begin{matrix} & N_A & & N_B & & N_C & & N_D & \\ E_D & & E_A & & E_B & & E_C & & E_D & & E_A \end{matrix} \qquad (11.3)$$

Thus $A = 0.5[N_A(E_B - E_D) + N_B(E_C - E_A) + N_C(E_D - E_B) + N_D(E_A - E_C)]$

$= 0.5[0 + 71(163) + 148(263 - 71) - 25(0 - 163)]$

$= 0.5[11\,573 + 28\,416 + 4075] = 22\,032\ \text{m}^2$

The stations must be lettered clockwise around the figure. If anticlockwise the result will be the same but has a negative sign.

11.2 PARTITION OF LAND

This task may be carried out by an engineer when sub-dividing land either for large building plots or for sale purposes.

11.2.1 To cut off a required area by a line through a given point

With reference to *Figure 11.7*, it is required to find the length and bearing of the line *GH* which divides the area *ABCDEFA* into the given values.

Method

(1) Calculate the total area *ABCDEFA*.
(2) Given point *G*, draw a line *GH* dividing the area approximately into the required portions.
(3) Draw a line from *G* to the station nearest to *H*, namely *F*.
(4) From coordinates of *G* and *F*, calculate the length and bearing of the line *GF*.
(5) Find the area of *GDEFG* and subtract this area from the required area to get the area of triangle *GFH*.
(6) Now area $GFH = 0.5HF \times FG \sin\theta$, difference *FG* is known from (4) above, and θ is the difference of the known bearings *FA* and *FG* and thus length *HF* is calculated.
(7) As the bearing *FH* = bearing *FA* (known), then the coordinates of *H* may be calculated.
(8) From coordinates of *G* and *H*, the length and bearing of *GH* are computed.

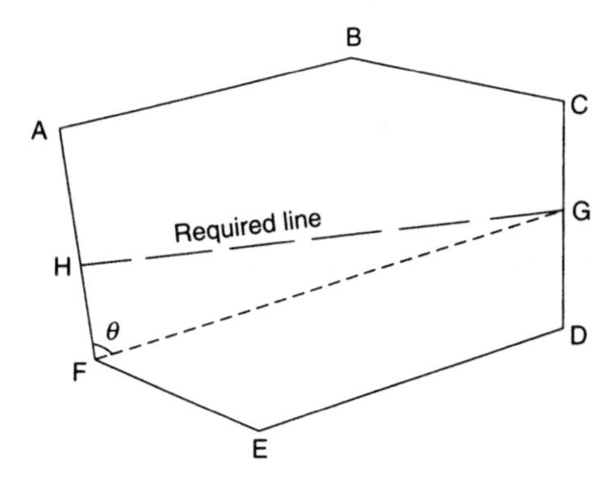

Fig. 11.7 *Divide an area by a line through a given point*

11.2.2 To cut off a required area by a line of given bearing

With reference to *Figure 11.8(a)*, it is required to fix line *HJ* of a given bearing, which divides the area *ABCDEFGA* into the required portions.

Method

(1) From any station set off on the given bearing a trial line that cuts off approximately the required area, say *AX*.
(2) Compute the length and bearing of *AD* from the traverse coordinates.
(3) In triangle *ADX*, length and bearing *AD* are known, bearing *AX* is given and bearing *DX* = bearing *DE*; thus the three angles may be calculated and the area of the triangle found.
(4) From coordinates calculate the area *ABCDA*; thus total area *ABCDXA* is known.

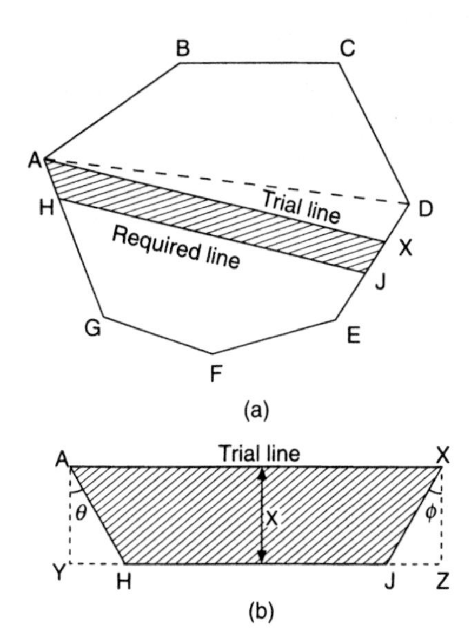

Fig. 11.8 *Divide an area by a line with a given bearing*

(5) The difference between the above area and the area required to be cut off, is the area to be *added* or *subtracted* by a line *parallel* to the trial line *AX*. Assume this to be the trapezium *AXJHA* whose area is known together with the length and bearing of one side (*AX*) and the bearings of the other sides.

(6) With reference to *Figure 11.8(b)*, as the bearings of all the sides are known, the angles θ and ϕ are known. From this $YH = x \tan\theta$ and $JZ = x \tan\phi$; now:

$$\text{Area of } AXJHA = \text{area of rectangle } AXZYA - (\text{area of triangle } AHY + \text{area of triangle } XZJ)$$

$$= AX \times x - \left(\frac{x}{2} \times x \tan\theta + \frac{x}{2} \times x \tan\phi\right)$$

$$= AX \times x - \left[\frac{x^2}{2}(\tan\theta + \tan\phi)\right] \quad (11.4)$$

from which the value of x may be found.

(7) Thus, knowing x, the distances *AH* and *XJ* can easily be calculated and used to set out the required line *HJ*.

11.3 CROSS-SECTIONS

Finding the areas of cross-sections is the first step in obtaining the volume of earthwork to be handled in route alignment projects (road or railway), or reservoir construction, for example.

In order to illustrate more clearly what the above statement means, let us consider a road construction project. In the first instance an accurate plan is produced on which to design the proposed route. The centre-line of the route, defined in terms of rectangular coordinates at 10- to 30-m intervals, is then set out in the field. Ground levels are obtained along the centre-line and also at right angles to the line (*Figure 11.9(a)*). The levels at right angles to the centre-line depict the ground profile, as shown in *Figure 11.9(b)*, and if the design template, depicting the formation level, road width, camber, side slopes, etc., is added, then a cross-section is produced whose area can be obtained by planimeter or computation. The shape of the cross-section is defined in terms of vertical heights (levels) at horizontal distances each side of the centre-line; thus no matter how complex the shape, these parameters can be treated as rectangular coordinates and the area computed using the rules given in *Section 11.1.2*. The areas may now be used in various rules (see later) to produce an estimate of the volumes. Levels along, and normal to, the centre-line may be obtained by standard levelling procedures, with a total station, or by aerial photogrammetry. The whole computational procedure, including the road design and optimization, would then be carried out on the computer to produce volumes of cut and fill, accumulated volumes, areas and volumes of top-soil strip,

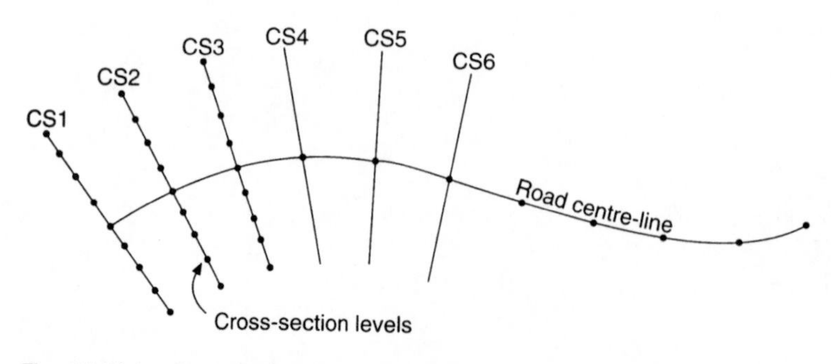

Fig. 11.9(a) *Cross-sections*

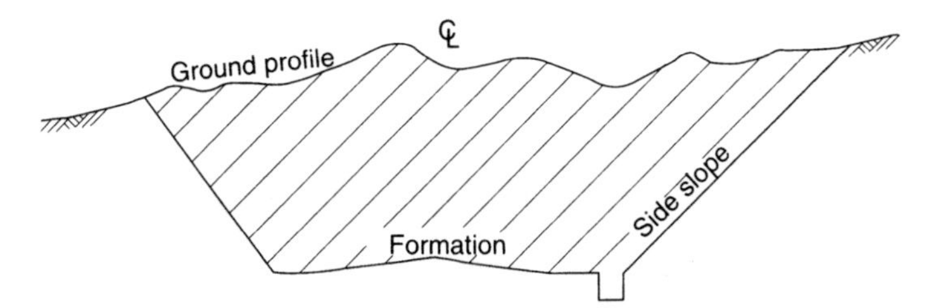

Fig. 11.9(b) *Cross-sectional area of a cutting*

side widths, etc. Where plotting facilities are available the program would no doubt include routines to plot the cross-sections for visual inspection.

Cross-sections may be approximated to the ground profile to afford easy computation. The particular cross-section adopted would be dependent upon the general shape of the ground. Typical examples are illustrated in *Figure 11.10.*

Whilst equations are available for computing the areas and side widths they tend to be over-complicated and the following method using 'rate of approach' is recommended (*Figure 11.11*).

Given: height x and grades AB and CB in triangle ABC.
Required: to find distance y_1.
Method: *Add* the two grades, using their absolute values, invert them and multiply by x.

I.e. $(1/2 + 1/5)^{-1}x = 10x/7 = y_1$ or $(0.5 - 0.2)^{-1}x = 10x/7 = y_1$

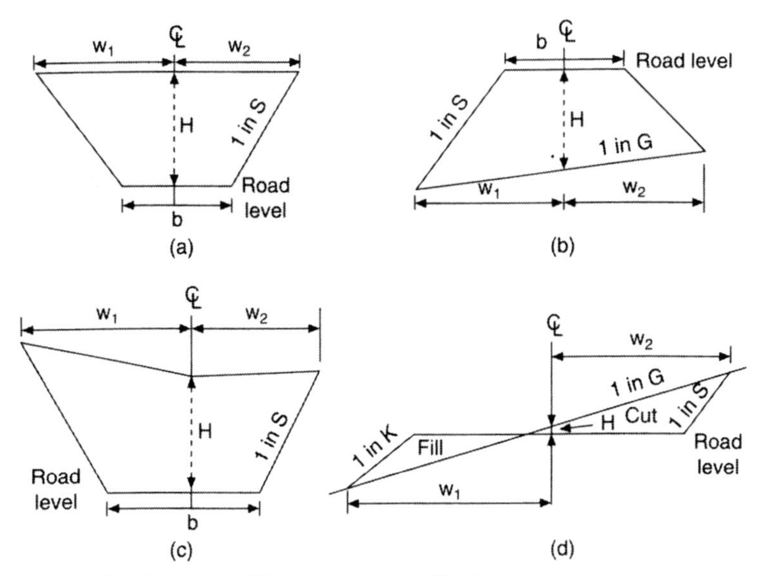

b = Finished road width at road or formation level
H = Centre height
W_1, W_2 = Side widths, measured horizontally from the centre-line and depicting the limits of the construction
1 in S = Side slope of 1 vertical to S horizontal (100 x 5^{-1}%)
1 in G = Existing ground slope

Fig. 11.10 *(a) Cutting, (b) embankment, (c) cutting and (d) hillside section*

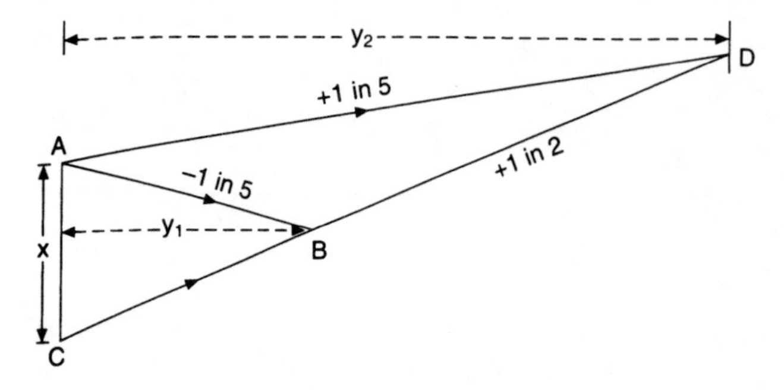

Fig. 11.11 *Rate of approach*

Similarly, to find distance y_2 in triangle *ADC*, subtract the two grades, invert them and multiply by x.

E.g. $(1/2 - 1/5)^{-1}x = 10x/3 = y_2$ or $(0.5 - 0.2)^{-1}x = 10x/3 = y_2$

The rule, therefore is:

(1) When the two grades are running in opposing directions (as in *ABC*), add (signs opposite + −).
(2) When the two grades are running in the same direction (as in *ADC*), subtract (signs same).

N.B. Height x must be *vertical* relative to the grades (*see Worked example 11.2*).

Proof

From *Figure 11.12* it is seen that 1 in 5 = 2 in 10 and 1 in 2 = 5 in 10, and thus the two grades diverge from *B* at the rate of 7 in 10. Thus, if $AC = 7$ m then $EB = 10$ m, i.e. $x \times 10/7 = 7 \times 10/7 = 10$ m.

Two examples will now be worked to illustrate the use of the above technique.

Worked examples

Example 11.2 Calculate the *side widths* and *cross-sectional area* of an embankment (*Figure 11.13*) having the following dimensions:

Road width = 20 m existing ground slope = 1 in 10 (10%)

Side slopes = 1 in 2 (50%) centre height = 10 m

As horizontal distance from centre-line to *AE* is 10 m and the ground slope is 10%, then *AE* will be 1 m greater than the centre height and *BD* 1 m less. Thus, $AE = 11$ m and $BD = 9$ m, area of $ABDE = 20 \times 10 = 200\,\text{m}^2$. Now, to find the areas of the remaining triangles *AEF* and *BDC* one needs

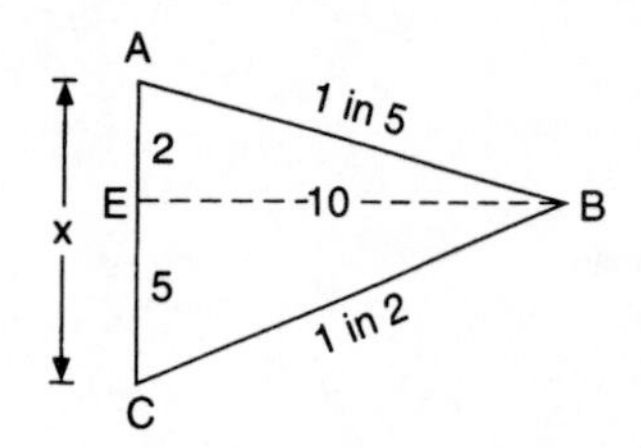

Fig. 11.12 *Rate of approach calculation*

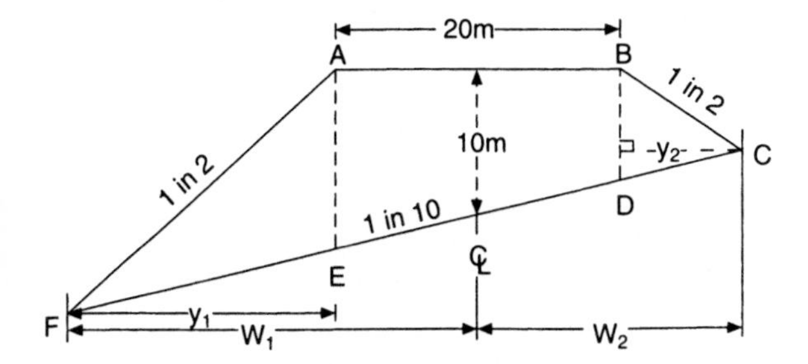

Fig. 11.13 *Rate of approach calculation*

the perpendicular heights y_1 and y_2, as follows:

(1) $1/2 - 1/10 = 4/10$, then $y_1 = (4/10)^{-1} \times AE = 11 \times 10/4 = 27.5\,\text{m}$
(2) $1/2 + 1/10 = 6/10$, then $y_2 = (6/10)^{-1} \times BD = 9 \times 10/6 = 15.0\,\text{m}$

$$\therefore \text{Area of triangle } AEF = \frac{AE}{2} \times y_1 = \frac{11}{2} \times 27.5 = 151.25\,\text{m}^2$$

$$\text{Area of triangle } BDC = \frac{BD}{2} \times y_2 = \frac{9}{2} \times 15.0 = 67.50\,\text{m}^2$$

$$\text{Total area} = (200 + 151.25 + 67.5) = 418.75\,\text{m}^2$$

$$\text{Side width } w_1 = 10\,\text{m} + y_1 = 37.5\,\text{m}$$

$$\text{Side width } w_2 = 10\,\text{m} + y_2 = 25.0\,\text{m}$$

Example 11.3 Calculate the side widths and cross-sectional areas of cut and fill on a hillside section (*Figure 11.14*) having the following dimensions:

Road width = 20 m	existing ground slope = 1 in 5 (20%)
Side slope in cut = 1 in 1 (100%)	centre height in cut = 1 m
Side slope in fill = 1 in 2 (50%)	

As ground slope is 20% and centre height 1 m, it follows that the horizontal distance from centre-line to B is 5 m; therefore, $AB = 5$ m, $BC = 15$ m. From these latter distances it is obvious that $AF = 1$ m and $GC = 3$ m.

$$\text{Now, } y_1 = (1/2 - 1/5)^{-1} \times AF = \frac{10}{3} \times 1 = 3.3\,\text{m}$$

$$y_2 = (1 - 1/5)^{-1} \times GC = \frac{5}{4} \times 3 = 3.75\,\text{m}$$

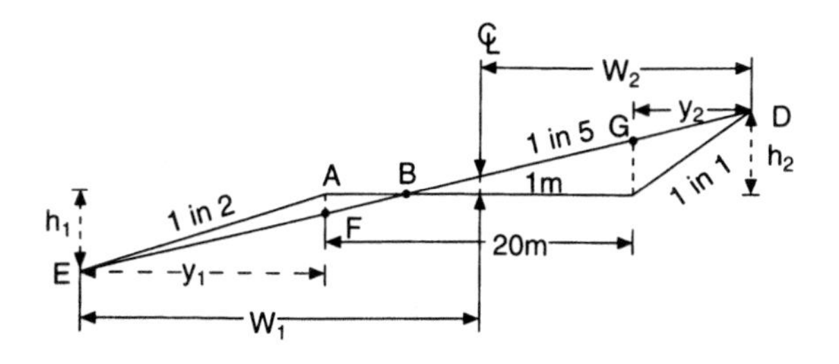

Fig. 11.14 *Cross-sectional area calculation*

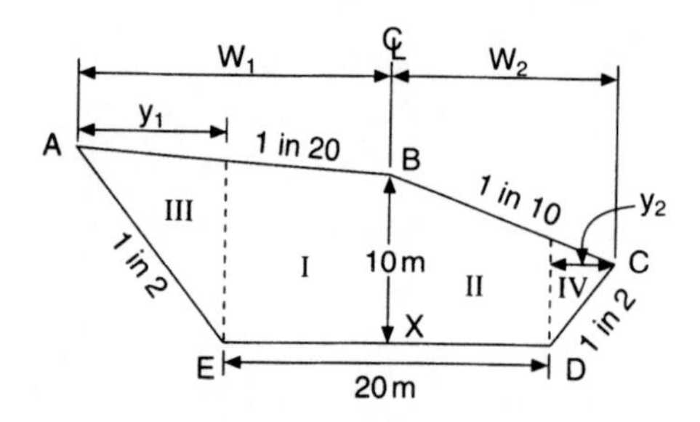

Fig. 11.15 *Cross-sectional area by coordinates*

$$\therefore \text{Side width } w_1 = 10\,\text{m} + y_1 = 13.3\,\text{m}$$

$$\text{Side width } w_2 = 10\,\text{m} + y_2 = 13.75\,\text{m}$$

Now, as side slope AE is 1 in 2, then $h_1 = y_1/2 = 1.65\,\text{m}$ and as side slope CD is 1 in 1, then $h_2 = y_2 = 3.75\,\text{m}$.

$$\therefore \text{Area of cut}\,(BCD) = \frac{BC}{2} \times h_2 = \frac{15}{2} \times 3.75 = 28.1\,\text{m}^2$$

$$\text{Area of fill}\,(ABE) = \frac{AB}{2} \times h_1 = \frac{5}{2} \times 1.65 = 4.1\,\text{m}^2$$

Example 11.4 Calculate the area and side-widths of the cross-section (*Figure 11.15*):

$$y_1 = \left(\frac{1}{2} - \frac{1}{20}\right)^{-1} \times 10.5 = \frac{20}{9} \times 10.5 = 23.33\,\text{m}$$

$$\therefore W_1 = 10 + y_1 = 33.33\,\text{m}$$

$$y_2 = \left(\frac{1}{2} + \frac{1}{10}\right)^{-1} \times 9 = \frac{10}{6} \times 9 = 15.00\,\text{m}$$

$$\therefore W_1 = 10 + y_2 = 25.00\,\text{m}$$

$$\begin{aligned}
\text{Area I} &= [(10 + 10.5)/2]10 &&= 102.50\,\text{m}^2\\
\text{Area II} &= [(10 + 9)/2]10 &&= 95.00\,\text{m}^2\\
\text{Area III} &= (10.5/2)23.33 &&= 122.50\,\text{m}^2\\
\text{Area IV} &= (9/2)15.00 &&= 67.50\,\text{m}^2\\
&\text{Total area} &&= 387.50\,\text{m}^2
\end{aligned}$$

11.3.1 Cross-sectional areas by coordinates

Where the cross-section is complex and the ground profile has been defined by reduced levels at known horizontal distances from the centre-line, the area may be found by coordinates. The horizontal distances may be regarded as Eastings (E) and the elevations (at 90° to the distances) as Northings (N).

Consider the previous example (*Figure 11.15*). Taking X on the centre-line as the origin, XB as the N-axis (reduced levels) and ED as the E-axis, the coordinates of A, B, C, D and E are:

$$\left\{ A = \frac{11.67}{-33.33} \quad B = \frac{10.00}{0} \quad C = \frac{7.50}{25.00} \right\} \quad \text{Ground levels}$$

$$\left\{ E = \frac{0}{-10.00} \quad D = \frac{0}{10.00} \right\} \quad \text{Formation levels}$$

(In the above format the denominator is the E-value.)

Point	*E*	*N*
A	−33.33	11.67
B	0	10.00
C	25.00	7.50
D	10.00	0
E	−10.00	0

From *equation (11.3)*:

$$\begin{array}{cccccccccccc} & N_A & & N_B & & N_C & & N_D & & N_E & & \\ E_E & & E_A & & E_B & & E_C & & E_D & & E_E & & E_A \end{array}$$

$$\text{Area} = \tfrac{1}{2}[N_A(E_B - E_E) + N_B(E_C - E_A) + N_C(E_D - E_B) + N_D(E_E - E_C) + N_E(E_A - E_D)]$$

$$= \tfrac{1}{2}[11.67\{0 - (-10)\} + 10\{25 - (-33.33)\} + 7.5(10 - 0)]$$

(As the N_D and N_E are zero, the terms are ignored.)

$$= \tfrac{1}{2}[116.70 + 583.30 + 75] = 387.50\,\text{m}^2$$

11.4 DIP AND STRIKE

On a tilted plane there is a direction of maximum tilt, such direction being called *the line of full dip*. Any line at right angles to full dip will be a level line and is called a *strike line* (*Figure 11.16(a)*). Any grade between full dip and strike is called *apparent dip*. An understanding of dip and strike is occasionally necessary for some earthwork problems. From *Figure 11.16(a)*:

$$\tan\theta_1 = \frac{ac}{bc} = \frac{de}{bc} = \left(\frac{de}{be} \times \frac{be}{bc}\right) = \tan\theta\cos\phi$$

i.e. tan (apparent dip) = tan (full dip) × cos (included angle) (11.5)

Worked example

Example 11.5 On a stratum plane, an apparent dip of 1 in 16 bears 170°, whilst the apparent dip in the direction 194° is 1 in 11; calculate the direction and rate of full dip.

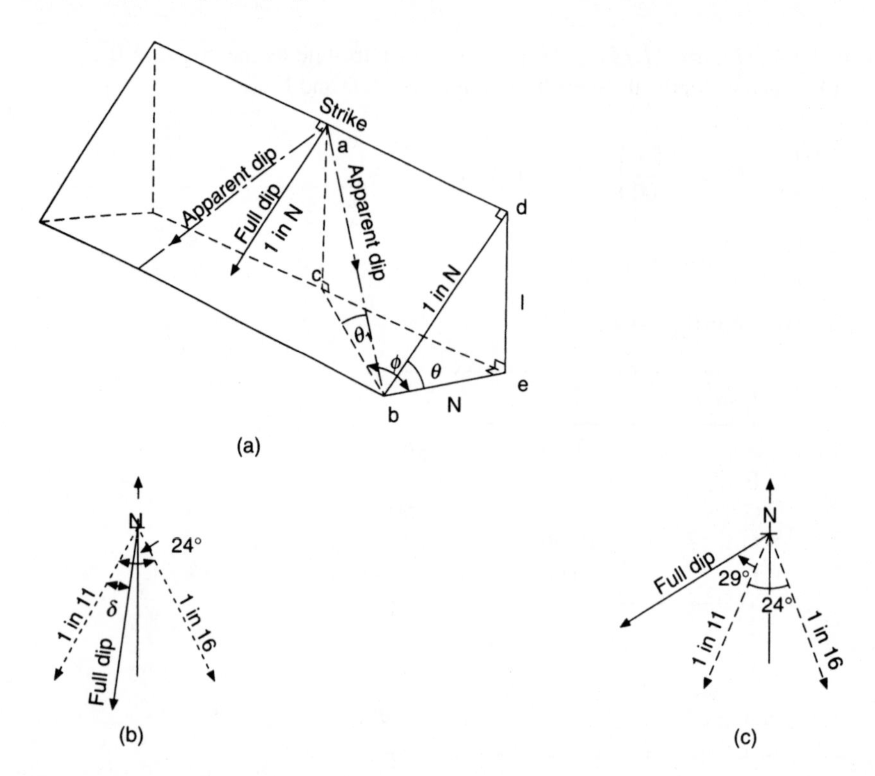

Fig. 11.16 *Dip and strike*

Draw a sketch of the situation (*Figure 11.16(b)*) and assume any position for full dip. Now, using *equation (11.5)*:

$$\tan\theta_1 = \tan\theta\cos\phi$$

$$\frac{1}{16} = \tan\theta\cos(24^\circ - \delta)$$

$$\tan\theta = \frac{1}{16\cos(24^\circ - \delta)}$$

Similarly, $\dfrac{1}{11} = \tan\theta\cos\delta$

$$\tan\theta = \frac{1}{11\cos\delta}$$

Equating (a) and (b):

$$16\cos(24^\circ - \delta) = 11\cos\delta$$

$$16(\cos 24^\circ\cos\delta + \sin 24^\circ\sin\delta) = 11\cos\delta$$

$$16(0.912\cos\delta + 0.406\sin\delta) = 11\cos\delta$$

$$14.6\cos\delta + 6.5\sin\delta = 11\cos\delta$$

$$3.6\cos\delta = -6.5\sin\delta$$

Cross multiply $$\frac{\sin\delta}{\cos\delta} = \tan\delta = -\frac{3.6}{6.5}$$

$$\therefore \delta = -29^\circ$$

N.B. The minus sign indicates that the initial position for full dip in *Figure 11.16*(*b*) is incorrect, and that it lies *outside* the apparent dip. As the grade is increasing from 1 in 16 to 1 in 11, the full dip must be as in *Figure 11.16*(*c*).

$\therefore$ Direction of full dip = 223°

Now, a second application of the formula will give the rate of full dip. That is

$$\frac{1}{11} = \frac{1}{x}\cos 29^\circ$$

$$\therefore x = 11\cos 29^\circ = 9.6$$

$\therefore$ Rate of full dip = 1 in 9.6

11.5 VOLUMES

The importance of volume assessment has already been outlined. Many volumes encountered in civil engineering appear, at first glance, to be rather complex in shape. Generally speaking, however, they can be divided into *prisms*, *wedges* or *pyramids*, each of which will now be dealt with in turn.

(1) Prism

The two ends of the prism (*Figure 11.17*) are equal and parallel, the resulting sides thus being parallelograms.

$$\text{Vol} = \text{AL} \tag{11.6}$$

(2) Wedge

$$\text{Volume of wedge (Figure 11.18)} = \frac{L}{6}(\text{sum of parallel edges} \times \text{vertical height of base})$$

$$= \frac{L}{6}[(a+b+c)\times h] \tag{11.7a}$$

when $a = b = c$: $$V = AL/2 \tag{11.7b}$$

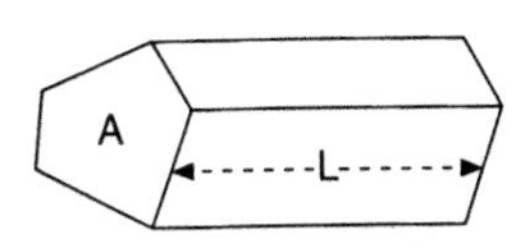

Fig. 11.17 *Prism*

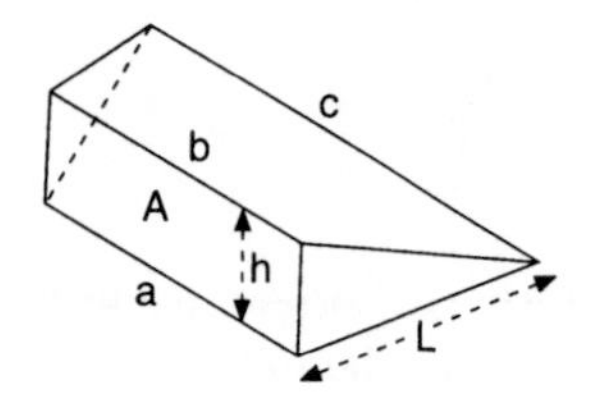

Fig. 11.18 *Wedge*

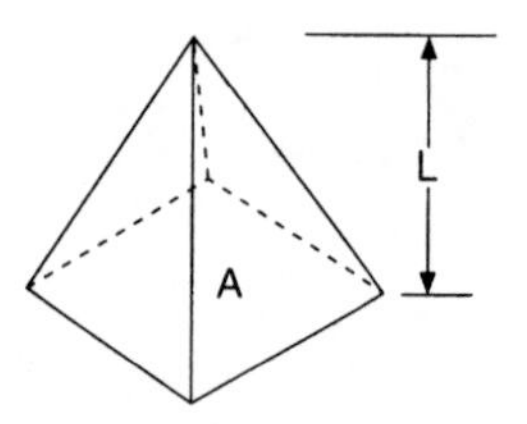

Fig. 11.19 *Pyramid*

(3) Pyramid

$$\text{Volume of pyramid (}\textit{Figure 11.19}\text{)} = \frac{AL}{3} \tag{11.8}$$

Equations (11.6) to *(11.8)* can all be expressed as the common equation:

$$V = \frac{L}{6}(A_1 + 4A_m + A_2) \tag{11.9}$$

where A_1 and A_2 are the end areas and A_m is the area of the section situated mid-way between the end areas. It is important to note that A_m is not the arithmetic mean of the end areas, except in the case of a wedge.

To prove the above statement consider:

(1) Prism

In this case $A_1 = A_m = A_2$:

$$V = \frac{L}{6}(A + 4A + A) = \frac{L \times 6A}{6} = AL$$

(2) Wedge

In this case A_m is the mean of A_1 and A_2, but $A_2 = 0$. Thus $A_m = A/2$:

$$V = \frac{L}{6}\left(A + 4 \times \frac{A}{2} + 0\right) = \frac{L \times 3A}{6} = \frac{AL}{2}$$

(3) Pyramid

In this case $A_m = \dfrac{A}{4}$ and $A_2 = 0$:

$$V = \frac{L}{6}\left(A + 4 \times \frac{A}{4} + 0\right) = \frac{L \times 2A}{6} = \frac{AL}{3}$$

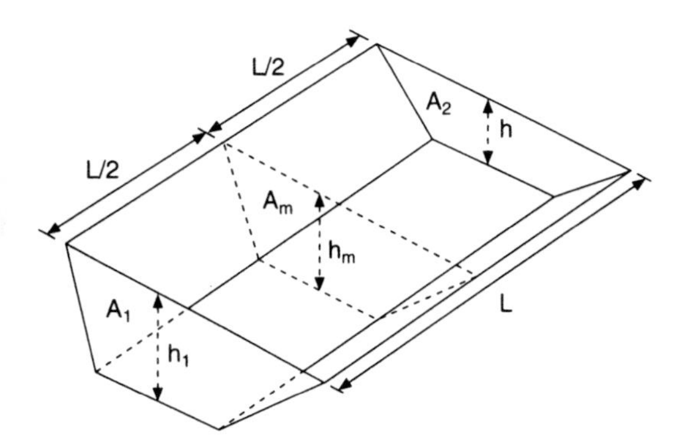

Fig. 11.20 *Prismoid*

Thus, any solid which is a combination of the above three forms and having a common value for L, may be solved using *equation (11.9)*. Such a volume is called a *prismoid* and the formula is called the *prismoidal equation*. It is easily deduced by simply substituting areas for ordinates in Simpson's rule. The prismoid differs from the prism in that its parallel ends are not necessarily equal in area; the sides are generated by straight lines from the edges of the end areas (*Figure 11.20*).

The prismoidal equation is correct when the figure is a true prismoid. In practice it is applied by taking three successive cross-sections. If the mid-section is different from that of a true prismoid, then errors will arise. Thus, in practice, sections should be chosen in order to avoid this fault. Generally, the engineer elects to observe cross-sections at regular intervals assuming compensating errors over a long route distance.

11.5.1 End-area method

Consider *Figure 11.21*, then

$$V = \frac{A_1 + A_2}{2} \times L \tag{11.10}$$

i.e. the mean of the two end areas multiplied by the length between them. The equation is correct only when the mid-area of the prismoid is the mean of the two end areas. It is correct for wedges but in the case of a pyramid it gives a result which is 50% too great:

$$\text{Vol. of pyramid} = \frac{A + 0}{2} \times L = \frac{AL}{2} \text{ instead of } \frac{AL}{3}$$

Although this method generally over-estimates, it is widely used in practice. The main reasons for this are its simplicity and the fact that the assumptions required for a good result using the prismoidal method are rarely fulfilled in practice. Strictly, however, it should be applied to prismoids comprising prisms and wedges only; such is the case where the height or width of the consecutive sections is approximately equal. It is interesting to note that with consecutive sections, where the height increases as the width decreases, or *vice versa*, the end-area method gives too small a value.

The difference between the prismoidal and end-area equations is called *prismoidal excess* and may be applied as a correction to the end-area value. It is rarely used in practice.

Summing a series of end areas gives:

$$V = L\left(\frac{A_1 + A_n}{2} + A_2 + A_3 + \cdots + A_{n-1}\right) \tag{11.11}$$

This formula is called the *trapezoidal rule* for volumes.

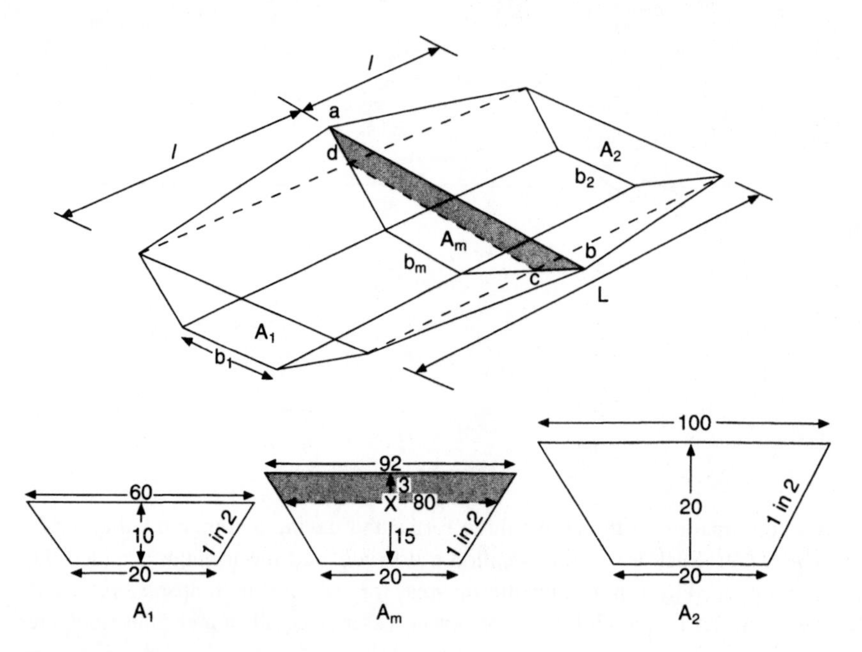

Fig. 11.21 *End-area calculation*

11.5.2 Comparison of end-area and prismoidal equations

In order to compare the methods, the volume of *Figure 11.21*, will be computed as follows:

Dimensions of *Figure 11.21*.

Centre heights: $h_1 = 10\,\text{m}, h_2 = 20\,\text{m}, h_m = 18\,\text{m}$

Road widths: $b_1 = b_2 = b_m = 20\,\text{m}$

Side slopes: 1 in 2 (50%)

Horizontal distance between sections: $l = 30\,\text{m}, L = 60\,\text{m}$

N.B. For a true prismoid h_m would have been the mean of h_1 and h_2, equal to 15 m. The broken line indicates the true prismoid, the excess area of the mid-section is shown tinted.

The true volume is thus a true prismoid plus two wedges, as follows:

(1) $$A_1 = \frac{60 + 20}{2} \times 10 = 400\,\text{m}^2$$

$$A_2 = \frac{100 + 20}{2} \times 20 = 1200\,\text{m}^2$$

$$A_m = \frac{80 + 20}{2} \times 15 = 750\,\text{m}^2$$

$$\text{Vol. of prismoid} = V_1 = \frac{60}{6}(400 + 4 \times 750 + 1200) = 46\ 000\,\text{m}^3$$

$$\text{Vol. of wedge 1} = \frac{L}{6}[(a + b + c) \times h] = \frac{30}{6}[(92 + 80 + 60) \times 3] = 3480\,\text{m}^3$$

$$\text{Vol. of wedge 2} = \frac{30}{6}[(92 + 80 + 100) \times 3] = 4080\,\text{m}^3$$

$$\text{Total true volume} = 53\,560\,\text{m}^3$$

(2) *Volume by prismoidal equation* (A_m will now have a centre height of 18 m)

$$A_m = \frac{92 + 20}{2} \times 18 = 1008\,\text{m}^2$$

$$\text{Vol.} = \frac{60}{6}(400 + 4032 + 1200) = 56\,320\,\text{m}^3$$

$$\text{Error} = 56\,320 - 53\,560 = +2760\,\text{m}^2$$

This error is approximately equal to the area of the excess mid-section multiplied by $L/6$, i.e. (Area $abcd \times L$)/6, and is so for all such circumstances; it would be negative if the mid-area had been smaller.

(3) *Volume by end area*

$$V_1 = \frac{400 + 1008}{2} \times 30 = 21\,120\,\text{m}^3$$

$$V_2 = \frac{1008 + 1200}{2} \times 30 = 33\,120\,\text{m}^3$$

$$\text{Total volume} = 54\,240\,\text{m}^3$$

$$\text{Error} = 54\,240 - 53\,560 = +680\,\text{m}^3$$

Thus, in this case the end-area method gives a better result than the prismoidal equation. However, if we consider only the true prismoid, the volume by end areas is $46\,500\,\text{m}^3$ compared with the volume by prismoidal equation of $46\,000\,\text{m}^3$, which, in this case, is the true volume.

Therefore, in practice, it can be seen that neither of these two methods is satisfactory. Unless the ideal geometric conditions exist, which is rare, both methods will give errors. To achieve greater accuracy, the cross-sections should be located in the field, with due regard to the formula to be used. If the cross-sections are approximately equal in size and shape, and the intervening surface roughly a plane, then end areas will give an acceptable result. Should the sections be vastly different in size and shape, with the mid-section contained approximately by straight lines generated between the end sections, then the prismoidal equation will give the better result.

11.5.3 Contours

Volumes may be found from contours using either the end-area or prismoidal method. The areas of the sections are the areas encompassed by the contours. The distance between the sections is the contour interval. This method is commonly used for finding the volume of a reservoir, lake or spoil heap (see *Exercise (11.3)*).

11.5.4 Spot heights

This method is generally used for calculating the volumes of excavations for basements or tanks, i.e. any volume where the sides and base are planes, whilst the surface is broken naturally (*Figure 11.22(a)*). *Figure 11.22(b)* shows the limits of the excavation with surface levels in metres at A, B, C and D. The sides are vertical to a formation level of 20 m. If the area $ABCD$ was a plane, then the volume of excavation would be:

$$V = \text{plan area } ABCD \times \text{mean height} \tag{11.12}$$

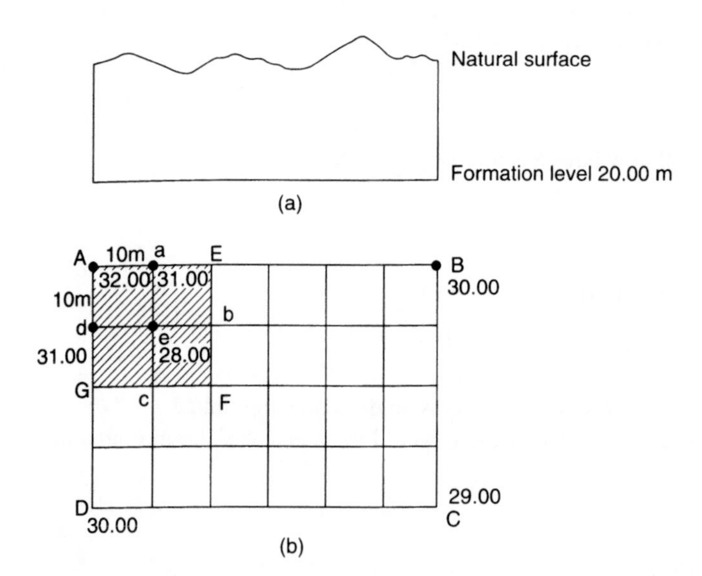

Fig. 11.22 *(a) Section, and (b) plan*

However, as the illustration shows, the surface is very broken and so must be covered with a grid such that the area within each 10-m grid square is approximately a plane. It is therefore the ruggedness of the ground that controls the grid size. If, for instance, the surface *Aaed* was not a plane, it could be split into two triangles by a diagonal (*Ae*) if this would produce better surface planes.

Considering square *Aaed* only:

$$V = \text{plan area} \times \text{mean height}$$

$$= 100 \times \frac{1}{4}(12 + 11 + 8 + 11) = 1050\,\text{m}^3$$

If the grid squares are all equal in area, then the data is easily tabulated and worked as follows:

Considering *AEFG* only, instead of taking each grid square separately, one can treat it as a whole.

$$\therefore V = \frac{100}{4}[h_A + h_E + h_F + h_G + 2(h_a + h_b + h_c + h_d) + 4h_e]$$

If one took each grid separately it would be seen that the heights of *AEFG* occur only once, whilst the heights of *abcd* occur twice and h_e occurs four times; one still divides by four to get the mean height.

This approach is adopted by computer packages and by splitting the area into very small triangles or squares (*Figure 11.23*), an extremely accurate assessment of the volume is obtained.

The above formula is also very useful for any difficult shape consisting entirely of planes, as the following example illustrates (*Figure 11.24*).

Vertical height at *A* and *D* is 10 m.

As $AB = 40$ m and surface slopes at 1 in 10, then vertical heights at *B* and *C* must be 4 m greater, i.e. 14 m.

Consider splitting the shape into two wedges by a plane connecting *AD* to *HE*.

In $\Delta ABB'$ (*Figure 11.24(a)*):

$$\text{By rate of approach: } y = \left(1 - \frac{1}{10}\right)^{-1} \times 14 = 15.56\,\text{m} = B'H = C'E$$

$$\therefore HE = 20 + 15.56 + 15.56 = 51.12\,\text{m}$$

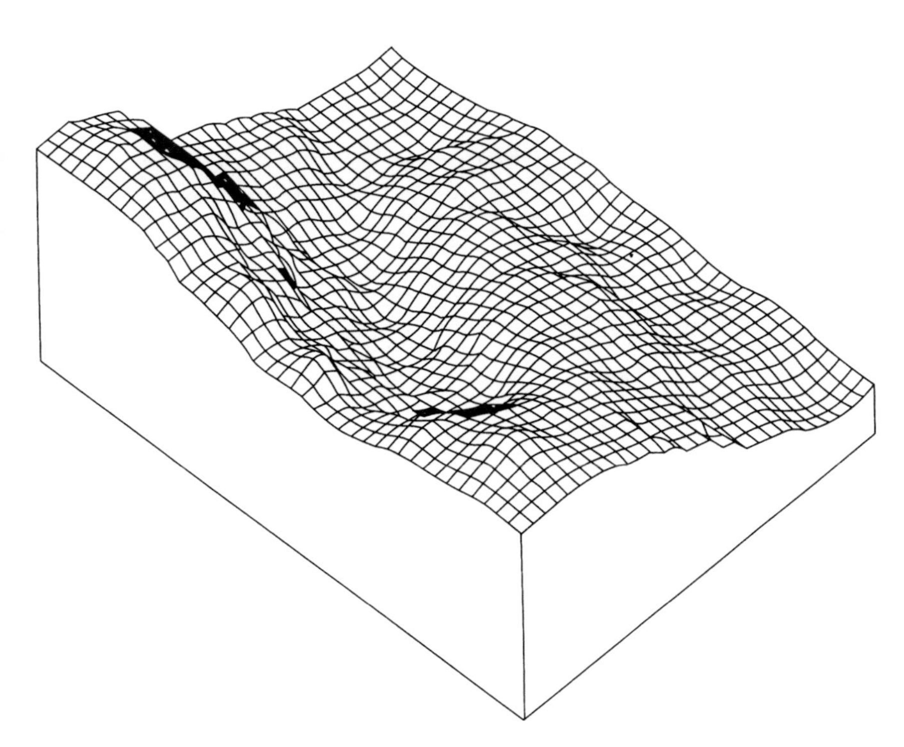

Fig. 11.23 *Ground model*

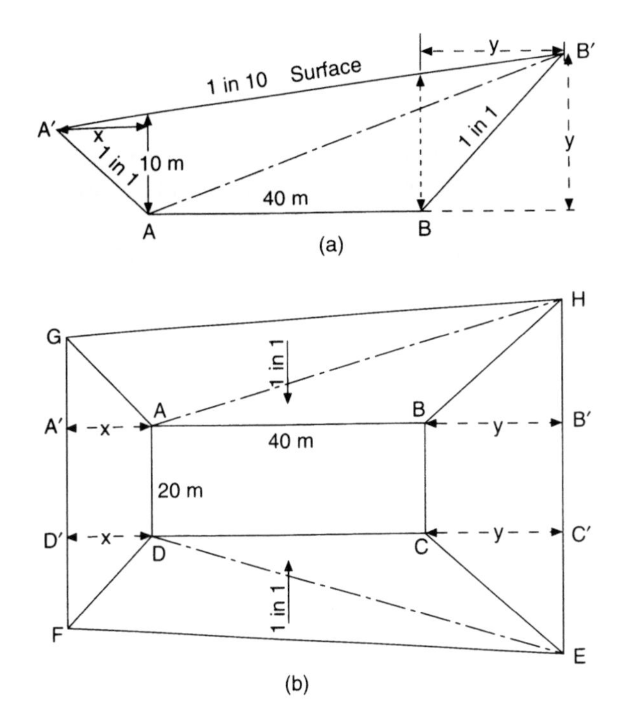

Fig. 11.24 *Calculate volume of shape definded by planes*

Area of $\Delta ABB'$ normal to AD, BC, HE

$$= \frac{40}{2} \times 15.56 = 311.20\,\text{m}^2$$

$$\therefore \text{Vol} = \text{area} \times \text{mean height} = \frac{311.20}{3}(AD + BC + HE)$$

$$= 103.73(20 + 20 + 51.12) = 9452\,\text{m}^3$$

Similarly in $\Delta AA'B'$:

$$x = \left(1 + \frac{1}{10}\right)^{-1} \times 10 = 9.09\,\text{m} = A'G = D'F$$

$$\therefore GF = 20 + 9.09 + 9.09 \times 38.18\,\text{m}$$

Area of $\Delta AA'B'$ normal to AD, GF, HE

$$= \frac{(x + AB + y)}{2} \times 10 = \frac{64.65}{2} \times 10 = 323.25\,\text{m}^3$$

$$\therefore \text{Vol} = \frac{323.25}{2}(20 + 38.18 + 51.12) = 11\,777\,\text{m}^3$$

Total vol $= 21\,229\,\text{m}^3$

Check

$$\text{Wedge } ABB' = \frac{40}{6}[(20 + 20 + 51.12) \times 15.56] = 9452\,\text{m}^3$$

$$\text{Wedge } AA'B' = \frac{64.65}{6}[(20 + 38.18 + 51.12) \times 10] = 11\,777\,\text{m}^3$$

11.5.5 Effect of curvature on volumes

The application of the prismoidal and end-area formulae has assumed, up to this point, that the cross-sections are parallel. When the excavation is curved (*Figure 11.25*), the sections are radial and a *curvature correction* must be applied to the formulae.

Pappus's theorem states that the correct volume is where the distance between the cross-sections is taken along the path of the centre of mass.

Consider the volume between the first two sections of area A_1 and A_2

Distance between sections measured along centre-line $= X'Y' = D$

Angle δ subtended at the centre $= D/R$ radians

Now, length along path of centre of mass $= XY = \delta \times$ mean radius to path of centre of mass where mean radius

$$= R - (d_1 + d_2)/2 = (R - d)$$

$$\therefore XY = \delta(R - d) = D(R - d)/R$$

$$\text{Vol. by end areas} = \frac{1}{2}(A_1 + A_2)XY = \frac{1}{2}(A_1 + A_2)D(R - d)R$$

$$= \frac{1}{2}(A_1 + A_2)D(1 - d/R)$$

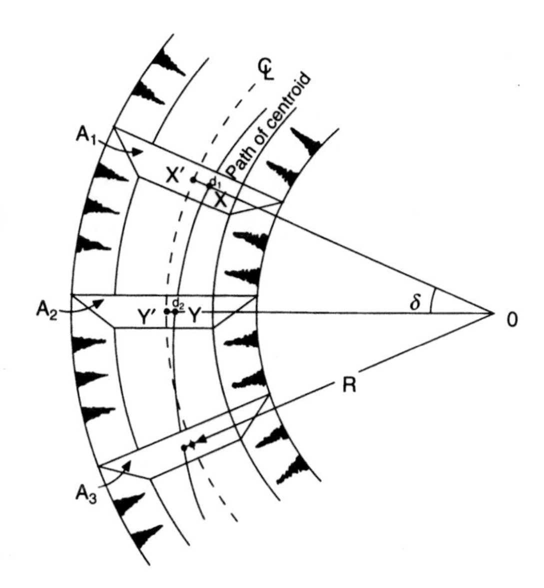

Fig. 11.25 *Curved excavation*

In other words, one corrects for curvature by multiplying the area A_1 by $(1 - d_1/R)$, and area A_2 by $(1 - d_2/R)$, the corrected areas then being used in either the end-area or prismoidal formulae, in the normal way, with D being the distance measured along the centre-line. If the centre of mass lies beyond the centre-line, as in section A_3, then the correction is $(1 + d_3/R)$.

This correction for curvature is, again, never applied to earthworks in practice. Indeed, it can be shown that the effect is very largely cancelled out on long earthwork projects. However, it may be significant on small projects or single curved excavations. (Refer to *Worked example 11.10*)

Worked examples

Example 11.6 Figure 11.26 illustrates a section of road construction to a level road width of 20 m, which includes a change from fill to cut. From the data supplied in the following field book extract, calculate the volumes of cut and fill using the end-area method and correcting for prismoidal excess. (KU)

Chainage	*Left*	*Centre*	*Right*
7500	$-\frac{10.0}{36.0}$	$-\frac{20.0}{0}$	$-\frac{8.8}{22.0}$
7600	$\frac{0}{10}$	$-\frac{6.0}{0}$	$-\frac{14.0}{24.6}$
7650	$\frac{16.0}{22.0}$	$\frac{4.0}{0}$	$\frac{0}{10}$
7750	$\frac{13.5}{24.0}$	$\frac{22.0}{0}$	$\frac{8.6}{26.0}$

N.B. (1) The method of booking and compare it with the cross-sections in *Figure 11.27*.
(2) The method of splitting the sections into triangles for easy computation.

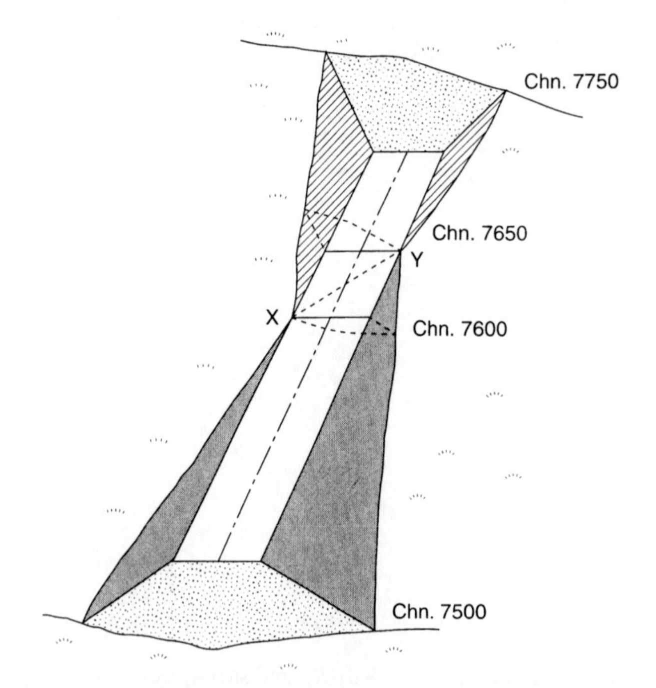

Fig. 11.26 *Road excavation with cut and fill*

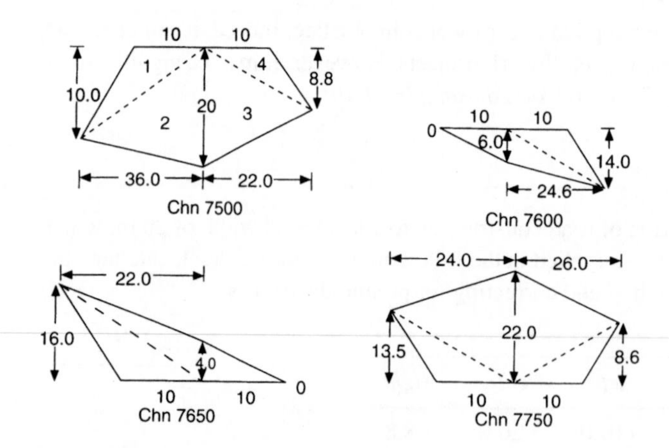

Fig. 11.27 *Cross-sections*

Area of cross-section 75 + 00

$$\text{area } \Delta 1 = \frac{10 \times 10}{2} = 50\,\text{m}^2$$

$$\text{area } \Delta 2 = \frac{36 \times 20}{2} = 360\,\text{m}^2$$

$$\text{area } \Delta 3 = \frac{22 \times 20}{2} = 220\,\text{m}^2$$

$$\text{area } \Delta 4 = \frac{8.8 \times 10}{2} = 44\,\text{m}^2$$

$$\textit{Total area} = 674\,\text{m}^2$$

Similarly, area of cross-section $76 + 00 = 173.8\,\text{m}^2$

$$\text{Vol. by end area} = \frac{674 + 173.8}{2} \times 100 = 42\,390\,\text{m}^3$$

The equation for prismoidal excess varies with the shape of the cross-section. In this particular instance it equals

$$\frac{L}{12}(H_1 - H_2)(W_1 - W_2)$$

where L = horizontal distance between the two end areas
H = centre height
W = the sum of the side widths per section, i.e. $(w_1 + w_2)$

$$\text{Thus, prismoidal excess} = \frac{100}{12}(20 - 6)(58 - 34.6) = 2730\,\text{m}^3$$

$$\textit{Corrected volume} = 39\,660\,\text{m}^3$$

Vol. between $76 + 00$ and $76 + 50$

Line XY in *Figure 11.26* shows clearly that the volume of fill in this section forms a pyramid with the cross-section $76 + 00$ as its base and 50 m high. It is thus more accurate and quicker to use the equation for a pyramid.

$$\text{Vol} = \frac{AL}{3} = \frac{173.8 \times 50}{3} = 2897\,\text{m}^3$$

$$\therefore \textit{Total vol. of fill} = (39\,660 + 2897) = 42\,557\,\text{m}^3$$

Now calculate the volume of cut for yourself.
(*Answer:* $39\,925\,\text{m}^3$)

Example 11.7 The access to a tunnel has a level formation width of 10 m and runs into a plane hillside, whose natural ground slope is 1 in 10 (10%). The intersection line of this formation and the natural ground is perpendicular to the centre-line of the tunnel. The level formation is to run a distance of 360 m into the hillside, terminating at the base of a cutting of slope 1 vertical to 1 horizontal (100%). The side slopes are to be 1 vertical to 1.5 horizontal (67%).

Calculate the amount of excavation in cubic metres. (LU)

Figure 11.28 illustrates the question, which is solved by the methods previously advocated.

By rate of approach

$$x = (1 - 0.1)^{-1} \times AB = 36/0.9 = 40\,\text{m} = DD'$$

As the side slopes are 67% and D' is 40 m high, then $D'G = 40/0.67 = 60\,\text{m} = E'H$. Therefore $GH = 130\,\text{m}$.

Area of $\Delta BCD'$ (in section above) normal to GH, DE, CF

$$= \frac{360}{2} \times 40 = 7200\,\text{m}^2$$

$$\therefore \text{Vol} = \frac{7200}{3}(130 + 10 + 10) = 360\,000\,\text{m}^3$$

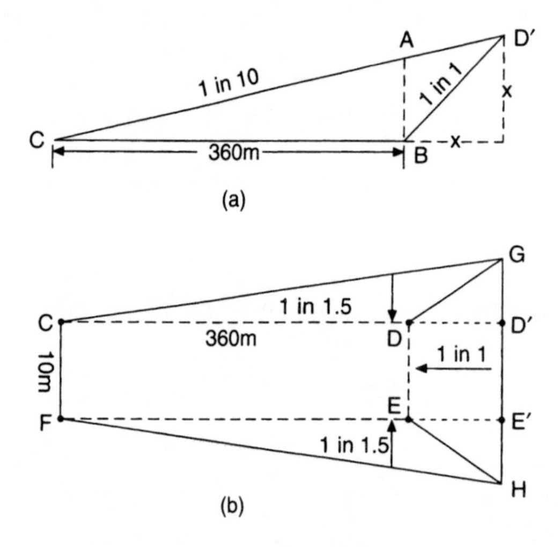

Fig. 11.28 *(a) Section, and (b) plan*

Check

$$\text{Wedge } BCD' = \frac{360}{6}[(130 + 10 + 10) \times 40] = 360\,000\,\text{m}^3$$

Example 11.8 A solid pier is to have a level top surface 20 m wide. The sides are to have a batter of 2 vertical in 1 horizontal and the seaward end is to be vertical and perpendicular to the pier axis. It is to be built on a rock stratum with a uniform slope of 1 in 24, the direction of this maximum slope making an angle whose tangent is 0.75 with the direction of the pier. If the maximum height of the pier is to be 20 m above the rock, diminishing to zero at the landward end, calculate the volume of material required. (LU)

Figure 11.29 illustrates the question. Note that not only is the slope in the direction of the pier required but also the slope at right angles to the pier.

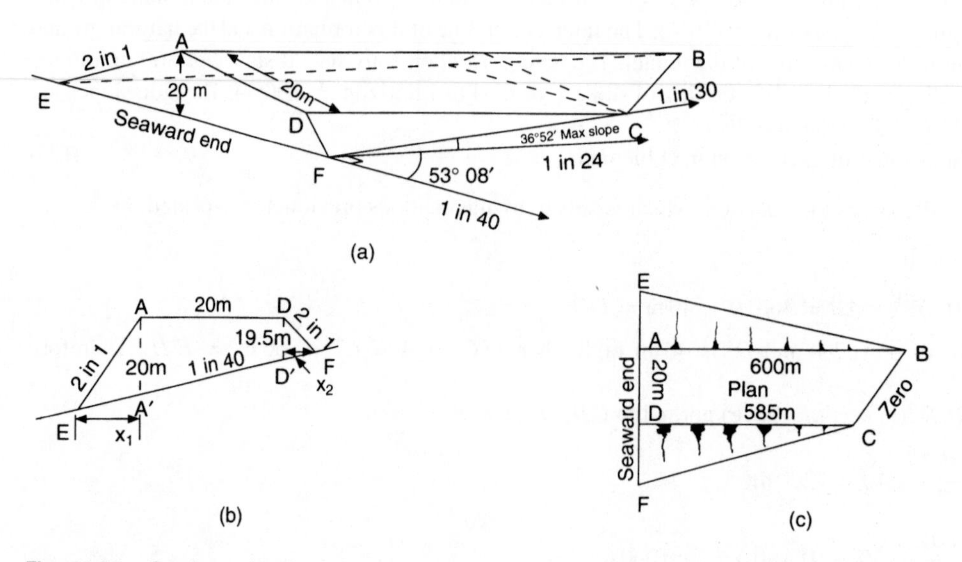

Fig. 11.29 *Sections of a pier*

By dip and strike:

tan apparent slope = tan max slope × cos included angle

$$\frac{1}{x} = \frac{1}{24}\cos 36^\circ\, 52' \qquad \text{where } \tan^{-1} 0.75 = 36^\circ\, 52'$$

$$x = 30$$

$\therefore$ Grade in direction of pier = 1 in 30, $\quad \therefore AB = 20 \times 30 = 600\,\text{m}$

Grade at right angles: $\dfrac{1}{y} = \dfrac{1}{24}\cos 53^\circ\, 08'$

$y = 40$. Grade = 1 in 40 as shown on *Figure 11.29(a)*

$\therefore DD' = 19.5$ m and $DC = 19.5 \times 30 = 585$ m

From *Figure 11.29(b)*:

$$x_1 = \left(2 - \frac{1}{40}\right)^{-1} \times 20 = 10.1\,\text{m}$$

$$x_2 = \left(2 + \frac{1}{40}\right)^{-1} \times 19.5 = 9.6\,\text{m}$$

$$\therefore \text{Area } \Delta EAA' = \frac{20 \times 10.1}{2} = 101\,\text{m}^2$$

$$\text{Area } \Delta DFD' = \frac{19.5 \times 9.6}{2} = 93.6\,\text{m}^2$$

Now, Vol. $ABCD$ = plan area × mean height

$$= \left(\frac{600 + 585}{2} \times 20\right) \times \frac{1}{4}(20 + 19.5 + 0 + 0) = 117\,315\,\text{m}^3$$

$$\text{Vol. of pyramid } EAB = \frac{\text{area } EAA' \times AB}{3} = \frac{101 \times 600}{3}$$

$$= 20\,200\,\text{m}^3$$

$$\text{Vol. of pyramid } DFC = \frac{\text{area } DFD' \times DC}{3} = \frac{93.6 \times 585}{3}$$

$$= 18\,252\,\text{m}^3$$

Total vol = $(117\,315 + 20\,200 + 18\,252) = 155\,767\,\text{m}^3$

Alternatively, finding the area of cross-sections at chainages 0, 585/2 and 585 and applying the prismoidal rule plus treating the volume from chainage 585 to 600 as a pyramid, gives an answer of $155\,525\,\text{m}^3$.

Example 11.9 A 100-m length of earthwork volume for a proposed road has a constant cross-section of cut and fill, in which the cut area equals the fill area. The level formation is 30 m wide, transverse ground slope is 20° and the side slopes in cut-and-fill are $\frac{1}{2}$ horizontal to 1 vertical and 1 horizontal to 1 vertical, respectively. Calculate the volume of excavation in the 100-m length. (LU)

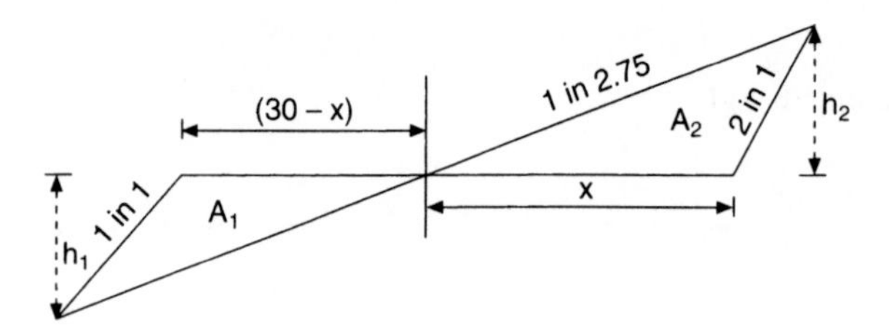

Fig. 11.30 *Road cross-section*

Imagine turning *Figure 11.30* through 90°, then the 1-in-2.75 grade (36°) becomes 2.75 in 1 and the 2-in-1 grade becomes 1 in 2, then by rate of approach:

$$h_1 = (2.75 - 1)^{-1}(30 - x) = \frac{30 - x}{1.75}$$

$$h_2 = \left(2.75 - \frac{1}{2}\right)^{-1} x = \frac{x}{2.25}$$

$$\text{Now, area } \Delta A_1 = \frac{30 - x}{2} \times h_1 = \frac{(30 - x)^2}{3.5}$$

$$\text{area } \Delta A_2 = \frac{x}{2} \times h_2 = \frac{x^2}{4.5}$$

But area A_1 = area A_2

$$\frac{(30 - x)^2}{3.5} = \frac{x^2}{4.5}$$

$$(30 - x)^2 = \frac{3.5}{4.5}x^2 = \frac{7}{9}x^2 \text{ from which } x = 16\,\text{m}$$

$$\therefore \text{Area } A_2 = \frac{16^2}{4.5} = 56.5\,\text{m}^2 = \text{area } A_1$$

$$\therefore \text{Vol in 100 m length } = 56.5 \times 100 = 5650\,\text{m}^3$$

Example 11.10 A length of existing road of formation width 20 m lies in a cutting having side slopes of 1 vertical to 2 horizontal. The centre-line of the road forms part of a circular curve having a radius of 750 m. For any cross-section along this part of the road the ground surface and formation are horizontal. At chainage 5400 m the depth to formation at the centre-line is 10 m, and at chainage 5500 m the corresponding depth is 18 m.

The formation width is to be increased by 20 m to allow for widening the carriageway and for constructing a parking area. The whole of the widening is to take place on the side of the cross-section remote from the centre of the arc, the new side slope being 1 vertical to 2 horizontal. Using the prismoidal rule, calculate the volume of excavation between the chainages 5400 m and 5500 m. Assume that the depth to formation changes uniformly with distance along the road. (ICE)

From *Figure 11.31*, it can be seen that the centre of mass of the increased excavation lies $(20 + x)$ m from the centre-line of the curve. The distance x will vary from section to section but as the side slope is 1 in 2, then:

$$x = 2 \times \frac{h}{2} = h$$

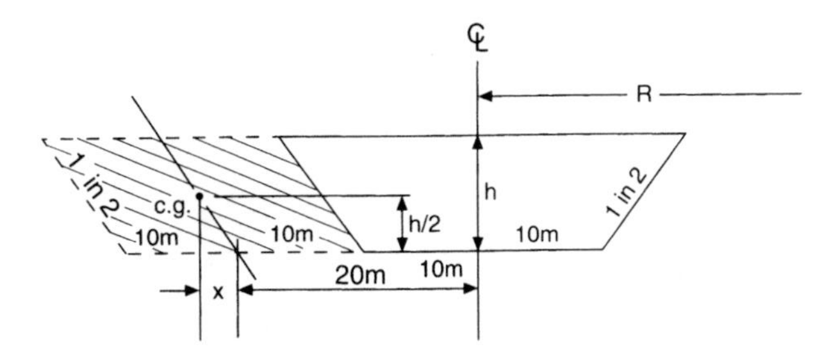

Fig. 11.31 *Road cross-section*

horizontal distance of centre of mass from centre-line $= (20 + h)$.

$$\text{At chainage } 5400\,\text{m}, \; h_1 = 10\,\text{m} \quad \therefore (20 + h) = 30\,\text{m} = d_1$$

$$\text{At chainage } 5450\,\text{m}, \; h_2 = 14\,\text{m} \quad \therefore (20 + h) = 34\,\text{m} = d_2$$

$$\text{At chainage } 5500\,\text{m}, \; h_3 = 18\,\text{m} \quad \therefore (20 + h) = 38\,\text{m} = d_3$$

$$\text{Area of extra excavation at } 5400\,\text{m} = 10 \times 20 = 200\,\text{m}^2 = A_1$$

$$\text{Area of extra excavation at } 5450\,\text{m} = 14 \times 20 = 280\,\text{m}^2 = A_2$$

$$\text{Area of extra excavation at } 5500\,\text{m} = 18 \times 20 = 360\,\text{m}^2 = A_3$$

The above areas are now corrected for curvature: $A\left(1 + \dfrac{d}{R}\right)$.

$$\text{At chainage } 5400\,\text{m} = 200\left(1 + \frac{30}{750}\right) = 208\,\text{m}^2$$

$$\text{At chainage } 5450\,\text{m} = 280\left(1 + \frac{34}{750}\right) = 292.6\,\text{m}^2$$

$$\text{At chainage } 5500\,\text{m} = 360\left(1 + \frac{38}{750}\right) = 378\,\text{m}^2$$

$$\therefore \text{Vol} = \frac{100}{6}(208 + 4 \times 292.6 + 378) = 29\,273\,\text{m}^3$$

Exercises

(*11.1*) An access road to a quarry is being cut in a plane surface in the direction of strike, the full dip of 1 in 12.86 being to the left of the direction of drive. The road is to be constructed throughout on a formation grade of 1 in 50 dipping, formation width 20 m and level, side slopes 1 in 2 and a zero depth on the centre-line at chainage 0 m.

At chainage 400 m the direction of the road turns abruptly through a clockwise angle of 40°; calculate the volume of excavation between chainages 400 m and 600 m. (KU)

(*Answer:* $169\,587\,\text{m}^3$)

(*11.2*) A road is to be constructed on the side of a hill having a cross fall of 1 in 50 at right angles to the centre-line of the road; the side slopes are to be 1 in 2 in cut and 1 in 3 in fill; the formation is 20 m wide and level. Find the position of the centre-line of the road with respect to the point of intersection of the formation and the natural ground.

(a) To give equality of cut and fill.
(b) So that the area of cut shall be 0.8 of the area of fill, in order to allow for bulking. (LU)

(*Answer:* (a) 0.3 m in cut, (b) 0.2 m in fill)

(*11.3*) A reservoir is to be formed in a river valley by building a dam across it. The entire area that will be covered by the reservoir has been contoured and contours drawn at 1.5-m intervals. The lowest point in the reservoir is at a reduced level of 249 m above datum, whilst the top water level will not be above a reduced level of 264.5 m. The area enclosed by each contour and the upstream face of the dam is shown in the table below.

Contour (m)	*Area enclosed* (m^2)
250.0	1 874
251.5	6 355
253.0	11 070
254.5	14 152
256.0	19 310
257.5	22 605
259.0	24 781
260.5	26 349
262.0	29 830
263.5	33 728
265.0	37 800

Estimate by the use of the trapezoidal rule the capacity of the reservoir when full. What will be the reduced level of the water surface if, in a time of drought, this volume is reduced by 25%? (ICE)

(*Answer:* 294 211 m^3; 262.3 m)

(*11.4*) The central heights of the ground above formation at three sections 100 m apart are 10 m, 12 m, 15 m, and the cross-falls at these sections are respectively 1 in 30, 1 in 40 and 1 in 20. If the formation width is 40 m and sides slope 1 vertical to 2 horizontal, calculate the volume of excavation in the 200-m length:

(a) If the centre-line is straight.
(b) If the centre-line is an arc of 400 m radius. (LU)

(*Answer:* (a) 158 367 m^3, (b) 158 367 $\pm$ 1070 m^3)

11.6 MASS-HAUL DIAGRAMS

Mass-haul diagrams (MHD) are used to compare the economy of various methods of earthwork distribution on road or railway construction schemes. By the combined use of the MHD plotted directly below the longitudinal section of the survey centre-line, one can find:

(1) The distances over which cut and fill will balance.
(2) Quantities of materials to be moved and the direction of movement.
(3) Areas where earth may have to be borrowed or wasted and the amounts involved.
(4) The best policy to adopt to obtain the most economic use of plant.
(5) The best use of plant for the distances over which the volumes of cut and fill are to be moved.

11.6.1 Definitions

(1) *Haul* refers to the volume of material multiplied by the distance moved, expressed in 'station metres'.
(2) *Station metre* (stn m) is 1 m^3 of material moved 100 m
Thus, 20 m^3 moved 1500 m is a haul of $20 \times 1500/100 = 300$ stn m.
(3) *Waste* is the material excavated from cuts but not used for embankment fills.
(4) *Borrow* is the material needed for the formation of embankments, secured not from roadway excavation but from elsewhere. It is said to be obtained from a 'borrow pit'.
(5) *Limit of economical haul* is the maximum haul distance. When this limit is reached it is more economical to waste and borrow material.

11.6.2 Bulking and shrinkage

Excavation of material causes it to loosen, and thus its excavated volume will be greater than its *in situ* volume. However, when filled and compacted, it may occupy a less volume than when originally *in situ*. For example, light sandy soil is less by about 11% after filling, whilst large rocks may bulk by up to 40%. To allow for this, a correction factor is generally applied to the cut or fill volumes.

11.6.3 Construction of the MHD

A MHD is a continuous curve, whose vertical ordinates, plotted on the same distance scale as the longitudinal section, represent the algebraic sum of the corrected volumes (cut +, fill −).

11.6.4 Properties of the MHD

Consider *Figure 11.32(a)* in which the ground XYZ is to be levelled off to the grade line $A'B'$. Assuming that the fill volumes, after correction, equal the cut volumes, the MHD would plot as shown in *Figure 11.32(b)*. Thus:

(1) Since the curve of the MHD represents the algebraic sums of the volumes, then any horizontal line drawn parallel to the base AB will indicate the volumes that balance. Such a line is called a *balancing line* and may even be represented by AB itself, indicating that the total cut equals the total fill.
(2) The rising curve, shown broken, indicates cut (positive), the falling curve indicates fill (negative).
(3) The maximum and minimum points of a MHD occur directly beneath the intersection of the natural ground and the formation grade; such intersections are called *grade points*.
(4) As the curve of the MHD rises above the balance line AB, the haul is from left to right. When the curve lies below the balance line, the haul is from right to left.
(5) The total cut volume is represented by the maximum ordinate CD.
(6) In moving earth from cut to fill, assume that the first load would be from the cut at X to the fill at Y, and the last load from the cut at Y to the fill at Z. Thus the haul distance would appear to be from a

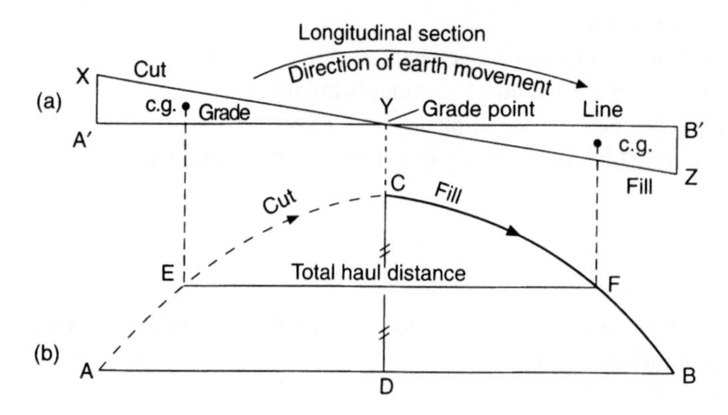

Fig. 11.32 *Mass haul diagram*

point mid-way between X and Y, to a point mid-way between Y and Z. However, as the section is representative of volume, not area, the haul distance is from the centre of mass of the cut volume to the centre of mass of the fill volume. The horizontal positions of these centres of mass may be found by bisecting the total volume ordinate CD with the horizontal line EF.

Now, since haul is volume × distance, the *total haul* in the section is total vol × total haul distance = $CD \times EF/100$ stn m.

Most if not all muckshifting contractors have long abandoned the traditional concepts of freehaul and overhaul. The muckshifter takes the haul length into account when pricing the job but does not measure or value the freehaul or overhaul as such any more. The price is per m^3 and depends upon the type of material being excavated. The Civil Engineering Method of Measurement Edition 3 (CESMM3) specifies five excavation material types but in the Method of Measurement of Highway Works which forms volume 4 of the Manual of Contract Documents for Highway Works there are over 40 classes of soil plus hard material which is measured and paid for 'extra over' the soft muck price elsewhere in the Bill of Quantities. Today, the volumes in the different cuttings and embankments come from a CAD program and in the case of MMHW these are then split into different soil classifications by the designer and shown in a schedule for the contractors pricing the work at tender stage. *Figure 11.33* shows a simplified time-chainage/time-location programme; such documents are often used as the major working document on site. There are also computer-based time-chainage/time-location programs that do all the hard work, especially when the programme of works needs to be revised, as it often has to be where the New Engineering Contract Edition 3 (NEC3) is used.

Worked examples

Example 11.11 The following notes refer to a 1200-m section of a proposed railway, and the earthwork distribution in this section is to be planned without regard to the adjoining sections. The table below shows the stations and the surface levels along the centre-line, the formation level being at an elevation above datum of 43.5 m at chainage 70 and thence rising uniformly on a gradient of 1.2%. The volumes are recorded in m^3, the cuts are plus and fills minus.

(1) Plot the longitudinal section using a horizontal scale of 1 : 1200 and a vertical scale of 1 : 240.
(2) Assuming a correction factor of 0.8 applicable to fills, plot the MHD to a vertical scale of 1000 m^3 to 20 mm.

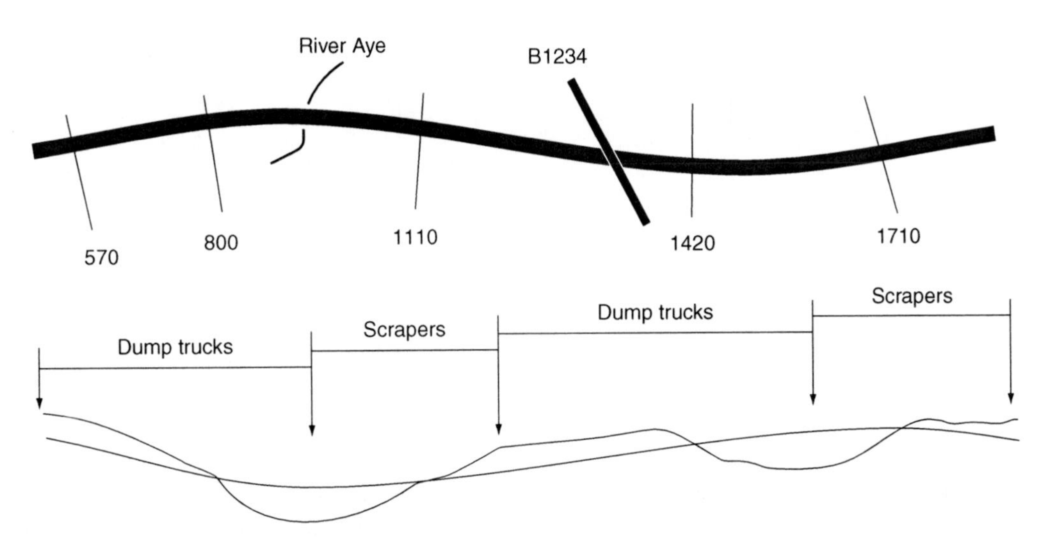

Chainage	570–800	800–1110	1110–1420	1420–1710
Topsoil	4130	4560	5590	5170
Excavate suitable	23450	540	34980	620
Excavate rock	2450	50	7890	260
Suitable fill		19650		16450
Excavate unsuitable	2150	2760	980	390

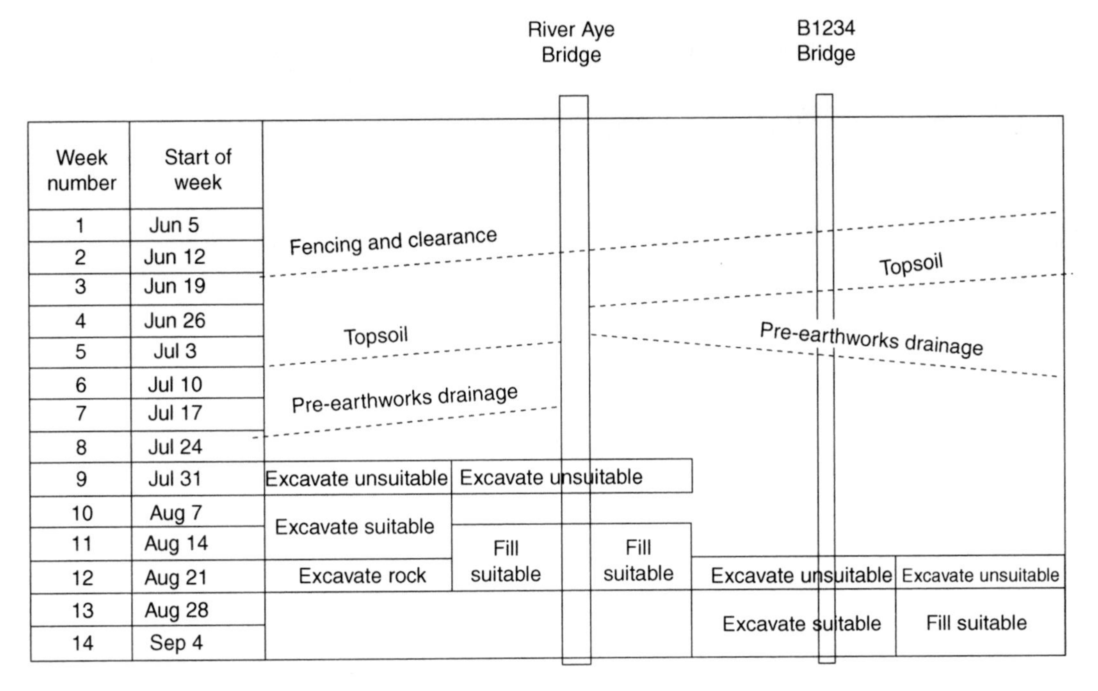

Fig. 11.33 *Simplified time-chainage time-location programme*

(3) Calculate *total haul* in stn m and indicate the haul limits on the curve and section. (LU)

Chn	*Surface level*	*Vol*	*Chn*	*Surface level*	*Vol*	*Chn*	*Surface level*	*Vol*
70	52.8		74	44.7		78	49.5	
		+1860			−1080			−237
71	57.3		75	39.7		79	54.3	
		+1525			−2025			+362
72	53.4		76	37.5		80	60.9	
		+547			−2110			+724
73	47.1		77	41.5		81	62.1	
		−238			−1120			+430
74	44.7		78	49.5		82	78.5	

For answers to parts (1) and (2) see *Figure 11.34* and the values in *Table 11.2*.

N.B. (1) The volume at chainage 70 is zero.
(2) The mass ordinates are always plotted at the station and not between them.
(3) The mass ordinates are now plotted to the same horizontal scale as the longitudinal section and directly below it.
(4) Check that maximum and minimum points on the MHD are directly below grade points on the section.

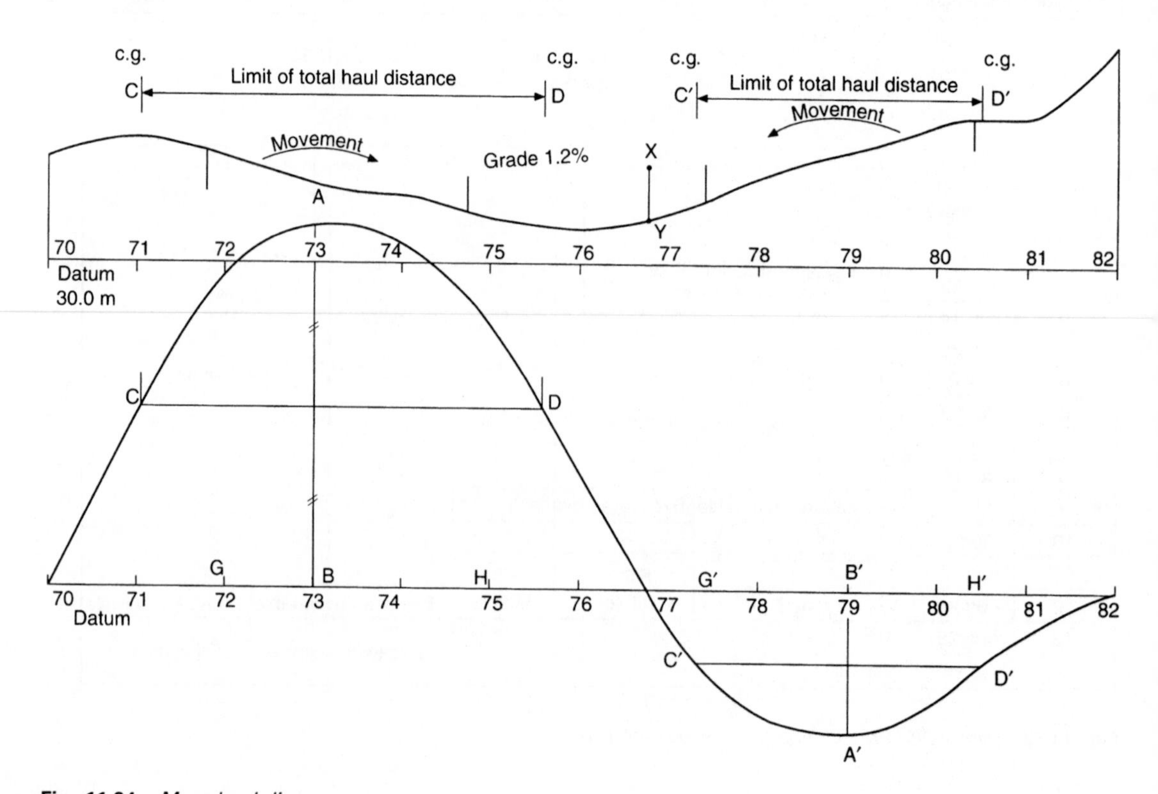

Fig. 11.34 *Mass haul diagram*

Table 11.2

Chainage	*Volume*	*Mass ordinate (algebraic sum)*
70	0	0
71	+1860	+1860
72	+1525	+3385
73	+547	+3932
74	$-238 \times 0.8 = -190.4$	+3741.6
75	$-1080 \times 0.8 = -864$	+2877.6
76	$-2025 \times 0.8 = -1620$	+1257.6
77	$-2110 \times 0.8 = -1688$	−430.4
78	$-1120 \times 0.8 = -896$	−1326.4
79	$-237 \times 0.8 = -189.6$	−1516
80	+362	−1154
81	+724	−430
82	+430	0

(5) Using the datum line as a balancing line indicates a balancing out of the volumes from chainage 70 to XY and from XY to chainage 82.
(6) The designation of chainage uses the convention that the chainage stated is in 100s of metres. Thus chainage 82 means 8200 m.

Total haul (taking each loop separately) = total vol × total haul distance. The total haul distance is from the centre of mass of the total cut to that of the total fill and is found by bisecting AB and $A'B'$, to give the distances CD and $C'D'$.

$$\text{Total haul} = \frac{AB \times CD}{100} + \frac{A'B' \times C'D'}{100}$$

$$= \frac{3932 \times 450}{100} + \frac{1516 \times 320}{100} = 22\,545 \text{ stn m}$$

N.B. All the dimensions in the above solution are scaled from the MHD.

Example 11.12 The volumes between sections along a 1200-m length of proposed road are shown below, positive volumes denoting cut, and negative volumes denoting fill:

Chainage (m)	0	100	200	300	400	500	600	700	800	900	1000	1100	1200
Vol. between sections ($m^3 \times 10^3$)	+2.1	+2.8	+1.6	−0.9	−2.0	−4.6	−4.7	−2.4	+1.1	+3.9	+43.5	+2.8	

Plot a MHD for this length of road to a suitable scale and determine suitable positions of balancing lines so that there is

(1) A surplus at chainage 1200 but none at chainage 0.
(2) A surplus at chainage 0 but none at chainage 1200.
(3) An equal surplus at chainage 0 and chainage 1200. (ICE)

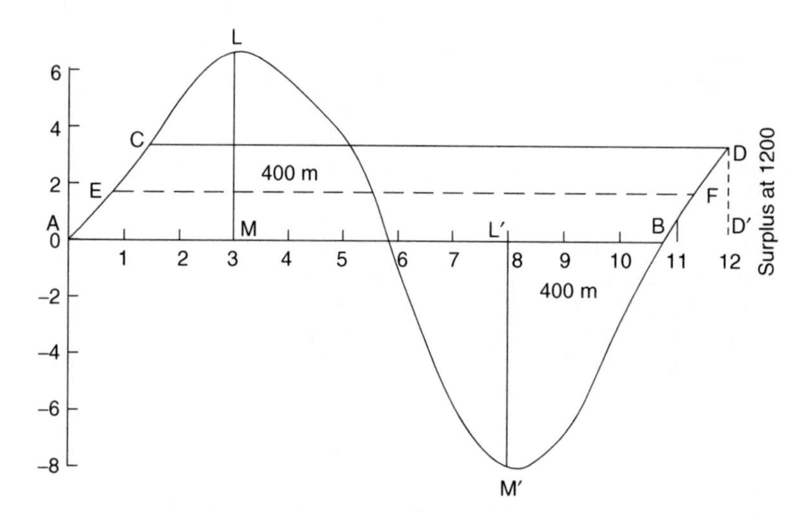

Fig. 11.35 *Mass haul diagram*

For plot of MHD see *Figure 11.35*.

Mass ordinates = 0, +2.1, +4.9, +6.5, +5.6, +3.6, −1.0, −5.7, −8.1, −7.0, −3.1, +0.4, +3.2 (obtained by algebraic summation of volumes)

(1) Balance line *AB* gives a surplus at chainage 1200 but none at 0.
(2) Balance line *CD* gives a surplus at chainage 0 but none at 1200.
(3) Balance line *EF* situated mid-way between *AB* and *CD* will give equal surpluses at the ends.

Example 11.13 Volumes in m^3 of excavation (+) and fill (−) between successive sections 100 m apart on a 1300-m length of a proposed railway are given.

Section	0		1		2		3		4		5		6		7
Volume (m^3)		−1000		−2200		−1600		−500		+200		+1300		+2100	

Section	7		8		9		10		11		12		13
Volume (m^3)		+1800		+1100		+300		−400		−1200		−1900	

Draw a MHD for this length. If earth may be borrowed at either end, which alternative would give the least haul? (LU)

Adding the volumes algebraically gives the following mass ordinates

Section	0		1		2		3		4		5		6		7
Volume (m^3)		−1000		−3200		−4800		−5300		−5100		−3800		−1700	

Section	7		8		9		10		11		12		13
Volume (m^3)		+100		+1200		+1500		+1100		−100		−2000	

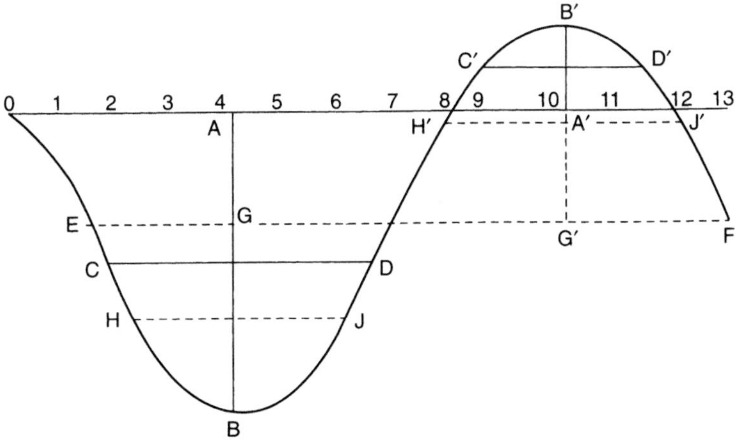

Fig. 11.36 *Mass haul diagram*

These are now plotted to produce the MHD of *Figure 11.36*. Balancing out from the zero end permits borrowing at the 1300 end.

(1)
$$\text{Total haul} = \frac{(AB \times CD)}{100} + \frac{(A'B' \times C'D')}{100} \text{ stn m}$$

$$= \frac{(5300 \times 475)}{100} + \frac{(1500 \times 282)}{100} = 29\,405 \text{ stn m}$$

Note: CD bisects AB and $C'D'$ bisects $A'B'$.

(2) Balancing out from the 1300 end (EF) permits borrowing at the zero end.

$$\text{Total haul} = \frac{(GB \times HJ)}{100} + \frac{(G'B' \times H'J')}{100}$$

$$= \frac{(3300 \times 385)}{100} + \frac{(3500 \times 430)}{100} = 27\,755 \text{ stn m}$$

Thus, borrowing at the zero end requires the least haul.

11.6.5 Auxiliary balancing lines

A study of the material on MHD plus the worked examples should have given the reader an understanding of the basics. It is now appropriate to illustrate the application of auxiliary balancing lines.

Consider in the first instance a MHD as in *Worked examples 11.11*. In *Figure 11.37*, the balance line is ABC and the following data are easily extrapolated:

Cut AD balances fill DB Vol moved $= DE$
Cut CJ balances fill BJ Vol moved $= HJ$

Consider now *Figure 11.38*; the balance line is AB, but in order to extrapolate the above data one requires an auxiliary balancing line CDE parallel to AB and touching the MHD at D:

Cut AC balances fill EB Vol moved $= GH$
Cut CF balances fill FD Vol moved $= FG$
Cut DJ balances fill JE Vol moved $= JK$

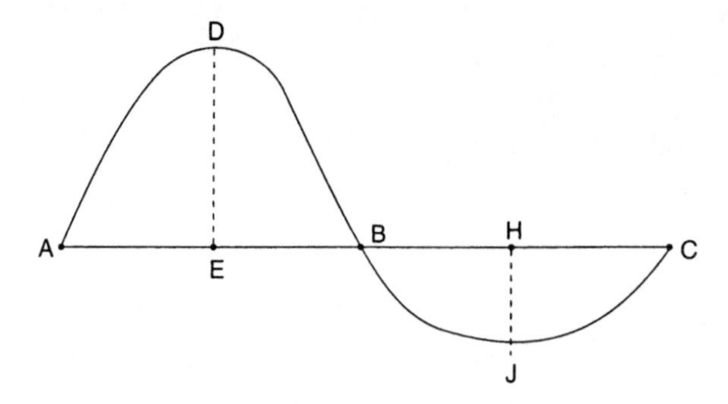

Fig. 11.37 *Mass haul diagram*

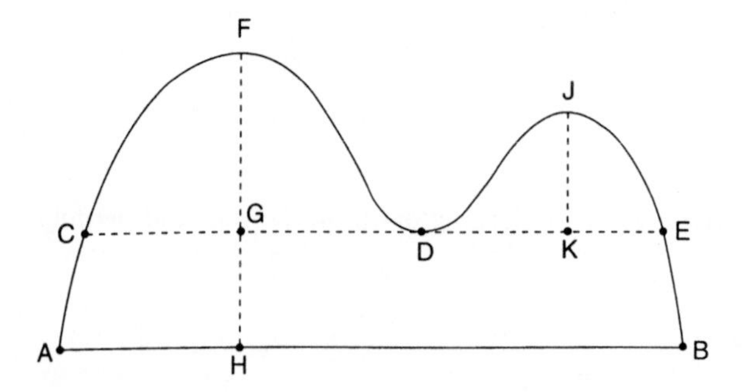

Fig. 11.38 *Mass haul diagram with auxiliary balancing lines*

Thus the total volume moved between A and B is $FH + JK$

Finally, *Figure 11.39* has a balance from A to B with auxiliaries at CDE and FGH, then:

Fill AC balances cut BE	Vol moved $= JK$
Fill CF balances cut DH	Vol moved $= KL$
Fill FM balances cut GM	Vol moved $= LM$
Fill GO balances cut HO	Vol moved $= NO$
Fill DQ balances cut EQ	Vol moved $= PQ$

The total volume moved between A and B is $JM + NO + PQ$.

The above data become apparent only when one introduces the auxiliary balancing lines.

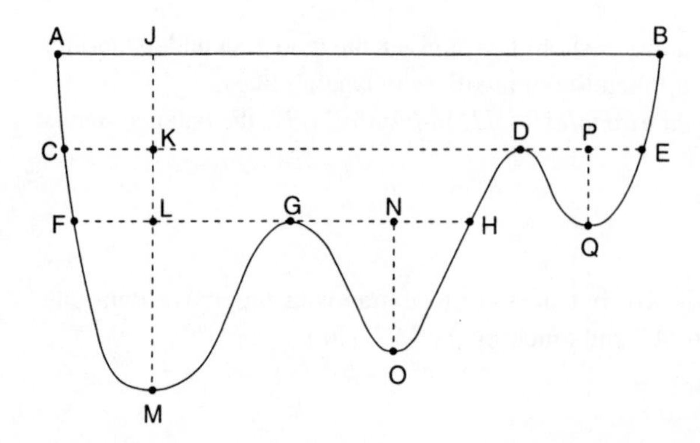

Fig. 11.39 *Mass haul diagram with auxiliary balancing lines*

Exercise

(*11.5*) The volumes in m^3 between successive sections 100 m apart on a 900-m length of a proposed road are given below (excavation is positive and fill negative):

Section	0		1		2		3		4		5		6		7
Volume (m^3)		+1700		−100		−3200		−3400		−1400		+100		+2600	

Section	7		8		9
Volume (m^3)		+4600		+1100	

Determine the maximum haul distance when earth may be wasted only at the 900-m end. (LU)

(*Answer:* 558 m)

12
Setting out (dimensional control)

In engineering the production of an accurate large-scale plan is usually the first step in the planning and design of a construction project. Thereafter the project, as designed on the plan, must be set out on the ground in the correct absolute and relative position and to its correct dimensions. Thus, surveys made in connection with a specific project should be planned with the setting-out process in mind and a system of three-dimensional control stations conveniently sited and adequate in number should be provided to facilitate easy, economical setting out.

It is of prime importance that the establishment and referencing of survey control stations should be carried out at such places and in such a manner that they will survive the construction processes. This entails careful choice of the locations of the control stations and their construction relative to their importance and long- or short-term requirements. For instance, those stations required for the total duration of the project may be established in concrete or masonry pillars with metal plates or bolts set in on which is punched the station position. Less durable are stout wooden pegs set in concrete or driven directly into the ground. A system of numbering the stations is essential, and frequently pegs are painted different colours to denote the particular functions for which they are to be used.

In the UK it is the Institution of Civil Engineers (ICE) Conditions of contract that are generally used. At the time of writing the ICE7th is the current edition but the New Engineering Contract Edition 3 (NEC3) has just been published and is rapidly replacing the ICE7th on many of the larger types of project.

12.1 RESPONSIBILITY ON SITE

Responsibility with regard to setting out is defined in Clause 17 of the 5th, 6th and 7th editions of the ICE Conditions of Contract:

> The contractor shall be responsible for the true and proper setting out of the works, and for the correctness of the position, levels, dimensions, and alignment of all parts of the works, and for the provision of all necessary instruments, appliances, and labour in connection therewith. If, at any time during the progress of the works, any error shall appear or arise in the position, levels, dimensions, or alignment of any part of the works, the contractor, on being required so to do by the engineer, shall, *at his own cost*, rectify such error to the satisfaction of the engineer, unless such error is based on incorrect data supplied in writing by the engineer or the engineer's representative, in which case the cost of rectifying the same shall be borne *by the employer*. The checking of any setting out, or of any line or level, by the engineer or the engineer's representative, shall not, in any way, relieve the contractor of *his* responsibility for the correctness thereof, and the contractor shall carefully protect *and preserve* all bench-marks, sight rails, pegs, and other things used in setting out the works.

The clause specifies three persons involved in the process, namely, the employer, the engineer and the agent, whose roles are as follows:

The *employer*, who may be a government department, local authority or private individual, requires to carry out and finance a particular project. To this end, he/she commissions an *engineer* to investigate and design the project, and to take responsibility for the initial site investigation, surveys, plans, designs, working drawings, and setting-out data. On satisfactory completion of the engineer's work the employer lets the contract to a contractor whose duty it is to carry out the work.

On site the employer is represented by the engineer or their representative, referred to as the *resident engineer* (RE), and the contractor's representative is called the *agent*.

The engineer has overall responsibility for the project and must protect the employer's interest without bias to the contractor; however the engineer is not a party to the contract which is between the client and the contractor. The agent is responsible for the actual construction of the project.

12.2 RESPONSIBILITY OF THE SETTING-OUT ENGINEER

The setting-out engineer should establish such a system of work on site that will ensure the accurate setting out of the works well in advance of the commencement of construction. To achieve this, the following factors should be considered.

(1) A complete and thorough understanding of the plans, working drawings, setting-out data, tolerances involved and the time scale of operations. Checks on the setting-out data supplied should be immediately implemented.
(2) A complete and thorough knowledge of the site, plant and relevant personnel. Communications between all individuals is vitally important. Field checks on the survey control already established on site, possibly by contract surveyors, should be carried out at the first opportunity.
(3) A complete and thorough knowledge of the survey instrumentation available on site, including the effect of instrumental errors on setting-out observations. At the first opportunity, a base should be established for the calibration of tapes, EDM equipment, levels and theodolites.
(4) A complete and thorough knowledge of the stores available, to ensure an adequate and continuing supply of pegs, pins, chalk, string, paint, timber, etc.
(5) Office procedure should be so organized as to ensure easy access to all necessary information. Plans should be stored flat in plan drawers, and those amended or superseded should be withdrawn from use and stored elsewhere. Field and level books should be carefully referenced and properly filed. All setting-out computations and procedures used should be clearly presented, referenced and filed.
(6) Wherever possible, independent checks of the computation, abstraction and extrapolation of setting-out data and of the actual setting-out procedures should be made.

It can be seen from this brief list of the requirements of a setting-out engineer, that such work should never be allocated, without complete supervision, to junior, inexperienced members of the site team. Consider the cost implications of getting it right or getting it wrong. If the most junior and therefore the cheapest surveyor is used a few £/$/€ may be saved on the setting-out work, but if he/she gets it wrong the cost of putting the error right is likely to be many times the money 'saved'.

All site engineers should also make a careful study of the following British Standards, which were prepared under the direction of the Basic Data and Performance Criteria for Civil Engineering and Building Structures Standards Policy Committee:

(1) BS 5964-1:1990 (ISO 4463–1, 1989) Building setting out and measurement. Part 1. Methods of measuring, planning and organization and acceptance criteria.
(2) BS 5964-3:1996 (ISO 4463–3, 1995) Building setting out and measurement. Part 3. Checklists for the procurement of surveys and measurement services.
(3) BS 5606:1990 Guide to Accuracy in Building.
(4) BS 7307:1990 (ISO 7976:1989) Building tolerances. Measurement of buildings and building products. Part 1. Methods and instruments. Part 2. Position of measuring points.

(5) BS 7308:1990 (ISO 7737:1986) Method for presentation of dimensional accuracy data in building construction.
(6) BS 7334:1990 (ISO 8322:1989) Measuring instruments for building construction. Methods for determining accuracy in use: Part 1 theory, Part 2 measuring tapes, Part 3 optical levelling instruments, Part 4 theodolites, Part 5 optical plumbing instruments, Part 6 laser instruments, Part 7 instruments when used for setting out, Part 8 electronic distance-measuring instruments up to 150 m.

These documents supply important information coupled with a wealth of excellent, explanatory diagrams of various setting-out procedures.

For instance BS 5964 provides acceptance criteria for the field data measured. The acceptance criteria is termed permitted deviation (PD), where PD $= 2.5$ (SD) (Standard Deviation). It is based on the assumption that a theodolite which reads directly to $10''$ is used and that if tapes are used then no more than two tape lengths are employed.

If a primary system of control points has been established as stage one of the surveying and setting-out process, its acceptability (or otherwise) is based on the difference between the measured distance and its equivalent, computed from the adjusted coordinates, that is the residual as described in Chapter 7; this difference should not exceed $\pm 0.75(L)^{\frac{1}{2}}$ mm, with a minimum of 4 mm, where L is the distance in metres. For angles it is $\pm 0.045(L)^{-\frac{1}{2}}$ degrees.

For a secondary system of control points the acceptance criteria is:

For distance, $\pm 1.5(L)^{\frac{1}{2}}$ mm, with a minimum of 8 mm.

For angles, $\pm 0.09(L)^{-\frac{1}{2}}$ degrees. No minimum value is stated. However, if the line length is 1000 m then the tolerance is $10''$ which would be hard to achieve with a $10''$ theodolite, therefore this criteria should only be applied to shorter lines.

For levelling between benchmarks the general acceptance criterion is ± 5 mm, with slight variations for different situations.

This small sample of the information available in these standards indicates their importance to all concerned with the surveying and setting-out on site.

12.3 PROTECTION AND REFERENCING

Most site operatives have little concept of the time, effort and expertise involved in establishing setting-out pegs. For this reason the pegs are frequently treated with disdain and casually destroyed in the construction process. A typical example of this is the centre-line pegs for route location which are the first to be destroyed when earth-moving commences. It is important, therefore, that control stations and BMs should be protected in some way (usually as shown in *Figure 12.1*) and site operatives, particularly earthwork personnel, impressed with the importance of maintaining this protection.

Where destruction of the pegs is inevitable, then referencing procedures should be adopted to relocate their positions to the original accuracy of fixation. Various configurations of reference pegs are used and the one thing that they have in common is that they must be set well outside the area of construction and have some form of protection, as in *Figure 12.1*.

A commonly-used method of referencing is from four pegs (A, B, C, D) established such that two strings stretched between them intersect to locate the required position (*Figure 12.2*). Distances AB, BC, CD, AD, AC, BD should all be measured as checks on the possible movement of the reference pegs, whilst distances from the reference pegs to the setting-out peg will afford a check on positioning. Ideally TP1 should be in line with DB and AC. Intersecting lines of sight from theodolites at, say, A and B may be used where ground conditions make string lining difficult.

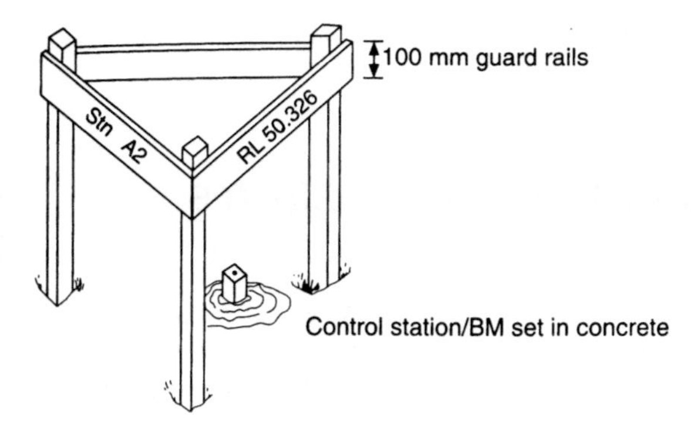

Fig. 12.1 *Control point protection*

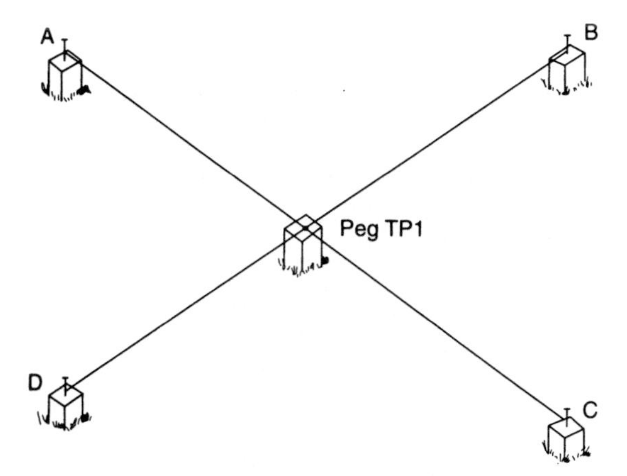

Fig. 12.2 *Control point reference pegs*

Although easy to construct, wooden pegs are easily damaged. A more stable and precise control station mark that is easily constructed on site is shown in *Figure 12.3*. A steel or brass plate with fine but deeply engraved lines crossing at right angles is set with Hilti nails into a cube of concrete cast into a freshly dug hole. To avoid any possible movement of the plate there should be a layer of epoxy resin between it and the concrete.

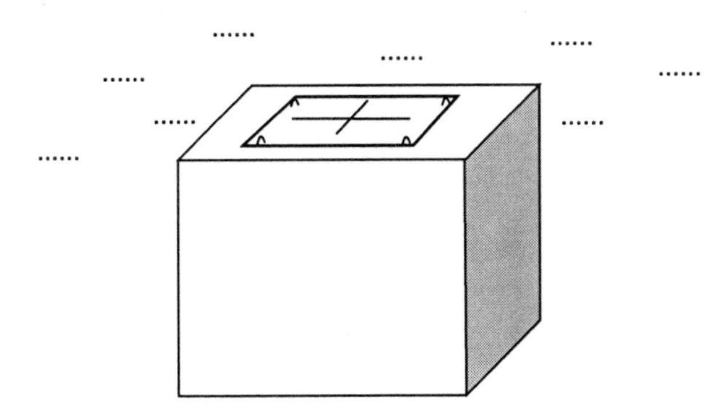

Fig. 12.3 *Control point*

All information relating to the referencing of a point should be recorded on a diagram of the layout involved.

12.4 BASIC SETTING-OUT PROCEDURES USING COORDINATES

Plans are generally produced on a plane rectangular coordinate system, and hence salient points of the design may also be defined in terms of rectangular coordinates on the same system. For instance, the centre-line of a proposed road may be defined in terms of coordinates at, say, 30-m intervals, or alternatively, only the tangent and intersection points may be so defined. The basic methods of locating position when using coordinates are either by polar coordinates or by intersection.

12.4.1 By polar coordinates

In *Figure 12.4*, *A*, *B* and *C* are control stations whose coordinates are known. It is required to locate point *IP* whose design coordinates are also known. The computation involved is as follows:

(1) From coordinates compute the bearing *BA* (this bearing may already be known from the initial control survey computations).
(2) From coordinates compute the horizontal length and bearing of $B - IP$.
(3) From the two bearings compute the setting-out angle *AB(IP)*, i.e. β.
(4) Before proceeding into the field, draw a neat sketch of the situation showing all the setting-out data. Check the data from the plan or by independent computation.

Alternatively the coordinate geometry functions in a total station can be used to avoid most of the computations. The field work involved is as follows:

(1) Set up theodolite at *B* and backsight to *A*, note the horizontal circle reading.
(2) Add the angle β to the circle reading *BA* to obtain the circle reading $B - IP$. Set this reading on the theodolite to establish direction $B - IP$ and measure out the horizontal distance *L*.

If this distance is set out by steel tape, careful consideration must be given to all the error sources such as standardization, slope, tension and possibly temperature if the setting-out tolerances are very small. It should also be carefully noted that the sign of the correction is reversed from that applied when measuring a distance. For example, if a 30-m tape was in fact 30.01 m long, when measuring a distance the recorded length would be 30 m for a single tape length, although the actual distance is 30.01 m; hence a *positive* correction of 10 mm is applied to the recorded measurement. However, if it is required to set out 30 m, the actual distance set out would be 30.01 m; thus this length would need to be reduced by 10 mm, i.e. a *negative* correction.

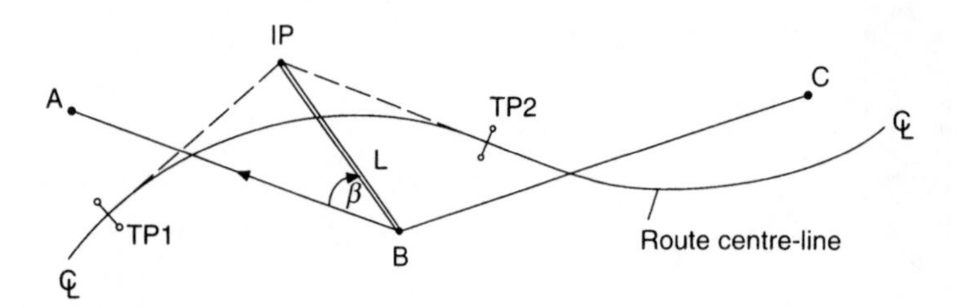

Fig. 12.4 *Setting out from control points*

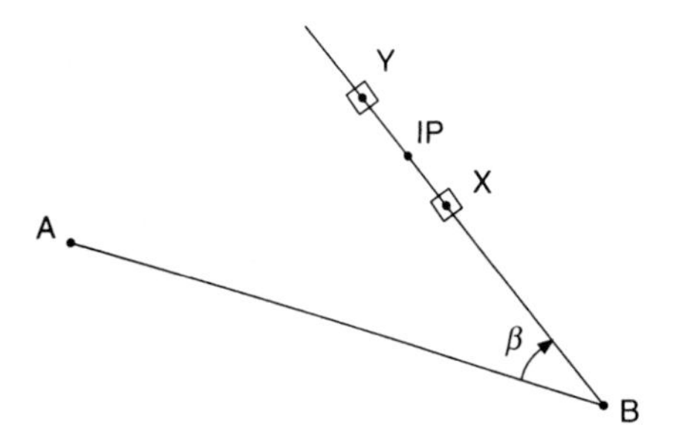

Fig. 12.5 *Setting out with a steel tape*

The best field technique when using a steel tape is carefully to align pegs at X and Y each side of the expected position of IP (*Figure 12.5*). Now carefully measure the distance BX and subtract it from the known distance to obtain distance $X - IP$, which will be very small, possibly less than one metre. Stretch a fine cord between X and Y and measure $X - IP$ along this direction to fix point IP.

A total station may be used to display horizontal distance, so the length $B - IP$ may be ranged direct to a reflector fixed to a setting-out pole.

12.4.2 By intersection

This technique, illustrated in *Figure 12.6*, does not require linear measurement; hence, adverse ground conditions are immaterial and one does not have to consider tape corrections. The technique is applicable if a total station is not available or if its batteries are flat.

The computation involved is as follows:

(1) From the coordinates of A, B and IP compute the bearings AB, $A - IP$ and $B - IP$.
(2) From the bearings compute the angles α and β.

The relevant field work, assuming two theodolites are available, is as follows:

(1) Set up a theodolite at A, backsight to B and turn off the angle α.
(2) Set up a theodolite at B, backsight to A and turn off the angle β.

The intersection of the sight lines $A - IP$ and $B - IP$ locates the position of IP. The angle δ is measured as a check on the setting out.

If only one theodolite is available then two pegs per sight line are established, as in *Figure 12.5*, and then string lines connecting each opposite pair of pegs locate position IP, as in *Figure 12.2*.

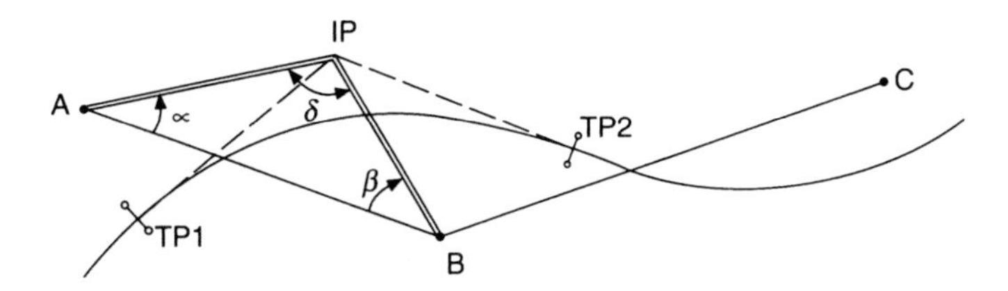

Fig. 12.6 *Setting out by intersection*

12.5 USE OF GRIDS

Many structures in civil engineering consist of steel or reinforced concrete columns supporting floor slabs. As the disposition of these columns is inevitably that they are at right angles to each other, the use of a grid, where the grid intersections define the position of the columns, greatly facilitates setting out. It is possible to define several grids as follows:

(1) *Survey grid*: the rectangular coordinate system on which the original topographic survey is carried out and plotted (*Figure 12.7*).
(2) *Site grid*: defines the position and direction of the main building lines of the project, as shown in *Figure 12.7*. The best position for such a grid can be determined by simply moving a tracing of the site grid over the original plan so that its best position can be located in relation to the orientation of the major units designed thereon.

In order to set out the site grid, it may be convenient to translate the coordinates of the site grid to those of the survey grid using the well-known transformation formula:

$$E = \Delta E + E_1 \cos\theta - N_1 \sin\theta$$

$$N = \Delta N + N_1 \cos\theta + E_1 \sin\theta$$

where $\Delta E, \Delta N$ = difference in easting and northing of the respective grid origins
E_1, N_1 = the coordinates of the point on the site grid
θ = relative rotation of the two grids
E, N = the coordinates of the point transformed to the survey grid

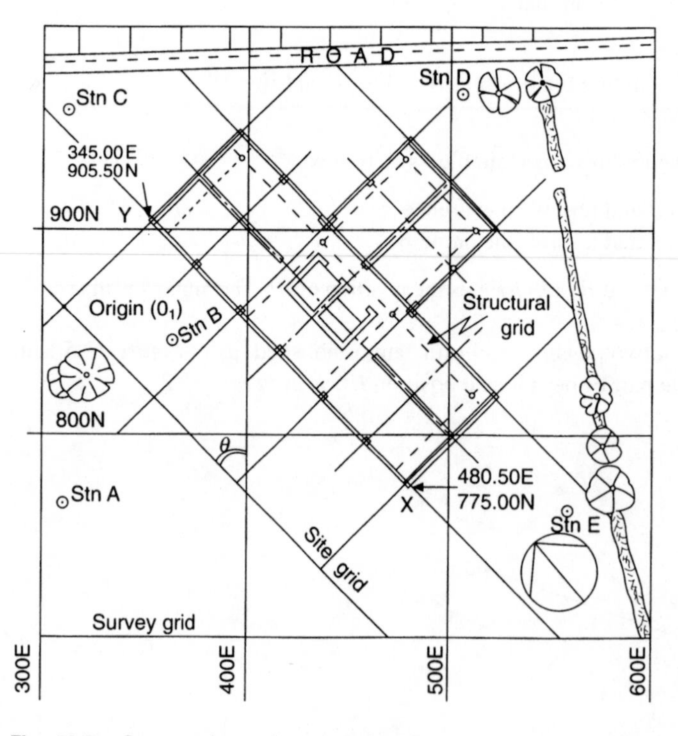

Fig. 12.7 *Survey, site and structural grids*

Thus, selected points, say X and Y (*Figure 12.7*) may have their site-grid coordinate values transformed to that of the survey grid and so set-out by polars or intersection from the survey control. Now, using XY as a baseline, the site grid may be set out using theodolite and steel tape, all angles being turned off on both faces and grid intervals carefully fixed using the steel tape under standard tension.

When the site grid has been established, each line of the grid should be carefully referenced to marks fixed clear of the area of work. As an added precaution, these marks could be further referenced to existing control or permanent, stable, on-site detail.

(3) *Structural grid*: used to locate the position of the structural elements within the structure and is physically established usually on the concrete floor slab (*Figure 12.7*). It may be used where the relative positions of points are much more important than the absolute positions, such as for the holding down bolts of a steel frame structure. The advantages of such a grid are that the lines of sight are set out in a regular pattern and so can be checked by eye even for small errors and that there is more check on points set out from the grid than if those points were set out individually by bearing and distance or by coordinates from a total station.

12.6 SETTING OUT BUILDINGS

For buildings with normal strip foundations the corners of the external walls are established by pegs located directly from the survey control or by measurement from the site grid. As these pegs would be disturbed in the initial excavations their positions are transferred by total station on to profile boards set well clear of the area of disturbance (*Figure 12.8*). Prior to this their positions must be checked by measuring the diagonals as shown in *Figure 12.9*.

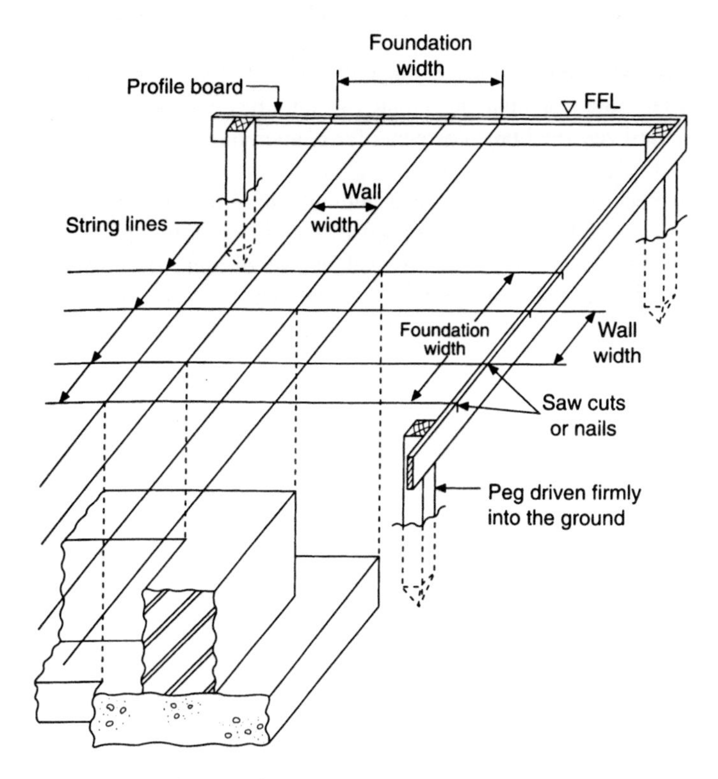

Fig. 12.8 *Profile boards*

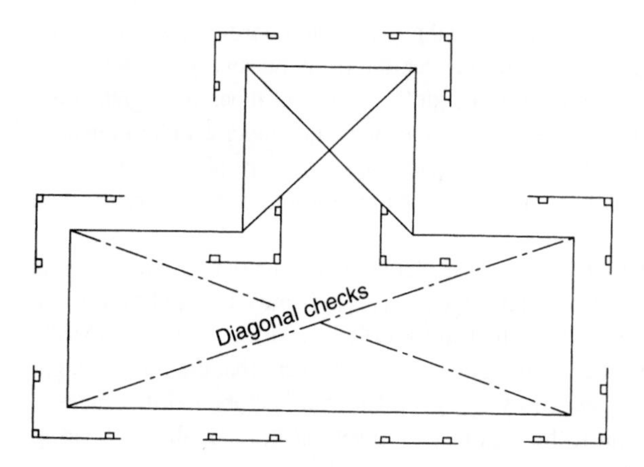

Fig. 12.9 *Check measurements*

The profile boards must be set horizontal with their top edge at some predetermined level such as *damp proof course* (DPC) or *finished floor level* (FFL). Wall widths, foundation widths, etc., can be set out along the board with the aid of a steel tape and their positions defined by saw-cuts. They are arranged around the building as shown in *Figure 12.9*. Strings stretched between the appropriate marks clearly define the line of construction.

In the case of buildings constructed with steel or concrete columns, a structural grid must be established to an accuracy of about ±2 to 3 mm or the prefabricated beams and steelwork will not fit together without some distortion.

The position of the concrete floor slab may be established in a manner already described. Thereafter the structural grid is physically established by Hilti nails or small steel plates set into the concrete. Due to the accuracy required a 1″ theodolite and standardized steel tape corrected for temperature and tension may be preferable to a total station.

Once the bases for the steel columns have been established, the axes defining the centre of each column should be marked on and, using a template orientated to these axes, the positions of the holding-down bolts defined (*Figure 12.10*). A height mark should be established, using a level, at a set distance (say, 75 mm) below the underside of the base-plate, and this should be constant throughout the structure. It is important that the base-plate starts from a horizontal base to ensure verticality of the column.

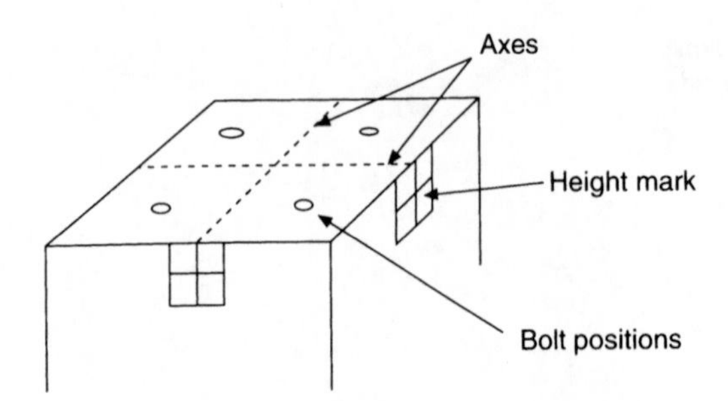

Fig. 12.10 *Holding down bolt positions*

12.7 CONTROLLING VERTICALITY

12.7.1 Using a plumb-bob

In low-rise construction a heavy plumb-bob (5 to 10 kg) may be used as shown in *Figure 12.11*. If the external wall were perfectly vertical then, when the plumb-bob coincides with the centre of the peg, distance *d* at the top level would equal the offset distance of the peg at the base. This concept can be used internally as well as externally, provided that holes and openings are available. The plumb-bob should be large, say 5 kg, and both plumb-bob and wire need to be protected from wind. The motion of the plumb-bob may need to be damped by immersing the plumb-bob in a drum of water. The considerations are similar to those of determining verticality in a mine shaft (*Chapter 13*) but less critical. To ensure a direct transfer of position from the bottom to the top floor, holes of about 0.2 m diameter will need to be left in all intermediate floors. This may need the agreement of the building's designer.

12.7.2 Using a theodolite

If two centre-lines at right angles to each other are carried vertically up a structure as it is being built, accurate measurement can be taken off these lines and the structure as a whole will remain vertical. Where site conditions permit, the stations defining the 'base figure' (four per line) are placed in concrete well clear of construction (*Figure 12.12*(*a*)). Lines stretched between marks fixed from the pegs will allow offset measurements to locate the base of the structure. As the structure rises the marks can be transferred up onto the walls by theodolite, as shown in *Figure 12.12*(*b*), and lines stretched between them. It is important that the transfer is carried out on both faces of the instrument.

Where the structure is circular in plan the centre may be established as in *Figure 12.12*(*a*) and the radius swung out from a pipe fixed vertically at the centre. As the structure rises, the central pipe is extended by adding more lengths. Its verticality is checked by two theodolites (as in *Figure 12.12*(*b*)) and its rigidity ensured by supports fixed to scaffolding.

The vertical pipe may be replaced by laser beam or autoplumb, but the laser would still need to be checked for verticality by theodolites.

Steel and concrete columns may also be checked for verticality using the theodolite. By string lining through the columns, positions *A–A* and *B–B* may be established for the theodolite (*Figure 12.13*); alternatively, appropriate offsets from the structural grid lines may be used. With instrument set up at *A*, the outside face of all the uprights should be visible. Now cut the outside edge of the upright at ground level with the vertical hair of the theodolite. Repeat at the top of the column. Now depress the telescope back to ground level and make a fine mark; the difference between the mark and the outside edge of the column is

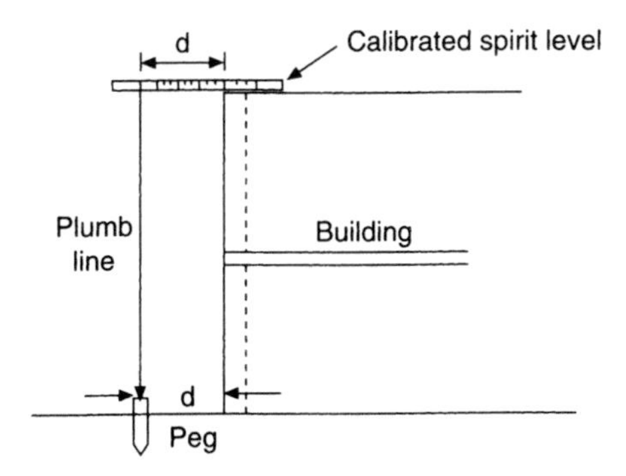

Fig. 12.11 *Plumb-bob for verticality*

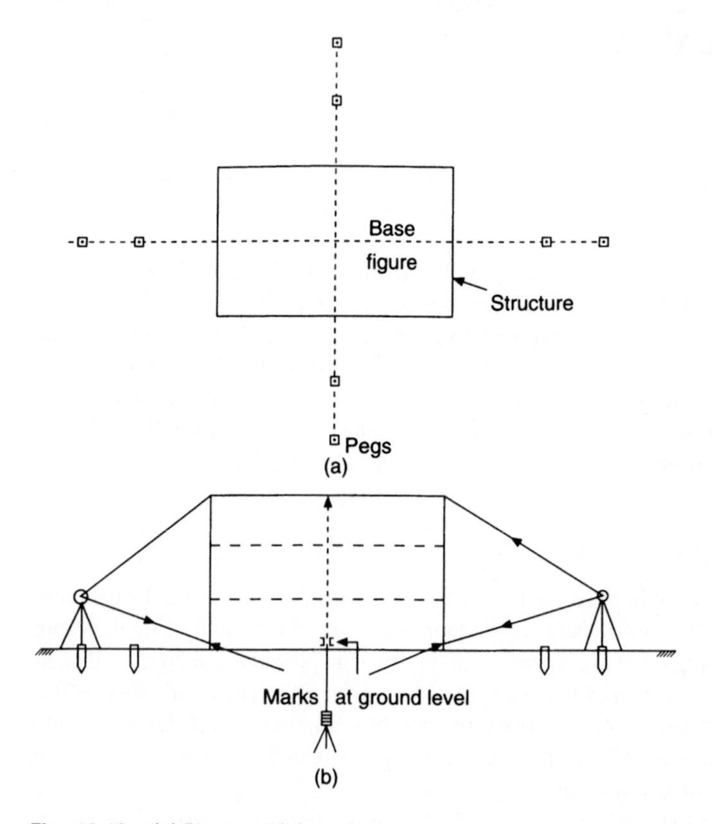

Fig. 12.12 *(a) Plan, and (b) section*

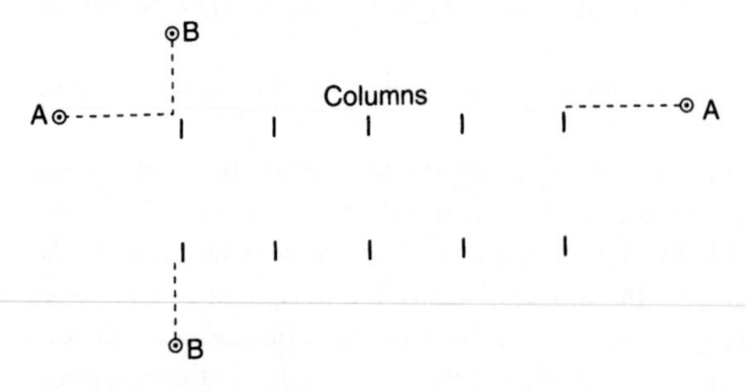

Fig. 12.13 *String lines*

the amount by which the column is out of plumb. Repeat on the opposite face of the theodolite. The whole procedure is now carried out at *B*. If the difference exceeds the specified tolerances the column will need to be corrected.

12.7.3 Using optical plumbing

For high-rise building the instrument most commonly used is an autoplumb (*Figure 12.14*). This instrument provides a vertical line of sight to an accuracy of ±1 second of arc (1 mm in 200 m). Any deviation from the vertical can be quantified and corrected by rotating the instrument through 90° and observing in all four quadrants; the four marks obtained would give a square, the diagonals of which would intersect at the correct centre point.

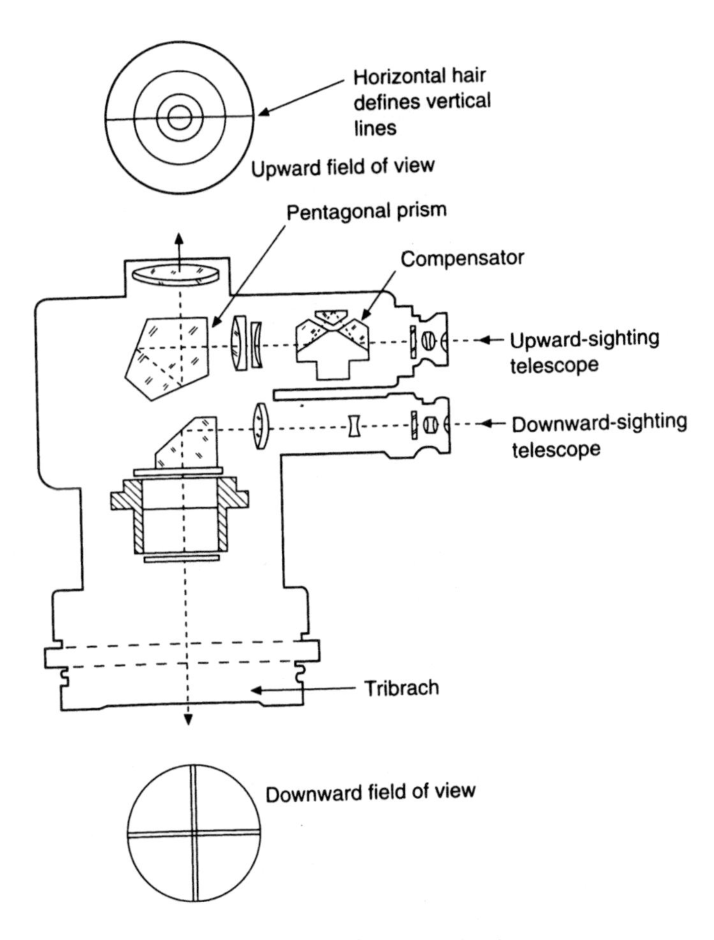

Fig. 12.14 *The optical system of the autoplumb*

A base figure is established at ground level from which fixing measurements may be taken. If this figure is carried vertically up the structure as work proceeds, then identical fixing measurements from the figure at all levels will ensure verticality of the structure (*Figure 12.15*).

To fix any point of the base figure on an upper floor, a Perspex target is set over the opening and the centre point fixed as above. Sometimes these targets have a grid etched on them to facilitate positioning of the marks.

The base figure can be projected as high as the eighth floor, at which stage the finishing trades enter and the openings are closed. In this case the uppermost figure is carefully referenced, the openings filled, and then the base figure re-established and projected upwards as before.

The shape of the base figure will depend upon the plan shape of the building. In the case of a long rectangular structure a simple base line may suffice but *T* shapes and *Y* shapes are also used.

12.8 CONTROLLING GRADING EXCAVATION

This type of setting out generally occurs in drainage schemes where the trench, bedding material and pipes have to be laid to a specified design gradient. Manholes (MH) will need to be set out at every change of

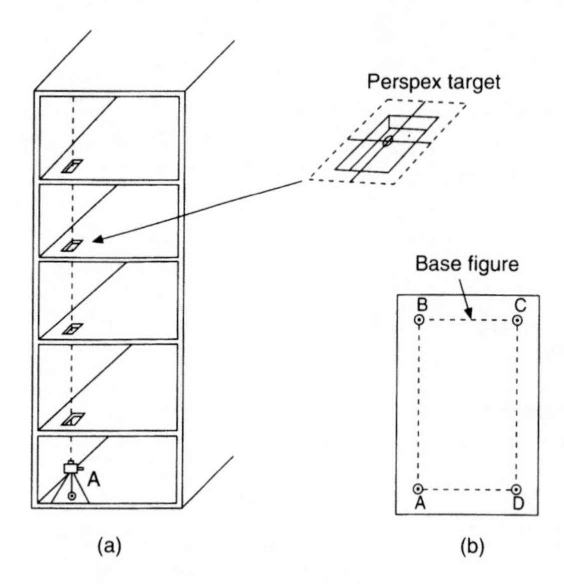

Fig. 12.15 *(a) Elevation, and (b) plan*

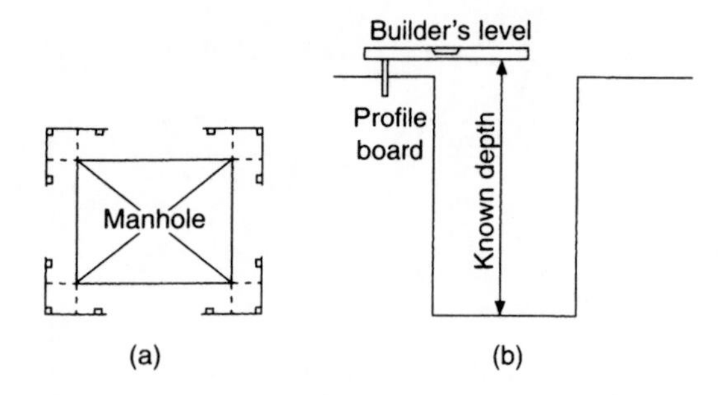

Fig. 12.16 *Setting out a manhole*

direction or at least every 100 m on straight runs. The MH (or inspection chambers) are generally set out first and the drainage courses set out to connect into them.

The centre peg of the MH is established in the usual way and referenced to four pegs, as in *Figure 12.2*. Alternatively, profile boards may be set around the MH and its dimensions marked on them. If the boards are set out at a known height above formation level the depth of excavation can be controlled, as in *Figure 12.16*.

12.8.1 Use of sight rails

Sight rails (SRs) are basically horizontal rails set a specific distance apart and to a specific level such that a line of sight between them is at the required gradient. Thus they are used to control trench excavation and pipe gradient without the need for constant professional supervision.

Figure 12.17 illustrates SRs being used in conjunction with a boning rod (or traveller) to control trench excavation to a design gradient of 0.5% (rising). Pegs *A* and *B* are offset a known distance from the centre-line of the trench and levelled from a nearby TBM.

Assume that peg *A* has a level of 40 m and the formation level of the trench at this point is to be 38 m. It is decided that a reasonable height for the SR above ground would be 1.5 m, i.e. at a level of 41.5;

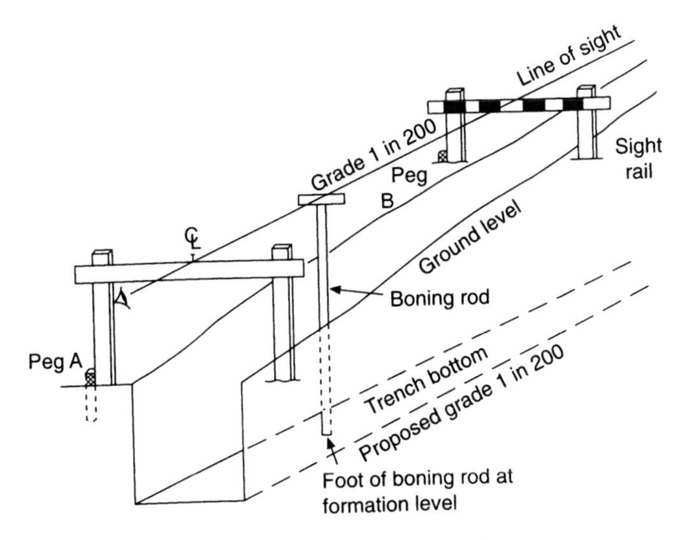

Fig. 12.17 *Trench excavation*

thus the boning rod must be made (41.5 − 38) = 3.5 m long, as its cross-head must be on level with the SR when its toe is at formation level.

Consider now peg *B*, with a level of 40.8 m at a horizontal distance of 50 m from *A*. The proposed gradient is 0.5%, which is 0.25 m in 50 m, and thus the formation level at *B* is 38.25 m. If the boning rod is 3.5 m, the SR level at *B* is (38.25 + 3.5) = 41.75 m and is set (41.75 − 40.8) = 0.95 m above peg *B*. The remaining SRs are established in this way and a line of sight or string stretched between them will establish the trench gradient 3.5 m above the required level. Thus, holding the boning rod vertically in the trench will indicate, relative to the sight rails, whether the trench is too high or too low.

Where machine excavation is used, the SRs are as in *Figure 12.18*, and offset to the side of the trench opposite to where the excavated soil is deposited. Before setting out the SRs it is important to liaise with the plant foreman to discover the type of plant to be used, i.e. will the plant straddle the trench as in *Figure 12.18* or will it work from the side of the trench and where the spoil will be placed to ensure the SRs will be useful.

Knowing the bedding thickness, the invert pipe level, the level of the inside of the bottom of the pipe, may be calculated and a second cross-head added to the boning rod to control the pipe laying, as shown in *Figure 12.19*.

Due to excessive ground slopes it may be necessary to use double sight rails with various lengths of boning rod as shown in *Figure 12.20*.

12.8.2 Use of lasers

The word *laser* is an acronym for Light Amplification by Stimulated Emission of Radiation and is the name applied to an intense beam of highly monochromatic, coherent light. Because of its coherence the light can be concentrated into a narrow beam and will not scatter and become diffused like ordinary light.

In controlling trench excavation the laser beam simply replaces the line of sight or string in the SR situation. It can be set up on the centre-line of the trench, over a peg of known level, and its height above the peg measured to obtain the reduced level of the beam. The instrument is then set to the required gradient and used in conjunction with an extendable traveller set to the same height as that of the laser above formation level. When the trench is at the correct level, the laser spot will be picked up on the centre of the traveller target, as shown in *Figure 12.21*. A levelling staff could just as easily replace the traveller; the laser spot being picked up on the appropriate staff reading.

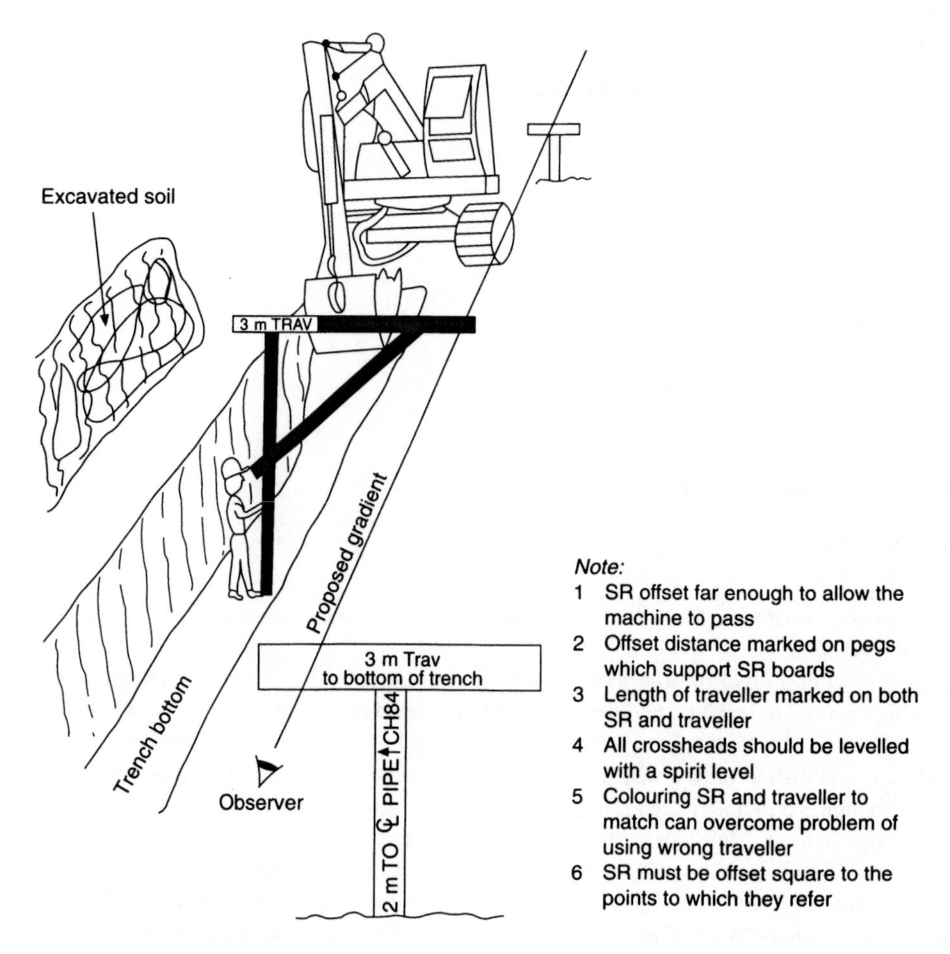

Fig. 12.18 *Use of offset sight rails (SR)*

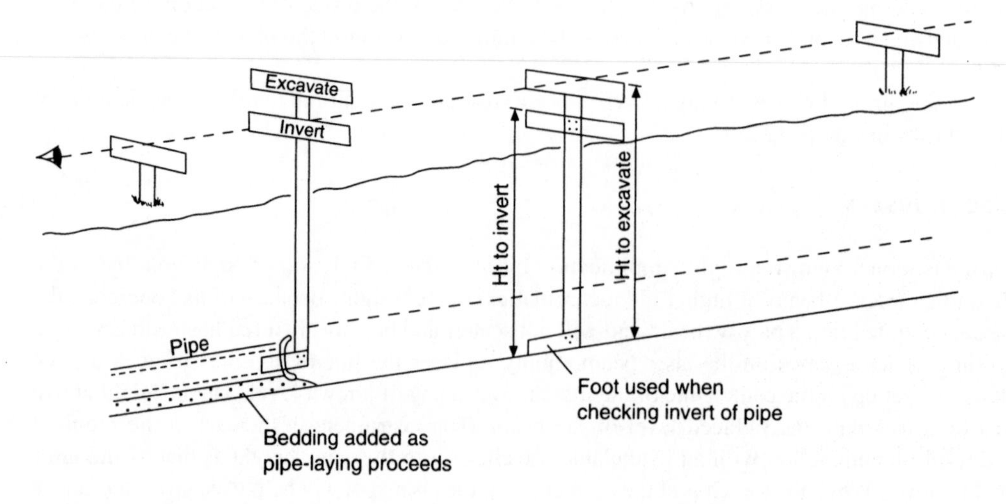

Fig. 12.19 *Pipe laying*

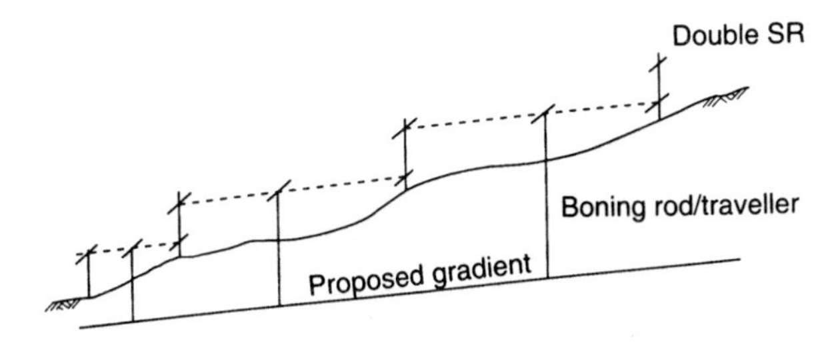

Fig. 12.20 *Double sight rails*

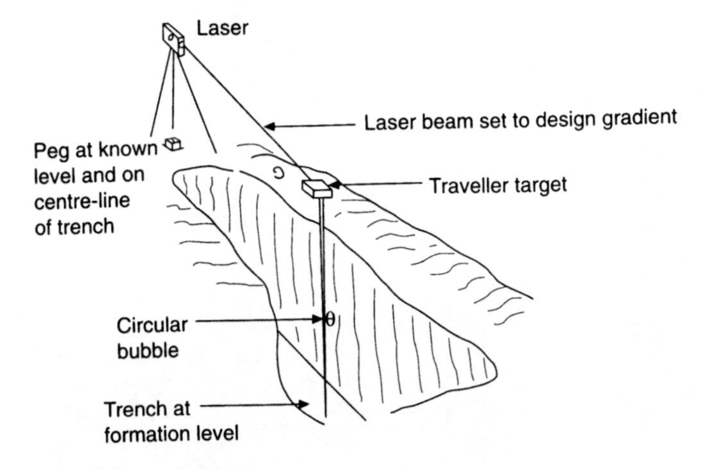

Fig. 12.21 *Laser level*

Where machine excavation is used the beam can be picked up on a photo-electric cell fixed at the appropriate height on the machine. The information can be relayed to a console within the cabin, which informs the operator whether he/she is too high or too low (*Figure 12.22*).

At the pipe-laying stage, a target may be fixed in the pipe and the laser installed on the centre-line in the trench. The laser is then orientated in the correct direction (by bringing it on to a centre-line peg, as in *Figures 12.23*(*a*, *b*)) and depressed to the correct gradient of the pipe. A graduated rod, or appropriately-marked ranging pole, can also be used to control formation and sub-grade level (*Figure 12.23*(*a*)). For large-diameter pipes the laser is mounted inside the pipe using horizontal compression bars (*Figure 12.23*(*c*)). If the laser is knocked off line or off level it may cease to function, or work intermittently, so indicating that it can no longer be relied on.

Where the MH is constructed, the laser can be orientated from within using the system illustrated in *Figure 12.24*. The centre-line direction is transferred down to peg *B* from peg *A* and used to orientate the direction of the laser beam.

12.9 ROTATING LASERS

Rotating lasers (*Figure 12.25*(*a*)) are instruments which are capable of being rotated in both the horizontal and vertical planes, thereby generating reference planes or datum lines.

When the laser is established in the centre of a site over a peg of known level, and at a known height above the peg, a datum of known reduced level is permanently available throughout the site.

Fig. 12.22 *Machine excavation*

Using a vertical staff fitted with a photo-electric detector, levels at any point on the site may be instantly obtained and set out, both above and below ground, as illustrated in *Figure 12.25(b)*.

Since the laser reference plane covers the whole working area, photo-electric sensors fitted at an appropriate height on earthmoving machinery enable whole areas to be excavated and graded to requirements by the machine operator alone.

Other uses of the rotating laser are illustrated in *Figure 12.26*.

From the above applications it can be seen that basically the laser supplies a reference line at a given height and gradient, and a reference plane similarly disposed. Realizing this, the user may be able to utilize these properties in a wide variety of setting-out situations such as off-shore channel dredging, tunnel guidance, shaft sinking, etc.

12.10 LASER HAZARDS

The potential hazard in the use of lasers is eye damage. There is nothing unique about this form of radiation damage; it can also occur from other, non-coherent light emitted, for example, by the sun,

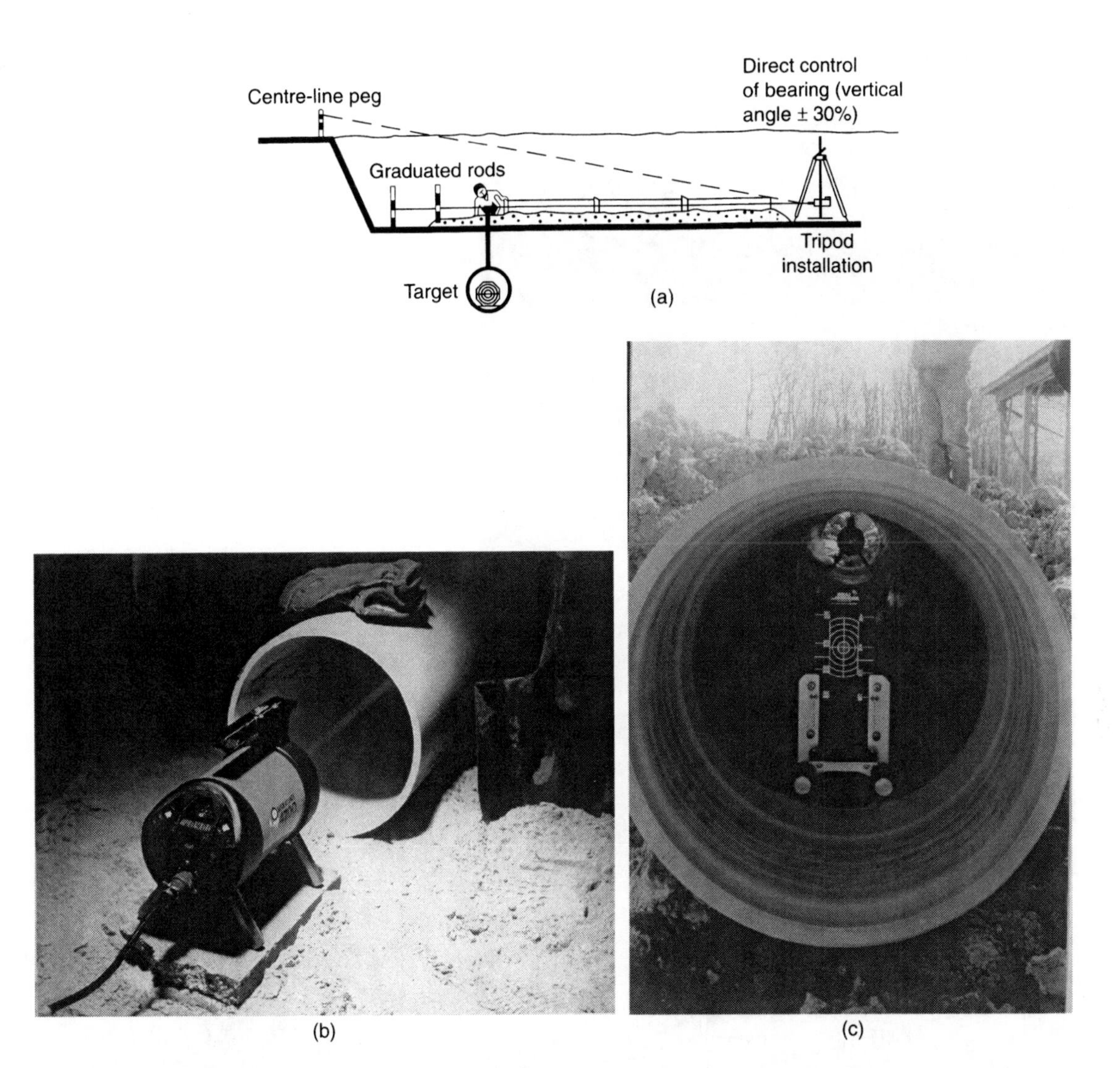

Fig. 12.23 *(a) Pipe work, (b) pipe laser, (c) laser alignment in large-diameter pipe*

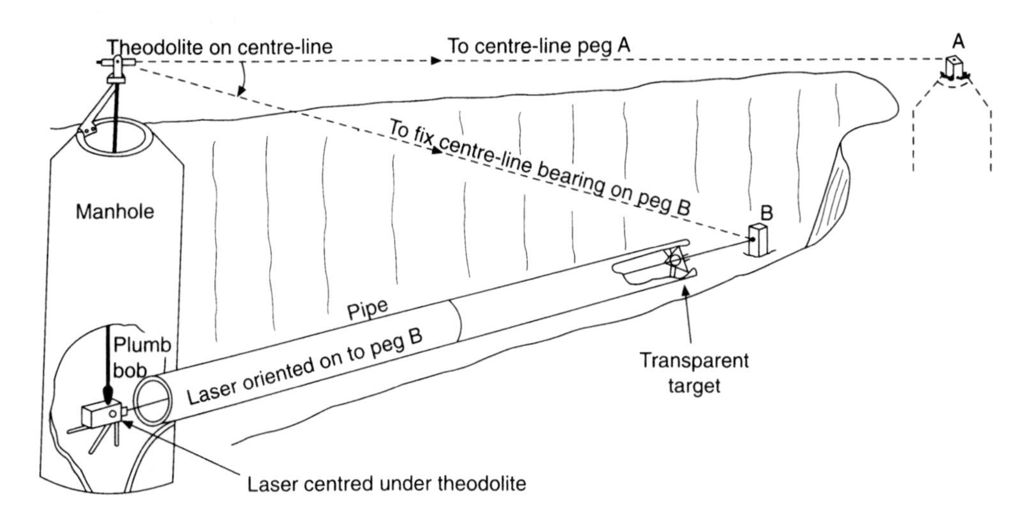

Fig. 12.24 *Pipe laser in a manhole*

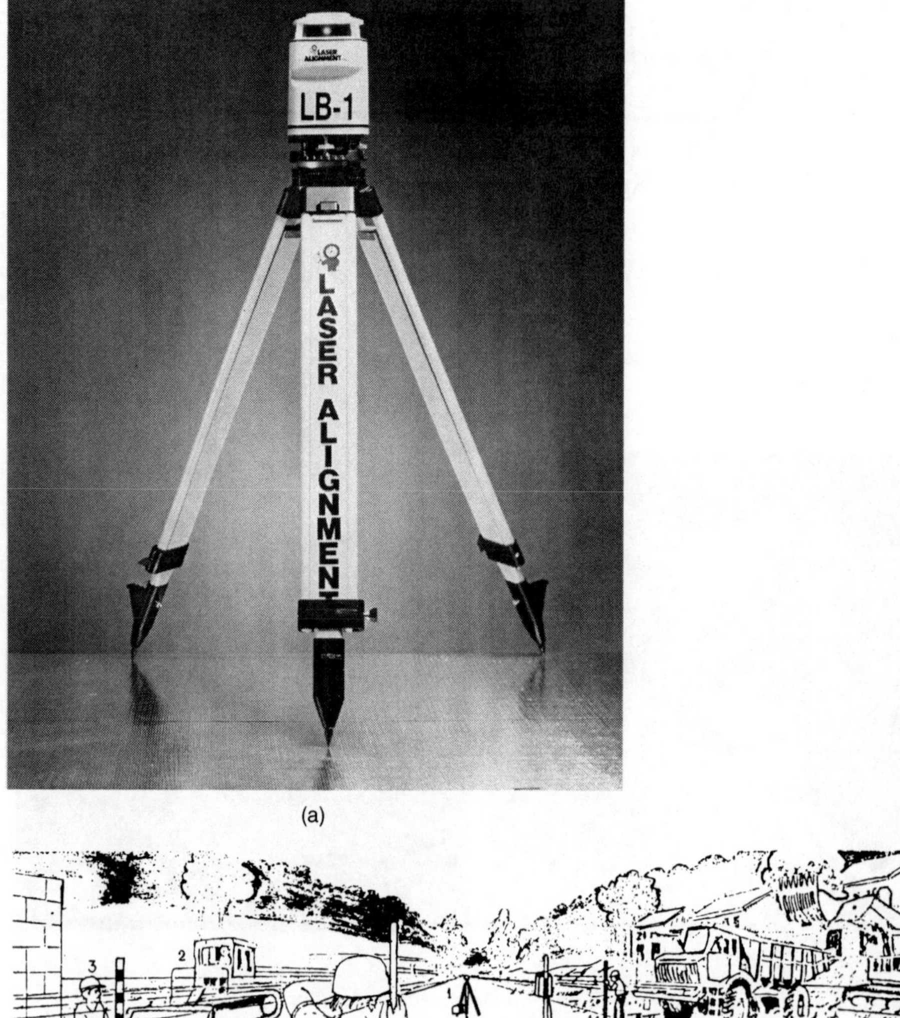

(a)

(b)

Fig. 12.25 *(a) Rotating laser, (b) 1 – Rotating laser; 2 – Laying sub-grade to laser control; 3 – Checking formation level; 4 – Fixing wall levels; 5 – Taking ground levels; 6 – Staff with detector used for fixing foundation levels; 7 – Laser plane of reference*

arc lamps, projector lamps and other high-intensity sources. If one uses a magnifying lens to focus the sun's rays onto a piece of paper, the heat generated by such concentration will cause the paper to burst into flames. Similarly, a laser produces a concentrated, powerful beam of light which the eye's lens will further concentrate by focusing it on the retina, thus causing an almost microscopic burn or blister which can cause temporary or permanent blindness. When the beam is focused on the macula (critical area of the

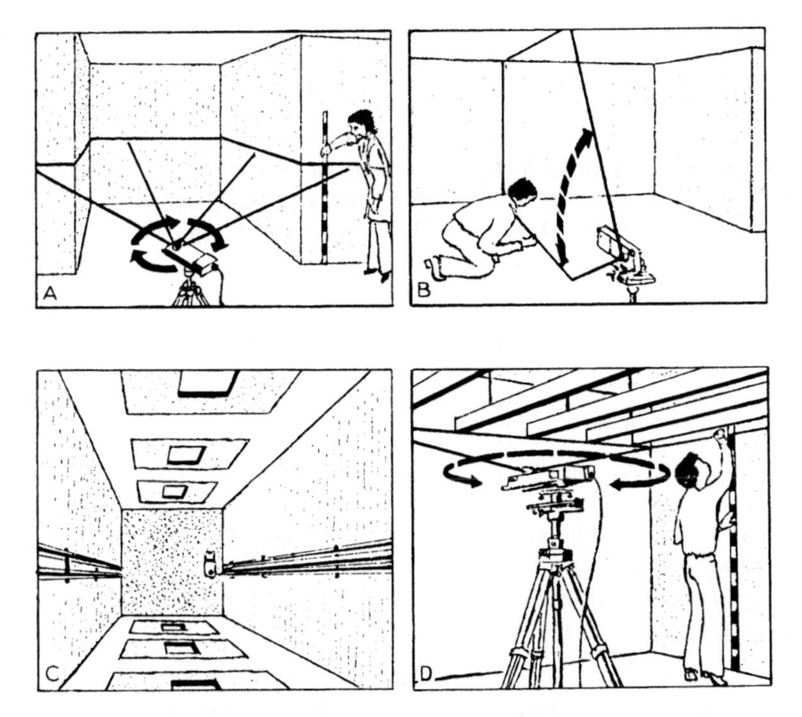

Fig. 12.26 *A – Height control; B – Setting out dividing walls; C – Use of the vertical beam for control of elevator guide rails, and slip-forming structures; D – Setting out and control of suspended ceiling*

retina) serious damage can result. Since there is no pain or discomfort with a laser burn, the injury may occur several times before vision is impaired. A further complication in engineering surveying is that the beam may be acutely focused through the lens system of a theodolite or other instrument, or may be viewed off a reflecting or refracting surface. It is thus imperative that a safety code be adopted by all personnel involved with the use of lasers.

Formerly the British Standards Institution published a guide to the safe use of lasers (BS 4803:Part 3: 1983) in which there were classified five types of laser, but only three of these were relevant to on-site working:

Class 2 A visible radiant power of 1 mW. Eye protection is afforded by the blink-reflex mechanism.

Class 3A Has a maximum radiant power of 1 to 5 mW, with eye protection afforded by the blink-reflex action

Class 3B Has a maximum power of 1 to 500 mW. Eye protection is not afforded by blink-reflex. Direct viewing, or viewing of specular reflections, is highly dangerous.

For surveying and setting-out purposes the BSI recommended the use of Classes 2 and 3A only. Class 3B may be used outdoors, if the more stringent safety precautions recommended are observed.

The most significant recommendation of the BSI document was that on sites where lasers are in use there should be a laser safety officer (LSO), who the document defined as 'one who is knowledgeable in the evaluation and control of laser hazards, and had the responsibility for supervision of the control of laser hazards'. Whilst such an individual was not specifically mentioned in conjunction with the Class 2 laser, the legal implications of eye damage might render it advisable to have an LSO present. Such an individual would not only require training in laser safety and law, but would have needed to be fully conversant with the RICS' Laser Safety Code produced by a working party of the Royal Institution of Chartered Surveyors.

The RICS code was produced in conjunction with BS 4803 and dealt specifically with the helium–neon gas laser (He–Ne) as used on site. Whilst the manufacturers of lasers would no doubt have complied with the classification laid down, the modifications to a laser by mirrors or telescopes may completely alter such specifications and further increase the hazard potential. The RICS code presented methods and computations for assessing the possible hazards which the user could easily apply to their working laser system, in both its unmodified and modified states. Recommendations were also made about safety procedures relevant to a particular system from both the legal and technical aspects. The information within the RICS code enabled the user to compute such important parameters as:

(1) The safe viewing time at given distances.
(2) The minimum safe distance at which the laser source may be viewed directly, for a given period of time.

Such information is vital to the organization and administration of a 'laser site' from both the health and legal aspects, and should be combined with the following precautions.

(1) Ensure that all personnel, visitors to the site, and where necessary members of the public, are aware of the presence of lasers and the potential eye damage involved.
(2) Erect safety barriers around the laser with a radius greater than the minimum safe viewing distance.
(3) Issue laser safety goggles where appropriate.
(4) Avoid, wherever possible, the need to view the laser through theodolites, levels or binoculars.
(5) Where possible, position the laser either well above or well below head height.

At the time of writing the 'BS 4803:Part 3:1983 Radiation safety of laser products and systems. Guidance for users' has been withdrawn. However, the European Union's Council of Ministers has recently approved a 'Physical Agents (Optical Radiation)' directive aimed at protecting workers from dangerous levels of optical radiation.

Guidelines have been drawn up by the International Commission on Non-Ionizing Radiation Protection (ICNIRP) which recommends limits for exposure to non-ionizing radiation in 'Revision of the Guidelines on Limits of Exposure to Laser radiation of wavelengths between 400 nm and 1.4 μm, *Health Physics*, Vol. 79, No. 4, pp. 431–440, 2000, and available at http://www.icnirp.de/documents/laser400nm+.pdf. Employers will be required to assess the risks posed by lasers and introduce measures to control exposure and provide employees with information and training as well as carry out health surveillance.

As yet there is no UK legislation that deals specifically with occupational exposure to optical radiation, but it is reported that the Health and Safety Executive (HSE) does not believe that the directive will place additional burdens on industry.

It is anticipated that the directive may soon become law but Member States will have several years to implement it.

12.11 ROUTE LOCATION

Figure 12.27 shows a stretch of route location for a road or railway. In order to control the construction involved, the pegs and profile boards shown must be set out at intervals of 10 to 30 m along the whole stretch of construction.

The first pegs located would be those defining the centre-line of the route (peg E), and the methods of locating these on curves have been dealt with in *Chapter 10*. The straights would be aligned between adjacent tangent points.

The shoulder pegs C and D, defining the road/railway width, can be set out by appropriate offsets at right angles to the centre-line chords.

Pegs A and B, which define the toe of the embankment (fill) or top edge of the cutting, are called *slope stakes*. The side widths from the centre-line are frequently calculated and shown on the design drawings or

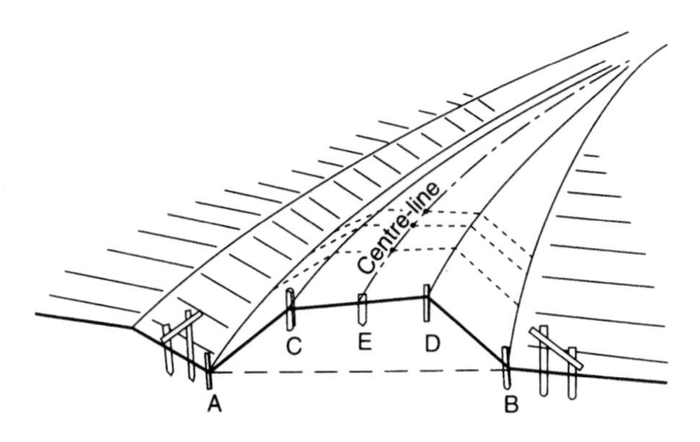

Fig. 12.27 *Pegging out a route*

computer print-outs of setting-out data. This information should be used only as a rough check or guide; the actual location of the slope stake pegs should always be carried out in the field, due to the probable change in ground levels since the information was initially compiled. These pegs are established along with the centre-line pegs and are necessary to define the area of top-soil strip.

12.11.1 Setting-out slopes stakes

In *Figure 12.28*(*a*) and (*b*), points A and B denote the positions of pegs known as *slope stakes* which define the points of intersection of the actual ground and the proposed side slopes of an embankment or cutting. The method of establishing the positions of the stakes is as follows:

(1) Set up the level in a convenient position which will facilitate the setting out of the maximum number of points therefrom.
(2) Obtain the height of the plane of collimation (HPC) of the instrument by backsighting on to the nearest TBM.
(3) Foresight onto the staff held where it is thought point A may be and obtain the ground level there.
(4) Subtract 'ground level' from 'formation level' and multiply the difference by N to give horizontal distance x.
(5) Now tape the horizontal distance $(x + b)$ from the centre-line to the staff. If the *measured distance* to the staff equals the *calculated distance* $(x + b)$, then the staff position is the slope stake position. If not move the staff to the *calculated distance* $(x + b)$, level again and re-compute the *calculated distance*. Repeat as necessary.

The above 'trial-and-error' approach should always be used on site to avoid errors of scaling the positions from a plan, or accepting, without checking, a computer print-out of the dimensions.

For example, if the side slopes of the proposed embankment are to be 1 vertical to 2 horizontal, the formation level 100 m OD and the ground level at A, say, 90.5 m OD, then $x = 2(100 - 90.5) = 19$ m, and if the formation width = 20 m, then $b = 10$ m and $(x + b) = 29$ m. Had the staff been held at A_1 (which had exactly the same ground level as A) then obviously the calculated distance $(x + b)$ would not agree with the measured distance from centre-line to A_1. They would agree only when the staff arrived at the slope stake position A, as x is dependent upon the level at the toe of the embankment, or top of the cutting.

If the ground is nearly level then it should only take one or two iterations to find the correct point for the slope stake. If the ground slope is steep then unless the first trial point is close to the correct point it may take many attempts to find the correct point. In this case one could use the 'rate of approach' concept

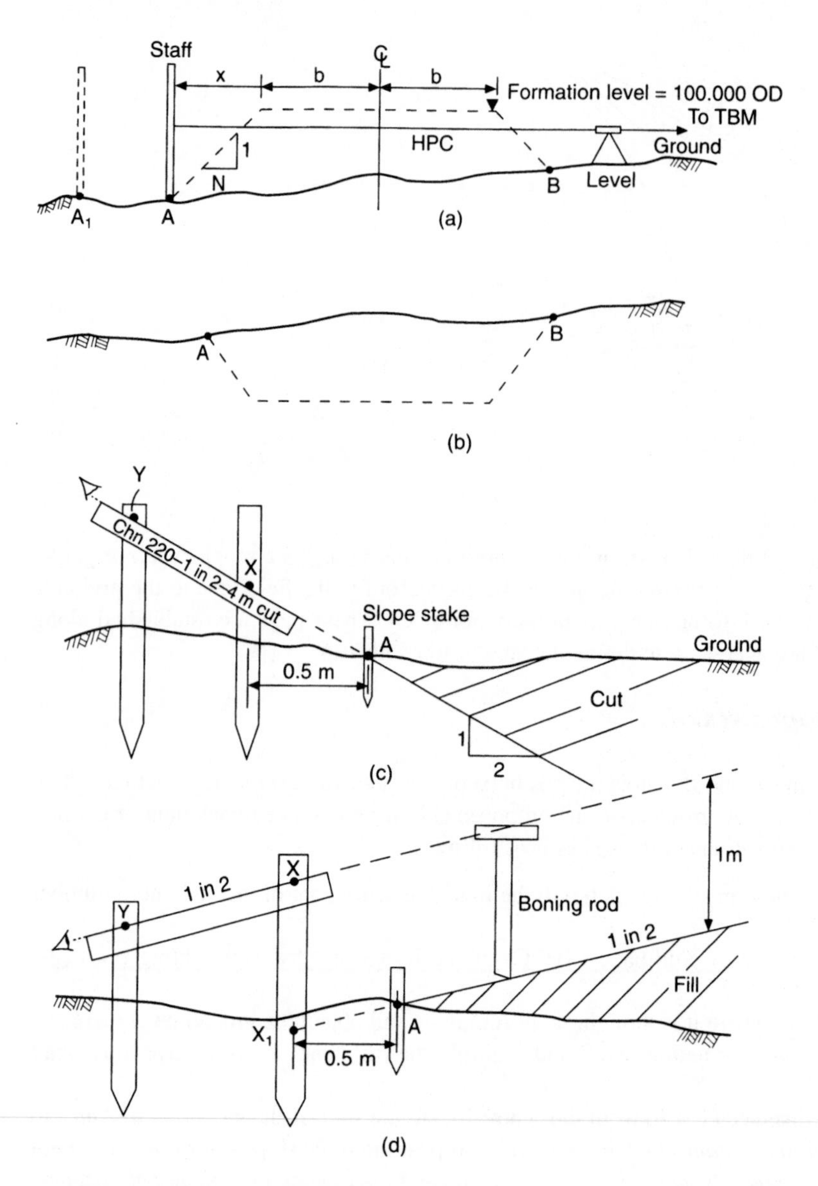

Fig. 12.28 *Setting out slope stakes*

developed in *Section 11.3* to get a better first estimate for the slope stake. For example, in *Figure 12.29*, if the embankment is to have a slope of 20% (= 0.2 = 1 in 5) and the ground slopes downwards from the slope stake towards the road centre-line at 10% (= 0.1 = 1 in 10) and the height of the top of the side slope (from the road design) is 5 m above the ground height of the centre-line (from survey) then:

calculated distance $(x + b)$ to the left of the centre-line is found as follows

Estimated height of ground below top of slope = 5 m − 10% of $b = 5 - 0.1b$

b of course is known from the design of the road

$$x + b = (0.2 + 0.1)^{-1}(5 - 0.1b) + b$$

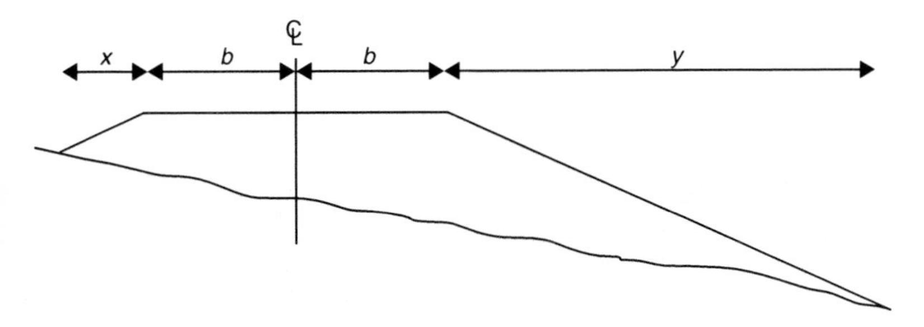

Fig. 12.29 *Setting out slope stakes by rate of approach*

and *calculated distance* $(y + b)$ to the right of the centre-line is

$$\text{Estimated height of ground below top of slope} = 5 \text{ m} + 10\% \text{ of } b = 5 + 0.1b$$

$$y + b = (0.2 - 0.1)^{-1}(5 + 0.1b) + b$$

If b is 10 m then the values of the calculated distance are $x + b = 23.33$ m and $y + b = 70.00$ m; very different values.

12.11.2 Controlling earthworks

Batter boards, or *slope rails* as they are sometimes called, are used to control the construction of the side slopes of a cutting or embankment (see *Figure 12.28*(*c*) and (*d*), and *Figure 12.27*).

Consider *Figure 12.28*(*c*). If the stake adjacent to the slope stake is set 0.5 m away, then, for a grade of 1 vertical to 2 horizontal, the level of point X will be 0.25 m higher than the ground level at A. From X, a batter board is fixed at a gradient of 1 in 2 (50%), using a 1 in 2 template and a spirit level. Stakes X and Y are usually no more than 1 m apart. Information such as chainage, slope and depth of cut are marked on the batter board.

In the case of an embankment (*Figure 12.28*(*d*)), a boning rod is used in the control of the slope. Assuming that a boning rod 1 m high is to be used, then as the near stake is, say, 0.5 m from the slope stake, point x_1 will be 0.25 m lower than the ground level at A, and hence point x will be 0.75 m above the ground level of A. The batter board is then fixed from x in a similar manner to that already described.

The formation and sub-base, which usually have setting-out tolerances in the region of ±25 mm, can be located with sufficient accuracy using profiles and travellers. *Figure 12.30* shows the use of triple profiles for controlling camber, whilst different lengths of traveller will control the thickness required.

Laying of the base course (60 mm) and wearing course (40 mm) calls for much smaller tolerances, and profiles are not sufficiently accurate; the following approach may be used.

Pins or pegs are established at right angles to the centre-line at about 0.5 m beyond the kerb face (*Figure 12.31*). The pins or pegs are accurately levelled from the nearest TBM and a coloured tape placed

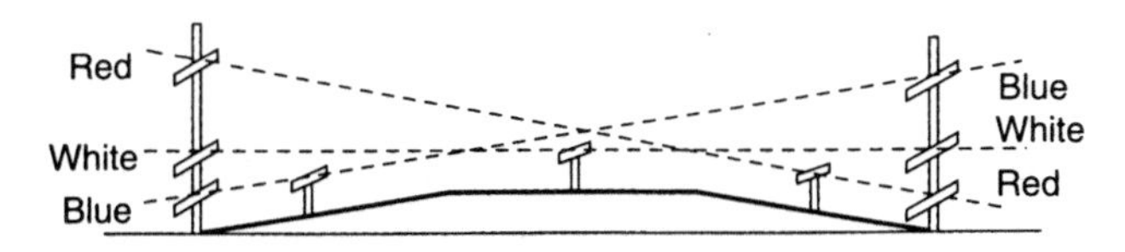

Fig. 12.30 *Controlling camber with triple profiles*

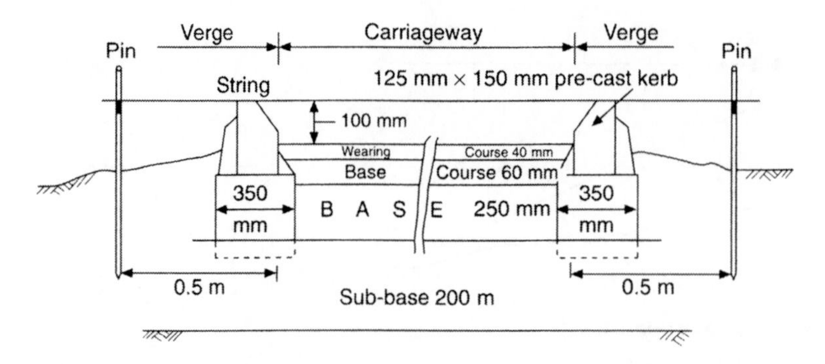

Fig. 12.31 *Pins used to control setting out of base and wearing courses*

around them at 100 mm above finished road level; this will be at the same level as the top of the kerb. A cord stretched between the pins will give kerb level, and with a tape the distances to the top of the sub-base, top of the base course and top of the wearing course can be accurately fixed (or dipped).

The distance to the kerb face can also be carefully measured in from the pin in order to establish the kerb line. This line is sometimes defined with further pins and the level of the kerb top marked on.

13 Underground surveying

Underground surveying is quite different from surveying on the surface. In tunnelling or mining operations it may be hot, wet, dark, cramped, dusty, dirty and dangerous, and usually most of these.

The essential problem in underground surveying is that of orientating the underground surveys to the surface surveys, the procedure involved being termed a *correlation*.

In an underground transport system, for instance, the tunnels are driven to connect inclined or vertical shafts (points of surface entry to the transport system) whose relative locations are established by surface surveys. Thus the underground control networks must be connected and orientated into the same coordinate system as the surface networks. To do this, one must obtain the coordinates of at least one underground control station and the bearing of at least one line of the underground network, relative to the surface network.

If entry to the underground tunnel system is via an inclined shaft, then the surface survey may simply be extended and continued down that shaft and into the tunnel, usually by the method of traversing. Extra care would be required in the measurement of the horizontal angles due to the steeply inclined sights involved (see *Chapter 5*) and in the temperature corrections to the taped distances due to the thermal gradients encountered.

If entry is via a vertical shaft, then optical, mechanical or gyroscopic methods of orientation are used.

13.1 OPTICAL METHODS

Where the shaft is shallow and of relatively large diameter the bearing of a surface line may be transferred to the shaft bottom by theodolite (*Figure 13.1*).

The surface stations A and B are part of the control system and represent the direction in which the tunnel must proceed. They would usually be established clear of ground movement caused by shaft sinking or other factors. Auxiliary stations c and d are very carefully aligned with A and B using the theodolite on both faces and with due regard to all error sources. The relative bearing of A and B is then transferred to $A'B'$ at the shaft bottom by direct observations. Once again these observations must be carried out on both faces with the extra precautions advocated for steep sights.

If the coordinates of d are known then the coordinates of B' could be fixed by measuring the vertical angle and distance to a reflector at B'.

It is important to understand that the accurate orientation bearing of the tunnel is infinitely more critical than the coordinate position. For instance, a standard error of $1'$ in transferring the bearing down the shaft to $A'B'$ would result in a positional error at the end of 1 km of tunnel drivage of 300 mm and would increase to 600 mm after 2 km of drivage.

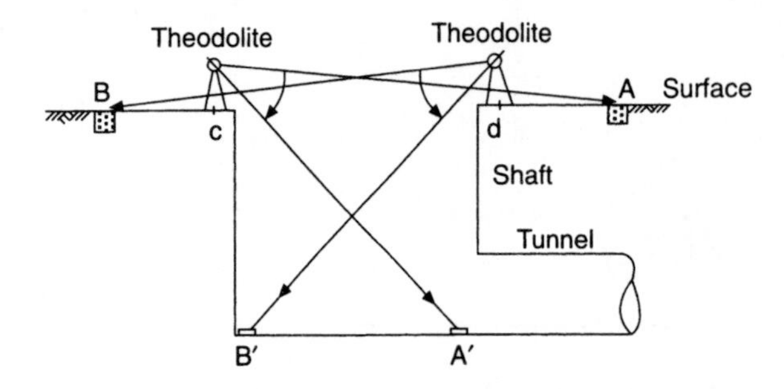

Fig. 13.1 *Transfer alignment to bottom of a wide shaft*

13.2 MECHANICAL METHODS

Although these methods, which involve the use of wires hanging vertically in a shaft, are rapidly being superseded by gyroscopic methods, they are still widely used and are described herewith.

The basic concept is that wires hanging freely in a shaft will occupy the same position underground that they do at the surface, and hence the bearing of the wire plane will remain constant throughout the shaft.

13.2.1 Weisbach triangle method

This appears to be the most popular method in civil engineering. Two wires, W_1 and W_2, are suspended vertically in a shaft forming a very small base line (*Figure 13.2*). The principle is to obtain the bearing and coordinates of the wire base relative to the surface base. These values can then be transferred to the underground base.

In order to establish the bearing of the wire base at the surface, it is necessary to compute the angle $W_sW_2W_1$ in the triangle as follows:

$$\sin \hat{W}_2 = \frac{w_2}{w_s} \sin \hat{W}_s \tag{13.1}$$

As the Weisbach triangle is formed by approximately aligning the Weisbach station W_s with the wires, the angles at W_s and W_2 are very small and *equation (13.1)* may be written:

$$\hat{W}_2'' = \frac{w_2}{w_s} \hat{W}_s'' \tag{13.2}$$

(The expression is accurate to seven figures when $\hat{W}_s < 18'$ and to six figures when $\hat{W}_s < 45'$.)

From *equation (13.2)*, it can be seen that the observational error in angle W_s will be multiplied by the fraction w_2/w_s.

Its effect will therefore be reduced if w_2/w_s is less than unity. Thus the theodolite at W_s should be as near the front wire W_1 as focusing will permit and preferably at a distance smaller than the wire base W_1W_2.

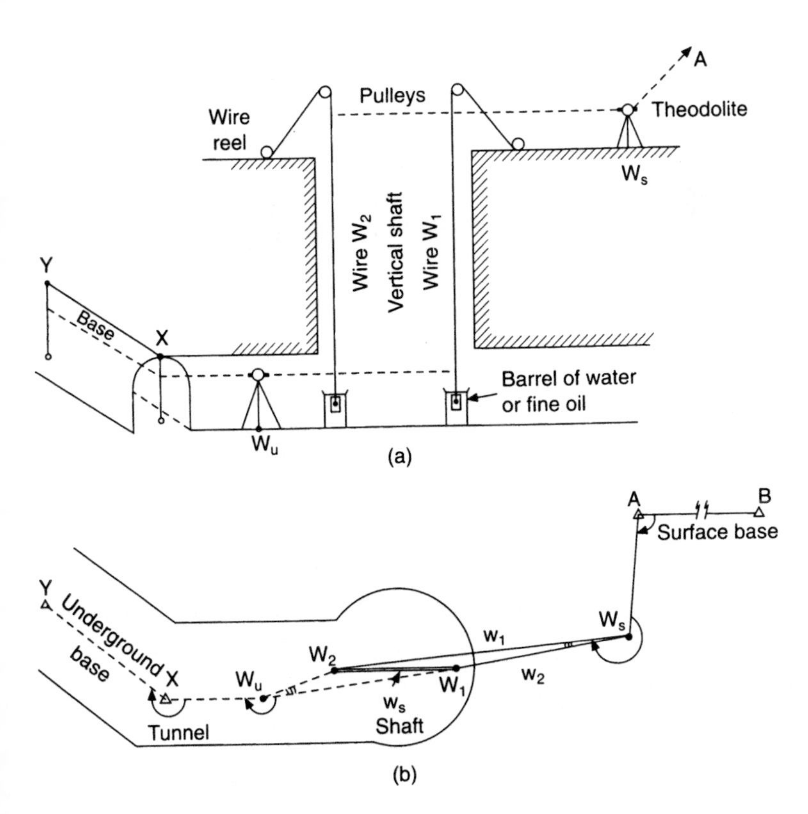

Fig. 13.2 *(a) Section, and (b) plan*

The following example, using simplified data, will now be worked to illustrate the procedure. With reference to *Figure 13.2(b)*, the following field data are obtained:

(1) Surface observations

Angle $BAW_s = 90° 00' 00''$	Distance $W_1W_2 = w_s = 10.000$ m
Angle $AW_sW_2 = 260° 00' 00''$	Distance $W_1W_s = w_2 = 5.000$ m
Angle $W_1W_sW_2 = 0° 01' 20''$	Distance $W_2W_s = w_1 = 15.000$ m

(2) Underground observations

Angle $W_2W_uW_1 = 0° 01' 50''$	Distance $W_2W_u = y = 4.000$ m
Angle $W_1W_uX = 200° 00' 00''$	Distance $W_uW_1 = x = 14.000$ m
$W_uXY = 240° 00' 00''$	

Solution of the surface Weisbach triangle:

$$\text{Angle } W_sW_2W_1 = \frac{5}{10} \times 80'' = 40''$$

Similarly, underground

$$\text{Angle } W_2W_1W_u = \frac{4}{10} \times 110'' = 44''$$

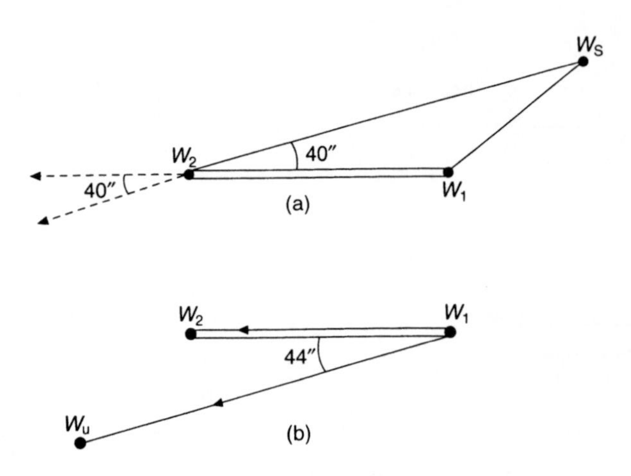

Fig. 13.3 *Part of Weisbach triangle*

The bearing of the underground base XY, relative to the surface base AB is now computed in a manner similar to a traverse:

Assuming	WCB of $AB = 89^\circ 00' 00''$	
then,	WCB of $AW_s = 179^\circ 00' 00''$	(using angle BAW_s)
	Angle $AW_sW_2 = 260^\circ 00' 00''$	
	Sum $= 439^\circ 00' 00''$	
	-180°	
	WCB of $W_sW_2 = 259^\circ 00' 00''$	see *Figure 13.3(a)*
	Angle $W_sW_2W_1 = +0^\circ 00' 40'$	
	WCB $W_1W_2 = 259^\circ 00' 40''$	
	Angle $W_2W_1W_u = -0^\circ 00' 44''$	see *Figure 13.3(b)*
	WCB $W_1W_u = 258^\circ 59' 56''$	
	Angle $W_1W_uX = 200^\circ 00' 00''$	
	Sum $= 458^\circ 59' 56''$	
	-180°	
	WCB $W_uX = 278^\circ 59' 56''$	
	Angle $W_uXY = 240^\circ 00' 00''$	
	Sum $= 518^\circ 59' 56''$	
	-180°	
	WCB $XY = 338^\circ 59' 56''$	(underground base)

The transfer of bearing is of prime importance; the coordinates can be obtained in the usual way by incorporating all the measured lengths AB, AW_s, W_uX, XY.

13.2.2 Shape of the Weisbach triangle

As already indicated, the angles W_2 and W_s in the triangle are as small as possible. The reason for this can be illustrated by considering the effect of accidental observation errors on the computed angle W_2.

From the basic equation: $\sin \hat{W}_2 = \dfrac{w_2}{w_s} \sin \hat{W}_s$

Differentiate W_2 with respect to each of the measured quantities in turn:

(1) with respect to W_s

$$\frac{dW_2}{dW_s} = \frac{w_2 \cos W_s}{w_s \cos W_2}$$

(2) with respect to w_2

$$\frac{dW_2}{dw_2} = \frac{\sin W_s}{w_s \cos W_2}$$

(3) with respect to w_s

$$\frac{dW_2}{dw_s} = \frac{-w_2 \sin W_s}{w_s^2 \cos W_2}$$

then:

$$\sigma_{W_2} = \left[\frac{w_2^2 \cos^2 W_s}{w_s^2 \cos^2 W_2}\sigma_{W_s}^2 + \frac{\sin^2 W_s}{w_s^2 \cos^2 W_2}\sigma_{w_2}^2 + \frac{w_s^2 \sin^2 W_s}{w_s^4 \cos^2 W_2}\sigma_{w_s}^2\right]^{\frac{1}{2}}$$

$$= \frac{w_s}{w_s \cos W_2}\left[\cos^2 W_s \sigma_{W_s}^2 + \sin^2 W_s \frac{\sigma_{w_2}^2}{w_2^2} + \sin^2 W_s \frac{\sigma_{w_s}^2}{w_s^2}\right]^{\frac{1}{2}}$$

but $\cos W_s = \dfrac{\sin W_s \cos W_s}{\sin W_s} = \sin W_s \cot W_s$, which on substitution gives:

$$\sigma_{W_2} = \frac{w_2}{w_s \cos W_2}\left[\sin^2 W_s \cot^2 W_s \sigma_{W_s}^2 + \sin^2 W_s \frac{\sigma_{w_2}^2}{w_2^2} + \sin^2 W_s \frac{\sigma_{w_s}^2}{w_s^2}\right]^{\frac{1}{2}}$$

$$= \frac{w_2 \sin W_s}{w_s \cos W_2}\left[\cot^2 W_s \sigma_{W_s}^2 + \frac{\sigma_{w_2}^2}{w_2^2} + \frac{\sigma_{w_s}^2}{w_s^2}\right]^{\frac{1}{2}}$$

by sine rule $\dfrac{w_2 \sin W_s}{w_s} = \sin W_2$, therefore substituing

$$\sigma_{W_2} = \tan W_2 \left[\cot^2 W_s \sigma_{W_s}^2 + \left(\frac{\sigma_{w_2}}{w_2}\right)^2 + \left(\frac{\sigma_{w_s}}{w_s}\right)^2\right]^{\frac{1}{2}} \tag{13.3}$$

Thus to reduce the standard error (σ_{W_2}) to a minimum:

(1) tan W_2 must be a minimum; therefore the angle W_2 should approach 0° .
(2) As W_2 is very small, W_s will be very small and so cot W_s will be very large. Its effect will be greatly reduced if σ_{W_s} is very small; the angle W_s must therefore be measured with maximum precision.

13.2.3 Sources of error

The standard error of the transferred bearing σ_B, is made up from the effects of:

(1) Uncertainty in connecting the surface base to the wire base, σ_s.
(2) Uncertainty in connecting the wire base to the underground base, σ_u.

(3) Uncertainty in the verticality of the wire plane, σ_p giving:

$$\sigma_B = (\sigma_s^2 + \sigma_u^2 + \sigma_p^2)^{\frac{1}{2}} \tag{13.4}$$

The uncertainties σ_s and σ_u can be obtained in the usual way from an examination of the method and the types of instrument used. The uncertainty in the verticality of the wire plane σ_p is vitally important in view of the extremely short length of the wire base.

Assuming that the relative uncertainty of wire W_1 with respect to wire W_2, at right angles to the line W_1W_2, is 1 mm and that the wires are 2 m apart, then $\sigma_p = 100''$ ($\sigma_p^2 = 10\,000$). If σ_B is required to be less than $2'$ ($= 120''$), then from *equation (13.4)*, and assuming $\sigma_s = \sigma_u$:

$$120 = (2\sigma_s^2 + 10\,000)^{\frac{1}{2}}$$

$$\therefore \sigma_s = \sigma_u = 47''$$

If, therefore, the uncertainty in the connection to the wire base is to be $47''$ and to be achieved over a baseline of 2 m then each wire must be sighted with a precision, σ_w, of

$$\sigma_w = 2 \sin 47''(2)^{-\frac{1}{2}}\ \text{m} = 0.0003\ \text{m} = 0.3\ \text{mm}$$

These figures serve to indicate the great precision and care needed in plumbing a shaft and sighting onto each wire, in order to minimize orientation errors.

13.2.4 Verticality of the wire plane

The factors affecting the verticality of the wires are:

(1) *Ventilation air currents in the shaft*
All forced ventilation should be shut off and the plumb-bob protected from natural ventilation.

(2) *Pendulous motion of the shaft plumb*
The motion of the plumb-bob about its suspension point can be reduced by immersing it in a barrel of water or fine oil. When the shaft is deep, complete elimination of motion is impossible and clamping of the wires in their mean swing position may be necessary.
The amplitude of wire vibrations, which induce additional motion to the swing, may be reduced by using a heavy plumb-bob, with its point of suspension close to the centre of its mass, and fitted with large fins.

(3) *Spiral deformation of the wire*
Storage of the plumb wire on small-diameter reels gives a spiral deformation to the wire. Its effect is reduced by using a plumb-bob of maximum weight. This should be calculated for the particular wire using a reasonable safety factor.

These sources of error are applicable to all wire surveys.

13.2.5 Co-planing

The principles of this alternative method are shown in *Figure 13.4*. The triangle of the previous method is eliminated by aligning the theodolite at W_s exactly with the wires W_1 and W_2. This alignment is easily achieved by trial and error, focusing first on the front wire and then on the back. Both wires can still be seen through the telescope even when in line. The instrument should be set up within 3 to 4 m of the nearer wire. Special equipment is available to prevent lateral movement of the theodolite affecting its level, but if this is not used, special care should be taken to ensure that the tripod head is level.

The movement of the focusing lens in this method is quite long. Thus for alignment to be exact, the optical axis of the object lens should coincide exactly with that of the focusing lens in all focusing positions.

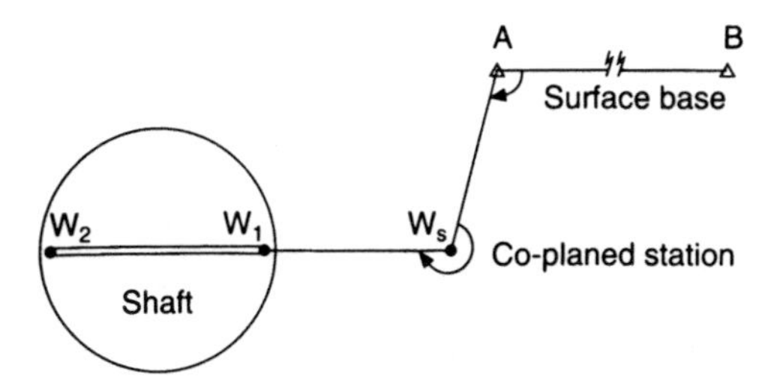

Fig. 13.4 *Co-planing*

If any large deviation exists, the instrument should be returned to the manufacturer. The chief feature of this method is its simplicity with little chance of gross errors.

13.2.6 Weiss quadrilateral

This method may be adopted when it is impossible to set up the theodolite, even approximately, on the line of the wire base W_1W_2 (*Figure 13.5*). Theodolites are set up at C and D forming a quadrilateral CDW_1W_2. The bearing and coordinates of CD are obtained relative to the surface base, the orientation of the wire base being obtained through the quadrilateral. Angles 1, 2, 3 and 8 are measured directly, and angles 4 and 7 are obtained as follows:

$$\text{Angle } 4 = \left[180^\circ - \left(\hat{1} + \hat{2} + \hat{3}\right)\right]$$

$$\text{Angle } 7 = \left[180^\circ - \left(\hat{1} + \hat{2} + \hat{8}\right)\right]$$

The remaining angles 6 and 5 are then computed from

$$\sin\hat{1}\sin\hat{3}\sin\hat{5}\sin\hat{7} = \sin\hat{2}\sin\hat{4}\sin\hat{6}\sin\hat{8}$$

thus

$$\frac{\sin\hat{5}}{\sin\hat{6}} = \frac{\sin\hat{2}\sin\hat{4}\sin\hat{8}}{\sin\hat{1}\sin\hat{3}\sin\hat{7}} = x \qquad (a)$$

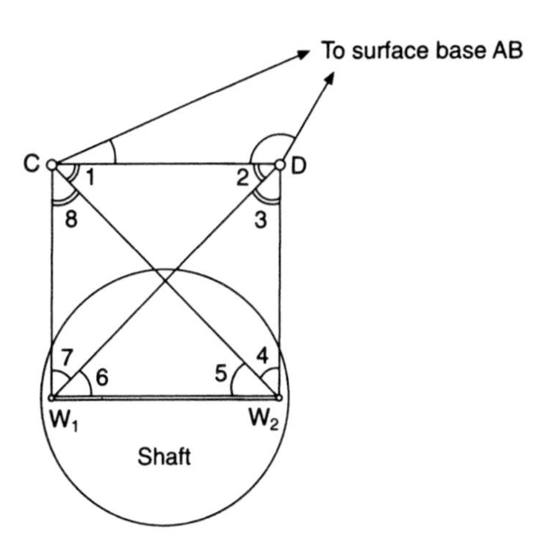

Fig. 13.5 *Weiss quadrilateral*

and $\quad \left(\hat{5}+\hat{6}\right)=\left(\hat{1}+\hat{2}\right)=\hat{y}$

$$\therefore \hat{5}=\left(\hat{y}-\hat{6}\right) \qquad (b)$$

from (a) $\sin\hat{5}=x\sin\hat{6} \quad \therefore \sin\left(\hat{y}-\hat{6}\right)=x\sin\hat{6}$

and $\quad \sin\hat{y}\cos\hat{6}-\cos\hat{y}\sin\hat{6}=x\sin\hat{6}$

from which $\quad \sin y\cot\hat{6}-\cos\hat{y}=x$

and

$$\cot\hat{6}=\frac{x+\cos\hat{y}}{\sin\hat{y}} \qquad (13.5)$$

Having found angle 6 from *equation (13.5)*, angle 5 is found by substitution in (b).
Error analysis of the observed figure indicates

(1) The best shape for the quadrilateral is square.
(2) Increasing the ratio of the length *CD* to the wire base increases the standard error of orientation.

13.2.7 Single wires in two shafts

The above methods have dealt with orientation through a single shaft, which is the general case in civil engineering. Where two shafts are available, orientation can be achieved via a single wire in each shaft. This method gives a longer wire base, and wire deflection errors are much less critical.

The principles of the method are outlined in *Figure 13.6*. Single wires are suspended in each shaft at *A* and *B* and coordinated from the surface control network, most probably by multiple intersections from as many surface stations as possible. From the coordinates of *A* and *B*, the bearing *AB* is obtained.

A traverse is now carried out from *A* to *B* via an underground connecting tunnel (*Figure 13.6(a)*). However, as the angles at *A* and *B* cannot be measured it becomes an open traverse on an *assumed* bearing

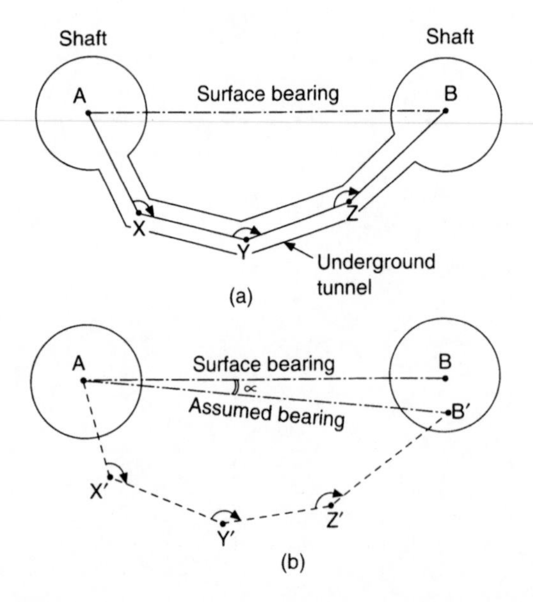

Fig. 13.6 *Alignment using single wires in two shafts*

for AX. Thus, if the assumed bearing for AX differed from the 'true' (but unknown) bearing by α, then the whole traverse would swing to apparent positions X', Y', Z' and B' (*Figure 13.6(b)*).

The value of α is the difference of the bearings AB and AB' computed from surface and underground coordinates respectively. Thus if the underground bearings are rotated by the amount α, this will swing the traverse almost back to B. There will still be a small misclosure due to linear error and this can be corrected by multiplying each length by a scale factor equal to length AB/length AB'. Now, using the corrected bearings and lengths the corrected coordinates of the traverse fitted to AB can be calculated. These coordinates will be relative to the surface coordinate system.

Alternatively, the corrected coordinates can be obtained directly by mathematical rotation and translation of AB/AB'; the corrected coordinates are obtained from

$$E_i = E_0 + K(E_i' \cos\alpha - N_i' \sin\alpha) \tag{13.6a}$$

$$N_i = N_0 + K(N_i' \cos\alpha + E_i' \sin\alpha) \tag{13.6b}$$

where E_0, N_0 = coordinates of the origin (in this case A)
E_i', N_i' = coordinates of the traverse points computed on the assumed bearing
E_i, N_i = transformed coordinates of the underground traverse points
K = scale factor (length AB/length AB')

There is no doubt that this is the most accurate and reliable method of surface-to-underground orientation. The accuracy of the method is dependent upon:

(1) The accuracy of fixing the position of the wires at the surface.
(2) The accuracy of the underground connecting traverse.

The influence of errors in the verticality of the wires, so critical in single-shaft work, is practically negligible owing to the long distance separating the two shafts. Provided that the legs of the underground traverse are long enough, then a total station could be used to achieve highly accurate surface and underground surveys, resulting in final orientation accuracies of a few seconds. As the whole procedure is under strict control there is no reason why the final accuracy cannot be closely predicted.

13.2.8 Alternatives

In all the above methods the wires could be replaced by autoplumbs or lasers.

In the case of the autoplumb, stations at the shaft bottom could be projected vertically up to specially arranged targets at the surface and appropriate observations taken directly to these points.

Similarly, lasers could be arranged at the surface to project the beam vertically down the shaft to be picked up on optical or electronic targets. The laser spots then become the shaft stations correlated in the normal way.

The major problems encountered by using the above alternatives are:

(1) Ensuring the laser beam is vertical.
(2) Ensuring correct detection of the centre of the beam.
(3) Refraction in the shaft (applies also to autoplumb).

In the first instance a highly sensitive spirit level or automatic compensator could be used; excellent results have been achieved using arrangements involving mercury pools, or photo-electric sensors.

Detecting the centre of the laser is more difficult. Lasers having a divergence of 10″ to 20″ would give a spot of 10 mm and 20 mm diameter respectively at 200 m. This spot also tends to move about due to variations in air density. It may therefore require an arrangement of photocell detectors to solve the problem.

Turbulence of air currents in shafts makes the problem of refraction less important.

Worked examples

Example 13.1 The national grid (NG) bearing of an underground base line, *CD* (*Figure 13.7*), is established by co-planning at the surface onto two wires, W_1 and W_2, hanging in a vertical shaft, and then using a Weisbach triangle underground.

The measured field data is as follows:

NG bearing *AB*	74° 28′ 34″
NG coords of *A*	E 304 625 m, N 511 612 m

Horizontal angles:

BAW_s	284° 32′ 12″
AW_sW_2	102° 16′ 18″
$W_2W_uW_1$	0° 03′ 54″
W_1W_uC	187° 51′ 50″
W_uCD	291° 27′ 48″

Horizontal distances:

W_1W_2	3.625 m
W_uW_2	2.014 m

Compute the bearing of the underground base. (KU)

The first step is to calculate the bearing of the wire base using the measured angles at the surface:

$$\begin{aligned}
\text{Grid bearing of } AB &= 74^\circ\, 28'\, 34'' \quad \text{(given)}\\
\text{Angle } BAW_s &= 284^\circ\, 32'\, 12''\\
\hline
\text{Grid bearing of } AW_s &= 359^\circ\, 00'\, 46''\\
\text{Angle } AW_sW_2 &= 102^\circ\, 16'\, 18''\\
\hline
\text{Sum} &= 461^\circ\, 17'\, 04''\\
&\ -180^\circ\\
\hline
\text{Grid bearing } W_1W_2 &= 281^\circ\, 17'\, 04''
\end{aligned}$$

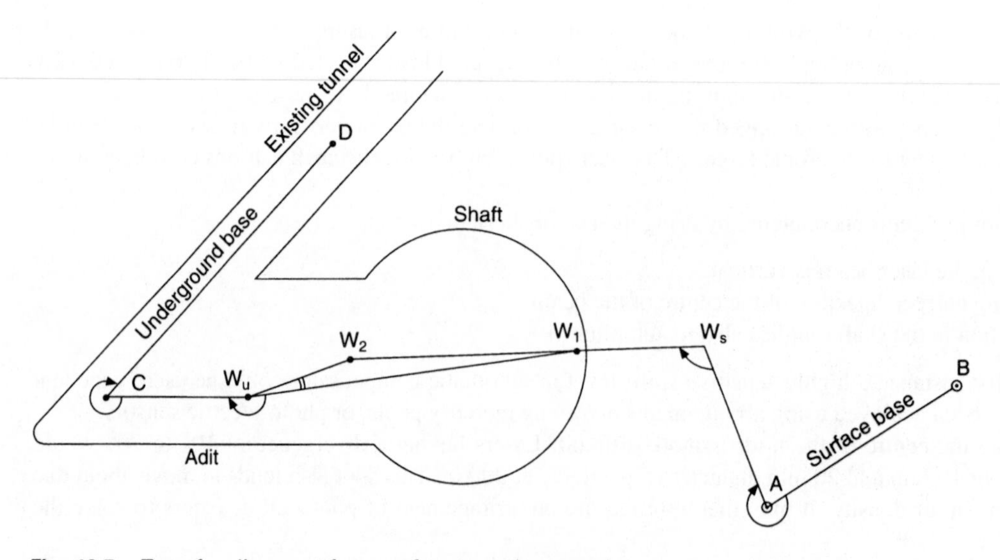

Fig. 13.7 *Transfer alignment from surface to underground*

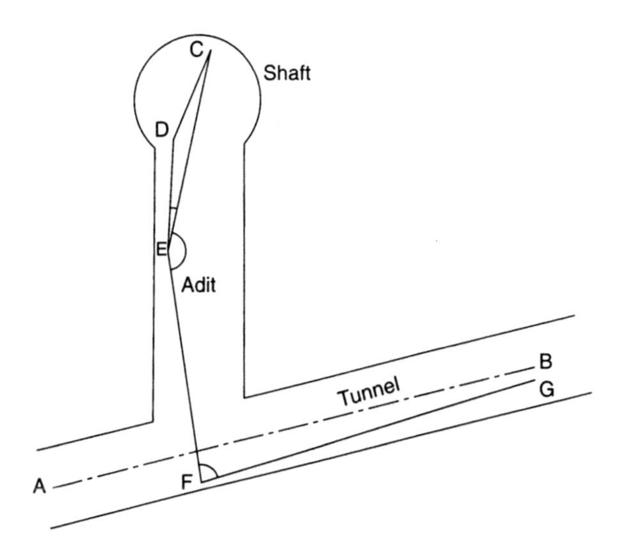

Fig. 13.8 *Tunnel, shaft and adit*

Now, using the Weisbach triangle calculate the bearing of the underground base from the wire base:

From a solution of the Weisbach triangle

$$\text{Angle } W_2W_1W_u = \frac{234'' \times 2.014}{3.625} = 130'' = 0°\,02'\,10''$$

$$\begin{aligned}
\text{Grid bearing } W_1W_2 &= 281°\,17'\,04'' \\
\text{Angle } W_2W_1W_u &= 0°\,02'\,10'' \\
\hline
\text{Grid bearing } W_1W_u &= 281°\,14'\,54'' \\
\text{Angle } W_1W_uC &= 187°\,51'\,50'' \\
\hline
\text{Sum} &= 469°\,06'\,44'' \\
& -180° \\
\hline
\text{Grid bearing } W_uC &= 289°\,06'\,44'' \\
\text{Angle } W_uCD &= 291°\,27'\,48'' \\
\hline
\text{Sum} &= 580°\,34'\,32'' \\
& -540° \\
\hline
\text{Grid bearing } CD &= 40°\,34'\,32'' \qquad \text{(underground base)}
\end{aligned}$$

Example 13.2 The centre-line of the tunnel *AB* shown in *Figure 13.8* is to be set out to a given bearing. A short section of the main tunnel has been constructed along the approximate line and access is gained to it by means of an adit connected to a shaft. Two wires *C* and *D* are plumbed down the shaft, and readings are taken onto them by a theodolite set up at station *E* slightly off the line *CD* produced. A point *F* is located in the tunnel, and a sighting is taken on to this from station *E*. Finally a further point *G* is located in the tunnel and the angle *EFG* measured.

From the survey initially carried out, the coordinates of *C* and *D* have been calculated and found to be E 375.78 m and N 1119.32 m, and E 375.37 m and N 1115.7 m respectively.

Calculate the coordinates of *F* and *G*. Without making any further calculations describe how the required centre-line could then be set out. (ICE)

Given data: $CD = 3.64$ m $\quad DE = 4.46$ m
$EF = 13.12$ m $\quad FG = 57.5$ m
Angle $DEC = 38''$
Angle $CEF = 167° 10' 20''$
Angle $EFG = 87° 23' 41''$

Solve Weisbach triangle for angle ECD

$$\hat{C} = \frac{ED}{DC}\hat{E} = \frac{4.46}{3.64} \times 38'' = 47''$$

By coordinates

$$\text{Bearing of wire base } CD = \tan^{-1}\frac{-0.41}{-3.62} = 186° 27' 19''$$

$$\begin{aligned}
\therefore \text{WCB of } CE &= 186° 27' 42'' - 47'' = 186° 26' 55'' \\
\text{WCB of } CE &= 186° 26' 55'' \\
\text{Angle } CEF &= 167° 10' 20'' \\
\text{WCB of } EF &= 173° 37' 15'' \\
\text{Angle } EFG &= 87° 23' 41'' \\
\text{WCB of } FG &= 81° 00' 56''
\end{aligned}$$

Line	*Length* (m)	*WCB*	*Coordinates*		*Total coordinates*		
			ΔE	ΔN	*E*	*N*	*Station*
					375.78	1119.32	*C*
CE	8.10	186° 26′ 55″	−0.91	−8.05	374.87	1111.27	*E*
EF	13.12	173° 37′ 15″	1.46	−13.04	376.33	1098.23	*F*
FG	57.50	81° 00′ 56″	56.79	8.99	433.12	1107.22	*G*

Several methods could be employed to set out the centre-line; however, since bearing rather than coordinate position is critical, the following approach would probably give the best results.

Set up at *G*, the bearing of *GF* being known, the necessary angle can be turned off from *GF* to give the centre-line. This is obviously not on centre but is the correct line; centre positions can now be fixed at any position by offsets.

Example 13.3 Two vertical wires *A* and *B* hang in a shaft, the bearing of *AB* being 55° 10′ 30″ (*Figure 13.9*). A theodolite at *C*, to the right of the line *AB* produced, measured the angle *ACB* as 20′ 25″. The distances *AC* and *BC* were 6.4782 m and 3.2998 m respectively.

Calculate the perpendicular distance from *C* to *AB* produced, the bearing of *CA* and the angle to set off from *BC* to establish *CP* parallel to *AB* produced.

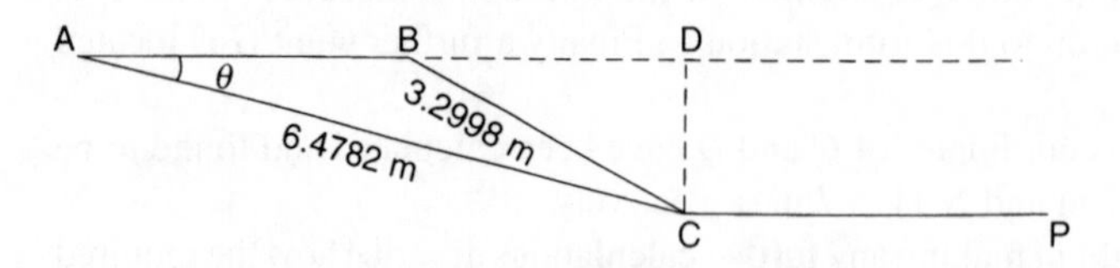

Fig. 13.9 *Wires in a shaft*

Describe how you would transfer a line AB above ground to the bottom of a shaft. (LU)

$$AB \approx AC - BC = 3.1784 \text{ m}$$

$$\text{Angle } BAC = \frac{3.2998}{3.1784} \times 1225'' = 1272'' = 21'\,12'' = \theta$$

$$\text{By radians } CD = AC \times \theta \text{ rad} = \frac{6.4782 \times 1272}{206\,265} = 0.0399 \text{ m}$$

$$\begin{aligned} \text{Bearing } AB &= 55^\circ\,10'\,30'' \\ \text{Angle } BAC &= 21'\,12'' \\ \text{Bearing } AC &= 55^\circ\,31'\,42'' \\ \therefore \text{Bearing } CA &= 235^\circ\,31'\,42'' \end{aligned}$$

$$\begin{aligned} \text{Angle to be set off from } BC = ABC &= 180^\circ - (21'\,12'' + 20'\,25'') \\ &= 179^\circ\,18'\,23'' \end{aligned}$$

13.3 GYRO-THEODOLITE

An alternative to the use of wire methods is the gyro-theodolite. This is a north-seeking gyroscope integrated with a theodolite, and can be used to orientate underground base lines relative to true north.

There are two main types of suspended gyroscope currently available, the older Wild GAK1 developed in the 1960s–70s, which requires careful manual handling and observing to obtain orientation and the more modern Gyromat 3000 which is very much more automated. There are still many GAK1s available on the second-hand market. The gyroscope will be explained with respect to the Wild GAK1 and at the end of this section the Gyromat 3000 will be described. The essential elements of the suspended gyro-theodolite are shown in *Figure 13.10*.

A theodolite is an instrument that enables the user to observe the difference in bearing, i.e. the angle, between two distant stations. Although angles are observed, it is often a bearing (relative to grid north) or azimuth (relative to true north) which is actually required. The suspended gyroscope is a device that may be attached to a theodolite to allow observations of azimuth rather than angle, to be taken, for example, to check an unclosed traverse in mining or tunnelling work.

A gyroscopic azimuth is the azimuth determined with a gyrotheodolite. If the gyrotheodolite has been calibrated on a line of known astronomic azimuth then the gyroscopic azimuth is effectively the same as the astronomic azimuth because astronomic and gyroscopic north are both defined in terms of the local vertical and the instantaneous Earth rotation axis.

13.3.1 GAK1 gyro attachment

Before the GAK1 can be used, a special mount or bridge must be fixed by the manufacturer to the top of the theodolite. The gyro attachment fits into the bridge so that three studs on the base of the gyro fit into the three grooves in the bridge with the gyro scale viewed from the same position as the theodolite on face left. *Figure 13.11*.

The GAK1 consists of a spinner that is mounted in a mast, which in turn is suspended by a fine wire, or tape, from a fixed point near the top of the gyro frame. The spinner is simply a cylinder of metal mounted on an axle which, in turn, is held by the mast. The mast is merely a carriage for the spinner. The spinner is driven by a small electric motor at a design angular rate of 22 000 rpm. To avoid damage during transit

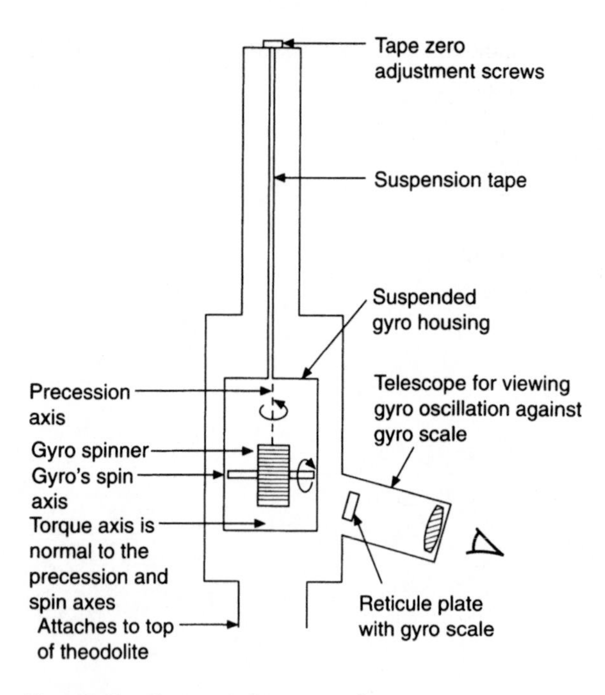

Fig. 13.10 *Suspended gyro attachment*

and during running up and slowing down of the spinner, the mast is held against the body of the gyro housing by a clamp at the gyro base. When the gyro is operating the clamp is released and the mast and spinner as a complete unit hang suspended. This system then oscillates slowly about its vertical axis. The amount of movement is detected by observing the shadow of a mark in a part of the optical train in the mast. The shadow is projected onto a ground glass scale that may be read directly. *Figure 13.11* shows a GAK1 mounted on a Wild T2 theodolite.

The scale is viewed through a detachable eyepiece. The scale is centred at zero with divisions extending from +15 on the left to −15 on the right (*Figure 13.12*(*b*)). The position of the moving mark may only be estimated to the nearest 0.1 of a division at best. Alternatively the scale may be viewed from the side of the eyepiece. This allows an instructor to monitor the observations of a student or two observers to work with the same gyro and observations.

When the observer uses the side viewing eyepiece the image is reversed. The precision of reading may be improved with the aid of a parallel plate micrometer attachment that allows coincidence between the image of the moving mark and a scale division to be achieved.

The position of the moving mark is found as the algebraic sum of the integer number of scale divisions, plus the micrometer scale reading, e.g. if the scale reading is −5 divisions and the micrometer reading +0.50 divisions, the result is −4.5 divisions.

The gyro motor is powered through a converter that contains a battery. The gyro may be powered by the converter's internal battery or by an external battery. The converter ensures that there is a stable power supply even when the external power supply or internal battery voltage starts to run down.

To run up the gyro, first make sure that the gyro is clamped up to ensure that no damage is done to the gyro mechanism during acceleration of the spinner. Next, see that the external power supply is correctly connected. Turn the switch to 'run'. The 'measure' display turns from green to white and the 'wait' display turns from white to red. When the spinner is running at full speed the displays on the converter change and the gyro is ready for use.

Fig. 13.11 *GAK1 mounted on Wild T2*

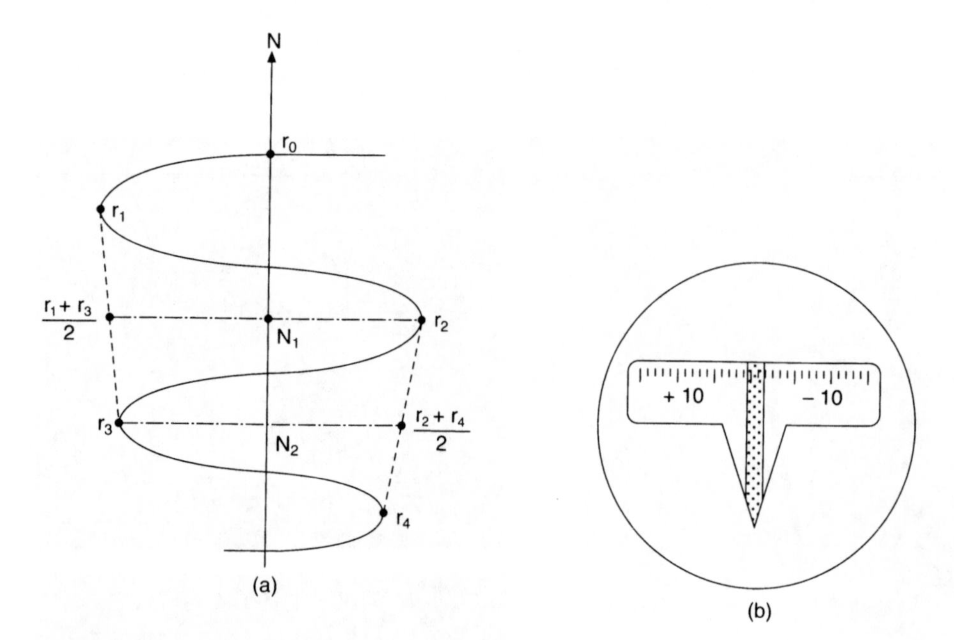

Fig. 13.12 *(a) Gyro precession, (b) gyro scale*

Now very carefully unclamp the mast so that the gyro hangs suspended only by the tape. This is a tricky operation requiring a steady controlled hand and a little bit of luck, to get a satisfactory drop without wobble or excessive swing. The clamp should be rotated until it meets the stop. A red line is now visible on the clamp. Pause for a few seconds to allow any unwanted movement to die down and then lower the clamp. The gyro is now supported only by the tape and is free to oscillate about its own vertical axis. Observations may now be made of the gyro scale.

If the gyro is badly dropped then there may be an excessive 2 Hz wobble which will make observations difficult and inaccurate. If this happens then clamp up and try again. Alternatively, wait a little while and the wobble will decay exponentially.

Also, with a bad drop, the moving mark may go off the scale; if this happens re-clamp the gyro and try again. Do not allow the moving mark to go off the scale as this may damage the tape, and anyway, no observations can be made.

When observations with the spinning gyro are complete the gyro must be clamped up and the spinner brought to rest.

13.3.2 Basic equations

There are two basic equations that govern the behaviour of the gyrotheodolite. The first is concerned with the motion of the moving mark as it appears on the gyro scale and in particular the midpoint of that motion. The second is concerned with the determination of north from the observed midpoints of swing of the moving mark on the gyro scale and other terms. The equations will be stated here, justified below and then applied.

The midpoint of motion of the moving shadow mark may be found from the equation of motion of the moving mark. This is an equation of damped harmonic motion:

$$Me^{-Dt}\cos(\omega(t-p)) + K - y = 0 \qquad (13.7)$$

where: M = magnitude of the oscillation
e = the base of natural logarithms = 2.718 281 8
D = damping coefficient
t = time
ω = frequency of the oscillation
p = time at a 'positive turning point', see below
K = midpoint of swing as measured on the scale
y = scale reading of the moving mark

The second equation, or north finding equation, finds the reading on the horizontal circle of the theodolite that is equivalent to north. It relates the K of *equation (13.7)*, determined when the spinner is spinning and also, separately, when the spinner is not spinning. The equation is:

$$N = H + sB(1 + C) - sCd + A \tag{13.8}$$

where: N = the horizontal circle reading of the theodolite equivalent to north
H = the fixed horizontal circle reading when the theodolite is clamped up ready for observations of the gyro
s = the value of one scale unit in angular measure
B = the centre of swing in scale units when the spinner is spinning
C = the 'torque ratio constant'
d = the centre of swing in scale units when the spinner is not spinning
A = the additive constant

Like astronomical north, the north determined by a gyro is also defined by gravity and by the Earth's rotation.

13.3.3 The equation of motion

The equation of motion will first be established rigorously and in fairly advanced mathematical terms. The effect will then be described again but in a rather looser and more physical sense.

Figure 13.13(a) shows the axis of the gyro spinner placed with respect to the surface of the Earth. A set of orthogonal axes is established with the origin at O, the i direction in the direction of gravity, the j and k directions are in the local horizontal plane with the k direction making a small angle, α, with the local direction of north. j is at right angles to i and k such that the axes form a right-handed orthogonal set. ϕ is the latitude of O. $\boldsymbol{i}$, $\boldsymbol{j}$ and $\boldsymbol{k}$ are unit vectors in the directions i, j and k. $\boldsymbol{\Omega}$ is the rotational velocity vector of the Earth.

The velocity vector of the earth may be broken into its component parts in the i, j and k directions so:

$$\boldsymbol{\Omega} = \Omega(\boldsymbol{i} \sin\phi - \boldsymbol{j} \cos\phi \sin\alpha + \boldsymbol{k} \cos\phi \cos\alpha) \tag{13.9}$$

If this triad of vectors is now rotated about the i axis with an angular velocity $-\dot{\alpha}$ then the angular velocity of the axes may be described in vector terms by

$$\boldsymbol{\omega}_a = -\dot{\alpha}\boldsymbol{i} \tag{13.10}$$

Substituting *equation (13.9)* into *equation (13.10)* gives

$$\boldsymbol{\omega}_a = \boldsymbol{i}(\Omega \sin\phi - \dot{\alpha}) - \boldsymbol{j}\Omega \cos\phi \sin\alpha + \boldsymbol{k} \cos\phi \cos\alpha \tag{13.11}$$

If the spinner is now placed so that it rotates about the k axis with an angular velocity of n, then the total angular velocity of the spinner is

$$\boldsymbol{\omega} = \boldsymbol{i}(\Omega \sin\phi - \dot{\alpha}) - \boldsymbol{j}\Omega \cos\phi \sin\alpha + \boldsymbol{k}(\Omega \cos\phi \cos\alpha + n)$$

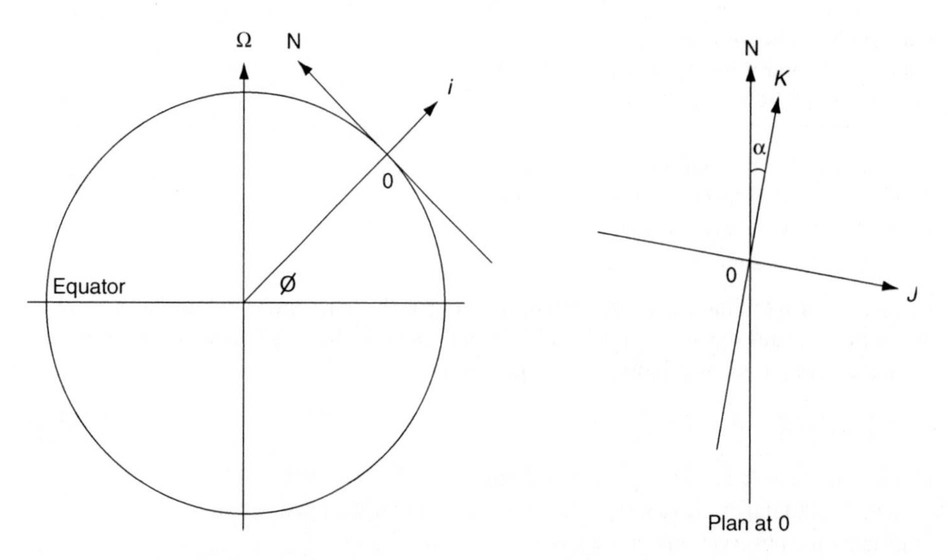

Fig. 13.13 *(a) Meridian section through the earth, (b) Plan view at 0*

If A is the moment of inertia of the spinner about the i or j axis (they will be the same because of rotational symmetry) and B is the moment of inertia of the spinner about the k axis then the total angular momentum of the spinner $\mathbf{L}$ is

$$\mathbf{L} = \boldsymbol{i}A(\Omega \sin \phi - \dot{\alpha}) - \boldsymbol{j}A(\Omega \cos \phi \sin \alpha) + \boldsymbol{k}B(\Omega \cos \phi \cos \alpha + n) \tag{13.12}$$

A couple (a twisting force) is now applied about the j axis to stop the spinner axis from wandering out of the horizontal plane. The magnitude of the couple is M so the couple is fully described by $M\boldsymbol{j}$, where

$$M\boldsymbol{j} = \dot{\boldsymbol{L}} \tag{13.13}$$

But from the theory of rotating axes $\dot{\boldsymbol{L}} = \dot{\mathbf{L}} + \boldsymbol{\omega}_{\mathrm{a}} \times \mathbf{L}$ (13.14)

where: $\dot{\boldsymbol{L}}$ = rate of change of momentum of the spinner in the ijk coordinate system
$\dot{\mathbf{L}}$ = absolute rate of change of angular momentum
$\boldsymbol{\omega}_{\mathrm{a}}$ = absolute angular velocity of the ijk coordinate system
$\times$ = the vector cross product symbol
$\mathbf{L}$ = the absolute angular momentum of the spinner

If values from *equations (13.11)*, *(13.12)* and *(13.13)* are now substituted into *equation (13.14)* then

$$M\boldsymbol{j} = -\boldsymbol{i}A\ddot{\alpha} + \boldsymbol{i}[(-\Omega \cos \phi \sin \alpha)B(\Omega \cos \phi \cos \alpha + n) + (\Omega \cos \phi \cos \alpha)A(\Omega \cos \phi \sin \alpha)]$$
$$+ \boldsymbol{j}\,(\text{other terms}) + \boldsymbol{k}\,(\text{other terms}) \tag{13.15}$$

Only the terms in $\boldsymbol{i}$ in *equation (13.15)* are now considered

$$A\ddot{\alpha} = -\Omega \cos \phi \sin \alpha Bn + \Omega^2(A - B)\cos^2 \phi \cos \alpha \sin \alpha \tag{13.16}$$

The Earth rotates once every day so $\Omega = 0.000\,0729$ radians/sec and the spinner in the GAK1 is kept at a nominal angular velocity of 22 000 rpm so $n \approx 2300$ radians/sec therefore $\Omega \approx 3.210^{-8}\,n$ so the second half of the right hand side of *equation (13.16)* is negligible compared with the first and so the equation may be simplified to

$$\ddot{\alpha} = -BA^{-1}n\Omega \cos \phi \sin \alpha \tag{13.17}$$

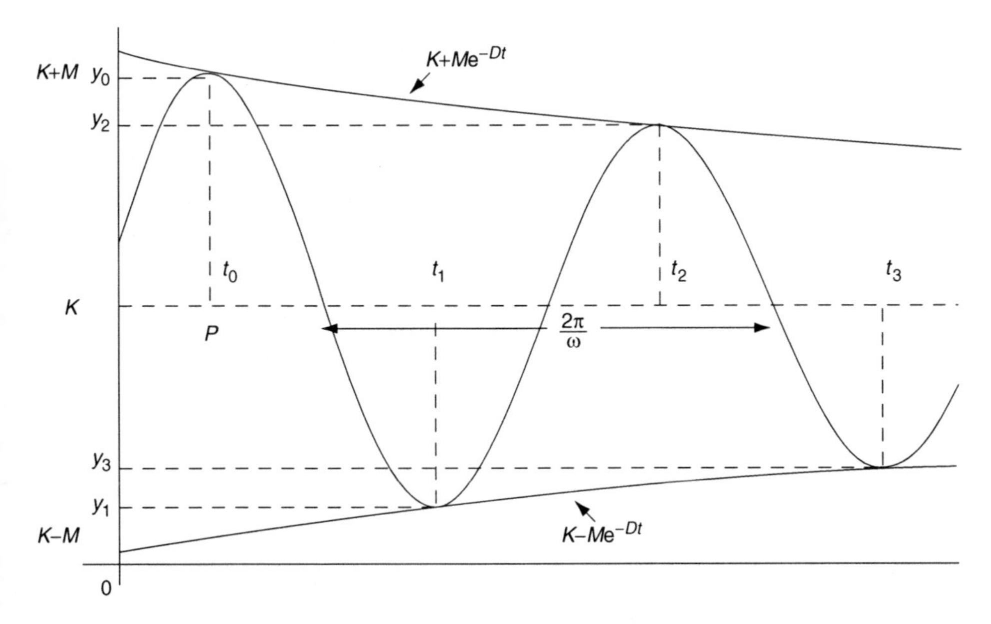

Fig. 13.14 *Damped sinusoidal oscillation of the moving mark*

If α is small, then, to a first approximation, $\alpha = \sin\alpha$ so that *equation (13.17)* becomes

$$\ddot{\alpha} = -\alpha BA^{-1}n\Omega\cos\phi \tag{13.18}$$

which is simple harmonic motion of period $= 2\pi A^{\frac{1}{2}}(Bn\Omega\cos\phi)^{-\frac{1}{2}}$ and the solution of *equation (13.18)*, a second order differential equation is

$$\alpha = M\cos(\omega(t-p)) \tag{13.19}$$

where $\omega^2 = BA^{-1}n\Omega\cos\phi$ and M and p are constants of integration. For those unfamiliar with differential equations the solution may be verified by differentiating *equation (13.19)* twice with respect to time. In practice friction and air resistance lightly damps the oscillation so a damping factor e^{-Dt} is introduced. Also, the actual centre of oscillation is offset from the scale zero by an amount K so that the final equation of motion is *equation (13.7)*.

The terms in the above equation are shown graphically in *Figure 13.14*. Notice that the moving mark oscillates about the centre of its swing K by an amount M when time equals 0. As time increases the magnitude of the swing reduces to Me^{-Dt} at time t. If the system were allowed to swing for ever, the magnitude of the oscillation would eventually decay to 0 and the moving mark would come to rest at K. The practical problem is to find the observations that can be made of the moving mark such that the value of K may be computed without having to wait, literally, for ever. This problem will be addressed later.

The behaviour of the spinner may also be considered in more descriptive terms. See *Figure 13.15*. If a spinner is set in motion, then, provided that there are no forces applied to the axle, the spinner will continue to rotate with the spinner axis maintaining the same direction in inertial space, that is, with respect to the stars. The spinner of a gyro, however, is suspended from a point which in turn is rotating with respect to the rotation axis of the Earth. In other words the direction of gravity, in inertial terms, changes as the earth completes one revolution on its own axis every 24 hours. If at the same time the mast has a rotation about the local vertical, the spinner axis continues to point to north. If, also, the spinner axis is not pointing to north when the mast is released then, as the earth turns, the apparent direction of gravity, as viewed from the spinner, will vary. This will have a component in the plane containing the axis of the spinner and

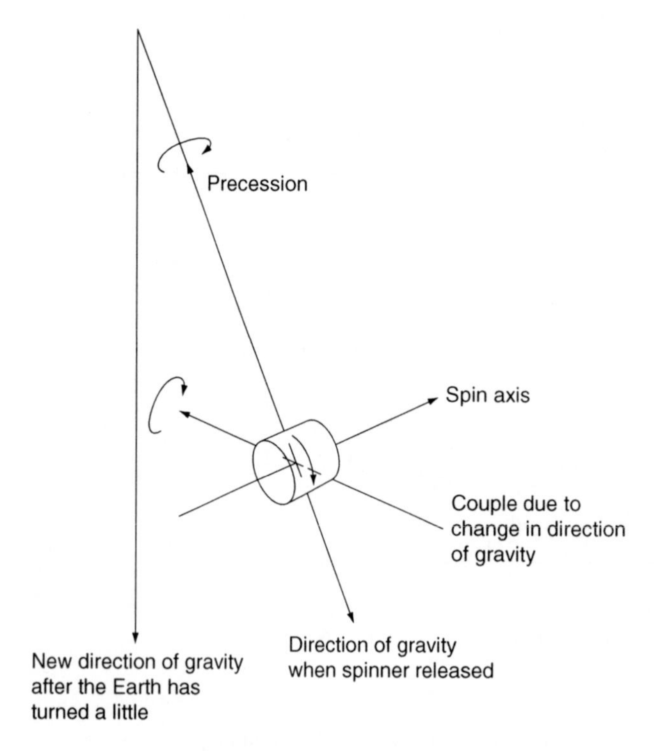

Fig. 13.15 *Motion of the spinner*

the point of suspension. There is, therefore, a twisting force about the ends of the spinner axis, which is mutually at right angles to the spinner axis and the direction of gravity.

If a couple is applied to the axis of the spinner then the result is a precessional torque about the axis which is at right angles to the plane containing the couple and spin axis. That is, the direction of the precessional torque rotation vector is defined by the vector cross product of the spin and the couple rotation vectors. It is possible to verify this by removing a wheel from a bicycle and holding it between outstretched hands. Get someone to spin the wheel as fast as possible and then try to turn around and face the opposite direction. By turning round you are applying a couple about the ends of the axle. The result is that the wheel tips over because of the precessional torque. Care must be taken to avoid personal injury. In this situation consider the axes of spin, couple and precessional torque and see that they are mutually at right angles. Similar effects may be seen with a child's toy gyroscope.

As the system of the mast and spinner are in precession the couple changes direction when the axis of the spinner passes through north and the precessional torque becomes applied in the opposite direction. The overall result is that the mast and spinner axis is always accelerating towards the local direction of north.

13.3.4 North finding equation

In *Figure 13.16* points of interest on the scale of the GAK1 Gyroscope are described. Scale zero is the centre of the scale. The moving shadow mark, as seen on the gyro scale, is at position y as measured on the gyro scale. In *Figure 13.16* it is shown as a double line because that is how it appears on the gyro scale. d is the position that the shadow mark would take if the mast with the spinner, not spinning, was allowed to hang until the mast had stopped moving. Ideally this would be at the scale zero position, but because the system cannot be adjusted that precisely it is unlikely to be so. B is the position that the shadow mark

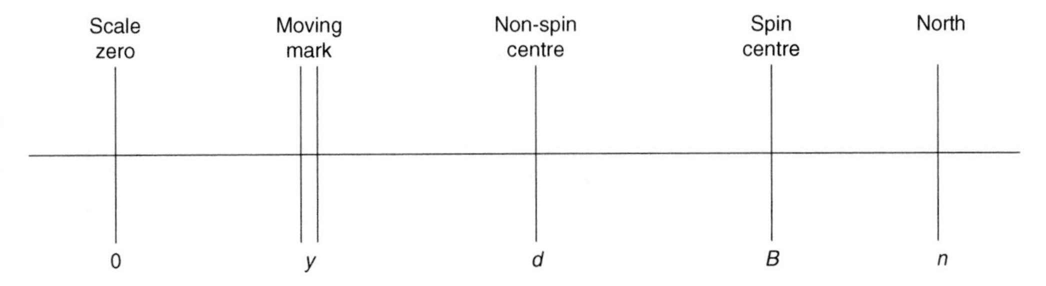

Fig. 13.16 *Points of interest on the gyro scale*

would take if the mast with the spinner, spinning, was allowed to hang until the mast had stopped moving. Finally n is the position of true north on the gyro scale. All these terms, y, d, B, and n, are referenced to the visible scale in the gyro and therefore are in scale units.

In *equation (13.8)*, H was the fixed horizontal circle reading when the theodolite was clamped up ready for observations of the gyro and s was the value of one scale unit in angular measure, then:

$$N = H + sn \tag{13.20}$$

At any one time the torque on the gyro is the sum of the torque due to precession and the torque due to the twisting of the suspending tape.
The torque due to precession $= k(n - y)s$ where k = torque per arc second of rotation due to precession.
The torque due to twisting $= k_1(d - y)s$ where k_1 = torque per arc second of rotation due to twisting.
The total torque is the sum of these two separate torques so that:

$$\text{Total torque} = k(n - y)s + k_1(d - y)s$$

and on rearranging becomes:

$$\text{Total torque} = s(k + k_1)\left[\frac{kn + k_1 d}{(k + k_1)} - y\right] \tag{13.21}$$

At the centre of the oscillation, i.e. when $y = B$, the total torque is zero, so, from *equation (13.21)*:

$$\frac{kn + k_1 d}{(k + k_1)} - B = 0$$

$$\text{so} \quad B = \frac{kn + k_1 d}{(k + k_1)} = \frac{n + Cd}{1 + C} \tag{13.22}$$

where $C = k_1/k$ is the ratio of the torque due to precession and the torque due to the twisting of the tape. C is usually referred to as the 'torque ratio constant'. *Equation (13.22)* may be rearranged to get

$$n = B(1 + C) - Cd$$

and this may be substituted into *equation (13.20)* to get

$$N = H + sB(1 + C) - sCd$$

A further term, A, is added to the equation to allow for the fact that the interface between the gyro and the theodolite is not perfectly matched. See *equation (13.8)* This may be thought of as a calibration constant or scale zero error.

13.3.5 The practical solution of the equation of motion

The equation of motion, *equation (13.7)*, is exactly the same whether the spinner is spinning or not. The only difference is in the values of the terms in the equation. When the spinner is spinning then the term K is the same as B in *equation (13.8)*. When the spinner is not spinning then the term K is the same as d in *equation (13.8)*. Observations may be made, either of time as the moving shadow mark passes scale divisions, or of the extent of the swing of the moving mark on the gyro scale. This is known as a turning point as it is the point at which the moving mark changes direction. In this latter case precise observations can only be made if the GAK1 is fitted with a parallel plate micrometer that allows the image of the moving shadow mark to be made coincident with a scale division. The reading on the parallel plate micrometer is then algebraically added to the value of the observed scale division. There are a number of methods available for finding K (i.e. B or d). Two are presented below.

13.3.5.1 Amplitude method

The scale value is observed at three or four successive turning points. With the motion of the moving mark slightly damped by hysteresis in the tape and air resistance, the damped harmonic motion of the moving mark is of the form described by *equation (13.7)* which can be rearranged as:

$$y = K + Me^{-Dt}\cos(\omega(t - p)) \tag{13.23}$$

In *Figure 13.17* the times t_0, t_1, t_2 and t_3 are successively half the period of the oscillation apart so that:

$$t_1 = t_0 + \frac{\pi}{\omega} \qquad t_2 = t_0 + \frac{2\pi}{\omega} \qquad t_3 = t_0 + \frac{3\pi}{\omega}$$

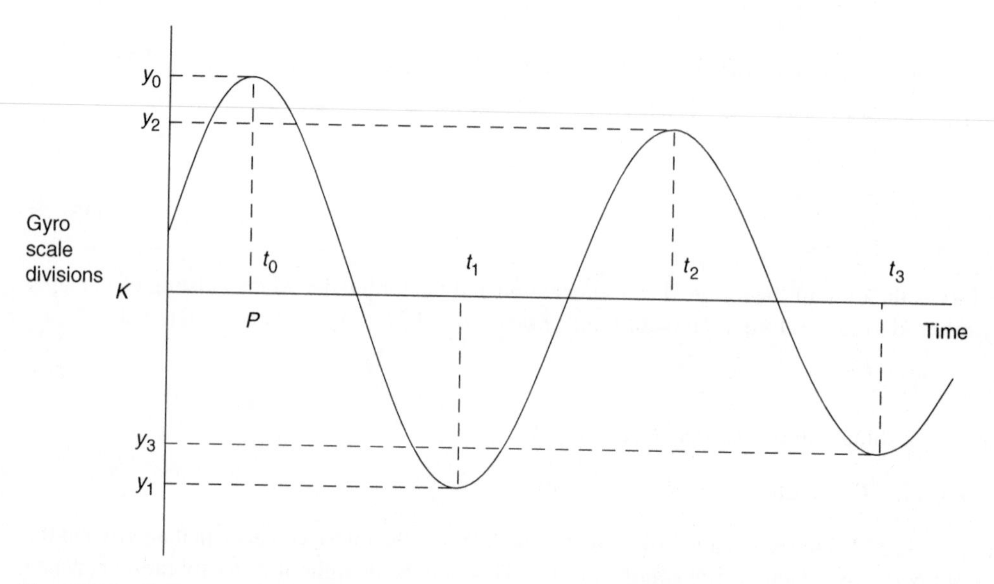

Fig. 13.17 *Amplitude method*

and so at successive turning points the scale readings will be:

$$y_0 = K + M e^{-Dt_0}\cos(\omega t_0 - \omega p) \qquad y_1 = K + M e^{-D(t_0+\pi/\omega)}\cos(\omega t_0 + \pi - \omega p) \tag{13.24}$$

$$y_2 = K + M e^{-D(t_0+2\pi/\omega)}\cos(\omega t_0 + 2\pi - \omega p) \quad y_3 = K + M e^{-D(t_0+3\pi/\omega)}\cos(\omega t_0 + 3\pi - \omega p)$$

If $\quad \cos(\omega t_0 - \omega p) = k \quad$ then $\quad \cos(\omega t_0 + \pi - \omega p) = -k$

$$\cos(\omega t_0 + 2\pi - \omega p) = k \qquad \cos(\omega t_0 + 3\pi - \omega p) = -k$$

and if $\ e^{-Dt_0} = g$ then *equation (13.34)* may be rewritten as

$$y_0 = K + Mgk \qquad y_1 = K + M e^{-D\pi/\omega} gk \tag{13.25}$$

$$y_2 = K + M e^{-2D\pi/\omega} gk \qquad y_3 = K + M e^{-3D\pi/\omega} gk$$

In practice D is in the region of 0.000 004 so the term $D\pi/\omega$ is about 10^{-3} $\sec^{-1}$ because, at UK latitudes the period is about 460 seconds when the spinner is spinning. Also $e^x = 1 + x + x^2/2! + x^3/3! + \cdots$. If the weighted mean of the scale readings is computed as follows:

$$\text{'Mean'} = \frac{y_0 + 2y_1 + y_2}{4}$$

then in terms of *equation (13.35)*:

$$\text{'Mean'} = \frac{y_0 + 2y_1 + y_2}{4} = K + \frac{3Mgk(D\pi)^2}{4\omega} + \cdots \tag{13.26}$$

but since, in practice, g and k are approximately 1, and in practice M will not be more than 9000″, then the error in the determination of K, as the midpoint of swing, using *equation (13.26)*, will be in error by

$$\frac{3Mgk(D\pi)^2}{4\omega} \quad \text{which is approximately } 0.006''$$

This is far less than the observational error. This simple weighted mean is perfectly acceptable for the determination of K when the spinner is spinning ($K = B$), but is not precise enough when the spinner is not spinning ($K = d$). This is because the damping coefficient is much larger. In this case the 'mean' is determined from four turning points as follows:

$$\text{'Mean'} = \frac{y_0 + 3y_1 + 3y_2 + y_3}{8} \tag{13.27}$$

To be successful the amplitude method requires an element of skill and experience. The observer must have light and nimble fingers to make the observations with the micrometer while not disturbing the smooth motion of the suspended gyroscope. The slightest pressure on the gyro casing may set up a 2 Hz wobble on the shadow mark making reading difficult and inaccurate. For ease of reading, at the next turning point, the micrometer should be returned to zero. This is easily forgotten especially in the nonspin mode when there will be at most 25 seconds before the next reading.

13.3.5.2 Transit method

In the transit method the observations are of time. The effect of the damping term is assumed to be negligible. Four observations of time are made as the moving shadow mark crosses specific divisions on the gyro scale. The parallel plate micrometer must of course be set to zero throughout these observations. Three of the times, t_0, t_1 and t_2, are when the moving shadow mark passes a scale division, y_0, near the midpoint of swing and the other, t_r, is when the moving shadow mark passes a scale division, y_r, near a turning point.

To do this in practice, first drop the mast, and see the approximate extent of the swing in both the positive and negative directions. Find the scale division nearest the centre of swing and a scale division close to the turning point. This division must not be so close to the turning point that there is a chance that the moving mark's swing might decay, so that the moving mark does not go far enough on subsequent swings. The first part of the swing may be to either side of the 0 scale division.

Having chosen the scale divisions which are to be y_0 and y_r, wait until the moving mark approaches the y_0 scale division in the direction of y_r. Take the time t_0. Next, time the moving mark as it crosses the y_r scale division, t_r. It does not matter whether the moving mark is on its way out, or on the return towards y_0 for this time. The formulae, to be seen later, are valid for both cases. The final two times, t_1 and t_2, are as the moving mark successively crosses y_0. See *Figure 13.18*.

Without damping, the equation of motion is:

$$y = K + M\cos(\omega(t - p))$$

where, from inspection of the diagram, it can be seen that:

$$\omega = \frac{2\pi}{t_2 - t_0} \text{ and} \tag{13.28}$$

$$p = \frac{t_0 + t_1}{2} \tag{13.29}$$

so that at times t_0 and t_r the equations become:

$$y_0 = K + M\cos(\omega(t_0 - p)) \tag{13.30}$$

$$y_r = K + M\cos(\omega(t_r - p)) \tag{13.31}$$

Now subtract *equation (13.31)* from *(13.30)* and making M the subject of the equation leads to:

$$M = \frac{y_0 - y_r}{\cos(\omega(t_0 - p)) - \cos(\omega(t_r - p))}$$

Substituting this back into *equation (30)* leads to:

$$K = y_0 - \frac{(y_0 - y_r)\cos(\omega(t_0 - p))}{\cos(\omega(t_0 - p)) - \cos(\omega(t_r - p))} \tag{13.32}$$

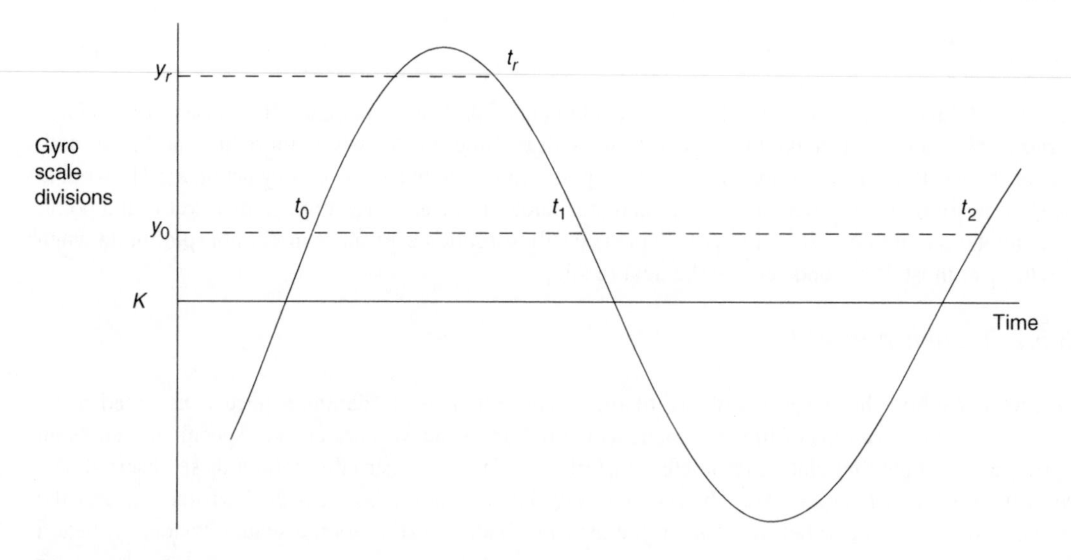

Fig. 13.18 *Transit method*

Substituting for ω and p from *equations (13.28)* and *(13.29)* into *equation (13.32)* leads to:

$$K = y_0 - \left[\frac{(y_0 - y_r)\cos\left(\dfrac{\pi(t_0 - t_1)}{t_2 - t_0}\right)}{\cos\left(\dfrac{\pi(t_0 - t_1)}{t_2 - t_0}\right) - \cos\left(\dfrac{\pi(2t_r - t_0 - t_1)}{t_2 - t_0}\right)} \right]$$

Although the above formula may look complicated it may be broken down into a computing routine as follows:

$$t_x = (t_0 - t_1)$$

$$t_y = (t_2 - t_0)$$

$$t_z = (2t_r - t_0 - t_1)$$

$$c_1 = \cos\left(\frac{180t_x}{t_y}\right) \quad (13.33)$$

$$c_2 = \cos\left(\frac{180t_z}{t_y}\right)$$

$$K = y_0 - \frac{(y_0 - y_r)c_1}{c_1 - c_2}$$

Note that the 180 in the c_1 and c_2 terms is to allow the operating mode of the calculator to be in degrees rather than radians.

The advantage of this method is that the observer does not need to touch the instrument during observations as all observations are of time. This is a considerable advantage as the gyro is a very sensitive instrument and the motion of the moving shadow mark will easily be upset by the slightest disturbance.

A further improvement of this method is to take the time at every scale division and use all the data in a least squares solution. Clearly taking each time with a stopwatch every few seconds is not practical but if the swing is recorded on video tape and played back to identify the frames where the moving mark passes each scale division then, at 25 frames per second, each time can be recorded to the nearest 0.02 seconds.

13.3.6 The practical solution of the north finding equation

To solve the north finding equation, *equation (13.8)*, each of the terms must be solved for.

In the above section, on the practical solution of the equation of motion, two methods were presented for the solution of B and d, the scale readings for the centres of swing in the spin and nonspin modes respectively. In practice d will be determined twice, before and after the determinations of B. The two solutions might vary a little because of heating effects in the gyro and stretching and twisting of the tape. If the two determinations of the midpoint of swing in the nonspin mode are d_1 and d_2 then the value of d is taken as:

$$d = \frac{d_1 + 3d_2}{4} \quad (13.34)$$

The weighting in *equation (13.34)* is purely arbitrary but reflects best experience.

To find the torque ratio constant, C, two sets of observations are made with the theodolite pointing a little either side of north. *Equation (13.8)* may be applied to each set of observations as follows:

$$N = H_1 + sB_1(1 + C) - sCd + A \quad \text{one side of north} \quad (13.35)$$

$$N = H_2 + sB_2(1 + C) - sCd + A \quad \text{the other side of north} \quad (13.36)$$

If *equation (13.36)* is taken from *equation (13.35)* then:

$$0 = (H_1 - H_2) + s(B_1 - B_2)(1 + C)$$

and so, on making C the subject of the equation:

$$C = \frac{(H_1 - H_2)}{s(B_2 - B_1)} - 1 \qquad (13.37)$$

Therefore, to determine the torque ratio constant for the instrument, the horizontal circle reading of the theodolite is taken for each pointing of the theodolite, a little east and a little west of north, when the observations for the determination of the midpoint of swing in the spin mode are made. Once the torque ratio constant C has been found then the value will hold for all observations over a limited latitude range of about $10'$ of arc, or about 20 km in a north–south direction. Alternatively if C is observed and computed, C_1 at latitude ϕ_1, its value, C_2 at latitude ϕ_2, may be found from:

$$C_2 = C_1 \frac{\cos \phi_1}{\cos \phi_2}$$

There is no means of determining s, the value of one division of the gyro scale in angular measure. The GAK1 handbook states that it is $10'$ or $0.19^g (=615.6'')$. $600''$/div is therefore taken as a working, though approximate value.

The additive constant, A, is found by calibrating the gyro on a line of known bearing. The bearing should be an astronomical azimuth, and if the gyro is subsequently to be used on stations where projection coordinates are known or are to be found then corrections for convergence and '$t - T$' must also be applied.

13.3.7 Practical observations

Firstly, an approximate direction of north is required. This may be obtained from known approximate coordinates of the instrument and another position, from a good magnetic compass provided that the local magnetic deviation and individual compass error are well known, an astronomic determination of north, or most easily from the use of the gyro in the 'unclamped method'.

In the unclamped method the mast is dropped in the spin mode. Instead of the theodolite remaining clamped and the moving mark being observed, the theodolite is unclamped and the observer attempts to keep the moving mark on the 0 division by slowly rotating the theodolite to follow the moving mark. At the full extremities of this motion the theodolite is quickly clamped up and the horizontal circle is read. The mean of two successive readings gives a provisional estimate of north.

If the difference between the two readings is greater than a few degrees it is best to repeat the process starting with the theodolite at the previous best estimate of north before dropping the mast. The precision of this as a method of finding north is, at best, a few minutes of arc. That is provided that a steady hand is used, the moving mark is kept strictly on the 0 division and that the value of d, the centre of swing on the gyro scale when the spinner is not spinning, is strictly 0. The whole process takes about 10 to 15 minutes.

Having achieved pre-orientation, the user is then ready to make sufficient observations to find the horizontal circle reading equivalent to north, or more likely, the azimuth of a Reference Object (RO).

The suggested order of observations, irrespective of the method of finding K (amplitude or transit), is as follows:

Point to RO. Horizontal circle reading (RO)	H_{RO}
Point the theodolite about half a degree to the west of north	
Horizontal circle reading (fix)	H_1
Observations for centre of swing, *non-spin mode*	$K = d_1$
Observations for centre of swing, *spin mode*	$K = B_1$
Point the theodolite about half a degree to the east of north	
Observations for centre of swing, *spin mode*	$K = B_2$
Observations for centre of swing, *non-spin mode*	$K = d_2$
Horizontal circle reading (fix)	H_2
Point to RO. Horizontal circle reading (RO)	H_{RO}

The solution is then found from *equations (13.44)* or *(13.45)*, and *(13.46)*.

The first and last observations are to the RO. This gives two independent estimates of the same horizontal circle reading and ensures that the theodolite has not been disturbed during the lengthy observational process. This may occur if the instrument is mounted on a tripod in damp or sunny conditions in which a wooden tripod may twist. A tripod may easily be knocked during operations.

The two determinations of $K = d$ are, time wise, either side of both the determinations of $K = B$. This means that the spinner only needs to be run up once for the two separate determinations of $K = B$. However, it is good practice to allow the spinner to hang spinning for about 20 minutes before any observations of either $K = d$ or $K = B$. This will allow the internal temperature of the instrument to stabilize and to remove any residual twisting of the tape that may have occurred while travelling. Between the determinations of $K = B$, east and west of north, the spinner must be clamped up when the theodolite is turned.

If pre-computed values of C and d already exist and can be relied upon at the level of precision that the operator is working to, then the observational process may be reduced to:

Point to RO. Horizontal circle reading (RO)	H_{RO}
Point the theodolite as close to north as possible	
Horizontal circle reading (fix)	H
Observations for centre of swing, *spin mode*	$K = B$
Point to RO. Horizontal circle reading (RO)	H_{RO}

The solution is then found from *equation (13.8)*.

Worked examples

Example 13.4 Amplitude method

From the following set of observations and data find the azimuth of the Reference Object (RO).

Observations

Horizontal circle reading (RO) 333° 45′ 16″ H_{RO}

Observations with the theodolite pointing about half a degree to the west of north:

Horizontal circle reading (fix) 338° 34′ 41″ H_1

Non-spin mode	$K = d_1$	Successive turning points at −8.16, +10.35, −8.00 and +10.27 scale divisions
Spin mode	$K = B_1$	Successive turning points at −0.30, +6.61 and −0.26 scale divisions

Observations with the theodolite pointing about half a degree to the east of north

Spin mode	$K = B_2$	Successive turning points at −6.74, +4.24 and −6.72
Non-spin mode	$K = d_2$	Successive turning points at +11.24, −8.88, +11.20 and −8.76

Horizontal circle reading (fix) 339° 34′ 52″ H_2
Horizontal Circle Reading (RO) 333° 45′ 16″ H_{RO}

Data $A = +01' 23''$ $s = 600''/\text{div}$
Computations
Compute midpoints of swing using *equations (13.36)* for B and *(13.37)* for d
$d_1 = 1.145$ $B_1 = 3.165$ $B_2 = -1.245$ $d_2 = 1.180$
Find the 'mean' value of d from *equation (13.33)* $d = 1.171$
Find the torque ratio constant, C, from *equation (13.36)* $C = 0.3647$
Use the north finding equation, *equation (13.8)*, to find N, the horizontal circle reading of the theodolite equivalent to north. Either the values observed to the east or to the west of north may be used. Both sets of data will produce exactly the same answer. Both computations may be done as a check upon the arithmetic of this computation and that of finding C. $N = 339° 15' 00''$
Azimuth of RO $= H_{RO} - 339° 15' 00'' = 354° 30' 16''$

Example 13.5 Transit method
From the following set of observations and data find the bearing of the Reference Object (RO).

Observations
Horizontal circle reading (RO) 333° 45′ 16″ H_{RO}
Observations with the theodolite pointing about half a degree to the west of north
Horizontal circle reading (fix) 338° 31′ 05″ H_1

Non-spin mode $K = d_1$ $y_0 = 0$ $y_r = +5$ $t_0 = 0.00^s$ $t_r = 23.32^s$ $t_1 = 33.45^s$ $t_2 = 58.34^s$
Spin mode $K = B_1$ $y_0 = +2$ $y_r = +8$ $t_0 = 0^m\,0.00^s$ $t_r = 1^m\,2.32^s$ $t_1 = 4^m\,3.47^s$ $t_2 = 7^m\,11.06^s$
Observations with the theodolite pointing about half a degree to the east of north
Spin mode $K = B_2$ $y_0 = -2$ $y_r = +3$ $t_0 = 0^m\,0.00^s$ $t_r = 1^m\,41.99^s$ $t_1 = 3^m\,43.41^s$ $t_2 = 7^m\,12.59^s$
Non-spin mode $K = d_2$ $y_0 = 0$ $y_r = +5$ $t_0 = 0.00^s$ $t_r = 21.48^s$ $t_1 = 33.90^s$ $t_2 = 58.01^s$
Horizontal circle reading (fix) 339° 41′ 39″ H_2
Horizontal circle reading (RO) 333° 45′ 16″ H_{RO}

Data $A = 01' 23''$ $s = 600''/\text{div}$
Computations
Compute midpoints of swing using *equation (13.33)*

d_1	$t_x = -33.45$	$t_y = 58.34$	$t_z = 13.19$	$c_1 = -0.2284$	$c_2 = 0.7582$	$d_1 = 1.158$
B_1	$t_x = -243.47$	$t_y = 431.06$	$t_z = -118.83$	$c_1 = -0.2022$	$c_2 = 0.6478$	$B_1 = 3.427$
B_2	$t_x = -223.41$	$t_y = 432.59$	$t_z = -19.43$	$c_1 = -0.0516$	$c_2 = 0.9901$	$B_2 = -1.752$
d_2	$t_x = -33.90$	$t_y = 58.01$	$t_z = 9.06$	$c_1 = -0.2620$	$c_2 = 0.8820$	$d_2 = 1.145$

Find the 'mean' value of d from *equation (13.33)* $d = 1.148$
Find the torque ratio constant, C, from *equation (13.36)* $C = 0.3624$
Use the north finding equation, *equation (13.8)*, to find N, the horizontal circle reading of the theodolite equivalent to north. Either the values observed to the east or to the west of north may be used. Both sets of data will produce exactly the same answer. Both computations may be done as a check upon the arithmetic of this computation and that of finding C. $N = 339° 15' 00''$
Azimuth of RO $= H_{RO} - 339° 15' 00'' = 354° 30' 34''$

13.3.8 Sources of error

The main sources of error associated with the gyro, other than those errors that apply just to the theodolite observations, are:

(1) The assumed value of 1 scale unit in angular measure

There is uncertainty in the value of s of about 5% of its true value. The term s appears twice in *equation (13.8)* and once in *equation (13.37)*. When the value of C has been determined with *equation (13.37)* it may be substituted into *equation (13.8)* and rearranged to give:

$$N = H_1 + \frac{(B_1 - d)(H_1 - H_2)}{(B_2 - B_1)} + sd + A$$

The effect of an error in s may therefore be minimized by minimizing the size of d. This may be achieved with very careful use of a screwdriver by adjusting the point of suspension of the tape. The resulting value of d can then be found by the amplitude or a transit method. The effect of the error in s upon the computed value of N will be $\delta N = d\delta s$. If δs is assumed to be $20''$, and this can be no more than an assumption, then its effect will be reduced to $2''$ if d is reduced to 0.1 of a scale division. This is not easily achieved, but it can be done.

(2) Other oscillations

So far it has been assumed that the suspended gyroscope is a single degree of freedom system. That is, the only way the mast can move is a simple rotation about its own axis. It can be shown that there are in fact five degrees of freedom. The top of the mast may swing like a pendulum both in a north–south and an east–west direction. The bottom of the mast may also swing like a pendulum both in a north–south direction and an east–west one with respect to the top of the mast. In practice, with a well set up gyro these swings should be very small, but when the spinner is spinning the effect upon the moving shadow mark is to add four further damped harmonic motion terms to the equation of motion. *Equation (13.7)* now becomes:

$$\sum_{i=1}^{i=5} \left[M_i e^{-D_i t} \cos(\omega_i (t - p_i))\right] + K - y = 0$$

This equation is of course quite unmanageable, there are 21 unknowns that need to be solved for. Fortunately the magnitudes of all the oscillations are quite small, and provided the gyro is allowed to settle down before the readings start the effects of the oscillations become negligible. This is because the damping terms associated with the other oscillations are larger than that of the main oscillation. One oscillation has a period of half a second. This oscillation is often seen on the scale as a 2 Hz wobble when the mast is released. It takes a steady hand to be able to release the mast so that this movement is not significant. If a small 2 Hz wobble is present then it is best to let the gyro oscillate until the 2 Hz wobble has decayed and is not visible to the naked eye.

(3) Tape zero drift

When the GAK1 is in transit, with the clamp tightened, the mast is held in position so that there is no tension in the tape. When the instrument is in operation, in either the spin or nonspin mode, the tape supports the full weight of the mast. The tape is a very fine piece of metal, designed so that the torsion in it is a practical minimum. The result must be that the tension in the tape is approaching the point where the tape may pass its elastic limit when in use. During operation, therefore, the tape may experience some plastic or permanent deformation. Although the main effect will be in the direction of gravity there may also be some rotational distortion.

The mast contains an electric motor to drive the spinner. As the instrument warms up during operation there may be temperature effects upon the tape. Although the instrument is designed so that local magnetic fields do not have a significant effect, there will be some residual effect, especially if the instrument is operated near electrical or magnetic sources.

(4) Change in the spin rate

The period of an oscillation depends, among other things, on the angular velocity of the spinner. The spinner is kept at a constant angular velocity by a power/control unit. As the battery in the power/control unit runs down during a set of observations there must be an effect upon the velocity of the spinner. It has been shown that a 0.1 volt change in the power produces a 0.1 second change in the period. It will not matter if the voltage is not the same for every set of observations, but changes in voltage during observations may upset the results. It has also been shown that changes in temperature will also affect the spin rate.

(5) Irregularities in the engraving of the gyro scale

The effect of errors in the engraving of a glass theodolite circle is reduced by comparing opposite sides of the circle in a single observation and by using different parts of the circle in different observations. This is not possible with the gyro and so errors must be assumed to be negligible. There is no practical test that the user can perform to determine the standard error of a scale division. In this respect the only quality assessment that can be made is in terms of the manufacturer's reputation.

(6) Dislevelment in the prime vertical

When the theodolite is perfectly levelled, the mast in its rest position would hang vertically from its point of suspension. If the theodolite was now dislevelled the mast would still hang vertically but would be in a different spatial position with respect to the body of the gyro housing and thus to the gyro scale. If the dislevelment is in the prime vertical then all subsequent readings on the gyro scale will be displaced left or right. This will be a constant error for each setup of the instrument. One second of dislevelment produces about one second of error in the computed azimuth. As the plate bubble on the theodolite may have a nominal scale value of 20″/division then dislevelment in the prime vertical would appear to be a significant source of error. The only practical solution to the problem is to level the instrument as precisely as possible. If further determinations are to be made so that a mean result is taken then the instrument should be dislevelled and re-levelled between determinations.

(7) Additive constant drift

The additive constant A is a function of the alignment of the axis of the spinner and the system of projection of the moving mark. The additive constant will be affected by rough handling of the gyro and scratches or knocks to the interfacing surfaces of gyro and theodolite. It will also be affected by overtightening the ring clamp. Even with good handling the additive constant can change slowly over a period of time, presumably because of wear, especially on the bearings of the spinner axle. For accurate work the gyro should be calibrated frequently on lines of known azimuth to monitor the drift of the additive constant.

13.3.9 Error analysis

There is little in the published literature that gives a definitive statement of the precision that may be achieved with a gyrotheodolite. The manufacturers' literature states that an azimuth good to 20″ may be achieved in 20 minutes of time with the amplitude method. This assumes that the torque ratio constant, C, and the midpoint of swing in the nonspin mode, d, are already known.

A formal error analysis of the amplitude method, *equations (13.36)* and *(13.37)*, using the realistic standard error of 0.04 scale divisions for an observation leads to uncertainties for d and B as $\sigma_B = 0.024$ and $\sigma_d = 0.019$ scale divisions respectively. For the transit method with a standard error of time of 0.2 seconds the values are $\sigma_B = 0.011$ and $\sigma_d = 0.038$. These values are for guidance only as the errors will vary with the size of the swings in the transit methods and with any disturbance of the instrument when the micrometer is moved in the amplitude method. In computing N from *equation (13.8)* using realistic values of σ_N around 22″ may be obtained for the amplitude method and 31″ for the transit method.

B is best determined by the transit method because it gives a marginally better result and the observer does not have to disturb the instrument by touching it. d is best determined by the amplitude method. In all cases the size of the swing should be minimized. In practical terms, if the swing appears to be large then the gyro should be clamped up and the spinner released again so that the swing does not exceed a few divisions.

The value of d should be adjusted to be less than 0.1 of a scale division, any larger and the standard errors described above will be significantly increased.

13.3.10 The gyroscopic azimuth

A gyroscopic azimuth is essentially an astronomical azimuth and therefore to find the instrumental constant A the gyrotheodolite must be calibrated on a line of known astronomical azimuth. To find the astronomical azimuth of a line:

(1) Astronomical observations to sun or stars must be made (outside the scope of this book).

(2) Or the deviation of the vertical must be calculated from the slope of the geoid which in turn must be derived from a geoid model related to the ellipsoid used in the determination of geodetic position and the values put into the Laplace equation:

$$A_A = A_G + (\lambda_A - \lambda_G)\sin\phi$$

where A_A = astronomic azimuth
A_G = geodetic azimuth
λ_G = geodetic longitude
λ_A = astronomic longitude. Apply the east–west deviation of the vertical (geoid slope) to λ_G
ϕ = latitude

(3) Or ignore the difference between astronomic and geodetic azimuth. The difference is not likely to be more than a few tens of seconds of arc at most.

13.3.11 Gyromat 3000

The Gyromat 3000 from Deutsche Montan Technologie GmbH is a fully automated gyrotheodolite which requires very little interaction with the observer once it is set up correctly (*Figure 13.19*). It takes about 10 minutes for the instrument to give a readout which is claimed to be accurate to 3″, or to 30″ in 2 minutes. The solution quality may be given as part on the data output. The measuring sequence of determining the reversal points is fully automatic. The instrument works at an optimum temperature and although there is internal software that corrects for variations in temperature it is usually a good idea to let the instrument stabilize in temperature before use. For precise work this is always good practice with any survey instrument. The operating temperature range of −20°C to +50°C should accommodate most work. For tunnelling work in winter, where the outside temperatures are low but the underground temperatures are high, a substantial portion of this range may be used. Like the GAK1 and any other suspended gyroscope,

Fig. 13.19 *Gyromat 3000 – courtesy of Deutsche Montan Technologie GmbH*

the Gryomat 3000 becomes less sensitive at higher latitudes and is effectively inoperative within 10° of latitude of the poles. There is a serial interface to connect to a data storage device or to an electronic theodolite and wireless remote control and data transmission via Bluetooth. The instrument is quite heavy at 11.5 kg plus theodolite, tripod and case. More details at http://www.gyromat.de.

13.4 LINE AND LEVEL

13.4.1 Line

The line of the tunnel, having been established by wire survey or gyro observations, must be fixed in physical form in the tunnel. For instance, in the case of a Weisbach triangle (*Figure 13.20*) the bearing W_uW_1 can be computed; then, knowing the design bearing of the tunnel, the angle θ can be computed and

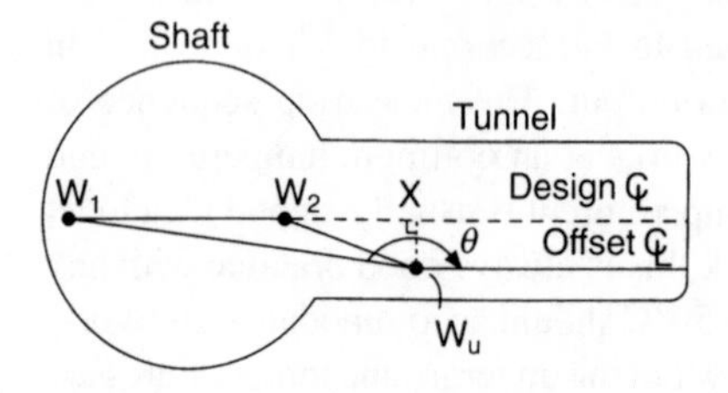

Fig. 13.20 *Plan view*

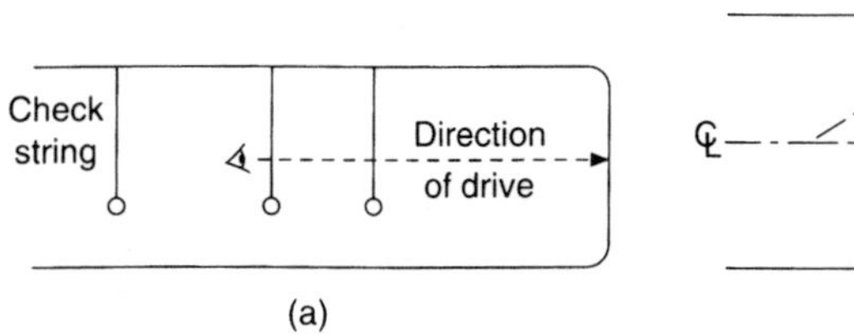

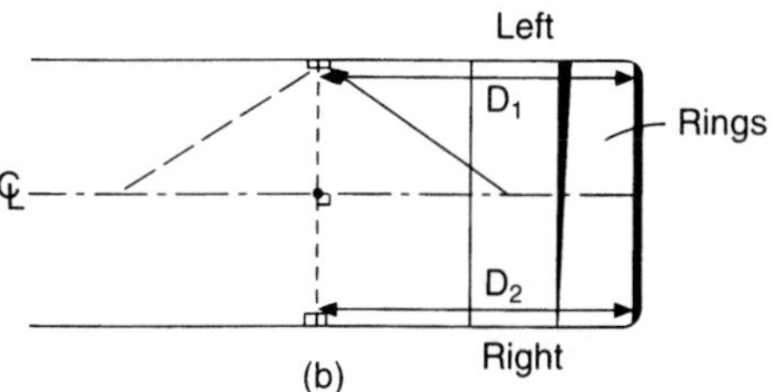

Fig. 13.21 *(a) Section, and (b) plan*

turned off to give the design bearing, offset from the true by the distance XW_u. This latter distance is easily obtained from right-angled triangle W_1XW_u.

The line may then be physically established by carefully lining in three plugs in the roof from which weighted strings may be suspended as shown in *Figure 13.21(a)*. The third string serves to check the other two. These strings may be advanced by eye for short distances but must always be checked by theodolite as soon as possible.

The gradient of the tunnel may be controlled by inverted boning rods suspended from the roof and established by normal levelling techniques.

In addition to the above, 'square marks' are fixed in the tunnel by taping equilateral triangles from the centre-line or, where dimensions of the tunnel permit, turning off 90° with a theodolite (*Figure 13.21(b)*). Measurements from these marks enable the amount of lead in the rings to be detected. For instance, if $D_1 > D_2$ then the difference is the amount of left-hand lead of the rings. The gap between the rings is termed *creep*. In the vertical plane, if the top of the ring is ahead of the bottom this is termed *overhang*, and the reverse is *look-up*. All this information is necessary to reduce to a minimum the amount of 'wriggle' in the tunnel alignment.

With a total station using conventional survey instrument and target mounts may be permanently attached to the wall of the tunnel to ensure forced centring. If that is the case then if all the mounts are on the same side of the tunnel then either it will not be possible to see round the bend if the curvature is to the same side as the mounts or there may be systematic horizontal refraction effects. Therefore where possible target and instrument mounts should be on alternate sides of the tunnel.

Where tunnel shields are used for the drivage, laser guidance systems may be used for controlling the position and attitude of the shield. A laser beam is established parallel to the axis of the tunnel (i.e. on bearing and gradient) whilst a position-sensing system is mounted on the shield. This latter device contains the electro-optical elements which sense the position and attitude of the shield relative to the laser datum. Immunity to vibrations is achieved by taking 300 readings per second and displaying the average. Near the sensing unit is a monitor which displays the displacements in mm automatically corrected for roll. Additionally, roll, lead and look-up (*Figure 13.22*) are displayed on push-button command along with details of the shield's position projected 5 m ahead. When the shield is precisely on line a green light glows in the centre of the screen. All the above data can be relayed to an engineers' unit several hundred metres away. Automatic print-out of all the data at a given shield position is also available to the engineers.

When a total station with automatic target recognition is used then it can lock on to and track a reflector on the tunnel boring machine (TBM). A steering control unit may be used to interface electronically between the guidance system, the push ram extensiometers and the TBM steering controls. The system briefly described here is the one designed and manufactured by ZED Instruments Ltd, London, England.

In general the power output of commercial lasers is of the order of 5 mW and the intensity at the centre of a 2-cm diameter beam approximately 13 mW/cm^2. This may be compared with the intensity of sunlight received in the tropics at noon on a clear day, i.e. 100 mW/cm^2. Thus, as with the sun, special precautions should be taken when viewing the laser (see *Section 12.9*).

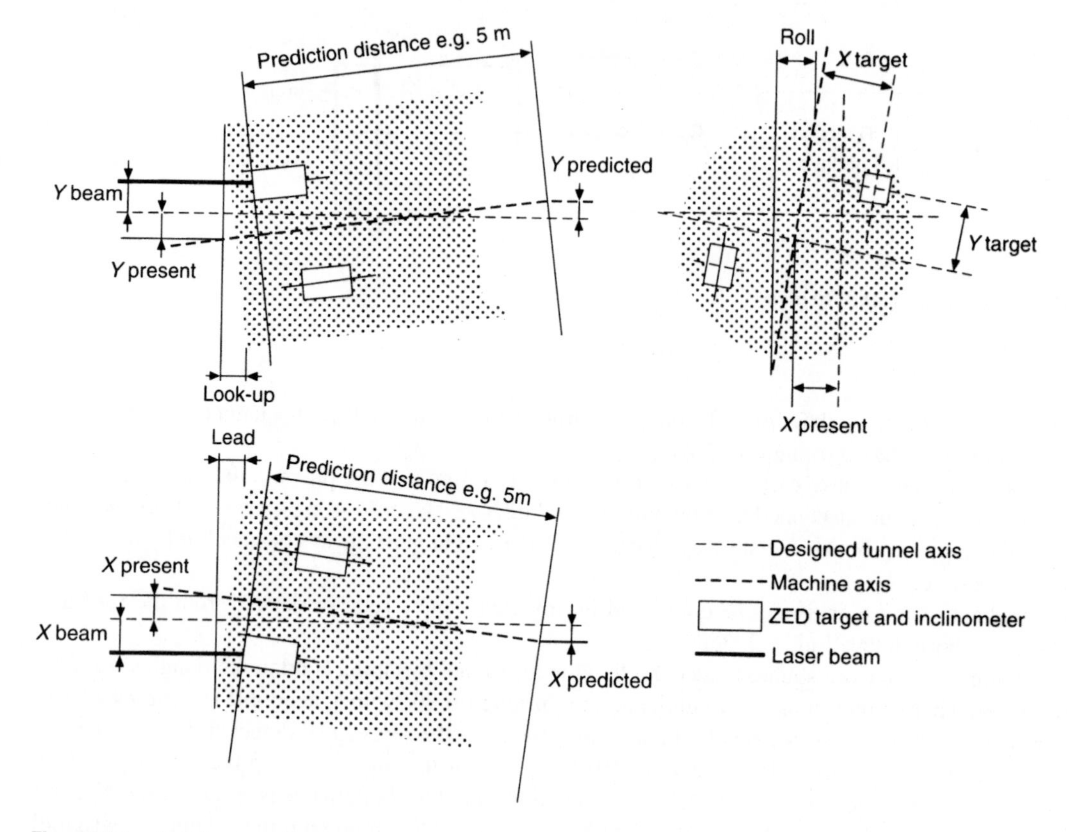

Fig. 13.22 *Roll, lead and look-up*

Practically all the lasers used in tunnelling work are wall- or roof-mounted, and hence their setting is very critical. This is achieved by drilling a circular hole in each of two pieces of plate material, which are then fixed precisely on the tunnel line by conventional theodolite alignment. The laser is then mounted a few metres behind the first hole and adjusted so that the beam passes through the two holes and thereby establishes the tunnel line. Adjustment of the holes relative to each other in the vertical plane would then serve to establish the grade line.

An advantage of the above system is that the beam will be obscured should either the plates or laser move. In this event the surveyor/engineer will need to 'repair' the line, and to facilitate this, check marks should be established in the tunnel from which appropriate measurements can be taken.

In order to avoid excessive refraction when installing the laser, the beam should not graze the wall. Earth curvature and refraction limit the laser line to a maximum of 300 m, after which it needs to be moved forward. To minimize alignment errors, the hole furthest from the laser should be about one-third of the maximum beam distance of the laser.

13.4.2 Level

In addition to transferring bearing down the shaft, height must also be transferred.

One particular method is to measure down the shaft using a 30-m standardized steel band. The zero of the tape is correlated to the surface BM as shown in *Figure 13.23*, and the other end of the tape is precisely located using a bracket fixed to the side of the shaft. This process is continued down the shaft until a level reading can be obtained on the last tape length at *B*. The standard tension is applied to the bottom of the

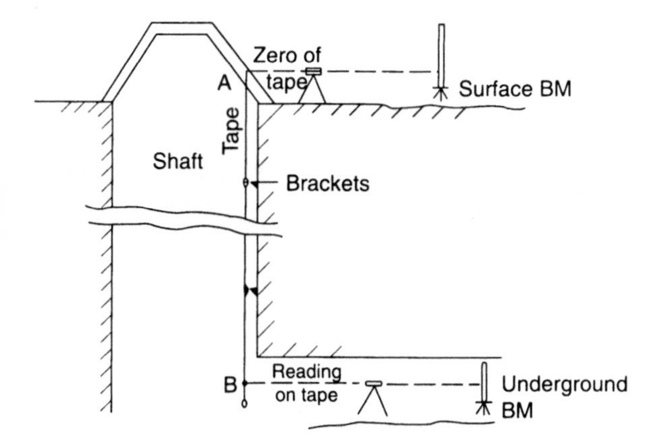

Fig. 13.23 *Transfer height down a shaft*

tape and tape temperature recorded for each bay. A further correction is made for elongation of the tape under its own weight using:

$$\text{Elongation (m)} = WL/2AE$$

where E = modulus of elasticity of steel (N/mm^2)
L = length of tape (m)
A = cross-sectional area of tape (mm^2)
W = mass of tape (N)

Then the corrected distance AB is used to establish the value of the underground BM relative to the surface BM.

If a special shaft tape (1000 m long) is available the operation may be carried out in one step. The operation should be carried out at least twice and the mean value accepted. Using the 30-m band, accuracies of 1 in 5000 are possible; the shaft tape gives accuracies of 1 in 10 000.

Total stations have also been used to measure shaft depths. A special reflecting mirror at the top of the shaft is aligned with the EDM instrument and then rotated in the vertical plane until the measuring beam strikes a corner-cube prism at the shaft bottom. In this way the distance from the instrument to the reflector is obtained and subsequently adjusted to give the distance from mirror to prism. By connecting the mirror and prism to surface and underground BMs respectively, their values can be correlated.

Exercises

(*13.1*) (a) Describe fully the surveying operations which have to be undertaken in transferring a given surface alignment down a shaft in order to align the construction work of a new tunnel.

(b) A method involving the use of the three-point resection is often employed in fixing the position of the boat during offshore sounding work.

Describe in detail the survey work involved when this method is used and discuss any precautions which should be observed in order to ensure that the required positions are accurately fixed. (ICE)

(*13.2*) Describe how you would transfer a surface bearing down a shaft and set out a line underground in the same direction.

Two plumb lines A and B in a shaft are 8.24 m apart and it is required to extend the bearing AB along a tunnel. A theodolite can only be set up at C 19.75 m from B and a few millimetres off the line AB produced.

If the angle *BCA* is 09′ 54″ what is the offset distance of *C* from *AB* produced? (ICE)

(*Answer*: 195 mm)

(*13.3*) The following observations were taken with a gyrotheodolite. Find the horizontal circle reading equivalent to north. The observations in the spin mode are by the transit method and the observations in the non-spin mode are by the amplitude method.

Horizontal circle reading (RO) 33° 5′ 6″ H_{RO}
Observations with the theodolite pointing about half a degree to the west of north
Horizontal circle reading (fix) 227° 23′ 30″ H_1
Non-spin mode $K = d_1$ Turning points: −3.19, +3.46, −3.03, +3.38
Spin mode $K = B_1$, $y_0=0$, $y_r=+7$, $t_0=0^m\,0.00^s$, $t_r=1^m\,3.02^s$, $t_1=4^m$, 3.24^s, $t_2=7^m\,10.90^s$
Observations with the theodolite pointing about half a degree to the east of north
Spin mode $K = B_2$, $y_0=-3$, $y_r=+2$, $t_0=0^m\,0.00^s$, $t_r=1^m\,41.69^s$, $t_1=3^m\,43.34^s$, $t_2=7^m\,11.65^s$
Non-spin mode $K = d_2$ Turning points: +4.14, −3.88, +4.06, −3.76
Horizontal circle reading (fix) 228° 23′ 58″ H_2
Horizontal circle reading (RO) 33° 5′ 6″ H_{RO}
$A = +02' 12''$ $s = 600''$/div

(*Answer*: $d = 0.1325$ $B_1 = 1.6418$ $B_2 = -2.7382$ $C = 0.3804$ $N = 227° 47' 52''$ Azimuth of RO = 165° 17′ 14″)

(*13.4*) The following set of observations, using the transit method has been made. Find the horizontal circle reading of the theodolite that is equivalent to north.

Horizontal circle reading (RO) 47° 34′ 29″ H_{RO}
Observations with the theodolite pointing about half a degree to the west of north
Horizontal circle reading (fix) 9° 30′ 23″ H_1
Non-spin mode $K=d_1$, $y_0=0$, $y_r=-3$, $t_0=2^m\,36.21^s$, $t_r=2^m\,44.50^s$, $t_1=3^m\,3.50^s$, $t_2=3^m\,31.64^s$,
Spinmode $K=B_1$, $y_0=+3$, $y_r=-7$, $t_0=10^m\,43.46^s$, $t_r=12^m\,2.60^s$, $t_1=14^m\,20.83^s$, $t_2=17^m\,56.99^s$
Observations with the theodolite pointing about half a degree to the east of north
Spin mode $K=B_2$, $y_0=-1$, $y_r=+8$, $t_0=9^m\,57.60^s$, $t_r=11^m\,10.48^s$, $t_1=13^m\,27.77^s$, $t_2=17^m\,10.64^s$
Non-spin mode $K=d_2$, $y_0=0$, $y_r=+2$, $t_0=6^m\,15.51^s$, $t_r=6^m\,23.49^s$, $t_1=6^m\,43.77^s$, $t_2=7^m\,10.93^s$
Horizontal circle reading (fix) 10° 30′ 19″ H_2
Horizontal circle reading (RO) 47° 34′ 31″ H_{RO}
$A = +2' 31''$ $s = 600''$/div

(*Answer*: $d_1 = +0.091$ $B_1 = +2.952$ $B_2 = -1.489$ $d_2 = +0.078$ $d = 0.081$ $C = 0.3494$ $N = 10° 12' 27''$ Azimuth of RO = 37° 22′ 3″)

14
Mass data methods

14.1 INTRODUCTION

Almost all the survey methods mentioned in this book so far relate to individual survey observations and their subsequent use in survey computations. For example, in a conventional resection, observations are taken at one point to three other points and those observations along with the coordinates of the targets are used to find the coordinates of the instrument point. If a number of points were required the productivity rate would be only a few points per day at best and so to find the coordinates of many points by such a method would be slow and tedious in the extreme. Finding a series of control points with a traverse, or by using the least squares adjustment of a network, would be more productive.

The surveyor could then use a total station and data logger to collect a series of detail points more rapidly. Even so the data capture rate would depend upon the speed at which the surveyor could travel from point to point and even in the most favourable environment the data capture rate would not exceed a few hundred points per day. If it is not necessary for the surveyor to physically identify each point on the ground and/or it is not necessary to remain static at each point during data capture then the productivity rate can be improved.

Reflectorless EDM makes it possible to measure to a point without that point being visited, though details of the point still need to be recorded to make the measurement useful. If stringlines of features or ground profiles are followed then data may be captured on the fly, for example with GPS. With ground profiles, such as for DTMs, the data rate may be limited by the speed with which the vehicle mounted GPS receiver can travel over the ground. The density of such data is therefore likely to be variable and will depend on the ability of the vehicle to maintain a constant speed, to cover the area without large or irregular gaps between runs and the ability of the GPS to maintain lock onto the satellites. Although relatively large amounts of data may be collected by a vehicle mounted GPS the overall quality of the data may be variable. With such a method several thousands of points a day could easily be captured. However, higher data capture rates are not really possible where it is necessary for instruments, whether GPS or corner cube prisms, to visit every point of detail.

This chapter is concerned with some of those data capture methods that allow the surveyor to identify points and collect their coordinates in a remote way and do so at data capture rates that far exceed those possible by the techniques described in the last paragraph.

14.2 PHOTOGRAMMETRY

As the word 'photogrammetry' implies, it means measurements from photographs, and in the case of aerial photogrammetry it is measurements from aerial photographs.

The major use of aerial photogrammetry is in the preparation of contoured plans from the aerial photographs. With the aerial camera in the body of the aircraft, photographs are taken along prearranged

Fig. 14.1 *Distortion; an oblique photograph of a rectangular door*

flight paths, with the optical axis of the camera pointing vertically down. Such photographs are termed vertical photographs.

A single photograph contains a lot of detailed information about the subject. However, a photograph cannot be a map of the ground or a plan of a building. At best, a photograph shows the view through the camera with distortions due to the optics of the camera, the atmosphere and the position and orientation of the camera. Consider *Figure 14.1* which shows a picture of a garage door. The door is rectangular but because of the position of the camera, the left and right edges appear with different heights, likewise the top and the bottom have different lengths. Therefore even if one of the dimensions had been measured with a tape, say, it would not be possible to take scaled measurements of the other dimensions from the photograph. Essentially there is insufficient information available.

Photogrammetry enables the user to derive metric information about a building façade, the terrain, or any three-dimensional object by making measurements on photographs of that object. Photogrammetry may be used with many forms of imagery taken within the visible electromagnetic spectrum and beyond. The mathematics involved in photogrammetry is relatively simple, but does involve many simultaneous calculations. Detail within a photograph may be given two-dimensional XY coordinates within the plane of the image and by the conformal transformations of shift, scale and rotation these may be turned into object XY coordinates in a plane parallel to the image. *Figure 14.2* shows the relationship between an object, the ground, the negative in the camera and an equivalent positive image. In conventional photography the positive image could be made by contact printing. In a digital camera the positive image may be derived with software.

The third dimension of the object, i.e. in the Z plane, can be found from the relative displacement of elements of the image from pairs of photographs. A point's depth within the object field will affect its displacement from its true orthogonal position. The image scale will dictate the magnitude of the linear and radial displacements in the image. The change in the position of an object from one photograph to the next is known as stereoscopic parallax or x-parallax. The amount of parallax depends on the object's height. Higher points have greater parallax. See *Figure 14.3*.

Given any two overlapping images, a stereo pair, it is possible to view both images simultaneously to replicate a stereo binocular view. Look at the top pair of photographs in *Figure 14.4*. The photographs

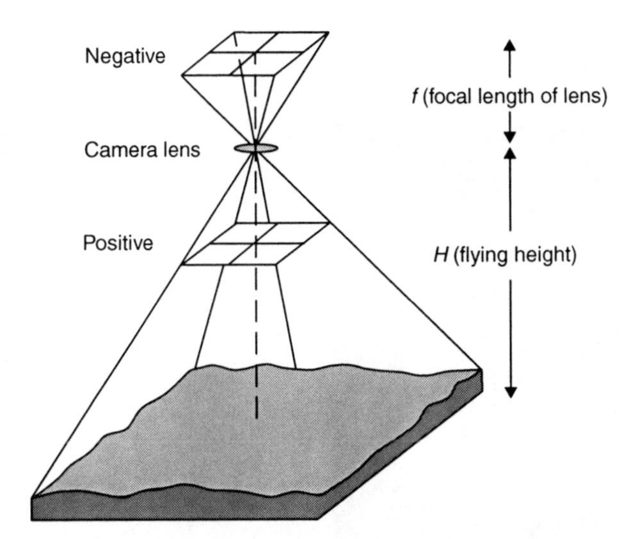

Fig. 14.2 *The photo/ground relationship*

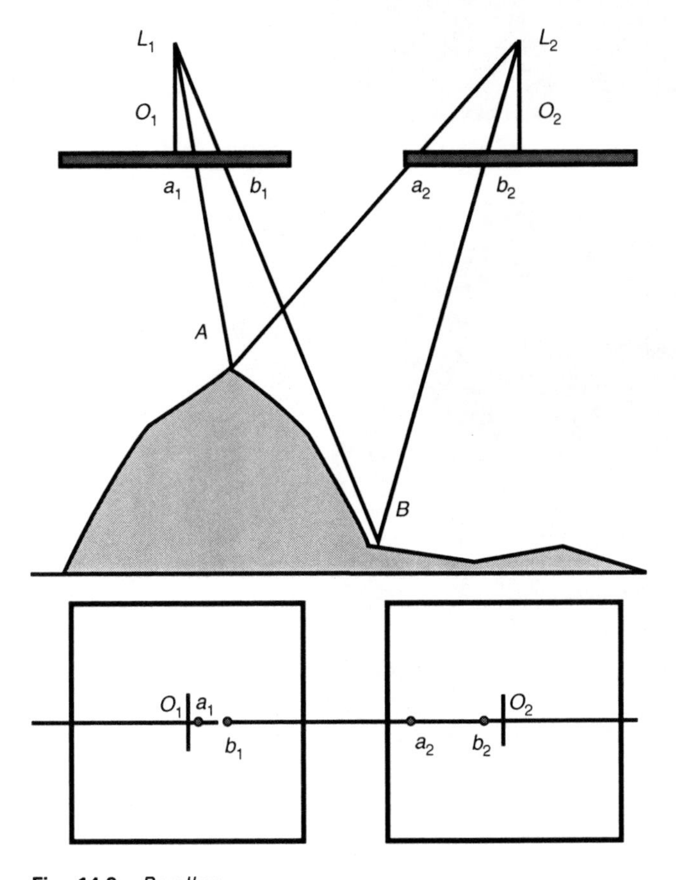

Fig. 14.3 *Parallax*

Fig. 14.4 *Stereo images*

were taken from two positions close to each other. In the second pair the photographs were taken from positions further away from each other. Hold the page up, focus on a distant object over the top of the page and quickly bring your gaze down so that you are looking at the left image with your left eye and the right image with your right eye. You should now see three images, with the central one appearing to be a three-dimensional image. Of the other images, the left one is now seen with the right eye and the right one with the left eye. Now look down to the lower set of images. Notice how the apparent depth of the second pair of images is greater than that of the first pair.

If you can form the three-dimensional image with the top pair comfortably you may have difficulty with the bottom pair because the depth exaggeration is much greater. When you look at any object you get a three-dimensional image because your two eyes see slightly different images; your eyes are a few centimetres apart. With the photographs in *Figure 14.4* the top pair of photographs were taken with the camera positions about a metre apart and the bottom pair much more than that.

The simplest practical viewing aid is a stereoscope, although this does not necessarily allow for accurate measurement. The two images can be brought into a clear three-dimensional stereo view at any point by bringing them closer together or further apart in the X direction, i.e. parallel to the human eye base. A displacement in the X direction is proportional to the magnitude of the *relief*, i.e. the Z dimension in the stereo model and so the third dimension of the object may be measured.

Developing the technology of stereoscopic viewing and measurement has presented manufacturers with many challenges over the last century. The major photogrammetric equipment manufacturers were those that could combine the high quality mechanical engineering needed for analogue stereo restitution instruments with high precision optics.

For many years photogrammetry was used almost exclusively for mapping using aerial photographs because of the opto-mechanical limitations of the instrumentation. Photogrammetry was the only practical way of mapping large areas and for producing digital elevation models (DEM). It has been used for mapping Mars and the Moon.

Photogrammetry has a number of advantages over conventional ground survey techniques using total stations or GPS. It is more cost effective over large areas. There is a much shorter time on site although much more work is later required in the office. Most of the data capture can be done without access to the site. The photographs form a permanent record of the site at the instant that the photographs were taken and this source data can be revisited and further information extracted if required.

Although photographs were taken for terrestrial measurements long before the invention of aircraft it was the military application of aerial survey in World War One that created the environment for aerial imaging and mapping by government agencies for decades after and terrestrial applications did not come to the fore until relatively recently. There are now available software systems that run on desktop PCs, the websites of the major manufacturers such as Leica, Zeiss and Rollei indicate the current level of equipment and software sophistication.

Prior to the other techniques described later in this chapter digital photogrammetry was the only way of capturing mass data at successive epochs to enable monitoring of change in a relatively short time frame. With digital output, direct comparisons are possible with statistical rigour, in days rather than in the weeks or months necessary to capture and process large projects with many points observed by conventional means. All the data in the photograph is captured instantaneously so no change can take place during the period of data capture. The photograph literally gives a snap shot in time. As all the information is contained with the photograph then it is possible to revisit the source data and verify that measurements taken from the photographs are correct. This is not possible with other survey techniques.

In monitoring, it is likely that coordinates from each epoch will be compared. Distances, areas or volumes may be determined from the coordinates if required. Absolute and relative displacement vectors with their associated confidence levels may be determined. In such systems, control may be arbitrary or fixed. If the control is arbitrary then the target itself provides unambiguous features that can be measured. If the target has no distinct features then stick on targets may be used. With arbitrary control there will be unique coordinate systems for each epoch and for each set of images. Lengths, areas and volumes may be derived for each epoch and these derived quantities compared. It may be necessary to include a scale bar in the images to give consistent scale to the derived quantities of each epoch. With fixed control there is a reference framework of control targets outside the object to be measured.

14.2.1 Basic principles

The essential processes involved in the production of a contoured plan or digital ground model from aerial photographs are:

(a) Photography
(b) Control
(c) Restitution

In the following sections some important relationships are stated. In a number of sections the proofs have been omitted for the sake of brevity. Those who wish to examine these proofs are directed to the second edition of this book.

14.2.2 Photography

Conventional photographs may be taken using dimensionally-stable film in precision-built cameras (*Figure 14.5*). However, increasingly, the photographs are taken with digital cameras. It is important

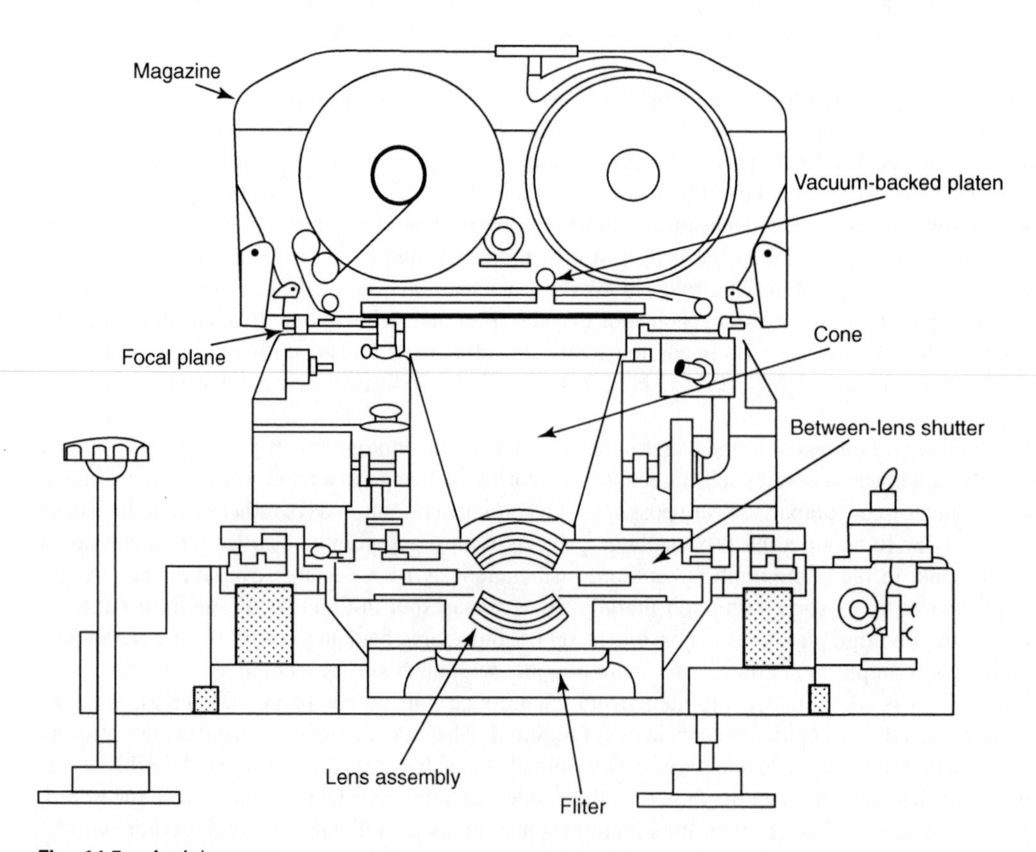

Fig. 14.5 *Aerial camera*

that all topographic detail must be clearly reproduced and therefore recognizable on the photograph, and that the geometric relationships between the ground objects and the photo images are rigorously maintained. These conditions are governed largely by the atmospheric conditions prevailing at the time of photography, aircraft movement, the characteristics of the camera and the scale used. If conventional film is used then it and its processing will also affect the quality of the final product.

Cameras used for air survey, as with all other survey equipment, are precision-built, and their lenses are of such high quality that aberrations are practically negligible. Lenses used in air survey cameras may be classified broadly as wide-angle (90°) with a principal distance (f) of 152.4 mm used for standard mapping or super-wide-angle (125°) with a principal distance of 305.0 mm used for small-scale mapping. From the engineering point of view the most popular lens is the wide-angle combined with a photograph format size of 230 mm×230 mm (see *Figure 14.6*). The air photo has a central perspective projection with the lens as origin.

After the lens system is a shutter which should be capable of exposing the whole film format for the required interval at the same instant of time. In addition, image movement, caused largely by the apparent movement of the ground relative to the aircraft, must also be reduced to negligible proportions, possibly by a *forward compensation mechanism* which moves the film slightly during exposure to compensate for the movement of the aircraft. The film may be kept flat by a low-pressure vacuum system or by pressure pads acting against the film.

The so-called *camera constants* obtained from the calibration process are:

(a) The position of the principal point
(b) The focal length of the lens
(c) The pattern and magnitude of distortion over the effective photographic field.

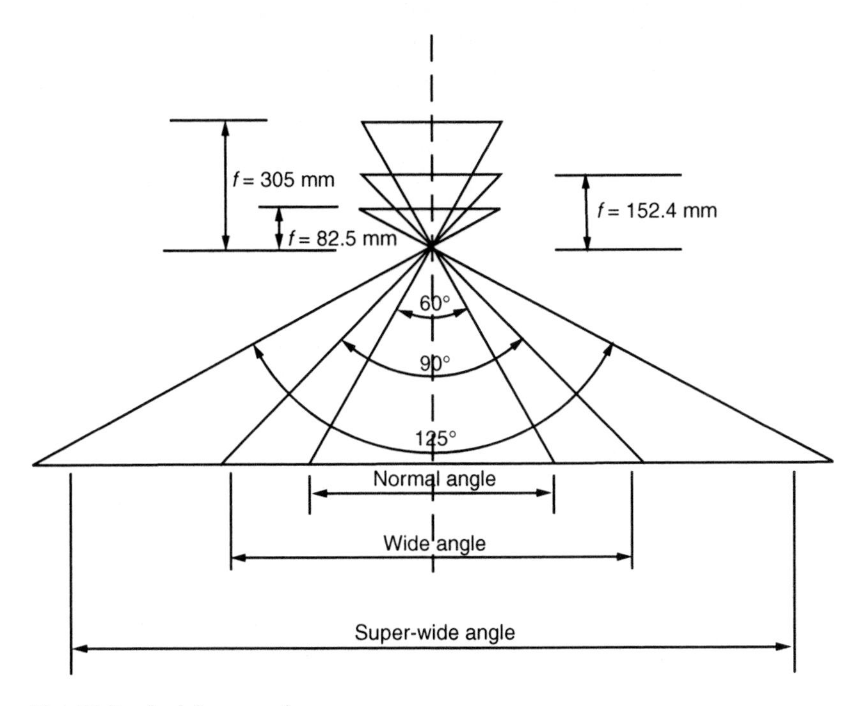

Fig. 14.6 *Aerial camera lenses*

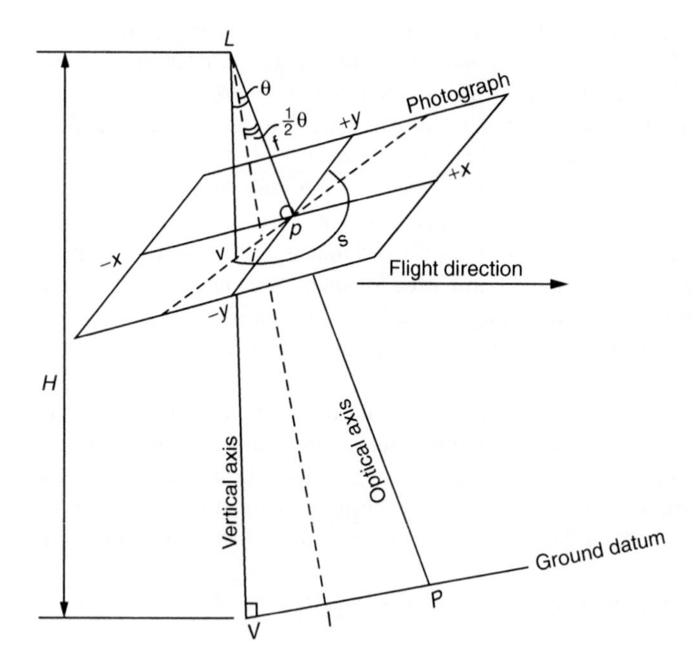

Fig. 14.7 *Near vertical photograph*

14.3 GEOMETRY OF THE AERIAL PHOTOGRAPH

First we need to consider the errors in an air photograph. These are largely caused by *tilt* in the plane of the film at the instant of exposure and also displacement of object position due to ground relief. A photograph is not a plan, except where the terrain is absolutely flat and level and the photograph axis is truly vertical. This of course never happens in practice.

14.3.1 Definitions

Because of the pitch and roll of the aircraft in flight it is rare for a truly vertical photograph to be taken. *Figure 14.7* shows a near vertical photograph with the optical axis tilted at θ to the vertical. In practice θ is usually less than 3°. The definitions of some commonly used terms are as follows.

Photo axis: the right-angled x–y-axis formed by joining the opposite fiducial marks of the photograph. This is the axis from which photo coordinates are measured. The x-axis approximates to the direction of flight.

Optical axis: the line LpP from the lens centre at 90° to the plane of the photograph.

Principal distance: the distance $Lp = f$, from the lens to the plane of the photograph. The principal distance may be referred to as the *focal length*.

Vertical axis: the line LvV in the direction of gravity, so 90° to a level datum plane.

Tilt: the angle θ between the vertical and optical axes (see also *principal line*).

Principal point (*PP*): the point p where the optical axis cuts the photograph, and coincides with the origin of the photo axes.

Plumb point: the point v where the vertical axis cuts the photograph.

Isocentre: the point i, where the bisector of the angle of tilt cuts the photograph.

Principal line: the line *vip* in the plane of the photograph giving the direction of maximum tilt of the photograph. It is therefore at the angle θ to the horizontal.

Plate parallels: the lines at 90° to the principal line; they are level lines.

Isometric parallel: the plate parallel passing through the isocentre and forming the axis of tilt of the photograph.

Flying height: the vertical height of the *lens* above ground at exposure. It is the height of the lens above datum (e.g. MSL) minus the mean height of the terrain.

Swing: the angle s measured in the plane of the photograph, clockwise from the $+y$ axis to the plumb point. It defines the direction of tilt relative to the photo axes.

The main sources of error in the air photo will now be outlined.

14.3.2 Scale and its variation due to ground relief

In *Figure 14.8* the scale of a photograph is the ratio of the distance on the ground to its imaged distance on the photograph. Hence, by similar triangles:

$$\text{Scale} = ab/AB = f/H$$

At point C, it is obvious that the scale $S_c = f/(H - h_c)$. Thus scale S varies with relief throughout the photograph and for any elevation (h) is given by:

$$S = f/(H - h) \tag{14.1}$$

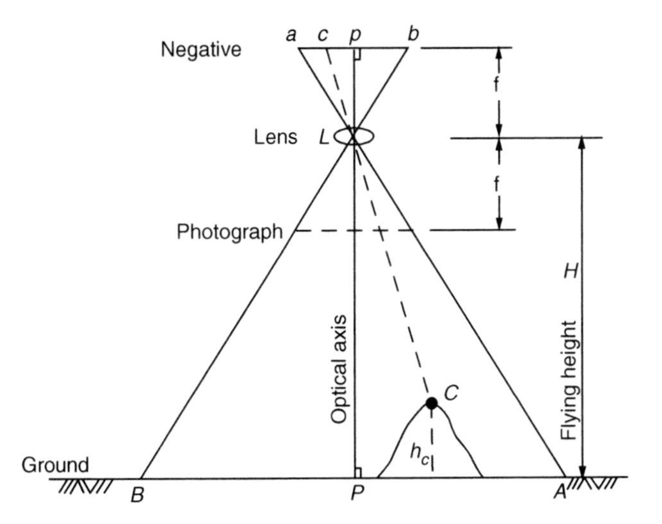

Fig. 14.8 *Photo scale*

14.3.3 Scale and its variation due to tilt

Figure 14.9 assumes flat terrain and indicates the axis of tilt through i; thus the scale at the isocentre is common to both a tilted and a truly vertical photograph

(a) *Scale at isocentre* $S_I = Li/LI = Lv_1/LV = f/H$
(b) *Scale at principal point* $S_p = Lp/LP = f/(H \sec \theta)$
(c) *Scale at plumb point* $S_v = Lv/LV = (f \sec \theta)/H$
(d) *Scale at random point* a $S_a = La/LA = Lv_2/LV = (Lv_1 + id_2)/LV$
$= (f + ai \sin \theta)/H = (f + y_a \sin \theta)H$
where $y_a = ai$, the distance from the isocentre
(e) *Scale at random point* b $S_b = (f - y_b \sin \theta)/H$
b is on the opposite side of i from a

Thus it can be seen that the scale continually varies along the principal line with distance from the isocentre. By definition, however, the scale along a plate parallel at a particular point will be constant providing the ground is level. The basic equation considering ground relief h is therefore

$$S = (f \pm y_b \sin \theta)/(H - h) \tag{14.2}$$

and $$\frac{dS}{dy} = \pm \sin \theta / H$$

14.3.4 Image displacement due to ground relief

Figure 14.10 shows an un-tilted photograph of undulating terrain. Point A, if projected orthogonally onto a plan, would appear at B. Its true position on the photograph is therefore at b, and distance ab is the displacement resulting from the height of A above the datum. By a consideration of similar triangles it can be shown that

$$ab = va\Delta h/H \tag{14.3}$$

From *equation (14.3)* it can be seen that any increase in flying height would reduce the amount of displacement ab, which in turn is directly proportional to the height Δh of the object. It can also be seen that if Δh and H remain constant, displacement will increase with distance va from the plumb point.

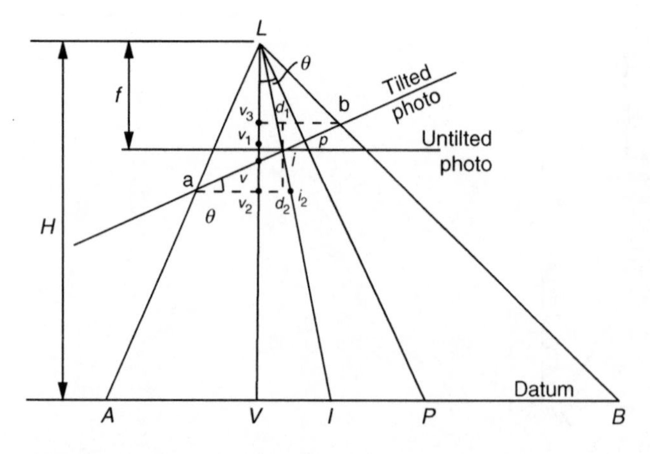

Fig. 14.9 *Tilted photograph*

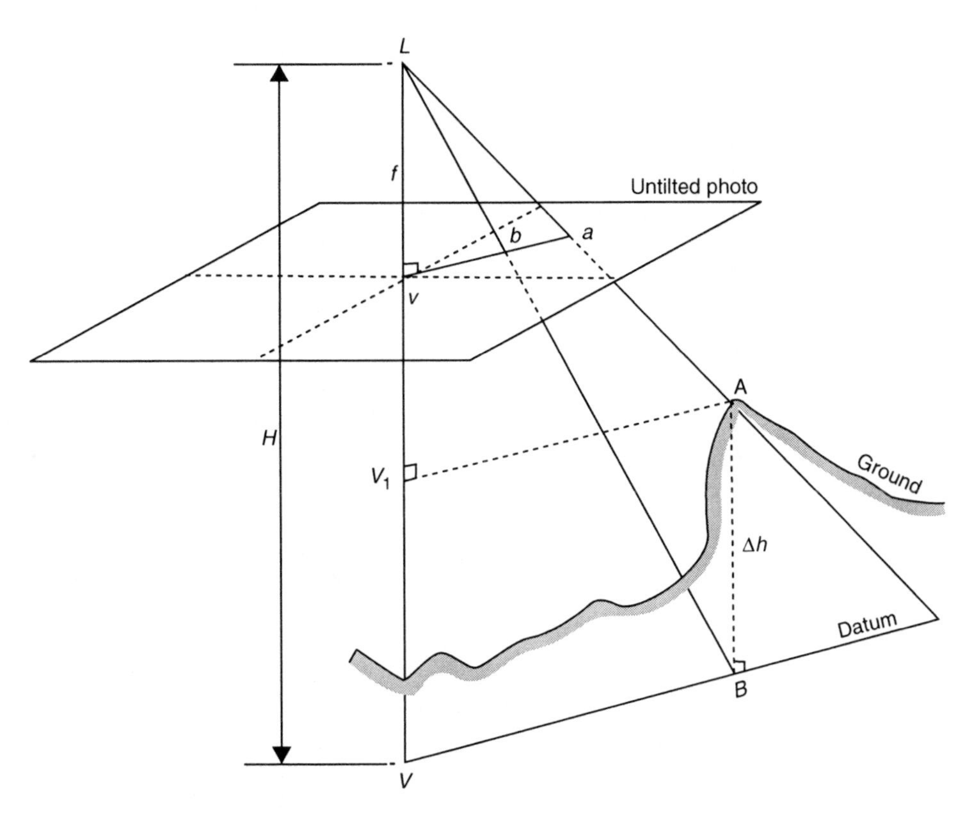

Fig. 14.10 *Untilted photograph*

From *Figure 14.10* it can be clearly seen that LV is parallel to AB and both are vertical. Thus $LABV$ forms a plane containing v, b and a, showing the displacement ba as being radial from the vertical LV at v, which in a vertical photograph is the plumb point.

14.3.5 Image displacement due to tilt

Considering point a in *Figure 14.11*, whose distance from the isocentre on the tilted photograph is ia, and on the un-tilted photograph is ia_1, then the displacement Δt due to tilt is given by

$$\Delta t = ia - ia_1$$

From consideration of the geometry of *Figure 14.11* it can be shown that

$$\Delta t = y^2 \sin\theta/(f + y\sin\theta)$$

where y is the distance from the isocentre measured along the principal line in the direction of downward slope. For any point off the principal line, the situation is as shown in *Figure 14.12*.

It can be seen that $ia = id\cos\varphi$ and $ia_1 = id_1\cos\varphi$. Therefore the displacement $dd_1 = \Delta t_1$, projected on to the principal line gives

$$\Delta t = \Delta t_1\cos\varphi = (y\cos\varphi)^2\sin\theta/(f + y\cos\varphi\sin\theta)$$

so the general equation becomes

$$\Delta t_1 = y^2\cos\varphi\sin\theta/(f + y\cos\varphi\sin\theta) \qquad (14.4)$$

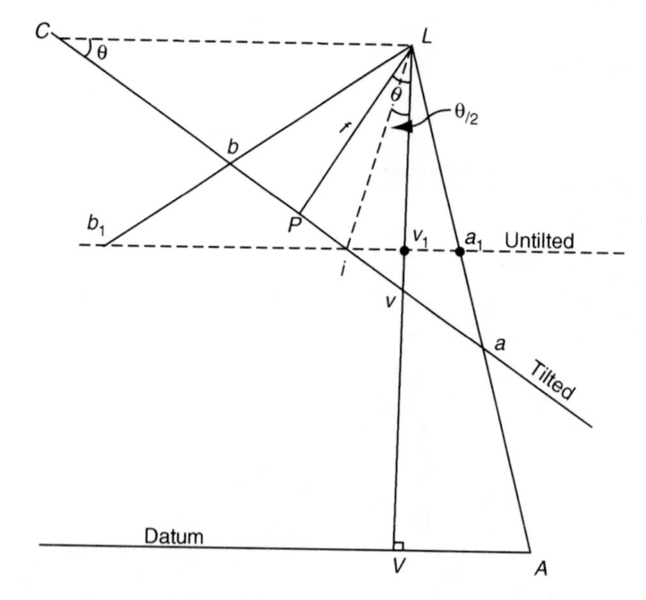

Fig. 14.11 *Untilted photograph*

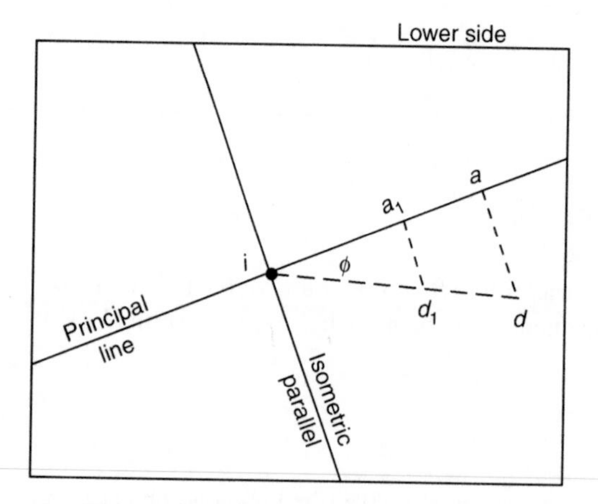

Fig. 14.12 *Tilt and image displacement*

and if θ is small then $y \cos \phi \sin \theta$ is negligible, compared with f, so

$$\Delta t_1 = y^2 \cos \varphi \sin \theta / f \tag{14.5}$$

The equation shows that displacement is proportional to the distance from the isocentre squared, and will therefore be greatest at the edges of the photograph. It shows also that increasing the focal length of the camera will help to reduce the displacement. It can be shown that image displacement due to tilt must be radial from the isocentre.

In addition to the complications already outlined, further displacements may result due to variation in flying height, refraction of the rays of light (particularly near the body of the aircraft), camera and photographic errors, etc. It can now be clearly seen that a photograph is not a plan, except where the axis of the photograph is truly vertical and the ground is flat and level.

14.3.6 Combined effect of tilt and relief

As already shown, displacement due to tilt and relief are not radial from any one point on the photograph. *Figure 14.13* shows ab as the top and bottom of a tall structure. The height displacement ab is radial from the plumb point v, whilst the tilt displacements aa_1 and bb_1 are radial from the isocentre. Note the reverse direction of displacement on the upper side of the photograph.

The treatment of such effects is to: (i) eliminate tilt displacement by a mathematical or optical rectification of the photograph, i.e. reduce the tilted photograph to its horizontal equivalent; (ii) consider height displacement on the rectified photograph – for instance, after rectification the equivalent position of the plumb point v is at v_1 (see also *Figure 14.11*) from which the height displacement a_1b_1 is radial.

14.3.7 To find the x and y tilts of a photograph

Knowing the focal length of the camera and the coordinates of the plumb point, the x and y tilts are given by

$$\sin\theta_y = y_v \cos\theta / f \qquad \sin\theta_x = x_v \cos\theta / f \tag{14.6}$$

where θ is the tilt of the photograph and θ_y and θ_x are the component parts of the tilt in the y and x directions respectively. Similarly y_v and x_v are the component parts of the vector pv in the y and x directions. Therefore, all lines parallel to the principal line have the same maximum tilt and all lines at 90° to the principal line (i.e. plate parallels) are horizontal.

14.3.8 Ground coordinates from a tilted photograph of flat terrain

In *Figure 14.14* consider point a whose photo coordinates x_a and y_a are measured about the fiducial axes. The flying height is H, the focal length is f, the tilt is θ and the swing is S. It is first necessary to obtain the coordinates of a relative to the principal line axes, with the isocentre as origin (tilt displacement radial from isocentre), i.e. x'_a, y'_a. *Figure 14.14(b)* illustrates the rotational effects where the angle between the respective axes is $\alpha = 180° + S$. The amount of translation necessary is $pi = f\tan(\frac{1}{2}\theta)$. Assuming a

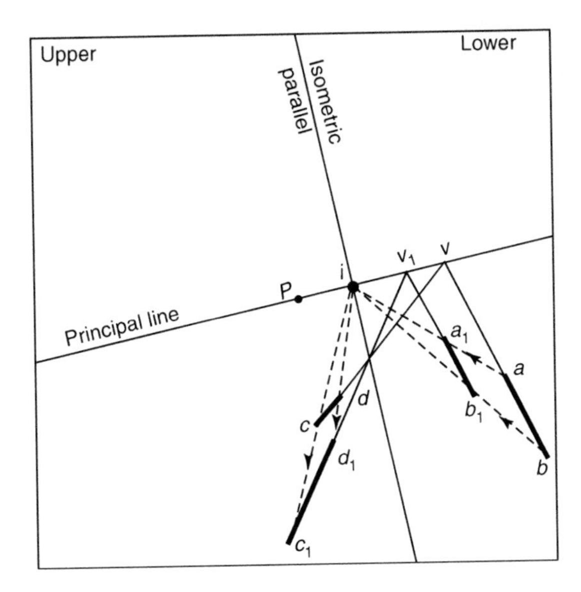

Fig. 14.13 *Effect of tilt and relief*

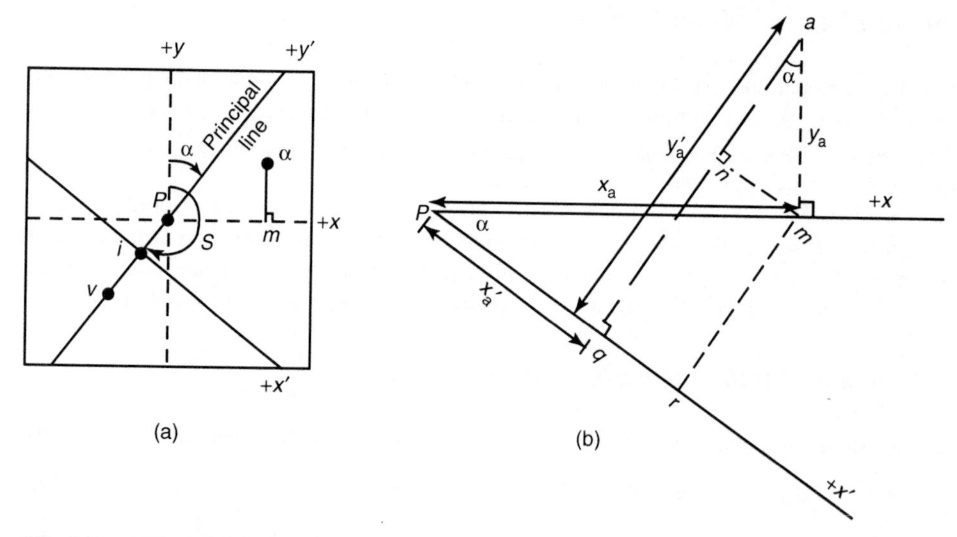

Fig. 14.14 *Tilted photograph of flat terrain*

parallel through p, an examination of *Figure 14.14(a)* shows that the angle between the x-axes is a. Then from *Figure 14.14(b)*:

$$x'_a = pr - qr = x_a \cos\alpha - y_a \sin\alpha \quad \text{and} \quad y'_a = an + mr = y_a \cos\alpha + x_a \sin\alpha$$

To obtain the general form for these equations, substitute $(180° + S)$ for α and add the translation amount $pi = f\tan(\frac{1}{2}\theta)$ to obtain the new origin at i, then

$$x' = -x\cos S + y\sin S \tag{14.7}$$

$$y' = -x\sin S - y\cos S + f\tan\left(\tfrac{1}{2}\theta\right) \tag{14.8}$$

where x and y are the photo coordinates measured about the fiducial axes.

Equation (14.1), for scale on a tilted photograph, can now be applied to the reduced coordinates to give the ground coordinates X and Y as follows:

$$X = Kx' \quad \text{and} \quad Y = Ky' \quad \text{where } K = H/(f - y\sin\theta)$$

14.3.9 Ground coordinates from a tilted photograph of rugged terrain

The data in this case are exactly the same as in *Section 14.4*, plus the elevations h (heights above/below MSL) of the points in question.

As the effect of ground relief is radial from the plumb point, then the rotation and translation is this time relative to v, where $pv = f\tan\theta$, then

$$x' = -x\cos S + y\sin S \tag{14.9}$$

$$y' = -x\sin S - y\cos S + f\tan\theta \tag{14.10}$$

Thus, from *Figure 14.15*, the new coordinates of a are

$$y'_a = vr \quad \text{and} \quad x'_a = ra$$

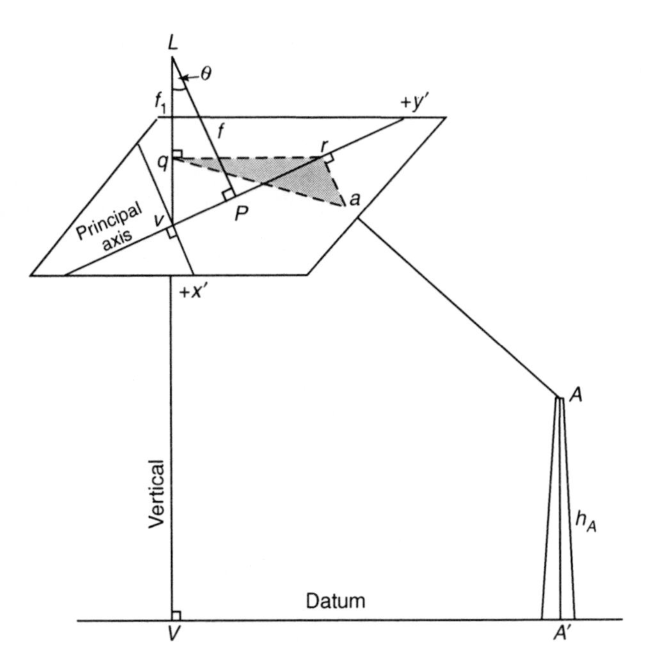

Fig. 14.15 *Tilted photograph of rugged terrain*

Mathematical rectification for each point is now carried out by considering horizontal planes passing through the plate parallels of the points in question. For example, in the case of point a, the horizontal plane through it is qra, and the rectified coordinates are therefore

$$y''_a = qr = y'_a \cos\theta \quad \text{and} \quad x''_a = ra = x'_a$$

The new focal length appropriate to the plane of rectification is now

$$f_1 = Lq = Lv - qv = f \sec\theta - y'_a \sin\theta$$

We now have an un-tilted photograph of ground point A with a new focal length f_1. It now only remains to multiply the photo coordinates by their appropriate scale as in *Section 14.3.2*, giving

$$X_A = x''_a(H - h_a)/f \quad \text{and} \quad Y_A = y''_a(H - h_a)/f$$

Worked example

Example 14.1 Two points A and B situated 10 and 40 m, respectively, above datum, are imaged on a near-vertical aerial photograph, taken from an altitude of 2000 m with a camera of focal length 152 mm. The photo coordinates of the points about the fiducial axes are measured by as follows:

	x (mm)	y (mm)
a	+50.00	+100.00
b	−100.00	+80.00

If the tilt and swing of the photograph are 2° and 20°, respectively, calculate the horizontal ground distance AB.

$$x'_a = -50\cos 20° + 100\sin 20° = -12.78 \text{ mm}$$

$$y'_a = -50\sin 20° - 100\cos 20° + 152\tan 2° = -105.76 \text{ mm}$$

$$x'_b = 100\cos 20° + 80\sin 20° = +121.33 \text{ mm}$$

$$y'_b = 100\sin 20° - 80\cos 20° + 152\tan 2° = -35.66 \text{ mm}$$

Rectified coordinates:

$$x''_a = x'_a = -12.78 \text{ mm}$$

$$y''_a = y'_a \cos 2° = -105.70 \text{ mm}$$

$$x''_b = x'_b = +121.33 \text{ mm}$$

$$y''_b = y'_b \cos 2° = -35.64 \text{ mm}$$

New focal length per point:

$$f_{1a} = f\sec\theta - y'_a \sin\theta = 152\sec 2° + 105.76\sin 2° = 155.78 \text{ mm}$$

$$f_{1b} = f\sec\theta - y'_b \sin\theta = 152\sec 2° + 35.66\sin 2° = 153.33 \text{ mm}$$

Ground coordinates:

$$X_A = x''_a(H - h_a)/f_{ia} = -12.78(2000 - 10.00)/155.78 = -163.26 \text{ m}$$

$$Y_A = y''_a(H - h_a)/f_{ia} = -105.70 \times 1990/155.78 = -1350.26 \text{ m}$$

$$X_B = x''_b(H - h_b)/f_{ib} = 121.33(2000 - 40.00)/153.33 = +1550.95 \text{ m}$$

$$Y_B = y''_b(H - h_b)/f_{ib} = -35.64 \times 1960/153.33 = -455.58 \text{ m}$$

Ground distance:

$$D = (\Delta X^2 + \Delta Y^2)^{\frac{1}{2}} = (1714.21^2 + 849.68^2)^{\frac{1}{2}} = 1933.64 \text{ m}$$

The above example serves to illustrate the need for a restitution system to correct photo measurements for the effects of tilt and relief displacement but would never be used for any purpose in practice.

14.4 GROUND CONTROL

The establishment of ground control points, which are clearly distinguishable on the air photographs, is very important to the photogrammetric process.

The minimum number of points required per photograph comprises two plan points to control position, scale and orientation and three height points to control level in the spatial model. Ground control, fixed by normal survey methods, should be more accurate than that attainable by the photogrammetric restitution system used.

Control points must consist of detail already clearly and sharply visible on the photographs and which can be well defined on the ground. Similarly, for height control the points chosen should lie in flat, horizontal ground free from vegetation. Steep slopes or peaks should be avoided to reduce the large height errors that would result from bad positioning within the photogrammetric process. The amount of control required depends largely on the scale and accuracy of the finished plan.

14.4.1 Pre-marked control

In the production of large scale engineering plans the control points are generally pre-marked targets. Their locations are indicated from initial photography and then, when established, the area is re-flown. A popular type of target used consists of a large white cross of durable material with arms in the region of 2 m long and 0.25 m wide. The size, however, is very much a function of photo scale, as already shown, and must be large enough to be clearly visible on the photographs but small for the centre to be precisely defined. Although pre-marking is more expensive than the use of existing detail, pre-marks may be so constructed as to be used for the control of setting-out, at a later stage.

14.4.2 Accuracy requirements

General rules quoted for the accuracy of fixing ground control are:

(a) for large-scale engineering plans $\pm 0.0002\,H$
(b) for medium-scale engineering plans $\pm 0.0003\,H$
(c) for small-scale topographic plans $\pm 0.0005\,H$

where H is the flying height.

The specifications apply to both plan position and height control. Thus for 1:10 000 photography using a wide-angle camera ($f = 150$ mm), H would be 1500 m and ground control fixed to an accuracy of ± 0.3 m in case (a) above. Based on this, an appropriate survey procedure could be instituted.

Normal survey procedures would be used for field completion of the final product to complete detail on the plan which may have been obscured on the photographs by cloud, glare, shadow, trees or other obstructions.

14.4.3 Aerial triangulation

For mapping at smaller scales and lower accuracies, the process of aerial triangulation may be used. This procedure provides control directly from the photographs, thereby reducing the amount, and thus cost, of ground control fixed by normal survey techniques. Aerial triangulation may be used to establish two- or three-dimensional control points, formerly by analogue methods in precision plotters, or by purely analytical processes.

In the analogue process, each stereo model is connected to the next, thus forming a strip of model coordinates. Each strip is then connected to the next; ultimately forming a set of block coordinates. Due to error propagation in the process, strip and block adjustments of the coordinates are necessary before they are transformed to fit the ground control.

Aerial triangulation forms a very important aspect of photogrammetry but is mentioned here only very briefly as it is beyond the scope of this book. Interested readers will find more detail in *Elements of Photogrammetry* by Paul Wolf and Bon de Witt (McGraw-Hill Publishing Co, 2000), *Introduction to Modern Photogrammetry* by Edward Mikhail *et al* (John Wiley & Sons, 2001) or *Digital Photogrammetry* by Yves Egels and Michel Kasser (Taylor & Francis, 2001).

14.4.4 Flight planning

The flight specifications for a particular project will vary with the type of project. For instance, photography required for interpretation purposes will not require the same detailed planning as that required for large-scale mapping.

The main factors to consider are the directions of the flight lines, the overlaps, scale and flying height. Some of the factors cannot be obtained until the flight has commenced. For instance, the heading direction and the time interval between exposures can only be calculated when the wind velocity at the time of flight is known. One also needs some idea of the number of photographs required in order to decide on the number of magazines or films to take or the storage requirement for digital photographs. The flying height of the aircraft is dependent on many factors ranging from aircraft capabilities, terrain conditions, and survey requirements. Flight planning is thus a skilled procedure requiring careful planning at all its various stages.

14.4.5 Direction of flight lines

Generally the area is flown parallel to the longest side to give the minimum number of strips. In this way the number of turns and run-ins, which are non-productive, are reduced to a minimum.

If large areas having different levels exist, such as mountain ranges or plateaus, the area may be flown parallel to these in order to avoid rapid variation in scale.

Each photograph in a strip normally overlaps the previous one by 60%, thus the new ground covered on each photograph is 40%. The purpose of the overlap is to permit stereoscopic viewing of the area. Each strip overlaps the previous one by 20 to 30% (*Figure 14.16*), thus complete coverage of the area is obtained.

The overlapping, which is automatically controlled on the air camera, is illustrated more clearly in *Figure 14.17*. The distance B between each photograph in the air is called the air base, while its equivalent on the photograph, b, is called the photo base. Due to the overlap, both of the principal points of the adjacent photographs will appear on the central photograph. The photo bases are in fact the direction of flight of the aircraft.

Care must be exercised when flying over steadily-rising ground, as failure to do so may result in the complete loss of the overlap required in both forward and lateral directions (*Figure 14.18*). The loss of forward overlap may be overcome by decreasing the exposure interval, whilst to ensure lateral overlap the flight lines must be based on the minimum lateral overlap over the highest ground.

The flight may also be affected by cross-winds (*Figure 14.19*) causing the aircraft to drift off the planned course. The triangle ABC may be solved to give the value of θ, and the craft is corrected on to course AD at a specific *air speed*, the wind velocity causing it to 'crab' the planned course AC at a different speed called the *ground speed*. Thus if the camera is not squared to the direction of flight the photographs will be crabbed, as shown, with resultant gaps in the coverage. This adjustment is carried out by the 'drift-ring

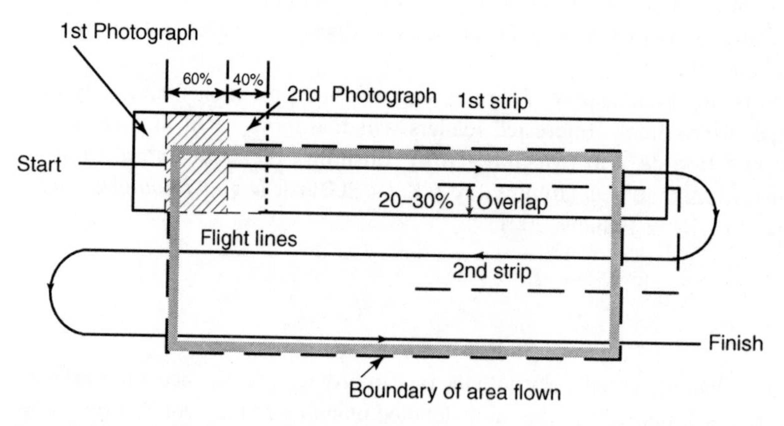

Fig. 14.16 *Flight lines*

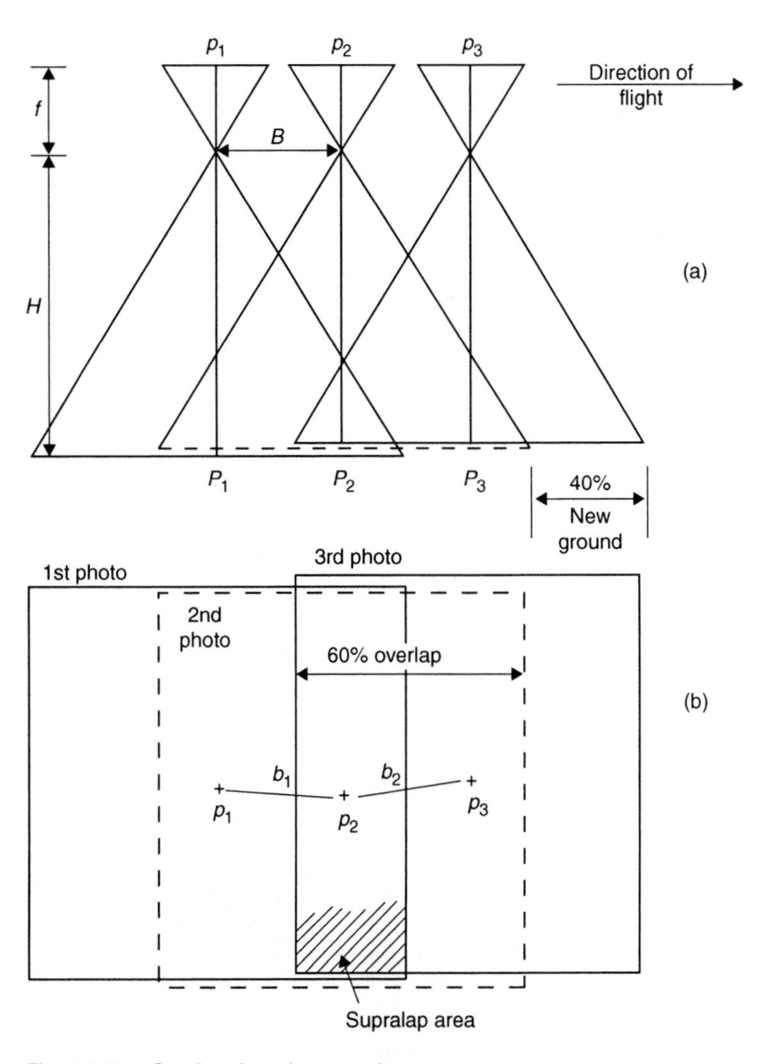

Fig. 14.17 *Overlapping photographs*

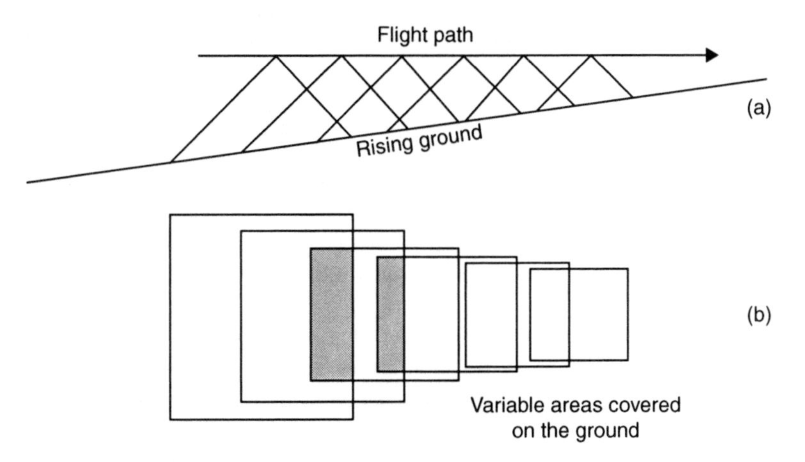

Fig. 14.18 *The effect of rising ground*

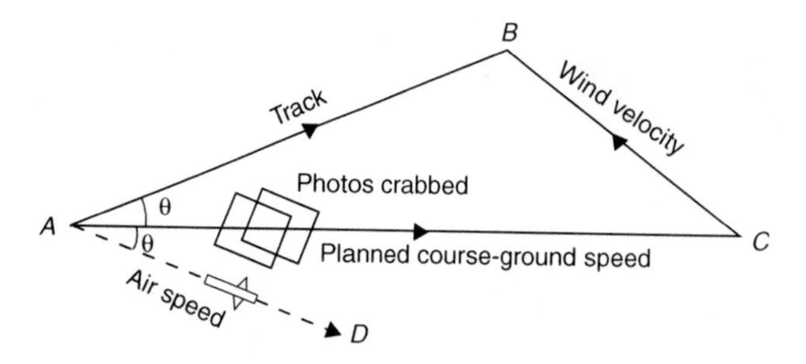

Fig. 14.19 *The effect of crab*

setting', which rotates the camera through θ about the vertical axis of the camera mount. Modern viewers and air cameras have largely eliminated this problem.

14.4.6 Scale and flying height

The scale of the photography will depend upon the map compilation techniques used. For example, 1 : 12 500 photography is frequently used to produce 1 : 2500 plans. As already shown, the flying height H is a function of the scale, thus using a normal wide-angle camera ($f = 152$ mm) for 1 : 12 500 photography gives:

$$f/H = 1/12\,500 = 0.152/H \quad \therefore\ H = 1900\text{ m}$$

Where there is great variation in ground relief or the area contains many tall buildings, the flying height may need to be increased.

Image movement, caused by the camera being in motion at the instant of exposure, can greatly affect the quality of the photograph. It can be reduced by flying higher, at slower speeds, by using faster shutter speeds or by *forward motion compensation* described earlier. As there is a limit to acceptable image movement, it will have an effect on the value of the flying height.

Where very hilly ground is encountered (*Figure 14.20*), or high urban construction with narrow streets, the use of a wide-angle lens at flying height H may result in much ground detail being obscured, i.e. dead ground at B and C. It may also be difficult to handle this photography stereoscopically. However, the use of a narrow-angle lens ($2f$) at twice the flying height ($2H$) would produce equivalent photography at the same scale. Also, all three points A, B and C are imaged on the normal-angle photography, whereas B and C are not imaged on the wide-angle.

14.4.7 Costing the project

The number of photographs needed to cover the project area will be required in order to cost the work and to estimate the amount of film required if a digital camera is not used. With conventional film, the points at which the film magazines should be changed can also be calculated. If possible, the magazines should be changed in the turns.

Consider an area 200 km by 100 km to be flown at an average scale of 1:10 000. At this scale the area covered by the photographs would be is 20 m by 10 m. If the photography has the standard format size of 230 mm × 230 mm, of which 60% is overlapped, so the new ground covered at this scale by each photograph is 40% of 230 mm = 92 mm.

Therefore the number of photographs per strip = 20 000 ÷ 92 mm = 218, plus say 2 each end to ensure complete coverage, i.e. a total of 222. In the same way the number of strips, assuming a 30% lateral overlap

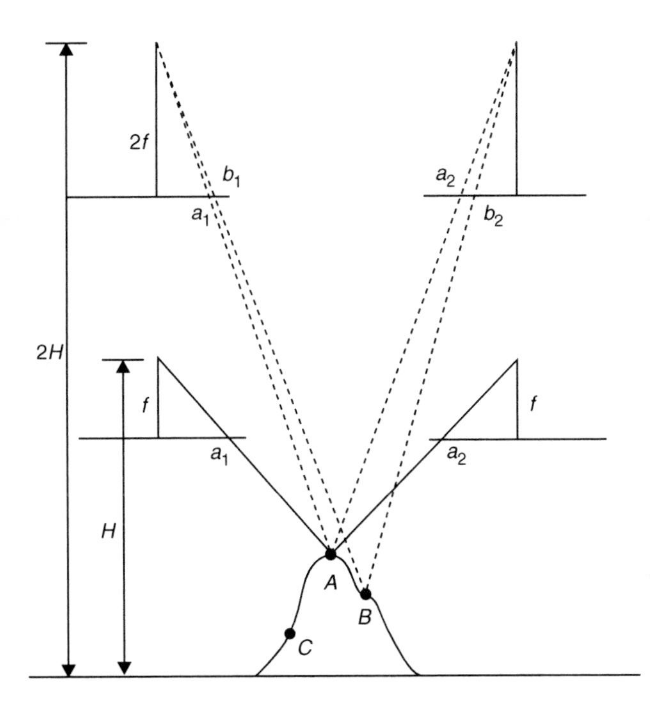

Fig. 14.20 *Flying height*

$= 10\,000 \div (70\%$ of 230 mm$) = 63$ strips.

$\therefore$ Total number of photographs $= 222 \times 63 = 13\,986$

The coordinates of the ends of the flight lines can be put into the GPS as waypoints to enable the aircraft to be flown along each flight line. In all, it can be seen that careful planning is needed to minimize the cost of a project. Look at *Figure 14.4* again.

14.5 STEREOSCOPY

So far, only the production of plan detail has been dealt with; *stereoscopy*, the process of seeing in three dimensions, enables the vertical dimension to be obtained. The application of stereoscopy to air survey will now be illustrated by relating the human sight processes to that of the air camera producing overlapping pairs of photographs.

14.5.1 Stereoscopy in air survey

Consider first a simplified explanation of the visualization processes when looking down at a survey arrow sticking in the ground (*Figure 14.21*). The arrow is viewed simultaneously from two different positions, the two images fusing together to form a three-dimensional image in the mind. The angles α_1 and α_2 are termed the *angles of convergence*, and exist because of the ability of the eyes to rotate simultaneously in their sockets. The ability to focus for varying distances is called *accommodation*, while the aperture control variation of the pupil of the eye is called *adaptation*. The angles β_1 and β_2 are called the *parallactic angles*

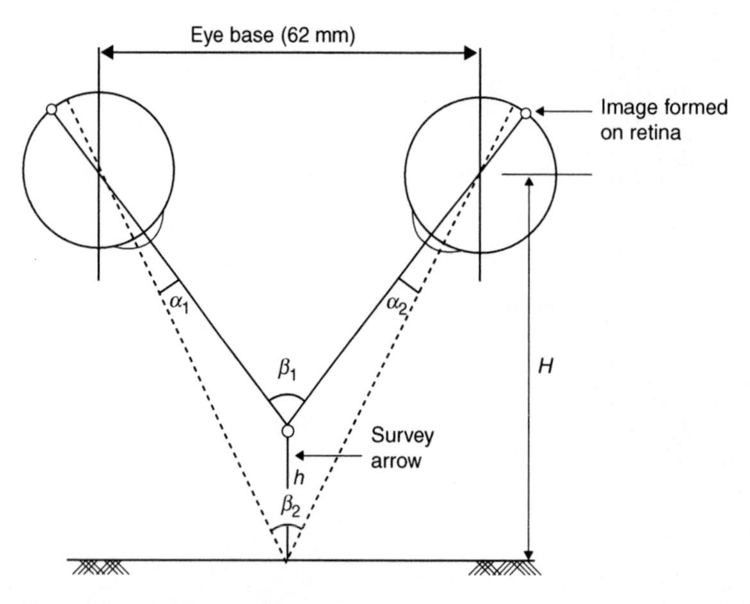

Fig. 14.21 *Eye base view*

and are a function of the stereoscopic perception of height, i.e.

$$h = f(\beta_1 - \beta_2) = H^2(\alpha_1 - \alpha_2)/e$$

where e is the eye base, about 62 mm for most adults

Commonsense tells us that if a person was taken to a height of, say, 2000 m, the parallactic angles would be so small as to render height perception impossible. The 'horizontal parallaxes' of the survey arrow are shown on the retina of the eye. That these are a function of the parallactic angles is obvious from the diagram.

In the determination of height there is a definite relationship of the eye or air base to the flying height. Thus when flying, the camera separation must be greatly increased as shown in *Figure 14.22*. The comparison between human viewing and air survey can now be clearly seen.

If the negatives are now printed as photographs (positives), and viewed simultaneously, so that the left eye sees only the left photograph and the right eye only the right photograph, then a three-dimensional image will form in the mind. The above condition can be most easily obtained by viewing the photographs under a stereoscope, as in *Figure 14.23*. The three-dimensional image formed is termed a *stereo model*, and the two photographs used are termed a *stereo pair*. The stereo model is usually exaggerated and this can be useful in the heighting process, particularly where the terrain is relatively flat. This effect can be increased or reduced when planning the photography. If the value of f is fixed, then from the base/height ratio, it can be seen that to halve the flying height would double the impression of height. It can also be shown that increasing the viewing distance of the stereoscope produces a proportionate increase in the impression of height.

14.5.2 Parallax

As already shown, stereoscopic height is a function of the parallactic angles, which are in turn a function of the horizontal parallaxes. As the angles occur in space, they cannot be measured on an aerial photograph. However, the horizontal parallaxes can be used to ascertain vertical heights.

Figure 14.24 illustrates a stereo pair of photographs in plan and elevation, on which it is intended to measure the parallax of A (P_A). By definition the parallax of a point is its apparent movement, parallel to

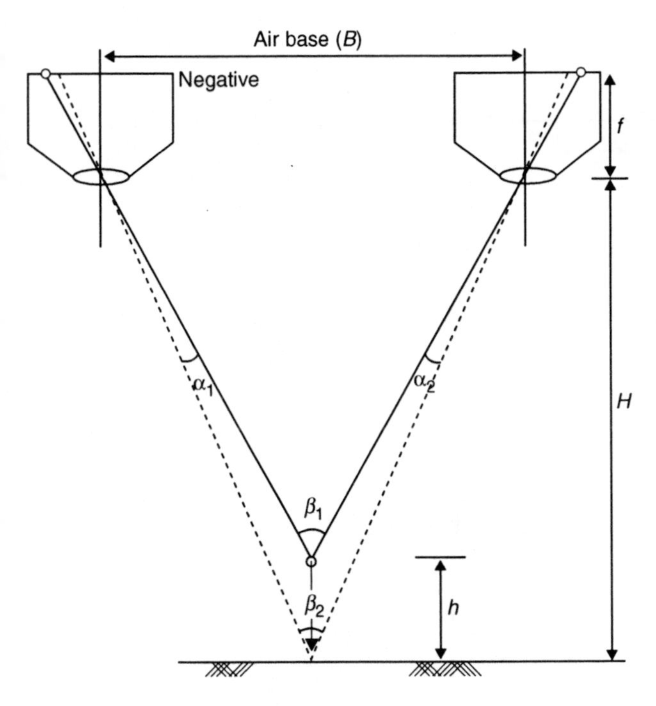

Fig. 14.22 *Air base view*

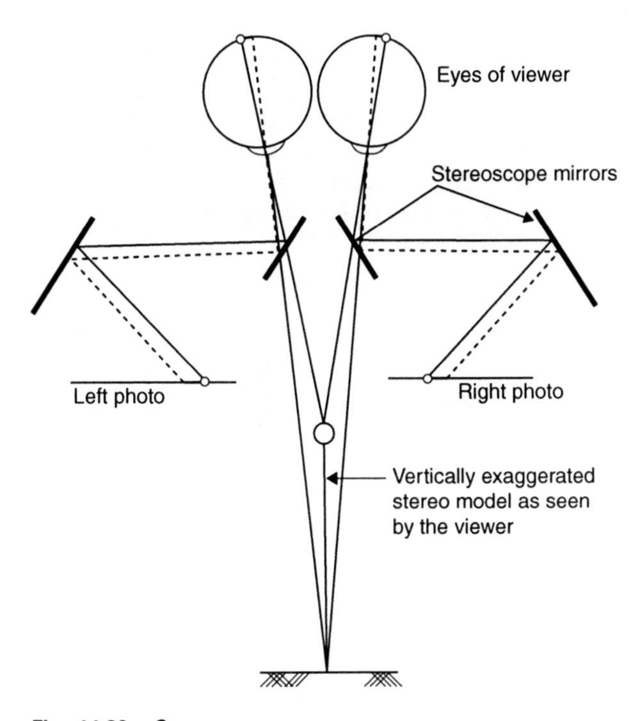

Fig. 14.23 *Stereoscope*

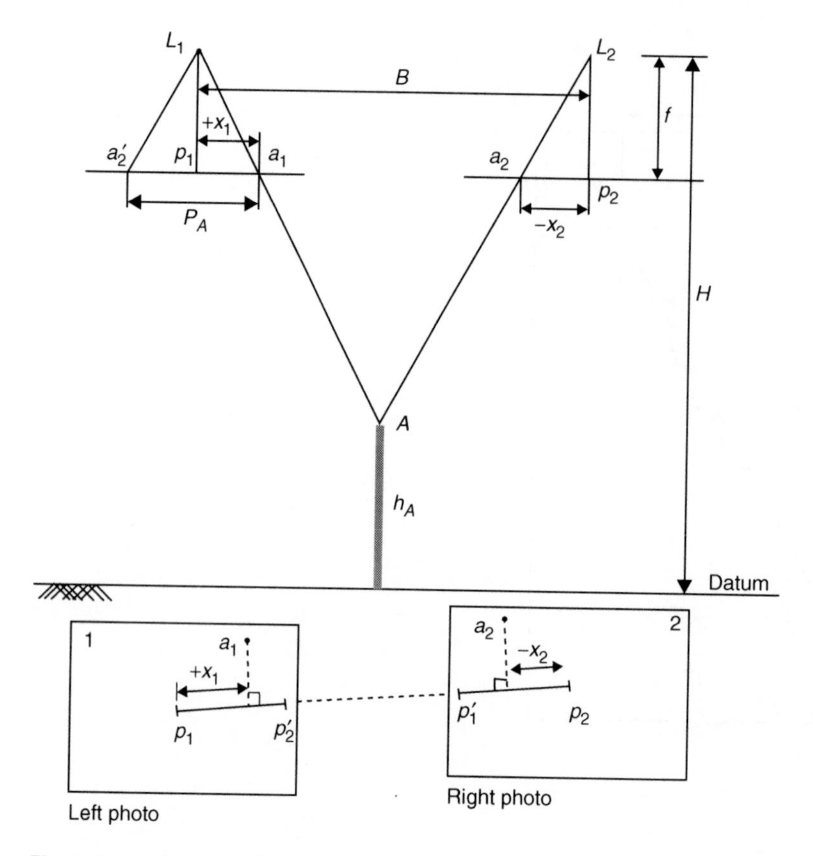

Fig. 14.24 *Photo parallax*

the eye base, when viewed from two different positions. Thus A appears at a_1 when viewed from L_1; and at a_2 when viewed from L_2. By overlapping the two photographs, the apparent movement of A is shown as a_1a_2', i.e. L_1a_2' is parallel to L_2a_2. It is thus shown that the parallax of A is the algebraic difference of the x-ordinates.

$$\therefore\ P_A = a_1a_2' = [x_1 - (-x_2)] = (x_1 + x_2)$$

The x-ordinates are always measured parallel to the photo base and not the fiducial axes. Whilst this indicates that the parallax of a point could easily be measured on the photograph using a simple ruler, in fact it is the difference in parallax between points which is measured, as will be shown later.

14.5.3 Basic parallax equation

The basic parallax equation is easily deduced from *Figure 14.24* in which triangles L_1L_2A and $a_2'L_1a_1$ are similar:

$$a_2'a_1/L_1p_1 = L_1L_2/(H - h_A)$$

but $a_2'a_1 = P_A \quad L_1p_1 = f$ and $L_1L_2 = B$

$$\therefore\ P_A = fB/(H - h_A) \tag{14.11}$$

As shown in *Figure 14.24, equation (14.11)* assumes absolutely vertical photographs taken at exactly the same flying height. This rarely occurs, so heights obtained using this formula are approximate.

The principal point on one photograph of a stereo pair may be plotted on the other, and *vice versa*. The distance between the principal point of a photograph and the plotted principal point of the other photograph is called the photo base. The photo base may be plotted on each of the pair of the photographs. For a variety of reasons, including imperfections in the plotting process, the two bases may be slightly different. From the scale of the photograph it is known that:

$$\text{Scale} = b/B = f/(H - h)$$

where b is the mean photo base $\frac{1}{2}(b_1 + b_2)$, and h is the mean height of the terrain.

$$\therefore\ B = b(H - h)/f$$

On substitution into *equation (14.11)* this gives

$$P_A = b(H - h)/(H - h_A)$$

As $(H - h)$ is the mean flying height, the equation is frequently written as

$$P_A = bH_0/(H - h_A) \tag{14.12}$$

However, as previously stated, it is normal practice to measure the difference in parallax (ΔP) between two points. This can be done using an instrument called a *parallax bar* (*Figure 14.25*). Consider two points A and C:

$$P_A = fB/(H - h_A) \quad \text{and} \quad P_C = fB/(H - h_C)$$

$$\therefore\ (H - h_A) = fB/PA \quad \text{and} \quad (H - h_C) = fB/P_C$$

$$\therefore\ (H - h_C) - (H - h_A) = h_A - h_C = \Delta h_{AC} = fB/(P_C^{-1} - P_A^{-1}) = fB/(P_A - P_C)P_A^{-1}P_C^{-1}$$

but $P_A - P_C = \Delta P_{AC} =$ difference in parallax between A and C.

$$\therefore\ \Delta h_{AC} = fBP_A^{-1} \times \Delta P_{AC} P_C^{-1}$$

But since $P_C = P_A + \Delta P_{AC}$

$$\Delta h_{AC} = fBP_A^{-1} \times \Delta P_{AC}(P_A + \Delta P_{AC})^{-1}$$

so

$$\Delta h_{AC} = (H - h_A)\Delta P_{AC}(P_A + \Delta P_{AC})^{-1} \tag{14.13}$$

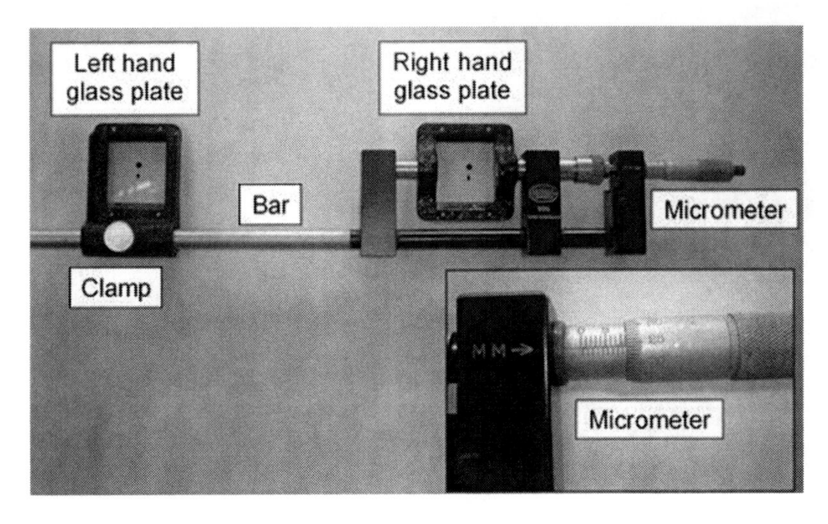

Fig. 14.25 *Parallax bar*

In relatively flat terrain ΔP_{AC} is negligible compared with P_A so

$$\Delta h_{AC} = (H - h_A)\Delta P_{AC} P_A^{-1} \qquad (14.14)$$

An inspection of the basic *equation (14.11)* shows that as h_A increases, then P_A must also increase: thus an important rule of parallax heighting is: the higher the point, the greater its parallax.

14.5.4 Measurement of parallax

Heighting may be carried out with the aid of a parallax bar (see *Figure 14.27*). This instrument is essentially a rod, carrying two glass plates with fine dots etched on them. The smaller pair of dots is used when the stereo model is viewed under magnification. The left plate can be moved anywhere along the bar and clamped in position; the right plate can be moved only by manipulation of the micrometer. Parallax measurements may be made to an accuracy of 0.01 mm.

In the heighting procedure the photographs are set with their bases co-planar for viewing under the stereoscope (*Figure 14.26*). It is important therefore that the bar is kept parallel to the base (p_1p_2) when measuring. Consider now the measurement of height *AC* in *Figure 14.27*. The bar is set to the mid-run of the micrometer and the right-hand dot (RHD) is placed over the image a_2, the left-hand plate is unclamped and the left-hand dot (LHD) placed over a_1. When viewed through the stereoscope the two dots will have fused into one, and appear to be resting on the point *A* in the stereo model. The parallax bar reading is M_A. This is *not* a measure of the distance a_1a_2, for the micrometer could have been set to any reading prior to the operation. The LHD now remains clamped in this position on the bar for all future heighting operations on this pair of photographs. It is now moved to c_1, and as the separation has not yet been altered, the RHD will be at *d*, causing the fused image to appear floating in space at *D*. While looking through the stereoscope the RHD is moved by manipulating the micrometer until the floating dot appears resting on the ground at *C*, in which case the RHD will, as *Figure 14.27* shows, be over c_2 on the photograph; the reading M_C is noted. As parallax has already been defined it should be obvious that the individual readings are meaningless. However, as L_1a_2' and L_1c_2' are parallel to L_2a_2 and L_2c_2, respectively, it can be seen that the 'difference' in the bar readings ($M_C - M_A$) is equal to the 'difference' in parallax ($P_C - P_A = \Delta P_{AC}$),

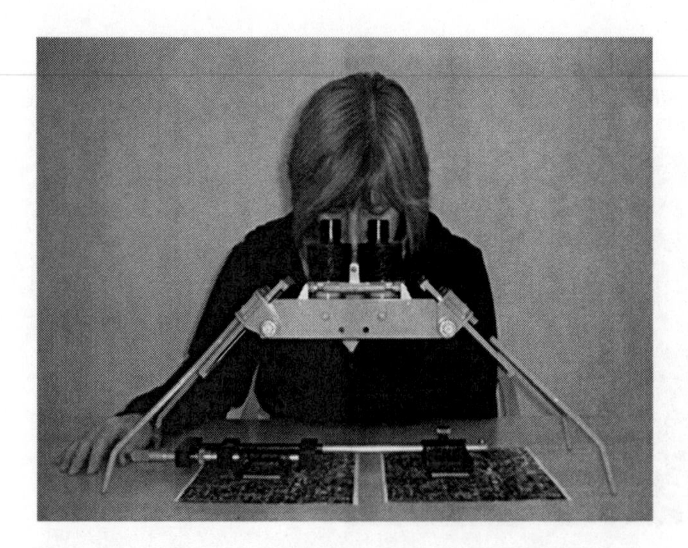

Fig. 14.26 *Stereoscope with air photographs and parallax bar*

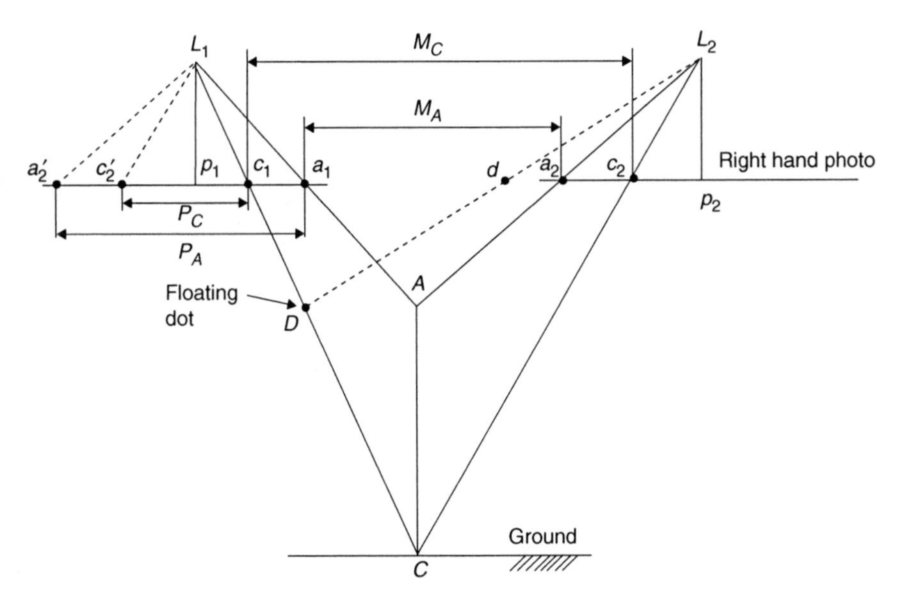

Fig. 14.27 *Parallax bar measurements*

which in turn is a function of the 'difference' in height of A and C (Δh_{AC}) and can be computed using *equation (14.13)*.

14.5.5 Basic procedure

Assuming that it is required to find the levels of a grid of points in the stereoscopic overlap of a pair of photographs, one must commence from a ground control point of known level, as follows:

(a) Using the basic parallax formula, $P_A = fB/(H - h_A)$, calculate the parallax P_A of ground control point A whose level h_A is known. f, B and H will also be known. (Refer to *Section 14.5.3*.)
(b) Obtain a parallax bar reading on the image points a_1 and a_2 of ground control point A.
(c) Now obtain a bar reading on point C. The difference between the readings on A and C will be equal to ΔP_{AC}.
(d) As P_A is known, then $P_C = P_A + \Delta P_{AC}$ (ΔP_{AC} will be positive if C is higher than A and negative if lower). Whether or not C is higher than A, may be detected from an examination of the stereo model and/or the bar readings.
(e) Now calculate the level of point C, i.e. h_C, from the basic formula $P_C = fB/(H - h_C)$. Alternatively, one may calculate Δh_{AC} from *equation (14.13)* and knowing the level of A thereby obtain the level of C.
(f) This process is now continued. For example, a bar reading on point D will give ΔP_{AD}, from which h_D or Δh_{AD} can be obtained as shown in (d) and (e) above.

14.5.6 Parallax height corrections

So far it has been assumed that when taking each photograph the aircraft is flying straight and level. Only by coincidence would this ever be the case. Therefore in practice it would be necessary to correct for the tilts of the aircraft and hence the camera.

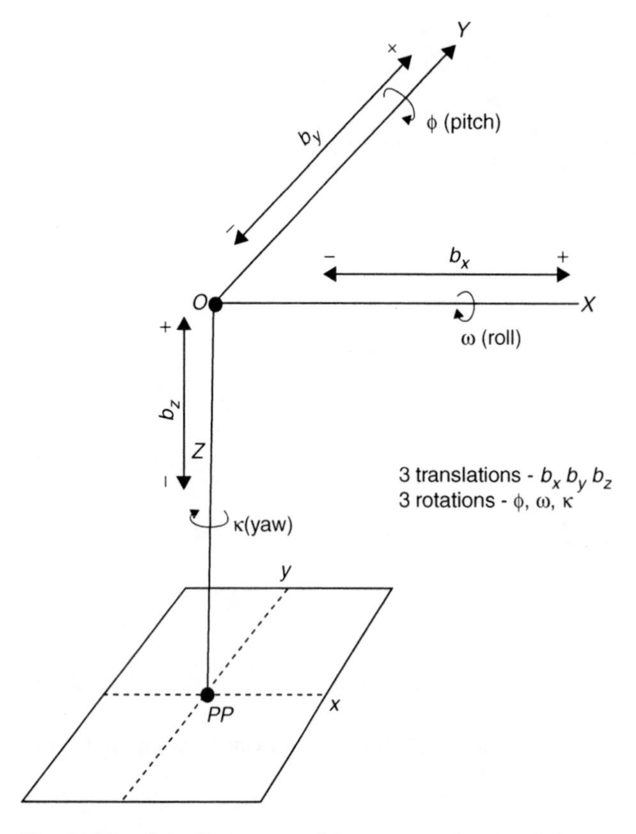

Fig. 14.28 *Coordinate axes of the camera showing 6 degrees of freedom*

The convention adopted for tilts is to denote rotation of each camera about three perpendicular axes, X, Y and Z, as ω, ϕ and κ, respectively, as shown in *Figure 14.28*. Translations along each axis are b_x, b_y and b_z, respectively.

Pitching of the aircraft from nose to tail would cause rotation about the y-axis ($\delta\phi$) as shown in *Figure 14.29*. The result of this tilt is to image points 3 and 4 at $3'$ and $4'$ with resultant error in the parallax measurement of $3b$ and $4a$. As shown, parabolic deformation of the stereo model takes place and the error in ground heights of any point can be shown to be equal to $(X^2 + Z^2)\delta\varphi/B$.

Rolling the aircraft about its longitudinal axis would cause rotation about the x-axis ($\delta\omega$), as shown in *Figure 14.30*. The result of this roll is to displace points 2 and 3 radially from PP1. The displacement of point 2 is in the y-direction, hence its parallax measurement is unaffected. The parallax measurement of point 3 will be in error by distance $3b$ which will distort the height by an amount equal to $XY\delta\omega/B$.

Yaw error is eliminated by careful baselining of the photographs. Error in this process will result in a raising and lowering of each half of the stereo model, as shown in *Figure 14.31*.

Variation in the heights of the camera at adjacent exposures tilts the air base and so tilts the stereo model about the y-axis, as shown in *Figure 14.32*.

The combined effect of all the above errors in each photograph comprising a stereo pair is to transform the stereo model. It may be shown to first order, that if the tilts are small and ground relief not excessive, the error in parallax heights may be expressed in terms of the photo coordinates (x, y) of the image point of the left-hand photograph as

$$\delta h = a_0 + a_1x + a_2y + a_3xy + a_4x^2 \tag{14.15}$$

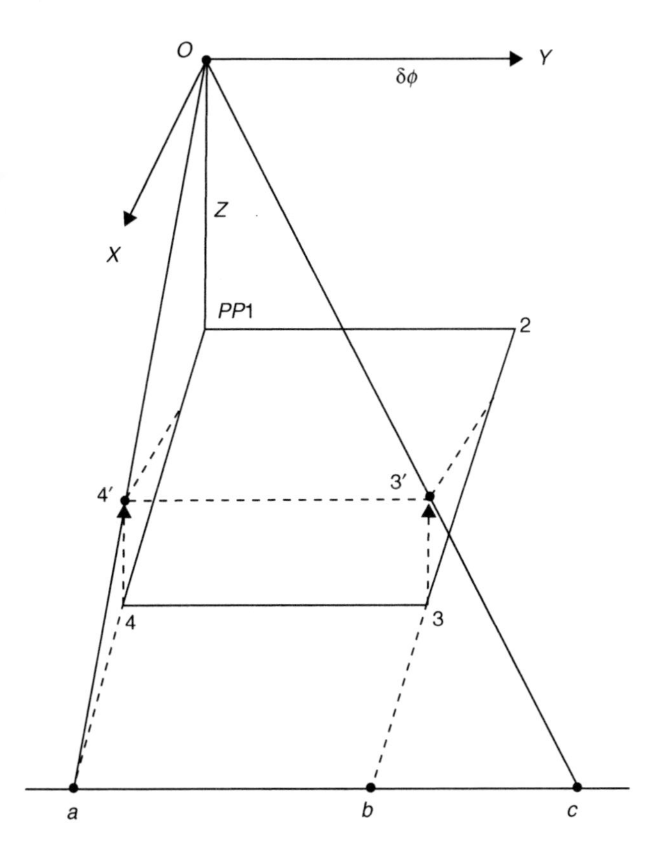

Fig. 14.29 *Pitch of the aircraft*

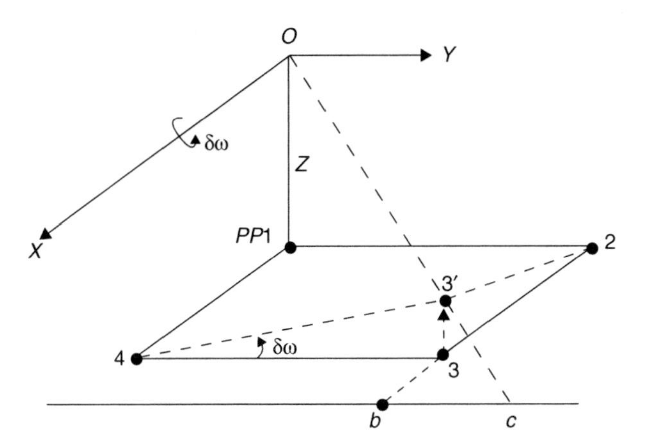

Fig. 14.30 *Roll of the aircraft*

To correct crude heights requires five ground control points whose levels (h_i) are known, distributed throughout the overlap area, as shown in *Figure 14.33*. The parallax heights (h'_i) of the five points are found taking, say, point 1 as datum and $h_i - h'_i = \delta h_i$. The centre of the photo base $p_1 p_2$ is taken as the origin for the x, y coordinate system and the coordinates of all five points scaled from the left-hand

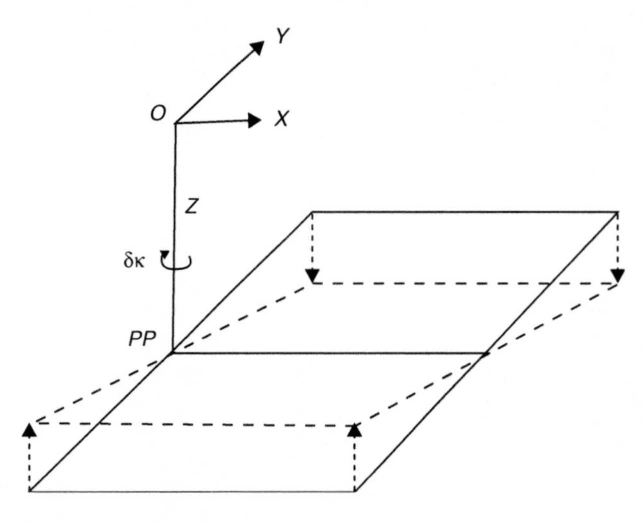

Fig. 14.31 *Yaw of the aircraft*

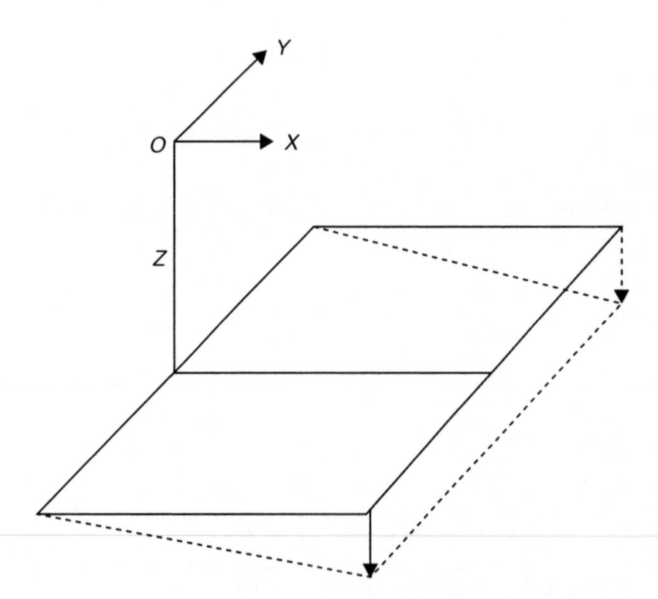

Fig. 14.32 *Tilt of the stereo model about the y-axis*

photograph and inserted in the five equations, i.e.

$$h_1 - h'_1 = \delta h_1 = a_0 + a_1x_1 + a_2y_1 + a_3x_1y_1 + a_4x_1^2$$

$$\vdots \qquad \vdots \qquad \vdots$$

$$h_5 - h'_5 = \delta h_5 = a_0 + a_1x_5 + a_2y_5 + a_3x_5y_5 + a_4x_5^2$$

The equations are then solved for the coefficients $a_0, a_1, \ldots, a_4$. Thereafter, the crude height (h') of any point in the overlap may be corrected by δh, using the above coefficients and its photo coordinates in *equation (14.15)*.

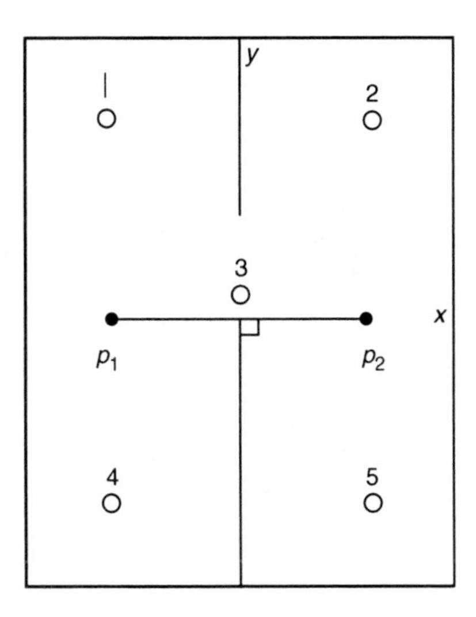

Fig. 14.33 *Control points for crude height corrections*

14.6 RESTITUTION SYSTEMS

Restitution is the fundamental problem in photogrammetry and involves establishing the positions and orientations the photographs had at the time of flight, and thereafter relating the correctly-formed stereo model to ground control.

In the past, restitution methods were either analogue, using universal or precision plotters, or analytical. With modern digital photogrammetry, they are now entirely analytical.

The problem may be illustrated as follows. *Figure 14.34* shows two photographs in a horizontal plane projecting images of a single point, *A*, onto a parallel plane. Because the projection system has not been

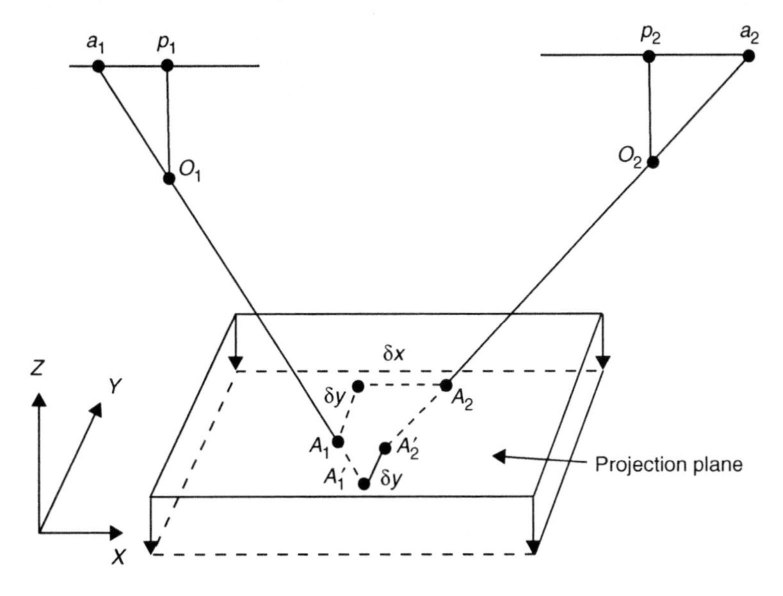

Fig. 14.34 *Projected images*

properly oriented, a_1 and a_2 intersect at A_1 and A_2 instead of coinciding at A. The discrepancy between A_1 and A_2 may be expressed in coordinate terms as δx and δy, and are called the x and y parallax.

The δx value is eliminated by lowering the projection plane in the Z direction (b_z) until the two points are imaged at A'_1 and A'_2, separated only by δy.

y-parallax is eliminated by orienting and moving the photographs through by b_z, b_y, ϕ, ω and κ as shown in *Figure 14.28*. This procedure, known as relative orientation, is carried out over six standard points distributed throughout the stereo model, and when complete it establishes the photographs in their correct relative positions. Failure to achieve correct *relative orientation* will result in a distorted model and may have a significant effect on the accuracy of height measurements.

The above model must now be correctly scaled and oriented to the ground coordinate system. This process is called *absolute orientation* and can only be achieved by the use of ground control.

First scale the model based, in its simplest form, on the known distance between two ground control points. This is achieved by altering the separation of the photographs in the x-direction, i.e. b_x translation.

Next level the spatial model by rotating it about its X- and Y-axis, i.e. ϕ and ω, until it conforms to the height data of at least three ground control points.

When this process is completed, detail, spot heights and contours may be plotted. Alternatively, the three-dimensional coordinates of points can be measured to produce a digital ground model (DGM).

The accuracy of restitution and thus the final plan is dependent on the accuracy of the ground control and its correct identification.

14.6.1 Orthophotomaps

Although the principal end product of photogrammetry, as far as the engineer is concerned, is a plan or DGM, other types of plans are available in the form of mosaics or orthophotomaps.

Mosaics are formed by matching up the photographs to get the best possible fit. No account is taken of displacements due to tilt and relief, and the assembly may or may not be fitted to any form of ground control.

An orthophoto is one that has been corrected for tilt and relief displacements. In this way the correctly oriented stereo model is plotted at true map position and correct scale (*Figure 14.35*). The end product may be a contoured photograph to correct scale containing very small errors of position and height. Although not quite as accurate as the line drawn plan, it can be produced much more quickly and will contain all the land form detail not usually shown on a plan and is the reason why a mosaic or orthophoto may be preferred, in some instances, to a plan. For example, in flood control, geological investigation or irrigation works the ability to see the areas involved could be extremely useful.

14.6.2 Commissioning vertical air photography and derived products

Air photography and products derived from it, although expensive, on a large project may be cheaper than ground based survey methods. Before commissioning such work it may be wise to consult the RICS's excellent guidance on the subject *Vertical Aerial Photography and Derived Digital Imagery – Client Specification Guidelines* written by the RICS's Mapping and Positioning Panel and available from RICS Books.

14.6.3 Applications of photogrammetry in engineering

Apart from its many applications in civil engineering, photogrammetry is widely used in forestry, town planning, architecture and even dentistry and medicine. Only some applications to civil engineering will be considered here.

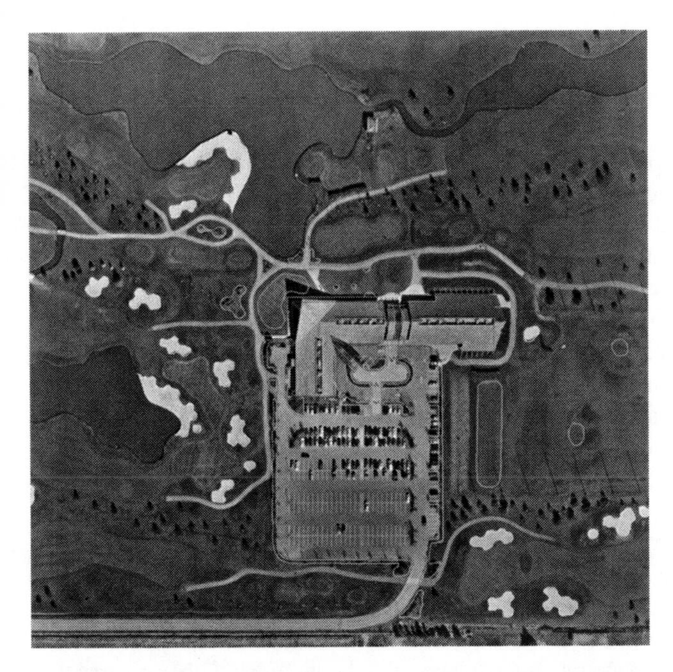

Fig. 14.35 *Orthophotomap – courtesy of Delta Aerial Surveys Ltd*

14.6.3.1 Highway optimization

UK is well-mapped so it is possible to reduce the area to be investigated for a new route into a relatively small band. Aerial photographic interpretation will serve as a very useful aid in this initial decision. The examination of even stereo pairs of photographs can supply an enormous amount of information to the trained eye, such as the geology of the area, main soil types, faults, land-slip areas, areas presenting drainage problems, location of borrow and quarry sites, major obstacles, expensive land, best grades, etc.

With restitution a digital terrain model may be formed and from this data software may be used to select the best route with earthwork quantities, plots of cross-sections and longitudinal sections along the centre line of the proposed route and produce mass-haul diagrams. The DTM may be used to generate flythrough models for visualization purposes.

14.6.3.2 Traffic engineering

Photographs may be used in land use studies to enable travel patterns to be estimated and predicted.

Time-lapse photography of a traffic route can provide information such as traffic speeds and density, concentration of traffic for selected time periods, en route travel time and the relative productivity of the various route segments. In this *sky-count* technique the tedium of counting the cars on the photographs has been eliminated by software applications using infra-red photography.

Traffic management, broadly aimed at improving traffic flow, can be aided in many ways by air photographs. The photographs provide a visual inspection of a large area at a glance and can be taken to show on- and off-peak flows, normal and congested routes, non-utilization of streets, parking characteristics, junction studies, effect of public transport on traffic flows, etc. They may also be used to produce traffic density contour maps and to provide a permanent inventory of roads, streets and car parks.

14.6.3.3 False colour imagery

A photograph, whether taken on the ground, from the air, or from space such as Landsat imagery (http://www.landsat.org/index.html) are able to detect the nature of an object without actually touching it. All photographs portray detail by a comparison of the visible light reflected from various objects. This light comprises electromagnetic energy with wavelengths from 0.4 μm to 0.7 μm. Energy whose wavelength is less than 0.4 μm is called *ultra-violet*, and above 0.7 μm *infra-red*. The camera is capable of recording energy within the 0.3 μm to 1.2 μm range, but above this upper value special equipment is required.

In colour photography, each distinctive colour is a function of the light reflected by objects, which in turn is a function of the energy absorption and reflection characteristics. Thus, as blue has different reflection characteristics to red, it is possible to distinguish between them. However, by sensing in the infra-red spectrum it is possible to distinguish different objects having the same colour, due to the variable energy reflection characteristics. This is particularly impressive in the field of ecology, where healthy and diseased vegetation will appear as different colours on infra-red colour film (false colour), even though to the human eye apparently identical.

To record energy in the 1–20 μm band, thermal infra-red devices are used. These devices record variation in energy due to variation in temperature. The terrain is sensed from the air and a thermal image built up. A typical example of its use is in the detection of river pollution where pollutants having different temperatures and may be recorded as shades varying from white through to black. Correlation with the various origins of these pollutants enables a more detailed analysis of the river to be made. This form of sensing can operate day or night, but cannot penetrate atmospheric cloud conditions. Other prospective uses of this technique are the detection of various rock and soil types, and assessment of the moisture content of soils.

14.7 TERRESTRIAL PHOTOGRAMMETRY

This form of photogrammetry utilizes photographs taken from a ground station. A camera may be combined with a theodolite which allows the position and orientation of the camera to be defined. The theodolite enables the direction of the principal axis of the camera to be found, relative to a base line.

14.7.1 Principle

At each station the camera is carefully centred and levelled such that the principal axis of the camera is horizontal and the plane of the photograph vertical. The plan position of a ground point can then be fixed from the terrestrial photograph. *Figure 14.36* indicates the position of a point A relative to the fiducial axes of the photograph. The horizontal axis x is called the *horizon line*, while the vertical axis y is called the *principal line*.

The horizontal and vertical angles, θ and ϕ, respectively, may be defined as follows:

$$\tan\theta = x_a/f$$

$$\tan\phi = -y_a/La' \quad \text{but} \quad La' = x_a/\sin\theta \quad \text{or} \quad La' = f/\cos\theta$$

$$\therefore \tan\phi = -y_a \sin\theta/x_a = -y_a \cos\theta/f$$

Figure 14.37 shows a terrestrial photogrammetric camera mounted on a theodolite.

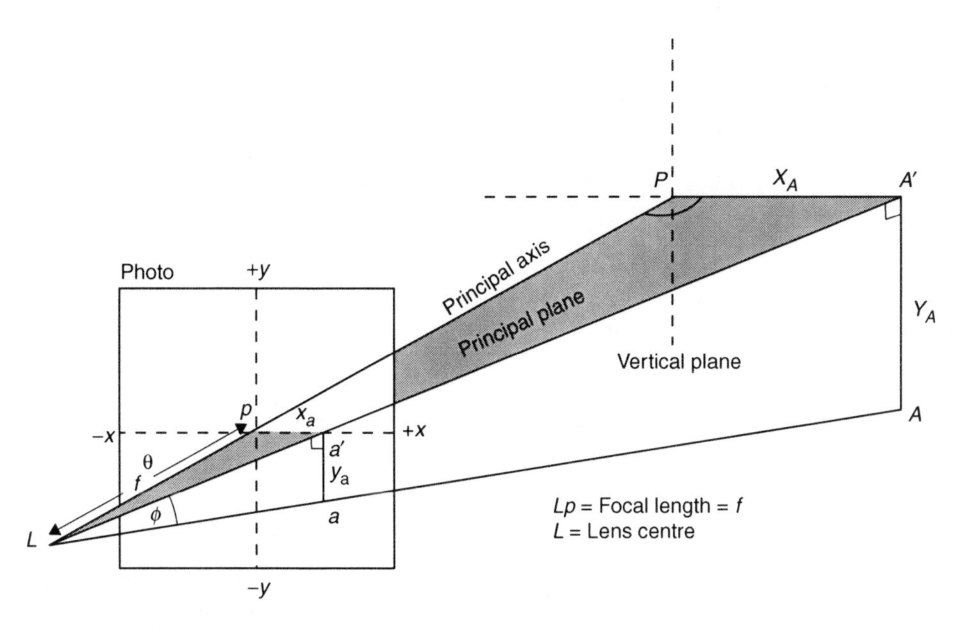

Fig. 14.36 *Terrestrial photograph*

Fig. 14.37 *Terrestrial photogrammetric camera*

14.7.2 Intersection

In this method (*Figure 14.38*) the camera axis is oriented at any angle to the base line L_1L_2 and the photographs are taken from both ends of the base line. The position of a point A may be fixed by intersection as in *Figure 14.38(a)*. The level of the point relative to the principal plane can be found by similar triangles

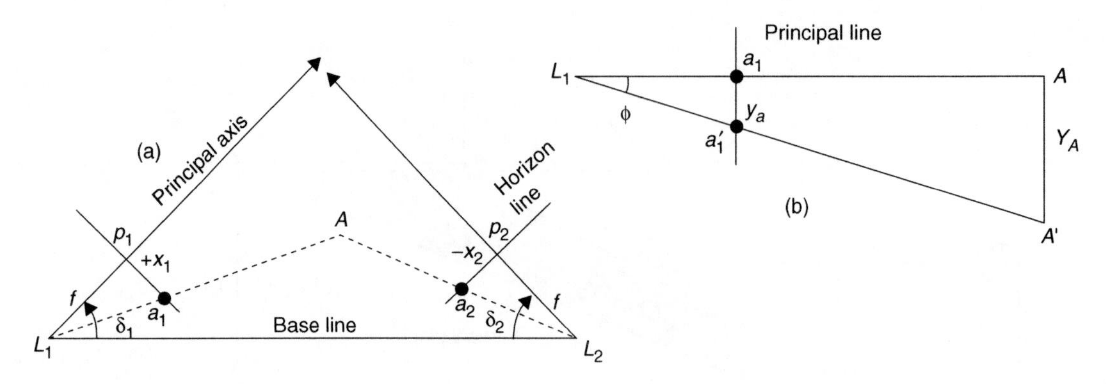

Fig. 14.38 *(a) Plan view and (b) elevation*

(*Figure 14.3(b)*)

$$Y_A/L_1A = y_a/L_1a_1$$

but from *Figure 14.38(a)* $L_1a_1 = (x_1^2 + f^2)^{\frac{1}{2}}$

$$\therefore Y_A = L_1A \times y_a/(x_1^2 + f^2)^{\frac{1}{2}}$$

These simple techniques have largely been replaced by stereoscopic methods.

14.7.3 Stereoscopic methods

To facilitate stereoscopic viewing the photographs are taken from each end of a base line with the principal axis at 90° to the base (*Figure 14.39*). The base should be of such length as to give a well-conditioned intersection of rays, and accurately measured to reduce the propagation of errors from this source. From *Figure 14.39*, triangles L_1AL_2 and $a_2'L_1a_1$ are similar:

$$\therefore Z_A/L_1L_2 = L_1p_1/a_1a_2'$$

$$\therefore Z_A = fB/P_A \tag{14.16}$$

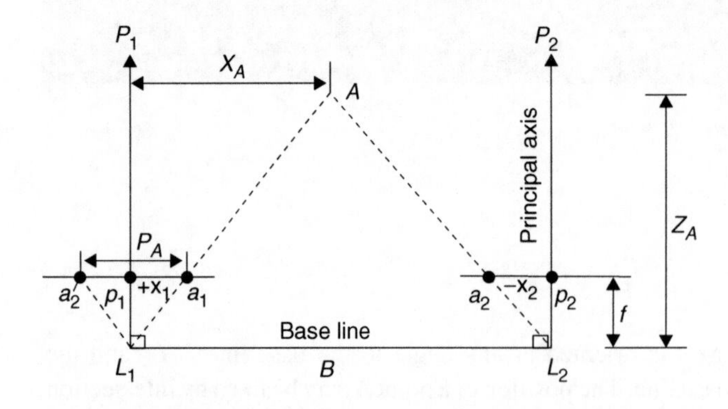

Fig. 14.39 *Plan view*

where P_A is the parallax of A. Similarly:

$$X_A/Z_A = x_1/f \ \therefore \ X_A = Z_A x_1/f$$

From *Figure 14.38(b)* $Y_A/y_a = L_1A/L_1a_1 = Z_A/f$ (from *Figure 14.39*)

$$\therefore \ Y_A = Z_A y_a/f$$

Note that in *equation (14.16)*, $P_A = [x_1 - (-x_2)] = (x_1 + x_2)$; if A was to the right of P_2 then $P_A = (x_1 - x_2)$, whilst if it was to the left of P_1, then $P_A = (x_2 - x_1)$.

14.7.4 Application

The method was originally devised for topographic surveys of very rugged terrain, and, as such, was widely utilized in Switzerland. The following instances of its use will serve to indicate present-day applications:

(a) Survey of sheer rugged faces in quarries, dam sites, etc.
(b) Short-base methods are used to make road-accident plans.
(c) Wriggle surveys in tunnels
(d) Recording architectural details for the restoration of ancient buildings.
(e) Scientific projects, such as stereoscopic photographs of intensely hot or other hazardous objects.

14.8 DIGITAL 'SOFTCOPY' PHOTOGRAMMETRY

In the application of photogrammetry, analogue and digital data is derived from traditional analogue photographs. The processing of that data involves manual applications from developing archiving and cataloguing the photographs to setting them up in photogrammetric machines and plotting features and extracting height information from the stereo model. Although much faster than detail capture by traditional ground survey methods, even by GPS, the process is still slow and labour intensive. The overall cost of the process is largely related to the man-hours necessary to complete the survey task rather than the cost of the equipment concerned, even the aircraft used to fly the photography.

The development of digital or softcopy photogrammetry has been largely driven by advances in the technology of data capture and of computing. Digital photogrammetry obviously needs digital data. Initially this was derived from traditional analogue photographs which were placed in a scanner to convert them into a digital product. If a 23 centimetre square standard photograph is scanned at 1200 dpi then approximately 120 mega-pixels are created. If these are stored using a greyscale of 256 levels, i.e. as an 8-bit byte, then each photo takes 120 Mb of storage. If scanning is at a greater level of dpi or colour images are used or a finer graduation of greyscale, such as 12-bit, then the storage for each image increases proportionately. However, the cost of storing an image in this digital form is less than the cost of storing its analogue equivalent. The resolution at which an image needs to be scanned is ultimately dictated by the resolution or scale of the final product.

Scanning traditional photographs is merely the intermediate stage in the development of digital photogrammetry. The ideal situation is for all data to be in digital form from start to finish of the process. This requires that digital air cameras are used so that, after data capture, all processes can be carried out by computer.

The digital air camera shown in *Figure 14.40* has three lenses so image data is captured looking forwards, downwards and backwards. The camera which has forward motion compensation also contains an inertial measuring unit (IMU) so that digital imagery, captured in the air can be rectified using GPS/IMU data. The camera may be placed on a stabilized mount, interfaced to IMU, for which the orientation may be

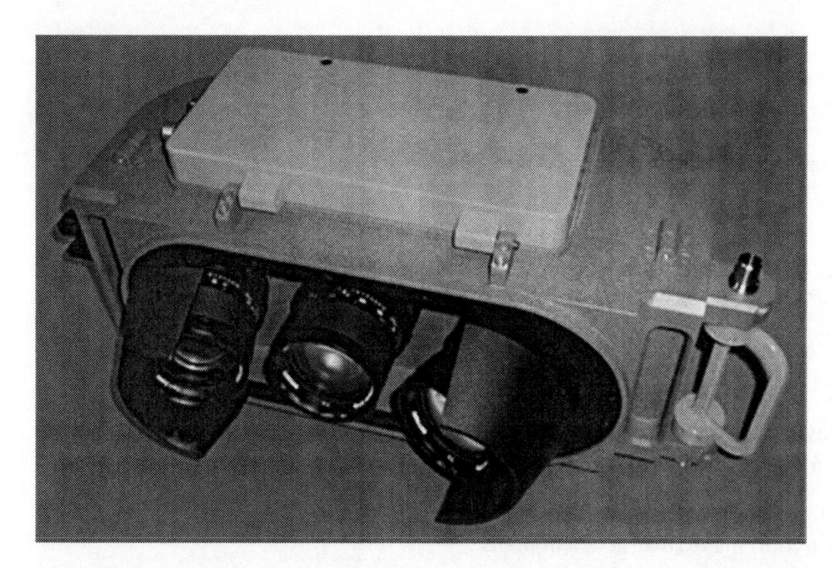

Fig. 14.40 *Wehrli/Geosystem 3-DAS-1 Digital Aerial Camera – courtesy of Wehrli & Associates Inc.*

determined to 0.02°. Output can be in various image formats with the continuous image strip cut into image frames.

Digital data capture has implications for flight planning. With traditional photogrammetry part of the usual criteria is to minimize the number of photographs taken because the overall costs of a large mapping project are largely proportional to the number of photographs used and hence the number of man-hours of work. With the traditional fore and aft overlap of 60% and sideways overlap of 20% most of the ground is only covered by one stereo model which in turn means that each point is defined by the intersection of a minimum of two rays. If digital images can be processed by computer in a completely automated manner then the number of images captured would be entirely immaterial as far as the cost of data processing was concerned.

If a point is defined by the intersection of two rays there can be no independent quality assessment of that point. With much greater fore and aft overlap of say 80% each point on the ground would appear on five images and so the point could be defined by the intersection of up to five rays. This leads to much greater redundancy and therefore greater precision in the determination of the three-dimensional position of a given point. It could also enable the selection of only those images that show the point clearly. With only two photographs a given ground point may be obscured by trees or buildings. With more images this problem will occur significantly less often.

If the lateral overlap is increased from 20% to 60% then the redundancy also increases, however this would involve flying double the number of flight lines and hence a significant increase in the cost of data acquisition.

Just as with traditional photogrammetry there is still the need for flight planning and the collection of other data such as ground control and camera information. Digital imagery may also be obtained from satellites such as ASTER, ENVISAT, EROS, ERS, IKONOS, IRS, Landsat, OrbView, QuickBird, SPOT and TopSat. The resolution varies with the product but sub-metre pixel sizes are now available. Commercial imagery is available from a number of sources such as http://www.infoterra-global.com which gives details of most of the aforementioned satellites and their products.

Stereo imaging of digital images may be achieved in a number of ways. Anaglyph systems with red and green images viewed through red and green glasses have been around for decades and in the modern application the images are viewed on one computer screen. Alternatively polarized images viewed through polarized glasses can be used. If special glasses are used in which the lenses are rapidly but alternately obscured using liquid crystal shutters in synchronization with alternating images on the screen the same effect can be achieved.

The products of digital photogrammetry are similar to those of conventional photogrammetry, i.e. orthophotos, DEMs and line or block features used to create maps, usually in vector format. Contours can then be derived from a DEM and orthophotomaps made by overlaying the orthophoto with the vector product. DEMs can be 'draped' with selected resampled elements of the original images to produce three-dimensional models.

In the ideal digital photogrammetric process all elements are automated. One area where this has been less successful is with matching images. Modern processes are semi-automated but an operator is still required to confirm acceptance and resolve problems. The main difficulties arise with highly three-dimensional environments such as steep valleys and urban canyons. Bodies of water present difficulties because of the variability of the surface due to waves and reflections being mistaken for hard features. Undoubtedly software associated with feature extraction will get better.

Although orthophotos are a major product of digital photogrammetry practical problems still exist with the depiction of sharply three-dimensional features such as buildings. Without very detailed manual work or sophisticated software it is not possible to show the outline of the roof to be coincident with the outline of the footprint of the building and so wall façades still appear.

Details of current systems may be found on manufacturers' websites such as those at:

DVP Complete W/S from DVP-GS, http://www.dvp-gs.com

Geomatica® from PCI Geomatics, http://www.pcigeomatics.com

Leica Photogrammetry Suite™ from Leica Geosystems, http://gis.leica-geosystems.com

SoftPlotter® from Boeing, http://sismissionsystems.boeing.com

Summit Evolution from Datem Systems International, http://www.datem.com

UltraMap™ Worksuite from Vexcel, http://www.askism.com

VirtuoZo from Supersoft Inc, http://www.supresoft.com

Z/I Imaging from Intergraph, http://www.intergraph.com/earthimaging/

If photogrammetry is fully digital then the technology lends itself to direct online solutions, i.e. where the images from the camera are processed in near real time. Such a system allows video images to be processed to find near real time change in a structure and has application for near real time deformation monitoring. Photogrammetry now extends to become videogrammetry. If the object to be measured does not have easily identifiable points then points can be artificially created using a laser spot.

14.9 LASER SCANNERS

A laser scanner combines a laser distance measuring device, two dimensional orientation measuring devices with a scanning mechanism. In its airborne application it is often called LiDAR (Light Detection and Ranging). The laser may work as a pulsed laser, which measures the time for the signal to travel from the laser to the target and back again, or as a continuous signal in which case it will use the phase difference method where the difference in phase between the outgoing and returning signals is measured in a manner similar to EDM. The phase difference method, which is preferred for surveying equipment that measures single points, leads to solutions that have high accuracy, but the range is usually limited and the instrument is slow. Pulsed lasers however are of lower accuracy but have ranges up to several kilometres and so are more suitable for laser scanners.

The advantages of laser scanning over conventional satellite survey are claimed to be that, point for point, it is close to the accuracy of conventional survey but very much quicker. It is a non-contact process so can be used for some of the more difficult and dangerous projects. Everything visible from the instrument can be observed at one go and large areas can be covered quickly. As with a total station three-dimensional coordinates may be generated directly.

On the other hand the equipment is expensive to buy or hire and since operators must be skilled there will also be an investment in training required. Only the points that can be directly seen from the instrument can be measured. In the case of ground based laser scanners it will probably be necessary to take measurements from several different positions, but even so it is likely that there will be pockets of detail that are missed and this will not become apparent until after the data is processed. In the case of airborne data capture (LiDAR) it may be necessary to make more than one pass over the area of interest. The data processing of the large volumes of data captured is quite complex and relies upon control data derived by other survey techniques and it is difficult to detect errors in the data caused by errors in the sensors. Measurement through water to river or sea bed is difficult and limited, and for longer ranges the accuracy may be affected by unmodelled meteorological conditions.

The range of a scanner will depend upon the quality of the return signal which in turn depends on the nature of the reflecting surface. For example the reflectivity of snow is 80–90%, white masonry about 85%, limestone and clay up to 75%, deciduous and coniferous trees typically 60% and 30% respectively, beach sand and bare desert about 50%, smooth concrete about 24%, asphalt about 17% and lava 8%.

As well as measurements for position the intensity of the return signal may also be recorded. The intensity may be used to interpret the type of surface that is being measured. An individual intensity value is unlikely to give much information but rapid changes in value may indicate changes in surface texture, such as the edge between wood and brickwork in a building.

What makes laser scanners particularly useful are their high rates of data capture; in the order of 10 000 points per second.

14.9.1 Terrestrial static laser scanner

A laser scanner has pan and tilt mechanisms that allow the scanner to be pointed in the desired direction. A rotating or oscillating mirror enables the scanning process to take place between the desired range of tilt and the pan mechanism allows the scanner to be reoriented for a new line of scan. If the mirror oscillates, the scan moves up and down with respect to the Z-axis and the data recording is near continuous. If the mirror rotates then each scan is always in the same direction, up or down, with the next scan starting afresh. The oscillating mirror has a very flexible adjustment of the viewing angle but there may additional errors caused by the mechanics wearing out due to accelerations of the moving parts and so regular calibration may be required. With the rotating mirror there is flexible adjustment of the viewing angle but misalignment of the mirrors may also occur due to wear. This way the moving parts are kept light and to the minimum. See *Figure 14.41*.

The simplest configuration for a laser scanner is when it is placed on a tripod. In this case the X-, Y- and Z-axes are defined by the initial orientation of the scanner and the three measurements of α, θ and s. See *Figures 14.42* and *14.43*.

The slope distance is s and if the instrument has been properly levelled then α is a vertical angle and θ a horizontal angle. From *Figure 14.44* the coordinates of any point, P, in the instrument reference frame are:

$$x = s \sin\theta \cos\alpha$$

$$y = s \cos\theta \cos\alpha$$

$$z = s \sin\alpha$$

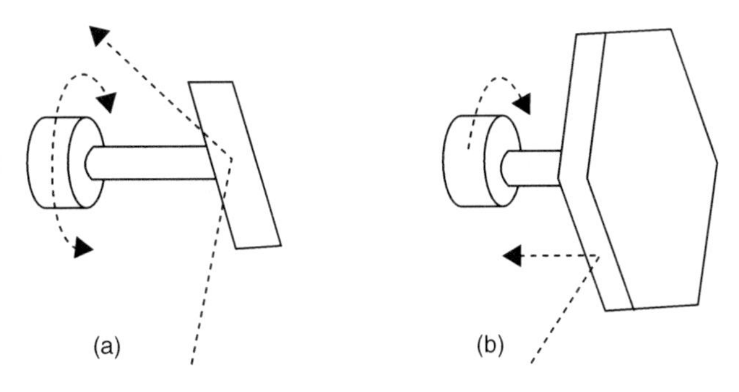

Fig. 14.41 *Laser scanner mirrors. (a) Oscillating. (b) Rotating*

Fig. 14.42 *Laser scanner – courtesy of Riegl UK Ltd*

Fig. 14.43 *Laser scanner on railway trolley and point cloud – courtesy of ABA Surveying*

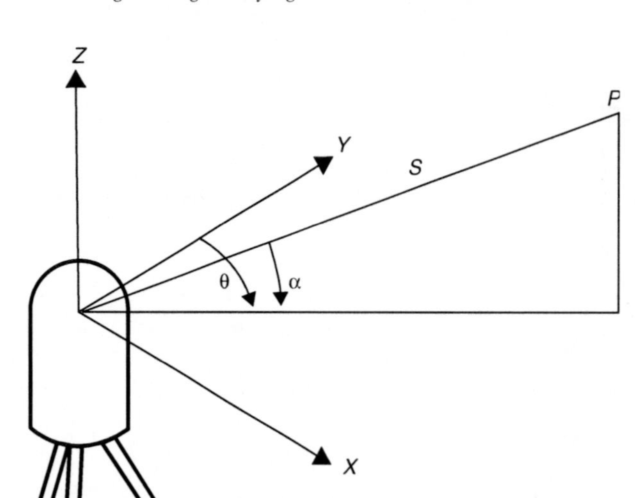

Fig. 14.44 *Laser scanner coordinates*

This simple arrangement is complicated by the fact that the scan and pan axes (vertical and horizontal axes in *Figure 14.44*) do not intersect; there is an offset between them. If the instrument axis origin is taken as the point where the plane defined by $\alpha = 0$ intersects with the pan axis, α_d is the error in α, i.e. α_d is the recorded value of α when it should be 0, and the scan axis is offset from the pan axis by r, then the coordinates of P relative to the instrument's $X_iY_iZ_i$-axes are:

$$x_i = s\sin\theta\cos(\alpha - \alpha_d) + r\sin\theta$$

$$y_i = s\cos\theta\cos(\alpha - \alpha_d) + r\cos\theta$$

$$z_i = s\sin(\alpha - \alpha_d)$$

If the instrument is static during the collection of data then the relationship between coordinates in the instrument's $X_iY_iZ_i$ frame may be related to the local mapping reference frame $X_mY_mZ_m$ by the application of an appropriate translation vector $x_Ty_Tz_T$ and a rotation matrix, M_i^m.

$$\begin{bmatrix} x_m \\ y_m \\ z_m \end{bmatrix} = \begin{bmatrix} x_T \\ y_T \\ z_T \end{bmatrix} + M_i^m \begin{bmatrix} x_i \\ y_i \\ z_i \end{bmatrix}$$

If M_i^m is defined by rotations about the three axes in order X_i, Y_i and Z_i by amounts χ, ψ and ω respectively then it can be shown that:

$$M_i^m = \begin{bmatrix} \cos\omega\cos\psi & (\sin\omega\cos\chi + \cos\omega\sin\psi\sin\chi) & (\sin\omega\sin\chi - \cos\omega\sin\psi\cos\chi) \\ -\sin\omega\cos\psi & (\cos\omega\cos\chi - \sin\omega\sin\psi\sin\chi) & (\cos\omega\sin\chi + \sin\omega\sin\psi\cos\chi) \\ \sin\psi & -\cos\psi\sin\chi & \cos\psi\cos\chi \end{bmatrix}$$

To find the six parameters x_T, y_T, z_T, χ, ψ and ω needed to calculate the coordinates of all points in the mapping frame from the instrument coordinates, several well spaced, coordinated points common to both frames are required. This will lead to a least squares solution for the parameters and so their quality, and the quality of all computed points can be assessed.

Although the technology has now matured, it is anticipated that many advances may still be made. However as a benchmark, at the time of writing in 2006, the Riegl LMS-Z420i has a class 1 laser with a measurement range to natural targets, with 80% reflectivity of up to 800 m and 10% reflectivity up to 250 m with a minimum range of 2 m. The measurement accuracy is ±10 mm for a single shot and ±5 mm averaged with a measurement resolution 5 mm. With an oscillating mirror the scan rate is up to 12 000 points/second and 8000 points/second with a rotating mirror. The laser wavelength is in the near infrared region and the beam divergence is 50″. The vertical scanning range is 0° to 80° and the scanning rate can vary from 1 to 20 scans/second. The minimum angle of step width is 30″ with an angular resolution 7″. The horizontal scan (pan) is through a full circle at a scanning rate of 35″/second to 15°/second with a minimum angle of step width of 35″. The angular resolution is 9″.

The output of laser scanning is often referred to as a *point cloud*. The term comes from the visualization of the output coordinate list where the density of points appears as a rendering of the project scene. The points are often coloured according to the intensity of the laser return signal and as the resultant image on the computer screen appears as many unconnected but closely spaced dots it is often referred to as a *point cloud; Figure 14.45.*

The above describes how a laser scanner may be used in a static environment. The concept may be extended to a *stop and go* mode where the laser scanner is mounted on a vehicle and moved from point to point around a project so that the whole project is observed and no, or at least few, holes are left in the observed data. The data is then most easily patched together if there are sufficient control points also observed by the scanner. As with photogrammetry, it is best if control targets are placed around the project to be picked up by the laser scanning, rather than after scanning trying to identify points to observe by conventional means from within the data *point cloud.*

14.9.2 Terrestrial mobile laser scanner

If the scanner is mounted on a vehicle and the pan fixed then with the Z-axis near vertical and the scan plane to the side of the vehicle, a continuous set of data profiling buildings, street furniture or power cables along the route may be recorded. Alternatively it could be mounted on top of the vehicle with the Z-axis near horizontal and the scan plane pointing forward covering the ground ahead of the vehicle or pointing upward to scan overhead power lines. In this configuration a profile of the route in front of the vehicle may be recorded and the technique is often referred to as *push broom*. In either configuration the axes of the laser scanner are continuously changing in orientation as the vehicle moves and therefore, the relatively simple method of finding the coordinates in the mapping frame from the coordinates in the instrument frame, by using a single translation vector and rotation matrix, is not possible.

It is therefore necessary to track the changing position and orientation of the laser scanner and this may be done using GPS in conjunction with an inertial measuring unit (IMU). Precise synchronization of timing between the GPS, IMU and laser scanner data is essential. GPS/IMU systems are described in *Section 14.11.*

14.9.3 Airborne laser scanner

When the laser scanner is mounted in an aircraft or helicopter the push broom technique, using GPS/IMU must be used, as shown in *Figure 14.46.* Given the technical parameters of the laser scanner, the desired density of measured points on the ground will dictate the flying altitude, swath width, and speed of the aircraft.

The size of the footprint will depend on the divergence of the beam, the flying height and the off-nadir angle. The greater the off-nadir angle, the more eccentric the footprint ellipse. In practice footprints may be from 0.1 m to several metres, depending on the application therefore there may be a number of distances to more than one object that could be measured. Early laser scanners could only record the first return pulse, off the nearest object, but modern scanners may record several pulses. Usually the first and last pulses are the most useful. When flying over a forest the first pulse is returned by the top of the canopy and the last may be returned by the ground providing there is at least a small hole through the trees. If the divergence of the beam is small there is less chance of getting a return from both canopy and ground.

Fig. 14.45 *3D point cloud of an excavation (above) and a building (below) – courtesy of terra international surveys ltd*

The quality of the final product of aerial laser scanning depends on the quality of all the instruments and processes that have been used. These include the errors in the hardware, i.e. GPS, IMU and the laser; errors in the data caused by the density of the ground points, flying height, the scan and off-nadir angles; errors caused by the target such as ground irregularities in the footprint, roughness of the terrain, multiple return signals from trees and vegetation; errors of filtering out blunders and unwanted objects, smoothing and interpolation created in the data processing.

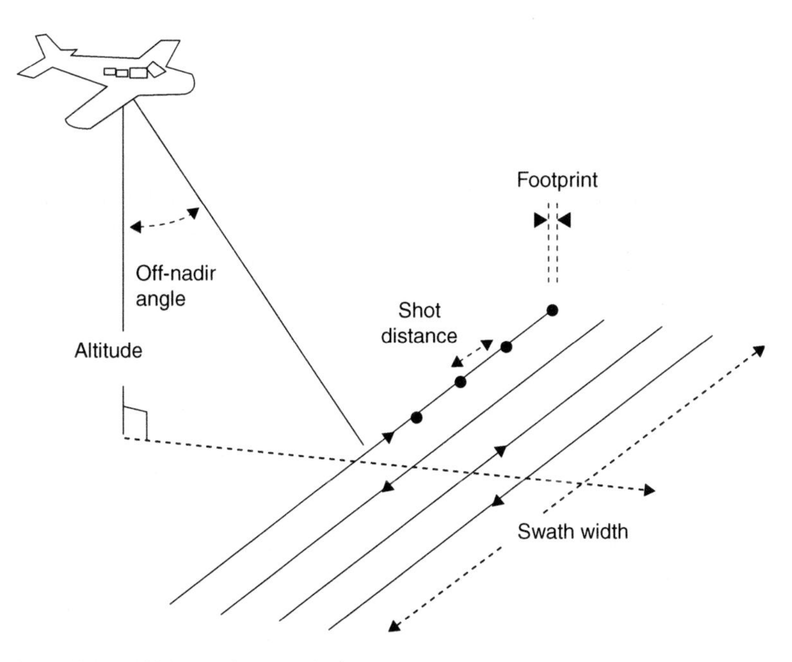

Fig. 14.46 *Airborne laser scanning*

The errors due to the GPS/IMU can be largely removed in a way similar to the concepts of strip and block adjustment in photogrammetry where the model derived from the data is fitted to sparse ground control.

Limited hydrographic survey is possible with a laser scanner. If signals of two different frequencies are used, one that penetrates water and one that does not, then the depth can be assessed by measuring the difference in time between the two return pulses. The refractive index of the water will alter the path of the signal as well as slowing it down and both these effects must be taken into account. See *Figure 14.47*.

A conventional near infra-red of pulse 1.06 μm is reflected by the surface but a green pulse of 0.53 μm may penetrate the water. However the depth to which the pulse may travel and return with a measurable signal is limited by the turbidity of the water. In clear water depths down to 70 metres may be measured, where there is mud in suspension, such as the River Humber, it may be only a few metres. The roughness of the water surface may reduce the quality of results. On land with a highly reflective flat surface the

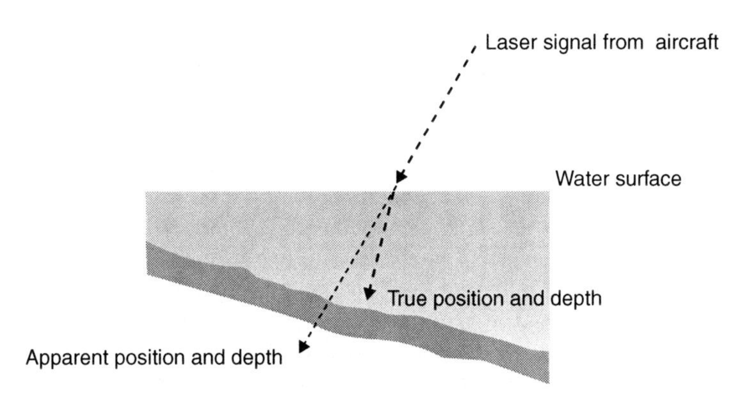

Fig. 14.47 *Refraction effects in water*

distance may be measured to a few centimetres; through water to the bottom and back the resolution is of the order of worse than 0.2 m.

14.9.4 Data processing

Processing laser scanning data can be complex and therefore it is essential to have suitable software for handling the raw scanning data and presenting the results. Apart from the scan data it will also be necessary to have the project data including coordinates of control and tie points, and hence the computed transformation matrices required to transform the data from individual scans into the common project coordinate system. If at least four, and preferably more, reflective targets for which the coordinates have been established by conventional survey, e.g. with a total station, are included in the scan then it may be possible for the processing software to compute and apply the translation vector and transformation matrix automatically. Alternatively, and at a lower precision, the position of the scanner may be found with GPS. If the scanner is levelled then only the orientation, in azimuth, of the scanner is required. The project coordinate system may be arbitrary or may be in the national or a global system such as WGS84.

A project point cloud is merely a collection of points in a common system. This data may be used to create meshes which can then be used to construct objects which in turn may be rendered with high-resolution elements of digital photography. However, prior to processing, the raw data may be visualized in two or three dimensions by colour coding according to the strength of the laser's return signal or the height or range of the point.

The large quantity of scanned data will inevitably contain some errors and inconsistencies. The software should have automatic anomaly detection systems to aid cleaning the data. The massive data files may need to be reduced by removing redundant data. For example if a wall is a plane surface it could be fully defined by the series of points joining the straight lines that delineate its edge.

With the aid of camera images it may be possible to construct orthophotos which in turn could be exported to render objects created within CAD software. If photographs have been taken from different positions it will be necessary to select the best one for each surface element taking account of brightness, contrast, texture, colour and shadows. Some processing of the selected image elements to ensure consistent overall images will be necessary.

14.10 COMPARISON OF PHOTOGRAMMETRY AND LASER SCANNING

Aerial photographs require the subject to be illuminated by the sun; lasers do not, so are not affected by cloud shadow or shadows from nearby trees or buildings. Laser scanning may be undertaken by night, which might be the best time to work in the vicinity of a busy airport.

Both systems see all that is in view. A traditional photograph is an analogue product and the resolution of detail is limited only by the quality of the camera, film and processing. Therefore, there is continuous information across the format and individual high contrast features, such as the intersection of white lines on a tarmac road can be identified and measured. A digital photograph has similar properties except that each pixel only has the average return signal over a small area. The information from a laser scanner is more discrete in that three-dimensional coordinates are derived for a small point or footprint but no information is available for the gap between one point and the next. A series of neighbouring points may be used to define a surface, such as a wall in terrestrial laser scanning. The intersection of two surfaces may define a line such as the join of two walls and the intersection of three surfaces, a point. So if the aim of the project is to detect movement from one epoch to another, such as in deformation monitoring then the movement could only be detected if perpendicular to a surface or in three dimensions with respect to a point.

A colour photograph will have information in a range of colours. A laser scanned point cloud only has coordinates of the point and the return signal intensity.

Detection and modelling of buildings from aerial or terrestrial laser scanning data present a number of challenges. The existence of buildings in aerial laser scanning data may be identified by sudden changes in height. The planes defining the building need to be established and the intersection of the planes then defines their limits and hence the outlines of the building.

The processing of the outputs of both systems is relatively complex but is most complex for laser scanning when the scanner is mobile during data capture, such as with aerial or motorized laser scanning.

The main advantage of photogrammetry is in the blanket nature of the data and the main advantage of laser scanning is in the precision of the data. Hybrid systems that take the best of both systems, for example to render a point cloud with photographic imagery, will give the most useful products but will also be the most costly. In this case, to be most effective, the image date should also be digital.

14.11 INERTIAL SYSTEMS

Inertial systems have been developed for surveying but their main limitations were those of reliability and cost. Today inertial measuring units used in conjunction with GPS are highly necessary and effective devices for giving position and orientation for aerial laser scanning and aerial photogrammetry.

The early inertial systems were developed for military use, in particular the German V2 rockets, but after World War II the principles were applied to aircraft and military vehicle navigation. It was not until the 1960s and 1970s that research was applied for geodetic and survey use.

An inertial system consists of a set of three accelerometers, mounted with their axes at right angles to each other, and two or three gyroscopes. The accelerometers detect acceleration in any direction and the gyroscopes, rotation or rate of rotation of the inertial unit. By doubly integrating the output of the accelerometers, change in position in the direction of each axis may be determined.

If a vehicle starts with zero velocity, i.e. it is standing still, and then accelerates at a constant rate in a straight line on level ground, its velocity and distance from the start point may be determined from the following simple well known formulae:

$$v = at \qquad \text{and} \qquad d = \frac{1}{2}at^2$$

where a = acceleration
v = velocity
d = distance
t = time

When the start position is not at the coordinate origin, the start velocity is not zero and the acceleration is variable, then the position and velocity of the vehicle at the end of a period of time may be more fully given by:

$$v_2 = v_1 + \int_{t_1}^{t_2} a\,dt \quad \text{and} \quad d_2 = d_1 + \int_{t_1}^{t_2} \left(v_1 + \int_{t_1}^{t_2} a\,dt \right) dt$$

Thus by continuously monitoring the output of an accelerometer aligned to the direction that a vehicle is moving it is possible, by doubly integrating that output, to find how far the vehicle has moved. This is fine in one dimension. For real movement in a three-dimensional world three mutually orthogonal accelerometers are required.

Inertial systems for navigation need real time processing because the navigator needs to know where he/she is at any time, *now*. For surveying purposes it is usually more appropriate to sacrifice the timeliness of a real time solution for the higher accuracy of a post-processed solution. This will allow better use of the information recorded when the platform is at rest, i.e. during a zero velocity update (ZUPT).

The continuously recorded data that is used for post-processed solutions should be the raw data and not that which may have been (Kalman) filtered for the optimum navigation solution.

Calibration surveys will need to be completed along north–south and east–west lines to scale the accelerometers. There is a practical time limit for an inertial survey before the un-modelled errors, i.e. non-linear drifts of the accelerometers, become excessive.

14.11.1 Coordinate transformations

When using inertial systems there are a number of coordinate systems that may be involved and it is necessary to be able to express coordinates in any systems that are linked. The most common coordinate systems that are used are Celestial Coordinates, Satellite Coordinates, Earth Centred Inertial Coordinates, Cartesian and Geographical Coordinates, Vehicle coordinates, Local Tangent plane coordinates. Some of these coordinate systems have already been covered in earlier chapters. The others will be explained here.

The notation for coordinate transformations used here is that $\mathbf{C}_B^A$ is a coordinate transformation matrix to transform coordinates in system A to system B. For example, $\mathbf{C}_{NED}^{ECI}$ is a coordinate transformation matrix to transform coordinates in an Earth Centred Inertial (*ECI*) coordinate system to an earth fixed North-East-Down (*NED*) coordinate system. $\mathbf{C}_{ENU}^{RPY}$ is a coordinate transformation matrix to transform coordinates in a Roll-Pitch-Yaw (*RPY*) of vehicle body coordinate system to an earth fixed East-North-Up (*ENU*). These coordinate frames will be described below.

Successive transformations may be achieved by successive pre-multiplication of individual transformation matrices, for example:

$$\mathbf{C}_{ECI}^{NED}\mathbf{C}_{NED}^{RPY} = \mathbf{C}_{ECI}^{RPY}$$

If $\mathbf{C}_{ENU}^{RPY}$ converts coordinates in the *RPY* frame to the *ENU* frame then

$$\begin{bmatrix} v_E \\ v_N \\ v_U \end{bmatrix} = \mathbf{C}_{ENU}^{RPY} \begin{bmatrix} v_R \\ v_P \\ v_Y \end{bmatrix}$$

where v_R is the component of velocity in the direction of the R coordinate axis, etc.

It can be shown that:

$$\mathbf{C}_{ENU}^{RPY} = \begin{bmatrix} \cos\theta_{ER} & \cos\theta_{EP} & \cos\theta_{EY} \\ \cos\theta_{NR} & \cos\theta_{NP} & \cos\theta_{NY} \\ \cos\theta_{UR} & \cos\theta_{UP} & \cos\theta_{UY} \end{bmatrix}$$

where θ_{ER} is the angle between the E- and the R-axes, etc.

14.11.2 Coordinate systems

In inertial space a non-rotating coordinate frame may be defined by two orthogonal directions.

- The direction of the earth's polar axis and the direction defined by the intersection of the equatorial plane with the plane of the ecliptic, the plane of the earth's rotation about the sun, define an earth centred inertial (*ECI*) frame. This last direction is called the *Gamma* or the *first point of Aries*. Actually such a frame is not strictly inertial but rotates every 26 000 years, a factor that is unlikely to concern a surveyor working over a period of hours at most. The direction of the earth's polar axis and the direction defined by the intersection of the Greenwich meridian with the equatorial plane define an earth centred terrestrial (*ECT*) coordinate system.
- Celestial coordinates are expressed in an *ECI* frame in terms of right ascension and declination. Right ascension is the arc measured on the celestial equator from Gamma, positive eastward in units of time, to the declination circle of the celestial body. Declination is the arc of the declination circle measured positive north, negative south, from the equator to the celestial body.

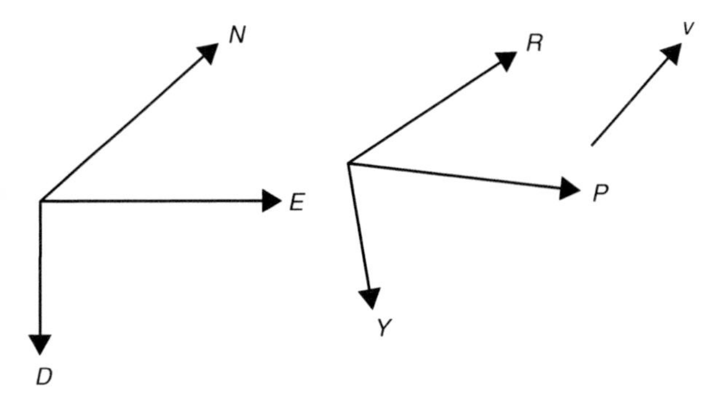

Fig. 14.48 *RPY and NED axes*

- Satellite coordinates when expressed in Keplerian elements are in an ECI but can be converted to *ECT* coordinates using Greenwich Apparent Sidereal Time.
- Cartesian and Geographical Coordinates have already been covered in *Chapter 8.*
- Vehicle coordinates are expressed in Roll Pitch Yaw (*RPY*) coordinates with respect to the axes of the vehicle where the roll axis is in the normal direction of forward motion, the pitch axis is from right to left across the vehicle and the yaw axis points upwards to make up the orthogonal set. The yaw axis on a yacht is about the mast.
- Local Tangent plane coordinates are the directions of East North Up (*ENU*) or North East Down (*NED*) at the vehicle's position.

To convert from *RPY* to *NED* can be achieved with three specific rotations defined in order (*Figure 14.48*). This may be achieved, for example, by rotation through *Euler angles* of Y about yaw axis, until P is parallel to the *DE* plane, then P about pitch axis, until R is parallel to the N-axis, then R about roll axis, until the *RPY* and the *NED* axes are parallel.

In this case the transformation matrix can be shown to be:

$$\begin{bmatrix} v_R \\ v_P \\ v_Y \end{bmatrix} = \mathbf{C}_{RPY}^{NED} \begin{bmatrix} v_N \\ v_E \\ v_D \end{bmatrix}$$

where

$$\mathbf{C}_{RPY}^{NED} = \begin{bmatrix} \cos P \cos Y & \cos P \sin Y & -\sin P \\ \sin R \sin P \cos Y - \cos R \sin Y & \cos R \cos Y + \sin R \sin P \sin Y & \sin R \cos P \\ \sin R \sin Y + \cos R \sin P \cos Y & \cos R \sin P \sin Y - \sin R \cos Y & \cos R \cos P \end{bmatrix}$$

The reverse transformation is:

$$\begin{bmatrix} v_N \\ v_E \\ v_D \end{bmatrix} = \mathbf{C}_{NED}^{RPY} \begin{bmatrix} v_R \\ v_P \\ v_Y \end{bmatrix}$$

where

$$\mathbf{C}_{NED}^{RPY} = \begin{bmatrix} \cos P \cos Y & \sin R \sin P \cos Y - \cos R \sin Y & \sin R \sin Y + \cos R \sin P \cos Y \\ \cos P \sin Y & \cos R \cos Y + \sin R \sin P \sin Y & \cos R \sin P \sin Y - \sin R \cos Y \\ -\sin P & \sin R \cos P & \cos R \cos P \end{bmatrix}$$

Because $\mathbf{C}_{NED}^{RPY} = \mathbf{C}_{RPY}^{NED\,-1} = \mathbf{C}_{RPY}^{NED\,\mathrm{T}}$

If angles R, P and Y are small then $\mathbf{C}_{NED}^{RPY}$ reduces to

$$\mathbf{C}_{NED}^{RPY} = \begin{bmatrix} 1 & -\delta Y & \delta P \\ \delta Y & 1 & -\delta R \\ -\delta P & \delta R & 1 \end{bmatrix}$$

To convert form *NED* to *ENU* coordinates

$$\begin{bmatrix} v_E \\ v_N \\ v_U \end{bmatrix} = \mathbf{C}_{ENU}^{NED} \begin{bmatrix} v_N \\ v_E \\ v_D \end{bmatrix}$$

where

$$\mathbf{C}_{ENU}^{NED} = \begin{bmatrix} 0 & 1 & 0 \\ 1 & 0 & 0 \\ 0 & 0 & -1 \end{bmatrix}$$

14.11.3 Methods for coordinate transformation

There are several ways in which coordinates in a Cartesian frame may be converted. They include Euler angles, rotation vectors and direction cosines. Each method has its own application.

14.11.3.1 Euler angles

These have already been covered above, where, for example:

$$\mathbf{C}_{RPY}^{NED} = \begin{bmatrix} \cos\theta_{RN} & \cos\theta_{RE} & \cos\theta_{RD} \\ \cos\theta_{PN} & \cos\theta_{PE} & \cos\theta_{PD} \\ \cos\theta_{YN} & \cos\theta_{YE} & \cos\theta_{YD} \end{bmatrix}$$

$$= \begin{bmatrix} \cos P\cos Y & \cos P\sin Y & -\sin P \\ \sin R\sin P\cos Y - \cos R\sin Y & \cos R\cos Y + \sin R\sin P\sin Y & \sin R\cos P \\ \sin R\sin Y + \cos R\sin P\cos Y & \cos R\sin P\sin Y - \sin R\cos Y & \cos R\cos P \end{bmatrix}$$

And therefore $\cos\theta_{RN} = \cos P\cos Y$, etc.

Euler angles give a good representation of vehicle attitude and may be used for driving ship and aircraft cockpit displays including compass cards, which are about the yaw axis, and for aircraft artificial horizons, which are about the pitch and roll axes.

Euler angles are not good for vehicle dynamics. For example if the vehicle points up as in the case of a rocket a small change in pitch would give a very big change in orientation, perhaps of the order of $\pm 180°$. Sensed vehicle body rates are a complicated function of Euler angle rates.

14.11.3.2 Rotation vectors

Any three-dimensional orthogonal coordinate system can be converted to any other three-dimensional orthogonal coordinate system with the same origin by a single rotation about a defined axis.

The rotation vector is defined by the magnitude of the rotation angle and the direction of the rotation axis. For example, to rotate from the *NED* system to the *ENU* system as *Figure 14.49* it is necessary to rotate by π about the north-east direction.

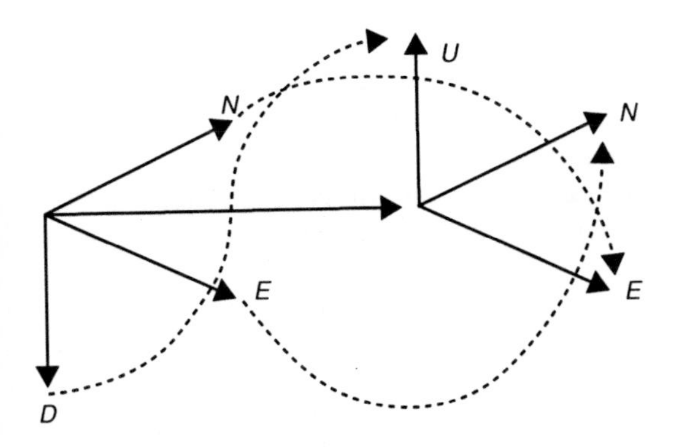

Fig. 14.49 *Rotation vector*

The rotation vector may be expressed as:

$$\rho_{ENU}^{NED} = \begin{bmatrix} \pi/\sqrt{2} \\ \pi/\sqrt{2} \\ 0 \end{bmatrix}$$

The $1/\sqrt{2}$ component in the above vector is because the vector lies half way between the east and the north directions. The magnitude of the vector is π. Note that $\boldsymbol{\rho}_{ENU}^{NED} = \boldsymbol{\rho}_{NED}^{ENU}$.

The difficulty with rotation vectors is that there is no change to a rotation vector if you add 2π to its magnitude and that there is a non-linear representation. For example, two successive rotations a and b may be represented by rotation c but c is a complicated function of a and b.

14.11.3.3 Direction cosines

Direction cosines are concerned with the angles between axes of the two systems. We have already seen that

$$\mathbf{C}_{RPY}^{NED} = \begin{bmatrix} \cos\theta_{RN} & \cos\theta_{RE} & \cos\theta_{RD} \\ \cos\theta_{PN} & \cos\theta_{PE} & \cos\theta_{PD} \\ \cos\theta_{YN} & \cos\theta_{YE} & \cos\theta_{YD} \end{bmatrix}$$

where θ_{RN} is the angle between the R- and the N- axes, etc. The inverse transformation is

$$\mathbf{C}_{NED}^{RPY} = \begin{bmatrix} \cos\theta_{NR} & \cos\theta_{NP} & \cos\theta_{NY} \\ \cos\theta_{ER} & \cos\theta_{EP} & \cos\theta_{EY} \\ \cos\theta_{DR} & \cos\theta_{DP} & \cos\theta_{DY} \end{bmatrix}$$

but since $\cos\theta_{RN} = \cos\theta_{NR}$, etc., then $\mathbf{C}_{RPY}^{NED} = \mathbf{C}_{NED}^{RPY_T}$.

14.11.4 Inertial technology

Inert means without power to move or act and so *inertia* is the property of matter by which it remains in a state of rest, or if in motion, continues moving in a straight line, unless acted upon by an external force. The main components of inertial technology are accelerometers, gyroscopes, the platform on which they are mounted and the software that controls them.

14.11.4.1 Inertial platforms

There are three possible types of inertial platform, space stable, local level and strap-down.

The space stable system uses the gyroscopes to keep the inertial platform oriented in inertial space. This requires very high quality mechanical gyroscopes and is expensive. Its application is particularly for long distance flight and spacecraft applications.

The local level systems use the gyroscopes to keep the platform oriented with respect to the local gravity vector (down) and North. This system usually also uses mechanical gyroscopes and torque motors to maintain the desired orientation of the inertial platform. The controlling and measuring software is more complex.

Strap-down systems are literally strapped down to the body of the host vehicle and use the gyroscopes to measure change of orientation. The continuously monitored output of the accelerometers and gyroscopes is processed through complex formulae to find the changes in position of the vehicle. Typically, they are integrated with other navigation systems such as GPS. The gyroscopes are usually non-mechanical, less precise and much cheaper. They measure and integrate over much shorter periods, sometimes at the sub-second level, but the software requirements are much more complex.

The function of the gyroscopes is to monitor the changes in the orientation of the inertial platform, i.e. the equipment frame.

14.11.4.2 Inertial sensors

All sensors are imperfect and have greater or lesser errors of bias, scale factor error, asymmetry, non-linearity, a dead zone where no response is detected and in digital systems there is quantization when continuous data is turned into discrete data.

For a cluster of three gyroscopes or accelerometers individual scale errors and input axis misalignments can be modelled by:

$$\mathbf{x}_{\text{out}} = S\{\mathbf{I} + \mathbf{M}\}\mathbf{x}_{\text{in}} + \mathbf{b} \tag{14.17}$$

where $\mathbf{x}_{\text{out}}$ are sensor outputs
$\mathbf{x}_{\text{in}}$ are sensed accelerations or gyro rates
S is a nominal sensor scale factor
$\mathbf{M}$ is a matrix of scale factor errors and input axis misalignments
$\mathbf{b}$ is a vector of sensor output biases.

14.11.4.3 Accelerometers

The accelerometer works by detecting a force along the axis of the accelerometer. Unfortunately, from the concepts of relativity, it is impossible to distinguish between acceleration and a displacing force and therefore any accelerometer's output will be a function of the effects of the local gravity field and the accelerating force on the vehicle. In other words if an accelerometer's axis is assumed to be horizontal, but in fact is not, then the accelerometer's output will be corrupted by small errors due to the accelerating force of the component of the gravity vector in the direction of the accelerometer's axis.

Accelerometers may sense acceleration from a strain under load, an electromagnetic force or a gyroscopic precession using measurement of angular displacement, torque rebalance, drag cups, a piezoelectric or piezoresistive effect or an electromagnetic effect.

All accelerometers work on the principle of Newton's second law of motion, $F = ma$, which connects force, mass and acceleration and measure the force to find the desired acceleration. Of particular recent interest are the new Micro-Electro-Mechanical Systems (MEMS) in which mechanical elements, sensors, actuators, and electronics are integrated on a common silicon substrate. See *Figure 14.50.*

Fig. 14.50 *Two-axis accelerometers*

14.11.4.4 Gyroscopes

The gyroscopes may be mechanical or laser. The mechanical gyro works on the principle of conservation of angular momentum. If a twisting force (a torque) is applied that attempts to change the orientation of the spin axis (axis of rotation) of a spinner then there is a precession (a rotation) of the gyroscope about an axis which is defined by the vector cross product of the inertial vector of the spinner and the applied torque vector.

This last sentence is a little wordy but it is possible to verify the gyroscopic effect by removing a wheel from a bicycle and holding it between outstretched arms and spinning the wheel as fast as possible. On trying to turn around and face the opposite direction a couple is applied to the end of the axle. The result is that the wheel seems to try to tip over, because of the precessional torque. Care must be taken to avoid personal injury. In this situation, the axes of spin, applied couple and precessional torque are all mutually at right angles. Similar effects can be seen with a child's toy gyroscope.

Gyroscopes may sense rotation or rotation rate from the Coriolis effect, angular momentum or light path properties using measurements of angular displacement, torque rebalance, vibration, rotation, ring lasers or fibre optic lasers.

Mechanical gyroscopes work on the principle of undisturbed momentum wheels that maintain orientation in inertial space. They measure either rotation of the platform relative to a gimballed gyro as in *local-level* and *space-stable* systems or the torque required to force a gyroscope to new orientation as in *strap-down systems*.

With a practical gyroscope, it is easier to calibrate than to calculate the scale factor that relates torque input to precession output (*local-level*) or precession input to torque output (*strap-down*).

Torque and precession have two dimensions so a momentum wheel gyroscope is a two-axis gyroscope.

The rotating Coriolis effect gyroscope, named after Gaspard Gustave de Coriolis 1792–1843, who discovered the effect for which he is named, uses an accelerometer mounted off the gyro rotational axis but with its sensitive axis parallel to rotation axis. When the system is rotated about an axis normal to the gyroscope Z-axis, such as the Y-axis in *Figure 14.51*, the accelerometer senses a sinusoidal Coriolis acceleration.

One tine of a tuning fork acts like a rotating Coriolis effect gyroscope when the tuning fork is rotated along its longitudinal axis. Very small vibrating gyroscopes act on the same principle.

The ring laser gyroscope (RLG) works on very different principles. Light from a single source is made to travel in opposite directions around a triangle with a mirror at each corner. If the triangle is stationary, i.e. not rotating about its own centre then the relative phase of the light waves from each path remains unchanged. However, if the triangle rotates, then because the light takes a finite amount of time to travel around the

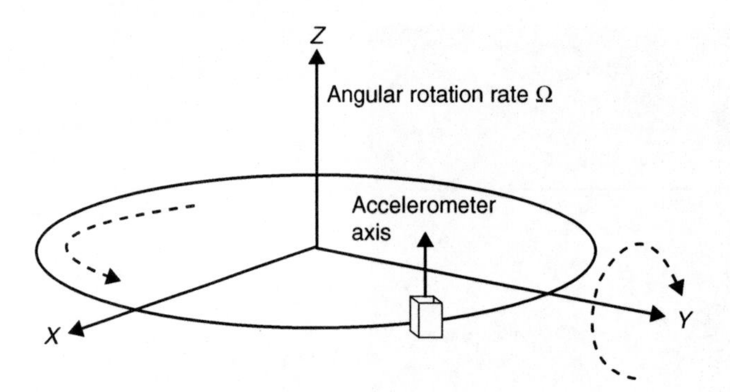

Fig. 14.51 *Coriolis effect gyroscope*

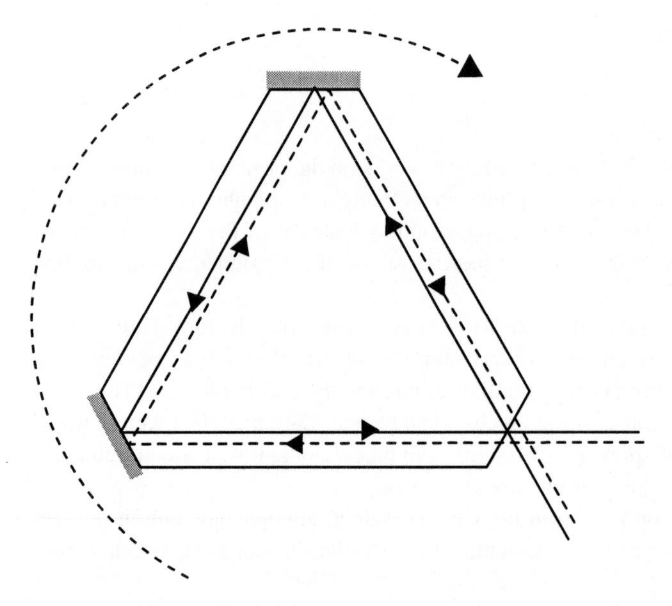

Fig. 14.52 *Ring laser gyroscope*

triangle, it will take different times depending upon which way it is going around the rotating triangle. The result is that there will be a relative phase change compared with the stationary case. The amount of phase change gives a measure of the rate of rotation. The advantage of the laser gyro over the mechanical gyro is that the laser gyro has no moving parts. The disadvantage is that it is less precise. RLGs are rate integrating gyroscopes and measure rotation not rotation rate. See *Figure 14.52*.

Fibre optic gyroscopes (FOG) use the same principle but instead of using a simple square or triangular path, the light is carried in a fibre optic coil of many turns. The main problem with this approach is strain distribution in the optical fibre caused by temperature changes and accelerations of the platform. See *Figure 14.53*.

14.11.5 Inertial navigation systems

Figure 14.54 illustrates a simple one-dimensional navigation system and so only applies in an ideal flat, non-rotating, linear world. The limitation of such a model is that accelerometers cannot measure gravitational

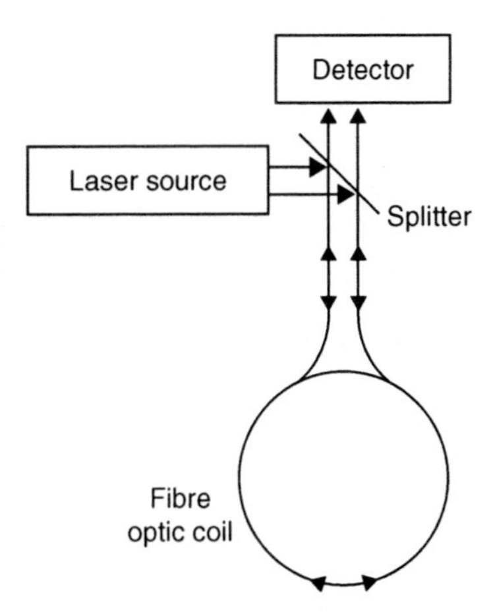

Fig. 14.53 *Fibre optic gyroscope*

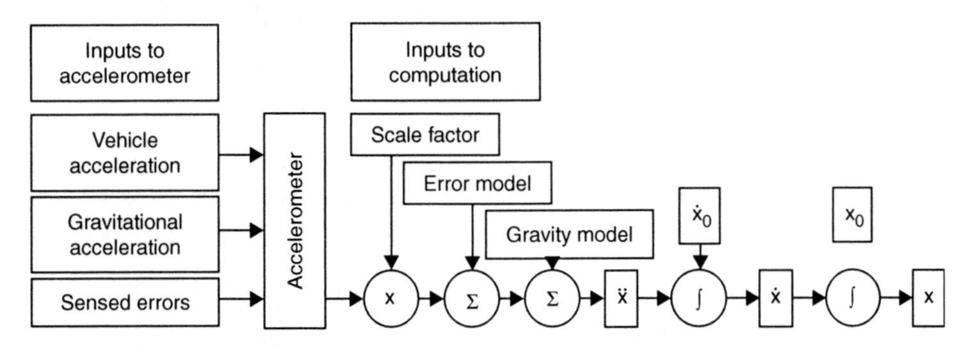

Fig. 14.54 *One-dimensional navigation system*

acceleration unless vehicle acceleration is zero. In practice, this is most easily achieved by stopping the vehicle so that the only sensed acceleration is that of gravity. Accelerometers have scale factor errors such as linear, bias and non-linearity errors and electronic and quantization noise errors. A model for gravitational accelerations will be required. Initial values of velocity and position are required for integration.

Three-dimensional navigation requires three accelerometers and two gyroscopes. Therefore there will be much more data processing. With *local-level* and *space stable* systems, sensors will need to be isolated from rotations.

Three-dimensional navigation systems may be gimballed, as in the *space stable* and *local-level* systems or may be *strap-down*. The gimbals isolate the platform from the vehicle. Gimballed systems are more precise but also more expensive than *strap-down* systems. A *local-level* system also has torque motors to realign the platform.

Although gimballed systems are high precision systems they are difficult to manufacture and therefore expensive. However, there are fewer error sources because the inertial platform is isolated from the vehicle. The main applications of gimballed systems are where updating is not possible by GPS such as in submarines, in space, underground, and in pipelines.

Strap-down systems, such as those used in IMUs for laser scanning, have higher rotation rates for all the sensors and therefore sensors that are subject to large rotation errors cannot be used, e.g. gyroscopic accelerometers. *Strap-down* systems may need error compensation for attitude rate. In *strap-down* systems, the gyroscopes maintain coordinate transformations from *RPY* to *ENU* or *NED*.

Initialization is the process of finding starting values for position, velocity and orientation and needs external sources of data such as GPS, transfer of motion information from a host vehicle or a zero velocity update (ZUPT).

Alignment is the name for attitude initialization and its purpose for gimballed systems is to align the platform axes (*RPY*) with the local tangent plane coordinate axes (*NED*). For *strap-down* systems, the purpose is to find initial values of coordinate transformation parameters.

The use of a gyrocompass for alignment is the only method that does not require external input and may be used for finding the attitude of a stationary vehicle as part of a ZUPT. Gravity defines the local vertical and so aligns the platform Y-axis with the local tangent plane D-axis. The process takes some time in a static location.

For an aircraft in the air GPS gives precise position but poor alignment and so the system will take several minutes before error models can correct for errors in alignment.

Alignment of a gimballed INS gyrocompass may take place on a stationary platform. The process is to tilt the platform until two accelerometers read zero. The platform Y-axis is now aligned with U-axis of navigation coordinates because the platform is now horizontal. Then the system rotates about the Y-axis until one horizontal gyroscope has zero input because the gyroscope axis is pointing East–West. In practice, there needs to be a filter to reduce the effects of disturbance to the vehicle due to wind, refuelling and loading.

14.11.6 Kalman filtering

The Kalman Filter (KF) is a way of using noisy sensor data to estimate a system's *state* where there are uncertain dynamics. In short, it is a process that can be used to get information from a disparate set of sensors so that the solution for position, for example, is significantly better than that which could be obtained from any one or subset of the sensors.

On a moving platform, data may come from GPS, accelerometers, gyroscopes, clocks, and air speed sensors. The KF process can be used to compute useful parameters such as position, velocity, acceleration, attitude and attitude rate but may also have to compute nuisance parameters such as sensor scale factor and bias. The KF process may also have to accommodate disturbances to the vehicle such as that due to wind buffeting, the vehicle being driven on bumpy road or aircraft crab.

14.11.6.1 Kalman filter equations

The Kalman Filter computes the terms of an estimated *state vector* and its associated covariance matrix.

The estimated *state vector* may have useful parameters of position and velocity but will also have nuisance parameters which are anything else that has to be estimated but is not ultimately required. The KF state variables are all the system variables that can be measured by the sensors, e.g. acceleration for accelerometers, angular rate for rate gyroscopes. State variables do not need to be aligned with body coordinate axes.

The KF is a *predictor-corrector* process in two steps. *Prediction* estimates the state vector and its covariance from one time epoch to the next and *correction* makes corrections to an estimated state using data from sensors.

In a simple navigation solution, at some point you will know your position, e.g. by GPS, and you may also know the uncertainty of that position. Sensors on your vehicle, such as compass and odometer give the bearing and distance travelled over the next short leg of the journey, again with uncertainty. On arrival, you can predict your new position using the information available so far, and its uncertainty. You can then correct that position by the addition of a new GPS fix and, in combination with the predicted position, compute a corrected position that is better than a position obtained from a single GPS fix. The process is repeated for each successive new position with each new position better than the last, but only up to a practical limit. See *Figure 14.55.*

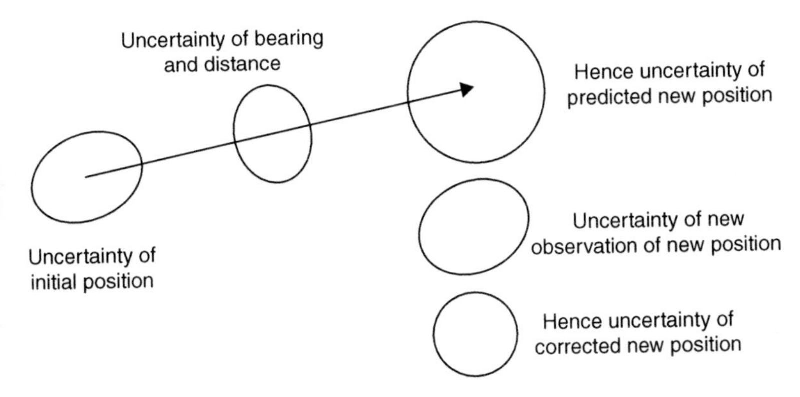

Fig. 14.55 *Kalman filter as a predictor-corrector process*

14.11.6.2 Correction equation

The correction equation corrects an estimate of the state vector using new measurements from sensors. The correction equation (without proof) is:

$$\mathbf{x}(+) = \mathbf{x}(-) + \mathbf{K}[\mathbf{z} - \mathbf{H}\mathbf{x}(-)]$$

where $\mathbf{x}(+)$ is the state vector after correction
$\mathbf{x}(-)$ is the state vector before correction
$\mathbf{K}$ is the Kalman gain matrix
$\mathbf{z}$ is the measurement vector
$\mathbf{H}$ is the measurement sensitivity matrix

The product $\mathbf{H}\mathbf{x}(-)$ is the predicted measurement based on the known behaviour of the system. The Kalman gain matrix applies a correction based on the difference between the actual measurements and the predicted measurements to update the state vector. The measurement $\mathbf{z} = \mathbf{H}\mathbf{x} +$ noise. Therefore, the measurement vector $\mathbf{z}$ is a linear function of state vector $\mathbf{x}$ and noise with statistical properties that are known.

The Kalman gain matrix, $\mathbf{K}$ is given by:

$$\mathbf{K} = \mathbf{P}(-)\mathbf{H}^{\mathrm{T}}(\mathbf{R} + \mathbf{H}\mathbf{P}(-)\mathbf{H}^{\mathrm{T}})^{-1}$$

where $\mathbf{P}(-)$ is the covariance of the state matrix before measurements
$\mathbf{H}$ is the measurement sensitivity matrix
$\mathbf{R}$ is the covariance matrix of the measurements

and $\mathbf{P}(+) = (\mathbf{I} - \mathbf{K}\mathbf{H})\mathbf{P}(-)$

where $\mathbf{P}(+)$ is the covariance of the state matrix after measurements

14.11.6.3 Prediction equation

The process is most easily seen through a simple navigation example. A vehicle moves with constant velocity so that:

$$\dot{E} = v_E \quad \dot{N} = v_N \quad \dot{v}_E = 0 \quad \dot{v}_N = 0$$

In matrix terms this is

$$\begin{bmatrix} \dot{E} \\ \dot{N} \\ \dot{v}_E \\ \dot{v}_N \end{bmatrix} = \begin{bmatrix} 0 & 0 & 1 & 0 \\ 0 & 0 & 0 & 1 \\ 0 & 0 & 0 & 0 \\ 0 & 0 & 0 & 0 \end{bmatrix} \begin{bmatrix} E \\ N \\ v_E \\ v_N \end{bmatrix}$$

With the addition of noise on the data this becomes:

$$\frac{d\mathbf{x}}{dt} = \mathbf{F}\mathbf{x} + \mathbf{w}$$

where $\mathbf{x}$ is the state vector
$\mathbf{F}$ is the dynamic coefficient matrix
$\mathbf{w}$ is the dynamic disturbance vector

The above is a continuous process, but real navigation processes are more usually discrete processes and so can be represented by:

$$\begin{bmatrix} E \\ N \\ v_E \\ v_N \end{bmatrix}_k = \begin{bmatrix} 0 & 0 & \Delta t & 0 \\ 0 & 0 & 0 & \Delta t \\ 0 & 0 & 0 & 0 \\ 0 & 0 & 0 & 0 \end{bmatrix} \begin{bmatrix} E \\ N \\ v_E \\ v_N \end{bmatrix}_{k-1}$$

which, with the addition of noise, becomes:

$$\mathbf{x}_k = \mathbf{\Phi}_{k-1}\mathbf{x}_{k-1} + \mathbf{w}_{k-1}$$

where $\mathbf{x}$ is the state vector
$\mathbf{\Phi}$ is the state transition matrix
$\mathbf{w}$ is the zero mean white Gaussian noise vector

In the zero mean white Gaussian noise process, $\mathbf{w}$ has an associated covariance matrix $\mathbf{Q}$ which is a diagonal matrix and models the unknown random disturbances. In the above formula $\mathbf{\Phi}$ models the system's known dynamic behaviour.

One source of noise may affect more than one component of the state vector. For example an applied force may affect a gyroscope as well as the accelerometers.

The discrete process variables can be computed from $\mathbf{\Phi}$ and $\mathbf{Q}$ where $\boldsymbol{\mu}$ is the expected value of $\mathbf{x}$ which is the mean vector of the distribution and $\mathbf{P}$ is the expected value of $[\mathbf{x} - \boldsymbol{\mu}][\mathbf{x} - \boldsymbol{\mu}]^T$ which is the $n \times n$ covariance matrix of the distribution.

The KF equations and process may be represented as in *Figure 14.56*.

In *Figure 14.56* the terms are:

$\hat{\mathbf{x}}$ is the state vector
$\mathbf{\Phi}$ is the state transition matrix
$\mathbf{P}$ is the state vector covariance matrix
$\mathbf{Q}$ is the matrix of unknown random disturbances to the state vector – the dynamic disturbance matrix
$\mathbf{K}$ is the Kalman Gain matrix

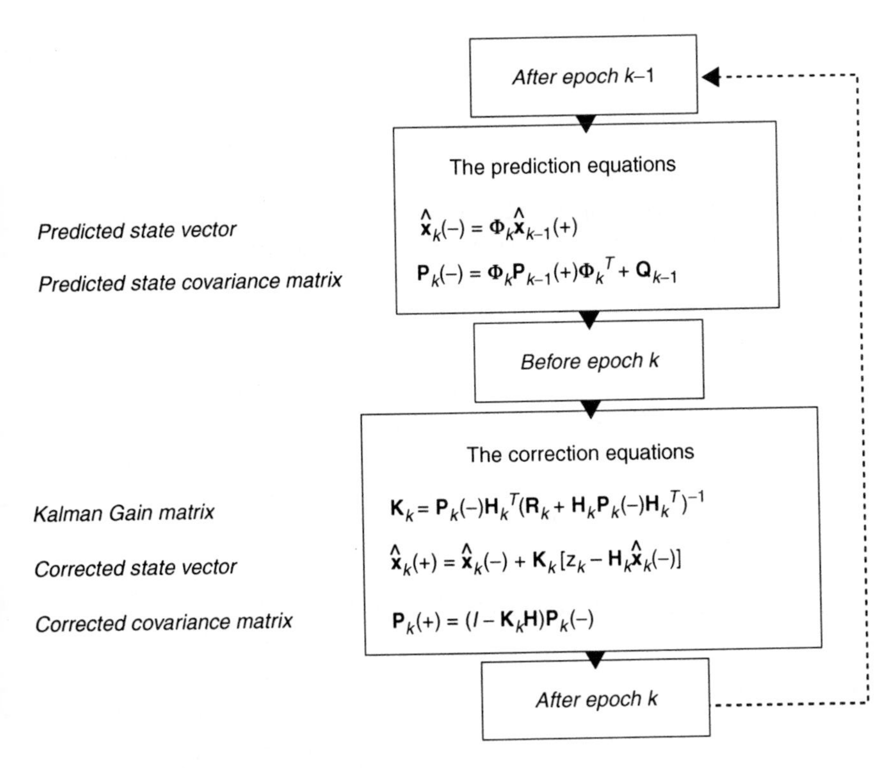

Fig. 14.56 *Kalman filter equations and process*

$\mathbf{H}$ is the measurement or observation sensitivity matrix
$\mathbf{R}$ is the measurement covariance matrix
z is the measurement or observation vector
$(-)$ is predicted
$(+)$ is corrected
$\mathbf{H}_k\hat{\mathbf{x}}_k(-)$ is the predicted measurement
$[\mathbf{z}_k - \mathbf{H}_k\hat{\mathbf{x}}_k(-)]$ is the innovations vector, the difference between the measurement vector and predicted measurement vector

The inputs to the process are $\mathbf{\Phi}_k, \mathbf{Q}_{k-1}, \mathbf{H}_k, \mathbf{R}_k$ and $\mathbf{z}_k$. The outputs from the process are $\hat{\mathbf{x}}_k(+)$ the corrected state vector and $\mathbf{P}_k(+)$ the state vector covariance matrix.

14.12 INTEGRATION OF GPS AND INERTIAL SYSTEMS

GPS and INS have complementary characteristics. GPS has low sample rates of pseudo-range measurements, typically less than 20 Hz. GPS gives position and velocity with bounded estimation error so the errors are reasonably easy to determine. There will be a short-term loss of signal if the lines of sight to the satellites are blocked.

INSs have high output rates, typically greater than 100 Hz. However, the computational method and the hardware limit the useable output rate. The state estimates are calculated from inertial sensor outputs but the sensor outputs do not depend on any external sources such as satellites. However, the major limitation is that position and velocity errors are unbounded in time.

GPS cannot respond suddenly to changing dynamics. If the vehicle accelerates, a GPS-only solution cannot recognize acceleration until the next GPS measurement. The KF uses the current state of the filter to weight the next GPS measurement and so the filter state slowly converges to true GPS state. The convergence rate depends on the accuracy of **P**, the state covariance, relative to **R**, the measurement covariance. A fast convergence implies noisy estimates.

A better performance could be achieved by adding inertial sensors. One solution would be to put inertial and GPS measurements into a KF. The state vector in the KF model then contains position, velocity, acceleration, attitude, angular rates and accelerometer and gyroscope errors. KF then uses inertial sensor measurements at high rates and GPS measurements at low rates.

There are three problems with this simple solution.

- The KF would have to be iterated at the high rate of the inertial measurements but the covariance equations are computationally intensive. This in turn would reduce the rate that inertial measurements could be incorporated into the process.
- Arbitrary models may have to be used in $\mathbf{\Phi}$, the state transition matrix. For example in GPS High Dynamic Receivers there are terms that describe the vehicle acceleration correlation time, that is the dependence of acceleration at one epoch with acceleration at the previous epoch. These parameters are generally unknown and vary with time.
- The filter is estimating the whole navigation state, which may change rapidly.

14.12.1 The complementary filter

The principle here is that the GPS is aiding the INS. Inertial measurements are processed separately and the inertial solution provides a reference trajectory to the KF.

At the GPS measurement epoch, the INS state vector is saved. The INS state vector is compared with the GPS data. A KF is used to estimate the navigation error state using the error between the GPS and the INS data.

Covariance update equations are used at the low rate of GPS updates. This reduces the computations and so the INS rate is not limited by the need for covariance propagation.

Filter design is now based on an error model so that all the model parameters are now properly defined. How the navigation system behaves now depends mainly on the update rate of the INS system and the KF reduces the errors on the GPS aiding signal because it is estimating slowly varying error quantities.

14.12.2 Coupling between GPS and INS

The flow of information in the integrated system depends upon whether the coupling between the systems is loose or tight.

If the GPS and INS are *uncoupled* the GPS and INS give position separately. If the time of interest is at a GPS epoch then the GPS position is accepted and the INS position is reset to that of the GPS. If the time of interest is after the last GPS epoch, then the INS position is accepted. However, no attempt is made to estimate the errors in the IMU of the INS. In this case, the INS errors are bounded at each GPS epoch but the rates of INS errors are unbounded. Therefore, the form of the navigation error will be as in *Figure 14.57*.

If the GPS and INS are *loosely coupled*, the GPS state vector will aid the INS. Both the INS and the GPS systems provide estimated vehicle position and these positions are used to estimate the errors in each navigation or calibration variable. The main advantage of this is that it gives a relatively simple integration approach. There is no need to process basic GPS observables and so this solution can use *off the shelf* GPS and INS systems' outputs, processed and combined in a single computational step. The disadvantage of such a system is that it will not give optimum performance.

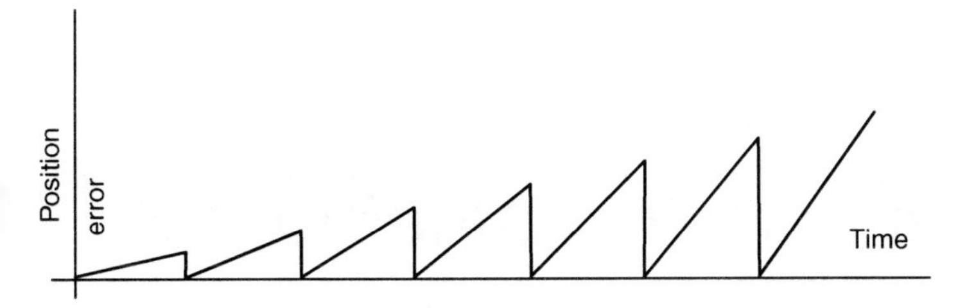

Fig. 14.57 *INS errors in an uncoupled system*

If the GPS and INS are *tightly coupled*, both the INS and GPS systems provide estimated vehicle-to-satellite pseudo-ranges. To do this it is necessary to use basic GPS observables (pseudo-ranges) and data from GPS navigation messages. A higher system performance may be possible with this. The GPS receiver carrier tracking loops are aided by acceleration data from the integrated navigation solution. However tight coupling is not easy to use because it becomes necessary to have access to the GPS hardware and variables used to implement the carrier tracking loops.

Appendix A
An introduction to matrix algebra

Matrix algebra is the only realistic and practical way to deal with the subject of least squares as applied to the solution of survey control networks. This appendix aims to be a primer for those new to the subject and presents the minimum necessary information to make good sense of the material in *Chapter 7*. For others it will be a handy reminder or reference.

A.1 THE MATRIX AND ITS ALGEBRA

A matrix is nothing more than an array of numbers set in a regular rectangular grid. To be useful the array must have some mathematical meaning, reflecting in survey, a physical purpose. The algebra that defines the properties of matrices must be one for which the rules of matrix manipulation lead to useful results. This appendix presents a summary of the rules. It is not written to be an authoritative mathematical text but more as a useful aide mémoire of the collected rules of matrix algebra as they apply to survey.

Matrix methods lend themselves to the solution of large sets of simultaneous equations. Solutions by matrix methods involve repetitive numerical processes that, in turn, lend themselves to computer manipulation. It is this that makes matrix methods particularly useful to the surveyor.

An element of a matrix may be referred to by its position in the matrix in terms of row and column numbers. The matrix as a whole is represented in upper case in bold type whereas the elements are represented in lower-case in ordinary type, sometimes with subscripts to represent the row and column numbers. In the example matrix **A** has three rows and two columns. Element a_{12} is in the first row and second column and has the value 6.

$$\mathbf{A} = \begin{bmatrix} a_{11} & a_{12} \\ a_{21} & a_{22} \\ a_{31} & a_{32} \end{bmatrix} = \begin{bmatrix} 2 & 6 \\ 5 & 4 \\ 3 & 7 \end{bmatrix}$$

Sometimes, to make a point about the dimensions of a matrix the matrix may be written with its dimensions with row subscript before and column subscript after. The above example is ${}_3\mathbf{A}_2$.

When a matrix has only one row or only one column it is referred to as a row or column vector. In this case the vector is written in bold lower-case and the elements have only one subscript to indicate their column or row respectively. **b** is a row vector and **c** is a column vector.

$$\mathbf{b} = \begin{bmatrix} b_1 & b_2 & b_3 \end{bmatrix} \qquad \mathbf{c} = \begin{bmatrix} c_1 \\ c_2 \\ c_3 \\ c_4 \end{bmatrix}$$

If a matrix has only one row *and* one column then the matrix contains a single element and may be treated as a single number or scalar.

Matrices may be added, subtracted and multiplied but **not** divided. Matrices may be inverted and this allows for an equivalent process to conventional division to take place.

A.2 ADDITION AND SUBTRACTION

For two matrices to be added or subtracted their dimensions must be the same, that is they must both have the same number of rows and the same number of columns. The result of the addition or subtraction of the matrices has the same dimensions as both the original matrices. The individual elements in the result are formed from the addition or subtraction of the respective elements from the two original matrices. In the following example:

$$\mathbf{A} = \begin{bmatrix} a_{11} & a_{12} \\ a_{21} & a_{22} \\ a_{31} & a_{32} \end{bmatrix} = \begin{bmatrix} 2 & 6 \\ 5 & 4 \\ 3 & 7 \end{bmatrix} \qquad \mathbf{B} = \begin{bmatrix} b_{11} & b_{12} \\ b_{21} & b_{22} \\ b_{31} & b_{32} \end{bmatrix} = \begin{bmatrix} 1 & 8 \\ 6 & 3 \\ 2 & 7 \end{bmatrix}$$

If $\mathbf{C} = \mathbf{A} + \mathbf{B}$ and $\mathbf{D} = \mathbf{A} - \mathbf{B}$

then $$\mathbf{C} = \begin{bmatrix} c_{11} & c_{12} \\ c_{21} & c_{22} \\ c_{31} & c_{32} \end{bmatrix} = \begin{bmatrix} a_{11} + b_{11} & a_{12} + b_{12} \\ a_{21} + b_{21} & a_{22} + b_{22} \\ a_{31} + b_{31} & a_{32} + b_{32} \end{bmatrix} = \begin{bmatrix} 3 & 14 \\ 11 & 7 \\ 5 & 14 \end{bmatrix}$$

and $$\mathbf{D} = \begin{bmatrix} d_{11} & d_{12} \\ d_{21} & d_{22} \\ d_{31} & d_{32} \end{bmatrix} = \begin{bmatrix} a_{11} - b_{11} & a_{12} - b_{12} \\ a_{21} - b_{21} & a_{22} - b_{22} \\ a_{31} - b_{31} & a_{32} - b_{32} \end{bmatrix} = \begin{bmatrix} 1 & -2 \\ -1 & 1 \\ 1 & 0 \end{bmatrix}$$

It follows from this that for any dimensionally compatible matrices:

$$\mathbf{X} + \mathbf{Y} = \mathbf{Y} + \mathbf{X}$$

$$\mathbf{X} + (\mathbf{Y} + \mathbf{Z}) = (\mathbf{X} + \mathbf{Y}) + \mathbf{Z}$$

A.3 MULTIPLICATION

The rules for matrix multiplication are a little more complex. For it to be possible to multiply two matrices the number of columns of the first matrix must equal the number of rows of the second matrix. The dimensions of the product are from the rows of the first matrix and columns of the second matrix. For example:

$${}_4\mathbf{F}_2 = {}_4\mathbf{E}_3\,{}_3\mathbf{A}_2$$

Note that $\mathbf{AE} \neq \mathbf{EA}$. That fact that **EA** may be possible does not mean that **AE** will also be possible. The order of multiplication is all important. In this example **A** is *pre-multiplied* by **E**, which is the same as saying that **E** is *post-multiplied* by **A**. The element f_{ij} is computed as:

$$f_{ij} = \sum_{x=1}^{x=n} (e_{ix} a_{xj})$$

where n is the number of columns in the first matrix and the number of rows in the second matrix. In this case:

$$f_{ij} = e_{i1}a_{1j} + e_{i2}a_{2j} + e_{i3}a_{3j}$$

If $$\mathbf{E} = \begin{bmatrix} e_{11} & e_{12} & e_{13} \\ e_{21} & e_{22} & e_{23} \\ e_{31} & e_{32} & e_{33} \\ e_{41} & e_{42} & e_{43} \end{bmatrix} = \begin{bmatrix} 1 & 4 & 6 \\ 5 & 6 & 0 \\ 3 & 7 & 2 \\ 3 & 4 & 2 \end{bmatrix}$$

then, for example:

$$f_{32} = e_{31}a_{12} + e_{32}a_{22} + e_{33}a_{32} = 3 \times 6 + 7 \times 4 + 2 \times 7 = 60$$

and so:

$$\mathbf{F} = \begin{bmatrix} f_{11} & f_{12} \\ f_{21} & f_{22} \\ f_{31} & f_{32} \\ f_{41} & f_{42} \end{bmatrix} = \begin{bmatrix} 1 & 4 & 6 \\ 5 & 6 & 0 \\ 3 & 7 & 2 \\ 3 & 4 & 2 \end{bmatrix} \begin{bmatrix} 2 & 6 \\ 5 & 4 \\ 3 & 7 \end{bmatrix} = \begin{bmatrix} 40 & 64 \\ 40 & 54 \\ 47 & 60 \\ 32 & 48 \end{bmatrix}$$

If two vectors are multiplied then the result is either a full matrix or a scalar. For example, if:

$$\mathbf{a} = \begin{bmatrix} a_1 \\ a_2 \\ a_3 \end{bmatrix} \quad \text{and} \quad \mathbf{b} = \begin{bmatrix} b_1 & b_2 & b_3 \end{bmatrix}$$

$$\text{then:} \quad \mathbf{ab} = \begin{bmatrix} a_1b_1 & a_1b_2 & a_1b_3 \\ a_2b_1 & a_2b_2 & a_2b_3 \\ a_3b_1 & a_3b_2 & a_3b_3 \end{bmatrix} \quad \text{and} \quad \mathbf{ba} = [b_1a_1 + b_2a_2 + b_2a_3]$$

It can be shown that for any dimensionally compatible matrices:

$$\mathbf{X(YZ) = (XY)Z}$$

$$\mathbf{X(Y + Z) = XY + XZ}$$

$$\mathbf{(W + X)(Y + Z) = WY + WZ + XY + XZ}$$

A.4 NULL MATRIX

The null matrix is a matrix or vector that contains only zeros. It may be added, subtracted or multiplied in the normal way. The product of any matrix and a null matrix is always a null matrix. A null matrix may be formed as the product of other non-null matrices. For example:

$$\begin{bmatrix} 1 & -2 \\ -1 & 2 \end{bmatrix} \begin{bmatrix} 1 & 1 \\ 0.5 & 0.5 \end{bmatrix} = \begin{bmatrix} 0 & 0 \\ 0 & 0 \end{bmatrix}$$

A.5 TRANSPOSE OF A MATRIX

The transpose of a matrix is one in which the rows and columns of the original matrix have been interchanged. Therefore if **G** is the transpose of **A** then:

$$\mathbf{G} = \mathbf{A}^T \quad \text{and} \quad g_{ij} = a_{ji}$$

In matrix notation:

$$\mathbf{G} = \mathbf{A}^T = \begin{bmatrix} a_{11} & a_{12} \\ a_{21} & a_{22} \\ a_{31} & a_{32} \end{bmatrix}^T = \begin{bmatrix} a_{11} & a_{21} & a_{31} \\ a_{12} & a_{22} & a_{32} \end{bmatrix} = \begin{bmatrix} g_{11} & g_{12} & g_{13} \\ g_{21} & g_{22} & g_{23} \end{bmatrix}$$

It can be shown that for any dimensionally compatible matrices:

Transpose of sum of matrices equals the sum of the transposes $(\mathbf{X} + \mathbf{Y})^T = \mathbf{X}^T + \mathbf{Y}^T$
Transpose of product of matrices equals the product of the transposes in reversed order $(\mathbf{XY})^T = \mathbf{Y}^T\mathbf{X}^T$

A.6 IDENTITY MATRIX

An identity matrix is a square matrix where all the elements are zero except for the elements on the leading diagonal which are 1. The leading diagonal is the line from top left to bottom right. Pre- or post-multiplication by an identity matrix does not change the matrix being multiplied. For example:

$$\mathbf{I} = \begin{bmatrix} 1 & 0 & 0 & 0 \\ 0 & 1 & 0 & 0 \\ 0 & 0 & 1 & 0 \\ 0 & 0 & 0 & 1 \end{bmatrix}$$

For multiplication to take place the dimensions of the identity matrix must be correct. The notation for an identity matrix is a bold upper case **I**. The following relationships follow from this definition.

$\mathbf{AI} = \mathbf{IA}$ Note that the dimensions of these **I**s are not the same unless **A** is square.

$\mathbf{I} = \mathbf{I}^{-1} = \mathbf{I}^T = \mathbf{II} = \mathbf{I}^2$. See below for an explanation of the inverse $(^{-1})$ of a matrix.

A.7 DIAGONAL MATRIX

A diagonal matrix is a square matrix for which all the elements are 0 except those on the leading diagonal. The identity matrix is an example of a diagonal matrix. If **A** and **B** are diagonal matrices where:

$$\mathbf{A} = \begin{bmatrix} a_{11} & 0 & 0 \\ 0 & a_{22} & 0 \\ 0 & 0 & a_{33} \end{bmatrix} \quad \text{and} \quad \mathbf{B} = \begin{bmatrix} b_{11} & 0 & 0 \\ 0 & b_{22} & 0 \\ 0 & 0 & b_{33} \end{bmatrix} \quad \text{and} \quad \mathbf{C} = \begin{bmatrix} c_{11} & c_{12} \\ c_{21} & c_{22} \\ c_{31} & c_{32} \end{bmatrix}$$

then:

$$\mathbf{AB} = \begin{bmatrix} a_{11}b_{11} & 0 & 0 \\ 0 & a_{22}b_{22} & 0 \\ 0 & 0 & a_{33}b_{33} \end{bmatrix} = \mathbf{BA}$$

$$\mathbf{AC} = \begin{bmatrix} a_{11}c_{11} & a_{11}c_{12} \\ a_{22}c_{21} & a_{22}c_{22} \\ a_{33}c_{31} & a_{33}c_{32} \end{bmatrix}$$

A.8 THE DETERMINANT OF A MATRIX

The determinant of a matrix is a scalar, a single number, no matter what the size of the matrix. The matrix must be square. In notation, the determinant of **A** is written $|\mathbf{A}|$. The easiest way to show the rules for the calculation of the determinant of a matrix is by example.

For a 1×1 matrix:

$$|{}_1\mathbf{A}_1| = |a_{11}| = a_{11}$$

For a 2×2 matrix:

$$|{}_2\mathbf{A}_2| = \begin{vmatrix} a_{11} & a_{12} \\ a_{21} & a_{22} \end{vmatrix} = a_{11}a_{22} - a_{12}a_{21}$$

This is the fundamental building block for computation of all determinants. Note the order of multiplication; top left times bottom right minus top right times bottom left.

For a 3 × 3 matrix:

$$|{}_3\mathbf{A}_3| = \begin{vmatrix} a_{11} & a_{12} & a_{13} \\ a_{21} & a_{22} & a_{23} \\ a_{31} & a_{32} & a_{33} \end{vmatrix}$$

Write out the elements of the first row of ${}_3\mathbf{A}_3$ with alternate changed signs starting with a_{11} taking its correct sign, a_{12} opposite and a_{13} correct. Multiply each element by the determinant of what is left of the original 3 × 3 matrix once the complete row and column containing the subject element have been removed. The pattern is the same for larger matrices.

$$|{}_3\mathbf{A}_3| = a_{11}\begin{vmatrix} a_{22} & a_{23} \\ a_{32} & a_{33} \end{vmatrix} - a_{12}\begin{vmatrix} a_{21} & a_{23} \\ a_{31} & a_{33} \end{vmatrix} + a_{13}\begin{vmatrix} a_{21} & a_{22} \\ a_{31} & a_{32} \end{vmatrix}$$

Now form the 2 × 2 determinants as above.

$$|{}_3\mathbf{A}_3| = a_{11}(a_{22}a_{33} - a_{23}a_{32}) - a_{12}(a_{21}a_{33} - a_{23}a_{31}) + a_{13}(a_{21}a_{32} - a_{22}a_{31})$$

For a 4 × 4 matrix:

$$|{}_4\mathbf{A}_4| = \begin{vmatrix} a_{11} & a_{12} & a_{13} & a_{14} \\ a_{21} & a_{22} & a_{23} & a_{24} \\ a_{31} & a_{32} & a_{33} & a_{34} \\ a_{41} & a_{42} & a_{43} & a_{44} \end{vmatrix}$$

Write out the elements of the first row of ${}_4\mathbf{A}_4$ with alternate changed signs starting with a_{11} taking its correct sign, a_{12} opposite, a_{13} correct and a_{14} opposite. Multiply each element by the determinant of what is left of the original 4 × 4 matrix once the complete row and column containing the subject element have been removed.

$$|{}_4\mathbf{A}_4| = a_{11}\begin{vmatrix} a_{22} & a_{23} & a_{24} \\ a_{32} & a_{33} & a_{34} \\ a_{42} & a_{43} & a_{44} \end{vmatrix} - a_{12}\begin{vmatrix} a_{21} & a_{23} & a_{24} \\ a_{31} & a_{33} & a_{34} \\ a_{41} & a_{43} & a_{44} \end{vmatrix} + a_{13}\begin{vmatrix} a_{21} & a_{22} & a_{24} \\ a_{31} & a_{32} & a_{34} \\ a_{41} & a_{42} & a_{44} \end{vmatrix}$$
$$- a_{14}\begin{vmatrix} a_{21} & a_{22} & a_{23} \\ a_{31} & a_{32} & a_{33} \\ a_{41} & a_{42} & a_{43} \end{vmatrix}$$

Now form the 3 × 3 determinants as above.

By now the pattern for finding the determinant of larger matrices should be clear.

The determinants above were developed from the first row of the matrix. They could just as well have been developed using the first column. Since changing the order of the rows or columns in the matrix only affects the sign but not the absolute value of the determinant then the determinant could have been developed using *any* row or column. An even number of exchanges of rows or columns does not affect the sign of the determinant.

If all the elements of a row or column are 0 then the determinant is 0. If the elements in one row are a combination of multiples of the elements of other rows the determinant is 0. The same applies to columns. For example, in the following matrix, the elements of row 3 equal the elements of row 1 plus 2 times the elements of row 2.

$$\begin{vmatrix} 1 & 2 & 3 \\ 2 & 3 & 1 \\ 5 & 8 & 5 \end{vmatrix} = 1(3 \times 5 - 1 \times 8) - 2(2 \times 5 - 1 \times 5) + 3(2 \times 8 - 3 \times 5) = 0$$

It can be shown that:

$$|\mathbf{A}||\mathbf{B}| = |\mathbf{B}||\mathbf{A}| \quad \text{because determinants are scalars}$$

$$|\mathbf{A}||\mathbf{B}| = |\mathbf{AB}| = |\mathbf{BA}| \quad \text{if } \mathbf{A} \text{ and } \mathbf{B} \text{ are dimensionally correct}$$

$$|\mathbf{A}^T| = |\mathbf{A}|$$

A.9 THE INVERSE OF A MATRIX

The inverse of a 2×2 matrix may easily be expressed as follows.

If: $$\mathbf{A} = \begin{bmatrix} a_{11} & a_{12} \\ a_{21} & a_{22} \end{bmatrix}$$

then $$\mathbf{A}^{-1} = \frac{1}{a_{11}a_{22} - a_{12}a_{21}} \begin{bmatrix} a_{22} & -a_{12} \\ -a_{21} & a_{11} \end{bmatrix}$$

This may be verified by forming the product $\mathbf{A}^{-1}\mathbf{A}$ and showing that is an identity matrix.

The inverse of a larger matrix is not so easily expressed. The computational process is as follows.

Form the determinant $|\mathbf{A}|$ as described above.
Form the adjoint of the matrix.
Divide the terms of the adjoint by the determinant.

The adjoint of the matrix has the same dimensions as the original matrix but its cofactor or 'signed minor' replaces each term of the original matrix. The whole is then transposed. The minor is the value of the determinant of the original matrix with the row and the column of the subject element removed. To become a *signed* minor, the minor is multiplied by -1 to the power of the sum of the row and column numbers of the subject element. When the element is in the ith row and the jth column $(-1)^{(i+j)}$ is $+1$ if $(i+j)$ is even and -1 if $(i+j)$ is odd. When i and j are both 1, i.e. in the top left-hand corner of the matrix, $(-1)^{(i+j)}$ is $+1$. On moving through the matrix, therefore, the value of $(-1)^{(i+j)}$ takes up the following pattern.

$$\begin{bmatrix} +1 & -1 & +1 & -1 & +1 & \dots \\ -1 & +1 & -1 & +1 & -1 & \dots \\ +1 & -1 & +1 & -1 & +1 & \dots \\ -1 & +1 & -1 & +1 & -1 & \dots \\ +1 & -1 & +1 & -1 & +1 & \dots \\ \dots & \dots & \dots & \dots & \dots & \dots \end{bmatrix}$$

The inverse of the matrix is then found as the adjoint scaled by the inverse of the determinant. The derivation is illustrated for a 3×3 matrix.

If: $$\mathbf{A} = \begin{bmatrix} a_{11} & a_{12} & a_{13} \\ a_{21} & a_{22} & a_{23} \\ a_{31} & a_{32} & a_{33} \end{bmatrix}$$

the determinant of $\mathbf{A}$, from above, is:

$$|\mathbf{A}| = a_{11}(a_{22}a_{33} - a_{23}a_{32}) - a_{12}(a_{21}a_{33} - a_{23}a_{31}) + a_{13}(a_{21}a_{32} - a_{22}a_{31})$$

The matrix of minors is:

$$\begin{bmatrix} (a_{22}a_{33}-a_{23}a_{32}) & (a_{21}a_{33}-a_{23}a_{31}) & (a_{21}a_{32}-a_{22}a_{31}) \\ (a_{12}a_{33}-a_{13}a_{32}) & (a_{11}a_{33}-a_{13}a_{31}) & (a_{11}a_{32}-a_{12}a_{31}) \\ (a_{12}a_{23}-a_{13}a_{22}) & (a_{11}a_{23}-a_{13}a_{21}) & (a_{11}a_{22}-a_{12}a_{21}) \end{bmatrix}$$

The adjoint or the transpose of the matrix of signed minors is:

$$\text{Adj}\,\mathbf{A} = \begin{bmatrix} (a_{22}a_{33}-a_{23}a_{32}) & -(a_{12}a_{33}-a_{13}a_{32}) & (a_{12}a_{23}-a_{13}a_{22}) \\ -(a_{21}a_{33}-a_{23}a_{31}) & (a_{11}a_{33}-a_{13}a_{31}) & -(a_{11}a_{23}-a_{13}a_{21}) \\ (a_{21}a_{32}-a_{22}a_{31}) & -(a_{11}a_{32}-a_{12}a_{31}) & (a_{11}a_{22}-a_{12}a_{21}) \end{bmatrix}$$

and therefore the inverse of **A** is:

$$\mathbf{A}^{-1} = \frac{\text{Adj}\,\mathbf{A}}{|\mathbf{A}|}$$

Worked example

Example A.1 Find the inverse of matrix A where:

$$\mathbf{A} = \begin{bmatrix} 1 & 2 & 3 \\ 2 & 1 & 4 \\ 3 & 2 & 5 \end{bmatrix}$$

The determinant of **A** is:

$$\begin{aligned} |\mathbf{A}| &= 1(1\times 5 - 4\times 2) - 2(2\times 5 - 4\times 3) + 3(2\times 2 - 1\times 3) \\ &= -3 - (-4) + 3 \\ &= 4 \end{aligned}$$

The matrix of minors is:

$$\begin{bmatrix} -3 & -2 & 1 \\ 4 & -4 & -4 \\ 5 & -2 & -3 \end{bmatrix}$$

The adjoint or the transpose of the matrix of signed minors is:

$$\text{Adj}\,\mathbf{A} = \begin{bmatrix} -3 & -4 & 5 \\ 2 & -4 & 2 \\ 1 & 4 & -3 \end{bmatrix}$$

and upon dividing throughout by the determinant the inverse of **A** is:

$$\mathbf{A}^{-1} = \begin{bmatrix} -0.75 & -1 & 1.25 \\ 0.5 & -1 & 0.5 \\ 0.25 & 1 & -0.75 \end{bmatrix}$$

Finally check to ensure that there are arithmetic errors by confirming that $\mathbf{AA}^{-1} = \mathbf{I}$:

$$\mathbf{AA}^{-1} = \begin{bmatrix} 1 & 2 & 3 \\ 2 & 1 & 4 \\ 3 & 2 & 5 \end{bmatrix} \begin{bmatrix} -0.75 & -1 & 1.25 \\ 0.5 & -1 & 0.5 \\ 0.25 & 1 & -0.75 \end{bmatrix} = \begin{bmatrix} 1 & 0 & 0 \\ 0 & 1 & 0 \\ 0 & 0 & 1 \end{bmatrix}$$

The inverse of a diagonal matrix is easily found because its inverse is also a diagonal matrix and the elements on the leading diagonal of the inverse are the inverse of the elements on the leading diagonal of the original matrix. Therefore if:

$$\mathbf{A} = \begin{bmatrix} a_{11} & 0 & 0 \\ 0 & a_{22} & 0 \\ 0 & 0 & a_{33} \end{bmatrix} \quad \text{then} \quad \mathbf{A}^{-1} = \begin{bmatrix} a_{11}^{-1} & 0 & 0 \\ 0 & a_{22}^{-1} & 0 \\ 0 & 0 & a_{33}^{-1} \end{bmatrix}$$

It can be shown that for any dimensionally compatible matrices:

Inverse of the transpose equals the transpose of the inverse $(\mathbf{X}^T)^{-1} = (\mathbf{X}^{-1})^T$
The product of a matrix and its inverse is an identity matrix $\mathbf{X}\mathbf{X}^{-1} = \mathbf{X}^{-1}\mathbf{X} = \mathbf{I}$
The inverse of a product is the product of inverses but in the reverse order $(\mathbf{XYZ})^{-1} = \mathbf{Z}^{-1}\mathbf{Y}^{-1}\mathbf{X}^{-1}$
The inverse of a sum is the sum of the inverses $(\mathbf{X} + \mathbf{Y})^{-1} = \mathbf{X}^{-1} + \mathbf{Y}^{-1}$
The determinant of the inverse is the inverse of the determinant $|\mathbf{X}^{-1}| = |\mathbf{X}|^{-1}$

A.10 SINGULARITY, ORDER, RANK AND DEGENERACY

A *singular matrix* is one for which the determinant equals 0. A singular matrix cannot have an inverse because the process for the determination of the inverse involves dividing by the determinant.

If a singular matrix is of the *order* n, i.e. it is an $n \times n$ matrix, and there is at least one sub-matrix of order $n - 1$ which is not singular then the matrix is said to have a *degeneracy* of 1. If the largest non-singular sub-matrix that can be formed is of order $n - m$ then the degeneracy is of order m. *Order* minus *degeneracy* equals *rank*.

If the matrix is not square then the *row-rank degeneracy* is the number of rows that are not independent. Similarly, *column-rank degeneracy* is the number of columns that are not independent.

A.11 ORTHOGONAL MATRICES

A square matrix is defined to be orthogonal only if its inverse equals its transpose, that is $\mathbf{A}^{-1} = \mathbf{A}^T$. Orthogonal matrices have particular application in surveying as rotation matrices which transform the coordinates of a point in one system to the coordinates of the same point in another system which is rotated with respect to the first.

In an orthogonal matrix the sum of the products of corresponding elements in any two rows and any two columns is zero. The sum of the squares of the elements in every row and in every column is 1. An orthogonal matrix is non-singular. If a matrix is orthogonal, so is its transpose. The product of any number of orthogonal matrices is another orthogonal matrix.

The *rotation matrix* below is an example of an orthogonal matrix and meets all the criteria of the preceding paragraph.

$$\begin{bmatrix} \cos x & \sin x \\ -\sin x & \cos x \end{bmatrix}$$

It can be shown that for orthogonal matrices:

$$\begin{aligned} \mathbf{X}\mathbf{X}^T &= \mathbf{X}^T\mathbf{X} = \mathbf{I} \\ \mathbf{X}^T &= \mathbf{X}^{-1} \\ |\mathbf{X}^T| &= |\mathbf{X}| = \pm 1 \\ |\mathbf{X}\mathbf{X}^T| &= 1 \\ (\mathbf{XY})^{-1} &= \mathbf{Y}^{-1}\mathbf{X}^{-1} = \mathbf{Y}^T\mathbf{X}^T = (\mathbf{XY})^T \end{aligned}$$

A.12 ROTATION MATRICES

Rotation matrices are most often used for coordinate transformations. If two coordinate systems A and B have the same origin and are rotated with respect to each other by an angle α, as in *Figure A.1* and a single point has coordinates in the two systems of:

$$\begin{bmatrix} x_A \\ y_A \\ z_A \end{bmatrix} \text{ in system } A \text{ and } \begin{bmatrix} x_B \\ y_B \\ z_B \end{bmatrix} \text{ in system } B$$

then the coordinates of the point in system B are related to the coordinates in system A by:

$$\begin{bmatrix} x_B \\ y_B \\ z_B \end{bmatrix} = \begin{bmatrix} 1 & 0 & 0 \\ 0 & \cos\alpha & \sin\alpha \\ 0 & -\sin\alpha & \cos\alpha \end{bmatrix} \begin{bmatrix} x_A \\ y_A \\ z_A \end{bmatrix}$$

this can be seen from *Figure A.1*.

$x_B = x_A$ because the rotation is about the x-axis. The equivalent rotations of β about y and γ about z are:

$$\begin{bmatrix} x_B \\ y_B \\ z_B \end{bmatrix} = \begin{bmatrix} \cos\beta & 0 & -\sin\beta \\ 0 & 1 & 0 \\ \sin\beta & 0 & \cos\beta \end{bmatrix} \begin{bmatrix} x_A \\ y_A \\ z_A \end{bmatrix}$$

$$\begin{bmatrix} x_B \\ y_B \\ z_B \end{bmatrix} = \begin{bmatrix} \cos\gamma & \sin\gamma & 0 \\ -\sin\gamma & \cos\gamma & 0 \\ 0 & 0 & 1 \end{bmatrix} \begin{bmatrix} x_A \\ y_A \\ z_A \end{bmatrix}$$

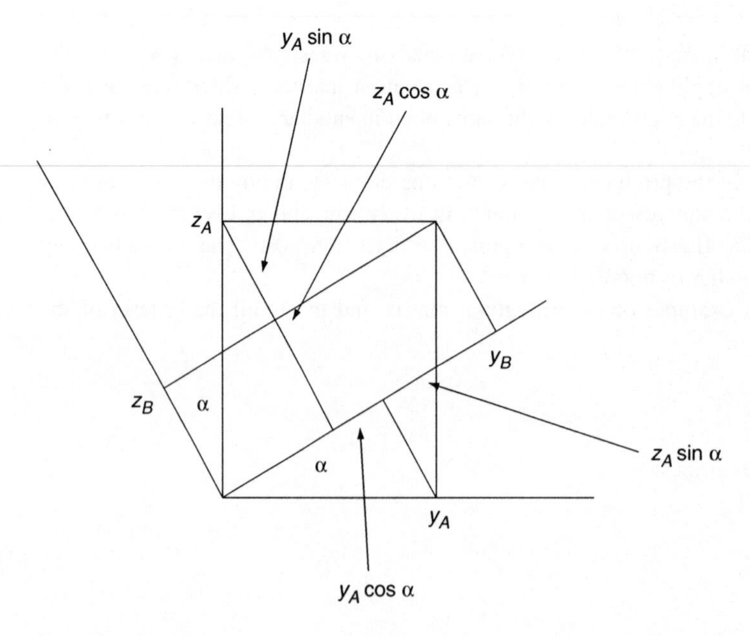

Fig. A.1 *Rotated axes*

A.13 EIGENVALUES AND EIGENVECTORS

If **N** is a square, non-singular matrix of order n, then in the equation

$$(\mathbf{N} - \lambda\mathbf{I})\mathbf{x} = \begin{bmatrix} n_{11} - \lambda & n_{12} & \dots & n_{1n} \\ n_{21} & n_{22} - \lambda & \dots & n_{2n} \\ \dots & \dots & \dots & \dots \\ n_{n1} & n_{n2} & \dots & n_{nn} - \lambda \end{bmatrix} \begin{bmatrix} x_1 \\ x_2 \\ \dots \\ x_n \end{bmatrix} = 0$$

there will be n values of λ for which the matrix $(\mathbf{N} - \lambda\mathbf{I})$ is singular. That is, there are n values of λ which make the determinant of $(\mathbf{N} - \lambda\mathbf{I})$ zero. If the determinant of $(\mathbf{N} - \lambda\mathbf{I})$ is formed and equated to zero it can be expressed in the form of its *characteristic equation* or *characteristic polynomial.*

$$a_n\lambda^n + a_{n-1}\lambda^{n-1} + \dots + a_1\lambda + a_0 = 0$$

All values of λ may not be distinct. These values are called *characteristic roots*, *latent roots* or *eigenvalues*. With solutions for λ, the equation

$$(\mathbf{N} - \lambda_i\mathbf{I})\mathbf{x} = 0$$

rewritten as $\mathbf{N}\mathbf{x} = \lambda_i\mathbf{x}$ will have as a solution

$$\mathbf{x} = c \begin{bmatrix} e_1 \\ e_2 \\ e_3 \\ \dots \\ e_n \end{bmatrix}$$

where c can take any value. The vector is then the *characteristic vector*, *latent vector* or *eigenvector* associated with the eigenvalue λ_i. If c is chosen such that:

$$\sum (ce_i) = 1$$

then **x** is said to be *normalized*. If **N** is singular then at least one of the eigenvalues is zero.

Worked example

Example A.2 Find the eigenvalues and eigenvectors of the 2 × 2 matrix **A** where:

$$\mathbf{A} = \begin{bmatrix} 3 & 2 \\ 4 & 5 \end{bmatrix}$$

then the eigenvalues are derived from the characteristic equation as follows:

$$|\mathbf{A} - \lambda\mathbf{I}| = \begin{vmatrix} 3 - \lambda & 2 \\ 4 & 5 - \lambda \end{vmatrix} = (3 - \lambda)(5 - \lambda) - 8 = 0$$

So: $\lambda^2 - 8\lambda + 7 = 0$
for which the solutions are $\lambda_1 = 1$ and $\lambda_2 = 7$.
If the eigenvalues are used in the expressions $\mathbf{A}\mathbf{x}_1 = \lambda_1\mathbf{x}_1$ and $\mathbf{A}\mathbf{x}_2 = \lambda_2\mathbf{x}_2$ then the solutions of:

$$(\mathbf{A} - \lambda_1\mathbf{I})\mathbf{x}_1 = \begin{bmatrix} 3-\lambda_1 & 2 \\ 4 & 5-\lambda_1 \end{bmatrix} \begin{bmatrix} x_1 \\ x_2 \end{bmatrix} = \begin{bmatrix} 0 \\ 0 \end{bmatrix} \quad (\mathbf{A} - \lambda_2\mathbf{I})\mathbf{x}_2 = \begin{bmatrix} 3-\lambda_2 & 2 \\ 4 & 5-\lambda_2 \end{bmatrix} \begin{bmatrix} x_1 \\ x_2 \end{bmatrix} = \begin{bmatrix} 0 \\ 0 \end{bmatrix}$$

are: $\begin{bmatrix} x_1 \\ x_2 \end{bmatrix} = \begin{bmatrix} k \\ -k \end{bmatrix}$ $\begin{bmatrix} x_1 \\ x_2 \end{bmatrix} = \begin{bmatrix} k \\ 2k \end{bmatrix}$

where k is any number.

Eigenvalues and eigenvectors are particularly useful in the determination of error ellipses, figures of positional uncertainty as described in *Chapter 7* and are the only practical solution for the determination of the three-dimensional equivalent, error ellipsoids.

Index

Lightning Source UK Ltd.
Milton Keynes UK
08 January 2010

148295UK00001B/50/P